# Flavourings

Edited by
Herta Ziegler

## 1807–2007 Knowledge for Generations

Each generation has its unique needs and aspirations. When Charles Wiley first opened his small printing shop in lower Manhattan in 1807, it was a generation of boundless potential searching for an identity. And we were there, helping to define a new American literary tradition. Over half a century later, in the midst of the Second Industrial Revolution, it was a generation focused on building the future. Once again, we were there, supplying the critical scientific, technical, and engineering knowledge that helped frame the world. Throughout the 20th Century, and into the new millennium, nations began to reach out beyond their own borders and a new international community was born. Wiley was there, expanding its operations around the world to enable a global exchange of ideas, opinions, and know-how.

For 200 years, Wiley has been an integral part of each generation's journey, enabling the flow of information and understanding necessary to meet their needs and fulfill their aspirations. Today, bold new technologies are changing the way we live and learn. Wiley will be there, providing you the must-have knowledge you need to imagine new worlds, new possibilities, and new opportunities.

Generations come and go, but you can always count on Wiley to provide you the knowledge you need, when and where you need it!

*William J. Pesce*
President and Chief Executive Officer

*Peter Booth Wiley*
Chairman of the Board

# Flavourings

Production, Composition,
Applications, Regulations

Edited by
Herta Ziegler

Second, Completely Revised Edition

WILEY-VCH Verlag GmbH & Co. KGaA

**The Editor**
Dr. Herta Ziegler
Erich Ziegler GmbH
Am Weiher 133
91347 Aufsess
Germany

**Cover**
Pictures were used with courtesy by Erich Ziegler GmbH.

Wiley Bicentennial Logo: Richard J. Pacifico

All books published by Wiley-VCH are carefully produced. Nevertheless, authors, editors, and publisher do not warrant the information contained in these books, including this book, to be free of errors. Readers are advised to keep in mind that statements, data, illustrations, procedural details or other items may inadvertently be inaccurate.

**Library of Congress Card No.:**
applied for

**British Library Cataloguing-in-Publication Data**
A catalogue record for this book is available from the British Library.

**Bibliographic information published by the Deutsche Nationalbibliothek**
The Deutsche Nationalbibliothek lists this publication in the Deutsche Nationalbibliografie; detailed bibliographic data is available in the Internet at <http://dnb.d-nb.de>.

© 2007 WILEY-VCH Verlag GmbH & Co. KGaA, Weinheim

All rights reserved (including those of translation into other languages). No part of this book may be reproduced in any form – by photoprinting, microfilm, or any other means – nor transmitted or translated into a machine language without written permission from the publishers. Registered names, trademarks, etc. used in this book, even when not specifically marked as such, are not to be considered unprotected by law.

**Composition** Strassner ComputerSatz, Leimen
**Printing** Strauss GmbH, Mörlenbach
**Bookbinding** Litges & Dopf GmbH, Heppenheim

Printed in the Federal Republic of Germany
Printed on acid-free paper

**ISBN** 978-3-527-31406-5

**Dedicated to Mr. Erich Ziegler
on the Occasion of his 80$^{th}$ Birthday**

# Preface to the Second Edition

## Flavourings – a tradition in the family

With the present $2^{nd}$ edition, this joint project of 41 authors has been updated and enlarged to include and reflect the recent changes and developments, which, also in the sector of flavourings and their technologies, occur at a breathtaking pace.

After laying the foundation for the first edition, Erich Ziegler has been able to pass on the editorship within the family, sharing his ongoing passion for the world of flavours. This $2^{nd}$ edition had initially been scheduled as homage on the occasion of his $80^{th}$ birthday in 2005, a target the large pool of authors could, however, not fulfil completely.

On behalf of all authors, I would like to dedicate this edition to Mr. Erich Ziegler, whose initiative and efforts were instrumental in gathering the first group of authors.

Edition 2 again represents a compendium which in its entirety is intended to familiarise the reader with the complex subject of flavourings, from raw materials to application methods and technology.

In addition to the numerous articles revised by their original authors, a considerable number of new authors and co-authors have joined our effort ensuring continuity and up-to-date contributions.

The preface to the $1^{st}$ edition, also intended as a summary to guide the reader through the book, has in the majority retained its relevance for the present edition. The already extensive survey of our field of work is complemented by a number of new topics. Prof. W. Grosch provides the reader with a comprehensive survey of aroma analysis with a special emphasis on key odourants. Contributors from multinational food companies introduce a focus on final products in the section on applications. Additionally, the sector on non-natural flavors has been expanded to include the current state of the European chemical group classifications.

This $2^{nd}$ edition today already possesses a 'historical' element for me, as a revision had originally been projected for the fifth year after the initial publication. However, with a team of authors as large as ours, the comparison with a ship - fully loaded, difficult to manoeuvre – may not be inadequate and I am, therefore grateful today that there have been 'only minor deviations' from the original schedule.

Unfortunately, a few authors did not succeed in submitting their revisions on time, but the publisher forged on, also to guarantee the topicality of those revisions which were submitted early.

The creation of such a collection of manuscripts is the result of that inner, mysterious urge to communicate, inherent to each and every author. To encourage, to revive this force is the small stimulus - sometimes gentle, sometimes more pronounced - provided by the editor in order to foster the conclusion that we all contribute towards making the magical world of flavours more accessible.

Dear co-authors, I do hope that you will not only pardon my persistence in trying to motivate you to write – in a world where time is more than scarce, especially for all those still tied up in the daily routines of companies or institutes – but will permit the light of positive retrospective to transform all these heights and depths into amusing anecdotes and commit negatives into the realm of oblivion.

I do also hope that you will share my pleasure and pride that we have succeeded in forming the majority of the manuscripts into a coherent whole and that the struggles of each and every 'comrade-in-pens' were in the end rewarded by the final outcome. I would like to again express my deep gratitude to all of you and also all those who participated in the prior edition. Without the support of all the companies and institutes, which made the participation of employees and access to their knowledge possible, this edition would not have been possible, a kindness for which I would like to express my appreciation.

I am also indebted to my sister-in-law Silvia Ziegler, whose untiring support as lector and translator made the book in its current form possible.

Additionally I would like to thank Wiley-VCH and especially Ms. Sora and Ms. Wüst for the constructive cooperation and their patience with our team of authors.

Just as in the first edition, acknowledging the support and help I received from so many sources is again a great pleasure. I would, therefore, like to express my deep gratitude to all those, who in personal or written form, offered assistance and encouragement. In this second edition, I could again rely on the valuable advice of Dr. George Clark, but I am also indebted to all those who provided a multitude of useful information and detailed insights into our industry.

Last – but not least, I would like to thank Dr. Salzer, who in addition to his numerous contributions as an author, has been invaluable for this second edition with his advice, understanding and support.

In remembrance of our co-author Mr. Herman Olsman, who passed away at the end of 2004 and whose contribution is no longer included in this book.

Although I certainly hope that every reader will come across interesting and innovative aspects concerning the world of flavourings within this collection of articles, it is certainly possible that one or the other aspect has been neglected, omitted or dealt with incompletely. All authors are entirely responsible for form and content of their respective contribution and will be pleased to receive questions, suggestions and any other scientific comments at the respective addresses.

Bayreuth / Aufsess, December 2006                                        Dr. Herta Ziegler

# Preface to the First Edition

The book **"Die natürlichen und künstlichen Aromen"** was first published by Erich Ziegler in German in 1982 as a collection of 21 articles written by authors who are experts on their respective subjects.

This first edition, an overview of this interesting and diverse field of work intended for those involved in food flavouring application, has been completely revised in order to take the manifold changes into consideration. The present expanded collection of 37 different contributions is certainly still only selective; it features enlarged versions of all previous chapters and also includes articles on a number of newly emerging topics and developments.

To open up the new edition to an international readership, English has been selected as the language of publication.

**"Flavourings"** intends to grant its readership an insight into the production, processing and application of various food flavourings and also focuses on the basic and new analytical methods employed in this field.

The book draws on the expert knowledge of contributors with backgrounds both in industry and academia.

*The following summary will guide you through the book:*

The book starts with a short overview of the industry, including historical and economic aspects as well as current trends and future perspectives.

The next chapter describes the basic physical and biotechnological processes which are today available for the production of flavourings and flavour extracts. These range from more traditional methods such as extraction and distillation to more recent developments, e.g. supercritical fluid extraction, spray and freeze drying as well as microencapsulation, and include the rapidly increasing field of biotechnology.

Chapter 3 deals with the raw materials which are of interest for the flavour sector. The topics range from chemically defined flavouring substances, both of natural and synthetic origin, to flavouring preparations and source materials, such as complex natural extracts, essential oils and juices. Furthermore, process flavourings and non-flavouring compounds which are important for food technology are also presented.

The next chapter focuses on the interesting area of blended flavourings, often regarded as an artistic field of work.

Beverages, confectioneries, dairy products and industrial food products are today important sectors for the application of flavourings and are therefore described in section 5.

In the following, quality control via sensory, analytical and microbiological methods is dealt with. As the quality of foods is, today more than ever, in the focus of the public interest, the methods available for standardised quality evaluation have undergone an enormous improvement and specification process. The recent analytical progress in the determination of natural origin of different food matrices will therefore be depicted in detail.

The last chapter focuses on questions dealing with legislative concerns, taking both geographic and ethical guidelines into consideration.

*Acknowledgements:*

The present book is a result of the efforts of 37 co-workers and the editors would like to express their gratitude to all those who contributed to this book and the respective companies and institutes which made this possible. In retrospect we have to admit that the co-operation of such a diverse crew of authors was not always easy to co-ordinate. In this context, we would like to express our deep gratitude to all those patient contributors who, despite the delays caused by those who were either late in submitting or who completely refrained from doing so, did not hesitate to keep their articles up-to-date. It is especially gratifying for us to see that our joint efforts have been rewarded by the present collection of articles which tries to capture the current state of knowledge.

The work of the last few years has been made considerably easier for the senior author through the constant help of his daughter-in-law, Dr. Herta Ziegler, who increasingly took over the tasks of an editor.

Moreover, we would like to express our gratitude to Dr. Uwe-Jens Salzer, who, apart from his contributions as author, was always very helpful and Mr. Kurt Roßbach, who kindly accepted the important and time-consuming office of the English technical reader.

We are also indebted to Mrs. Silvia Ziegler, who not only translated a number of articles but also took over all the tasks that an edition in English required.

Thanks also go to Mr. Günter Ziegler who always gave us new strength when our spirits faltered.

We would also like to thank Dr. T. Pillhofer for his technical advice and his expertise which was very helpful for the article on extraction.

Mr. J. Flores, FMC Corporation, kindly provided illustrations which appear in the chapters on fruit juices and citrus oils and we are also indebted to Dr. George Clark for his valuable advice.

In remembrance of Dr. R. Emberger, who was a contributor to the first edition, the editors would like to acknowledge his kindness and helpfulness also with this edition, where he provided the sensory evaluations for the citrus chapter.

Last but not least we would like to thank the Hüthig Verlag that made this book possible.

All authors are entirely responsible for their contributions and will be pleased to answer any questions concerning their subject or to receive constructive comments - please feel free to use the author's addresses listed in the Index of Authors.

Aufseß, September 1997

Erich Ziegler
Herta Ziegler

# Contents

Preface . . . . . . . . . . . . . . . . . . . . . . . . . . . . . . . . . . . . . . . . . . . . . . . . . . . . . . . . VII
Index of Authors . . . . . . . . . . . . . . . . . . . . . . . . . . . . . . . . . . . . . . . . . . . . . . . XVII
List of Companies . . . . . . . . . . . . . . . . . . . . . . . . . . . . . . . . . . . . . . . . . . . . . XXI

| | | |
|---|---|---|
| **1** | **Introduction** . . . . . . . . . . . . . . . . . . . . . . . . . . . . . . . . . . . . . . . . . | 1 |
| | A Dynamic Business With Taste – | |
| | The Flavour Industry *(Herta Ziegler)* . . . . . . . . . . . . . . . . . . . . . . . | 1 |
| **2** | **Manufacturing Processes** . . . . . . . . . . . . . . . . . . . . . . . . . . . . . . . | 15 |
| 2.1 | Physical Processes . . . . . . . . . . . . . . . . . . . . . . . . . . . . . . . . . . . . . | 17 |
| 2.1.1 | Extraction *(Manfred Ziegler)* . . . . . . . . . . . . . . . . . . . . . . . . . . . . | 17 |
| 2.1.1.1 | Introduction . . . . . . . . . . . . . . . . . . . . . . . . . . . . . . . . . . . . . . . . . . | 17 |
| 2.1.1.2 | Solid-Liquid Extraction . . . . . . . . . . . . . . . . . . . . . . . . . . . . . . . . . . | 17 |
| 2.1.1.3 | Liquid-Liquid Extraction . . . . . . . . . . . . . . . . . . . . . . . . . . . . . . . . | 24 |
| 2.1.2 | Supercritical Fluid Extraction (SFE) | |
| | *(Karl-Werner Quirin, Dieter Gerard)* . . . . . . . . . . . . . . . . . . . . . . | 49 |
| 2.1.2.1 | Solvent Evaluation . . . . . . . . . . . . . . . . . . . . . . . . . . . . . . . . . . . . . | 49 |
| 2.1.2.2 | Near Critical Gas Solvents . . . . . . . . . . . . . . . . . . . . . . . . . . . . . . . | 50 |
| 2.1.2.3 | Solvent Character of $CO_2$ . . . . . . . . . . . . . . . . . . . . . . . . . . . . . . . | 52 |
| 2.1.2.4 | Selectivity . . . . . . . . . . . . . . . . . . . . . . . . . . . . . . . . . . . . . . . . . . . . | 52 |
| 2.1.2.5 | $CO_2$-Extraction Process . . . . . . . . . . . . . . . . . . . . . . . . . . . . . . . . . | 54 |
| 2.1.2.6 | Extraction of Flavourings . . . . . . . . . . . . . . . . . . . . . . . . . . . . . . . . | 58 |
| 2.1.2.7 | Economic Considerations . . . . . . . . . . . . . . . . . . . . . . . . . . . . . . . . | 63 |
| 2.1.2.8 | Other Applications . . . . . . . . . . . . . . . . . . . . . . . . . . . . . . . . . . . . . | 64 |
| 2.1.3 | Distillation *(Manfred Ziegler, Martin Reichelt)* . . . . . . . . . . . . . . | 66 |
| 2.1.3.1 | Introduction . . . . . . . . . . . . . . . . . . . . . . . . . . . . . . . . . . . . . . . . . . | 66 |
| 2.1.3.2 | Fundamental Considerations . . . . . . . . . . . . . . . . . . . . . . . . . . . . . | 66 |
| 2.1.3.3 | Thermodynamic Fundamentals of Mixtures . . . . . . . . . . . . . . . . . | 71 |
| 2.1.3.4 | Equipment . . . . . . . . . . . . . . . . . . . . . . . . . . . . . . . . . . . . . . . . . . . . | 83 |
| 2.1.4 | Spray Drying and Other Methods for Encapsulation of Flavourings | |
| | *(Uwe-Jens Salzer)* . . . . . . . . . . . . . . . . . . . . . . . . . . . . . . . . . . . . . | 97 |
| 2.1.4.1 | General Introduction . . . . . . . . . . . . . . . . . . . . . . . . . . . . . . . . . . . | 97 |
| 2.1.4.2 | Spray Drying and 'Complementary' Procedures . . . . . . . . . . . . . . | 97 |
| 2.1.4.3 | New Methods for Encapsulation . . . . . . . . . . . . . . . . . . . . . . . . . . | 105 |
| 2.1.4.4 | Outlook . . . . . . . . . . . . . . . . . . . . . . . . . . . . . . . . . . . . . . . . . . . . . . | 108 |
| 2.1.5 | Freeze Drying *(Karl Heinz Deicke)* . . . . . . . . . . . . . . . . . . . . . . . | 109 |
| 2.1.5.1 | General Remarks on Drying . . . . . . . . . . . . . . . . . . . . . . . . . . . . . | 109 |
| 2.1.5.2 | The Freeze Drying Process . . . . . . . . . . . . . . . . . . . . . . . . . . . . . . | 110 |
| 2.1.5.3 | The Quality of Freeze-dried Products . . . . . . . . . . . . . . . . . . . . . . | 112 |
| 2.2 | Biotechnological Processes *(Joachim Tretzel, Stefan Marx)* . . . . . . . . | 120 |
| 2.2.1 | Introduction . . . . . . . . . . . . . . . . . . . . . . . . . . . . . . . . . . . . . . . . . . | 120 |
| 2.2.2 | Flavour Generation by Fermentation of Food Raw Materials . . . . . . . . | 121 |
| 2.2.3 | Flavour Generation in Bioreactors . . . . . . . . . . . . . . . . . . . . . . . . . | 123 |

| 2.2.4 | Surface Fermentation | 124 |
|---|---|---|
| 2.2.5 | Submerged Fermentation | 125 |
| 2.2.6 | Downstream Processing | 126 |
| 2.2.7 | Enzyme Reactors | 127 |
| 2.2.8 | Cell Culture Reactors | 129 |
| 2.2.9 | Genetic Engineering | 131 |
| **3** | **Raw Materials for Flavourings** | **135** |
| 3.1 | Introduction (*Günter Matheis*) | 137 |
| 3.2 | Flavouring Ingredients | 138 |
| 3.2.1 | Chemically Defined Flavouring Substances | 140 |
| 3.2.1.1 | Natural Flavouring Substances | 140 |
| 3.2.1.2 | Nature-Identical and Artificial Flavouring Substances (*Gerhard Krammer*) | 158 |
| 3.2.2 | Flavouring Preparations and Some Source Materials | 166 |
| 3.2.2.1 | Fruit Juices and Fruit Juice Concentrates (*Martin Simon*) | 166 |
| 3.2.2.2 | Citrus Oils (*Herta Ziegler, Wolfgang Feger*) | 187 |
| 3.2.2.3 | Herbs, Spices and Essential Oils (*Maren D. Protzen, Jens-Achim Protzen*) | 214 |
| 3.2.2.4 | Flavouring Preparations Based on Biotechnology (*Joachim Tretzel, Stefan Marx*) | 260 |
| 3.2.3 | Process Flavourings (*Christoph Cerny*) | 274 |
| 3.2.3.1 | Introduction | 274 |
| 3.2.3.2 | Definition and General Guidelines | 274 |
| 3.2.3.3 | Process Flavour Chemistry | 276 |
| 3.2.3.4 | Industrial Process Flavourings | 288 |
| 3.2.3.5 | Outlook | 292 |
| 3.2.4 | Smoke Flavourings | 298 |
| 3.2.4.1 | Nature, Preparation and Application (*Gary Underwood, John Shoop*) | 298 |
| 3.2.4.2 | Situation in Europe (*Uwe-Jens Salzer*) | 309 |
| 3.3 | Non-flavouring Ingredients | 314 |
| 3.3.1 | Extraction Solvents (*Howard D. Preston, Robert F. van Croonenborgh*) | 314 |
| 3.2.3 | Permitted Carriers and Carrier Solvents (*Howard D. Preston*) | 317 |
| 3.3.3 | Emulsifiers – Stabilisers – Enzymes (*William J. N. Marsden*) | 322 |
| 3.3.3.1 | Emulsifiers | 322 |
| 3.3.3.2 | Stabilisers | 330 |
| 3.3.3.3 | Enzymes | 335 |
| 3.3.4 | Flavour Modifiers (*Günter Matheis*) | 351 |
| 3.3.4.1 | Definition and Classification | 351 |
| 3.3.4.2 | Monosodium Glutamate, Purine 5'-Ribonucleotides, and Related Substances | 352 |
| 3.3.4.3 | Maltol and Ethyl Maltol | 362 |
| 3.3.4.4 | Furanones and Cyclopentenolones | 366 |
| 3.3.4.5 | Vanillin and Ethyl Vanillin | 368 |

| | | |
|---|---|---|
| 3.3.4.6 | Other Flavour Modifiers | 369 |
| 3.3.4.7 | Final Remarks | 370 |
| 3.3.5 | Antioxidants and Preservatives | 373 |
| 3.3.5.1 | Antioxidants (*Dieter Gerard, Karl-Werner Quirin*) | 373 |
| 3.3.5.2 | Preservatives (*Howard D. Preston*) | 377 |
| | | |
| **4** | **Blended Flavourings** (*Willi Grab*) | **391** |
| 4.1 | Introduction | 395 |
| 4.1.1 | Flavour Analysis | 395 |
| 4.1.2 | Flavour Profiles | 395 |
| 4.1.3 | Flavouring Raw Materials | 397 |
| 4.1.4 | Flavouring Composition | 398 |
| 4.1.5 | Flavour Release | 402 |
| 4.1.6 | Application of Flavourings | 404 |
| 4.1.7 | Flavouring Production | 405 |
| 4.1.8 | Flavouring Stability | 409 |
| 4.2 | Fruit Flavours | 411 |
| 4.2.1 | The Flavour of Natural Fruits | 412 |
| 4.2.2 | World of Commercially Important Fruit Flavours | 413 |
| 4.2.3 | Sensorially Interesting Fruit Flavours | 419 |
| 4.3 | Other Blended Flavourings | 425 |
| 4.3.1 | Processed Flavourings | 425 |
| 4.3.2 | Fermented Flavourings | 429 |
| 4.3.2.1 | Alcoholic Flavourings | 429 |
| 4.3.2.2 | Dairy Flavourings | 430 |
| 4.3.3 | Vegetable Flavourings | 431 |
| 4.3.4 | Vanilla Flavourings | 432 |
| 4.3.5 | Fantasy Flavourings | 432 |
| | | |
| **5** | **Application** | **435** |
| 5.1 | Introduction (*Günter Matheis*) | 437 |
| 5.1.1 | Flavour Binding and Release | 437 |
| 5.1.2 | Interactions with Carbohydrates | 438 |
| 5.1.3 | Interactions with Proteins | 445 |
| 5.1.4 | Interactions with Free Amino Acids | 449 |
| 5.1.5 | Interactions with Lipids | 450 |
| 5.1.6 | Interactions with Inorganic Salts, Fruit Acids, Purine Alkaloids, Phenolic Compounds and Ethanol | 454 |
| 5.1.7 | Interactions with Fat Replacers | 455 |
| 5.1.8 | Interactions with Complex Systems and with Foodstuffs | 458 |
| 5.1.9 | Conclusion | 462 |
| 5.2 | Flavourings for Beverages | 466 |
| 5.2.1 | Soft Drinks (*Matthias Sass*) | 466 |
| 5.2.1.1 | Clear Soft Drinks | 468 |
| 5.2.1.2 | Cloudy Soft Drinks | 469 |
| 5.2.1.3 | Colours for Soft Drinks | 472 |

| 5.2.1.4 | Sweeteners for Soft Drinks | 474 |
| --- | --- | --- |
| 5.2.1.5 | Healthy Nutrition – Functional Drinks | 478 |
| 5.2.1.6 | Flavour Systems for Convenient and Efficient Soft Drink Production | 482 |
| 5.2.1.7 | Use of Flavour Systems at the Consumer Level | 483 |
| 5.2.2 | Alcoholic Beverages (*Günther Zach*) | 487 |
| 5.2.2.1 | Introduction | 487 |
| 5.2.2.2 | Generic Distilled Spirits | 487 |
| 5.2.2.3 | Liqueurs | 496 |
| 5.2.2.4 | Wine | 503 |
| 5.2.2.5 | Beer | 506 |
| 5.2.2.6 | Malt/Beer-based Alcoholic Beverages | 511 |
| 5.2.2.7 | Cocktails/Spirit-based Alcoholic Beverages | 512 |
| 5.2.2.8 | Production of Flavours for Alcoholic Beverages | 512 |
| 5.3 | Flavourings for Confectioneries, Baked Goods, Ice-cream and Dairy Products | 515 |
| 5.3.1 | Flavourings for Confectioneries (*Jan-Pieter Miedema*) | 515 |
| 5.3.1.1 | Introduction | 515 |
| 5.3.1.2 | High Boilings | 516 |
| 5.3.1.3 | Fondants | 518 |
| 5.3.1.4 | Jellies and Gums | 520 |
| 5.3.1.5 | Caramel and Toffee | 521 |
| 5.3.1.6 | Chewing Gum | 522 |
| 5.3.1.7 | Compressed Tablets | 524 |
| 5.3.1.8 | Panned Work | 525 |
| 5.3.1.9 | Chocolate and Cocoa Products | 526 |
| 5.3.2 | Flavourings for Baked Goods (*Jan-Pieter Miedema*) | 531 |
| 5.3.2.1 | Introduction | 531 |
| 5.3.2.2 | What are Fine Bakery Products and what is a Dough? | 531 |
| 5.3.2.3 | Flavourings for Fine Bakery Products | 532 |
| 5.3.3 | Flavourings for Ice-Cream (*Alan D. Ellison, John E. Jackson, Jan-Pieter Miedema, Jürgen Brand*) | 535 |
| 5.3.3.1 | Introduction | 535 |
| 5.3.3.2 | Classification | 535 |
| 5.3.3.3 | Ingredients | 536 |
| 5.3.3.4 | Flavourings | 537 |
| 5.3.3.5 | The Influence of Other Ingredients on the Ice-cream | 537 |
| 5.3.3.6 | Sensory Aspect of Ice-cream | 539 |
| 5.3.3.7 | The Production of Ice-cream | 539 |
| 5.3.4 | Flavourings for Dairy Products and Desserts (*Jan-Pieter Miedema, Eckhard Schildknecht*) | 542 |
| 5.3.4.1 | Introduction | 542 |
| 5.3.4.2 | Production of Flavoured Dairy Products | 542 |
| 5.3.4.3 | Flavouring Components for Dairy Products | 545 |
| 5.4 | Flavouring of Dehydrated Convenience Food and Kitchen Aids (*Chahan Yeretzian, Imre Blank, Stefan Palzer*) | 549 |

| | | |
|---|---|---|
| 5.4.1 | Introduction | 549 |
| 5.4.1.1 | Flavour World | 551 |
| 5.4.1.2 | Savoury Taste: Umami and Saltiness | 552 |
| 5.4.2 | Savoury Flavours | 556 |
| 5.4.2.1 | Base Notes: Flavour Body and Taste Enhancement | 557 |
| 5.4.2.2 | Middle Notes: Process and Reaction Flavours | 559 |
| 5.4.2.3 | Top Notes: Key Compounds of the Culinary Aroma | 561 |
| 5.4.3 | Manufacturing of Dehydrated Convenience Foods | 563 |
| 5.4.3.1 | Legal Aspects | 563 |
| 5.4.3.2 | Flavour Application during Manufacturing of Dehydrated Convenience Foods | 564 |
| 5.4.3.3 | Physical and Chemical Flavour Changes during Processing and Shelf-Life | 568 |
| 5.4.4 | Conclusions | 570 |
| | | |
| **6** | **Quality Control** | **573** |
| 6.1 | Sensory Analysis in Quality Control (*Helmut Grüb*) | 575 |
| 6.1.1 | Introduction | 575 |
| 6.1.2 | "Sensory" | 575 |
| 6.1.3 | The Senses | 576 |
| 6.1.4 | Man as a Measuring Instrument | 579 |
| 6.1.5 | Sensory Quality Measuring Methods | 579 |
| 6.2 | Analytical Methods | 587 |
| 6.2.1 | General Methods (*Dirk Achim Müller*) | 587 |
| 6.2.1.1 | Introduction | 587 |
| 6.2.1.2 | Determination of Colouring Principles by UV/VIS-Spectroscopy | 587 |
| 6.2.1.3 | Determination of Heavy Metal Contamination | 587 |
| 6.2.1.4 | Modern Chromatographic Techniques | 589 |
| 6.2.2 | Stable Isotope Ratio Analysis in Quality Control of Flavourings (*Hanns-Ludwig Schmidt, Robert A. Werner, Andreas Rossmann, Armin Mosandl, Peter Schreier*) | 602 |
| 6.2.2.1 | Correlations between (Bio-)Synthesis and Isotope Content or Pattern of Organic Compounds | 604 |
| 6.2.2.2. | Methods for the Determination of Average Isotope Abundances in Organic Compounds | 608 |
| 6.2.2.3 | Intermolecular, Site-specific and Positional Isotope Ratio Analysis | 613 |
| 6.2.2.4 | Results of Stable Isotope Ratio Measurements in Authenticity Checks and in the Verification of the Natural Origin of Flavouring Compounds | 619 |
| 6.2.2.5 | Perspectives and Future Aspects | 646 |
| 6.2.3 | Enantioselective Analysis (*Armin Mosandl*) | 664 |
| 6.2.3.1 | Chiral Resolution and Chromatographic Behaviour of Enantiomers | 664 |
| 6.2.3.2 | Sample Clean-up | 665 |
| 6.2.3.3 | Detection Systems | 666 |
| 6.2.3.4 | Stereodifferentiation and Quantification | 668 |
| 6.2.3.5 | Limitations | 669 |

| | | |
|---|---|---|
| 6.2.3.6 | Analysis of Individual Classes of Compounds | 670 |
| 6.2.3.7 | Special Measurements | 690 |
| 6.2.3.8 | Authenticity Assessment – Latest Developments | 696 |
| 6.2.4 | Key Odorants of Food Identified by Aroma Analysis (*Werner Grosch*) | 704 |
| 6.2.4.1 | Introduction | 704 |
| 6.2.4.2 | Aroma Analysis | 704 |
| 6.2.4.3 | Applications | 711 |
| 6.3 | Microbiological Testing (*Heinz-Jürgen Lögtenbörger*) | 744 |
| **7** | **Legislation / Toxicology** (*Uwe-Jens Salzer*) | **753** |
| 7.1 | Introduction / Definitions | 755 |
| 7.2 | Toxicological Considerations | 758 |
| 7.3 | FAO/WHO and Council of Europe | 761 |
| 7.4 | European Union | 762 |
| 7.4.1 | General | 762 |
| 7.4.2 | Flavour Legislation | 763 |
| 7.4.3 | Food Additive Legislation | 767 |
| 7.4.4 | Other Rules Concerning Flavourings | 768 |
| 7.5 | America | 771 |
| 7.5.1 | Introduction | 771 |
| 7.5.2 | The NAFTA Countries | 772 |
| 7.5.3 | South American Countries | 779 |
| 7.6 | Asia | 784 |
| 7.6.1 | General | 784 |
| 7.6.2 | The "Middle East" | 784 |
| 7.6.3 | The "Far East" | 785 |
| 7.7 | South Africa, Australia and New Zealand | 800 |
| 7.8 | Religious Dietary Rules | 801 |
| 7.8.1 | Introduction | 801 |
| 7.8.2 | "Kosher" | 801 |
| 7.8.3 | "Halal" | 804 |
| 7.8.4 | Comparison of Kosher and Halal Requirements | 805 |
| **Index** | | **811** |

# Index of Authors

(in alphabetical order)

**Imre Blank**
Sentier de Courtaraye 2
1073 Savigny
Switzerland
imre.blank@rdor.nestle.com

**Jürgen Brand**
Grafenstrasse 9d
64331 Weiterstadt
Germany
juergen.brand@unifine.de

**Christoph Cerny**
Firmenich SA
Rue de la Bergère 7
PO Box 148
1217 Meyrin 2, Geneva
Switzerland
christoph.cerny@firmenich.com

**Karl-Heinz Deicke**
Ricarda-Huch-Strasse 35
48161 Münster
Germany

**Alan D. Ellison**
Quest International Inc.
5115 Sedge Boulevard
Hoffman Estates IL 60192
USA
alan.d.ellison@questintl.com

**Wolfgang Feger**
Erich Ziegler GmbH
Am Weiher 133
91347 Aufsess
Germany
wf@erich-ziegler.de

**Dieter Gerard**
Königsstrasse 2A
66780 Rehlingen-Siersburg
Germany
dg@flavex.de

**Willi Grab**
Givaudan Singapore Pte. Ltd.
1 Woodlands Avenue 8
Singapore 738972
Singapore
willi.grab@givaudan.com

**Werner Grosch**
Burgstrasse 3B
85604 Zorneding
Germany
GroschT13@t-online.de

**Helmut Grüb**
PO Box 1423
37594 Holzminden
Germany
hgrueb.HOL@web.de

**John E. Jackson**
Quest International Inc.
5115 Sedge Boulevard
Hoffman Estates Il 60192
USA
john.e.jackson@questintl.com

**Gerhard Krammer**
Symrise GmbH
Mühlenfeldstrasse 1
37603 Holzminden
Germany
gerhard.krammer@symrise.com

**Heinz-Jürgen Lötgenbörger**
Ringstrasse 27
64404 Bickenbach
Germany
heinz.loegtenboerger@doehler.com

**William J. N. Marsden**
Van Lenneplan 6a
1217 NC Hilversum
The Netherlands
Bill.Marsden@KerryBioScience.com

**Stefan Marx**
Stetteritzring 106
64380 Roßdorf
Germany
marx@n-zyme.de

**Günther Matheis**
Charlottenstrasse 34
37603 Holzminden
Germany

**Jan-Pieter Miedema**
Seekatzstrasse 27E
64285 Darmstadt
Germany
jan-pieter.miedema@doehler.com

**Armin Mosandl**
Institute of Food Chemistry
University of Frankfurt
Max-von Laue-Strasse 9
60438 Frankfurt am Main
Germany
mosandl@em.uni-frankfurt.de

**Dirk Achim Müller**
Takasago Europe GmbH
Industriestrasse 40
53909 Zülpich
Germany
dirk.mueller@takasago.de

**Stefan Palzer**
Stiegerstrasse 11
78377 Öhningen
Germany
stefan.palzer@rdsi.nestle.com

**Howard D. Preston**
140 Sandhurst Lane
Ashford
Kent, TN25 4NX
United Kingdom
howard.preston@tiscali.co.uk

**Jens-Achim Protzen**
Paul Kaders GmbH
Eschelsweg 27
22767 Hamburg
jens-achim.protzen@kaders.de

**Maren Protzen**
Paul Kaders GmbH
Eschelsweg 27
22767 Hamburg
Germany
maren.protzen@kaders.de

**Karl-Werner Quirin**
Haustadter-Tal-Strasse 52
66701 Beckingen-Haustadt
Germany
wq@flavex.com

**Martin Reichelt**
Erich Ziegler GmbH
Am Weiher 133
91347 Aufsess
Germany
mr@erich-ziegler.de

**Andreas Rossmann**
Isolab GmbH
Woelkestrasse 9/1
85301 Schweitenkir hen
Germany
Isolab_GmbH@t-online.de

Index of Authors

**Uwe-Jens Salzer**
Carl-Diem-Weg 34
37574 Einbeck
Germany
Uwe-Jens.Salzer@email.de

**Matthias Sass**
Rudolf Wild GmbH & Co. KG
Rudolf-Wild-Strasse 107-115
69214 Eppelheim
Germany
matthias.sass@wild.de

**Eckhard Schildknecht**
Neunkirchen 49a
64397 Modautal
Germany
Eckhard.Schildknecht@doehler.de

**Hanns-Ludwig Schmidt**
Prielhofweg 2
84036 Landshut
Germany
hlschmidt@web.de

**Peter Schreier**
Chair of Food Chemistry
University of Würzburg
Am Hubland
97074 Würzburg
Germany
schreier@pzlc.uni-wuerzburg.de

**John Shoop**
Red Arrow Products Company LLC
P.O. Box 1537
633 South 20th Street
Manitowok, WI 54221-1537
USA
j.shoop@redarrowusa.com

**Martin Simon**
Im Immental 17
53819 Neunkirchen-Seelscheid
Germany
Martin.Simon@t-online.de

**Joachim Tretzel**
In den Löser 3
64342 Seeheim-Jugenheim
Germany
J.u.S.T@t-online.de

**Garry Underwood**
Red Arrow Products Company LLC
P.O. Box 1537
633 South 20th Street
Manitowok, Wis 54221-1537
USA
g.underwood@redarrowusa.com

**Roland Werner**
Institute of Plant Science
Department of Grassland Science
ETH Zentrum, LFW C48.1
Universitätsstrasse 2
8092 Zürich
Switzerland
rwerner@ipw.agrl.ethz.ch

**Rob F. van Croonenborgh**
Quest International (Nederland) B.V.
28 Huizerstraatweg
1411 GP Naarden
The Netherlands
rob-van.croonenborgh@questintl.com

**Chahan Yeretzian, PhD, MBA**
Chemin du Miroir 14
1090 La Croix (sur Lutry)
Switzerland
Chahan.yeretzian@nespresso.com

**Günther Zach**
Gabeläcker 15
86720 Nördlingen
Germany
guenther.zach@symrise.com

**Herta Ziegler**
Rehleite 16
95445 Bayreuth
Germany
hz@erich-ziegler.de

**Manfred Ziegler**
Erich Ziegler GmbH
Am Weiher 143
91347 Aufsess
Germany
mz@erich-ziegler.de

# List of Companies / Institutes and Authors

**DöhlerGroup**
Riedstraße
64295 Darmstadt
Germany
www.doehler.com

*Brand J.*
*Lötgenbörger H.-J.*
*Miedema J.-P.*
*Schildknecht E.*
*Tretzel J.*

**Erich Ziegler GmbH**
Am Weiher 133
91347 Aufsess
Germany
www.erich-ziegler.com

*Feger W.*
*Reichelt M.*
*Ziegler H.*
*Ziegler M.*

**Firmenich SA**
Rue de la Bergère 7
P.O. Box 148
CH-1217 Meyrin 2
Switzerland
www.firmenich.com

*Cerny C.*

**FLAVEX Naturextrakte GmbH**
Nordstrasse 7
66780 Rehlingen
Germany
www.flavex.com

*Gerard D.*
*Quirin K.-W.*

**German Research Center for Food Chemistry**
Lichtenbergstrasse 4
85748 Garching
Germany
www.dfa.leb.chemie.tu-muenchen.de/

*Grosch W.*[*]

**Givaudan SA**
5, Chemin de la Parfumerie
1214 Vernier
Switzerland
www.givaudan.com

*Grab W.*[*]

---

[*] Address for correspondence see 'Index of Authors'.

**Ingenieurbüro für Verfahrenstechnik**  
Postfach 1160  
53810 Neunkirchen-Seelscheid  
Germany

*Simon M.*

**Institute of Biological Chemistry**  
TU Munich  
Vöttinger Strasse 40  
85350 Freising-Weihenstephan  
Germany  
www.lrz-muenchen.de/

*Rossmann A.*  
*Schmidt H.-L.*[*]  
*Werner R. A.*

**Institute of Food Chemistry**  
University of Würzburg  
Am Hubland  
97074 Würzburg  
Germany  
www.pzlc.uni-wuerzburg.de

*Schreier P.*

**Institute of Food Chemistry**  
University of Frankfurt  
Max-von-Laue-Strasse 9  
Room 3.07 / N210  
60438 Frankfurt am Main  
www.uni-frankfurt.de/fb/fb14/LMCFFM/en/index.html

*Mosandl A.*

**Isolab GmbH**  
Laboratory Schweitenkirchen  
Woelkestrasse 9/1  
85301 Schweitenkirchen  
Germany  
www.isolab-gmbh.de/index2.html

*Rossmann A.*  
*Schmidt H.-L.*

**Kerry Bio-Science Europe**  
Veluwezoom 62  
1327 AH Almere  
The Netherlands  
www.kerrygroup.com

*Marsden W. J. N.*

---

[*] Address for correspondence see 'Index of Authors'.

List of Companies / Institutes and Authors                                                    XXIII

**Nestlé Product Technology Center (PTC)**              *Blank I.*\*
Lange Strasse 21                                        *Palzer S.*\*
78221 Singen                                            *Yeretzian C.*\*
Germany
www.nestle.com


**N-Zyme BioTec GmbH**                                  *Marx S.*
Riedstrasse 7
64295 Darmstadt
www.n-zyme.de


**Paul Kaders GmbH**                                    *Protzen J.-A.*
Eschelsweg 27                                           *Protzen M.*
22768 Hamburg
Germany
www.paulkaders.de


**Quest International**                                 *Ellison A. D.*\*
Huizerstraatweg 28                                      *Jackson J. E.*\*
1411 GP Naarden                                         *Preston H. D.*\*
The Netherlands                                         *van Croonenborgh R. F.*
www.questintl.com


**Red Arrow Products Company LLC**                      *Underwood G.*
P.O. Box 1537                                           *Shoop J.*
Manitowoc, WI , 54220
USA
www.redarrowusa.com/redarrow.html


**Rudolf Wild GmbH & Co. KG**                           *Sass M.*
Rudolf-Wild-Straße 107-115
69214 Eppelheim / Heidelberg
Germany
www.wild.de


**Symrise GmbH & Co KG**                                *Grüb H.*\*
Mühlenfeldstrasse 1                                     *Krammer G.*
37603 Holzminden                                        *Matheis G.*\*
Germany                                                 *Salzer U.-J.*\*
www.symrise.com                                         *Zach G.*\*

---

\*   Address for correspondence see 'Index of Authors'.

**Takasago Europe GmbH**  *Müller A. D.*
Industriestr. 40
53909 Zülpich
Germany
www.takasago.com/

---

\* Address for correspondence see 'Index of Authors'.

# 1 Introduction

## A Dynamic Business With Taste – The Flavour Industry

*Herta Ziegler*

Humans are decisively influenced by their sense of taste and odour and human history is, therefore, closely tied to the development and usage of flavours. Whereas in prehistoric times, only herbs and spices could be employed for flavouring purposes, today a broad spectrum of flavourings is available, not only for use in the individual household, but especially for the production of food on an industrial scale.

The application of all products from the flavour and fragrance industry is solely aimed at enhancing the human striving for increased pleasure and sensual enjoyment. Hedonistic aspects, therefore, form the basis of our industry [1].

The roots of this industry date back to early Egyptian history, as this extraordinarily advanced civilisation was already thoroughly aware of and acquainted with perfumery and the embalming characteristics of certain spices and resins. Simple methods for the distillation and extraction of essential oils and resins were already known in pre-Christian times and subsequently elaborated by the Arabs. Balsamic oils produced by these methods were later on primarily used for pharmaceutical purposes; it was not before the times of the courtly baroque period that fragrance was an aspect of growing importance. In the medieval age, mostly monks were the pioneers in the art of capturing natural essences and transforming them into substances capable of flavouring food [2].

The onset of the industrial production of essential oils can be dated back to the first half of the $19^{th}$ century. After the importance of single aroma chemicals was recognised in the middle of the century, efforts were started to isolate such compounds from corresponding natural resources for the first time. This was soon followed by the synthesis of aroma chemicals. In this context, the most important pioneers of synthetic aroma chemicals have to be mentioned, such as methyl salicylate [1843]*, cinnamon aldehyde [1856]*, benzyl aldehyde [1863]* and vanillin [1872]*, as they constitute the precursors of a rapidly growing number of synthetically produced (nature-identical) aroma chemicals in the ensuing years.

From this starting point, the flavour and fragrance industry first developed in Europe, expanded to the USA and later reached an international scope. Today Western European companies have reconquered the leadership position in this market, which, after the $2^{nd}$ World War, was held by American companies.

Generally, the dynamics of the flavour and fragrance industry mirror the trend of many industrial sectors: the most important representatives of a large number of nationally oriented companies have through mergers, acquisitions and market expansion developed into globally operating multinational enterprises. As a result of this

---

\* year of the first synthesis

concentration process, the number of small and medium-size businesses decreased, a trend that will certainly result in a more uniform, less diverse market. Already an analysis of the year 1995 showed that approximately 65% of the total turnover of the flavour and fragrance industry is achieved by fewer than 10 firms (Fig. 1.1).

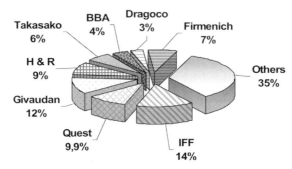

*Fig. 1.1:* Competitors' share of world market (1995) in aroma chemicals, fragrances and flavours (estimated by Haarmann & Reimer) [3]

Also, today analysts estimate the market share of the 'Top Ten' flavour houses at approximately 65% of the entire world market. The preceding decade, often described as the 'Age of Acquisitions', has for the Top Ten of the flavour and fragrance industry resulted in the current market shares depicted in Fig. 1.2.

Givaudan, IFF, Firmenich and Symrise are the contestants for the leadership positions, followed by Quest and Takasago in centre field, while Sensient, Hasegawa, Mane, Charabot and Danisco, with rather similar market share, compete every year to join the higher ranks of the Top Ten. However, it is of considerable importance in this context on which data the respective analysts base their evaluation. Therefore, in the data employed for 2005 [5], sales of non-flavour and fragrance industry items, included by some flavour and fragrance houses in their sales totals, have been subtracted or eliminated from the total sale figures (items eliminated include materials such as sugar, sunscreen chemicals, chemical intermediates, pharmaceutical chemicals, stabilisers, gums, etc.).

Comparison of the sales figures for the years 1995 and 2005 clearly reflect the ongoing changes in the corporate landscape. The merger of the two German flavour giants Haarmann & Reimer and Dragoco to form Symrise has strengthened the company's position in the top ranks. Names that are deeply rooted in and intertwined with the traditions and outstanding developments of the flavour and fragrance industry – such as the vanillin synthesis and the name Haarmann & Reimer (founded 1870) – today remain without contemporary counterpart. Analogously, with IFF's acquisition of Bush Boake & Allen in 2001, the name BBA, considered an invariable constant in Britain, ceased to exist. The pending merger of Givaudan with Quest in November 2006 marks another step towards further market consolidation. Givaudan´s current unrivalled market leadership will certainly be source and aim of other interesting developments in the industry.

The landscape of the big players of the flavour business is still centred on companies with European roots, which, however, all constitute global players.

# A Dynamic Business With Taste – The Flavour Industry

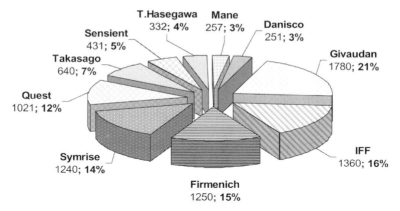

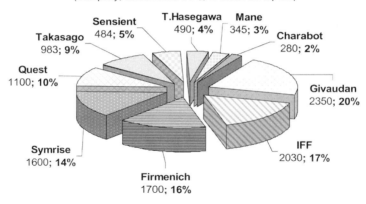

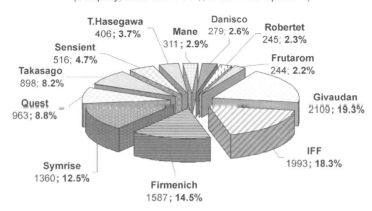

***Fig. 1.2:*** *Competitors' share of world market (2002, 2004 and 2005) in aroma chemicals, fragrances and flavours (calculated by www.leffingwell.com [4])*

These companies are closely followed by a considerable number of international and national manufacturers (not resellers) of flavours and fragrances with sales figures which are sometimes only slightly lower, but often not published as a result of private ownership. Danisco, Ungerer & Co., Robertet, Bell, Shiono, Chr. Hansen, Frutarom, Wild, McCormick, Treatt, Todd and Mastertaste (Kerry) deserve mentioning as examples of a long list of flavour and fragrance companies [4, 5].

These manufacturers are countered by the big purchasing companies, the multinational giants of the food and beverage industry as well as the household and consumer goods sector (Procter & Gamble, Unilever, Nestle, Kraft, Coca-Cola, Pepsi, General Mills, Danone, etc.).

In this context, an analysis of the flavour and fragrance sector along geographic regions and national boundaries is of considerable interest. As a single nation, the USA continues to be the world's largest consumer of flavour and fragrance products [6]. Together with Europe and Japan, the USA accounts for only 15% of the world population, but made up 71% of the overall demand for flavours and fragrances in the year 1999 and 66% in 2004 [www.leffingwell.com]. This clearly reflects the trend of increasing industrialisation usually coupled with a growing demand for flavours and fragrances in other parts of the world, especially Asia. The magical 'A' of Asia has to be granted as much importance in this context as the 'A' of acquisitions, as both 'A-words' decisively influence the investment trends of the flavour and fragrance industry in the beginning 21$^{st}$ century.

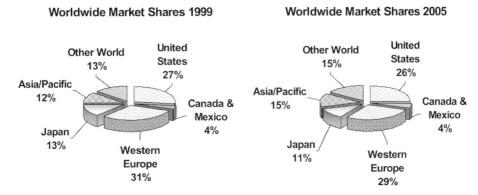

*Fig. 1.3: Worldwide market shares of the flavour industry for the years 1999 and 2005 (estimated by Freedonia; see: www.leffingwell.com/1372pr.pdf)*

The total market, valued at US$ 9.6 billion in 1995, has nearly doubled in the ensuing decade. The share of the typical flavour sector with its classic division into beverages, sweets, dairy and savoury, can only be estimated today and is usually valued at slightly over 40% of the total sales volume. Generally speaking, the global share of the flavour industry on the one hand and the fragrance industry on the other hand can be best approached with an approximate 50:50 ratio.

Since the 1960s both the usage of flavours and fragrances and their general acceptance in a broad array of consumer goods has been continually on the rise. This development in combination with the growing industrialisation in a number of coun-

# A Dynamic Business With Taste – The Flavour Industry

tries and, as a consequence, the predilection for flavours and fragrances does indeed portend well for the flavour and fragrance sector. This industry can realistically look forward to positive expectations and increasing turnover in the future. As far as fragrances are concerned, David J. Rowe has remarked with pleasant cynicism: 'This trend might perhaps suggest we have become afraid of smelling human' [7].

## The Flavour and Fragrance Industry – Sectors and Materials

Basically, three main subdivisions can be distinguished [6]:

- essential oils and natural extracts
- aroma chemicals
- formulated flavours and fragrances.

While **essential oils** and **natural extracts**, which are obtained from natural resources by various processes, mainly constitute complex mixtures, **aroma chemicals** are uniform compounds, which can be both of natural or synthetic origin. A number of representatives of frequently used *aroma chemicals* show an enormous discrepancy between synthetic and natural material. Raspberry ketone shall be used as an example here: for the year 1992, an estimated yearly worldwide consumption of 400 kg of natural material is countered by the 300-fold amount of synthetic material which found industrial usage [8].

*Formulated flavours and fragrances* are complex blends of aromatic materials such as essential oils, aroma chemicals and natural extracts. Depending on their intended usage and the type of flavour release envisioned by product design, they are available in concentrated form, diluted in solvents or bound to carriers.

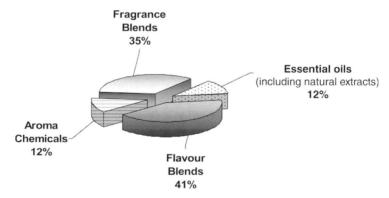

*Fig. 1.4:* Market share of the individual sectors of the flavour and fragrance industry (2002, estimated by Freedonia Group, C&EN estimates)

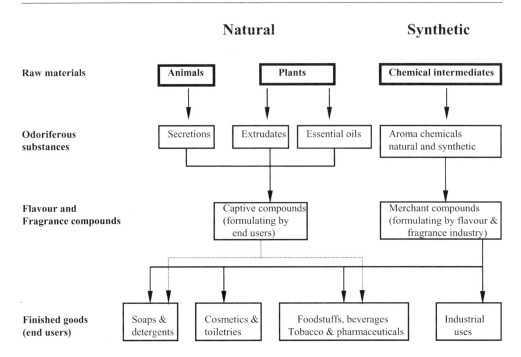

*Fig. 1.5: Industrial usage of flavour and fragrance materials [9]*

## The Flavour and Fragrance Industry – Trends, Expectations, Functionality

The demand for food flavourings has been constantly growing over the last 100 years as a result of the dramatic changes caused by our increasingly industrialised life-style. The shift of food production from the individual household to craftsmen and on to the food industry was accompanied by an increasing need for flavours.

Whereas earlier, technologically caused flavour losses were often the reason for the addition of flavourings, improved technology did not lead to a reduced demand for flavourings. This is a result of changed consumer expectations that went hand in hand with improved standard of living and changed life-styles and philosophy of life [1]. Today this trend can again be observed in new industrially developing countries.

In the 1950s and 1960s, consumers welcomed technological advances and were fascinated by and had a positive attitude towards progress. Better tasting, strongly flavoured food was just as acceptable as new convenience products, which often still required compromises in taste. The acceptance of synthetic materials was all-embracing; this was also the case in the flavour sector.

In the following decades, consumer attitudes changed dramatically: food and its quality evolved into a symbol of personality, expressed by the slogan 'you are what you eat'. Health, fitness and diet became the precursors of all current trends up to the turn of the century. Today, especially wellness, well-being and a well-balanced lifestyle have to be added. The fortification with vitamins and minerals results in products that implicate pharmacological benefits, a trend which is increasingly called for by consumers.

Demographics, therefore, play an increasingly important role in today's flavour industry [10].

The informed chemophobic consumer of the multi-media age of the 1990s was already rather demanding [10, 11]:

- natural, pure, whole
- freshness
- vegetarian products
- ethnic foods
- high fibre content
- high vitamin content
- low calories
- low fat
- low cholesterol
- low caffeine
- low nicotine

All these attributes and a number of others continue to characterise the current food trends. Additionally, **health, wellness, variety** and **anti-aging** are the major driving forces of today's **functional foods**. Never before has the consumer been so sensitive to the correlation between health consciousness, diet and long life, nutrition and fortification with a simultaneous acceptance and growing consumption of better tasting, ready-to-use convenience foods [12, 13].

While the unbroken strength of the focus on 'all natural, food-minus (especially low-fat) and food-plus' continues, we have to add the following aspects which drive our consumer trends today:

- healthy
- low sugar, low carbohydrate, low glycemic (with all aspects of the glycemic index (GI), and GI reference labelling)
- low sodium
- fortification with minerals (calcium on top) and vitamins
- functional
- wholegrain
- organic
- no additives and no preservatives – a very strong recent trend resulting from the discussions on allergies and intolerances
- gluten free
- portion control as an aspect of diet and daily requirements.

The results of all current trends are more and more convenient products which combine many of the actual tendencies (e.g. new soups classified as 'all natural, high fibre, wholegrain, cholesterol and additive-free, fortified with minerals') in products which possess a good window of opportunity for fast and successful market entry. Supported by skilful and clever sales promotion, it is suggested to consumers, especially the youngest ones, that 'it's cool to eat healthy'.

The aspects mentioned above certainly constitute important trends on a worldwide basis; however, it has to be taken into account that the individual trends are valued differently, depending on culture and geographic region. The evaluation of 'Food-Minus' and 'Food-Plus' in the different regions of the world market is depicted as an example in Tables 1.1 and 1.2.

*Table 1.1:* Trends in 'Food-Minus' in different markets (2004) [14]

| Latin America | 1. low calorie | 2. low fat | 3. low sugar | 4. no additives, no preservatives | 5. no cholesterol |
|---|---|---|---|---|---|
| North America | 1. low carbohydrate | 2. low fat | 3. no additives, no preservatives | 4. low sugar | 5. low calorie |
| Asia/Pacific | 1. no additives, no preservatives | 2. low fat | 3. low sugar | 4. low cholesterol | 5. low calorie |
| Europe | 1. low fat | 2. no additives, no preservatives | 3. low sugar | 4. low calorie | 5. low cholesterol |

*Table 1.2:* Trends in 'Food-Plus' in different markets (2004) [14] *

| Latin America | 1. Vit/Min* fortified | 2. all natural | 3. added fibre | 4. wholegrain | 5. added calcium |
|---|---|---|---|---|---|
| North America | 1. all natural | 2. organic | 3. Vit/Min* fortified | 4. vegetarian | 5. add calcium |
| Asia/Pacific | 1. Vit/Min* fortified | 2. add calcium | 3. all natural | 4. vegetarian | 5. functional |
| Europe | 1. Vit/Min* fortified | 2. vegetarian | 3. organic | 4. all natural | 5. gluten free |

Whereas the importance placed on the respective trend attributes varies considerably in different regions, the general tendencies are ubiquitous. Moreover, today's consumer focuses on an interesting, pleasurable, exiting or completely new taste experience. Within the flavour sectors, the developments for beverages took the lead in 2004 with 17% new introductions, followed by bakery products (12%), confectionery (11%), dairy (9%), sauces and seasonings (8%), snacks (8%), meals and meal centres (7%), processed fish, meat and egg products (6%), desserts and ice-creams (6%), side dishes (3%), fruits and vegetables (3%) [14].

---

\* Vitamins and minerals

The key categories of new flavour trends can be divided into three application directions:

- **Salty snacks** with mostly typical flavours (cheese, salt, chilli), hot and new flavours, which indicate potential growth segments (meaty flavours, ethnic flavours in new ways).
- **Juices** with orange being predominant (number one in all regions) or extremely fragmented flavour blends (orange plus other flavours (aloe vera, mango, hibiscus, vitamins fortified)).
- **Sugar confectionery** (strawberry on top in all regions) and regions with very specific flavours and generally a high geographic diversity (chocolate within the top ten of Asia, liquorice (Europe), tamarind (Latin America), sour (North America)).

Additionally, strong increases are predicted for ethnic offerings in meals. Seasonings remain spicy, new beverage flavours come from a variety of sources, and children's flavours continue to be popular.

A new trend is also to surprise consumers with flavours in unexpected categories (banana mayonnaise for children (Asia), or green tea cereals (Japan)); this trend is called *flavour migration*. 'Marrying of good flavour with nutrition' is also predicted.

Therefore, a balance of good taste combined with good nutrition, supplied in 'cool packaging' that appeals to children, seems to show the most effective way for product placement in the future. Additionally there seems to be a revival of comfort foods associated with 'nostalgia', which give the consumer the promise of basic security, familiar classics and casual lifestyle. Indulgence does play a considerable role in the sweets sector: to spoil oneself, easy-to-use small packaging units (e.g. drink desserts) and portion-controlled convenience mini meals which feature daily affordability, and possess considerable marketability *[15]*.

The consumer's expectations towards natural, creative products with sensational effects increase, while the tolerance threshold for accepting expensive brands in the food sector decreases dramatically, especially in Western Europe. This trend is actually a leading one: price restrictions constitute a decisive criterion in each and every product development.

This constitutes a great challenge, not only for the food industry but especially for the flavour and ingredients industry.

## The Flavour and Fragrance Industry – Challenges and Opportunities

In the course of the last decade, this enormous challenge led to nearly revolutionary structural changes, especially in the technological sector. This was the only way to answer the trends towards natural systems, while simultaneously increasing cost effectiveness.

This resulted also in the transferral of biotechnological basic knowledge into large areas of industrial production processes *[16]*. Additionally, gentle, modern technologies, such as reverse osmosis, ultra-filtration, column chromatography and cold extraction processes, were increasingly employed to obtain stable, final products with the utmost degree of naturalness – a driving force of the flavouring and fragrance industry.

Today broad analytical knowledge, the result of the rapid development in the analysis of different matrices, is, thanks to computer technology, omnipresent. From simple gas chromatography assistance up to the highly improved analytical technique of the electronic nose detector – as an example of a relatively new routine analytical approach – modern techniques are available for all areas of flavour creation, technological production and quality control. In the end, the composition of a flavour remains a creative act of art, despite the fact that today scientific knowledge of modern analytical methods is a prerequisite. Based on flavour science, the combination of flavour compositions and building blocks permits the creation of taste sensations tailored for the customer's delight. The recipes resulting from such compositions are today the last well-kept secrets of the flavour houses.

Moreover, it has to be mentioned that our industry was not spared from efforts to reduce costs via suppliers – the well-known 'Lopez Syndrome' of the 1990s *[11]*. As a consequence, the demands of the food industry on its ingredients and the respective suppliers intensified considerably.

This trend became increasingly demanding towards the turn of the century and culminated in the first decade of the $21^{st}$ century. Commercials that celebrate the coolness of greed have transferred this fixation on low price onto overall consumer attitude. However, the balance should not be lost here. As far as flavours are concerned, it should be kept in mind that as a percentage of the total product costs, flavour costs are usually rather low and it is often solely its flavour that accounts for victory or defeat of a product in the market place *[17]*.

For this reason, product design oriented towards the 'Da Vinci Principle' is today considered as the most effective method for creating an innovative new product endowed with optimised properties for market acceptance and penetration. The utilisation of a balance between art, science, logic and imagination, known as the 'Da Vinci Principle' can be utilised in every step of product development to reach higher efficiency through this 'whole-brain' development approach *[18]*.

The intelligent direct confluence of product development in flavour houses and application teams at the customer level constitutes another tool for achieving success and cost effectiveness *[19]*.

The slogan **multifunctionality** *[1]* plays an important role in the 'flavours of the future'.

*Multifunctionality* with regard to the single components will simultaneously lead to simplified process technology and cost reductions and is, therefore, increasingly called for today *[20]*. A lactobacillus culture, which on the one hand imparts a positive mouthfeel effect to a beverage while producing natural stabilisers through its metabolism on the other hand, is just as good an example as thickening agents, which simultaneously have positive effects on stabilisation.

Cooling agents that simultaneously strengthen the flavour of a product should be mentioned in this context. The usage of a variety of different spices can, apart from their flavouring properties, at the same time impart additional benefits to the product as far as preservation, colour and health are concerned. Especially for organically oriented consumers, such ingredients constitute a valued alternative to chemical preservatives and artificial colours [21].

The so-called ***intelligent flavours*** (flavours being liberated when food is prepared or when it is eaten, depending on different factors such as pH value and temperature) have been gaining increasing importance. These high-tech intelligent compounds give access to clearly defined product properties.

In this context, the potential of a number of diverse ingredients with significant potential as *flavour enhancers* or *masking agents* have to be mentioned. In particular, special minus-diets, e.g. low-carbohydrate or low-fat diets, change the taste, texture and sensory qualities of a product and therefore require corresponding alterations to endow the products with the properties called for by the consumer. Flavour enhancers are defined as: 'natural substances which are components of proteins or cell tissue. They have no typical taste or smell, but their presence potentiates other flavours present in the food.' In this field more and more studies are looking at the synergistic abilities of flavour-enhancing substances and the possibility of flavour masking.

Bitter blockers and sweetness potentiators are another field of current importance.

Additionally the new trend of '*kokumi*' has to be mentioned in this context. Special flavours, which add the kokumi taste, are declared to be the 'key to deliciousness'. The Japanese word *kokumi* apparently denotes 'a mixture of different taste or mouthfeel characteristics, including impact, mouthfulness, mildness and taste continuity' [22].

These research interests of the last decade are today partly available in the form of products and will certainly lead to further interesting developments.

The combination of scientific techniques such as genetic engineering, biotechnology, enzymology, physics and electronics will play an important role in the development of new, innovative flavours.

*Multifunctionality* with regard to the ingredients industry today means additional service, food innovations and product design, also from the flavour industry [11]. This part transferral of R&D costs from the food industry into the flavour and ingredient industry requires enormous additional efforts, but constitutes an extraordinary challenge with a high potential to guide the trends towards the favoured products of the flavour industry.

The possibility of gaining market shares for the flavour and fragrance industry by establishing new trend products or by expanding into areas which so far have remained 'unflavoured' constitute only the best known varieties of possible expansion prospects. As other examples from the beverage sector, the manifold new creations of flavoured coffees and ready-made milk drinks as well as the increasing demand for ice teas in Europe deserve mentioning [23]. Additionally, in the beverage sector new beverages borrow flavours from other categories (e.g. peppermint waters as well as brain-twist sensation drinks and '*think-drinks*' with omega 3-fortification).

In particular, the product developments in the sector of the *'free from certified allergens'* products, which guarantee the absence of a group of allergens, are examples of sophisticated foods, which certainly possess growing market potential. This places a double challenge on the flavour industry, as, for example, a tomato-free ketchup certainly has a considerable need for a substantial amount of flavour. Similarly, food additives such as the category *fat replacer* necessarily lead to a higher demand for flavourings in these products, as the fat's loss of taste has to be compensated.

Potential for growth and new perspectives are, therefore, for the flavour and fragrance industry mainly a question of imagination and ingenuity, market observation and skilful marketing. Opportunities abound.

*'Change is occurring in our industries at an ever faster pace. Fast progress is both exhilarating and painful, but the rewards for the company which thrives on the opportunities presented by change are often associated with an accelerated progress towards industry leadership'* [20].

Additionally, the expansion into emerging markets on an international level plays a fundamental role in this context, as saturated markets, such as the USA, only promise trend shifts with small growth rates. The improved standard of living in Eastern Europe and Asia continues to promise an enormous potential of new consumers, which decisively contributes to improved turnover and positive future perspectives *[11]*.

A look at the figures of new introductions in the beverage sector confirms the actual increase in the number of newly introduced products in the years between 2002 and 2004 at an annual worldwide average rate of 20%. In certain regions, such as Latin America, it is not uncommon that the number of innovative products is double that of the preceding year *[24]*.

The constantly falling barriers between cultures, which, sparked by ever increasing mass tourism, led to a boom in ethnic foods in Europe and America, now increasingly expected for developing countries.

Decisive political factors such as the creation of free trade zones with single currencies and shared legislative guidelines offer promising prospects also for the flavour industry with its pronounced orientation towards further globalisation *[16]*.

Within the scope of this book, this glimpse at the dynamic network between the flavour and fragrance industry and the sophisticated consumer of the 21$^{st}$ century illustrates the interesting perspectives for the future of the business with taste. Increasing client demands on flexibility and service will be countered by the flavour industry with improved customer support and by providing complete solutions ranging from 'concept-to-market' to 'creating brands'.

Today the leading flavour companies declare themselves as 'customer-focused and technology-driven' *[17]*.

In the future an all-embracing understanding of 'sensory intelligence, sensory creation, sensory technology and sensory science' will contribute to the success of the flavour and fragrance industry. *'Sensory expertise reveals today how much is still to discover and innovate in our industry'* *[25]*.

## REFERENCES

[1]  Willis B.J., Perfumer & Flavorist, 18 (4), 1-10 (1993)
[2]  Torrell F.M., Perfumer & Flavorist, 29 (3), 16-19 (2005)
[3]  Hartmann H., Perfumer & Flavorist, 21 (2), 21-24 (1996)
[4]  'The Top-Ten', Perfumer & Flavorist, 30 (5), 27-47 (2005), Perfumer & Flavorist, 28 (4), 32-38 (2003), and Perfumer & Flavorist, 31 (10), 22-32 (2006)
[5]  Clark G., personal communication, Dec. 2005
[6]  Somogyi L.P., Chemistry & Industry, 169-173, March 4, 1996
[7]  Rowe, D.J., Chemistry and Technology of Flavors and Fragrances, pp 5-11, Oxford, Blackwell, 2005
[8]  Clark G., Perfumer & Flavorist, 17 (4), 21-26 (1992)
[9]  Global Industry Analysts, Inc., 210 Fell Street, San Francisco, CA 94102, USA; A Global Business Report on Food Additives, October 1996; SBR-070
[10] Abderhalden H., Perfumer & Flavorist, 16 (6), 31-34 (1991)
[11] Hartmann H., Perfumer & Flavorist, 20 (5), 35-42 (1995)
[12] Blake A., Perfumer & Flavorist, 17 (1), 27-34 (1992)
[13] Furth, D.C., Food Technology, 58 (8), 30-34 (2004)
[14] www.mintel.com, IFT2005
[15] Symrise, IFT, Chicago 2003, Perfumer & Flavorist, 28 (5), 14 (2003)
[16] Leccini S.M.A., Perfumer & Flavorist, 19 (6), 1-6 (1994)
[17] Goldstein R.A., Perfumer & Flavorist, 29 (6), 20-24 (2004)
[18] Sucan M., Perfumer & Flavorist, 30 (3) 62-65 (2005).
[19] Bedford J., Perfumer & Flavorist, 30 (6), 32-37 (2005).
[20] Leccini S.M.A., speech presented at the Canadian European Beverage Seminar, Venice, October 19-20, 1989
[21] Raghavan S., Food Technology, 58 (8), 35-42 (2004)
[22] Pszszola D.E., Food Technology, 58 (8), 56-69 (2004)
[23] Sinki G., Perfumer & Flavorist, 19 (6), 19-23 (1994)
[24] www.mintel.com, Jago D., Dornblase L., IFT Tasting Sessions 2005
[25] Andrier G., Perfumer & Favorist, 30 (6), 14-18 (2005)

# 2 Manufacturing Processes

## 2.1 Physical Processes

### 2.1.1 Extraction

*Manfred Ziegler*

#### 2.1.1.1 Introduction

Extraction is one of the oldest techniques known for the production of aromatic mixtures and medicines from plants using water as an auxiliary agent. Extraction vessels excavated in Mesopotamia can be dated back to around 3500 BC. The products were mainly used for medical, religious and cosmetic purposes and the first written extraction procedures found in Egypt, known as the Papyrus Ebers, are famous. Apart from fats and oils, also alcoholic solutions (wine) were employed as solvents for the first time. In the Arab cultures, these extraction techniques were further developed and the first liquid-liquid extractions for decolourising oils or purifying sugar were established. Ethanol produced in medieval times from wine resulted in improved yield and selectivity of extraction processes. In the Renaissance, extraction was extended to a variety of commercial purposes, such as the purification of metals. The rapid development of thermodynamics in the second half of the 19$^{th}$ century led to the formulation of the distribution law for phases in equilibrium by Walter Ernst in 1891. At the end of the 19$^{th}$ century, important developments for continuous solid-liquid and liquid-liquid extraction processes were made. Since the 1930s, this technique has been employed on a large industrial scale in oil raffination and the petrochemical industry. In this context, aromatic compounds are obtained with glycolene and sulfolane via multi-stage continuous extraction from paraffins. Acetic acid concentration, purification of caprolactame and tar oil and tall oil processing constitute other fields of application [1, 2]. In the future, the focus of extraction will be on chemical and biotechnological applications as well as in the area of environmental protection [3]. Extractive processes that result from chemical reactions, e.g. interactions between solvent and carrier matrix, will not be dealt with in this context. Electrokinetic processes, the basis of electrophoresis, are employed for the separation of biomolecules and are of little importance for the flavour industry. The area of liquid-liquid extraction in combination with adsorption has, in contrast, found application in the flavour industry.

#### 2.1.1.2 Solid-Liquid Extraction

The principle of solid-liquid extraction consists in adding a liquid solvent to a solid matrix in order to selectively dissolve and remove solutes. The chosen extractant must be capable of preferentially dissolving the compound to be extracted, forming the miscella.

Solid-liquid extraction is applied on an industrial scale to produce oils and fats from oil-bearing seeds. In the food and flavour industry, extracts and resins, such as hop, chamomile, peppermint, valerian, vanilla, red pepper and liquorice, are obtained from herbs, roots, seeds and drugs. The technology has also found application in the pharmaceutical industry for the extraction of antibiotics, alkaloids and caffeine.

It is characteristic for solid-liquid extraction that no defined distribution coefficient for the distribution of solute in extract and feed exists. Practically, an equilibrium is never reached, as the solid matrix still contains adsorptively bound solute in the capillaries. A quasi-equilibrium is presumed to be achieved when the solution in the capillaries possesses the same concentration as the free solution.

The parameters for solid-liquid extraction are determined by the properties of the matrix to be extracted, such as moisture content, type and amount, reduction ratio, as well as selectivity and amount of solvent.

$$\text{degree of extraction} = \frac{\text{solute in raw material} - \text{solute in residue}}{\text{solute in raw material}} \cdot 100\,\%$$

It is advisable to carry out practical measurements in order to determine the amount of solvent required, yield and plate number for the process. See the literature [4-6] for further details. The selectivity of the solvent is of special importance with solid-liquid extraction. If the solute to be extracted is chemically uniform, the required solvent can be selected based on the similar polarity with the solutes. Table 2.1 provides further details.

Mixtures, such as herb and drug extracts, can be obtained selectively if the polarity of the solvent is similar to that of the solute. To facilitate this process, the solvent should additionally have a low surface tension for wetting and penetrating the solid's capillaries. The following steps can be described for the extraction of herbs [7, 8]:

- permeation of the solvent into the herb cells
- dissolving the solutes by diffusion
- elutriating the extract from the destroyed herb cells.

The section on liquid-liquid extraction provides further details on the considerations which should be taken into account when choosing a solvent.

### 2.1.1.2.1 Maceration

If the matrix to be extracted is only brought into contact with the solvents once, maceration occurs. If maceration proceeds at an enhanced temperature, the process is called digestion. These methods are relatively simple and can be carried out in a number of ways. For trial purposes on the laboratory scale, maceration is called for. The raw material is ground with special mills. Depending on feed material, crushing, grinding or cutting processes are appropriate [9]. Then a suitable solvent is chosen which should function selectively for the constituent to be extracted. In most cases, however, a mixture of solutes is the basis for isolation. After theoretical considerations of the solvent properties [10], tests can be performed. Apart from polarity, high diffusion rate, characterised by viscosity, and intensity of agitation also have to be taken into account. Maceration can be improved by agitating the extraction material, with a mixer in the laboratory and with a stirring device on an industrial scale. If a turbine mixer is employed as homogeniser, both a reduction of size and agitation are achieved, resulting in an improvement of the process.

Treatment with sound waves at the lower range of ultrasound by magnetostrictive frequency generators constitutes another source of agitation. When employing ultra-

# Extraction

**Table 2.1:** Extraction solvent properties

| Name or Trivial name | Formula | d.c. | D | $d_{20}$ | $n_D^{20}$ | $vp_{20}$ | bp | fp | i.t. | $sol_{20}$ | η | Expl. | $δ_{LV}H$ | MAK |
|---|---|---|---|---|---|---|---|---|---|---|---|---|---|---|
| n-Hexane | $C_6H_{14}$ | 1.9 | 0 | 0.66 | 1.3779 | 160 | 68.9 | -22 | 240 | unsol. | 0.31 | 1.1-7.4 | 335 | 50 |
| Cyclohexane | $C_6H_{12}$ | 2.0 | 0 | 0.78 | 1.4264 | 104 | 80.7 | -26 | 260 | unsol. | 0.94 | 1.2-8.3 | 360 | 300 |
| Trichloroethene | $C_2HCl_3$ | 3.4 | 0.94 | 1.46 | 1.477 | 77 | 87 | - | 410 | 1 | 0.57 | 7.9-90 | 240 | 50 |
| Diethyl ether | $(C_2H_5)_2O$ | 4.3 | 1.15 | 0.71 | 1.353 | 587 | 34 | -40 | 170 | 69 | 0.24 | 1.7-36 | 356 | 400 |
| n-Butyl acetate | $C_6H_{12}O_2$ | 5.0 | 1.74 | 0.88 | 1.395 | 13 | 126 | 22 | 370 | 7 | 0.73 | 1.2-7.5 | 311 | 200 |
| Ethyl acetate | $C_4H_8O_2$ | 6.0 | 1.82 | 0.90 | 1.373 | 103 | 77.2 | -4 | 460 | 79 | 0.44 | 2.4-11.5 | 368 | 400 |
| Methyl acetate | $C_3H_6O_2$ | 6.7 | 1.74 | 0.93 | 1.359 | 217 | 57 | -13 | 455 | 240 | 0.38 | 3.1-16 | 438 | 200 |
| Dichloromethane | $CH_2Cl_2$ | 8.9 | 1.58 | 1.32 | 1.424 | 453 | 40 | - | 605 | 20 | 0.43 | 13.0-22 | 330 | 100 |
| Pentanol, amyl alcohol | $C_5H_{12}O$ | 13.9 | 1.65 | 0.81 | 1.411 | 3 | 138 | 48 | 300 | 27 | 3.4 | 1.3-10.5 | 424 | 360 |
| Butan-2-ol | $C_4H_{10}O$ | - | 1.70 | 0.81 | 1.397 | 16.5 | 101 | 24 | 390 | 125 | 3.68 | 1.4-9.8 | 592 | 100 |
| Butan-1-ol | $C_4H_{10}O$ | 17 | 1.66 | 0.81 | 1.399 | 6.7 | 117.2 | 30 | 340 | 79 | 2.95 | 1.4-11.3 | 601 | 100 |
| Methyl ethyl ketone | $C_4H_8O$ | 18 | 2.75 | 0.80 | 1.379 | 103 | 80 | -4 | 514 | 260 | 0.42 | 1.8-11.5 | 441 | 200 |
| Propanol-2 | $C_3H_8O$ | 18.3 | 1.64 | 0.78 | 1.377 | 42 | 82.4 | 12 | 425 | sol. | 2.27 | 2.0-12 | 670 | 400 |
| Propanol-1 | $C_3H_8O$ | 20.1 | 1.66 | 0.80 | 1.386 | 19 | 97.4 | 15 | 405 | sol. | 2.2 | 2.1-13.5 | 697 | - |
| Acetone | $C_3H_6O$ | 21.5 | 2.81 | 0.79 | 1.359 | 232 | 56.5 | -20 | 540 | sol. | 0.32 | 2.5-13 | 525 | 1000 |
| Glycerine triacetate/Triacetin | $C_9H_{14}O_6$ | - | - | 1.16 | 1.43 | <0.1 | 258 | 138 | 433 | 64 | 23 | 1.1-7.7 | - | - |
| Glycerine diacetate/Diacetin | $C_7H_{12}O_5$ | - | - | 1.18 | 1.44 | <1 | 270 | 156 | 425 | sol. | - | - | - | - |
| Triethylene glycol | $C_6H_{14}O_4$ | 23 | - | 1.12 | 1.456 | 0.001 | 288 | 173 | 355 | sol. | - | 0.9-9.2 | 752 | - |
| Ethanol | $C_2H_6O$ | 24.5 | 1.69 | 0.79 | 1.361 | 58 | 78.5 | 9 | 425 | sol. | 1.2 | 3.5-15 | 846 | 1000 |
| 1,2-Propandiol/Propyleneglycol | $C_3H_8O_2$ | 32 | 2.28 | 1.04 | 1.433 | 0.11 | 188 | 99 | 371 | sol. | 55 | 2.4-17.4 | - | - |
| Methanol | $CH_4O$ | 33.7 | 1.69 | 0.79 | 1.329 | 127 | 65 | 12 | 455 | sol. | 0.55 | 5.5-31 | 1100 | 200 |
| 1,1,1,2 Tetrafluoroethane | $C_2H_2F_4$ | - | 2.06 | 1.20 | - | 6685 | -26.1 | - | - | unsol. | 0.202 | - | 218 | - |
| 1,2,3-Propantriol/Glycerin | $C_3H_8O_3$ | 42.5 | 2.67 | 1.261 | 1.474 | 0.01 | 290 | 160 | 400 | sol. | 1400 | 0.9 | 773 | - |
| Water | $H_2O$ | 80.4 | 1.85 | 1.00 | 1.333 | 23 | 100 | - | - | - | 1.0 | - | 2260 | - |

d.c. = Dielectric Constant, DIN 53483
D = Dipole Moment in Debye
$d_{20}$ = Density at 20°C in g·cm⁻¹
$n_D^{20}$ = Refractive Index nD at 20°C
$vp_{20}$ = Vapour Pressure mbar at 20°C
bp = Boiling Point °C at 1013 mbar
fp = Flash Point °C

$sol_{20}$ = Solubility in $H_2O$ at 20°C in g/l
i.t. = Ignition Temperature °C
η = Dynamic Viscosity in cP
Expl. = upper and lower Explosion Limit in Vol%
$δ_{LV}H$ = Heat of Evaporation in J/g at 1013 mbar
MAK = MAK Value in air, ppm = ml/m³

sound, special attention has to be paid to possible changes in the constituents. Piezo-electric donators (quartz pulsation) reach frequencies up to 2400 kHz. It has been reported *[11, 12]* that the range between 25 and 1000 kHz has been tested for extraction. Electric discharges with a broad frequency band can be helpful for accelerating and improving extraction *[13]*. Decomposition of and changes in the overall flavour profile also have to be considered when raising the temperature during extraction (digestion). A temperature of 40 to 50°C should not be exceeded. Finally, a change of the pH value of the extraction material can have an impact on yield and quality.

It is a disadvantage of maceration that no exhaustive extraction is achieved. Moreover, the concentration of the solute is low.

### 2.1.1.2.2 Countercurrent Extraction

In percolation, the solid matrix is repeatedly extracted with fresh solvent, up to depletion. This process is often used both in the laboratory and on an industrial scale, since the valuable solute can be removed to a high degree. In the laboratory, the process is carried out as Soxhlet extraction. The solvent is continuously evaporated and the condensed vapour is introduced into the raw material. The extract is allowed to flow into a solvent reservoir. After a certain extraction time, the feed is exhausted and the solute is enriched in the solvent. The multi-purpose extractor (Fig. 2.1) permits flow and Soxhlet extraction.

***Fig. 2.1:*** *Multi-purpose extractor*

# Extraction

If the solid matrix moves continuously towards the solvent, countercurrent extraction is performed. Depending on the techniques developed in percolation, three different types can be distinguished.

In relative continuous countercurrent extraction, the leaving feed from the extractor is extracted with fresh solvent and the miscella is used for the incoming feed. This process can be carried out in several stages. The solvent is, therefore, moving countercurrently towards the immobile feed. Several extractors operating on this principle have been developed, e.g. the Bollmann extractor and the Rotocel extractor *[14-16]*. They work according to the flow principle. This method requires a good solvent permeability of the feed, but possesses the advantage that the solid matrix is exposed to minimal mechanical stress. Moreover, self-filtration is achieved and the extract has a lower solid content.

In the Bollmann extractor the raw material is fed into buckets which are arranged on a moving belt, similar to a mill wheel. The descending buckets are sprayed partially with enriched solvent. As the buckets rise on the other side of the extractor, the feed leaving on top is sprayed with pure solvent. The enriched solvent flows through the buckets in a countercurrent stream. The exhausted solid matrix is dumped at the top of the moving belt and the buckets move again to the other side.

In the Rotocel extractor, several circularly arranged compartments move over a horizontal perforated plate. Countercurrent is achieved by spraying fresh solvent onto the matrix in the last compartment before dumping through a large opening occurs. The solid matrix is sprayed successively on each preceding compartment with the effluent from the succeeding one. To minimise the solid content, the full miscella sometimes leaves from the compartment preceding the one charged with raw material. The advantage of these extractors is their small space requirement.

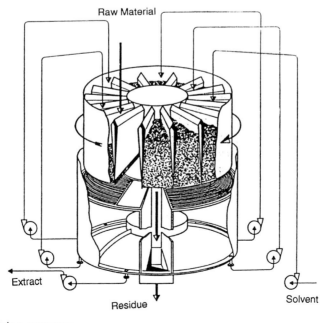

*Fig. 2.2: Revolving extractor*

In discontinuous countercurrent extraction, similar extractors are used as for maceration. However, several units are employed and set in such a way that the solvent has to pass successively in countercurrent mode through the extractors. Different designs for pot extractors have been developed, but all feature numerous lids for filling and dumping the solid feed and they all have considerable space requirements. The pure solvent is again sprayed onto the pot which is dumped in the next step. A variety of operation modes can be achieved through switching. For high enrichment of solute, the leaving miscella is sprayed over the percolator with the fresh raw material.

In contrast to the previously described modes of operation, solvent and feed move continuously towards each other in absolute countercurrent extraction. These extractors, e.g. the screw-conveyor extractor, the Bonotto extractor, the Kennedy extractor and extraction batteries with decanter, all move the solid material and are, therefore, mechanically stress objected. For the miscella this requires extended filtration for removing solids.

In the screw-conveyor extractor, the unit consists of a pressure-resistant cylinder with in-line screws. The solid matrix enters the extractor at the opposite side of the pure solvent. The feed is moved, under the effects of mixing and compaction, to the other end of the screw-conveyor.

In the Bonotto extractor, the extraction tower is equally divided by horizontal plates into cylindrical compartments. Each plate has a radial opening, staggered from the openings of the plates above and below. Each compartment is wiped by a rotating radial blade which allows the feed to enter from the plate above. Therefore, the solid matrix moves from plate to plate from the head of the tower towards the pure solvents at the bottom. At the sump the solids are discharged by a screw-conveyor.

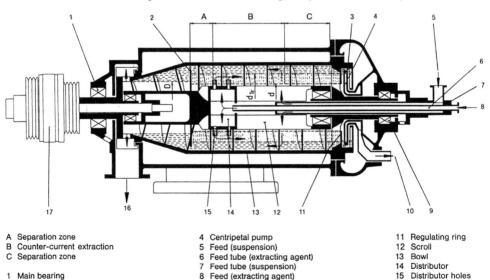

| A | Separation zone | 4 | Centripetal pump | 11 | Regulating ring |
| B | Counter-current extraction | 5 | Feed (suspension) | 12 | Scroll |
| C | Separation zone | 6 | Feed tube (extracting agent) | 13 | Bowl |
|   |   | 7 | Feed tube (suspension) | 14 | Distributor |
| 1 | Main bearing | 8 | Feed (extracting agent) | 15 | Distributor holes |
| 2 | Separating disc | 9 | Main bearing | 16 | Discharge (raffinate) |
| 3 | Centripetal pump chamber | 10 | Discharge (extract) | 17 | Cyclo gear |

*Fig. 2.3: Decanter*

The pulsed tower is of similar construction. Here the solid feed is moved from one disc to the next by pulsation.

In the extraction battery with decanter, several decanters are arranged in succession and the solvent moves countercurrently to the feed.

### 2.1.1.2.3 Work-Up Procedures

After the separation of extract and extraction material, a complete removal of solids and cloudy components has to ensue. This filtration process can be performed with continuous or discontinuous filters or by centrifugation with full-jacket, reciprocal pusher or sieve centrifuges *[17]*. If inflammable solvents are employed, protective measures against explosion have to be considered for the entire unit. The choice of extraction solvents permitted for usage in the food industry is limited by legislative guidelines. With the exception of an extraction with ethanol or water for usage in the flavour industry (e.g. vanilla extract), the solvent has to be recycled. If the extract and the solvent are present as a liquid matrix, this can be achieved with a distillation process by evaporating the solvent. Polar solvents are removed by thin-film evaporators under high vacuum at low temperatures. To prevent thermal stress solvent mixtures which form an azeotrope can be useful. With non-polar solvents, this is achieved by carrier distillation with steam, also possible under vacuum (see 2.1.3.3.3). Crystallisation of the extracted component may ensue (sugar industry).

It is often necessary to remove all traces of solvent; for further details see chap. 2.1.3. The content of solvent residue can be determined by gas chromatography, especially by head-space measurements with standardisation of the respective solvents.

### 2.1.1.2.4 Quantitative Considerations

Size and separation capacities of the extraction units have to be determined arithmetically. Knowledge of the required amount of solvent is not only important for processing, but also for the subsequent work-up of the extract. Since natural spices and drug extracts constitute complex mixtures, a significant constituent is here selected for calculation purposes and is determined in raw material, extract and residue.

Similar to the height of theoretical plates in distillation (see 2.1.3.3.2.), the number of theoretical extraction stages is important for extraction. Just as with distillation, these stages are equivalent to the number of theoretical solvent equilibria, which are necessary to reach certain concentration conditions. These stages are determined either graphically (Fig. 2.4) or numerically. A simplified method for calculating the theoretical extraction stages has been published by Schoenemann and Voeste *[18, 19]*. Similar to the McCabe-Thiele diagrams in distillation, an operating line is employed for determining the stages in a coordinate system. The following equation is used for establishing the operating line:

$$y_n = \frac{(E_{end} - e_a) + e_a}{(L_{end} - l_a) + l_n} \cdot x_n$$

E: amount of solute in extract
e: amount of solute in feed
L: amount of extract
l: amount of solvent in feed
a: state before extraction
end: state after extraction
n: arbitrary average plate number
y: E/L solute concentration in feed
x: e/l solute concentration in extract

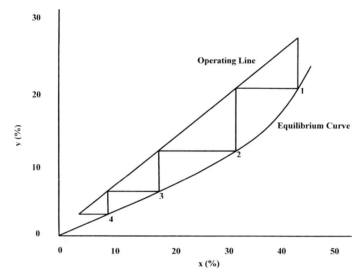

*Fig. 2.4:* Graphical depiction of the theoretical stages

### 2.1.1.3 Liquid-Liquid Extraction

In liquid-liquid extraction, a solvent is added to a liquid matrix (feed) to remove selectively transition components by the formation of two coexisting, immiscible liquid phases. The selected solvent (receiver phase) must be capable of preferentially dissolving the solutes to be extracted and be either immiscible or only partly miscible with the carrier (release phase). This process is, therefore, based on the different affinities of the solute distributing between the two coexisting liquid phases. Of the two phases, the solvent-rich solution containing the extracted solute is the extract and the solvent-lean, residual feed mixture is the raffinate. In the case of a closed miscibility gap, the correlation of the solute mole fraction in the extract and the raffinate phase is called the distribution coefficient (partition coefficient) K:

$K = Y_S/X_S$

In the case of an ideal or very narrow concentration system, this distribution coefficient is a constant in the liquid-liquid equilibrium diagram. See the literature *[20-23]* for further details.

## Fundamentals of Liquid-Liquid Equilibria

The thermodynamic equilibrium requires that the chemical potential of a component in two coexisting phases is equal:

$$\mu'_{i0} + RT \cdot \ln a'_i = \mu''_{i0} + RT \cdot \ln a''_i$$

$$a'_i = a''_i$$

with the activity coefficient defined as: $\gamma_i = a_i/x_i$

$$x'_i \cdot \gamma'_i = x''_i \cdot \gamma''_i$$

$$k_i = \frac{x'_i}{x''_i} = \frac{\gamma''_i}{\gamma'_i}$$

This leads to the relation that the distribution coefficient (partition coefficient) is only influenced by the activity coefficients for both phases and depends only on pressure, temperature and concentration. The relationship of the Gibbsche excess enthalpy to the activity coefficient was used for a variety of modern calculations (Wilson, UNIQUAC equations):

$$\sum x'_i \cdot \ln \gamma'_i = \frac{G^E}{RT}$$

The UNIFAC method considers the liquid phase as a combination of structural elements. It correlates interaction parameters from molecular group structures with the activity coefficient. As an incremental method with a large number of parameters, the UNIFAC method provides a means of calculating liquid-liquid phase equilibria and partition coefficients in multicomponent systems [24-26].

Liquid-liquid extraction has long been a powerful separation technique when the following characteristics are encountered in the system to be separated:

- the boiling points of the components are too close together
- components have very high or low boiling points
- the components are thermally labile
- separation of components belonging to the same compound class or complex mixtures with a large boiling point difference
- the components are present in very low concentration.

These features render the liquid-liquid extraction process a powerful tool for the food and flavour industries.

The suitability of a solvent for a given extraction problem can be assessed by the distribution coefficient as long as the evaluation refers to the separation of the solute. In order to elucidate the solubilities, it has to be determined whether the intermolecular forces present in the liquids are caused by polar molecules with their dipolar interaction, hydrogen-bridge bonding or by non-polar molecules with van der Waals interaction. The impact of such forces is discussed in the literature [27].

The dielectric constants and dipole moment can serve as first indicators for the solvent parameters. Table 2.1 shows that the dielectric constant is low for hydrocarbons and increases with increasing polarity. It also depicts the following data that are of interest for extraction:

- density and refractive index for solvent characterisation
- explosion limits, flash point, ignition temperature, vapour pressure at room temperature and MAK value are safety-relevant data
- water solubility has a considerable impact on solvent polarity
- boiling point and vaporisation enthalpy play an important role when recycling the solvent.

Chapters 3.3.1 and 7.4.4 provides details on legislative guidelines for extraction solvents. The solvent may be a single chemical species, but in some cases the active reagent known as extractant is dissolved in a liquid, the diluent. The diluent is not involved in specific interactions with the solute, but it can also influence the equilibrium and the maximum solute concentration.

Additional criteria for selecting the solvent in liquid-liquid extraction are:

- high extraction capacity for the solute (characterised by the distribution coefficient)
- high selectivity (characterised by the separation factor)
- low solubility with the carrier
- reasonable density difference (for high throughput without flooding)
- reasonable interfacial tension (too low a value leads to formation of emulsion, too high a value to schlieren formation)
- reasonable viscosity for a good mass transfer coefficient.

The selectivity of an extraction is defined as the concentration ratio of the solute and carrier components in the extract phase divided by the same ratio in the raffinate phase:

$$\alpha_{SC} = \frac{Y_S \cdot X_C}{X_S \cdot Y_C} = \frac{K_S}{K_C}$$

A prerequisite for extraction of solutes from one phase to the other is the intensive interface contact of solvent and feed, followed by gravitational separation of the two coexisting phases. In all industrial extraction equipment, one phase is dispersed as droplets in the continuous phase. Thus the density difference is used to facilitate phase separation. The mass transfer rate is defined as:

$$n_c = K \cdot a \cdot \Delta c$$

$n_c$: mass transfer rate of solute
$K$: mass transfer coefficient
$a$: interfacial area per unit volume
$\Delta c$: mean concentration difference

The two liquid phases are in equilibrium if no further changes in concentration occur and thus a theoretical stage is established. The interfacial area per unit volume depends directly on the fractional hold-up and is inversely proportional to the mean drop size of the dispersed phase. The former is influenced by the internals of the extractor, the latter by the physical properties of the two phases, such as interfacial tension and the degree of agitation. A large density difference and high interfacial tension prevent, on the one hand, emulsification and flooding, and, on the other, they obstruct dispersing and reduce the interfacial area. Considering the actual transport process of a single drop, three processes which promote the mass transfer can be distinguished:

- mass transfer during droplet formation (high internal circulation during the new interface formation; new interfaces are build by energy input)
- mass transfer during free drop movement (due to shear forces, turbulent internal circulations are formed when moving through the continuous phase)
- mass transfer during drop coalescence (drop interaction between coalescence and redispersion cycles promotes interface renewal).

Therefore, the following criteria should be taken into consideration for the assignment of dispersed phase:

- the transition component should be transferred from the continuous phase to the dispersed phase
- to reach the highest interfacial area, the phase with the highest flow rate should be dispersed
- the dispersed phase should be easiest dividable and possess the lowest surface tension
- to obtain the smallest droplets, the dispersed phase should not be wetted by the dispersing tool.

However, as mentioned before, chemical interactions are also often involved, which renders the description of the mass transfer more difficult. Hanson [23, 28] has performed investigations of the mass transfer processes at contact surfaces, of the coalescence processes of single drops at the boundaries, as well as of their consequences on practical operation.

Solvent recycling constitutes another important step of extraction. Different physical processes are available for this purpose; thermal separation methods are often chosen as they are cost-effective. Therefore, the solvent should possess a low evaporation energy, its boiling point should differ considerably and it should be non-toxic, non-inflammable and non-corrosive. No solvent fulfils all these requirements and a compromise has to be found by experimental trials.

### 2.1.1.3.1 Single-Stage Extraction

In the case that carrier and solvent are immiscible, the concentration of solute in extract and raffinate can be graphically depicted with the equilibrium curve in the loading diagram. Together with the volumes of feed and solvent, the mass balance for the solute leads to the amount of solute that can be recovered.

Whenever the miscibility of the two phases varies and is dependent on concentration, a triangular diagram is employed (Fig. 2.5). Here the three corners of the equilateral triangle stand for the pure components, the solvent S, the carrier T and the solute C. Each side of the triangle represents binary mixtures, each point within the triangle a ternary mixture. Since the sum of the perpendicular lines of any point in the triangle equals the height of the triangle, the length of these lines corresponds to the concentration of each component.

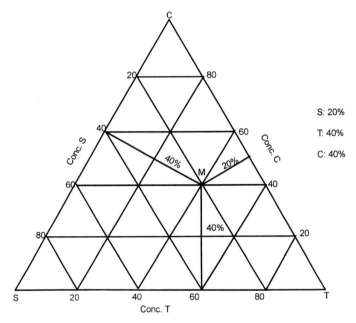

*Fig. 2.5:* Ostwald's triangle diagram

In liquid-liquid extraction at least one miscibility gap between solvent and feed is present. The binodal curve encloses the region of immiscibility (Fig. 2.6). In this area, a mixture with concentration M will separate into two equilibrium phases. The composition of the conjugate phases at equilibrium will lie on the binodal curve at either end of the tie line that passes through the average composition M of the total system. In a triangular diagram also the 'lever-arm rule' applies, where the lengths EM and MR correspond to the relative amounts of raffinate and extract. On the binodal curve, the plait point KP shows where the two conjugate phases disappear and approach each other in composition.

The tie lines and the miscibility gap are strongly influenced by changes in temperature.

A simplified representation of the phase equilibrium is the distribution diagram (Fig. 2.7). As demonstrated, the distribution equilibrium curve can be developed out of the triangular diagram. The slope of the equilibrium curve represents the distribution coefficient K. The position of the binodal curve and its tie lines in the liquid-liquid equilibrium is only determined by the activity coefficient.

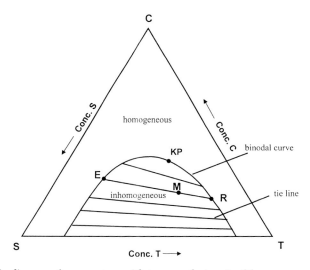

*Fig. 2.6:* Triangle diagram for a system with two partly immiscible components

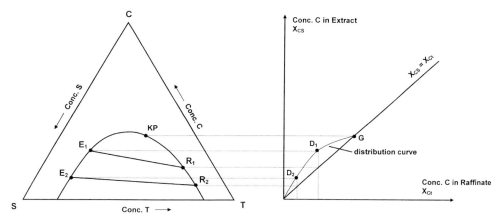

*Fig. 2.7:* Construction of the distribution diagram from the triangle diagram

For a single-stage extraction the following considerations can be made using the triangular diagram (Fig. 2.8). A binary mixture of carrier solvent T and solute C, denoted by the feed concentration F, is to be depleted in solute by an appropriate solvent S. The resulting heterogeneous mixture will separate at equilibrium into two coexisting phases E and R, the concentration of which is determined by the tie line through M.

The selectivity of the extraction can be determined graphically if the concentrations of extract D and raffinate phase G are converted on a solvent-free basis, the ratio of both concentrations represents the selectivity. If this distribution equilibrium is reached, a so-called theoretical stage is present. In reality, the achieved enrichment is far smaller than the theoretically possible one. When designing extractors, not only the theoretically required stages, but also a stage exchange degree, to be determined empirically, has to be taken into consideration. This is of special importance with multi-stage units.

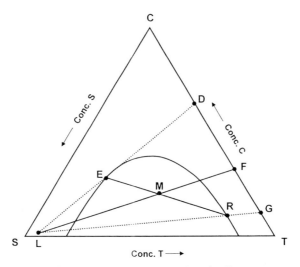

*Fig. 2.8:* Depiction of a single-stage extraction in a triangle diagram

### 2.1.1.3.2 Multi-Stage Liquid-Liquid Extraction

On an industrial scale, emphasis is put on good solute depletion and liquid-liquid extraction is, therefore, carried out in several stages.

In discontinuous cross-current extraction, the solvent is mixed with the feed and subsequently separated; the leaving raffinate is again extracted with fresh solvent. An arbitrary number of extraction stages can follow. The result of a cross-current extraction is obviously determined by the distribution coefficient as well as the solvent ratio. In the case of a high distribution coefficient, the required number of extraction stages is low and the obtained solute therefore has a high concentration. By using large amounts of fresh solvent, a good solute depletion in the raffinate can be achieved. On the other hand, in case of a low distribution coefficient many extraction stages are necessary and the obtained solute concentration decreases rapidly. The graphical determination by employing the triangular diagram will lead to a tie line through M for every stage (Fig. 2.9). The corresponding concentration of solute in extract and raffinate will lie again on the binodal curve.

For the extractive processes in the flavour industry, it is useful to determine an analytically identifiable constituent in the extract after each stage. Calculations then result in the number of actually employable stages.

In the laboratory, a multi-stage liquid-liquid extraction can be performed by a simultaneous distillation-extraction process according to Likens-Nickerson *[29]* (Fig. 2.10). Here, the liquid matrix with the solute in one flask is evaporated together with an immiscible solvent in a second flask. Extraction takes place in the vapour phase where an intensive distribution of both phases is ensured. The condensed vapours from the two phases are separated via a siphon using their different densities and their reintroduction into the original flasks. As the distillation process is continued, extraction is repeated until the solute is exhausted in the original matrix. This method is very useful when traces of non-volatile solutes are present, which are only partly miscible in the liquid matrix. Here carrier distillation lowers the boiling temperature of the solute considerably.

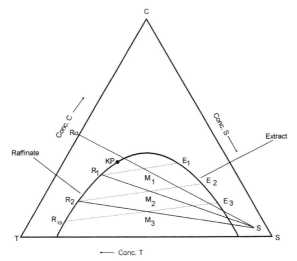

*Fig. 2.9:* Multi-stage cross-current extraction

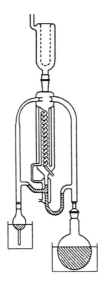

*Fig. 2.10:* Likens-Nickerson apparatus

Vacuum can be applied in order to reduce thermal exposure. The cooling funnel requires a deep-freezing mixture. This extraction method can easily be transferred onto an industrial scale. An important application is essential oils in water where steam distillation is carried out. For the distillative extraction process, different water-immiscible solvents are used. Thermal deterioration and retrieval ratio in the solvent have been studied intensively for fragrance materials *[30]*.

In multi-stage continuous countercurrent extraction, it is characteristic that solvent and feed are continuously moving countercurrently towards each other in the extractor. The fresh solvent first comes into contact with the leaving raffinate and on the opposite side the leaving extract with the introduced feed. This leads to a large loading capacity for both sides and, therefore, to high enrichment of solute in the extract and high depletion in the raffinate. Therefore, the concentration of solute in the extract is much higher compared to cross-current extraction and less solvent is necessary for the depletion of solute in raffinate. Multi-stage extraction is achieved by the addition of the successive single stages with countercurrent flow of feed and solvent.

Depending on the technical construction of these extractors operating in the countercurrent mode, two different classifications can be described: stage-wise or differential contacting of the two countercurrently flowing phases. In a stage-wise extractor, the concentration profile changes stepwise, since in each stage the separated layer of extract and raffinate are newly distributed in the following unit (e.g. mixer-settler battery, Robatel extractor).

In stage-wise liquid-liquid extraction, calculations can be performed from stage to stage. A convenient method for the determination of the necessary theoretical stages or minimum solvent ratio is again graphical depiction (Fig. 2.11). In the known loading diagram, the necessary stages can be determined by inserting the operating line. Together with the mass flow ratio of feed and solvent and the final concentration of solute in the raffinate, an operating line can be drawn into the loading diagram and the necessary steps to reach this value are counted.

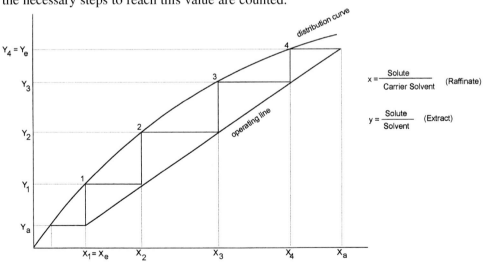

*Fig. 2.11: Loading diagram*

$$F/S = \tan \alpha$$

The slope of the operating line is defined by the solvent ratio. The minimum solvent ratio depicts the operating line in the loading diagram with a common point on the equilibrium curve. Indefinite theoretical stages on the operating line would be necessary to reach this position; therefore the solvent ratio is higher for operational use. Similar to distillation, extraction efficiency in stage-wise extraction units is expressed in HETS (height equivalent of a theoretical stage). The HETS is calculated from the theoretical stages and the total length of the extraction unit:

$$\text{HETS} = H/n_{th}$$

For the projection of an extraction unit, the practical theoretical plate number is determined by dividing the theoretical plate number by the plate efficiency value:

$$S = n_{th}/n_{pr}$$

The height of a single plate in the unit is then defined by the total height of the mass transfer zone and the practical plate number.

In more complicated ternary mixtures, the triangular diagram is again suitable for graphical description (Fig. 2.12). The mass balance for the determination of the point M, depicted previously for the single stage, is now required along the entire unit. The difference between mass flow of extract and raffinate in every cross-section is equal and corresponds to the net mass flow at either end of the apparatus. Since the sum of two amounts in a triangular diagram is represented by a point on a line between them, the difference may be represented by a point on an extended line through them. The corresponding lines for each cross-section originate from one single point which represents the net mass flow at one end. This point P is designated as difference point or pole. The location of this point P can be determined graphically if the inlet phase, the required purity of the raffinate phase and the ratio of raffinate and extract flow are known. Originating from the inlet feed F and solvent S, the point M is determined. As this mixture is also considered to be a mixture of raffinate and extract, outlet extract $E_\omega$ can be found as the intersection between the binodal curve and the line through point M and outlet raffinate $R_\omega$. The location of the difference point is found at the intersection of both lines $E_\omega F$ and $R_\omega S$.

The position of the difference point can also be situated on the other side of the triangular diagram. There are also some restrictions with respect to the necessary minimum solvent/feed ratio. Further details can be found elsewhere [20, 21].

After determination of the difference point, it is possible to determine the necessary number of theoretical stages. Starting from the extract $E_\omega$ the raffinate $R_1$ for the first stage is found by using the tie line through $E_\omega$. A line through P and $R_1$ intersects with the binodal curve and results in $E_2$. This procedure is repeated until the final raffinate point $R_\omega$ is reached.

The example results in three theoretical stages. By modifying the solvent/feed ratio, it is possible to change the number of theoretical stages. A larger number of stages means less solvent flow in relation to feed flow and consequently reduced cost of solvent recovery. Opposite results will be received for increased solvent flow.

In the differential extraction mode, the concentration profile changes continuously as the two phases have no exact stepwise phase separation and a continuous movement towards each other is present. Here, an ideal contacting pattern for the two phases corresponds to a perfect countercurrent plug flow (e.g. extraction towers, Podbielniak extractor). For the determination of extraction efficiency in these kinetic systems, the HTU-NTU (height of a transfer unit and number of a transfer unit) models were developed. The mass transfer model is based on a differential volume element in the column where a thin-film contact for the two phases results in the solute transfer into the extract. The HTU is again defined over the column length H:

HTU = H/NTU

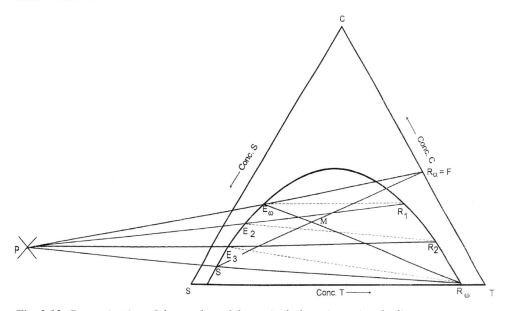

*Fig. 2.12: Determination of the number of theoretical plates in a triangle diagram*

The HTU values for the solute transfer of the releasing (raffinate) and receiving (extract) phase are calculated with the two following equations, where the integrals describe the NTU from the loading difference of the two respective phases:

$$HTU_R \cdot NTU_R = \frac{F}{AK_R a} \cdot \int \frac{dX}{X - X'} = H$$

F/S : flowrate of feed and solvent
K: mass transfer coefficient
$X - X'$: loading difference of the raffinate phase

$$HTU_E \cdot NTU_E = \frac{S}{AK_E a} \cdot \int \frac{dY}{Y' - Y} = H$$

a: interfacial area per unit volume
A: cross-section area
$Y' - Y$: loading difference of the receiving phase

The first HTU term contains the physical and fluid-dynamic parameter and the second NTU term expresses the number of theoretical stages as function of the solute concentration difference. The extractor-specific HTU value is, on the one hand, described by the quotient of flow rate and cross-sectional area of the column, and, on the other hand, it is characterised by the interfacial area per unit volume and the mass transfer coefficient. The former is mainly influenced by drop size and phase hold-up, the latter by the relative movement of the dispersed phase. These characteristic HTU values can be experimentally measured for a certain extractor type and are used for comparison with other extractors or for the projection of larger units.

The NTU values that characterise the concentration profile can be graphically determined if the operation line is parallel to the equilibrium curve in the loading diagram. In this case $NTU_R = NTU_E$ and reflects the theoretical stage number $n_{th}$. A deviation of the ideal plug flow in the continuous and dispersed phase occurs for the following reasons:

- eddy diffusion in axial and radial direction in the continuous phase
- different velocity spectrum over column cross-section
- carry along of small dispersed droplets in axial direction
- broad velocity spectrum of ascending droplets due to the different droplet size.

These phenomena are defined as axial dispersion which reduces the mass transfer. Therefore, additionally a term called HDU (height of diffusion rate) has to be taken into account for the measured HTU':

$$HTU' = HTU + HDU$$

The HDU can also be experimentally determined by measuring the residence time distribution of the two phases in the extractor unit.

### 2.1.1.3.3 Equipment

When performing single-stage extraction on the laboratory scale, the chemist employs a separating funnel as mixing and precipitating vessel *[23, 31]*.

On an industrial scale, the following requirements for extraction equipment are called for:

- generation of small droplets
- high turbulence in the mixing zone
- homogeneous droplet distribution
- generation of high mass transfer coefficient
- prevention of axial back-mixing
- fast phase separation after solute transfer.

In technical applications, mixing of the two phases is achieved by

- blenders, intensive mixers or high-speed mixers
- static blender of a centrifugal or jet pump
- mixing centrifuges
- or, until the distribution equilibrium is reached, with sound waves and electric discharge, similar to solid-liquid extraction *[32]*.

Gravitational forces are used for separating the two phases. Horizontal chambers with mixers are selected for dispersion purposes. In the subsequent settler, the drops coalesce forming a separate layer, resulting in the name mixer-settler. Good extraction efficiency is obtained, apart from high interfacial area, if a certain minimum residence time in the mixer is achieved. In the settler, the dispersed phase must coalesce and form a homogeneous phase layer. Most settlers consist of horizontal vessels, as experience has shown that phase separation efficiency is proportional to the interface area. Improvements in separation have been accomplished by installing settling aids that must be wetted by the dispersed phase. The throughput is therefore dominated by the settling process. The reinforcement ratio of a mixer-settler is in the range of 0.8 to 1. For detailed studies see *[31, 33]*.

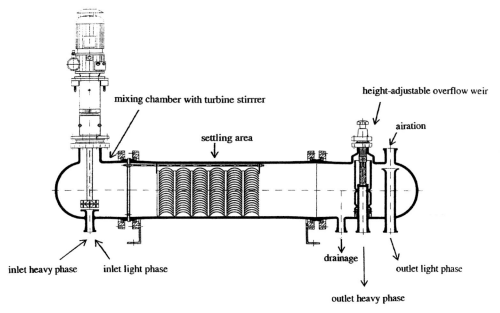

*Fig. 2.13: Mixer-settler extractor*

For high-stage efficiency and rapid phase separation, centrifugal forces constitute a suitable tool. Mixing and separation can be performed with a centrifugal mixer and separator. Self-cleaning or nozzle separators are employed to discharge any solids or sludges present. The former are equipped with a hydraulic bowl-opening mechanism for intermittent solids. Nozzle separators feature continuous sludge discharge.

#### 2.1.1.3.3.1 Extraction Batteries

For this technique, several units are arranged together in such a way that the effluents of the two phases flow countercurrently. All mixing and separation equipment of single-stage extraction can be employed in the corresponding countercurrent mode. Extraction batteries have large space and material requirements. They have, nevertheless, widespread application in the flavour industry, as they can be employed universally. They can also be used with larger extract requirements in multi-stage liquid-liquid extraction and, depending on arrangement, both continuously and discontinuously.

The earliest extraction units were mixer-settlers that involved separate mixing and settling vessels. The disadvantage of these units was their high space requirement, and soon different configurations were developed. In this context, box or tower mixer-settler extractors found a broad range of application *[34]*. The most common unit is the mixer-settler battery consisting of a mixer chamber and an integrated downstream settler chamber. The settler is separated from the mixer chamber by a slotted baffle. The separated phases are removed at the end of the settler chamber according to their densities and then pumped into the next unit. The advantages of mixer-settler units are the simple addition of further stages and the broad loading capacity with the possibility of extreme phase ratios.

### 2.1.1.3.3.2 Centrifugal Extractors

Here centrifugal force is used for mixing and separating the two phases at high flow rates and small density differences are already sufficient. The centrifugal extractors are arranged in such a way that countercurrent flow is achieved. The low volume hold-up as well as the short residence time are big advantages of these units. As described in single-stage extraction, centrifugal mixers and separators are set in series in such a way that countercurrent flow of feed and solvent is established. In the last decade some improved annular centrifugal contactor designs became commercially available *[35]*. This centrifugal extractor operates in a similar manner to a mixer-settler: the entering two liquids are rapidly mixed in the annular space between the housing and the spinning cylindrical rotor and are then pumped through the central opening of the rotor to the centrifugal separator. The mixed phases are accelerated to the rotor speed and separation begins as the liquids are displaced upwards by the self-pumping rotor. The interface between heavy and light phase is adjusted by a heavy phase weir ring and an optimum rotor speed. Good separation efficiency is maintained even with changes in flow rate or phase ratio due to the large dynamic interface zone. For multi-stage extraction these contactors can easily be set in series because the discharge port is higher than the inlets. The further development of the one-stage centrifugal extractors has resulted in technically elaborate, cost-intensive equipment *[36-38]*. As a result of the different design of these extractors, they vary in throughput and in their capacity to separate different density ratios.

In the Podbielniak extractor (Fig. 2.14) rotation is around a horizontal shaft, which is equipped with radial tubes for central and peripheral passage of the entering and leaving liquids. The body of the extractor is a cylindrical drum containing concentric perforated cylinders. The liquids are introduced through the rotating shaft with the help of special mechanical seals; the light liquid is led internally to the drum periphery and the heavy liquid to the axis of the drum. Rapid rotation (up to several thousand rpm, depending on size) causes radial counterflow of the liquids, which are then led out again through the shaft. During operation three zones are formed within the extractor: two narrow zones near the shaft and the rim and a large zone in which the extraction takes place. This principal interface position is adjusted by a back-pressure control of the light-phase outlet. Depending on the ratio of back pressure to light liquid inlet pressure, either the light or heavy phase can be continuous. These machines are particularly characterised by extremely low hold-up of liquid per stage but require a certain density difference.

The Quadronics extractor is a horizontally rotated device in which either fixed or adjustable orifices may be inserted radially as a package. These permit control of the mixing intensity as the liquids pass radially through the extractor.

The Alfa-Laval extractor contains a number of perforated cylinders revolving around a vertical shaft. The liquids follow a spiral path about 25 m long in countercurrent fashion radially and mix when passing through the perforations. Up to 20 theoretical stages can be achieved.

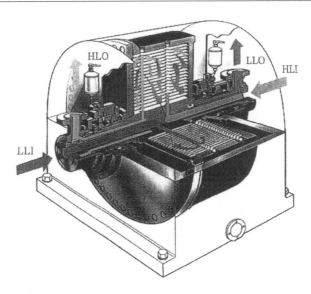

*Fig. 2.14:* Podbielniak extractor

The Robatel extractor (Fig. 2.15) is basically a mixer-settler set above each other, which uses the centrifugal force to reach fast phase separation in each stage. The extractor consists of a rotating bowl divided by baffles into horizontal compartments. Each compartment has connections to lead the separated phases to the next stage. The stationary central shaft has mixing discs and pumps the liquids into the settling part of the stage. The heavy and light liquids are introduced on the opposite side of the extractor to reach a countercurrent flow in the stages. Thereby the liquid volume of a multi-stage unit is reduced to a minimum. Models are available that reach up to seven stages with short contact time.

### 2.1.1.3.3.3 Extraction Towers

These towers are basically similar to those employed in countercurrent distillation (chapter 2.1.3.3.2). In the case of extraction towers, gravitational forces are used for the phase flow. The two immiscible phases enter at opposite ends of the tower. According to their different densities, the light phase is introduced at the bottom and the heavy phase at the head of the tower to realise a vertical countercurrent flow. The introduction of the dispersed phase into the whole cross-section of the column is achieved by distribution units such as nozzles and sprinklers. The dispersed phase is introduced into the tower as small droplets which coalesce again into a homogeneous phase on the opposite side. The entering fluid on the opposite side is called the continuous phase. The columns are characterised by their cross-section area and their height. The diameter of the column influences mainly the throughput capacity and the height influences mainly the extraction efficiency. The maximum throughput of a column is determined by the maintenance of continuous countercurrent flow without flooding. The internals of an extraction tower are constructed in such a way that high hold-up and large interface renewal for the dispersed phase are reached. All fluid elements of the phase should have a narrow residence time distribution in the apparatus.

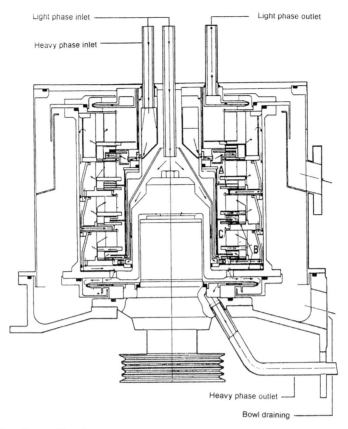

*Fig. 2.15: Robatel centrifugal extractor*

Extraction towers can be basically divided in two groups, with and without energy input.

Packed, sieve-tray and spray towers are used without agitation *[39-41]*.

The simplest extraction tower is the spray tower. The droplets of the dispersed phase are generated only once at the input. The extraction efficiency of these towers is very low, due to the broad range of droplet diameter and poor interface renewal. Additionally, the back-mixing effects increase dramatically by increasing the ratio between diameter and height of the column. The throughput is generally influenced by the density difference and the viscosity of the two phases.

Static sieve-tray columns (Fig. 2.16) have a number of applications. The sieve-tray column is designed according to the same principle as distillation columns with overflow weirs and downcomers. The droplets are reformed on every plate, since the dispersed phase will back-up on the trays, coalescing into continuous layers. By means of suitable drain pipes, a considerable cross-flow is generated in the column which results in additional hold-up for the dispersed phase. Due to the high interface renewal and the smaller back-mixing effects in the trays, the separation efficiency is high and mainly influenced by the height on the back-up layers and the tray spacing.

The throughput through these columns depends again on the density difference and the height of the back-up layer under the plates. The loading range is small.

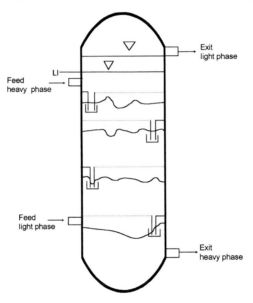

*Fig. 2.16:* Static sieve-tray column

Static packing columns also have an improved mass transfer with less axial dispersion; however, the dispersed phase should not wet the packings. These extraction towers, which depend on density differences, have fairly large throughput and are inexpensive extraction units.

Countercurrent columns with additional kinetic energy input have found a broad range of industrial applications [42-48]. Examples of extraction towers with energy input are pulsed towers, pulsed packed columns and pulsed perforated-plate towers. A number of units with some form of mechanical agitation are also used (Karr column, Scheibel column, Oldshue-Rushton column, Kühni column, RZE extractor, RDC and ARD extractor, Graesser contactor).

As discussed, the supply of mechanical energy reduces the droplet size and increases interfacial turbulence, resulting in a higher theoretical plate number. The energy for agitation can be introduced by pulsation or agitation.

Pulsed extraction towers have been used throughout the industry for many years. Pulsation is generated at the sump of the column using special piston pumps or compressed air. Extensions for phase settling are installed at the bottom and head of the column. The design of the column is rather difficult since back-mixing will occur with increasing diameters.

In pulsed sieve-plate towers, the entire column cross-section is occupied with trays, and thus the lighter phase passes through the holes in the upward stroke and the heavy phase in the downward stroke. This will continuously create new interfaces, which improves the mass transfer. By low pulsation intensities the dispersed phase is discontinuously moving through the holes (mixer-settler mode). The appropriate relation between pulsation intensities and throughput has to be determined empirically in order to operate the tower efficiently. The ideal operation mode where the phase is continuously dispersed throughout the column can be depicted graphically (Fig. 2.17). The product of frequency and amplitude (pulsation intensity) of the pulsator in relation to the throughput per free column cross-sectional area will lead to a flooding curve with a maximum. The area enclosed by the flooding curve will show the allowed operation range. On one side of the maximum, the enclosed area will give the mixer-settler mode and on the other side the dispersion mode. The pulsator frequency is between 30 and 150 strokes/min and the amplitude between 5 and 10 mm. The spacing of the trays and the hole diameter have to be adapted to the physical data in use and will influence throughput and separation efficiency.

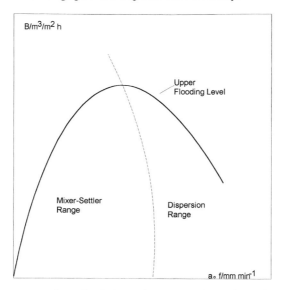

***Fig. 2.17:*** *Operation range of a pulsed sieve-plate extraction column for a certain free cross-sectional area*

Pulsed packed columns have internal packings which should neither have cavities nor be wetted by the dispersed phase. The throughput of these columns as well as the loading range is low but they reach high extraction efficiency. Packed columns also require a minimum density difference for operation. Operation of these towers at high throughput and the pulsation in the dispersion area lead to good extraction efficiency *[42]*. Disadvantages of these columns are the small loading range and the proneness to clogging with sticky products.

The disadvantage of all pulsation towers is the high energy input for moving the whole column content and use is therefore limited for larger units.

This led to the development of reciprocating plate towers which consist of a stack of perforated plates and baffle plates. The column developed by Karr (Fig. 2.18) reaches high maximum loads and is suitable for systems with low interfacial tension *[44]*. Here the perforated plates move up and down driven by an outside motor. The uniform distribution of the energy dissipation across the whole cross-section gives uniform droplet size and low axial mixing which leads to good HTU values.

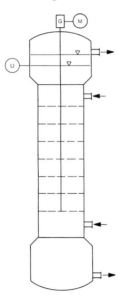

*Fig. 2.18: Karr column*

Agitated extraction towers have in common that the internals are formed into compartments by differently shaped horizontal baffles. Within each compartment, agitation is achieved by the rotation of discs or impellers on a shaft. Without the calming compartments, a high degree of agitation would cause complete back-mixing in all stages of the column. One of the oldest extraction towers with agitators is the Scheibel column. In this column double-bladed agitators are mounted at certain intervals on a vertical shaft in order to achieve phase mixing. In the separation zone between the agitators wire-mesh packings, which should wet the dispersed phase, are installed to improve coalescence. These columns work according to the mixer-settler principle. In the calming zones, three times the height of the mixing zone, the light phase passes upward to the mixing zone. The capacity of these columns is very sensitive to the interfacial properties; high throughput at low speed leads to low efficiency and vice versa. The maximum throughput is low due to the separation zones. At higher column diameters, the efficiency decreases as a result of the broad droplet spectrum generated by the impeller. This has been improved by the Scheibel-York design, where the impeller is surrounded by a shrouded baffle. In addition to lowering the compartment height, better HETS values were achieved.

Advances in development resulted in the Oldshue-Rushton and Kühni extractors *[43, 45]*. The Kühni extractor has shrouded turbine impellers for agitation to promote radial discharge characteristics (Fig. 2.19). The stator discs are made from perforated plates and the residence time can be varied by changing the distance and the hole diameter of these plates. These extractors can be used in the dispersion or mixer-settler mode. Therefore, they can be adapted to extreme phase ratios and reach high theoretical plate numbers. Throughput, which also depends on the mixer speed, is high. A disadvantage is their difficult and expensive installation.

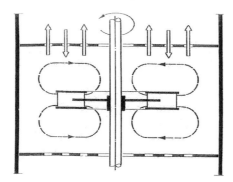

***Fig. 2.19:*** *Kühni extractor*

The RZE extractor (Fig. 2.20) is of similar construction *[47]*. Here, an impeller mounted on a central shaft is used for agitation and the stator discs are again arranged on rods at certain intervals. The small aperture in the centre forms a meander band on the inside ring. The modular design of these internals allows facile extension of the column. The maximum throughput is mainly influenced by the free cross-section of the stator discs. The loading range is broad as the impeller speed can be varied and they show a high theoretical plate number.

Probably the best-known agitated towers are the RDC and ARD contactors *[46, 48]*. In the former, rotating discs are mounted on a central shaft in the tower. Offset against the agitator disc, stator rings, with aperture greater than the disc diameter, are installed at the column walls. These extractors show a broad loading range since the disc's speed is adjustable. In practical applications, certain ratios between the baffle intervals and the column diameter were found but, depending on the physical properties of the phases, these proportions may be varied. Furthermore, they show high throughput but the rotating discs require a certain minimum viscosity of the liquids. The theoretical stage number is low since at higher disc speed the influence of axial mixing is high.

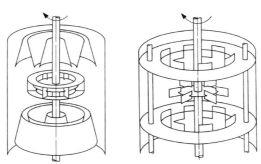

***Fig. 2.20:*** *RZE extractor*

Further improvement was achieved by the ARD extractor (Fig. 2.21). Asymmetric rotating discs are mounted on a shaft that is off-centre in the tower. The mixing zones are separated by horizontal stator rings and additional vertical metal sheets form settling zones. This leads, on the one hand, to lower HETS values but, on the other hand, reduces the maximum throughput. A minimum viscosity of the phase is also a prerequisite for this design. In contrast to the RDC, the ARD column can be easily extended by couplings of the shaft and is available in large diameters.

Finally, the Graesser contactor deserves mentioning. It employs a horizontal format where a series of discs is mounted on a central shaft, with C-shaped buckets mounted between the discs (Fig. 2.22). There is a peripheral gap between the discs and the interior of the shell and longitudinal flow of the phases is through this gap. The heavy phase outlet level is adjusted in such a way that it results in an interface level approximately on the centreline of the unit. In operation, the rotor assembly is slowly rotated and each phase is dispersed, in turn, in the other. The design is virtually unique in not having one phase dispersed and the other continuous throughout.

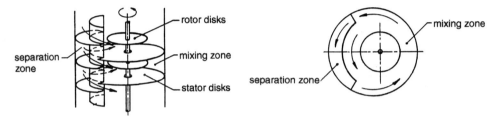

*Fig 2.21: Asymmetric rotating disc contactor*

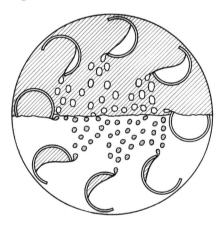

*Fig. 2.22: Graesser raining-bucket contactor*

In the membrane contactor, the interface of the two phases is immobilised on a porous membrane. This membrane must have the ability to separate both phases from each other. Therefore, it is a prerequisite that the membrane is wetted only from one phase and the other phase has a high surface tension towards the membrane *[49, 50]*. Since the pore size is <1.5 µm and the thickness of the wall with is very small (<100 µm), the membrane pores are completely filled with wetted phase due to capillarity. Therefore, the interface is between the two liquid phases on the pore entrance from the non-wetted phase. To prevent penetration of the wetted phase, a higher pressure on the non-wetted phase has to be applied. With this pressure, it is guaranteed that the interface is immobilised in the pores. For this reason, it is necessary to evaluate the suitability of the membrane material for the extraction system used.

In general, membrane-supported liquid-liquid extraction is offered as a micro-porous hollow fibre module (Fig. 2.23). The membrane contactor contains thousands of micro-porous hollow fibres knitted into an array that is wound around a distribution tube with a central baffle. The hollow fibres are arranged in a uniform open packing allowing the utilisation of the total membrane surface area. The liquid flows over the shellside (outside of the hollow fibre), is introduced through the distribution tube and moves radially across the array of hollow fibres and then around the baffle and is carried out by the collection tube.

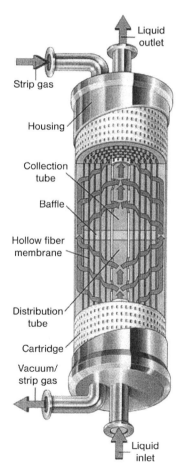

*Fig. 2.23:* Membrane contactor

The mass transfer is governed by molecular diffusion through the pores and the interface phase equilibrium. The significant difference to all previously described extraction units is that no dispersion between both phases is necessary, which leads to the following characteristics:

- no density difference is necessary
- no formation of emulsions
- no phase separation
- every phase ratio is possible
- the fluid dynamic is not restricted by flooding.

As a result of the construction of the membrane module, the interface area for mass transfer per volume is very high compared to extraction towers and is not influenced by flow volumes. The flow volumes are only restricted by the phase breakthrough into the other phase caused by the pressure loss along the contactor. HTU values are lower compared to other extraction units; therefore the membrane contactor has higher extraction efficiency. HTU values increase with higher loading of the membrane module. To reach a higher theoretical plate number, comparable to extraction towers, more module units have to be set in series.

With all extraction columns, especially those with movable internal fittings, success is mainly dependent on construction and, therefore, on the supplying manufacturers. Close cooperation is necessary for all experimental trials regarding the particular problem to be solved.

### 2.1.1.3.3.4 Selection of Companies Supplying Equipment

**Solid-Liquid Extraction**

Alfa Laval GmbH, Wilhelm-Bergner-Str. 1, D-21509 Glinde, Germany
Crown Iron Works Co., PO Box 1364, Minneapolis, MN 55440-1364, USA
De Dietrich Process Systems, SAS, PO Box 8, F-67110 Zinswiller, France
E & E Verfahrenstechnik GmbH, Düsternstr. 55, D-48231 Warendorf, Germany
Flottweg GmbH & Co. KGaA, Industriestraße 6-8, D-84137 Vilsbiburg, Germany
F.T. Industrial Pty. Ltd, 680 Pacific Highway, Killara, NSW-2071, Australia
i-Fischer Engineering GmbH, Dachdeckerstr. 2, D-97297 Waldbüttelbrunn, Germany
Innoweld Metallverarbeitung, Industriepark, A-8682 Mürzzuschlag-Hönigsberg, Austria
NORMAG LABOR- und PROZESSTECHNIK GmbH, Auf dem Steine 4, D-98683 Ilmenau, Germany
Normschliff Gerätebau Dr. Friedrichs – Dr. Matschke GmbH & Co. KG, Hüttenweg 3, D-97877 Wertheim, Germany
PRUESS Anlagentechnik GmbH, Blumenstr. 24, D-85283 Wolnzach, Germany
QVF Engineering GmbH, Hattenbergstr. 36, D-55122 Mainz, Germany
Rousselet Robatel, Avenue Rhin, F-07104 Annanay, France
Schrader Verfahrenstechnik GmbH, Schleebergstr. 12, D-59320 Ennigerloh, Germany
SITEC-Sieber Engineering AG, Aschbach 621, CH-8124 Maur/Zürich, Switzerland
Sulzer Chemtech AG, Hegifeldstr. 10, CH-8404 Winterthur, Switzerland
Uhde GmbH, Friedrich-Uhde-Str. 15, D-44141 Dortmund, Germany

Uhde High Pressure Technologies GmbH, Buschmühlenstr. 20, D-58093 Hagen, Germany
Westfalia Separator AG, Werner-Halbig-Str. 1, D-59302 Oelde, Germany

**Liquid-Liquid Extraction**

Alfa Laval GmbH, Wilhelm-Bergner-Str. 1, D-21509 Glinde, Germany
B & P Process Equipment, 1000 Hess Avenue, Saginaw, MI 48601, USA
De Dietrich Process Systems, SAS, PO Box 8, F-67110 Zinswiller, France
E & E Verfahrenstechnik GmbH, Düsternstr. 55, D-48231 Warendorf, Germany
i-Fischer Engineering GmbH, Dachdeckerstr. 2, D-97297 Waldbüttelbrunn, Germany
Flottweg GmbH & Co. KGaA, Industriestraße 6-8, D-84137 Vilsbiburg, Germany
Julius Montz GmbH, Hofstr. 82, D-40723 Hilden, Germany
Kühni AG, Gewerbestr. 28, CH-4123 Allschwill 2, Switzerland
NORMAG LABOR- und PROZESSTECHNIK GmbH, Auf dem Steine 4, D-98683 Ilmenau, Germany
Normschliff Gerätebau Dr. Friedrichs – Dr. Matschke GmbH & Co. KG, Hüttenweg 3, D-97877 Wertheim, Germany
QVF Engineering GmbH, Hattenbergstr. 36, D-55122 Mainz, Germany
Rousselet Robatel, Avenue Rhin, F-07104 Annanay, France
Schrader Verfahrenstechnik GmbH, Schleebergstr. 12, D-59320 Ennigerloh, Germany
Sulzer Chemtech AG, Hegifeldstr. 10, CH-8404 Winterthur, Switzerland
TOURNAIRE S.A., Route de la Paoute, F-06131 Grasse Cedex, France
Uhde GmbH, Friedrich-Uhde-Str. 15, D-44141 Dortmund, Germany
Uhde High Pressure Technologies GmbH, Buschmühlenstr. 20, D-58093 Hagen, Germany
Westfalia Separator AG, Werner-Halbig-Str. 1, D-59302 Oelde, Germany

**REFERENCES**

[1] Levey M., Chemistry and Chemical Technology in Ancient Mesopotamia, Amsterdam, Elsevier, 1959
[2] Blass E., Liebl T., Häberl M., Extraktion - ein historischer Rückblick, Chem. Ing. Tech., 69, 431-437 (1997)
[3] Villeumier H., Schlegel J., Burckhart R., Chem. Ing. Tech., 64, 899-904 (1992)
[4] Voeste T., Wesp K., Extraktion von Feststoffen – Ullmann's Enzyklopädie der Technischen Chemie Bd. 2, Weinheim, Verlag Chemie, 1972
[5] List P.H., Schmidt P.C., Technologie pflanzlicher Arzneimittelzubereitungen, Stuttgart, Wissenschaftliche Verlagsgesellschaft, 1984
[6] Ullmann's Encyclopedia of Industrial Chemistry, Weinheim, VCH Verlagsgesellschaft, 1988: Vol. B 3 Unit Operations II, 6-1 Liquid-Liquid Extraction; Vol. B-3 Unit Operations II, 7-1 Liquid-Solid Extraction
[7] Muravev I.A., Ponomarev V.D., Pshukov U.G., Pharm. Chem. J. (Engl. ed.), 5, 102 (1971)
[8] Melichar M., Studien über Gleichgewichtszustände, Pharmazie, 17, 290 (1962)
[9] Samans, H., Zerkleinerungstechnik in der Lebensmittelindustrie, Lebensmitteltechnik, 8, 424(1976)
[10] Riddick J.A., Bunger W.B., Organic Solvents, New York, Wiley Interscience, 1970
[11] Schultz O.E., Klotz J., Versuche zur Verbesserung von Extraktionsausbeuten, 3. Milt. Arzneimittel Forschung, 4, 325 (1954)
[12] Süss W., Hanke H., Pharmazie, 24, 270 (1969)
[13] Issaev I., Mitev D., Drogenextraktion durch elektrische Entladung, Pharmazie, 27, 236(1972)

[14] Vauck R.A., Müller H.A., Grundoperationen chemischer Verfahrenstechnik, Leipzig, VEB Deutscher Verlag, 1978
[15] Kirk-Othmer, Encyclopedia of Chemical Technology, 2nd edn, Vol. 8, p 761, New York
[16] Kehse W., Fest-Flüssig Extraktion mit dem Karusell-Extrakteur, Chemiker Ztg., 94, 56-62 (1970)
[17] Brunner K.H., Separatoren und Dekanter für die kontinuierliche Extraktion, Oelde, Westfalia Technisch-wissenschaftliche Dokumentation No. 4, 1982
[18] Schönemann K., Voeste T., Fette und Seifen, 54, 385-393 (1952)
[19] Figlmüller J.K., Die quantitativen Grundlagen, Oester. Chemiker Ztg., 56, 221(1956)
[20] Sattler K., Thermische Trennverfahren, Weinheim, VCH Verlag, 1995
[21] Treybal R.E., Liquid Extraction, 2nd edn, New York, McGraw-Hill, 1963
[22] Proceedings of the International Solvent Extraction Conference ISEC 74, London, Society of Chemical Industry, 1974
[23] Hanson C., Baird M., Lo T., Handbook of Solvent Extraction, New York, Wiley Interscience, 1983
[24] Sorensen J.M., Arlt W., Grenzheuser P., Vorausberechnung von Flüssig-Flüssig Gleichgewichten, Chem. Ing. Tech., 53, 519-528 (1981)
[25] Gmehling J., Rasmussen P., Fredensland A., Chem. Ing. Tech., 52, 724 (1980)
[26] Sorensen J.M., Arlt W., Liquid-Liquid Equlibrium Data Collection, Vol. V, 3 Bde, Frankfurt, Dechema, 1979, 1980
[27] Hampe M., Chem. Ing. Tech., 50, 647-55 (1978); Chem. Ing. Tech., 57, 669-681 (1985)
[28] Hanson C., Neuere Fortschritte der Flüssig-Flüssig Extraktion, Aarau, Sauerländer Verlag, 1974
[29] Likens S.T., Nickerson G.B., Proc. Am. Soc. Brew. Chem., 5, 5-13 (1964)
[30] Bartsch A., Hammerschmidt F., Perf. & Flav., 18, 41 (1993)
[31] Godfrey J.C., Slater M.J., Liquid-Liquid Extraction Equipment, Chichester, Wiley Interscience, 1994
[32] Bart H.-J., Gneist G., Zielgerichtetes Dispergieren mittels hochfrequenter elektrischer Felder in Flüssig/Flüssig Systemen, Chem. Ing. Tech., 73, 819-823 (2001)
[33] Jeffreys G.V., Davies G.A., Recent Advances in Liquid-Liquid Extraction, Oxford, Pergamon Press, 1971
[34] Mueller E., Flüssig-Flüssig Extraktion, Weinheim, Verlag Chemie, 1972
[35] Meikrantz D.H., Method for Separating Disparate Components in fluid Stream, Patent 4,959,158 (1990)
[36] Todd T., Chem. Eng. Prog., 62, 119-124 (1966)
[37] Hanson C., Chem. Eng., 75 (18), 98 (1968)
[38] Miachon J.P., La Technique Moderne, April 1974
[39] Bailes P., Hanson J., Hughes M.A., Chem. Eng., 19, 86-100 (1976)
[40] Blass E., Goldmann G., Hirschmann K., Mihailowitsch P., Pietzsch W., Chem. Ing. Tech., 57, 565-581 (1985)
[41] Pilhofer T., Goedel R., Chem. Ing. Tech., 49, 431 (1977)
[42] Wolf D., Bender E., Berger R., Leuckel W., Untersuchungen zur Betriebscharakteristik pulsierter Füllkörperkolonnen, Chem. Ing. Tech., 51, 192-199 (1979)
[43] Oldshue, J.Y., Rushton, H., Chem.Eng.Prog., 48, 297 (1952)
[44] Karr A.E., Lo T.C., Chem. Eng. Prog., 72, 68-70 (1976)
[45] Mögli A., Chem. Ing. Tech., 37, 210-213 (1965)
[46] Marr R., Moser F., Husung G., Chem. Ing. Tech., 49, 203-212 (1977)
[47] Pilhofer T., Auslegungskriterien von Rührzellenextraktoren, Wiesbaden, Firmenschrift QVF Glastechnik GmbH
[48] Mišek T., Marek J., Br. Chem. Eng., 15, 202-207 (1970)
[49] Ho W.S., Sirkar K.K., Membrane Handbook, New York, Kluwer Academic, 1992
[50] Schneider, J. et al., Microporous membrane contactors, reprints des 8. Aachener Membran Kolloqiums, 2001

## 2.1.2 Supercritical Fluid Extraction (SFE)

*Karl-Werner Quirin*
*Dieter Gerard*

Extraction and separation processes are basic industrial operations applied in many areas with considerable economic relevance. Supercritical fluids, especially carbon dioxide, are of increasing interest for new separation processes in the fields of foodstuffs, cosmetics and pharmaceuticals.

To take away the mystery from the word supercritical, it should be recognized that a supercritical fluid can be used like any other solvent for maceration and percolation processes. The only restriction is that it must be handled under high pressure, a fact that requires a special and expensive design of the extraction apparatus. This obvious disadvantage however is compensated by many benefits as demonstrated below. Supercritical $CO_2$-extraction of botanical materials is today a well-developed and reliable procedure applied on an industrial scale for about 20 years. The equipment is available turn-key from various suppliers in multi-purpose design or tailor made for special applications.

### 2.1.2.1 Solvent Evaluation

For the design of the extraction process the choice of the right solvent is most important. What are the properties of the ideal solvent? Table 2.2 lists some general criteria that must be observed for solvent evaluation.

*Table 2.2:* Criteria for solvent evaluation

| | |
|---|---|
| – selectivity | – evaporation enthalpy |
| – capacity | – specific heat |
| – stability | – combustibility |
| – reactivity | – flash point |
| – viscosity | – explosion limits |
| – surface tension | – maximal allowable working concentration |
| – boiling point | – environmental relevance |

The solvent selectivity should be as high as possible in order to avoid coextraction of useless or disturbing by-products. If the desired extract is rich in valuable ingredients the direct use is possible without further refining connected not only to additional processing costs but also to thermal stress and product losses. At the same time and mostly in contradiction to the previous point, a high capacity is required for a fast extraction and for limiting the amount of solvent flow that is necessary for quantitative extract recovery.

It is obvious that solvents should have a high stability and no reactivity towards the substances to be extracted since it is a precondition that the solvent can be used in many extraction cycles and since the genuine raw material ingredients should be recovered in unadulterated form.

The ideal solvent is characterized by low viscosity and surface tension that both enable a close contact with the material to be extracted by wetting the surface and by penetrating into small capillaries. The yield of extract is thus increased and the speed of extraction is accelerated.

Since a solvent is only an auxiliary medium, which has to be removed after the separation step, it needs to feature a low boiling point in order to avoid thermal degradation, the formation of off-flavours and the loss of top notes. Low values of the evaporation enthalpy and specific heat, physical properties which determine the energy consumption during solvent recovery are of similar importance.

The ideal solvent should not be flammable or at least should have a high flash point and the narrowest possible explosion interval of mixtures with air. This again is contrary to the requirement for boiling points and evaporation enthalpy. Combustible solvents not only require additional flame and explosion proofing, but bear the imminent risk of hazardous reactions if safety guidelines are not strictly observed.

The maximum allowable working concentration of the solvent in air to which employees may be exposed is regulated by law. Solvents with small toxic potential and health risk have high exposure values. The toxicity and other stability and reactivity aspects are important in terms of environmental relevance, e.g. the amount of solvent that is permitted to be vented into the atmosphere. As such the working concentration of the solvent has an impact on the investment in and operational costs of the solvent recovery system. It determines whether the process needs official permission and to what extent regular inspections are necessary. If the amount of solvent to be vented is not restricted this simplifies very much the design of the whole process, as the different steps do not need to be sealed completely.

### 2.1.2.2 Near Critical Gas Solvents

Supercritical fluid extraction – also referred to as dense gas extraction or near critical solvent extraction – means that the operational temperature of the process is in the vicinity of the critical temperature of the solvent. Since the extraction of herbal raw materials requires non-drastic gentle process temperatures the choice of suitable near critical solvents is limited to pure or partly halogenated $C_1$-$C_3$ hydrocarbons, dinitrogen monoxide and carbon dioxide. All these solvents, especially carbon dioxide, exhibit favourable properties in view of the afore-mentioned aspects.

*Table 2.3: General features of supercritical gas solvents*

| general | $CO_2$ |
|---|---|
| – adjustable selectivity<br>– moderate dissolving capacity<br>– stable and inert behaviour<br>– high diffusion rates and low viscosity<br>– low operation temperature<br>– exclusion of oxygen<br>– easy solvent recycling<br>– no solvent residues | – ideal critical temperature<br>– approved for food application<br>– high purity<br>– inexpensive and readily available<br>– bacteriostatic<br>– not flammable<br>– environmentally safe<br>– no waste stream |

Dense gases in this context have a strong lipophilic selectivity. This means on one hand a reduction towards special substance classes that can be separated, on the other hand it should be recognized that there is no need to replace polar hydroalcoholic solvents which are well established and accepted without restrictions. Moreover, the selectivity of dense gases can be adapted by density change to the separation problem to be dissolved whereas conventional lipophilic solvents have a strictly defined solvent power which cannot be influenced.

According to their high selectivity dense gases have a moderate solvent capacity. However this disadvantage is partly compensated by their favourable mass transport properties. Although their density is comparable to liquid organic solvents, their dynamic viscosity is nearer to the low values of normal gases and their diffusion coefficient is more than ten times that of a liquid *[1]*. These values allow supercritical gases to pass even through finely powdered materials with high mass flow rates. Thus reasonably short extraction times can be realized despite the small capacity.

Dense gases are stable and inert. They are non-reactive towards the extract and they can be recirculated in the process without changing their properties. The excess pressure in the equipment prevents the entry of oxygen and damage by oxidation and the closed extraction cycle excludes the loss of highly volatile top notes.

The solvent recovery takes place under gentle conditions and is in practice automatically included in the solvent circulation. The extract is precipitated simply by lowering the pressure while the temperature is kept constant or even adjusted to somewhat lower levels compared to the extraction stage. Thus there is no thermal strain that might lead to rearrangement or decomposition of delicate plant constituents.

Of course there are no solvent residues left in the products because gas residues disappear very quickly at atmospheric pressure.

In addition to the general benefits, carbon dioxide which is the only solvent of practical relevance for industrial scale processes, exhibits additional advantages. It is a solvent generally recognized as safe (GRAS-status) for the production of food ingredients by the FDA. It is bacteriostatic and not flammable. It is readily available at high purity and it is inexpensive, the price being largely independent from oil price movements.

Carbon dioxide is harmless to the environment and creates no waste products. Considerable expense is thus avoided. The use of carbon dioxide as extraction solvent does not contribute to the greenhouse effect because the gas is obtained as by-product of fermentation processes and chemical reactions. The world-wide $CO_2$-amount in the atmosphere is not increased by extraction processes but only by burning gas, mineral oil and coal.

The attractiveness of supercritical carbon dioxide extraction is shown by the already existing industrial applications of hop extraction, decaffeination of tea and coffee, defatting of cocoa powder, and extraction of herbs and spices and is also demonstrated by the large number of patent applications and scientific publications in recent years.

## 2.1.2.3 Solvent Character of $CO_2$

$CO_2$-extracts are by nature lipophilic products. To give an idea about the substance classes which can be separated, some rules of thumb have been derived from practical experience.

*Table 2.4: Solvent properties of supercritical $CO_2$*

| easily soluble | – small lipophilic molecules < 400 u<br>    hydrocarbons, ethers,<br>    esters, ketones, lactones,<br>    alcohols, i.e. mono- and<br>    sesquiterpenes |
|---|---|
| sparingly soluble | – depending on polarity, substances up to 2.000 u<br>    fatty oils, waxes, resins,<br>    steroids, some alcaloids,<br>    carotenoids, oligomers,<br>    water |
| insoluble | – polar substances<br>    sugars, glycosides,<br>    amino acids, saponins,<br>    tannins, phospholipids<br><br>– polymers and mineral salts<br>    proteins, polysaccharides,<br>    polyterpenes, plastics |

All flavouring and fragrance materials which are comparatively volatile are easily soluble (1-10% by weight), e.g. monoterpenes, phenylpropane derivatives and sesquiterpenes not only the hydrocarbons but also the oxygenated molecules like ethers, esters, ketones, lactones and alcohols. All of which are typical components of essential oils.

The solubility decreases with increasing molecular weight and polarity. Fatty oils, waxes, resins, steroids, alcaloids, carotenoids and oligomers are less soluble (0.1-1% by weight). Also water exhibits low solubility which mainly depends on temperature, e.g. 0.3% by weight at 50°C. Consequently lipophilic $CO_2$-extracts derived from dried plant materials with 10% residual moisture contain small amounts of water. This water however can simply be removed since it is not miscible in the lipophilic extract.

Polar substances like organic and inorganic salts, sugars, glycosides, amino acids, saponins, tannins and phospholipids are completely insoluble; so are all polymers like proteins, polysaccharides, polyterpenes and plastics. This offers the advantage that $CO_2$-extracts are virtually free of these substances, especially of inorganic salts and heavy metals. Also $CO_2$ can be useful for cleaning such insoluble materials i.e. by removing impurities from polymers.

## 2.1.2.4 Selectivity

Supercritical carbon dioxide offers the possibility to change the solvent power within a wide range by adjusting the gas density. This can be done by variation of the parameters for temperature and, more important, for pressure. Liquefied carbon dioxide in contrast is more similar to normal solvents without the possibility to influence the dissolving power.

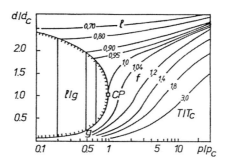

*Fig. 2.24:* Density isotherms as function of pressure (reduced values)
*l* liquid, *g* gas, *f* supercritical, CP critical point

This situation is illustrated in Figure 2.24 where the density is plotted versus the pressure and the lines inside are isotherms. All numbers indicated are reduced values, i.e. they are not absolute but divided by their value at the critical point.

Reduced temperatures below 1.0 are subcritical and the gas becomes liquefied with increasing pressure. The density changes from the low value of the gas phase to the high value of the liquid phase. Then it remains almost constant with rising pressure because the liquid is almost incompressible. As reduced temperatures approach unity the isothermal compressibility of the gas rises rapidly. At values above unity in the supercritical area there is no further liquefaction and the gas density can be adjusted continuously with increasing pressure, which offers the option to adjust the dissolving power.

This advantage together with the fact that supercritical gas at high densities is a better solvent than liquefied gas is the reason why modern extraction plants rather work with supercritical than with liquefied carbon dioxide although the supercritical plant design involves higher pressures and subsequently higher investment costs. There is even a tendency to increase the working pressure from 300 bar before to 500 bar in order to improve further the $CO_2$-solvent power.

Skilful treatment with supercritical carbon dioxide thus can yield extracts with high contents in active principles e.g. from pyrethrum flowers *[2]*, valarian roots *[3]* or chamomile flowers *[4]*. This method produces directly high grade products which do not need further refining after the classical extraction.

Two different types of $CO_2$-extracts can be produced in the field of herbs and spices. These can be characterized as selective extracts and as total extracts. The average data for their extraction are given in Table 2.5.

*Table 2.5:* Average extraction criteria for the production of selective and total $CO_2$-extracts

| type | selective | total |
|---|---|---|
| pressure (bar) | 90-120 | 250-500 |
| temperature (°C) | 30-60 | 40-80 |
| rel. amount of $CO_2$ (kg/kg) | 2-10 | 10-60 |
| yield (%) | 0.5-5 | 5-40 |

Selective extracts obtained in the pressure range around 100 bar contain only small volatile molecules like mono- and sesquiterpenes. Thus they are similar to conventional steam distillates (essential oils). Total extracts recovered in the pressure range around 300 bar contain in addition higher molecular weight lipophilic constituents like fatty oils, resins and waxes and thus are comparable to classical hexane extracts (oleoresins). Consequently supercritical $CO_2$-extraction is the only procedure which produces completely different extracts from one and the same raw material on the same equipment.

Entrainers are often recommended to modify the solvent power and selectivity of supercritical carbon dioxide, this especially before the background to open up the technology for the extraction of more polar components. The entrainer must be completely miscible with the supercritical gas. This is true for most of the conventional solvents, although only water and alcohol are considered to fit into the 'natural' process of $CO_2$-extraction.

It should be recognized that only small amounts of entrainer (up to 5%) are reasonable. Such low levels cannot change very much the polarity of carbon dioxide. If larger amounts are necessary the high pressure process is less viable and it should be replaced by direct alcoholic extraction. Even the small amounts destroy most of the advantages of the pure carbon dioxide as they leave solvent residues in the extract and raw material. Such residues would have to be removed in a second step. Additionally new problems are created in maintaining a strictly defined and constant entrainer concentration during the operation.

Subsequently the use of entrainers is restricted to rare and specific applications mainly to enhance the solubility of substances on the lower edge of extractability.

### 2.1.2.5 $CO_2$-Extraction Process

The process of supercritical $CO_2$-extraction is very simple. A closed gas circulation is divided into high and low gas density. At high gas densities, in the pressure range from 90-500 bar, marked in Figure 2.25 by the thick line, the carbon dioxide takes into solution the substances to be extracted. At low gas densities corresponding to pressures of 40-70 bar this dissolving power is lost, the extract precipitates and the gas is regenerated.

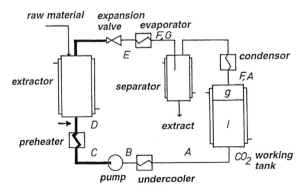

***Fig. 2.25:*** *Diagram of a dense gas extraction plant, letters A-G refer to Figure 2.26*

# Supercritical Fluid Extraction (SFE)

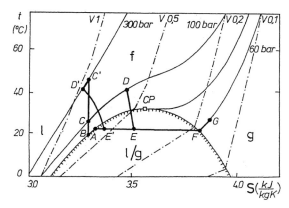

*Fig. 2.26:* Extraction circuit A-G in the t,s-diagram of carbon dioxide
CP critical point, V isochors with density indication, g gas, l liquid, f supercritical

In detail the extraction plant has a working tank that provides the carbon dioxide necessary for the process. In the tank at about 60 bar and ambient temperature the liquefied carbon dioxide is in equilibrium with the gas phase (A). Before entering the pump the liquid gas is cooled to avoid cavitation (B). The pump then isentropically increases the pressure to the extraction value of 100 bar (C) respectively 300 bar (C') for example. In the next step the extraction temperature is adjusted mostly to 40°C (D, D'). Then the dense $CO_2$ is passing through the material in the extractor and takes the lipophilic components into solution.

After leaving the extractor the pressure of the fluid is released by the expansion valve to the low level of 60 bar again by which the gas is cooled down and partly liquefied (E, E'). The liquid part is evaporated (F) and the temperature adjusted to a value near 30°C (G) by flowing through a heat exchanger. The $CO_2$ has now lost its solvent power, the solute is precipitated and collected in the separator. The gas coming out of the separator is regenerated and liquefied (F, A) again in the condenser before flowing back into the working tank and closing the cycle.

All parameters are exactly controlled and regulated in order to have a well standardized extraction procedure. It is obvious from the isochores in the t,s-diagram that under extraction conditions (D) gas densities from 0.6 to 0.9 are realized whereas in the separation stage (G) the density is less than 0.2.

## 2.1.2.5.1 Extraction of Solids

$CO_2$-extraction is best suitable for dry botanical materials, i.e. with a water content of about 10%. For fast and complete extraction the material needs conditioning which is achieved by cutting and powdering mills and by pelletisation which is recommended for materials with low bulk density, i.e. herbs and flowers.

For extraction the gas passes through a fixed bed of the raw material and removes the soluble components from the solid particles in the direction of the solvent flow. The yield vs. the specific solvent consumption in kg $CO_2$/kg substrate gives in the first period a straight line representing the maximum efficiency under the initial conditions. The extraction curve flattens asymptotically in a second phase of the extraction due to a reduced gas loading when the final yield is approached.

In order to increase the efficiency of such a batch process and to reduce the inconvenience of discontinuous operation when the extractor is decompressed and opened for replacing the spent material against fresh one the extraction volume is spread over three or four vessels. These are switched into the gas circulation in a battery-type sequence utilizing the countercurrent principle. This means the extractor containing already depleted material is first contacted with the fresh gas and the extractor with the fresh material containing the full extract concentration is contacted in the second or last position in order to benefit as much as possible from the dissolving capacity of the gas.

The handling can be simplified by providing special quick-acting closure types to the extraction vessels and in some cases by using baskets to bring the raw material into the extractors (Fig. 2.27). We still have the disadvantage of additional energy consumption for recompressing the freshly filled extractors.

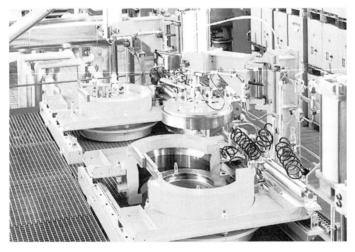

***Fig. 2.27:*** *Supercritical $CO_2$-extraction plant*

This batch mode operation for solids even if quasi-continuous and according to the countercurrent principle affects the economics and restricts the application to products which provide a certain added value or to separation cases which cannot be solved otherwise. For this reason large-scale operations like the production of vegetable oils is still the domain of hexane extraction, but there are many other examples where the $CO_2$-process is highly competitive or offers new unique possibilities and solvent-free products.

### 2.1.2.5.2 Extraction of Liquids

Truly continuous is the extraction of liquids if the extractor is replaced by a column. The liquid is pumped continuously onto the head of the pressurised column and flows down by gravity. Supercritical fluid extraction is normally operated in the so-called droplet regime and not in the film regime. This means the liquid in contact with the

# Supercritical Fluid Extraction (SFE)

supercritical gas breaks up into droplets. A large surface is created by coalescence and redispersion of the droplets by impingement. The type of column packing has no significant influence on the separation efficiency in the droplet regime but can influence the capacity. In most cases the column is equipped with a regular stainless steal wire mesh type of packing.

While the droplets are falling down in the continuous dense gas phase, which is moving in countercurrent mode from the bottom to the top, some components are dissolved and carried out as extract into the separator whereas the insoluble part is collected and withdrawn from the bottom of the column as raffinate. This is illustrated in Fig. 2.28 as trickle flow mode. Examples are the deterpenation of citrus oils, the deacidification of vegetable oils, the separation of alcohol and water and the enrichment of carotenes or of EPA- and DHA-fatty acid esters from fish oils.

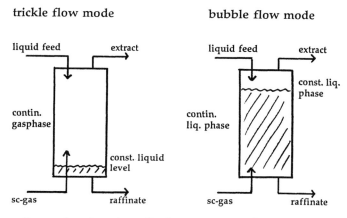

*Fig. 2.28: Operation modes of a column for dense gas extraction*

A different mode of operation, characterised as bubble flow, is possible if the continuous phase is the liquid to be extracted. The liquid level is kept constant at the top of the column and the supercritical gas which is introduced at the bottom is bubbled through the liquid for extraction. This method is mainly used if small amounts of dense gas are sufficient for complete extraction, i.e. if the solvent ratio of dense gas/liquid feed is around 1 kg/kg. Examples are the extraction of flavours from wine and fruit juice where only very small amounts of extract can be expected.

It is a precondition for both operations that the density of the liquid pumped onto the top of the column is higher than the density of the supercritical gas. The difference should be large enough in order to give a proper phase separation. If, at very high pressures, the gas density surpasses the liquid density a so-called barotropic phase inversion can be observed at which the liquid 'swims' on the dense gas.

The liquid feed can also be introduced into the column at an intermediate position which separates the column in an enriching and a stripping section. It is possible to create an internal reflux by temperature change in the upper section of the column or an external reflux for improving the separation efficiency. More details on column operation and some application samples are given in a recent review article *[5]*.

### 2.1.2.6 Extraction of Flavourings

The most important procedures for producing natural flavourings and volatile aromatic oils are steam distillation and selective $CO_2$-extraction, both of which will be compared more in detail.

Steam distillation is based on an aceotropic or carrier-gas distillation of two immiscible liquids. Due to the unfavourable ratio of vapour pressures and thus of mole fractions in the distillate, large amounts of water must be evaporated for the separation of small amounts of essential oils. This is connected to long distillation times at around 100°C and a considerable thermal stress leading to the formation of artefacts, oxidation and isomerisation to a certain extent. Moreover the water itself can be a reactant and hydrolyse terpene esters that make up the core of a flavour; terpene alcohols remain partially dissolved in the water and thus are lost from the essential oil. All this can modify the essential oil composition and change the original typical flavour impression.

Other disadvantages are enzymatic processes especially during the heating-up phase of the water giving some off-flavours and cooking notes. Last but not least some chemicals are often added to the water in order to keep the distillation chamber and heat exchanger clean or to prevent foaming. Also, some essential oils need solvent addition in the collecting vessel for a better phase separation which might be a problem for really pure and natural products.

The positive aspects of steam distillation are the simple method, the universal application and the inexpensive equipment. Also, steam distillation is suitable for fresh plant material and production in place or even in the field is possible.

$CO_2$-extraction, in comparison, is a more complex technology using more expensive equipment than steam distillation. The method has a limited suitability for fresh materials and is not recommended for floral fragrances which should be produced from fresh flowers immediately in place. The $CO_2$-oils are slightly more coloured than steam distillates and they can contain some wax traces.

$CO_2$-extraction is, however, a rapid and gentle procedure working under the exclusion of oxygen which better preserves the genuine composition of the flavouring components. There is no loss of top notes and the back notes which are made up of less volatile and sensitive oxygenated sesquiterpenes are fully preserved as well. This also applies to the core notes since the hydrolysis of terpene esters or the loss of terpene alcohols is largely avoided during $CO_2$-extraction. Thus a representative essential oil composition can be expected in the $CO_2$-extract near to the botanical raw material and with a typical fresh flavour impression. The difference in composition between steam distillates and selective $CO_2$-extracts are presented below based on some practical examples.

#### 2.1.2.6.1 Hydrolysis

The first example is clove bud oil which is composed mainly of caryophylene, eugenol and eugenylacetate. For oil production clove buds were cold milled and one part of the powder was $CO_2$-extracted, the other part steam distilled. The yield was 14.8% in both cases. The oils were analysed by GC-F1D and the composition of the

# Supercritical Fluid Extraction (SFE)

volatile ingredients was calculated by the 100% method. For comparison purposes a third clove bud oil which was steam distilled in origin was also analysed. The eugenol and eugenylacetate content of all three oils is given in Table 2.6.

*Table 2.6: Constituents of different clove bud oils*

|  | clove bud oil (same material) | | clove bud oil (different material) |
|---|---|---|---|
|  | $CO_2$-extracted | steam distilled in lab. | distilled in prod. scale |
| eugenol | 67.2 | 74.6 | 85.0 |
| eugenylacetate | 16.8 | 8.6 | 0.4 |
| total | 84.0 | 83.2 | 85.4 |

Whereas the sum of eugenol and eugenylacetate is fairly constant in all three oils, the eugenylacetate is hydrolysed into eugenol and acetic acid during steam distillation by 50% in the oil produced under gentle conditions on a laboratory scale and almost completely in the oil of different origin distilled on a production scale. The ester degradation is connected to a noticeable flattening of the clove bud flavour impression.

The clove bud oil might be a drastic example for ester degradation since the exceptionally high oil content needs long distillation times and stressing distillation conditions. There are, however, other samples where the same problem is found to a smaller extent, e.g. for linalylacetate which is a main constituent of lavender oil.

Terpene esters are also important in cardamom oil which depending on the raw material has 25-45% terpenylacetate and 3-6% linalylacetate. Cardamom fruits of IA green quality from Guatemala were steam distilled or $CO_2$-extracted and the oils were investigated by gas chromatography for their ester content. Quantification was done with menthol as internal standard and for comparison the values were calculated as milligrams of linalylacetate or terpinylacetate in 100 g of cardamom fruits. $CO_2$-extraction gave 354 mg linalylacetate in 100 g fruits, which was more than three times the amount of the steam distillate of 101 mg in 100 g. Hydrolysis was less pronounced for terpinylacetate giving values of 2147 mg/100 g and 1992 mg/100 g, respectively.

Another interesting example is the breakdown of sabinenehydrate acetate (Fig. 2.29), a typical and important flavour component of marjoram oil. This molecule is easily hydrolysed to sabinenehydrate which is subsequently transformed by dehydration into sabinene, a monoterpene hydrocarbon. Sabinene itself can be isomerised to the more stable terpinenes and terpinolene and even oxidised to terpinene-4-ol. It is clear that this whole reaction chain, which is quantified in Table 2.7 for marjoram oil, means a change in the olfactive property, especially since the sabinenehydrate and its ester are most important for the marjoram oil character.

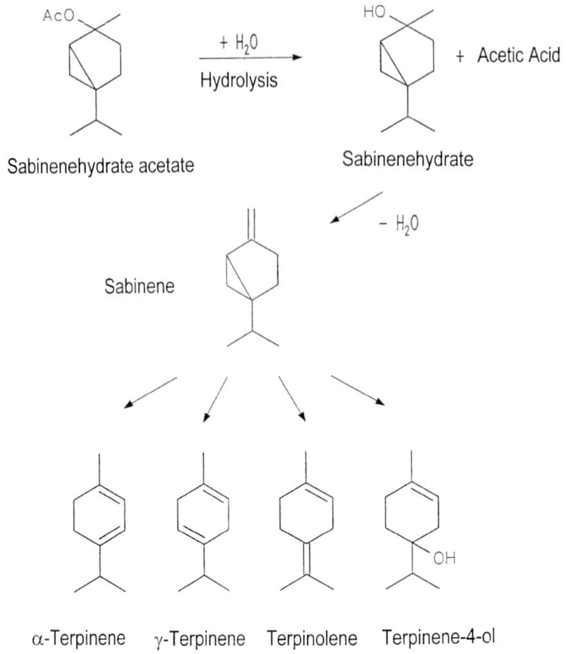

*Fig. 2.29: Schematic of the breakdown of sabinenehydrate acetate*

*Table 2.7: Typical composition of marjoram essential oil (%)*

| substance | $CO_2$ | steam |
|---|---|---|
| sabinenehydrate acetate | 20 | 3 |
| sabinenehydrate | 35 | 20 |
| terpinene-4-ol | 5 | 20 |

### 2.1.2.6.2 Isomerisatian and Oxidation

Fennel oils are composed mainly of phenylpropane derivatives, the different anethol isomers making up 60-80% of the oil, followed by fenchone, a bicyclic monoterpene lactone making up 10-35%, and monoterpene hydrocarbons, mainly limonene, with a content of up to 10%.

The essential oil of fennel seeds from the USA, Germany and India was separated by steam distillation or selective $CO_2$-extraction. Both procedures gave similar essential oil yields of about 6% for the German and US material (bitter fennel) and 2% for the Indian material (sweet fennel). Under stressing conditions the main component, *trans*-anethol (sweet taste), can be isomerised into *cis*-anethol (not sweet), or can be oxidised to anisaldehyde, as shown in Fig. 2.30. Both compounds can simply be detected by HPLC-UV as trace components in fennel oils. Their content should be as low as possible, since fennel oil with more than 2% of anisaldehyde is considered to be spoiled by oxidation and *cis*-anethol has a 15-times higher toxicity compared to the natural *trans*-isomer.

The result of the HPLC analysis is summarised in Table 2.8. It is significant that all three fennel varieties had 10-13 times more anisaldehyde in the steam-distilled than in the $CO_2$-extracted oil. Also, the *cis*-anethol content was 1.2-2.3 times higher in the steam distillates. This finally means a loss in essential oil quality during the steam distillation procedure.

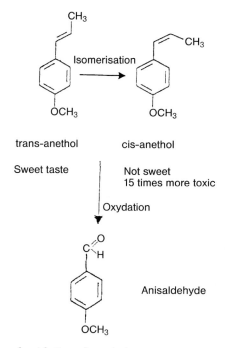

*Fig. 2.30: Isomerisation and oxidation of anethol*

*Table 2.8: Selected trace components in essential fennel oils (%)*

| compound | fennel (Germany) | | fennel (India) | | fennel (USA) | |
|---|---|---|---|---|---|---|
| | steam | $CO_2$ | steam | $CO_2$ | steam | $CO_2$ |
| *cis*-anethol | 0.08 | 0.07 | 0.25 | 0.11 | 0.14 | 0.09 |
| anisaldehyde | 0.19 | 0.02 | 4.15 | 0.30 | 0.89 | 0.06 |

In order to verify further these results the same investigations were carried out with anise seeds from Turkey and star anise from China which both contain *trans*-anethol as the main constituent. The formation of aldehyde during steam distillation was less pronounced compared to the fennel oils, although a 1.6-2.6 times higher content of anise aldehyde was detected in the steam distillates compared to the $CO_2$-extracts. The *cis*-anethol content was increased by 50% for the distillates, which confirms the results found for the fennel oils.

In summary, it can be said that there is less oxidation and isomerisation observed for $CO_2$-extraction compared to steam distillation, which is an important quality feature for $CO_2$-extracted essential oils even if this refers only to trace components in these oils.

### 2.1.2.6.3 Specific Artefacts

There are other cases where steam distillation is responsible for more specific problems of artefact formation which can be avoided if $CO_2$-extraction is applied for essential oil separation. One example is the blue colour of distilled German camomile oil. This is caused by the transformation of matricin, which is the genuine plant ingredient, into chamazulene [4]. Both substances have anti-inflammatory properties but the matricin, which is preserved during $CO_2$-extraction, is believed to have the better efficacy. Also, the sensitive fragrance of camomile flowers is better preserved in the $CO_2$-extract than in the steam distillate.

The degradation of the thermo-sensitive sesquiterpene ketones in calamus oil has also been well investigated. Under the influence of heat the genuine and characteristic acoragermacrones are decomposed into shyobunones [6].

### 2.1.2.6.4 Conclusion

The examples discussed above clearly demonstrate that $CO_2$-extraction is a more gentle procedure than steam distillation. The smaller processing stress widely avoids the formation of artefacts. Therefore $CO_2$-extracts often have a better efficacy or a richer aroma profile reflecting the complete flavour or fragrance spectrum of the herbal raw material. This is also confirmed in the literature where professional flavourists have compared the aroma profiles of $CO_2$-extracts, essential oils and oleoresins for a range of different spices [7]. Moreover $CO_2$-extraction is carried out under precisely standardised and controlled conditions which allow reproducible results. Since $CO_2$-extracts have their own character different from the usual distillates, they are new and powerful means for flavourists and food technologists to modify, improve or boost existing products or to create new premium flavour qualities.

The advantages of gentle process conditions apply also to the production of $CO_2$-total extracts as opposed to solvent oleoresins. Thus it is possible to extract unstable valepotriates from valerian root without decomposition [3] and to extract the bitter principle artabsin from wormwood leaves without transformation to dihydrochamazulenes [8]. Supercritical ginger extracts have a high content of flavour oils and of pungent gingerols but a very low shogaol content which is formed by dehydration of gingerols. Supercritical extraction of antioxidants from rosemary and sage gives a high-grade composition of active diterpene phenols with the genuine carnosolic acid as the main constituent [9]. During other extraction procedures this is more or less decomposed into carnosol, which nonetheless still has antioxidant properties.

Supercritical extracts therefore have a unique and concentrated spectrum of lipophilic ingredients. They have the general advantage of being free of solvents, inorganic salts and heavy metals. They are practically sterile [10] and they need no preservatives since they do not provide a base for germ growth due to the absence of water, proteins and polysaccharides. All this allows a safe application and simple declaration.

Supercritical extracts anticipate legal requirements and the consumer's expectation regarding safety, naturalness and purity in food production, and they set market trends. The advantages described above are summarised in Table 2.9.

# Supercritical Fluid Extraction (SFE)

***Table 2.9:*** *Advantages of $CO_2$-extracts*

| |
|---|
| – no thermal degradation |
| – no hydrolysis |
| – no loss of top notes |
| – full content of back notes |
| – high concentration of valuable ingredients |
| ↔ **superior organoleptic quality** |
| – no solvent residues |
| – no inorganic salts or heavy metals |
| – no microbial activity |
| ↔ **clean products** |
| – conformance with all regulatory requirements |
| – meet consumers' expectation |
| – no export restrictions and simple declaration |
| – compatible with kosher criteria |
| – compatible with certified organic criteria |
| ↔ **safe in the future** |

## 2.1.2.7 Economic Considerations

The extraction costs of solids depend on several factors. One is the specific solvent ratio, kg $CO_2$/kg raw material, which is required to achieve the intended extraction result under optimised process conditions. The usual aim is to separate 85-95% of the valuable extract components; a more complete extraction might not be economic. The solvent ratio is determined by the amount and the solubility of the extract to be separated from the starting material. In most cases the solvent ratio used for extraction is between 5 and 50 kg $CO_2$/kg material.

For economic reasons it is best to do the extraction as fast as possible by increasing the flow rate of the supercritical gas. This is, however, limited by the pressure drop from inlet to outlet of the extractor which should not exceed 10 bar in order to avoid compaction and channelling. The maximal flow depends on the geometry of the extractor and on the nature and particle size of the material. Other limiting factors especially towards the end of the process are diffusion and mass transfer kinetics which require time rather than large amounts of solvent.

For low solvent ratios and very fast extractions, e.g. if the raw material contains only 1% of essential oil to be removed, it should be considered that the exchange of spent vs. new material including de- and re-pressurisation takes 45-90 minutes depending on the size of the extractor even if baskets are used for material handling. In such cases were the extraction time is shorter than the handling time the gas flow can be reduced for energy savings.

Another parameter affecting the economics of the process is the bulk density of the material, i.e. the amount that can be filled in the existing high-pressure volume. Finally the size of the extractor and the total amount of material to be processed are important since it makes a difference in terms of capacity if there are more or less changes of product type in a certain period which all are connected to a thorough cleaning and the exchange of the gas filling in the working tank in order to avoid cross contamination.

A 3 × 500 litre multipurpose extraction plant which is operated 250 days, for 24 hours each day, and with a change of raw material every two weeks has a capacity of 500-600 tons of botanical material a year and the extraction expenses are estimated to be € 3 (±30%)/kg feed in total. Costs of the raw material and raw material conditioning have to be added, as well as the costs of analysis and marketing if applicable. This price can be considerably reduced to about € 1/kg solid feed or less if there are no product changes, if the hardware layout is adjusted to the feed material to be processed, if the extract separation from the gas is optimised and if the scale of the installation is increased.

A breakdown according to cost centres reveals that investment costs (interest and depreciation) are the major cost factor of production at about 40%, followed by personnel, energy, consumables, maintenance and administration expenses. A more detailed description including plant price indices, operating expenses and profitability as well as more details on supercritical extraction mechanisms and modelling of solid botanical matrices and a presentation of the Latin American scenario are given in a recent review article [11].

There are many examples demonstrating that SFE is competitive compared to other procedures for the extraction of solids as well as of liquids on an industrial scale. Supercritical extracts are today no exotic novelties; they are widely included in our daily food, food supplements and cosmetics.

### 2.1.2.8 Other Applications

Apart from the extraction of botanicals with the aim of obtaining a valuable extract (e.g. hops, herbs and spices) or a purified botanical feed material (e.g. decaffeination of green coffee beans and tea leaves, pesticide removal from ginseng roots), the same principle can also be applied to other substrates. Supercritical or liquefied gases are used for cleaning polymers, adsorbents, catalysts and electronic semiconductors, for working up grinding debris of glass or metallic type, for cleaning wastewater and contaminated soils and for refinement of spent mineral oils.

$CO_2$-extraction is in the course of being commercialised in dry cleaning machines for textiles, replacing harmful chlorinated solvents, and impregnation procedures have been developed for the colouring of textiles and plastics and for wood preservation in more eco-friendly processes, all these being huge fields of application. Pressurised $CO_2$ is applied for drying procedures, for tobacco expansion, in processes of pest control and for pasteurisation of fruit juices. Compressed $CO_2$ can be used in heat pumps and air conditioning machines and it is used for mobilising viscous dead oil in mineral oil production. Supercritical water oxidation is a recent procedure for destroying toxic and problematic waste materials.

Supercritical $CO_2$ is used in different procedures for the formation of small particles, and also as an antisolvent to precipitate substances out of a solution in conventional solvents. Such small particles improve the dissolving kinetics of pharmaceuticals and are a precondition for inhalative applications. Dense $CO_2$ is applied as a solvent for reactions and chemical synthesis, e.g. for hydrogenation of vegetable oils, and it is increasingly important for preparative-scale chromatographic separations which require large amounts of solvent that can be simply recycled in the supercritical process.

The driving forces for all these applications, which are partly under development and partly commercialised, are the improved cost effectiveness and legal and environmental pressure for sustainable and non-polluting processes and solvent-free products.

## REFERENCES

[1]  Schneider G.M., Angew. Chem. Ind. Ed., 17, 716 (1978)
[2]  Stahl E., Schütz E., Planta Med., 40, 12 (1980)
[3]  Stahl E., Schütz E., Planta Med., 40, 262 (1980)
[4]  Stahl E., Schütz E., Arch. Pharm., 311, 992 (1978)
[5]  Brunner G., J. Food Eng., 67, 21 (2005)
[6]  Stahl E., Keller E., Planta Med., 47, 75 (1983)
[7]  Hartmann G., ZFL, 42, 502 (1991)
[8]  Stahl E., Gerard D., Parfuem. Kosmet., 64, 237 (1983)
[9]  Gerard D., Quirin K.-W., Schwarz E., Int. Food Market. Technol., 9, 46 (1995)
[10] Manninen P., Häivälä, E, Sarimo S., Kallio H., Z. Lebensm. Unters. Forsch. A, 204, 202 (1997)
[11] del Valle J.M., de la Fuente J.C., Cardarelli D.A., J. Food Eng., 67, 35 (2005)

## 2.1.3 Distillation

*Manfred Ziegler*
*Martin Reichelt*

### 2.1.3.1 Introduction

Distillation and rectification are among the most common physical separation methods, both on a laboratory scale and for industrial production, which have encountered widespread application in the flavour and fragrance industry. It is the intent of the following discussion to familiarise the reader with the basic fundamentals necessary for the daily operation of such units; extensive literature is available on the theoretical and practical aspects of designing and operating such equipment *[1-9]*.

The development of this important thermal separation process can be dated back to ancient times. Until the end of the $18^{th}$ century, however, the structure of the distillation apparatus had remained almost unchanged. It consisted of an evaporation unit, a distillation flask heated by an oven, and a condensation unit with an air- and later water-cooled condenser. In the early $19^{th}$ century, progress in distillation techniques was spurred by the necessity to produce sugar in Europe; this politically motivated development resulted in numerous patents for the production of alcohol. Depending on the different starting materials, a number of distillation units and the first rectification columns were developed in various European countries *[10]*. The $19^{th}$ and $20^{th}$ centuries saw a rapid development of distillation technology prompted by increasing applications in the petrochemical, chemical and pharmaceutical industries.

### 2.1.3.2 Fundamental Considerations

Distillation is a process for the thermal separation of liquid mixtures, where the mixture is separated by boiling the liquids, condensing the vapour and collecting the components according to their boiling points. The fundamental principle of this separation is the fact that the liquid phase possesses a different composition than the corresponding vapour phase *[11]*. The distillation process can be carried out in discontinuous or continuous mode of operation. In discontinuous operation, the distillate is fractionated with an increasing ratio of less volatile components. In the continuous mode, a constant feed is introduced into an instant evaporator and the low-boiling constituents are separated from the high-boiling constituents in the mixture.

Distillation can be carried out as equilibrium, flash or carrier distillation. Equilibrium distillation proceeds under isobaric conditions, whereas flash distillation is achieved by heating the feed at higher pressure and expanding the vapour. In flash distillation, the overheated entering feed is expanded in the vapour-liquid separator where, under adiabatic conditions, the vapour cools down and partial separation with one theoretical stage is achieved. In carrier distillation, boiling of the liquid mixture is facilitated by the addition of a vaporised agent. The agent should not dissolve in the liquid feed in order to add up to the vapour pressure of the components in the mixture for the total operation pressure.

In multistage distillation, successive stages are employed, and in every stage the more volatile compound of the mixture will be present in higher concentration. In order to obtain the more volatile component in higher purity in one stage, countercurrent distillation (rectification) is employed. In rectification, the generated vapour is introduced into a column in which a part of the condensate is in countercurrent with the vapour. During the intensive contact of the countercurrent phases of vapour and reflux, a mass and energy transfer occurs and the component with the lowest boiling point is enriched towards the top of the column.

### 2.1.3.2.1 Influence of Pressure

Apart from temperature, distillative processes are dependent on the total operation pressure. The correlation between boiling point and total pressure can be depicted with vapour pressure curves (Fig. 2.31)

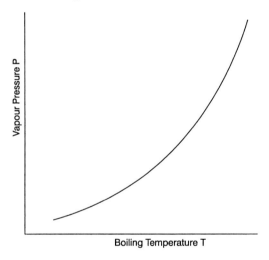

*Fig. 2.31:* Vapour pressure curve

Since many flavour compounds are sensitive to heat, oxygen and light, the application of gentle distillation methods is indispensable.

As the vapour pressure curve shows, the boiling point of a compound can be lowered by reducing the total pressure; this also has consequences on equipment design. According to the law of Boyle and Mariotte, $p \cdot V$ = constant, the volume of gases increases under reduced pressure. These higher values have to be taken into account when calculating evaporation and condensation areas.

**Pressure Range**

The following distinctions are made with distillative methods:

| | |
|---|---|
| overpressure | |
| normal pressure | at 1000 mbar |
| coarse vacuum | up to 1 mbar |

| | |
|---|---|
| medium high vacuum | up to $10^{-3}$ mbar |
| high vacuum | up to $10^{-5}$ mbar |

Most distillation equipment works at the coarse vacuum level. It is necessary to study the kinetic gas theory in order to understand the conditions under medium high or high vacuum. This theory maintains that a gas consists of a large number of molecules which move in straight lines at various speeds. Collisions between the molecules or with the wall can change the individual molecular speed, but the speed distribution remains the same. The average speed of the molecules depends on the type of gas and on temperature. At the same temperature, lighter gases move on average faster than heavier gases and, for example, air exhibits an average speed of about 450 m/s. A mole diameter of $10^{-8}$ to $10^{-7}$ cm results in

| | |
|---|---|
| coarse vacuum in | a particle number of appr. $10^{17}$ per cm$^3$ and an average free path length of appr. $10^{-4}$ cm. |
| high vacuum in | a particle number of appr. $10^{12}$ per cm$^3$ and an average free path length of above 5 cm. |

With pipe diameters of this size, the molecules do not collide with each other, but solely with the walls of equipment and tubes. Therefore, different flow laws apply here than at higher pressures where the molecules collide mostly with each other. With vapours that are sensitive to heat, the collision numbers at distillation temperatures are in the hundreds of thousands. This has a negative impact on the gentle handling of the substance to be distilled. All these factors have to be taken into account for the correct evaluation of the type and size of the vacuum pumps, the tubes and the other distillation equipment.

### 2.1.3.2.2 Vacuum Generation

Pumping capacity, the required forevacuum and the final vacuum have to be taken into consideration for the correct selection of a vacuum pump *[12]*. Pumping speed is defined as the gas volume L or m$^3$ fed per time unit. It also depends on the pressure at the pump's intake port: the more the pressure is reduced, the smaller the pumping speed. Vapour jet pumps, such as steam ejectors or oil ejectors, can compress the sucked-off air only up to approximately 1 mbar. This pressure limit is also called 'stability of forevacuum'. The final vacuum is defined by the vapour pressure of the propellant.

Single-stage or two-stage rotary vane pumps with gas ballast reach medium high vacuum, whereas vapour jet pumps, depending on the propellant, reach high vacuum. Frequently, volatile substances are present and their vapours, therefore, require a huge volume of space under vacuum. In order to increase pumping speed, the use of large pumps would be necessary. The use of condensers before entering the pumps is, therefore, called for. Usually water-cooled condensers suffice; cold traps, e.g. liquid nitrogen, are required under high vacuum.

Liquid seal pumps are frequently employed as mechanical pumps in the coarse vacuum range. During the pump's rotation, a water ring forms as a result of centrifugal forces. As the rotor is mounted eccentrically, a pump chamber forms on the pumping side which decreases on the pressure side. Air is introduced on the pumping side and thrust out on the pressure side. When liquid seal pumps are in tandem arrangement, pumping speed and final vacuum can be improved up to the vapour pressure of water, appr. 10 mbar.

Oil seal pumps, such as rotary vane or single-lobe pumps, are generally employed for medium high vacuum. In order to pump off small amounts of condensed vapour, a measured amount of fresh air is introduced into the pump's chamber. This allows the removal of vapours before condensation and improves the final vacuum. Ballasting is therefore performed if condensed vapours reduce the vapour pressure of the pump oil. Alternatively, fresh lubricated rotary vane vacuum pumps, which offer advantages when aggressive vapours are present, can be employed. Rotary vacuum pumps equipped with two stages and special oil can be used up to $10^{-3}$ mbar. In the medium high vacuum range, the pumping speed of these vacuum pumps does often not suffice. A rootspump can be inserted before the rotary vane pump in such cases. These are displacement pumps with two pistons which run counter-rotatingly with a high number of revolutions, resulting in large suction power.

Vapour jet pumps fulfil a similar purpose, as they also improve the pumping speed of rotary vacuum pumps. The vapour jet of a high-boiling propellant flows through nozzles and thus vapours and gases are sucked off. The propellant is condensed in a cooling funnel and flows back into the boiling flask.

In the high-vacuum range, fractionating vapour jet pumps are employed as diffusion pumps after oil seal vacuum pumps. These vapour jet diffusion pumps are equipped with especially constructed nozzles with a diffusion slot. The working range of this type only starts in medium high vacuum and leads to high vacuum or molecular distillation.

The following pump combinations result:

| | |
|---|---|
| coarse vacuum | liquid seal pump or water jet pump + steam ejector |
| medium high vacuum | oil rotary pump + oil ejector or rootspump |
| high vacuum | two-stage oil rotary pump + vapour jet diffusion pump |

The connection between pump and distillation unit also requires consideration. As large gas volumes are present under vacuum, pipe connections have to be short and wide. If the pipe diameter is too small, a pressure loss results and the pump's suction efficiency decreases. The pressure loss due to the flow conductance is the reciprocal sum of all connecting parts from the pump to the distillation unit. With short pipes at least the diameter of the pump's intake port has to be selected; longer pipes require correspondingly larger pipe diameters. Pump manufacturers (see 2.1.3.5) usually provide the necessary information.

### 2.1.3.2.3 Heat Generation

Since the process of distillation is based on the evaporation of components, heat has to be introduced. The heat required for vaporising a single component is the sum of the energy necessary for reaching the boiling point and of the energy for transferring the component into the gaseous phase. The latter constitutes the molar evaporation energy of a single component and is approximately proportional to the boiling point in kelvin.

The molar evaporation enthalpy is described by Trouton's law:

$$H_V = k \cdot T$$

$H_V$: molar evaporation energy
T: boiling point in kelvin
k: constant

For the majority of chemical compounds, Trouton's constant is between 80 and 105 kJ/(mol K).

When calculating the parameters for the design of a distillation unit, the assumption is made that Trouton's law applies. The energy necessary for heating up and evaporation can be transferred by

1. direct heat achieved by:
   - electrical resistance heating
   - electrical energy of low or high frequency
   - introduction of saturated or superheated steam

2. indirect heat achieved by:
   - low-pressure steam, up to 8 bar and 170°C
   - high-pressure steam, up to 30 bar and 230°C
   - superheated steam, up to 500°C
   - heat transfer fluid, up to 300°C.

The most common source of indirect heat is steam in combination with heat exchangers or double jackets. The heat capacity introduced into the apparatus depends on the heat transfer coefficients, the heat exchange area and the temperature difference between steam condensation temperature and product temperature.

The advantages of steam are the excellent heat transfer coefficient (>6000 W/(m² K)), the constant temperature over the whole exchange area and the fast and convenient regulation of heating by means of a throttle valve. Lowering the steam pressure leads to lower condensation temperatures [13].

The advantage of heat transfer fluids is the low operation pressure even at high temperatures [14]. Disadvantages of this heat medium are the temperature differences over the heat exchange area and the difficult heat regulation. Changes of the flow rate through valves have an immediate influence on the heat transfer coefficient. These variations can be reduced by employing a secondary heat transfer fluid cycle.

The heat transfer also depends on the heat conductivity of the material employed (glass, stainless steel or copper). Therefore the following points should be taken into account:

- chemical resistance
- catalytic influences
- pressure resistance
- visibility of the processes (glass as material).

# Distillation

## 2.1.3.3 Thermodynamic Fundamentals of Mixtures

### 2.1.3.3.1 Equilibria, ideal - nonideal

In order to describe the principles of distillation, the following thermodynamic concepts should be taken into consideration *[15-17]*. The liquid phase is in equilibrium with the vapour phase when their chemical potentials are equal. Since the chemical potential is dependent on temperature and pressure, one parameter is always kept constant for the thermodynamic description of distillation.

Mixtures of liquids exhibit ideal behaviour when their intermolecular forces are equal among and between themselves, their partial enthalpies are independent of concentration and equal to the molar enthalpies of the pure components. In this case they obey Raoult's law, which states that the partial vapour pressure of a component is proportional to its mole fraction in the liquid mixture:

$$p_i = p_{i0} \cdot x_i$$

$p_{i0}$: vapour pressure of the pure component
$x_i$: mole fraction of component in the liquid phase
$p_i$: partial vapour pressure of the component

As the partial vapour pressures of the individual components add up to the total pressure, the more volatile component accumulates in the vapour phase. This fact constitutes the basis of distillation. The ideal relative volatility factor then results from the vapour pressures of the pure components:

$$\alpha_{ij} = \frac{p_{i0}}{p_{j0}}$$

and provides information on the distillative separability of the components. Together with Dalton's law, the relationship between the mole fraction of the vapour phase and the mole fraction of the liquid phase is defined as

$$y_i = \frac{\alpha_{ij} \cdot x_i}{1 + x_i \cdot (\alpha_{ij} - 1)}$$

Graphical depiction of an ideal binary mixture results in the equilibrium phase diagram shown in Fig. 2.32.

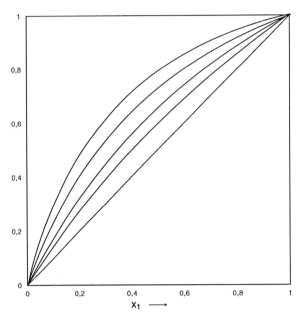

*Fig. 2.32: Phase diagram with equilibrium curves for various relative volatilities of a binary mixture*

The larger the relative volatility factor the more the curves deviate from the bisecting line of an angle, resulting in an improved distillative separability. Unfortunately, most mixtures do not exhibit ideal behaviour. The components do not act independently of each other and instead of Raoult's law the following correlation applies for the partial vapour pressure:

$$p_i = p_{i0} \cdot a_i$$

In this case, the real behaviour of the liquid phase is expressed by the activity $a_i$:

$$a_i = x_i \cdot \gamma_i$$

Here the activity coefficient $\gamma$ is also dependent on concentration:

$$\lim_{x_i \to 1} \gamma = 1$$

Applying Dalton's law, the mole fraction $y_i$ of a component in the vapour phase can be expressed as

$$y_i = \frac{p_{i0} \cdot \gamma_i \cdot x_i}{\varphi_i \cdot p_G}$$

Here the fugacity coefficient $\varphi$ demonstrates nonideal behaviour of the vapour phase:

$$\lim_{p \to 1} \varphi_i = 1$$

The pressure dependency of the fugacity coefficient leads to the following integral over the pressure p:

$$\ln \varphi_i = \frac{1}{RT} \cdot \int V_i - \frac{RT}{P} \, dp$$

In the gas law for real gases, the molar volume $V_i$ can be expressed with one or two virial coefficients according to the equations from Redlich-Kwong or Prausnitz [18, 19]. With low pressures, the dependency of the fugacity coefficient can be neglected.

The relationship between the composition of the vapour phase and the corresponding liquid phase can be depicted in equilibrium phase diagrams. These equilibrium curves are either measured at constant temperature or pressure.

As the activity coefficient can be smaller or larger than 1, real mixtures with vapour pressure maxima and minima exist.

vapour pressure minimum: $\gamma_i < 1$ maximum azeotrope, e.g. $HNO_3/H_2O$
vapour pressure maximum: $\gamma_i > 1$ minimum azeotrope, e.g. $C_2H_5OH/H_2O$

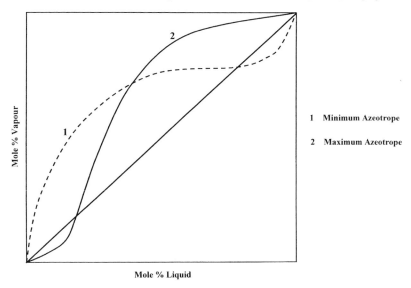

*Fig. 2.33:* *Equilibrium diagram*

In Fig. 2.33, the equilibrium curves intersect with the bisecting line of an angle, an azeotrope mixture is present and the vapour phase possesses the same composition as the liquid phase.

The free excess enthalpy takes the real behaviour of the components in a mixture into account:

$$G^E = RT \cdot \Sigma \, x_i \cdot \ln \gamma_i$$

At constant temperature and pressure, the concentration-dependent activity coefficient can be determined from the free excess enthalpy by differentiation through the mole fraction. These equations are the basis for the methods of Wilson and Prausnitz to calculate the activity coefficient [19, 20]. The Gibbs-Duhem equation is again a convenient method for checking the obtained equilibrium data:

$$\Sigma (X_i \cdot \partial \ln \gamma_i / x_i)_{T,P} = 0$$

This leads to the following equation for binary mixtures:

$$X_1 \cdot \frac{\partial \ln \gamma_1}{\partial x_1} + \frac{\partial \ln \gamma_2}{\partial x_2} \cdot X_2 = 0$$

Phase equilibrium data have been measured for many binary mixtures with a special apparatus and are available in compiled form. These McCabe Thiele diagrams show the mole fraction of volatiles in the liquid phase in relation to the mole fraction of volatiles in the vapour phase in equilibrium at constant pressure [21, 22].

The vapour pressure dependency on temperature for a component can be calculated with the Clausius-Clapeyron equation:

$$d\ln p = (H_V/RT^2)\, dT$$

### 2.1.3.3.2 Rectification

A column in which the ascending vapour is in contact with the refluxing liquid is used for this purpose. The reflux is generated by an overhead condenser. Due to the phase transfer in the column, a height equivalent of theoretical plate HETP can be defined where the two phases, liquid and vapour, are in thermodynamic equilibrium. The theoretical plate number is defined as

$$n_{th} = H/HETP$$

It is dependent on the mass transfer across the interface of a two-phase system. The maximum transfer rate is obtained when all three terms reach high values in the following equation:

$$N = K \cdot A \cdot \Delta c$$

N: mass transfer rate
K: mass transfer coefficient
A: interfacial area
$\Delta c$: concentration difference in the phases

The mass transfer coefficient depends on the flow condition of gas and liquid phases, the interface area is influenced by the geometry of the column internals and local velocity of the two phases. The largest driving force for the mass transfer is the concentration difference when the two phases are uniformly distributed over the entire flow area. This is achieved when a countercurrent flow pattern of the two phases without remixing is reached in a theoretical plate.

The two-phase flow is influenced by the interior construction of the column. The internals provide a large mass transfer rate through the intensive contact of gas and liquid due to the formation of dispersions. Depending on the vapour-liquid loading and the individual column construction, different flow regimes and residence times of the two phases occur. The main forms of dispersion are the bubble regime with continuous liquid phase, the drop regime with continuous vapour phase and the froth regime where no clear dispersed phase exists and the gas-liquid layer is intensively agitated. This leads to the exchange efficiency $S_q$ of the different columns and the actual plate number can be expressed as

$$S_q = n_{th}/n_{pr}$$

Also for calculation purposes, in every theoretical plate the ascending vapour is in thermodynamic equilibrium with the refluxing liquid. Therefore, together with the mass flow and the mole fractions, the calculations in the rectification unit are performed from plate to plate. The minimal theoretical plate number can be graphically and analytically solved by the method of Fenske, using the following assumptions:

- constant mass flow
- constant volatility coefficient
- indefinite reflux ratio.

The first method employs the phase diagram where steps are inserted between the equilibrium curve and the diagonal (Fig. 2.34).

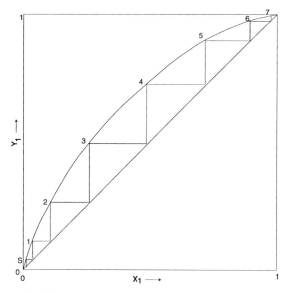

**Fig. 2.34:** *Continuous distillation process with large reflux*

The second one uses the mole fraction of the volatile component at the head and sump and the relative volatility factor $\alpha$:

$$n_{th\,min} = \frac{\lg \frac{x_D}{(1-x_D)} \cdot \frac{(1-x_B)}{x_B} - 1}{\lg \alpha}$$

$x_B$: mole fraction of volatile at bottom
$x_D$: mole fraction of volatile at head

Since the aim of distillation is to remove a distillate, a definite reflux ratio must be employed. The reflux ratio $V_R$ is defined by the mass flow of reflux R and distillate D:

$$V_R = R/D$$

In the case of discontinuous batch rectification, it has to be taken into account that the concentration of the volatile component decreases in the course of the rectification process. The calculations, therefore, have to be carried out up to the concentration which is to be left in the flask. The reflux ratio is usually increased during the rectification process in practical applications in order to remove the volatile component with a certain purity.

## Continuous Rectification

Here the mixture to be separated is continuously fed into the rectification column. At set parameters of the adiabatic rectification the mass flow of feed, distillate and bottom are therefore constant. Depending on the feed F, which can be situated at the beginning or at the end of the column, two basic separation phenomena can be described. When introduced at the beginning of the column, the more volatile component has the possibility to increase its purity towards the head. This method is called amplification column. In a stripping column, the more volatile component leaves at the bottom of the unit at the lowest concentration. Using the McCabe-Thiele diagram, a theoretical plate number can again be graphically determined by indefinite reflux. In the amplification column, the mass flow and the corresponding mole concentrations result in the following operating line:

$$Y = \frac{V_R}{V_R + 1} X + \frac{X_R}{V_R + 1}$$

$X_D$: mole fraction of volatile in distillate
$V_R$: reflux ratio

Every step shown in Fig. 2.35 will lead to increased purity towards the head of the column.

For the stripping column (Fig. 2.36), the operating line is defined as:

$$Y = \frac{V_{R'}}{V_{R'} - 1} \cdot X - \frac{X_B}{V_{R'} - 1}$$

$X_B$: mole fraction of volatile in sump

$$V_{R'} = L/B$$

L: mass flow in stripping column
B: mass flow bottom

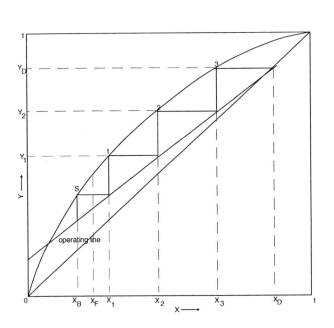
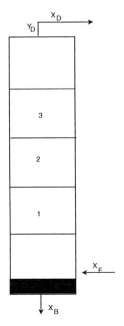

*Fig. 2.35: Amplification column*

# Distillation

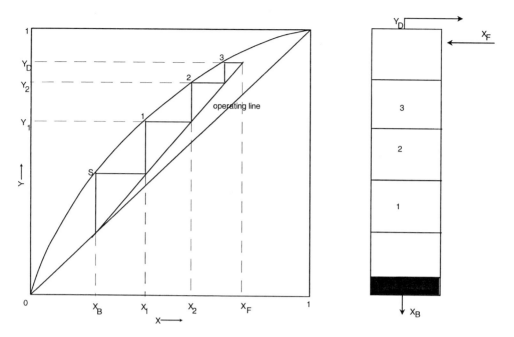

*Fig. 2.36: Stripping column*

To operate a rectification apparatus at economical cost, a minimum reflux ratio is calculated which can be depicted as a function of theoretical plates (Fig. 2.37).

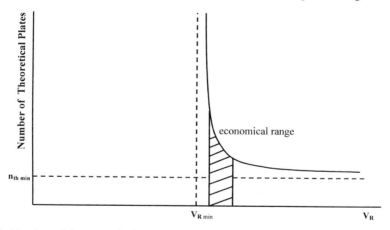

*Fig. 2.37: Number of theoretical plates as a function of reflux ratio*

The minimum reflux ratio $V_{R\,min}$ is reached by an indefinite number of theoretical plates. The effective reflux ratio is preferably between $1.1 V_{R\,min}$ and $1.2 V_{R\,min}$ Together with the envisioned final mole concentrations in both cases, the number of theoretical plates can be determined graphically using the McCabe-Thiele diagram by counting the necessary plates. When the feed is somewhere between head and bottom, the two operating lines for amplification and stripping are used for the determination of theoretical plates (Fig. 2.38).

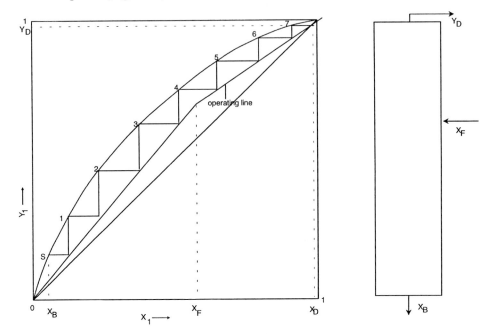

*Fig. 2.38:* Coupled stripping and amplification column

An accurate determination is achieved via iterative procedures by calculating from plate to plate. It is obvious that in the case of more than two components, (n–1) columns are necessary for continuous rectification *[23, 24]* (Fig. 2.39).

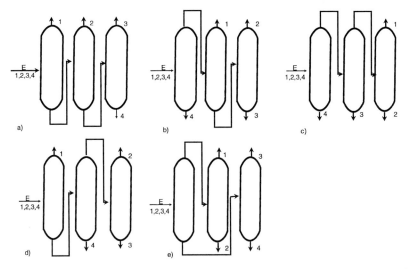

*Fig. 2.39:* Five possible networks for the separation of a quaternary mixture: a) to d) serial connection e) parallel connection

## 2.1.3.3.3 Carrier Distillation

Liquid phases that are only partly miscible or immiscible are graphically depicted by equilibrium curves which show miscibility gaps *[25]* (Fig. 2.40).

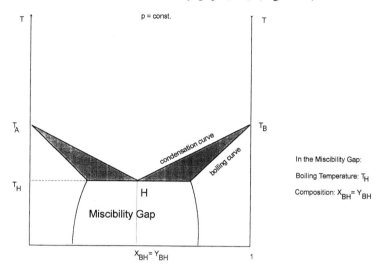

***Fig. 2.40:*** *Boiling point curve of partly immiscible liquid phases*

In the area of the miscibility gap, we have liquid phases which show vapour pressure maxima. In this immiscible liquid phase system, the condensation curve therefore has a common point with the boiling curve and is called a heteroazeotrope. In this miscibility gap, the boiling temperature $T_H$ will be lower than for the pure compounds and the vapour phase in equilibrium will have a constant composition.

$$Y_i = p_{i0}/p_G$$

$p_{i,0}$: pressure of the pure compound
$p_G$: total pressure

Dalton's law applies in the two liquid phase area where immiscible liquids are present. Here, the total pressure is the sum of the vapour pressure of the pure components:

$$p_G = p_{i0} + p_{j0}$$

The expression shows that the vapour pressures of liquid phases act independently of each other within the miscibility gap. This relationship is the basis for steam distillation. Here components with high boiling points can be distilled close to the boiling point of water with a vapour composition according to their vapour pressure:

$$p_i/p_S = (p_G - p_S)/p_S$$

In contrast, by varying the total pressure (vacuum), steam distillation can be carried out at lower temperatures. In this case, the Clausius-Clapeyron equation can be used for the determination of the temperature.

It has to be taken into account that the miscibility gaps are dependent on temperature and can vanish at certain temperatures.

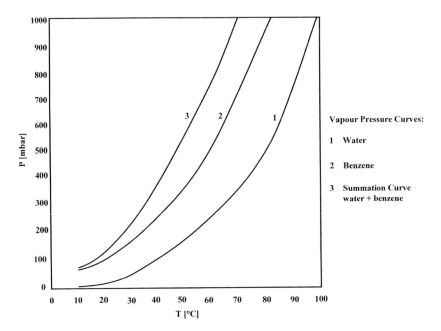

*Fig. 2.41: Vapour pressure curves*

For the flavour industry, steam distillation constitutes one of the most important techniques for the production of essential oils from various plants. One can distinguish between steam distillation, hydro-distillation and hydrodiffusion [26]. In steam distillation, the plant material is placed in a perforated basket and the steam is introduced through a grill at the bottom of the still. Hydro-distillation is mostly carried out with flowers. In a perforated basket, the flowers are heated in 2-3 times their weight of water with indirect steam. A volume of water equal to the weight of the flowers is distilled. The yield of separated oil is low and the water condensate is saturated with polar compounds. In hydrodiffusion, vegetable matter comes into contact with low-pressure steam (<0.1 bar) and the volatiles are replaced by osmotic action. In the hydrodiffusor, the low-pressure steam flows, according to the law of gravity, from the top through the plant material to the condenser at the bottom. As a consequence, the water condensate is more or less saturated with polar constituents.

In continuous steam distillation, an insulated conveying system with superheated steam as carrier is used for providing a countercurrent flow of steam and pulverised plant material. During transport, the oil is transferred into the vapour phase and exits the system with the steam. A cyclonic vessel separates the gas phase from the solid phase. In the last step the gas phase (steam and oil) is condensed, the oil is separated using a Florentine flask and the water recycled to the boiler [27].

Steam distillation is also a very simple and effective method for the purification or deodorisation of high-boiling organic substances (e.g. essential oils). In the food industry, fats or fatty acids are deodorised and decolourised with steam. Continuous steam distillation in a column is used for this purpose. In this operation, the heated oil is run as a thin film countercurrently to the steam; this reduces the amount of steam

# Distillation

for effective stripping. The low residence time and low required temperature avoid thermal degradation of sensitive products; however, the generation of emulsions or foams restricts the applications.

### 2.1.3.3.4 Azeotrope Distillation

If the equilibrium curves of the mixtures approach the diagonal sigmoidly in the lower or upper range of the equilibrium diagram, the point of intersection indicates an azeotrope point. As already set forth, the following distinction is made:

minimum azeotrope: vapour pressure maximum
positive deviation from the diagonal: boiling point minimum

maximum azeotrope: vapour pressure minimum
negative deviation from the diagonal: boiling point maximum

Thus, if an azeotrope point is reached during rectification, this mixture boils at constant temperature and the composition of the vapour is identical to that of the liquid in equilibrium: $y_i = x_i$

If a two-component mixture is present, the azeotrope can be characterised by the relative volatility factor. If $\alpha = 1$, an azeotrope point is present. For a two-component mixture, this relative volatility can be defined as follows, equating the fugacity coefficient with 1:

$$\alpha = \frac{\gamma_i \cdot p_{i0}}{\gamma_j \cdot p_{j0}}$$

This clearly indicates that with defined vapour pressures of the pure components, the relative volatility, and therefore the azeotrope, is only influenced by the activity coefficients. The basis of azeotrope distillation is the addition of a selected compound which forms a new azeotrope with the original mixture. In the ternary azeotrope, the so-called entrainer should additionally have a partial miscibility with one of the original compounds to form two liquid phases (heteroazeotrope). This miscibility gap is a prerequisite for the easy removal of the entrainer from the original compounds by phase separation. In case the present mixture exhibits a minimum azeotrope, the added compound usually again shows a new minimum azeotrope and allows the separation of one of the original compounds at the sump. If a maximum azeotrope is present, the selected compound again results in a new maximum azeotrope in the mixture, so that one original compound can be removed at the head. The formation of new azeotropes is used in technical applications to facilitate distillation of compounds which exhibit close boiling points or azeotropes. Furthermore, the formation of a minimum azeotrope with heat-sensitive natural constituents is often desired in the flavour industry. In these cases, the addition of a selected compound allows the separation of certain constituents in the original mixture at a lower boiling temperature. This method is used to gently remove solvents from natural extracts.

Due to the total pressure dependency of the chemical potential, the equilibrium curves with their azeotrope points can be shifted by applying different pressures. Azeotrope rectification can be performed either in batch or in continuous mode. On a technical scale, azeotrope rectification is often carried out continuously and the selected added compound is recycled and fed back to the original mixture. In both cases, columns are employed and, depending on minimum or maximum azeotrope, the feed is introduced at different points of the column [28-30].

For example, during the dehydration of ethanol with toluene a new ternary azeotrope is formed (Fig. 2.42). Due to the phase separation of the added toluene from water, toluene can be continuously recycled using rectification equipment with two columns.

Examples for the usage of azeotrope rectification on an industrial scale are:
- dehydration of solvents, such as alcohols, esters, ketones and acids
- purification of uniform flavour constituents from mixtures, e.g. of biotechnological origin
- isolation of rare natural components which occur only in the trace range.

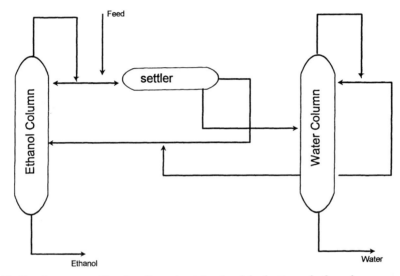

*Fig. 2.42:* Continuous rectification flow sheet for the dehydration of ethanol

### 2.1.3.3.5 Extractive Distillation

In contrast to azeotrope distillation, it is the aim of extractive distillation to find a component to be added, a so-called extracting agent, which dissolves an azeotrope or closely boiling mixtures by increasing the relative volatility considerably. This is achieved by the addition of a high-boiling extracting agent which strongly influences the activity coefficient of one constituent in the azeotrope mixture [31]. The extracting agent is introduced at the head of the column and removed together with the less volatile component from the sump, while the more volatile component is distilled solely from the head. The amount of added component which is introduced via the column has to be tailored to the requirements of the respective separation problem. The added component is also selected with regard to good separability from the forming mixture. It is the intent of extractive rectification to apply continuous processes which allow feedback of the added component. An example for the usage of extractive distillation is the separation of the closely boiling mixture acetone/methanol. The dehydration of isopropanol constitutes another field of application. In this case, the mixture is introduced in the middle of the column. The employed high-boiling extracting agent ethylene glycol is introduced in the correct ratio at the head of the column *[32]* (Fig. 2.43).

The selection of a suitable extracting agent can be supported by gas chromatographic head space analysis. This technique allows the fast determination of an extracting agent by measuring the difference in relative volatility via peak areas in the vapour phase. It permits the fast and reliable assessment of various components to be added with regard to suitability *[33]*.

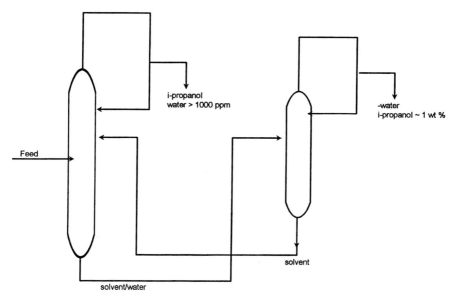

*Fig. 2.43: Flow sheet of an extractive distillation for the dehydration of isopropanol*

### 2.1.3.4 Equipment

### 2.1.3.4.1 Distillation

In the laboratory, discontinuous distillation is performed with glass equipment, e.g. Claisen flasks or micro-distillation kits *[34-36]*. With a number of mixtures, delay in boiling may cause problems. Overheating may occur, which can cause an explosive discharge. This can be prevented by the addition of boiling stones, which continuously create bubbles as a result of their capillary effect. Another option is the continuous introduction of small amounts of inert gas or agitation of the boiling mixture.

The application of vacuum technology allows evaporation at lower temperatures and, therefore, gentle product handling. Usually, coarse vacuum up to 1 mbar is used for distillation. Leakage, as a result of defective equipment, also has a negative impact on the final vacuum. These undesired gases can cause product oxidation and losses, as they function as carrier gases. Therefore, the leakage rate Q has to be determined for larger units:

$$Q = (\Delta p \cdot V)/t$$

where $\Delta p$ is defined as the increase in pressure in the time t with an equipment volume V.

The gas dissolved in the product is also released during distillation. Moreover, a chemical reaction (decomposition) may lead to a gas flow and thus to a sudden increase in pressure.

Generally, discontinuous flask distillation is employed; however, the long time of direct contact in the flask at high temperatures can be a disadvantage [37]. To perform continuous distillation in the laboratory, falling-film or circulation evaporators may be employed (Fig. 2.44). If made entirely of glass, good observation is ensured and the apparatus can also be used under vacuum.

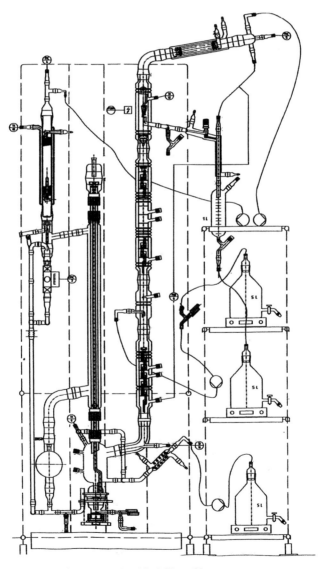

*Fig. 2.44:* Laboratory rectification unit with falling-film evaporator

For gentle evaporation of volatiles, the rotary evaporator is used in the laboratory (Fig. 2.45). The rotating distillation flask creates high turbulence and a new thin film in the upper part of the flask with every rotation. This allows a high heat and mass transfer rate and overheating of the liquid is prevented.

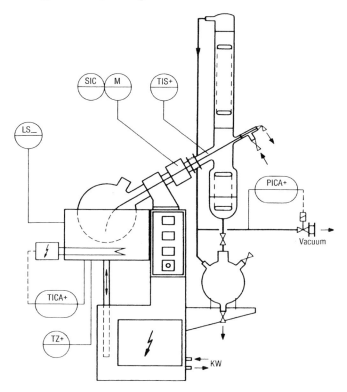

*Fig. 2.45:* Rotary evaporator

On the technical scale, different types of reboilers have been developed [38]. Depending on the distillation requirements, various reboiler constructions can be used. The simplest ones are heat exchangers or double jackets; evaporators constitute more elaborate technical constructions. The advantage of the latter systems is generally the short residence time and the handling of products with difficult physical properties, such as those with high viscosity or a tendency to crystallisation. These evaporators have very short direct contact times and, therefore, allow distillation of heat-sensitive products.

Employing a forced circulation evaporator for evaporating a mixture ensures considerable improvement (Fig. 2.46). Here, the liquid moves upwards between the heating tubes, either by upward thrust or by forced circulation, and is evaporated. The part that is not evaporated is fed back via a circulation tube, thus creating a cycle. The vapour bubbles, which form during boiling, ensure good heat transfer, while the liquid is not exposed to high temperatures for too long. A forced circulation evaporator can be used under vacuum up to 60 mbar. A further pressure reduction results in even lower temperatures and, therefore, gentle product handling.

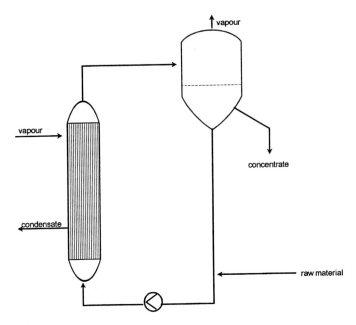

*Fig. 2.46: Forced circulation evaporator*

Falling-film evaporators eliminate the problem of a hydrostatic head with liquid phase. The feed enters at the head of the evaporator, is distributed evenly across the heating tubes and, as a result of gravity, flows downward as a thin film. The liquid-vapour separation takes place at the bottom. The falling-film evaporator is widely used for heat-sensitive materials, because the contact time is small and the liquid is not overheated during passage.

Thin-film evaporators are employed for evaporation purposes up to 1 mbar [39]. Agitated thin-film evaporators permit handling of highly viscous materials and have residence times of only a few seconds (Fig. 2.47). The feed is distributed evenly over the whole circumference above the heated surface. The rotor creates a homogeneous continuous thin film across the heating jacket, preventing overheating. The liquid phase exits at the bottom through gravity and the vapour flows countercurrently to the head of the evaporator. The LUWA evaporator is available in a vertical design. A rotor with rigid blades evenly distributes the highly viscous liquid on the heating jacket. The Sambay thin-film evaporator features movable wipers on the rotor. Centrifugal forces press them onto the heating jacket. This allows squeezing of residues and processing of products which tend to coat the surfaces. The wipers' frequency and thrust can be adjusted to fit the product optimally. The SAKO evaporator requires little space and can already form a continuous liquid film with small amounts of concentrated product for trial purposes. This apparatus features a conical construction and, therefore, the gaps between rotor and heating jacket can be adjusted continuously. This permits one to vary the thickness of the liquid film as well as the time of contact. Even wetting and turbulences create ideal conditions for heat and mass transfer.

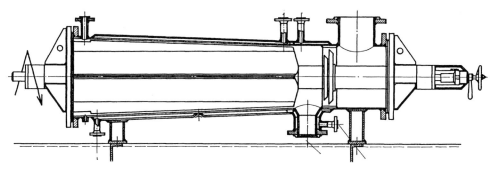

*Fig. 2.47: Thin-film evaporator*

A gas haze of introduced air covers the condenser in vacuum distillation. This has a negative impact on vapour condensation. The thickness of the gas haze is calculated according to the laws of diffusion. It is inversely proportional to the condensation heat which is released in unit time. In practical applications, values between 1 and 5 mm can be expected. As the gas haze has a particularly negative effect on medium high vacuum distillation, it is advisable to carry out pre-degasification, also with technical units with several stages. It is also possible to minimise the gas haze by employing pumps with high pumping speed. If all these details are taken into account, gentle product handling can be achieved while product yield can be increased.

If gentle product handling is called for during distillation and, therefore, a significant reduction of the boiling temperature is required, high-vacuum distillation is employed *[37]*. The pressure is reduced to such an extent that the molecules, during their thermal motion, do not collide with each other. Each molecule is only evaporated once and immediately reaches the condensation unit, which, therefore, has to be opposite to the evaporator at short distance, approximately equivalent to the distance of the average free path length of the molecules. The pressure range is around $10^{-3}$ mbar. This procedure is called molecular distillation (Fig. 2.48). Under these conditions, the evaporation rate is expressed by the Langmuir-Knudsen equation:

$$m = 1500 \cdot p \cdot \sqrt{M/T}$$

m: output (kg/(m² h))
M: molecular weight (g/mol)
T: temperature (K)
P: operation pressure (bar)

Irrespective of the product evaporated, when the conditions of molecular distillation are applied, an output of 1 kg/(h m²) is reached at 0.001 mbar. These small capacities restrict this distillation technique to applications in research and on the laboratory scale. If a larger product capacity is to be reached on an industrial scale, the area of free molecule diffusion has to be left and higher pressures have to be accepted.

The Langmuir-Knudsen correlation now results in 10 kg/(h m²) at a pressure of 0.01 mbar and in 1000 kg/(h m²) at a pressure of 1 mbar. This is the range of flash distillation. In this case, thin-film evaporators are employed where the exterior evaporation surface is closely opposite to the interior condensation area. One constructive example is the short path evaporator with wiped film. The product is dispersed onto a heated jacket with rotating plates at the circumference. A mechanical wiper system creates a homogeneous, continuous product film. The condenser is located in the centre of the evaporator. A cooled pipe bundle results in a good flow diameter, where gases can be sucked off from the empty interior. The distillate is removed via the condensate pipes, and the residue is collected in a loading cup. The time of direct heat contact is reduced to a few seconds. For degasification purposes, a preliminary step is introduced with a pressure of a few millibar. This also results in a pre-separation of the volatile first runnings.

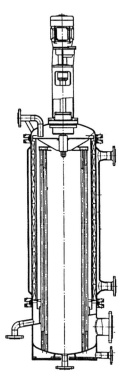

**Fig. 2.48:** *Shortway evaporator*

Centrifugal evaporators are used on an industrial scale for gentle distillation of temperature-sensitive materials (Fig. 2.49). Here, a thin film is evenly spread on a heated conical plate by centrifugal force. This technique reduces hold-up, contact time and foaming of the liquid on the heated surface considerably. Furthermore, the centrifugal force immediately throws the condensed steam away from the rotor's heating surface (dropwise condensation). This results in a uniformly heated rotor surface with a high overall heat transfer coefficient up to 30,000 kJ/(m² h K). This heated evaporation surface is located opposite to a cooled box which results in the distillate removal.

In another centrifugal evaporator construction, a nested stack of hollow conical discs rotates on a common spindle. The heating medium is supplied through the hollow spindle to the steam chamber surrounding the cone stack. Again due to the centrifugal forces a dropwise condensation of the steam on the heating surface is achieved. The feed enters the evaporator through a common tube at the top and injection nozzles distribute the liquid onto the underside of each rotating cone. Centrifugal force spreads the liquid as a thin film over the heating surface. The vapour is released through the centre of the cones to an external condenser.

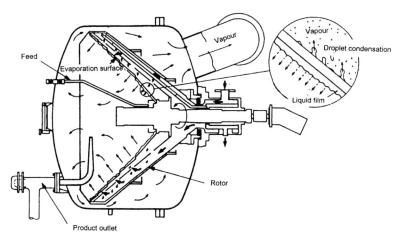

*Fig. 2.49: Centrifugal evaporator*

Flash distillation with its low residence time and high vapour velocities constitutes another gentle distillation method and has found widespread application in heat-sensitive juice concentration.

As molecular and flash distillation constitute simple 'one-way distillation methods', only products that possess considerably differing vapour pressures can be separated.

Based on a concept initially developed in the 1930s, the spinning cone column is a multistage centrifugal evaporator which achieves higher separation efficiency of volatiles under gentle conditions [40]. In a column, a series of alternate stationary and spinning cones are either assembled on the housing or attached on a centrally rotating shaft (Fig. 2.50). The preheated feed is pumped into the top of the column and falls onto the stationary cone where a liquid film flows across the upper surface towards the column shaft. The liquid then falls onto the rotating cone and due to the centrifugal force, is forced upward and outward across the cone surface. The turbulent thin film leaves the lip of the spinning cone, falls onto a stationary cone and the process is repeated. The vapour is pumped upwards through the column, thus creating a countercurrent flow. The addition of fins to the underside of the rotating cones creates high turbulences in the vapour and liquid flow resulting in a high mass transfer rate. Its ability to handle viscous fluids or slurries renders the spinning cone column into an interesting tool for the food industry.

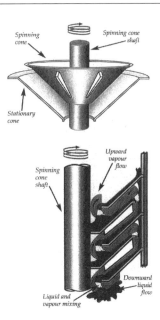

***Fig. 2.50:*** *Spinning cone column*

### 2.1.3.4.2 Countercurrent Distillation; Rectification

The equipment for continuous distillation can only separate one stage in the equilibrium diagram. Countercurrent distillation, also called rectification, has found widespread application with normal pressure and coarse vacuum distillation when complex mixtures or components with small relative volatility factor are to be separated. The fundamentals are discussed above (2.1.3.3.2); the technical side will be dealt with here *[41-45]*.

Rectification can be carried out:

- discontinuously
- semi-continuously
- continuously.

Laboratory equipment for the two important types is depicted in Fig. 2.51.

Discontinuous rectification is characterised by non-recurring feed and a separation column. To generate a reflux, an overhead condenser is inserted and a cold trap cools the obtained distillate. For analytical-preparative purposes in the laboratory, an apparatus with annulus columns has proved to be successful. This apparatus has a vacuum of up to 0.01 mbar, an output of 1-50 ml/h and low hold-up volume. Larger rectification units have tray columns, sieve plates and packing material with a vacuum of up to 0.1 mbar and outputs of several litres per hour.

On the one hand, the simple, easy to control mode of operation, which is especially called for in the laboratory or with trial runs, is of advantage. On the other hand, the long time of direct contact, the high energy requirements and the difficult automation, as product and distillate change continuously, have turned out to be of a disadvantage.

# Distillation

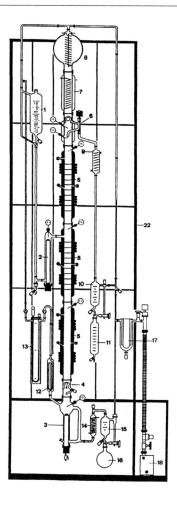

*Fig. 2.51: Laboratory rectification unit with circulating evaporator*

Therefore, partly continuous rectification is a mode for special applications. In this case, the mixture to be separated is introduced into the flask as distillate is removed. The mixture is fed into the flask or just above the flask and proceeds at boiling temperature. The condensation heat from the overhead condenser can be employed as a heat source. After a certain time when the flask is filled, the feed is stopped and the high-boiling constituents are separated discontinuously. This technology is used for removing the volatile first runnings or the solvent from high-boiling components. Again, the mode of operation is simple; the long thermal exposure of the flask contents is not desirable.

Continuous rectification does not suffer from these disadvantages. Here, a continuous feed between amplifier and stripping column exists. In industrial applications, the following criteria should be considered in operating a rectification unit at economical cost [46]:

- reduction of the operation pressure increases relative volatility and reduces the energy input
- optimisation of reflux ratio (minimum reflux)
- in vacuum rectification, the low pressure loss per plate permits energy savings
- the heat of the leaving distillate and especially of the bottom flow can be utilised for preheating the feed
- the usage of heating pumps in applications with high energy demand.

For heating pumps, employing a direct product stream, e.g. vapour recompression, results in a higher difference between head and sump temperature than the usage of an external compression fluid. After adjustment of the operating conditions, a head and a sump product form. The reduced thermal exposure is of advantage, as well as the low energy requirements and the high output. Methods developed on a laboratory scale can be transferred to semi-industrial and industrial units.

If multi-compound mixtures are present, it is necessary to introduce the bottom product into a further continuous unit. Mixtures with n components require n–1 separation columns. This technology is used for processes on a large industrial scale. In the flavour industry, mixtures with three or more components are first separated into two to three fractions and then subjected to a discontinuous separation process. The after-run may be separated from its high-boiling constituents by thin-film evaporators.

The essential parameters for the construction of a column are:

- vapour loading coefficient: this value is characteristic for the specific throughput
- the height of theoretical plates: this value shows the separation efficiency HETP
- the pressure loss per plate: important for usage under vacuum.

The columns can be structured into three basic types: tray, filling bodies and packed columns [47-49].

Tray columns are available in lengths up to 100 m and with very large diameters. The interfacial area in the column is generated by the vapour which permeates through the holes with high velocity and collides with the descending liquid to form dispersions. Tray columns show a broad range of vapour loading capacity with high separation efficiency, even with small loads. The number of theoretical plates is, depending on the technical construction, relatively high; the same applies to capacity. The loss of pressure with 2-5mbar per plate has a negative impact on the operation under vacuum.

A number of different constructions have been developed for tray columns; the essential ones are still bubble-cap, valve, sieve and grating plates. Bubble-cap plates are the oldest development and, due to the high production costs, they are rarely used today.

Sieve plates have also been known for a long time and possess fine drillings. The vapour flows through them and comes into exchange with the liquid, thus forming a bubble layer. This reduces the loss of pressure, but, as a result of their construction, they also have a lower vapour loading coefficient. The danger of encrustation is, however, high, as the drillings are blocked easily. Cross-flow sieve plates and cross-current sieve plates are available from a number of manufacturers and they are optimally designed to meet individual requirements.

Column fillings with irregular packings have found increasing application. The cylindrical Raschig rings have been replaced by Pall rings, Berl saddles, Intalox and grating rings. Due to their high surface, these columns possess good separation efficiency and show small loss of pressure. The disadvantage is the poor distribution of the liquid, especially with larger column diameters. This can be countered by the insertion of a special distribution device. Since the phase transfer should use the high surface area of the fillings, the wetting characteristics of these materials are important. Adequate packing material can ensure appropriate wetting. Apart from different metals, glass, porcelain, ceramics, carbon and plastics have been used [49].

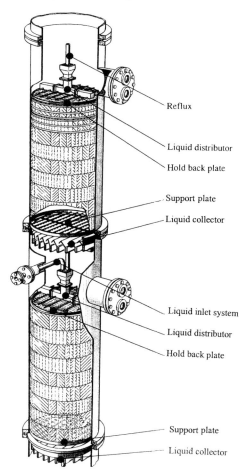

*Fig. 2.52: Rectification column with regular packings*

The ordered structure of the regular packings column with uniform flow canals allows a precise phase distribution and small loss of pressure [50] (Fig. 2.52). The very large surface area results in a high number of theoretical plates. Since regular packings ensure a good distribution of the liquid, a high vapour loading is encountered. Due to the capillary mechanism of the net structure, the packings also work at very low vapour loading. Again, columns with bigger diameters should have a good liquid distribution device and wetting of the packings also has to be taken into account. These regular packings are prone to encrustations and expensive to produce.

As a result of their positive characteristics, packings have found a broad range of application in the flavour industry [51]. Both regular and irregular packings result in a small HETP value with very low pressure loss in the column. These are important features for the use of these packings in the flavour industry.

## COMPANIES SUPPLYING EQUIPMENT

Alcatel Vacuum Technology, Avenue de Brogny 98, F-74009 Annecy, France
ANA-Verfahrenstechnik, Am Saalehang 4, D-06217 Merseburg, Germany
Anhydro A/S, Østmarken 7, DK-2860 Soeborg, Denmark
API Schmidt-Bretten GmbH & Co. KG, Pforzheimer Straße 46, D-75015 Bretten, Germany
Artisan Industries Inc., 73 Pond Street, Waltham, MA 02451-4594, USA
BOC Edwards, Manor Royal, Crawley, RH10 2LW, UK
Büchi AG, Geschwaderstr. 12, CH-8610 Uster, Switzerland
Busch GmbH, Postfach 1251, D-79689 Maulburg, Germany
Buss-SMS GmbH, Kaiserstr. 13–15, D-35510 Butzbach, Germany
Canzler GmbH, Kölner Landstr. 332, D-52351 Düren, Germany
De Dietrich Process Systems, SAS PO Box 8, F-67110 Zinswiller, France
E & E Verfahrenstechnik GmbH, Düsternstr. 55, D-48231 Warendorf, Germany
F.T. Industrial Pty. Ltd., 680 Pacific Highway Killara, NSW-2071, Australia
i-Fischer Engineering GmbH, Dachdeckerstr. 2, D-97297 Waldbüttelbrunn, Germany
GEA Wiegand GmbH, Einsteinstr. 9–15, D-76275 Ettlingen, Germany
Heraeus Quarzglas GmbH & Co. KG, Quarzstr. 8, D-63450 Hanau, Germany
Julius Montz GmbH, Hofstr. 82, D-40723 Hilden, Germany
Kühni AG, Gewerbestr. 25, CH-4123 Allschwill 2, Switzerland
Leybold Vakuum GmbH, Bonner Str. 498, D-50968 Köln, Germany
Norbert Karasek GmbH, Neusiedler Str. 15–19, A-2640 Gloggnitz, Austria
NORMAG LABOR- und PROZESSTECHNIK GmbH, Auf dem Steine 4, D-98683 Ilmenau, Germany
Normschliff Gerätebau Dr. Friedrichs – Dr. Matschke GmbH & Co. KG, Hüttenweg 3, D-97877 Wertheim, Germany
Pfeiffer Vacuum GmbH, Berliner Str. 43, D-35614 Aßlar, Germany
QVF Engineering GmbH, Hattenbergstr. 36, D-55122 Mainz
Raschig GmbH, Mundenheimer Str. 100, D-67061 Ludwigshafen, Germany
Rauschert Verfahrenstechnik GmbH, Paul-Rauschert-Str. 6, D-96349 Steinwiesen, Germany
Rietschle GmbH, Postfach 1260, D-79642 Schopfheim, Germany
Rosenmund VTA AG, Gestadeckplatz 6, CH-4410 Liestal, Switzerland
Schrader Verfahrenstechnik GmbH, Schleebergstr. 12, D-59320 Ennigerloh, Germany
Sterling SIHI GmbH, Lindenstraße 170, D-25524 Itzehoe, Germany

Sulzer Chemtech AG, Hegifeldstr. 10, CH-8404 Winterthur, Switzerland
Uhde GmbH, Friedrich-Uhde-Str. 15, D-44141 Dortmund, Germany
UIC GmbH, Am Neuen Berg 4, D-63755 Alzenau-Hörstein, Germany
Vereinigte Füllkörper Fabriken GmbH & Co. KG, Rheinstr. 176, D-56235 Ransbach-Baumbach, Germany
VTA Verfahrenstechnische Anlagen GmbH, Josef-Wallner-Str. 10, D-94469 Deggendorf, Germany

## REFERENCES

[1]  Stichlmair J., Fair J., Distillation Principles and Practice, New York, Wiley-VCH 1998
[2]  Kister H., Distillation Operation, New York, McGraw-Hill 1990
[3]  Billet R., Industrielle Destillation, Weinheim, Verlag Chemie 1973
[4]  Sattler K., Thermische Trennverfahren, Weinheim, Verlag Chemie 1995
[5]  Stichlmair J., Ullmanns Encyclopedia of Industrial Chemistry Bd. 3, 5. Aufl., Weinheim, Verlag Chemie 1988
[6]  Weiß S., Militzer K.-E., Gramlich K., Thermische Verfahrenstechnik, Stuttgart, Deutscher Verlag für Grundstoffindustrie 1993
[7]  Ricci L., Separation Techniques I: Liquid-Liquid Systems, New York, McGraw-Hill 1980
[8]  Van Winkle M., Distillation, New York, McGraw-Hill 1967
[9]  King C.J., Separation Processes, New York, McGraw-Hill 1971
[10] Deibele L., Die Entwicklung der Destillationstechnik, Chem. Ing. Tech., $\underline{66}$, 809-818 (1994)
[11] Holland C.D., Fundamentals and Modelling of Separation Processes, Englewood Cliffs, Prentice Hall 1975
[12] Jorisch W. (Ed.), Vakuumtechnik, Weinheim, Wiley-VCH 1998
[13] Nitsche M., Dampfförmig oder Flüssig, Verfahrenstechnik $\underline{36}$, 7-8 (2002)
[14] Wagner W. Wärmeträgertechnik, Würzburg, Vogel Fachbuch Verlag 2005
[15] Kortüm G., Lachmann H., Einführung in die Chemische Thermodynamik, Weinheim, Verlag Chemie 1981
[16] Redlich O., Thermodynamics, Fundamentals, Application, New York, Elsevier 1976
[17] Prausnitz J.M., Molecular Thermodynamics of Fluid Phase Equilibria, New York, Prentice Hall 1969
[18] Redlich O., Kwong J., Chem. Rev., $\underline{44}$, 223 (1949)
[19] Prausnitz J.M., Eckert C.A., Computer Calculations for Multicomponent Vapour/Liquid Equilibria, New York, Prentice Hall 1967
[20] Wilson G., J. Am. Chem. Soc., $\underline{86}$, 127 (1964)
[21] Gmehling J., Onken U., Arlt W., Vapour-Liquid Equilibrium Data Collection, Frankfurt, Dechema 1977 (Vol. 1)–1996
[22] Horsley L.H., Azeotropic Data III, Washington, American Chemical Society 1973
[23] Vogelpohl A., Definition von Destillationslinien bei der Trennung von Mehrstoffgemischen, Chem. Ing. Tech., $\underline{65}$, 512-522 (1993)
[24] Hollard C.D., Fundamentals of Multicomponent Distillation, New York, McGraw-Hill 1981
[25] Haase R., Thermodynamik der Mischphasen, Berlin, Springer Verlag 1956
[26] Boelens M., Proceedings 12th International Congress of Flavour, Fragrance and Essential Oils, 1-10, Vienna, October 1992
[27] Boucard G., Serth R., Continuous Steam Distillation of Essential Oils, Perf. & Flav., $\underline{23}$, 3/4 (1998)
[28] Düssel R., Stichlmair J., Zerlegung azeotroper Gemische durch Batch-Rektifikation unter Verwendung eines Zusatzstoffes, Chem. Ing. Tech., $\underline{9}$, 1061-64 (1994)
[29] Warter M., Düssel R., Batch-Rektifikation azeotroper Gemische in Verstärkungs- und Abtriebskolonnen, Chem. Ing. Tech., $\underline{72}$, 675-682 (2000)
[30] Schadler, N., Köhler, J., Haverkamp, H., Zur diskontinuierlichen Rektifikation azeotroper Gemische mit Hilfsstoffeinsatz, Chem. Ing., $\underline{67}$, 967-971 (1995)
[31] Berg L., Chem. Eng. Prog., $\underline{65}$, 52-57 (1969)
[32] Bauer M., Stichlmair J., Bestimmung der minimalen Lösungsmittelmenge bei der Extraktivrektifikation, Vortrag Chem. Ing. Tech., $\underline{68}$ (1996)

[33] Hachenberg H., Schmidt A., Charakterisierung von Zusatzstoffen für Extraktdestillation durch die GC-Dampfraummethode, Verfahrenstechnik, 8 (1974)
[34] Krell E., Handbuch der Laboratoriumsdestillation, Heidelberg, Hüthig Verlag 1976
[35] Zuiderweg F.J., Laboratory of Batch Distillation, New York, Interscience 1957
[36] Schröter J., Deibele L., Steude H., Miniplant Technik, Chem. Ing. Tech., 69, 623-631 (1997)
[37] Frank W., Kutsche D., Die schonende Destillation, Mainz, Krauskopf 1969
[38] Billet R., Verdampfung und ihre technischen Anwendungen, Weinheim, Verlag Chemie 1981
[39] Billet R., Trennleistung von Dünnschichtverdampfern, Chem. Ing. Tech., 72, 565-570 (2000)
[40] Sykes S., Casimir D., Prince R., Recent advances in spinning cone column technology, Food Australia, 44, 462-464 (1992)
[41] Kister H., Distillation Design, New York, McGraw-Hill 1992
[42] Rose M., Distillation design in practice, Amsterdam, Elsevier 1985
[43] Devoy R.H., Chilton C.H., Chemical Engineering Handbook, Tokyo, McGraw-Hill 1973
[44] Billet R., Optimierung in der Rektifiziertechnik, Mannheim, Bibliographisches Institut 1967
[45] Wolf D., Kaiser R., Eiden U., Schuch G., Scale-up von Destillationskolonnen, Chem. Ing. Tech., 67, 269-279 (1995)
[46] Blaß E., Pöllmann P., Köhler J., Berechnung des minimalen Energiebedarfs nichtidealer Rektifikationen, Chem. Ing. Tech., 65, 143-156 (1993)
[47] Billet R., Bewertung von Füllkörpern und die Grenzen ihrer Weiterentwicklung, Chem. Ing. Tech., 65, 157-166 (1993)
[48] Billet R., Stand der Entwicklung von Füllkörpern und Packungen und ihre optimale geometrische Oberfläche, Chem. Ing. Tech., 64, 401-410 (1992)
[49] Schultes M., Füllkörper oder Packungen? Wem gehört die Zukunft?, Chem. Ing. Tech., 70, 254-261 (1998)
[50] Steiner R., Becker O., Erprobung neuartiger Carbonfaser-Packungen, Chem. Ing. Tech., 67, 883-888 (1995)
[51] Guenther E., The Essential Oils, New York, Van Nostrand 1948

## 2.1.4 Spray Drying and Other Methods for Encapsulation of Flavourings

*Updated and enlarged by Uwe-Jens Salzer*
*Based on the manuscript by Abelone Nielsen and Jens Lütken Getler*

### 2.1.4.1 General Introduction

For a long time, flavourings were liquids which were added drop by drop to their intended application. However, the ever expanding industrial production of food started the need for flavourings in dry form. The most basic way to obtain a dry flavour is to plate a liquid concentrate onto a dry carrier as is done with 'spice salts', i.e. spice oleoresins mixed into salt. Such flavourings are very susceptible to the influences of air and light. Hence the industry has been looking for possibilities of not just drying, but encapsulating. The spray drying technique – originally developed for the coffee and dairy industries – proved to be successful for this purpose. It results in a dry, free-flowing encapsulated flavour powder with low water activity. Thus the flavour profile is stabilised, it is protected against decomposition and the release of the flavour itself can be controlled. The spray drying process is described below in more detail.

Spray dried flavourings have satisfied the demands of the food industry for quite some time. However, manufacturing technologies and application forms for food continue to develop. Modern flavourings have to keep up with these changes and with other special requirements put forward by producers. This has resulted in some further developments in the manufacture of encapsulated flavourings – the so-called 'complementary procedures' – which have come into use in recent years. Complementary procedures are refinements, later developments of the spray drying technique. These methods are dealt with later in this section *[1-3]*.

### 2.1.4.2 Spray Drying and 'Complementary' Procedures

#### 2.1.4.2.1 Introduction

Spray drying is recognised as an efficient method of converting liquid flavours into powders. By spray drying it is possible to produce a powder with controlled physical properties such as flowability, residual moisture content and bulk density. At the same time it is possible to control the properties of the active flavour component in terms of its controlled release, its organoleptic quality etc.

Flavourings are normally not spray dried as pure substances. The active, volatile flavour component is mixed and homogenised with a solution of carrier material. During and after the spray drying process, this carrier will form a protective layer on the surface of each formed particle. Whether the active ingredient is soluble or not, the protective carrier acts as facilitator for the spray drying process. At the same time the carrier acts as a microencapsulating agent for solid and liquid flavourings.

Spray drying is an extremely cost-effective and widely applicable process; the equipment has been in use for a long time and it has been optimised over many years. Its advantages are:

- existing well-proven equipment can be used to a large extent
- it is an economical process
- it ensures good results with a wide selection of carrier materials
- it can be used for a broad range of flavouring materials to be encapsulated.

*Fig. 2.53: Small-scale spray drying plant*

### 2.1.4.2.2 Principles of the Procedure and its Application in the Flavour Industry

Spray drying in general is achieved by pumping a solution or a homogenised emulsion to an atomizer and spray it as a fine mist of droplets into a drying chamber. Here the droplets are brought into contact with hot air in a co-current flow. The hot air provides the necessary energy to evaporate the water. The process air will be sucked in through filters, heated and distributed in a controlled flow throughout the drying chamber. Thus the temperature of the individual droplets and the resulting powder is kept low. The air is cooled almost instantaneously due to the very large liquid surface area created from the atomization. The droplet temperature and the temperature of the formed solid powder particles will be kept at a low level because of the evaporation of water. By adjusting the fan speeds for air input and air extraction, the drying chamber can be operated with defined pressure conditions. This results in controlled residing times for the powder in the chamber which, in the case of flavourings, should not exceed 30 seconds.

# Spray Drying and Other Methods for Encapsulation of Flavourings

Most of the powder is discharged directly from the bottom of the chamber. Spent drying air containing fine powder particles is cleaned in highly efficient cyclones and in some cases further cleaned in a bag filter or a wet scrubber. Powder from the cyclone may be discharged separately or it may be returned to the drying chamber in the vicinity of the atomizer where it will adhere to new droplets and form agglomerates.

After drying, the resulting powder is cooled and conveyed to further handling, storage and packing. The main components of a single-stage spray drying plant are shown in Fig. 2.54.

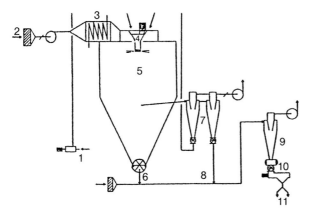

1 Dosing pump
2 Air filter
3 Air heating
4 Atomizer
5 Drying chamber
6 Spray dried product: main part
7 Cyclone
8 Residual part of spray dried product
9 Blender
10 Sieve
11 Blended spray dried product

*Fig. 2.54: Principle of a spray drying plant*

The quality of the spray dried flavour is mainly influenced by

- the particle forming at the atomiser
- the primary drying immediately after the formation of the droplet
- the secondary drying, i.e. diffusion of the residual moisture from the center of the droplet.

The liquid is dispersed into the chamber either by centrifugal force or with a nozzle. Today centrifugal atomising is used much more widely as it offers the following advantages:

- it is flexible, versatile and easy to operate
- the resulting powder quality is not very sensitive to variations in feed rate, feed viscosity, etc.

In a centrifugal atomiser, droplets are formed by the liquid being fed through a distributor into an accelerating chamber and from there they are forced through radial holes in a fast-rotating wheel. Medium-sized wheels are 16-40 cm in diameter and they normally rotate at 16,000 to 6,300 rpm. Atomisation takes place with a peripheral velocity of 130 m/s resulting in droplets of 50-80 µm in diameter. The diameter is controlled by the number of wheel revolutions and the properties of the original liquid.

The characteristics of the powder on the other side depend on the speed of the liquid feed and the design of the atomising wheel. A centrifugal atomiser and its spray mist are shown in Fig. 2.55 and Fig. 2.56 *[4]*.

***Fig. 2.55:*** *Centrifugal atomizer*

In case of use of pressure nozzles, the liquid feed is pumped through a small orifice under high pressure, typically 2-300 bars, and the liquid breaks up into droplets by friction with the atmosphere. Nozzles may produce powders with particularly narrow particle size distribution or relatively dense particles. The particle size from a given pressure nozzle is influenced by the feed rate. This is controlled by varying the feed pressure which again influences the flow capacity. Pressure nozzles are therefore less flexible than centrifugal atomizers with regard to ease of operation and control of product characteristics.

The materials to be spray dried, e.g. flavouring concentrates or fruit, have to be prepared into a solution or an emulsion. Flavour concentrates, e.g. essential oils, extracts and/or mixtures of these with other flavouring substances, are emulsified in water with gum arabic and then homogenised with a solution of the dry carrier. Useful carriers are modified starch products, maltodextrin, sugar, modified whey proteins, cellulose ether and certain forms of polymers. Fruits have to be milled thoroughly to a fine pulp, after which starch sugar is added.

*Fig. 2.56:* Spray mist from rotating wheel of centrifugal atomiser

The solids content of these preparations is important. It may range from 25 to 50%. The residual moisture in the spray dried powder is typically 3-5% and the content of the original liquid flavour is about 20%. The drying temperatures generally vary between 200 and 280°C at air entrance and 90-120°C at air exit.

In general, spray dried flavourings can be considered small capsules of an active volatile and/or tasting ingredient which is embedded in a matrix material. The spray dried capsules are almost spherical particles, each containing a high number of oil-globules or solid particles. Most liquid or solid materials can be encapsulated, whether they are hydrophobic or hydrophilic.

When the fine droplets of emulsion created by atomization come into contact with the hot air within the drying chamber, the moisture evaporates and the carrier material solidifies. Hereby particles of the flavour base are trapped within a dry shell formed by the carrier.

The surface of the droplets dries first and creates a diffusion-resistant boundary layer. The small water molecules diffuse fast through this barrier while larger flavouring molecules are retained. This selective permeability allows for rapid evaporation of water with a minimum loss of volatile flavouring components. At the same time the water evaporation will cool the surface of the particles. This protects heat-sensitive ingredients against overheating and thermal degradation.

The carrier material may also protect the active ingredient against external influences. Flavourings which tend to oxidize, e.g. citrus oils, require encapsulation in order to prevent contact with atmospheric oxygen. These encapsulated citrus flavourings, mainly consisting of essential oils, may be stabilised further by removing the residual amounts of surface oil from the capsule with an inert gas [6]. Further, hygroscopic products may be protected from atmospheric moisture by micro-encapsulation which enhances storage stability and shelf life.

Desired properties, e.g. a specific water solubility, can be designed into encapsulated flavourings through selection of the spray drying technology. And, finally, there is no dust or odour when these flavourings are processed.

However, the primary deciding factor for flavour release is the choice of encapsulation technology. Flavour release involves an extremely wide variety of requirements. We speak of solubility-driven release when a flavour capsule is dissolved in water and thus releases the flavour. The speed at which the capsule dissolves, and subsequently the speed at which the flavour is released, can be determined through the selection of the carrier material. It is also possible to design encapsulation systems that are not soluble in water. They keep the flavour locked up in aqueous products (e.g. sorbets) until the product is consumed. Temperature-driven flavour release can be achieved by coating an encapsulated flavouring with fats of specific melting points, e.g. in cake mixes.

Flavourings used in baked goods develop their flavour in the oven at temperatures of $\geq 70°C$ while the kind of kibbled flavours that are employed in teas, soups or candies are not released until the product is consumed. Instant soup mixes release some of their flavour – especially the highly volatile parts – after hot water is poured over the dry mix in order to produce the soup's characteristic aroma. Flavourings for tea are designed to behave similarly, i.e. the aroma starts to develop whilst the tea is steeping. In the case of chewing gum, the flavour should be released instantly with the start of the chewing process (= impact), but it should also be clearly perceivable after 10-20 minutes of chewing (= long-lasting effect).

These few examples may demonstrate how differently flavours can be released, either at the point of processing the food or at the point of consumption. No single encapsulation method can satisfy all these different requirements. Hence methods other than spray drying, i.e. spray chilling, compacting, agglomerating and fluidised spray drying, as described below, have been developed.

### 2.1.4.2.3 Environmental Considerations

When operating with flavourings, a spray drying system with low odour emission is desirable. This can be achieved with either a low-oxygen system or with an oxygen-free closed circuit using nitrogen as the drying medium. The closed circuit is the choice if organic and inflammable solvents are present.

The low-oxygen system (Fig. 2.57) is characterised by being self-supplying with almost inert drying air. Ambient air is supplied to the direct gas fired air heater at a rate which is sufficient for a complete combustion of the gas. When operating the burner with only little excess of combustion air, the oxygen content in the re-circulating drying air decreases and the system becomes self-inertizing. Usually the oxygen content will be about 4 volume %.

# Spray Drying and Other Methods for Encapsulation of Flavourings

In the spray drying process, the heated drying air, low in oxygen content, is led to the drying chamber, and after the drying process, powder is recovered in a cyclone.

The drying air is cooled in a scrubber/condenser where the water, evaporated in the spray dryer, is removed. The air is subsequently recycled to the direct gas-fired air heater. A vent will exhaust a volume of drying air equal to the volume of combustion products from the gas burner. However, most of the air is recycled in the system.

A low-oxygen system is ideal for many flavouring products: It reduces the risk of powder explosions and/or of oxidation of sensitive products, and the release of exhaust air to the environment is very small compared to the air volume needed in the drying process. The reduced airstream, leaving the plant, can be deodorized or detoxicated by incineration.

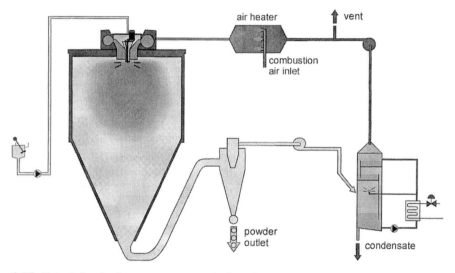

*Fig. 2.57: Principle of a low-oxygen spray drying plant*

## 2.1.4.2.4 Safety Aspects

Some spray dried flavourings are flammable and potential fires and dust explosions pose a considerable risk to operators and equipment.

Dust explosions occur when finely divided combustible solids are airborne in sufficient concentration and when they are subjected to an ignition source of sufficient energy – provided the oxygen level is sufficiently high.

When operating a spray drying plant for flavourings, the explosion hazard should be determined and the correct protective measures should be taken. The before-mentioned low-oxygen spray drying system and the closed-cycle system offer a high level of safety and the additional advantage of low emission.

Other ways to reduce the risk is the use of spark-free materials in the construction, explosion suppression systems or the use of explosion venting by rupture discs or pressure relief panels [5].

## 2.1.4.2.5 Spray Chilling

Spray chilling is performed in the same type of equipment as spray drying. The active flavouring is mixed with a molten wax or fat. This emulsion or suspension is atomized into cold air where the wax solidifies and forms spherical or nearly spherical particles. As carrier material, either vegetable oil or fat is used with air temperatures of 45-122°C ('spray cooling'), or hydrogenated or fractionated vegetable oil with air temperatures of 32-42°C ('spray chilling'). The flavour will be released when the coating melts. Such flavourings may be used in food products which are prepared through a cold process and which are heated before consumption like soups, sauces and deep-fried food.

## 2.1.4.2.6 Compaction and Agglomeration

These processes complement spray drying, especially when larger particles are needed. Compaction results in a compressed product with low porosity, i.e. more strength. Agglomeration, in contrast, produces a fluffy powder with high porosity, ideally suited for applications that which call for instant dissolving.

*Compaction.* The spray dried flavours are compressed under high pressure to lumps and subsequently crushed into small pieces ranging in size from 0.7 to 3.0 mm. This procedure is useful for applications where grainy structures are required. It ensures that the flavour will not separate from the final product and seep through bags with larger pores such as in tea applications.

*Agglomeration.* Spray dried flavourings are fluidised in hot air. The fluidisation process separates the individual powder particles and allows them to be sprayed from all sides. By spraying on a binder – such as water – the powder particles gradually stick together and form larger granules.

The principles of both procedures are shown in Fig. 2.58.

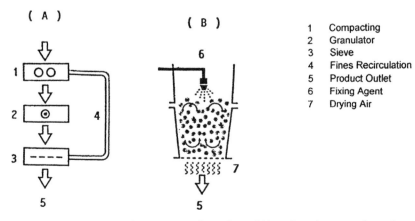

*Fig. 2.58:* Principle of a plant for compacted products (A) and agglomerated products (B)

## 2.1.4.2.7 Fluidised Spray Drying

A fluidised spray dryer is a spray dryer with an attached fluid bed which is integrated at the base of the drying chamber. Hot air enters the spray drying chamber through a roof-mounted air dispenser around the atomiser and leaves through its ceiling. The spray droplets travel downwards towards the fluid bed, whilst the hot exhaust air is led to either cyclones or bag filters. Product recovered from the dry particulate collectors is reintroduced into the process.

### 2.1.4.3 New Methods for Encapsulation

#### 2.1.4.3.1 Introduction

The 'classic' procedures mentioned above have their limitations, especially when it comes to encapsulating very volatile and/or reactive flavourings. In order to satisfy this demand other methods of encapsulation have been developed recently. Also, processes from the pharmaceutical industry have been adjusted to the technological and legislative requirements of the flavour industry.

An important part of this work is the selection of the right material for encapsulating and coating. They have to be neutral in taste and they need to be approved and safe for food use. By choosing the right materials one can also achieve a better control over the release of the flavour, with the trigger being temperature, pH or the solvent used.

On the process side, techniques like the fluid bed method, extrusion, coacervation, the submerged nozzle process and molecular inclusion have gained importance within the last 10-15 years [1-3, 7].

#### 2.1.4.3.2 Fluid Bed Methods: Spray Granulation and Coating

Specific temperature-controlled flavour release is a key function of this delivery system. Similar to agglomeration and fluidised bed spray drying 'fluidised bed spray granulation' is performed in a fluid bed, the difference being that an aqueous emulsion is used here from the very beginning. This technique yields very precise particle sizes of 0.2-1.2 mm and low porosities of the granules. It also allows specificity in size distribution within the resulting powder mix. By spraying repeatedly, re-applying and drying droplets in a fluid bed, it is possible to structure the granulate like the layers of an onion.

Continuous spray granulation starts off identical to spray drying, i.e. an aqueous emulsion. This granulation technology has the advantage of producing large flavoured particles of uniform size and shape without the need for any additional production steps.

An alternative batch spray granulation process utilises the fluidised bed rotor granulator for the manufacture of spherical flavour granules. In this case a flavouring emulsion is sprayed into a fluidised bed of core material.

Both methods allow the subsequent coating of the granules [8]. By selecting appropriate coating materials one can design specific properties into the encapsulated flavourings. Apart from water-soluble materials, it is also possible to coat with fat.

Flow diagrams for both technologies are shown in Fig. 2.59.

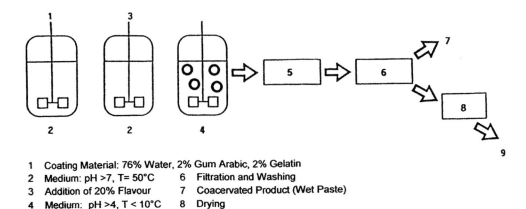

| | |
|---|---|
| 1 Coating Material: 76% Water, 2% Gum Arabic, 2% Gelatin | |
| 2 Medium: pH >7, T= 50°C | 6 Filtration and Washing |
| 3 Addition of 20% Flavour | 7 Coacervated Product (Wet Paste) |
| 4 Medium: pH >4, T < 10°C | 8 Drying |
| 5 Hartening | 9 Coacervated Product (Powder) |

*Fig. 2.59: Principle of a coacervation plant*

### 2.1.4.3.3 Extrusion

Extrusion processes have gained importance in recent years. Highly viscous carriers can be processed into glassy systems that are characterised by high stability and long shelf life. Water or other softeners/plasticisers are added to sugars (or other carbohydrates) which have been melted before the liquid flavour is added. The flavoured melt is forced with high pressure through the extruder's die plate. The extrudate is solidified quickly and so forms an amorphous glassy yet firm mass in pellets with the shape of small needles. The flavouring is completely entrapped in this matrix.

This process is especially well suited for encapsulating highly sensitive flavourings, e.g. all citrus types. The advantage is good protection against oxidation and therefore an extended shelf life compared to other methods.

This principle is shown together with the submerged nozzle process in Fig. 2.60.

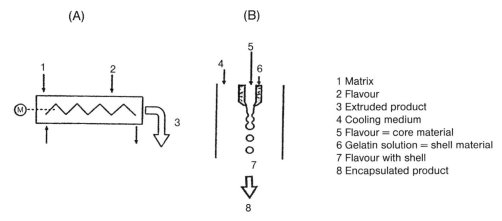

1 Matrix
2 Flavour
3 Extruded product
4 Cooling medium
5 Flavour = core material
6 Gelatin solution = shell material
7 Flavour with shell
8 Encapsulated product

*Fig. 2.60: Principle of an extruder (A) and a submerged nozzle plant (B)*

### 2.1.4.3.4 Encapsulation with Gelatin

Gelatin capsules with enclosed flavour droplets can be produced by either coacervation or the submerged nozzle process.

In *coacervation*, the capsule materials – usually gelatin and gum arabic – are dissolved in water. Afterwards the water-insoluble flavouring is added. By altering temperature or pH, the interfacial surface between the water phase and the flavour droplet forms a thin skin which envelops the droplet. To stabilise the skin, the gelatin has to be chemically treated to cross-link and curve it further after it has been separated from the surrounding water. For various applications the resulting pasty capsules have to be gently dried in a final step. The flow diagram is shown in Fig. 2.59.

Another method for enclosing flavourings in gelatin capsules is the *submerged nozzle process*. The resulting capsules are significantly larger compared to coacervation. The flavouring and the gelatin are forced simultaneously through a special coextrusion nozzle into a suitable medium such as vegetable oil, with the gelatin capsule curing and fully surrounding the flavour droplet. This process calls for the utmost precision and requires constant monitoring of the process steps. The principle is shown in Fig. 2.60 (together with that of an extruder plant).

### 2.1.4.3.5 Molecular Inclusion in β-Cyclodextrin

Molecular inclusion complexes are another technique for the encapsulation of flavouring substances. β-Cyclodextrin is particularly well suited for this method. It is a cyclic glucose oligomer of seven glucopyranosyl units which forms inclusion compounds with substances which, in terms of molecular structure, fit into the active centre and which are less polar than water. Encapsulation with β-cyclodextrin has been widely recognised as one of the most effective ways for protecting flavours against oxidation, evaporation, heat and light degradation. The outcome of this encapsulation, as reflected by the flavour load, product yield, efficiency, etc., is not only affected by intrinsic properties of the flavouring, but also by the preparation method. For instance, the actual flavour load in a finished product arises as the initial flavouring loading increases. However, it stops increasing beyond a saturation point whereas the overall flavour recovery declines [9].

Forming a β-cyclodextrin complex can be as simple as mixing the cargo into a water solution of the complex former, then drawing off the water by drying. The complex is so easily formed because the hydrophobic interior of the β-cyclodextrin drives out the water through thermodynamic forces. The hydrophobic portions of the cargo flavouring readily take the water's place. In general, it takes 10 parts of β-cyclodextrin to every part of the cargo flavouring. Once encapsulated, the flavour is protected from many of the same stresses as a traditionally encapsulated flavouring. The flavouring does not even need to be completely within the cavity to be protected. However, β-cyclodextrin does not offer much control over release. The complex releases as soon as it contacts water. As soon as the water dissolves, an equilibrium forms between what is encapsulated and what is free. Because the equilibrium is dynamic, it creates a reservoir for the release of the flavouring. Whatever stays complexed is still protected. As the free flavour molecules interact with taste receptors, more are freed from the complex to maintain the equilibrium. This can help to extend the time intensity curve of the flavouring *[10]*.

A typical application of this process is the protection of instable, highly volatile and high value-added flavour chemicals. Molecular inclusion can, for example, be used to achieve a 'long-lasting taste effect' in chewing gum. The relatively high price of β-cyclodextrin, however, does not allow a broad use in this context.

### 2.1.4.4 Outlook

All the important and relevant requirements for dry flavourings such as adjustable properties, easy handling, improved shelf life and controlled release can be achieved adequately with the described technologies. However, progress never stops and already other, newer technologies such as liposome and alginate encapsulation, cocrystallisation and boundary surface polymerisation are ready for use. Liposomes form capsules with one or more layers of phospholipids with a particle size from 25 nm up to several micrometres. Alginate pearls may trap flavourings within their gelatinous matrix; but, since flavour diffusion is somewhat restricted, the use of this technique may be limited. Cocrystallisation means the inclusion of flavourings in carbohydrate crystals and in boundary surface polymerisation, which takes place at the boundary surface between lipophilic flavourings and the water phase. However, at this point we are lacking approved effective polymers for food use. Furthermore the use of nanotechnology, which involves the study and use of materials at sizes of some nanometres, could increasingly be used in the creation and development of flavouring systems in the future *[11]*.

All these processes may develop into useful methods for future encapsulation of flavourings. Flavour technology remains an exciting business.

## REFERENCES

[1] Eckert M., Mikroverkapselte Aromen, Herstellung und Anwendung, Teil I ZFL, 47(5), 57-50 (1996); Teil 2 ZFL, 47(6), 63-65 (1996)

[2] Uhlemann J., Schleifenbaum B., Encapsulated Flavors for Intelligent Products, H&R Contact, 79, 3-8 (1999)

[3] Uhlemann J., Schleifenbaum B., Bertram H., Flavor Encapsulation Technologies: An Overview Including Recent Development, Perf. Flavor., 27(5), 52-60 (2002)

[4] Getler J., Spray Drying of Bioproducts in Granulation Technology for Bioproducts, Boca Raton, CRC Press 1990

[5] Skov O., Siwek R., Modellberechnungen zur Dimensionierung von Explosionsklappen auf der Basis von praxisnahen Explosionsversuchen, VDI-Berichte, 701(2), 569 (1988)

[6] Haarmann & Reimer GmbH/Symrise GmbH & Co KG, EP 1.099.385, 16 May 2001

[7] Salzer, Siewek, Gerhardt (Eds.), Handbuch Aromen und Gewürze, chap. 3B.1.3, Hamburg, B. Behr's Verlag 1999

[8] Dewettinek K., Huygebaert A., Fluidized Bed Coating in Food Technology, Trends Food Sci. Tech., 10, 163-168 (1999)

[9] Qi Z.H., Xu A. and Embuscado M.E., Methods and Techniques for Encapsulation of Flavors using β-Cyclodextrin, 1999 AACC Annual Meeting, Oct. 31-Nov. 3, Meeting Abstract 249, www.aacc-net.org/meetings/99mtg/abstracts/acabc51htm

[10] Hegenbarth S., Future Solutions Through Molecular Encapsulation, Food Product Design – New Technologies, April 1993, www.foodproductdesign.com

[11] http://www.foodnavigator.com/news/printNewsBis.asp?id=60741

## 2.1.5 Freeze Drying

*Karl Heinz Deicke*

### 2.1.5.1 General Remarks on Drying

Drying is one of the oldest methods of preservation employed by mankind. The various methods of drying have a drastic impact on the original character of the natural products. This has been acceptable as it always was the main aim to preserve the most important nutritive substances.

The methods which are basically available for dehydration will be explained by employing the phase diagram of water (Fig. 2.60). The starting point shall be within the liquid phase.

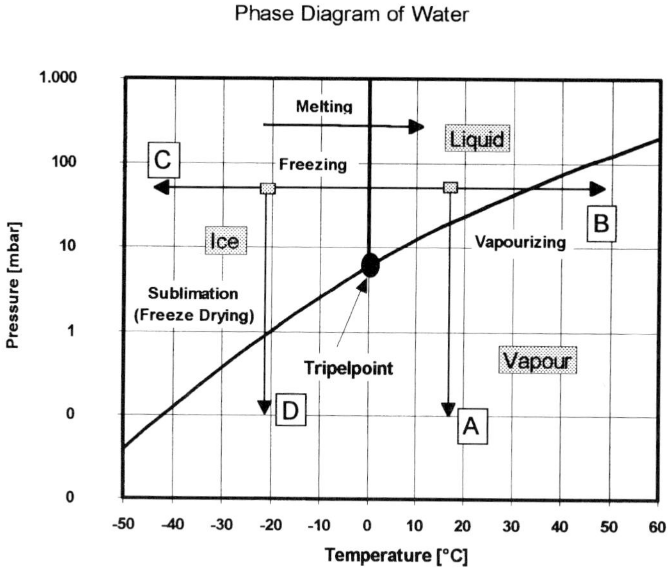

*Fig. 2.60: Phase Diagram of Water*

*Method A: Pressure reduction at constant temperature (vacuum drying)*

If vacuum is applied, the phase boundary between liquid and vapour is reached at a certain pressure and water starts to vaporize. To prevent a loss of temperature, heat – the so-called heat of evaporation – has to be introduced from the outside. If all water has been evaporated, a rise in temperature will ensue that can be taken as an indicator for the end of the drying process. Although this method is gentle, the material to be dried is puffed up and structural and cell deformations follow.

*Method B: Temperature increase at constant pressure (thermal drying)*

The phase boundary is reached by heating – although here at higher temperatures – and the water vaporizes. If the system is kept open, the pressure remains constant. The temperature remains at the evaporation point until all water has been vaporized. This leads to heavy strain on the biological material. Proteins and builders are denaturized, chemical reactions are accelerated.

*Method C: Temperature reduction at constant pressure (freezing)*

With this method, the phase boundary between water and ice is reached. The free water starts congealing below the freezing point. When all material is frozen, method D can then follow.

*Method D: Pressure reduction at constantly low temperature (freeze drying)*

If the pressure is now reduced, analogous to method A, the phase boundary between ice and vapour state is reached. Here, the ice starts to transform directly into vapour, it sublimates. Again, heat has to be introduced, in this case not only the heat of evaporation, but also the heat of sublimation inherent to ice. This does not lead to melting of the ice, as the melting point is skipped by the detour C-D. If the heat of sublimation is not introduced, the product will cool down further and the condition of constant temperature would not be maintained. If all water is evaporated, the temperature in the product increases as a result of the positive thermal balance and the drying process has come to an end. This process, a combination of methods C and D, is called freeze drying.

### 2.1.5.2 The Freeze Drying Process

#### 2.1.5.2.1 General Considerations

The sublimation process can also be observed in nature. If wet laundry is hung on the clothes-line in winter, it freezes and becomes stiff. After a certain period of time, the laundry will be dry, although the temperature was continuously below the freezing point. If such a 'freeze-dried' piece of clothing is compared with one which was dried over a heater, it can be ascertained that the freeze-dried piece of clothing feels softer. Thus, the method of drying must have had a considerable impact on the texture of the material.

Basically, all products which contain water can be dried in this manner. After freezing, the ice crystals remain at first at their original location in the structure. If vacuum is applied, the ice sublimates directly without melting. This results in a porous, dry product which has retained its original form. The volume has hardly changed, the cell walls have been little affected, flavour and other main constituents have been treated gently. If the product is to be restored to its former condition – to be reconstituted – only water has to be added. Water sorption in freeze dried materials proceeds faster than in conventionally dried products. For detailed studies on the fundamentals of freeze drying see the literature *[1-10]*.

#### 2.1.5.2.2 The Construction of Freeze Drying Plants

The core of a freeze drying plant (Fig. 2.61) consists of a drying chamber and a set of vacuum pumps. The pump should not be overexposed to water vapour, which possesses a very large volume in the low pressure range, and a cold trap is, therefore, always inserted before the pump – a so-called 'ice-condenser'. The material to be dried is positioned on special trays which are put onto heater plates. These heater plates can be heated with steam or electricity. Such a system represents a static drying method. With dynamic methods, the product is directly conveyed from one heater plate to the other; this can be achieved by scrapers or by vibration. See Willemer *[11]* and Kamps *[12]* for further details.

Other heating methods, such as IR and microwaves have also found application *[13-15]*.

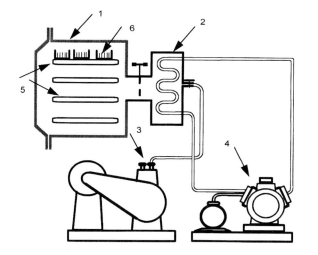

1 Drying chamber
2 Ice condenser
3 Vakuum pump
4 Freezing machine
5 Heater plates
6 Trays

***Fig. 2.61:*** *Sketch of a freeze drying plant*

***Fig. 2.62:*** *Industrial freeze drying unit*

Fig. 2.62 shows an industrial unit employed at Dr. Otto Suwelack, D-48727 Billerbeck, Germany: a tunnel type with inlet and outlet sluices for continuous operation.

### 2.1.5.3 The Quality of Freeze-dried Products

### 2.1.5.3.1 General Remarks

The quality of a freeze-dried product depends on a number of factors and questions of processing technology. The most important are:

- condition of the raw material
- pretreatment of the raw material
- freezing step
- drying program
- treatment with inert gas
- packaging and storage.

The characteristics and importance of these processing steps will subsequently be dealt with employing examples from the spice and flavour industry. A division into three aspects can be made:

- flavour production from natural products
- flavour preservation in natural products and extracts
- flavouring other products.

### 2.1.5.3.2 Production of Flavourings from Natural Products

Finished mixtures for the food sector are often prepared with powdered flavourings. Especially if natural finished flavourings are called for, the use of freeze-dried products should be taken into consideration. The dry matter of a natural extract contains carbohydrates, proteins and other nitrogen compounds, fats and waxes, minerals, vitamins, acids and flavouring substances, which all have an impact on the drying behaviour.

It is advisable to perform a concentration step before drying. The less water has to be removed during the actual freeze drying process, the more economical is the processing. On the other hand, concentration can only be applied within certain limits. Apart from problems with viscosity, the freezing point will for physical reasons decrease with increasing concentration to such an extent, that already the freezing step can become uneconomical. For an excellent overview of processing options for concentration purposes see Pala and Bielig *[16]*.

For the following discussion, we will chose the method of freeze-concentration, as it combines well with the ensuing freeze drying process. A solution of fructose in water will be selected as a practical example for illuminating the physical principles.

If a 20% fructose solution is selected, Young *[17]* shows in diagram (Fig. 2.63), that pure ice freezes out from this mixture at appr. -2.5°C; this is a result of the known phenomenon of freezing point reduction in solutions. The freezing-out of pure ice, however, causes an increase in the sugar concentration and thus a further freezing point depression. This continues until a mixture of ice and fructose dihydrate is present at -10°C. This lowest common solidification point is called 'eutectic point', the mixture is referred to as 'eutectic'. True freeze drying is only possible below the eutectic point, when all water is present as ice. A certain amount of pure ice will be present in the sugar solution before reaching the eutectic point, which can be removed by a simple separation process, such as centrifugation. This method is called ice- or freeze-concentration and can be performed in several stages. During the subsequent freeze drying less water has to be vaporized.

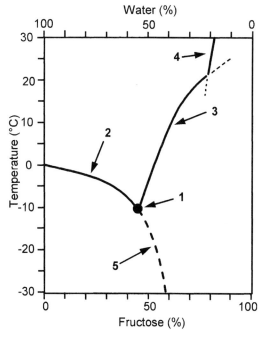

*Fig. 2.63: Phase diagram of the system D-fructose / water [17]*

The frozen material is then exposed to vacuum in a chamber. The ice can now sublimate as described. However, it has to be taken into consideration that, as mentioned, the heat of sublimation of 680 kcal per kg ice has to be applied. For practical applications this means that the material has to be heated to prevent a decrease in temperature and thus a reduction of the sublimation speed. At the end of the drying process a porous product is obtained, the macro structure of which is basically the same as in the preceeding frozen state.

Problems can occur with sugar-containing solutions and fruit juices if, during freezing the solution, the sugar delays in crystallizing below the eutectic point and a supersaturated solution forms. The seemingly crystallized sugar then tends to foaming and splashing: the dried layer breaks down – a phenomenon which is described as collapse in the literature *[1, 18-21]*. In practical applications, the addition of water or carbohydrates can facilitate drying.

It remains to be added that the described model fructose/water can only be taken as an illustration of the principles. Natural extracts possess a far more complicated composition and trials in the laboratory have to be performed to determine the drying behaviour.

Beke, Bartucz-Kovacs and Degen *[22]* report on the combination of the two methods for coffee. As a result of its success in the market-place, freeze-dried coffee has become the generic notion for freeze drying in general. Its production is the show-piece for the combination of highly modern technologies. For an excellent summary with a detailed overview of literature see Sylla *[19]*, Schweinfurt *[23]* and Kerkhof *[24]*. The results depicted therein can be applied to the majority of questions concerning the technology of flavour production from natural products.

Apart from the good flavour characteristics, freeze-dried coffee also possesses an interesting structure. It consists of rather coarse, spongy, irregularly formed bits which show good dissolution properties. The granulated material can be obtained by performing a foaming-up process in the cold. This process can be directed within certain limits to vary the bulk density in order to meet the individual demands. The expectations of the consumer set the standard for dosage. It is obvious that one teaspoon of granular material should yield one cup of beverage.

Tea has encountered increasing importance. The various concentration and drying methods for tea are discussed by von Bomben, Bruin, Thijssen and Merson *[25]*. With black tea, it is possible to dry the pure extract. The freeze-drying process is exhaustively treated in Deicke *[26]*.

In the majority, herbal teas require a carrier matrix. Maltodextrines of various qualities are used. These carbohydrates are capable of retaining volatile compounds to a certain extent, as will be discussed in 2.1.5.3.3. These results can also be transferred to other extracts.

### 2.1.5.3.3 Flavour Preservation in Natural Products and Extracts

#### (1) Natural Products

Experience has shown that freeze drying results in qualitatively superior products when compared to other drying and preservation methods. Herbs such as basil, chervil, dill, parsley, garlic, marjoram, oregano, rosemary, sage, tarragon, thyme and watercress are especially suitable. Economical reasons can also play a role. Freeze-dried products are, in contrast to fresh produce, constantly available all year round at rather stable prices *[27]*.

In the majority, good flavour preservation can be observed if the raw material's structure undergoes few changes. The volatile flavour constituents are well encapsulated and can hardly be perceived in the dried products; it possesses a hay-like smell. The natural flavour reappears with remarkable expressiveness only after rehydration and swelling. Moreover, the natural colour of the fresh products reappear. The colour of the dried product can be influenced by adjusting freezing temperature and pressure, as Poulsen and Nielsen demonstrate for parsley and chives *[28]*.

Detailed investigations on the behaviour of flavour components in herbs are described in Huopalathi and Kesaelathi *[29]*. Tschogowadse and Bakhtadze *[30]* have performed a comparison of thermal drying and sublimation drying for the constituents of coriander.

Green pepper is ideally suited for freeze drying. It possesses a different flavour profile than the usual white or black product. The fresh, unripe fruits constitute the raw material. During freeze drying, the berries hardly undergo any volume contraction and the green colour is largely retained. The freeze-dried berry can be easily reconstituted with water and it regains its original softness. On the other hand, it can also be well used when dry, as the peppercorns can be easily crumbled with the fingers, no peppermill is required.

To maximize the preservation of the volatile flavour constituents in natural products, it is advisable to freeze out as much free water as possible, as the flavour constituents then bind more strongly to the remaining structure. The single products show retention maxima at varying amounts of frozen-out free water. Therefore, the optimal processing and drying program for each product to be freeze-dried should to be determined individually through trials. This also applies to raw material composed of several varieties or coming from different growing areas.

Maelkki Nikkilae, Aalto and Heinonen *[31, 32]* stress the importance of the raw material quality of onions. Stieger *[33]* has investigated the suitablity of various strawberry varieties. Processes for flavour preservation have been thoroughly examined with cultivated mushrooms, leading to an illumination of the most important freeze drying parameters *[34]*.

Even the best freeze drying process can not compensate poor prior preparation.

## (2) Extracts

What applies to natural products, is also valid for extracts. The latter possess the advantage that the optimal matrix for freeze drying can be assembled independently. As mentioned above, good raw material, gentle extraction and an optimal freezing process are prerequisitive. Fruit juices have been investigated by Capella, Lercker and Lerici *[35]*.

It also has to be mentioned that certain flavour losses always occur in the course of freeze drying. In the majority, these are water vapour volatile constituents which escape during the removal of water. A number of authors had dealt with this topic and have investigated the possibility of flavour preservation through various absorbents *[36-40]*. Studies with model character on the flavour retention of terpenic and non-terpenic essential oils have been performed by Smyri and LeMaguer *[41]*.

Maltodextrines constitute a good carrier matrix. They are characterized through the range of starch degradation products with varying molecule size. A characteristic number in this context is the so-called DE-value = dextrose-equivalent. Generally the rule applies that the larger the molecules (low DE-value), the higher, and thus the more advantageous, is the eutectic temperature. Mixtures of simple, reduced (high DE-value) sugars are unsuitable. The investigations of Saint-Hilaire and Solms *[42-44]* of orange juice give an overview.

The preservation effect for flavours, however, proceeds oppositely. Kopelman, Meydav and Wilmersdorf *[45]* have demonstrated with freeze-dried citrus flavourings that the flavour preservation increases with rising DE-value. For practical applications, analysis of the raw material and pre-trials with various carrier matrixes have, therefore, to be employed to find the optimum. The groups of Thijssen as well as Flink and Karel *[45-56]* have performed exhaustive basic studies on this problem.

The structure of carbohydrates can also play an important role for flavour sorption. Studies of Niediek and Babernics *[58]* deal with the flavour sorption properties of amorphous saccharose and lactose. So far, they were able to confirm that the sorption capability in the amorphous state is considerably higher than in the crystalline state.

Grinding processes also play a role for product quality, as Grinberg et al. *[59-60]* report for apricots and apple purée.

### (3) Shelf Life

When stored in closed, aroma-tight containers, ideally in the absence of air, freeze-dried herbs possess good stability. Applying a nitrogen blanket results in further improvement.

The storage stability of dried products depends on a number of material characteristics. The water vapour sorption behaviour plays an important role *[61]*. It is characterized by sorption isotherms, which are typical for every material. The review article by Wolf and Jung *[61]* provides abundant references, also on a number of herbs.

### 2.1.5.3.4 Flavouring Other Products

The number of so-called convenience products is increasing steadily. For the consumer, they should be ready-to-use and ready-to-eat. With dried products, they mainly contain powdered flavourings and, mainly for optical reasons, fruit or other granular material. Such granules should, for reasons of quick reconstitutability, be present in freeze-dried form.

They can be easily enriched with flavourings before freezing or freeze drying. In this context, it is often advantageous to concentrate the defrosting-water of the ice-condensers, which contains volatile flavour constituents, and to add it to the granules.

Similarily, blanching water, an inevitable by-product of some production processes, can be used. An example for cultivated mushrooms is described by Wu et al. *[62]*, who compare various drying methods. They report that thermal drum drying has the best sensorical effect. In this case, freeze drying would only be of interest if the main emphasis is on the structure of the granular material.

Freeze drying can be of advantage with convenience products if the flavouring is mixed with the basic product and they are then dried together. With curd dessert with fruit content, curd dishes and spreads it has turned out that the addition of certain seasonings and flavourings has a stabilizing effect on the final product. This results from their antioxidative properties. The components of curd, such as lactose and milk protein, furthermore cause a good retention of the herb constituents. It is, however, prerequisitive for these effects to take place that the seasonings and herbs are mixed with the substrate when fresh and that they are subsequently frozen together and then freeze-dried.

For other product classes, it is characteristic that the flavour only forms or is changed significantly during production. Thermal, enzymatic and microbiological processes can be responsible for these changes. Examples are the ripening of salami and sour dough fermentation. In both cases, undesired bacteria can cause off-flavour formation.

In the case of salami, starter cultures can be cultivated on meat substrate and freeze-dried, where they remain active at a large rate. The storable culture substrate is added to the salami raw mixture to start the desired ripening process. Depending on process course, it can already contain flavour or flavour-precursors which contribute to accelerating the ripening process.

The latter plays an important role with freeze-dried sour dough. The microorganisms cultivated on flour substrate already form a number of flavour constituents, which stick well to the flour matrix. The subsequent freeze drying preserves both, flavour and microorganisms. The resulting product saves time in the bakery and improves the quality of bread.

### 2.1.5.3.5 Final Observations

The discussion of all these examples shows that, with the application of freeze drying, each problem is situated differently. A detailed discussion with an application or drying technologist can illuminate whether the technology is promising for the problem at hand. Often only trials give the desired answer and clarify further steps. Freeze drying of foods, spices and flavourings represents a high-tech method, which can be tailored to individual needs to yield modern products of the highest quality.

### REFERENCES

[1] Goldblith, S. A.; Rey, L.; Rothmayr, W. W.: Freeze-drying and advanced food technology.; (1975)
[2] Kessler, H. G.: Chem. Ing Techn 47 (18), 55-759 (1975)
[3] Kessler H. G.: Gordian 6, 174-178 (1976)
[4] Lorentzen, J.: Kälte, 157-169 (1970)
[5] Gutcho, M. H.: Freeze drying processes for the food industry.; Food Technology Review, Noyes Data CorporationNo. 41, xii (1977)
[6] Neumann, K.: Grundriß der Gefriertrocknung, Göttingen (1954)
[7] Griesbeck, G.: Ernährungsindustrie 11; 54-56; 12; 51-54 (1979)
[8] Willemer, H.; Lentges, G.; Honrath, M.: Developments in freeze-drying. Proceedings of the Institute of Refrigeration; 79, 11-15 (1982)
[9] Lentges, G.; Oetjen, G. W.; Willemer, H.; Wilmanns, J.: Problems of measurement and control in freeze-drying down to temperatures of -180 DEGC., Proceedings of the International Congress of Refrigeration (13th Washington); 3, 707-715 (1971)
[10] Mellor, J. D.: Fundamentals of freeze-drying. Academic Press Inc, London (1978)
[11] Verein Deutscher Ingenieure: Gefriertrocknung, Verfahren und Anwendungen in der Praxis. Handbuch zum Lehrgang, (Düsseldorf 1979)
[12] Kamps, H.: Industrielles Gefriertrocknen. Ernährungswirtschaft; No. 8, 379-381 (1977); und No. 9, 438-442 (1977)
[13] Attiyate, Y.: Microwave Vakuum Drying, Food Eng 51, 78-79 (1979)
[14] Aglio Gdall'; Gherardi, S.; Versitano, A.: Microwave, infra red and contact plate heating for freeze-drying of some vegetable products. Industria Conserve; 51, 4, 282-289 (1976)
[15] Huet, R.: Aroma retention in tropical fruit powders obtained in a vacuum microwave oven. Fruits; 29 (5), 399-405 (1974)

[16] Pala, M.; Bielig H.J.: Industrielle Konzentrierung und Aromagewinnung von fluessigen Lebensmitteln. Fortschritte in der Lebensmittelwissenschaft; No. 5, xiii386 (1978)

[17] Young, F. E., Jones, F. T., Lewis, A. J.: J. Phys.Chem 56, 1093-1096 (1952)

[18] Tsourouflis, S.; Flink, J. M.; Karel, M.: Loss of structure in freeze-dried carbohydrates solutions; Journal of the Science of Food and Agriculture; 27 (6), 509-519 (1976)

[19] Sylla, K. F.: Die Herstellung von Kaffee-Extrakt; Int. Kongr. "Kaffeechemie" Abidjan (1977)

[20] McKenzie, A. P.: Ann N Y Acad Sci 125, 522-547 (1965)

[21] McKenzie, A. P.: Bull Parenter Drug Assoc 20, 101-129 (1966)

[22] Beke, G.; Bartucz-Kovacs, O.; Degen, G.: Production of soluble coffee using the combination of freeze-concentration and freeze-drying. Bulletin de l'Institut International du Froid; 59 (4), 1120-1121 (1979) Abstr. C1-2/

[23] Schweinfurth, H.: Kaffee- und Kaffee-Extrakt, (in Kröll, K., Kast W., Trocknungstechnik 3, 186-190 (1989), Berlin, Springer Verlag 1989

[24] Kerkhof, P. J. A. M.: Preservation of aroma components during the drying process of extracts. (In '8th International Scientific Colloquium on Coffee'), 235-248 (1979)

[25] Bomben, J. L.; Bruin, S.; Thijssen, H. A. C. ; Merson, R. L.: Aroma recovery and retention in concentration and drying of foods. Advances in Food Research; 20, 1-111 (1973)

[26] Deicke, K. H.: Tee- und Tee-Extrakt, (in Kröll, K., Kast W., Trocknungstechnik 3, 86-197 (1989), Berlin, Springer Verlag 1989

[27] N.N.: Herbal essence comes to food industry with new IQF/freeze dry plant in California.; Quick Frozen Foods International; 35 (1), 126 (1993)

[28] Poulsen, K. P.; Nielsen, P.: Freeze-drying of chives and parsley – optimization attempts. Bulletin de l'Institut International du Froid; 59 (4), 1118-1119 (1979) Abstr. C1-77

[29] Huopalahti, R.; Kesaelahti, E.: Effect of drying and freeze-drying on the aroma of dill – Anethum graveolens cv. Mammut.(In 'Essential oils and aromatic plants', 179-184 (1985))

[30] Tschogowadse, SchK., Bakhtadze, D.M.: Untersuchungen zur Veraenderung der aromatischen Stoffe des Korianders durch Waerme- und Sublimationstrocknung. (Studies on changes in aromatic substances of coriander during heat-drying and freeze-drying.)Lebensmittel-Industrie; 24 (11), 513-515 (1977)

[31] Maelkki, Y.; Heinonen, S.: Freeze-drying of high aroma onions. Journal of the Scientific Agricultural Society of Finland; 50 (2), 125-136 (1978)

[32] Maelkki, Y.; Nikkilae, O. E.; Aalto, M.: The composition and aroma of onions and influencing factors.Journal of the Scientific Agricultural Society of Finland; 50 (2), 103-124 (1978)

[33] Stieger, M.: Untersuchung von Erdbeersorten im Hinblick auf ihre Eignung für die Gefriertrocknung. (Study of strawberry varieties with regard to their suitability for freeze-drying.) Erwerbsobstbau; 17 (2), 26-29 (1975)

[34] Kompany, E.; Rene, F.: Aroma retention of cultivated mushrooms (Agaricus bisporus) during the freeze-drying process. Lebensmittel-Wissenschaft und -Technologie; 26 (6), 524-528 (1993)

[35] Capella, P.; Lercker, G.; Lerici, C. R.: Aroma retention during freeze-drying of fruit juices: volatiles behaviour evaluated by head-space gas chromatography. IV International Congress of Food Science and Technology; 5b, 18 (1974)

[36] Kerkhof, P. J. A. M.; Thijssen, H.A.: Proc. Int. Syp. Aroma Research, Zeist , (Centre for Agr., Publ. and Dok. Wageningen 1975), 167-192 (1975)

[37] Maier H. G.: Lebensm Unters Forsch 141, 65 (1969) und 151, 384-386 (1973)

[38] Thijssen, H. A. C.: J Food Technol 5, 211-229 (1970)

[39] Maier, H. G.: Lebensm Unters Forsch 149, 65-69 (1972)

[40] Maier, H. G.; Hartmann, R. U.: Lebensm Unters Forsch 163, 251-254 (1977)

[41] Smyrl, T.G.; LeMaguer, M.: Retention of sparingly soluble volatile compounds during the freeze drying of model solutions. Journal of Food Process Engineering; 2 (2), 151-170 (1978)

[42] Saint-Hilaire, P.; Solms, J.: Ueber die Gefriertrocknung von Orangensaft. I. Der Einfluss der chemischen Zusammensetzung auf die Sublimationstemperatur. [Freeze-drying of orange juice. I. Effect of chemical composition of sublimation temperature.], Lebensmittel-Wissenschaft Technologie; 6 (5), 170-173 (1973)

[43] Saint-Hilaire, P.; Solms, J.: Ueber die Gefriertrocknung von Orangensaft. II. Der Einfluss der Einfriermethode auf die Gefriertrocknung. [Freeze-drying of orange juice. II. Effect of the freezing method.], Lebensmittel-Wissenschaft Technologie; 6 (5), 174-178 (1973)

[44]  Saint-Hilaire P; Solms J: [Some aspects of the freezing and freeze-drying of orange juice.] Mitteilungen aus dem Gebiete der Lebensmitteluntersuchung undHygiene 64 (1), 90-95 (1973)

[45]  Kopelman, I. J. ; Meydav, S.; Wilmersdorf, P.: Freeze drying encapsulation of water soluble citrus aroma. Journal of Food Technology; 12 (1), 65-72 (1977)

[46]  Flink, J.; Karel, M.: Effect of processing conditions on quality of freeze dried foods. IV International Congress of Food Science and Technology; 5b, 15-16 (1974)

[47]  Karel, M.; Flink, J.: Mechanisms of retention of organic volatiles in freeze-dried systems. Journal of Food Technology; 7 (2), 199-211 (1972)

[48]  Flink, J.; Gejl-Hansen, F.: Retention of organic volatiles in freeze-dried carbohydrate solutions: microscopic observations. Journal of Agricultural and Food Chemistry; 20 (3), 691-694 (1972)

[49]  Flink, J.; Karel, M.: Effects of process variables on retention of volatiles in freeze-drying. Journal of Food Science; 35 (4), 444-447 (1970)

[50]  Rulkens, W. H.; Thijssen, H. A. C.: The retention of organic volatiles in spray-drying aqueous carbohydrate solutions. Journal of Food Technology; 7 (1), 95-105 (1972)

[51]  Flink, J. M.; Gejl-Hansen, F.; Karel, M., Microscopic investigations of the freeze drying of volatile-containing model food solutions. Journal of Food Science; 38 (7), 1174-1178 (1973)

[52]  Karel, M.; Flink, J. M.: Influence of frozen state reactions on freeze-dried foods. Journal of Agricultural and Food Chemistry; 21 (1), 16-21 (1973)

[53]  Karel, M.; Flink, J. M.: Influence of frozen state reaction on freeze dried foods., Abstracts of Papers. American Chemical Society; 163: AGFD 12 (1972)

[54]  Thijssen, H. A. C.: Effect of process conditions in drying liquid foods on aroma retention.( „Proceedings of the 3rd Nordic Aroma Symposium.") 154, 5-38 (1972)

[55]  Thijssen, H. A. C.: Prevention of aroma losses during drying of liquid foods.Dechema-Monographien; 70, 353-366 (1972)

[56]  Menting, L. C.; Hoogstad, B.; Thijssen, H. A. C.: Aroma retention during the drying of liquid foods. Journal of Food Technology; 5 (2), 127-39 (1970)

[57]  Menting, L. C.; Hoogstad, B.; Thijssen, H. A. C.: Diffusion coefficients of water and organic volatiles in carbohydrate-water systems. Journal of Food Technology; 5 (2), 111-26 (1970)

[58]  Niediek, E. A.; Babernics, L.: Aromasorptionseigenschaften von amorpher Saccharose und Lactose Gordian; 79 (2), 35-36, 38-40, 42-44; (1979)

[59]  Grinberg, N. Kh.; Popovskii, V. G.; Kolesnichenko, A. I.: Investigation into the retention of aromatic components of freeze-dried apricot puree during granulation.Konservnaya i Ovoshchesushil'naya Promyshlennost'; No. 7, 38-40 (1977)

[60]  Grinberg, N.Kh.; Popovskii, V. G.: Aroma retention during freeze-drying of apple puree. Konservnaya i Ovoshchesushil'naya Promyshlennost'; No. 11, 41-45 (1976)

[61]  Wolf, W., Jung, G.: Wasserdampfsorptionsraten für die Lebensmitteltrocknung; ZFL Intern. Z. f. Lebensmittel-Technologie u. -Verfahrenstechnik; 35, (2), 61-126 (1985)

[62]  Wu, C. M.; Wu, J. L. P.; Chen, C.C.; Chou, C. C.: Flavor recovery from mushroom blanching water; The quality of foods and beverages. Chemistry and technology. Vol. I., 133-145 (1981)

## 2.2 Biotechnological Processes

*Joachim Tretzel*
*Stefan Marx*

### 2.2.1 Introduction

The previously described manufacturing processes for flavour chemicals and flavour extracts primarily regard the physical and physico-chemical isolation and purification of naturally occurring flavour chemicals derived from plant and animal tissue. The huge area of organic chemical synthesis of nature-identical and synthetical flavour chemicals is not within the scope of this book.

Table 2.10 shows that the isolation and purification of naturally occurring flavour chemicals and extracts from animal and plant raw materials is most important for the preparation of natural flavours. About 75% of the commercially used flavours come from such natural sources. Physico-chemical reactions of typical flavour precursors may also lead to natural flavouring substances when mild conditions ('kitchen technology') are applied. In addition, natural flavour chemicals may be prepared by biotechnological processes. This chapter outlines the most important biotechnical manufacturing techniques.

*Table 2.10: Manufacturing processes for natural flavourings with typical examples. Technical routes to natural flavourings*

| **Physical Processes** Isolation / Purification | **Chemical Processes** Modification | **Biochemical Processes** Biosynthesis |
|---|---|---|
| ◆ Extraction *Meat Extract* ◆ Distillation *Essential Oils* ◆ Chromatography *Beer Flavour* | ◆ Hydrolysis *NVP* ◆ Thermochemistry *Coffee Roasting* | ◆ In-Situ-Fermentation *Cheese* ◆ Technical Bioreactors *γ-Decalactone* |

All flavouring substances found in nature are exclusively products of biochemical reactions in Single cells (bacteria, yeasts, moulds) or in higher organisms. Most of these reactions are enzyme catalyzed and take place in the ideal 'bioreactor': the living cell.

In biotechnology man makes use of these multiple opportunities offered by nature for the synthesis of materials. In general, all biotechnical processes are based on the reaction of a living cell either directly with the raw material (fermentation with intact micro-organisms, e.g. mediated by starter cultures) or these reactions take place inside well controlled technical equipment, so-called bioreactors.

Therefore, there are in principle two different technical approaches to flavour production:

(1) In-situ-fermentation with intact micro-organisms

(2) 'Technical bioreactors' for

- The propagation of flavour producing micro-organisms
- The chemical transformation of flavour precursors by micro-organisms (biotransformation)
- The production and modification of flavour chemicals from precursors by enzymes
- The generation and modification of flavour materials by plant cell cultures.

The processes can be operated in open systems (unsterile fermentation) and in hermetically closed sterile equipment, respectively, depending on the sensitivity of the reaction system for microbiological contamination.

The generation of flavour chemicals by starter cultures or raw material borne enzymes occurs directly in the food raw material itself. The flavouring products do not have to be processed after reaction. In all other processes it is necessary to isolate, purify and process the flavour substances by more or less complicated separation of the reaction mixture. In the same way as desired flavouring compounds may be obtained by flavour intensification or modification, also the opposite effect is observed. Flavour loss and even formation of typical off-flavours may likewise be the consequence of reactions of food-borne enzymes or such being generated by contaminating micro-organisms in the food. Examples for such adverse biochemical reactions are off-flavours in soy beans or oat flakes imparted by plant-borne lipoxygenases, rancid taste in fat-containing food formed by microbial peroxidases and protein hydrolysis leading to bitter peptides by microbial proteases [1].

## 2.2.2 Flavour Generation by Fermentation of Food Raw Materials

The in-situ-generation of taste-giving materials directly in food by raw material borne or imparted cultures of micro-organisms – so-called starter cultures – is one of the most traditional and oldest food technological processes.

Table 2.11 gives an overview of traditionally produced foods by raw material borne micro-organisms via spontaneous fermentation. In all cases the fermentation produces or intensifies typical flavours. Starting materials in general are the usual building blocks of foods: carbohydrates (oligo- and polysaccharides), proteins, peptides and amino acids, fats and fatty acids, nucleic acids and minerals, organic and inorganic acids. Still today, in the time of modern biotechnology, these traditional processes have the biggest significance of all biotechnical processes in food technology. The food industry, however, together with the pharmaceutical industry, has become by far the most important employer of modern biotechnologists.

As result of a multitude of scientific investigations there is a very good theoretical understanding of the basic mechanisms of flavour formation in biochemical reactions. The application, however, of such processes is still more art and craftsmanship than exact natural science. On the other hand, the industrial evaluation of such manufacturing processes has led into better defined process conditions. There is significant progress, especially in the reproducibility and constancy of quality of the obtained products.

*Table 2.11: Spontaneous fermentation of plant-and animal-derived raw materials for generation of typical food flavours*

| Product | Raw Materials | Fermentation by |
|---|---|---|
| **Plant Derived** | | |
| *alcoholic* | | |
| Fruit derived: | | |
| wine | Grape juice | Saccharomyces cerevisiae, Oenococcus oenos |
| champagne | Wine | Saccharomyces cerevisiae |
| Cereal derived: | | |
| Beer | Barley malt | Saccharomyces cerevisiae |
| Kwass | Malt, Bread | Yeasts, Lactobacillae |
| Brandy | Wine | Saccharomyces cerevisiae |
| Rum | Sugar cane | Saccharomyces cerevisiae |
| Vodka | Potato, Corn | Saccharomyces cerevisiae |
| *non alcoholic* | | |
| Vinegar | Wine, Malt, Ethanol | Acetobacter aceti |
| Sauerkraut | Cabbage | Gluconobacter oxydans |
| | | Lactobacillus plantarum |
| | | L. brevis, Leuconostoc mesenteroides |
| Sour dough | Wheat flour | Lactobacillae, |
| | | Saccharomyces cerevisiae |
| Coffee | Coffee beans | Enterobacteria, Lactobacillae, Yeasts |
| Cocoa | Cocoa beans | Yeasts, Lactobacillae, Acetobacteria |
| Kombucha | Tea leaves, herbs | Yeast, Lactobacillae ('Kombucha fungus') |
| Tea | Tea leaves | plant derived enzymes |
| Tobacco | Tobacco leaves | plant derived enzymes |
| Soy sauce | Rice, Wheat, Soy beans | Aspergillus oryzae, Lactobacillae ssp., Pediococci, Zygosaccharomyces ssp., Torulopsis ssp. |
| **Animal Derived** | | |
| Emmentaler Cheese | Milk | Lactobacillus helveticus, Streptococcus salivarius |
| | | Propionibacterium freudenreichii |
| Blue Cheese | Milk | Penicillium roquefortii |
| Roquefort | | Penicillium caseicolum and camembertii |
| Salami | Meat | Vibrio costicola, Pen. nalgiovensis |
| Anchovies | Herring | Staphylococcus spp. |

An important step towards this was the further development of spontaneous fermentation by application of defined cultures of the desired micro-organisms which effect the transformations (so-called starter cultures) [2].

The parameters for effecting and influencing a biotechnological reaction are:
- Type and amount of the fermenting micro-organisms
- Preconditioning of the food raw materials
- Temperature are aeration
- Other physico-chemical conditions (pH, viscosity, ionic strength).

The application of well defined starter cultures enables the unequivocal determination of the main route of fermentation. The micro-organisms imparted to the food raw materials in the form of concentrated starter cultures quickly overgrow undesired germs in the raw material because of their high initial count and suppress deviating metabolic pathways which may give rise to misfermentations and off-flavours.

Starter cultures are mainly applied as liquid cultures with about $10^8$ to $10^{10}$ micro-organisms per ml. They are also available in a freeze-dried or deep frozen preparation with the advantage of very simple application.

To a limited extent a starter culture once established can be propagated on site (sour dough, yoghurt cultures, yeast in the brewery).

Besides the application of a well defined starter culture which has to be tested every time before use, the preconditioning of the food raw material is of critical significance. This preconditioning ranges from simple washing to complete sterilization. Besides, the supplementation of nutrients and the adjustment of a defined pH-value is frequently decisive for the result of a fermentation.

Apart from these measures, the application of spontaneous and starter culture mediated fermentative generation of flavour active substances does not require any further technical equipment. This fact and the traditional experience with such processes are the reasons for the still considerably high amount of food produced by this technology. In 1990, such products exhibited a value of € 23 billion compared to the total amount of € 91 billion in the same year for all food products [3].

Very recently an innovation in the field of soft drinks has been presented by the beverage industry derived from fermented food materials. Traditional processes have been used like the brewing of Kombucha, fermentation of malt wort and applewine and cider-making for obtaining beverage bases after de-alcoholisation, if necessary. In combination with fruit juices, purees and flavours, good but unconventionally tasting beverages are obtained which at the same time provide additional benefits in the health and well-being area, some of which have been proved in scientific assessments [4].

### 2.2.3 Flavour Generation in Bioreactors

When isolated biochemical catalysts (enzymes) or specially selected cultures of micro-organisms are to be used and more precisely defined reaction conditions are required, the biotechnological system demands a better controlled environment. Modern biotechnology, therefore, can be addressed as a refinement and further development of spontaneous fermentation with raw material inherent organisms.

It is the role of technical bioreactors in such advanced systems to create a specifically defined environment for the biochemical reaction system producing or modifying flavour substances. The most important purpose of such bioreactors is the well controlled combination of food raw material and flavour precursor, respectively, with the biological reaction centres. It also provides the means for survival and maintenance of the centre's metabolic activity. Presently there is laboratory and partially also industrial experience with mostly all kinds of different micro-organisms and isolated biocatalysts:

- Isolated purified enzymes, enzyme-mixtures, enzyme complexes
- Yeasts
- Moulds
- Bacteria
- Protozoa and algae
- Cell and tissue cultures of higher plants.

All biochemical reactions are enzyme-mediated. The rate of an enzyme reaction depends on the substrate concentration at the location of the enzyme and thereby on the diffusion rate of a substrate to the enzyme. It is therefore important to permanently obtain an intimate contact between a cell or enzyme and substrate molecules. Additionally, the product generated in the bioreactor has to be extracted because it may under certain conditions inhibit its own production. In some processes there may also be even a prepurification in the bioreactor itself. If living micro-organisms have to be applied, it is necessary to provide sufficient nutrition and respiration gases in case of aerobic fermentation. All other reaction parameters such as temperature, pH-value and reaction time have to be controlled precisely. In many cases (generally with modern processes) the maintenance of microbiological integrity (sterile process) is absolutely mandatory for a successful fermentation.

Besides that, fermentation can only be industrially attractive if the process provides highest yields and exhibits an efficient isolation and purification process (downstream processing) with only minimal product losses. Additionally, suitable substrates must be commercially available at low cost. Finally, the generation of flavours by fermentation in bioreactors will only be profitable if the desired product, be it a pure substance or a complex flavour extract, is not obtainable with comparable quality by inexpensive classical techniques.

By far the most critical economical parameter is the yield of a bioprocess, i.e. the amount of product obtained relative to the raw materials applied, the working volume of the bioreactor and the elapsed reaction time. The yield may be improved by the use of more efficient biocatalysts as a result of screening programmes or genetic engineering as well as optimisation of downstream separation and purification techniques.

## 2.2.4 Surface Fermentation

The oldest and a relatively simple method for production of complex flavour extracts is the fermentation of flavour raw materials by bacteria and moulds (aspergillus, mucor) on solid nutrient media. The media is placed in simple thermostatic boxes on baking-tray like plates to which the micro-organisms are inoculated. This technique which has originated in Japan for the production of Koiji seasonings is still used today for the production of soy sauce made from rough-ground cereals and soy beans which are inoculated with special moulds. After incubating the nutrient media for several days, it is finally extracted with water to isolate the fermented products contained in the media [5]. For many years the Koiji technique was also the preferred method for the production of enzymes which are secreted by bacteria and moulds into the extracellular environment (the nutrient media). Another example for a traditional surface fermentation process is vinegar production by acetic acid bacteria grown on the surface of wood chips.

## 2.2.5 Submerged Fermentation

Submerged fermentation is generally applicable for the manufacturing of cell products by propagation of micro-organisms and cell cultures in a fluid nutrient media. The submerged fermenter (diagram shown in Figure 2.64) is normally operated in a sterile manner i.e. all components have to be sealed against the environment by ports which are air-tight and impenetrable by bacteria. The whole fermenter consisting of an agitated tank with thermostatic mantle, a stirrer and several lines for respiratory gases, pH regulation agents, nutrient source etc. has to be autoclaved prior to reaction. Therefore, it must be able to withstand sterilization with overheated steam at 121°C. In the course of aerobe fermentations, the micro-organisms have to be supplied with respiratory gas in the fermenter by an intense aeration system. Depending on the heat balance of the reaction sometimes huge amounts of heat have to be removed by cooling registers which are built into the fermenter. With complex measuring and controlling devices the environmental conditions within the fermenter for pH, temperature, ionic strength and nutrient concentration are controlled with high accuracy.

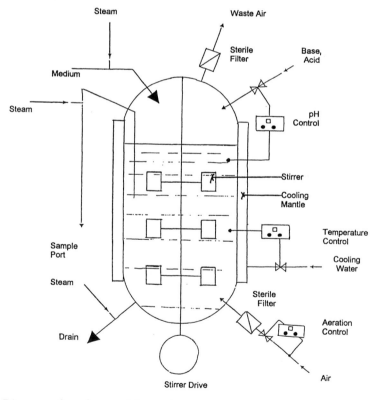

*Fig. 2.64: Diagram of a submerged fermenter*

As a consequence of the adverse operating conditions (sterilization heat; viscose, particle containing media with high dry matter content; protein containing solutions tending to sedimentation; aggressive cleaning aids; gas-bubble containing media) the sensors for retrieval of the control data are constructed in a complicated and expensive way. The on-line analysis of exhaust gas composition by specific sensors allows the continuous evaluation of the activity of the fermenting micro-organisms. In this way an on-line evaluation of the productivity rate of the micro-organism culture is achievable.

By combination of all data thus obtained the process may be controlled with sufficient precision by a process computer. The culture within the fermenter can be maintained over a considerable amount of time in the state of maximum productivity.

The initial propagation of a culture of micro-organisms for production fermenters with payloads up to 200 m³ is achieved via several steps of one order of magnitude each. One starts with the laboratory scale (1 l) via diverse inoculum fermenters (10, 100 l) to obtain a sufficient count and concentration of living micro-organisms for the fermentation. For any successful fermentation a limiting density of living micro-organisms of about $10^8$ per ml is required.

Submerged fermentations are mostly operated in batch processes but can also be run continuously in certain cases (continuous fermentation). Batch fermentations may last up to 10 days. Following the fermentation the flavour raw material is extracted from the fermentation broth. In industrial fermentations typically cell counts of 10-30 g/l are obtained. For a profitable cost/efficiency relation a product yield of 20-30 g/l has to be achieved. Aerobe fermentations require oxygen transfer rates to the fermentation broth of about 100 mmol/l per hour. Depending on the viscosity of the media 0.75-2.5 KW stirring power has to be applied for each m³ of fermentation broth.

## 2.2.6 Downstream Processing

Depending on the remainder of the product synthesized by the micro-organisms (inside the cell or secreted to the media) downstream processing commences either with a disruption of the cell walls or directly with the separation of biomass and particulate matter from the fermentation broth by centrifugation, decantation or filtration. Further processing and purification of the product is achieved by salting-out, fluid extraction, distillation or concentration by evaporation of the fermentation broth, depending on the chemical nature of the product. According to the desired product purity further purification steps may have to follow e.g. chromatography or fractionated precipitation *[6]*. All isolation and purification processes rely on the classical physico-chemical processes described in chapter 2.1. The diversity of aroma compounds from micro-organisms is apparently unlimited, but many of the processes used are currently not economically feasible. In particular, product inhibition can limit the yield of a bioreaction, as metabolism of most micro-organisms is only possible within a narrow range of metabolite concentration. As product inhibition often results in a fermentation broth with a low concentration of products, large reactor volumes are required to meet production demands and the broth has to undergo costly downstream processing to recover the diluted product. Product inhibition can, however, be reduced or avoided by withdrawing inhibiting substances from the fermentation. This concept, often referred to as 'in situ product removal (ISPR)' requires a biocompatible separation operation that is highly selective for the inhibiting substances. These requirements can be met by the use of organophilic pervaporation membranes *[7, 8]*. In general all recovery processes of flavour substances have to follow the principle of minimal processing, i.e. minimising excessive heat, pressure and mechanical forces, in order to maintain the quality of the flavours and the integrity of the biocatalyst (enzyme or micro-organism). Chromatographic and membrane techniques have been widely applied for this purpose. Figure 2.65 schematically depicts fermentation and downstream processing for a water soluble cell product.

## 2.2.7 Enzyme Reactors

In the case of utilizing isolated enzymes for generation and modification of flavour substances from precursors, enzyme reactors may be applied.

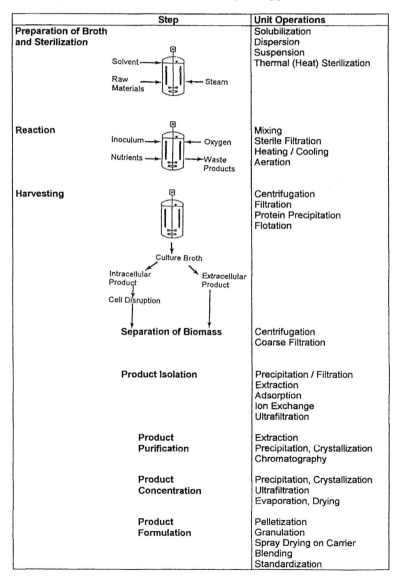

*Fig. 2.65:* Fermentation and processing for bioproduction of natural flavourings

To put it simple, the enzyme reactor is an agitated tank reactor in which the substrate is placed and stirred together with the enzyme for a certain period of time at a certain temperature in the liquid phase. The enzyme is added preferably in a concentration of 0.1 to 1% and is lost after completion of the reaction.

If expensive enzyme preparations have to be used it is possible to immobilize the enzyme and reuse it after the reaction is completed. For this purpose the enzyme may be coupled to a particulate carrier *[9]*. It is also possible to use an enzyme membrane reactor by including the high molecular weight enzyme proteins into a membrane compartment *[10]*. Reactions with carrier bound enzymes may be performed in special reactors (e.g. fixed bed, fluidized bed). When the reaction is completed the enzyme containing carrier beads are separated from the media and reused after gentle cleaning. In the enzyme membrane reactor the high molecular weight enzyme protein is rejected from the membrane which it cannot permeate. Educts and products of the enzyme reaction contained in the media flowing across the membrane can be transported through the membrane. With such membrane reactors (diagram in Figure 2.66) it is possible to transform precursors continuously into flavour substances. The obtained reaction mixture also has to be processed after reaction by the techniques mentioned above.

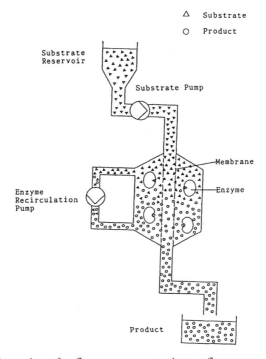

***Fig. 2.66:*** *Biotransformation of a flavour precursor into a flavour product by enzymes in a membrane reactor*

## 2.2.8 Cell Culture Reactors

The production of flavour substances by cell or tissue cultures is still a dream for the future in most cases. Today the extraction of product from intact living plants is still less expensive than the production by isolated cells and tissues. On the other hand, it is very attractive to make use of the secondary metabolism of plant cells for the synthesis of natural flavours in a controlled way to avoid contaminating by-products and thus considerably simplify downstream processing. Further advantages of such cell culture systems would be the independence from agriculture combined with the risk for possible shortage and variances in product quality, the ability to scale-up the process to create an inexhaustible source of well-defined product.

Secondary metabolism comprises the side paths of the ordinary metabolism, so-called primary metabolism which are activated in the cell in rest situations or under limiting conditions for nutrient and energy supply. In most cases, secondary metabolism is linked to the building blocks responsible for growth and reproduction which are products of primary metabolism and is hallmarked by a multitude of reactions, intermediates and final products. Starting materials for the secondary metabolism are e.g. amino acids, sugars and the co-enzymes of the primary metabolism. Only a very small fraction of formation mechanisms and the product variants of plant secondary metabolism have been characterized yet.

Fermentation of cell cultures is difficult and technically complicated *[11, 12]*. Contrary to single cell micro-organisms, higher cells require an environment similar to the one found in living organized tissue for survival and growth as well as for their metabolism. Critical environmental factors are light, nutrients, respiratory gases, hormones and growth factors. Metabolism waste materials and the products of synthesis from primary and secondary metabolism pathways have to be removed from the fermentation broth. Additionally, absolute sterility of the media is mandatory, because isolated cell tissue is by far more sensitive to microbial contamination than natural organisms.

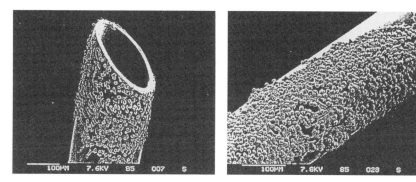

***Fig. 2.67:*** *Fibroblast cell culture grown on a dialysis membrane*

Isolated plant cells also tend to de-differentiate with progressing in-vitro cultivation time. Thus product generation can come to an absolute halt. In many cases de-differentiation can be delayed by the formation of a tissue-like structure of cells. This may be achieved by immobilizing cells in gels or by growing the cells to surfaces like polymer beads, textile fiber fabrics and membrane surfaces (Figure 2.67) to form a cell layer, a so-called callus culture *[13]*.

Preparing a plant cell culture one makes use of the pluripotency of plant cells which develop back to meristema cells after explanting them from the natural tissue (de-differentiation). Meristema cells may undergo unlimited segmentation and therefore can be propagated very easily. In culture such cells quickly lose their special tasks (piping, supporting, metabolic activity), which they had in the intact living plant. In many cases, however, their secondary metabolism pathways remain sufficiently active or may be reactivated by applying special culture conditions regarding nutrients, hormones or growth factors. As mentioned above, these pathways are also responsible for the formation of typical flavour substances in the plant.

In modern plant propagation techniques plant callus cultures are increasingly utilized. For this reason, there are considerable practical experiences with that type of cell culture methodology. Figure 2.68 outlines the preparation of a plant cell culture.

Plant cells or tissues may be fermented like micro-organisms in the submerged fermenter if grown on the surface of carrier beads or are kept in suspension. There is also experience in the operation of special membrane reactors for this purpose *[13]*.

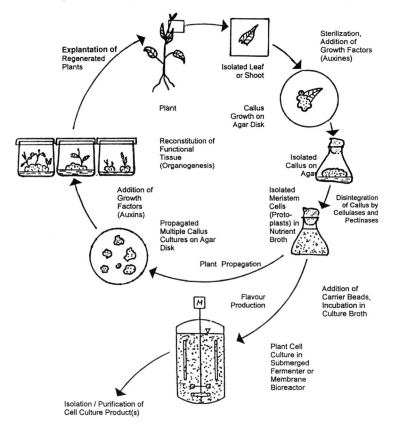

***Fig. 2.68:*** *Preparation of a plant cell (callus) culture*

Scientific results regarding the generation of flavouring materials from plant cell cultures on laboratory scale are available for quinine, capsaicin, quassin, vanillin, cocoa, citrus oils, peppermint oils, apple etc. *[14, 15]*.

The various pathways of secondary metabolism of plants leading mostly into complex flavour mixtures need further systematical experimental evaluation. After all it seems to be very attractive to make use of the synthetic performance of the natural organism for the production of flavours. In order to develop this technology into competitive and reliable industrial production, further optimization of process engineering is mandatory, e.g. identification and characterisation of the biosynthetic pathways involved or engineering of plant secondary metabolism to overcome limitations. Although huge advantages in understanding the metabolic production have been obtained by genomic, proteomic and metabolic approaches, this target still seems to be in the distant future *[16, 17]*.

## 2.2.9 Genetic Engineering

Genetic engineering comprises the modification of organisms in order to support traditional screening methods to alter metabolic pathways and products by manipulation of the DNA in the genes *[18]*. Figure 2.69 shows the cloning of a gene as example for a production process based on genetically engineered bacteria. The toolbox of a genetic engineer is depicted in Figure 2.70.

Genetic engineering provides organisms which may produce products of normal metabolism in considerably enhanced yields or which may also synthesize completely new products. The modifications achieved by genetic engineering are inherited by the following generations. Genetic engineering today is a routine technique for microorganisms. There is wide experience with a significantly enhanced production rate of enzyme synthesis in industrial production by multiple expression of genes responsible for enzyme synthesis in production organisms (self-cloning). The first results have been reported on industrial scale production with purposely modified higher organisms like plants *[19, 20]*. Genetic engineering offers other possibilities for further development to producers of biotechnical flavours.

It is possible to ferment ultra-pure enzymes without side activities by micro-organisms. Such products have only been accessible through isolation from animal tissue (e.g. chymotrypsin, a protease in cheese manufacturing). On the other hand, genetic engineering can also lead to the production of enzymes which are specifically tailored for their desired usage ('enzyme engineering'). Taking into account cell culture techniques instead of intact flavour producing plants, genetic engineering may make it feasible to significantly increase production of flavouring substances (e.g. flavr savr® – tomatoes, genetically modified by CALGENE, Inc. in California: Depressed synthesis for plant-borne polygalacturonases which liquefy the fruit tissue in the ripe state) *[21]*. State-of-the-art techniques of the 'directed evolution' approach and the 'genome shuffling' approach have led to dramatic progress by the directed and time-efficient development of novel enzymatic processes. These techniques in combination with high-throughput screening *[22]* enable the efficient production of enzymes which are specifically tailored for their desired application. A general overview is given by Bornscheuer and Pohl *[23]*.

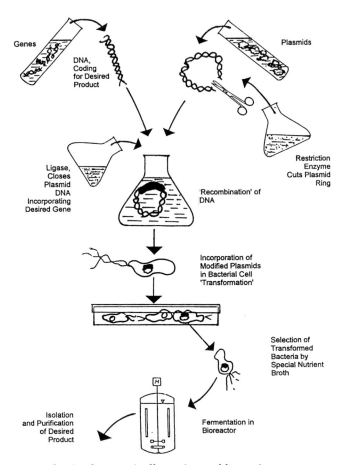

*Fig. 2.69:* Flavour production by genetically engineered bacteria

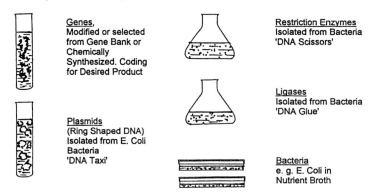

*Fig. 2.70:* Important tools for genetic engineering

Table 2.12 gives an overview of presently investigated genetically engineered flavours, processing aids and other food ingredients *[24]*.

*Table 2.12: Genetically engineered products for the food industry*

| Product | Biological Source | Use | Actual Situation |
|---|---|---|---|
| **Chymosin** | Calves | Cheese Making | Production Process by Fermentation of Yeast Available. Approval in the Final Stage. |
| **Thaumatin** | Plants | Sweetener | Production Process by Fermentation of Yeast Available. |
| **Protease Peptidase** | Various Microorganisms | Cheese Flavour | Various Proteases Cloned, Improvements by Enzyme Engineering (Focused on Use in Detergents) |
| **Lipases** | Various Microorganisms | Flavour Production (Cheese) | Various Enzymes Cloned |
| **Amylase** | Various Microorganisms; Plants | Flour Supplement | Various Enzymes Cloned |
| **Phospholipase** | Pig's Pancreas | Modified Egg Yolk; Emulsifying | Expression in Yeast; Protein Engineering |
| **α-Galactosidase** | Plant | Gum Modification | Production Process by Fermentation of Yeast Available |
| **Diacetyl Vitamin C Amino Acids** | Streptococcus | Flavour | Genetic Engineering under Study. Production of Precursor in E. Coli |
| **Terpenoids** | Moulds | Flavour | Genetic Engineering under Study |

All genetically engineered flavour production processes yield purely natural flavours. However, as to their labelling they have to be seen in the context of the EU Biotechnology Guideline. Presently, however, there is a widespread scepticism and reluctance among legislatures and non-governmental organisations as regards genetically modified organisms (GMO) and biotechnical processes which have led to a significant aversion among consumers. Incentives from the market place are therefore lacking to rapidly drive this technology.

## REFERENCES

[1] Belitz H.D., Scharf U., Lebensmitteltechnologie, in: Winnacher Küchler ed., Chemische Technologie, Vol. 7, pp 405-514, L. Aufl., München, Hanser 1986
[2] Luck T., Biotechnologie der Lebensmittelproduktion, in: Lebensmittelbiotechnologie, Czermak P., ed., Darmstadt, GIT-Verlag 1993
[3] Statistisches Jahrbuch über Ernährung, Landwirtschaft und Forsten, Münster, Landwirtschaftsverlag 1992
[4] Hugenholtz J., Smid E.J., Nutraceutical production with food-grade microorganisms, Curr. Opin. Biotechnol. 13, 497-507 (2002)
[5] Klappach G., Pilzlich fermentierte Lebensmittel, in: Lebensmittelbiotechnologie, Ruttloff H., ed., Berlin, Akademie Verlag 1991

[6]   Eisele A., Finn R.K., Samhaber W., Mikrobiologische und biochemische Verfahrenstechnik: eine Einführung, Weinheim, VCH Verlagsgesellschaft 1985
[7]   Stefer B. Bioprozesstechnische Charakterisierung eines organophilen Pervaporation-Bio-Hybridreaktors am Beispiel einer Aromasynthese. Fortschr. Ber. VDI Reihe 3 Nr. 814. Düsseldorf: VDI Verlag 2004
[8]   Lye G.J., Woodley J.M., Application of in situ product-removal techniques to biocatalytic processes, Trends Biotechnol. 17, 395-402 (1999)
[9]   Uhlig H., Enzyme arbeiten für uns: technische Enzyme und ihre Anwendung, München, Hanser 1991
[10]  Czermak P., König A., Tretzel J., Reimerdes E.H., Bauer W., Enzymkatalysierte Prozesse in Dialyse-Membranreaktoren, Forum Mikrobiologie, 11, 368-373 (1988)
[11]  Balandrin M.F., Klocke J.A., Natural Plant Chemicals: Sources of Industrial and Medicinal Materials, Science, 228, 1154-1160 (1985)
[12]  Böhm H., Hermersdörfer H., Möglichkeiten der Zellkulturtechnik zur Herstellung pflanzlicher Lebensmittelzusatzstoffe, in: Lebensmittelbiotechnologie, Ruttloff H., ed., Berlin, Akademie-verlag 1991
[13]  Mark U., Tretzel J., Verfahren und Vorrichtung zum Kultivieren von tierischen Zellen DE-OS3633891 (1988)
[14]  Robins R.J., Progress towards Producing Food Flavor Compounds by the Biotechnological Exploitation of Plant Cell Cultures, Abstracts Woerman Oslo, Flavor Symposium 1987
[15]  Westcott R., Progress and Prospects in Plant Biotechnology, in: Biotechnology Challenges for the Flavor and Food Industry, Lindsay R.C., Willis B.J. eds., London, Elsevier 1989
[16]  Chappell J., Valencene synthase – a biochemical magician and harbinger of transgenic aromas, Trends Plant Sci., 9, 266-269 (2004)
[17]  Guterman I., Shalit M., Menda N., Piestun D., Dafny-Yelin M., Shalev G., Bar E., Davydov O., Ovadis M., Emanuel M., Wang J., Adam Z., Pichersky E., Lewinsohn E., Zamir D., Vainstein A., Weiss D., Rose scent: genomics approach to discovering novel floral fragrance-related genes, The Plant Cell, 14, 2325-2338 (2002)
[18]  Gassen H.G., Martin A., Sachse G., Der Stoff, aus dem die Gene sind, München, Schweitzer 1986
[19]  Memelink J., Kijne J.W., van der Heijden R., Verpoorte R., Genetic modification of plant secondary metabolite pathways using transcriptional regulators, Adv. Biochem. Eng. Biotechnol., 72, 103-125 (2001)
[20]  Dudareva N., Pichersky E., Gershenzon J., Biochemistry of plant volatiles, Plant Physiol., 135, 1893-1902 (2004)
[21]  Overbeeke N., Genetic Engineering in Flavour Research, in: Biotechnology Challenges for the Flavor and Food Industry, Lindsay R.C., Willis B.J., eds., London, Elsevier 1989
[22]  Smit B.A., Engels W.J., Bruinsma J., van Hylckama Vlieg J.E., Wouters J.T., Smit G., Development of a high throughput screening method to test flavour-forming capabilities of anaerobic microorganisms, J. Appl. Microbiol., 97, 306-313 (2004)
[23]  Bornscheuer U.T., Pohl M., Improved biocatalysts by directed evolution and rational protein design, Curr. Opin. Chem. Biol., 5, 137-142 (2001)
[24]  Frankfurter Allgemeine Zeitung, Nr. 159, Seite 16 (12 July 1994)

# 3 Raw Materials for Flavourings

# 3.1 Introduction

*Günter Matheis*

During the early days of history, people used mainly herbs and spices (whole or ground) to impart flavour to, or modify the flavour of, foods. It was only in the Middle Ages that some extraction of plant materials started, followed by distillation of essential oils. The latter were predominantly used by pharmacists, and it was not until the 19th century that some people found out that essential oils can be used to impart flavour to foods. In the second half of the 19th century, chemists began to realize the flavouring potential of some synthetic chemicals (e.g. vanillin, which was synthesized from guaiacol). Thus was born the flavour industry around the middle of the 19th century.

The first raw materials for the flavour industry included extracts, tinctures, oleoresins, juice concentrates, essential oils, and a few synthetic chemicals (Tab. 3.1). Up to the 1950s, flavour research was concentrated on the isolation, structural analysis, and synthesis of just a few quantitatively outstanding natural materials (Tab. 3.1). The situation changed dramatically with the advent of gas chromatography as a means of analysis, especially in conjunction with mass spectrometry.

At the start of modern flavour research, the prime object of investigation was to find out how many individual components made up a flavour and which they were. That phase has today passed its peak. More than 4,000 chemicals have been identified as flavouring substances. There may well be more awaiting discovery.

*Table 3.1:* Chronology of flavour research and development [1]

| Period | Remarks |
|---|---|
| Before 1900 | Import of tropical fruits and spices<br>Identification of some flavouring substances<br>Synthesis of some flavouring substances |
| Ca. 1900-1950 | Start of structural analysis of flavouring substances<br>Improvement of extraction and distillation processes<br>Improvement of synthesizing techniques |
| Ca. 1950-1970 | Introduction of chromtography and spectral photometry for analytical characterization of flavouring substances<br>Evidence of hundreds of individual components in a single flavour<br>Development of flavourings with up to 30 individual components (predominantly nature-identical flavouring substances) |
| Since ca. 1970 | Identification of flavouring substances by gas chromatography coupled with mass spectrometry (GC/MS)<br>Development of flavourings with up to 80 individual components (predominantly nature-identical flavouring substances)<br>Biotechnological methods to produce natural flavourings and natural flavouring substances<br>Growing importance of natural flavourings |

Today, society demands that the food industry should provide the required quantity of good tasting, nourishing food to satisfy the demand at as low a price as possible. The modern trend is towards the consumption of more processed and convenience foods calling for a wide spectrum of flavouring effects. The flavour industry serves the food industry by providing a great variety of flavourings.

### *What is a flavouring?*

Various definitions in the literature emphasize the idea that a flavouring is any substance (single chemical entity or blends of materials of synthetic or natural origin) whose primary purpose is to impart flavour [2-4]. The definitions of the International Organization of the Flavour Industry (IOFI) and of the Council of the European Communities are given in Tab. 3.2. It is obvious from these definitions that flavourings consist of ingredients that contribute to the flavour of foods and of ingredients that do not contribute to the flavour. The latter are flavour adjuncts and include, for example, solvents, carriers, preservatives, and food additives.

*Table 3.2:* Definitions of flavouring according to the International Organization of the Flavour Industry [5] and to the Council of the European Communities [6]

| International Organization Flavour Industry, 1990 | Council of the European Communities, 1988 |
|---|---|
| Concentrated preparation, with or without flavour adjuncts[a], used to impart flavour, with the exception of only salty, sweet or acid tastes. It is not intended to be consumed as such. | Flavouring means flavouring substances, flavouring preparations, process flavourings, smoke flavourings or mixtures thereof.<br>Flavourings may contain foodstuff as well as other substances[b]. |
| a:<br>Food additives and food ingredients necessary for the production, storage and application of flavourings as far as they are nonfunctional in the finished food. | b:<br>Additives necessary for the storage and use of flavourings, products used for dissolving and diluting flavourings, and additives for the production of flavourings (processing aids) where such additives are not covered by other Community provisions. |

It should be noted here that, according to the recommendations of the IOFI [5], the terms flavour and flavouring should not be used as synonyms. **Flavour** should only be used to describe effects upon the senses, whereas *flavouring* means a preparation used to impart flavour.

## 3.2 Flavouring Ingredients

IOFI and Council of Europe definitions of flavouring ingredients that are permitted in flavourings are summarized in Tab. 3.3. They will be discussed in detail in the following paragraphs.

*Table 3.3:* IOFI and Council of Europe definitions of flavouring substance, flavouring preparation, process flavouring, and smoke flavouring [5, 6]

| | **IOFI, 1990** | **Council of Europe, 1988** |
|---|---|---|
| Flavouring substance | Defined chemical component with flavouring properties, not intended to be consumed as such | Defined chemical substance with flavouring properties |
| Natural flavouring substance | Defined substance obtained by appropriate physical, microbiological or enzymatic processes from a foodstuff or material of vegetable or animal origin as such or after processing by food preparation processes | Flavouring substance obtained by physical processes (including distillation and solvent extraction) or enzymatic or microbial processes from material of vegetable or animal origin either in the raw state or after processing for human consumption by traditional food preparation processes (including drying, torrefaction and fermentation) |
| Nature-identical flavouring substance | Flavouring substance obtained by synthesis or isolated through chemical processes from a natural aromatic raw material and chemically identical to a substance present in natural products intended for human consumption, either processed or not | Flavouring substance obtained by chemical synthesis or isolated by chemical processes and which is chemically identical to a substance naturally present in material of vegetable or animal origin |
| Artificial flavouring substance | Flavouring substance, not yet identified in a natural product intended for human consumption, either processed or not | Flavouring substance obtained by chemical synthesis but which is not chemically identical to a substance naturally present in material of vegetable or animal origin |
| Flavouring preparation | A preparation used for its flavouring properties, which is obtained by appropriate physical, microbial or enzymatic processes from a foodstuff or material of vegetable or animal origin, either as such or after processing by food preparation processes | A product, other than flavouring substances, whether concentrated or not, with flavouring properties which is obtained by appropriate physical processes (including distillation and solvent extraction) or by enzymatic or microbial processes from material of vegetable or animal origin, either in the raw state or after processing for human consumption by traditional by traditional food-preparation processes (including drying, torrefaction and fermentation) |

|  | **IOFI, 1990** | **Council of Europe, 1988** |
|---|---|---|
| Process flavouring | A product or mixture prepared for its flavouring properties which is produced from ingredients or mixtures of ingredients which are themselves permitted for use in foodstuffs, or are present naturally in foodstuffs, or are permitted for use in process flavourings, by a process for the preparation of foods for human consumption. Flavour adjuncts may be added. This definition does not apply to flavouring extracts, processed natural food substances or mixtures of flavouring substances | A product which is obtained according to good manufacturing practices by heating to a temperature not exceeding 180 °C for a period not exceeding 15 minutes a mixture of ingredients, not necessarily themselves having flavouring properties, of which at least one contains nitrogen (amino) and another is a reducing sugar |
| Smoke flavouring | Concentrated preparation, not obtained from smoked materials, used for the purpose of imparting a smoke type flavour to foodstuffs. Flavour adjuncts may be added | A smoke extract used in traditional food foodstuff's smoking processes |

## 3.2.1 Chemically Defined Flavouring Substances

A flavouring substance is a defined chemical component or substance with flavouring properties (Tab. 3.3). Various synonyms are in use, e.g. flavour substance, flavour chemical, flavour(ing) component, flavour(ing) compound, flavouring agent, aroma compound, aroma chemical, and others. Since flavour includes both taste and odour, a flavouring substance may be a substance that causes either taste or odour impressions, or both.

Flavouring substances that cause only taste impressions are defined as "substances that are usually non-volatile at room temperature. Therefore, they are only perceived by the taste receptors" [7]. Examples are sucrose (sweet) or caffeine (bitter). Flavouring substances causing odour impressions are "volatiles that are perceived by the odour receptors" [7]. Examples are ethyl butyrate or dimethyl sulfide. Some flavouring substances are perceived by taste and odour receptors (e.g. acetic acid, butyric acid).

Flavouring substances may be classified into natural, nature-identical, and artificial flavouring substances (Tab. 3.3). These will be discussed below in detail.

### 3.2.1.1 Natural Flavouring Substances

Natural flavouring substances may be obtained by physical, enzymatic, or microbial processes from materials as defined in Tab. 3.3. Enzymatic and microbial processes are also known as biotechnological processes.

Thirty years ago, relatively few natural flavouring substances were available. Today, more than two hundred natural flavouring substances of high purity are at our disposal. Among these are approx. one hundred esters. In addition, up to one hundred natural flavouring substances of less purity and mixtures of 2 to 5 natural flavouring substances are available, e.g. "green notes" (hexanal, isomeric hexenals, and corresponding $C_6$ alcohols), "pineapple enhancer" (allyl caproate and other substances), pyrazine mixture (2,5- and 2,6-dimethyl pyrazines), methyl ketone mixture (2-heptanone, 2-nonanone, 2-undecanone), sinensal fraction ex orange (20 % of α- and β-sinensal), and many more. These products belong to the class of "flavouring preparations" (see 3.2.2.).

### 3.2.1.1.1 Natural Flavouring Substances Manufactured by Physical Processes

Physical processes (see chapter 2) for isolation of natural flavouring substances include distillation, solvent extraction (including supercritical carbon dioxide), and chromatography. Major sources are essential oils. These may be derived from various parts of aromatic plants such as fruits (e.g. citrus, fennel), fruit parts (e.g. mace), flowers (e.g. safflower), flower parts (e.g. saffron), flower buds (e.g. clove), bulbs (e.g. onion), barks (e.g. cinnamon), leaves (e.g. basil), leaves and twigs (e.g. mandarin petitgrain), rhizomes (e.g. ginger), roots (e.g. angelica), and seeds (e.g. mustard).

Tab. 3.4 lists some natural flavouring substances that are isolated from essential oils by physical processes. Cinnamic aldehyde and benzaldehyde have been isolated as early as 1834 and 1837, respectively.

*Table 3.4:* Selection of flavouring substances isolated from essential oils by physical methods

| Flavouring substance | Odour [8] | Possible Source |
|---|---|---|
| Anethol | Herbaceous-warm, anisic | Anise (Pimpinella anisum) Fennel (Foeniculum vulgare) Staranise (Illicium verum) |
| Allyl isothiocyanate | Pungent, stinging | Black mustard (Brassica nigra) |
| Benzaldehyde | Bitter almond | Bitter almond (Prunus amygdalus var. amara) |
| D-Carvone | Warm-herbaceous, breadlike, spicy, floral, caraway, dill | Caraway (Carum carvi) |
| L-Carvone | Warm-herbaceous, breadlike, spicy, spearmint | Spearmint (Mentha spicata) |
| 1,8-Cineole | Fresh, camphoraceous-cool | Eucalyptus (Eucalyptus globulus) |
| Cinnamic aldehyde | Warm, spicy, balsamic | Cassia (Cinnamomum cassia) Cinnamon (Cinnamomum zeylanicum) |
| Citral | Lemon | Lemongrass (Cymbopogon citratus, C. flexuosus) Litsea cubeba |
| Citronellal | Fresh, green, citrus | Eucalyptus citriodora |
| Decanal | Orange peel | Orange (Citrus sinensis) |
| Dimethyl sulfide | Sharp, green radish, cabbage | Cornmint (Mentha arvensis) |
| Eugenol | Warm-spicy | Clove (Syzygium aromaticum) |

| Flavouring substance | Odour [8] | Possible Source |
|---|---|---|
| Geraniol | Floral, rose | Palmarosa (Cymbopogon martini) Citronella (Cymbopogon nardus) |
| Geranyl acetate | Sweet, fruity-floral, rose, green | Lemongrass (Cymbopogon citratus) |
| (Z)-3-Hexenol | Green, grassy | Cornmint (Mentha arvensis) |
| D-Limonene | Fresh, orange peel | Citrus (Citrus species) |
| Linalool | Refreshing, floral-woody | Basil (Ocimum basilicum) Bois de Rose (Aniba rosaeodora) Camphor tree (Cinnamomum camphora) |
| Linalyl acetate | Sweet, floral-fruity | Bergamot mint (Mentha citrata) |
| Massoia lactone | Coconut | Massoia tree (Cryptocaria massoia) |
| Methyl chavicol | Sweet-herbaceous, anise, fennel | Basil (Ocimum basilicum) |
| Methyl cinnamate | Fruity-balsamic | Eucalyptus campanulata |
| Methyl N-methyl anthranilate | Musty-floral, sweet | Mandarin (Citrus reticulata) |
| Nootkatone | Fruity, sweet, citrus, grapefruit peel | Grapefruit (Citrus paradisi) |
| Terpinenol-4 | Warm-peppery, earthy-musty | Tea tree (Melaleuca alternifolia) |
| Thymol | Sweet-medicinal, harbaceous, warm | Thyme (Thymus vulgaris) Origanum (Origanum vulgare) |
| 2-Undecanone | Fruity-rosy, orange-like | Rue (Ruta graveolens) |

Besides these volatiles, various non-volatile flavouring substances are also isolated from plant material. Examples are the bitter tasting substances amarogentin and naringin, the pungent components capsaicin and piperine, and the sweet tasting substances hernandulcin (from the Mexican herb *Lippia dulcis*), stevioside (from the leaves of *Stevia rebaudiana*), glycyrrhizin (from licorice root), osladin (from the rhizomes of *Podiophyllum vulgare*), phyllodulcin (from *Hydragea macrophylla*), thaumatins (from *Thaumatococcus danielli*), and monellin (from *Dioscoreophyllum cummiusii*).

### 3.2.1.1.2 Natural Flavouring Substances Manufactured by Biotechnological Processes

A selection of biotechnological processes that are used to produce natural flavouring substances is shown in Fig. 3.1 (see also chapter 2.2). Note that physical processes (see 2.1) are always involved in isolating the products from the fermentation broth.

*Plant homogenates* have been considered as potential sources of natural flavouring substances. Special interest has been devoted to the so-called "green notes" (hexanal, isomeric hexenals, and corresponding $C_6$ alcohols) *[9-10]*. In spite of our knowledge about the fundamental steps of biogenesis (Fig. 3.2), industrially interesting amounts have not been achieved as yet. Nevertheless, the process has been patented using strawberry leaves *[11]* and soy beans *[12]* as homogenates.

Flavouring Ingredients

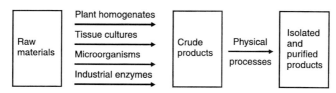

*Fig. 3.1:* Selection of biotechnological processes for the production of natural flavouring substances

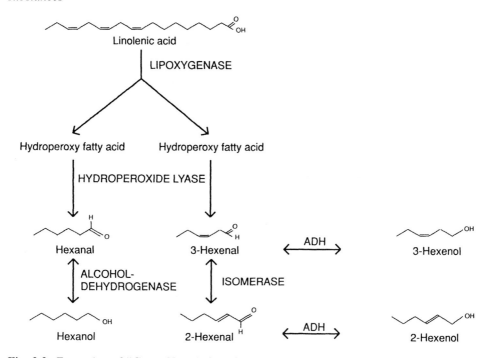

*Fig. 3.2:* Formation of "Green Notes" from lipids (ADH = alcohol dehydrogenase)

With mushroom homogenate, a process for the production of (R)-1-octen-3-ol (Fig. 3.3) has been developed *[13]*. To improve the yield of (R)-1-octen-3-ol, a lipid rich in linoleic acid and the commercially available enzyme lipase may be added to the mushroom homogenate.

A number of different types of *plant tissue cultures* (e.g. suspension cultures, differentiated cultures, immobilized cultures, and transformed cultures) have been studied for the production of flavouring substances *[14-19]*. As *de novo* biosynthesis has been found unsuccessful in most cases (exceptions are shown in Tab. 3.5.), biotransformation of added precursors has been studied extensively. Fig. 3.4 shows some examples of biotransformation of terpenes by suspension cultures. Tab. 3.6 lists some biotransformations by suspension or immobilized cultures.

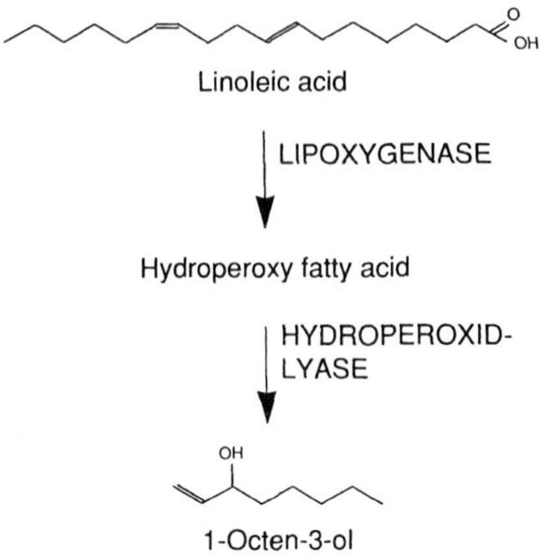

*Fig. 3.3:* Formation of (R)-1-octen-3-ol from lipids

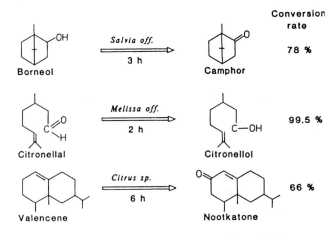

*Fig. 3.4:* Biotransformation of terpenses by suspension cultures [23]

*Table 3.5:* Examples of de novo biosynthesis of flavouring substances by tissue cultures [16, 20-22]

| Plant | Flavouring substance | Organoleptic properties |
| --- | --- | --- |
| Pimpinella anisum | Anethol | Herbaceous-warm, anisic |
| Coffea arabica | Caffeine | Bitter |
| Capsicum frutescens | Capsaicin | Pungent |
| Glycyrrhiza glabra | Glycyrrhizin | Sweet |
| Quassia amara | Quassin | Bitter |
| Cinchona ledgeriana | Quinine | Bitter |
| Vanilla fragrans | Vanillin | Vanilla |
| Rubus idaeus | Raspberry ketone | Raspberry |

*Table 3.6: Biotransformation of various substrates by suspension or immobilized cultures [16, 23-26]*

| Plant | Substrate | Product | Odour [8] |
|---|---|---|---|
| Cannabis sativa | Geraniol | Geranial | Lemon |
|  | Nerol | Neral | Lemon |
| Lavandula angustifolia | Geranial | Geraniol | Floral, rose |
|  | Neral | Nerol | Rose, sea shore |
|  | Citronellal | Citronellol | Fresh, rosy-leafy, petal-like |
| Nicotiana tabacum | Linalool | 8-Hydroxy linalool | Floral |
| Vanilla planifolia | Ferulic acid | Vanillin | Vanilla |
| Mentha canadensis, Mentha piperita | L-Menthyl acetate | L-Menthol | Refreshing, cooling, peppermint |

To date, most cultures have been unable to produce adequate yields of flavouring substances. The accumulation of larger amounts of flavouring substances in tissue cultures will continue to be a challenging scientific problem [20, 27].

There are two principal ways for utilization of *microorganisms* (yeasts, fungi, bacteria) for the production of flavouring substances, i.e. fermentation (*de novo* biosynthesis) and biotransformation (Tab. 3.7). Fermentation products are usually complex (see 3.2.2.4.). Nevertheless, there are some single flavouring substances that are produced by fermentation, such as acetic, butyric, and propionic acids and others (Tab. 3.8). For biotransformations by microorganisms, suitable substrates are necessary. Some examples are given in Tab. 3.9 and Fig. 3.5.

*Table 3.7: Characteristics of fermentation and microbial transformation. Modified from [17]*

| Parameter | Fermentation | Biotransformation |
|---|---|---|
| Microorganisms | Growing cells | Growing, permanent, or treated cells |
| Reaction | Complex reaction chain | Simple (one- or multistep) reaction |
| Reaction time | Long | Short |
| Substrate | Cheap carbon and nitrogen sources | Specific (sometimes expensive) |

*Table 3.8: Examples of de novo biosynthesis of flavouring substances by microbial fermentation [20, 28-36]*

| Microorganism | Flavouring substance | Organoleptic properties |
|---|---|---|
| Lactococcus species Leuconostoc species | Diacetyl | Buttery |
| Aspergillus niger | Citric acid | Sour |
| Pseudomonas species | 3-Isopropyl-2-methoxy pyrazine | Green pea |
| Streptococcus lactis | Methyl butanol | Malty |
| Trichoderma viride | 6-Pentyl-α-pyrone | Coconut |

| Microorganism | Flavouring substance | Organoleptic properties |
|---|---|---|
| Bacillus subtilis Corynebacterium glutamicum | Tetramethyl pyrazine | Nutty |
| Trametes sauvolens | Anise aldehyde | Anise |
| Trametes sauvolens | Benzaldehyde | Almond |
| Ceratocystis variospora, Trametes odorata | Citronellol | Rose |
| Ceratocystis variospora | Citronellyl acetate | Fruity, rose |
| Fusarium poae | γ-Dodecalactone | Fatty, buttery |
| Phellinus igniarius | Ethyl benzoate | Fruity |
| Lactobacillus casei, L. diacetylactis, Pseudomonas fragi | Ethyl butyrate | Fruity |
| Ceratocystis variospora | Geranial | Rose |
| Penicillium italicum, Ceratocystis variospora, Phellinus igniarius | Linalool | Floral |
| Phellinus tremulus | Methyl benzoate | Fruity |
| Phellinus igniarius | Methyl salicylate | Wintergreen |
| Lasiodiploida theobromae | 2-Octene-δ-lactone | |
| Mycoacia uda | ρ-Tolyl aldehyde | Almond |

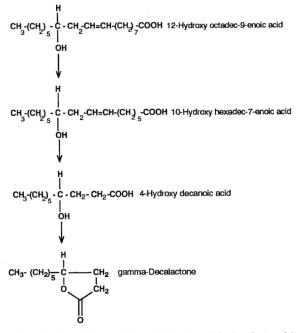

*Fig. 3.5: Oxidative degradation of ricinoleic acid by Candida lipolytica [40]*

*Table 3.9: Examples of biotransformation by microorganisms [20, 29, 40-67]*

| Product | Substrate | Microorganism | Organoleptic properties of product [8] |
|---|---|---|---|
| Acetaldehye | Ethanol | Candida utilis | Pungent, ethereal-nauseating |
| Benzyldehyde | Benzoic acid, benzyl alcohol or phenyl-alanine | Polyporus tuberaster, Pichia pastoris | Almond |
| γ-Decalactone | Ricinolic acid, oleic acid or 3-decen-4-olide | Candida lipolytica, Sporobolomyces odorus, baker's yeast | Peach |
| δ-Decalactone | Ricinolic acid, corrolic acid, Massoia lactone or 11-hydroxy hexadecanoic acid | Candida species, Clador-sporium suavolens, baker's yeast | Peach, buttery |
| 7-Decen-4-olide | Densipolic acid | Sporobolomyces odorus | Fatty, buttery, nutty |
| Ethyl acetate | Ethanol | Candida utilis | Ethereal-fruity |
| Ethyl isovalerate | L-Leucine | Geotrichum fragrans | Ethereal, vinous-fruity |
| Furfuryl-mercaptane | Furfural | Eubacterium limosum | |
| 2-Heptanone | Caprylic acid | Penicillium roqueforti | Herbaceous-green |
| 2,4-Hexadienal | Sorbic acid | Colletotrichum gleosporoides | Fruity-green |
| Hexanol | Hexanoic acid or 2-hexenal | Colletotrichum gleosporoides, baker's yeast | |
| 4-Hexen-1-ol | Sorbic acid | Colletotrichum gleosporoides | |
| 4-Hydroxy-2,5-dimethyl-3(2H)-furanone (Furaneol®) | Fructose-1,6-diphosphate | Zygosaccharomyces rouxii | Caramel, roasty |
| 3-Hydroxy-2-pentanone | 2,3-Pentandione | Baker's yeast | |
| Isobutyric acid | Isobutanol | Acetobacter species | Buttery, cheesy |
| L-Menthol | L-Menthone | Pseudomonas putida, Cullulomonas turbata | Refreshing, cooling, peppermint |
| L-Menthol (+D-menthyl acetate) | D, L-Menthyl acetate | Penicillium species, Rhizopus species, Trichoderma species | Refreshing, cooling, peppermint |
| Methyl anthranilate | Methyl N-methyl anthranilate | Trametes versicolor | Mosty-fruity, Concord grapes |

| Product | Substrate | Microorganism | Organoleptic properties of product [8] |
|---|---|---|---|
| 2-Methyl butyric acid | 2-Methylbutanol | Gluconobacter species | |
| 6-Methylhept-5-en-2-one | Nerol, citral | Penicillium digitatum | |
| δ-Octalactone | Jalap resin | Saccharomyces cerevisiae | Coconut |
| γ-Octalactone | Coconut oil fraction, ethyl caprylate | Polyporus durus, Mucor circinelloides | Coconut, Tonka bean |
| α-Terpineol | Limonene | Pseudomonas gladioli | Floral, lilac |
| Vanillin | Eugenol | Serratia species, Klebsiella species, Enterobacter species | Vanilla |
| Vanillin | Ferulic acid, eugenol, isoeugenol | Corynebacterium glutamicum, Pycnoporus cinnabarinus, Serratia species, Klebsella species, Enterobacter species, Pseudomonas species, Aspergillus niger | Vanilla |

Many (if not most) biotechnologically produced flavouring substances are produced in processes using *industrially available enzymes*. Enzymes are proteins with catalytic properties (biocatalysts). Conspicuous characteristics of an enzyme are:
(a) the ability to increase reaction velocity, (b) substrate specificity (= specificity with respect to the substances it acts upon), and (c) reaction specificity (= specificity with respect to the type of reaction that is catalyzed). Substrate specificity is more marked in some cases than in others. A particularly striking feature is the strong specificity with respect to stereoisomeric compounds. In the case of substances with chiral centres, only one enantiomer is converted. Specificity with respect to diastereomers, particularly cis-trans-isomers, is also widespread.

Reaction specificity is the most significant and most widespread classification of enzymes (Tab. 3.10). Another way of classifying enzymes is by their complexity (Tab. 3.11). As already pointed out, enzymes are proteins or at least consist predominantly of a protein portion. Some enzymes need cofactors (prosthetic groups or cosubstrates; see Tab. 3.11).

*Table 3.10: Classification of enzymes according to reaction specificity [68]*

| Main Class | Remarks | Examples |
| --- | --- | --- |
| (1) Oxidoreductases | Catalyze oxidation and reduction | Peroxidase, lipoxygenase, phenoloxidase |
| (2) Transferases | Transfer chemical groups from one molecule to another | Hexokinase |
| (3) Hydrolases | Catalyze hydrolytic cleavage | Proteinases, lipases, pectinesterase |
| (4) Lyases | Remove groups to form double or add groups to double bonds | Pectinlyase, alliinlyase |
| (5) Isomerases | Catalyze isomerization and interconversion within a molecule | Glucose phosphate isomerase |
| (6) Ligases | Synthesize compounds | |

*Table 3.11: Classification of enzymes according to their comlexity [68]*

| |
| --- |
| (1) Enzymes without cofactor |
| (2) Enzymes with prosthetic group (covalently bonded cofactor) |
| (3) Enzymes with cosubstrate (non covalently bonded cofactor) |
| (4) Multi-enzyme complexes |

More than 10,000 enzymes occur in nature. Of these, approx. 3,000 are characterized. Approx. 800 are commercially available, but only approx. 20 in industrial amounts (predominantly hydrolases and oxidoreductases). They are isolated from microorganisms, plants, or animals. Lipases (which belong to the class of hydrolases) and oxidoreductases catalyze, for example, the reactions depicted in Fig. 3.6. Note that all reactions are reversible. Examples of flavouring substances that are produced with lipases and oxidoreductases are shown in Tab. 3.12.

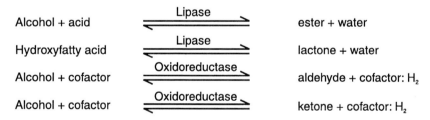

*Fig. 3.6: Reactions catalyzed by lipases and oxidoreductases [68]*

*Table 3.12: Selection of flavouring substances produced with various enzymes [18, 40, 68-72]*

| Product | Substrate(s) | Enzyme | Odour of product [8] |
|---|---|---|---|
| Butyric acid | Butter fat | Lipase | Sour, rancid |
| Caproic acid | Butter fat | Lipase | fatty-rancid, sweat-like |
| Caprylic acid | Butter fat | Lipase | Oily-rancid, sweat-like |
| Capric acid | Butter fat | Lipase | Sour-fatty, rancid |
| Ethyl butyrate[a] | Ethanol, butyric acid | Lipase | Ethereal-fruity |
| γ-Butyrolactone | γ-Hydroxy butyric acid | Lipase | Sweet-aromatic |
| Acetaldehyde | Ethanol | Alcohol dehydrogenase | Pungent, ethereal nauseating |
| Benzaldehyde | Benzyl alcohol | Alcohol dehydrogenase | Almond |
| 4-Hydroxy-2,5-dimethyl-3(2H)-furanone | Fructose-1,6-diphosphate, lactaldehyde | Aldolase, triosephosphate isomerase | Caramel, roasty |
| Geranial | Geraniol | Alcohol dehydrogenase | Lemon |
| Cinnamic alcohol | Cinnamic aldehyde | Alcohol dehydrogenase | Warm-balsamic, floral |
| Methanethiol | Methionine | Methioninase | Sulfureous |

[a] Ethyl butyrate is only one example of the many esters that are produced with lipases.

*Table 3.13: Advantages and disadvantages of using enzymes [30]*

| Parameter | Advantage | Disadvantage |
|---|---|---|
| Reaction medium | aqueous[a] | |
| Pressure | atmospheric | |
| Temperature | 20-50°C | |
| Specificity | high | |
| Product | natural[b] | |
| Reaction time | | relatively long |
| Stability of the enzyme | | low[c] |
| Availability of enzyme | | low |
| Cost of the enzyme | | high |
| Cost of enzyme co-factors | | high[d] |

[a] apart from a few exceptions
[b] if original material is natural
[c] can be increased by immobilization
[d] can be minimized by regeneration

Advantages and disadvantages of using enzymes are summarized in Tab. 3.13. The stability of enzymes can be increased by immobilization. Fig. 3.7 shows as an example the formation of mustard oils through immobilized myrosinase. Mustard seeds contain myrosinase naturally. The higher yields resulting from the use of additional, immobilized enzyme make the process more economical. Immobilized lipase has been used successfully for the production of esters *[69, 73, 74]*. The conversion efficiencies of various alcohols and acids to their corresponding esters are shown in Tab. 3.14.

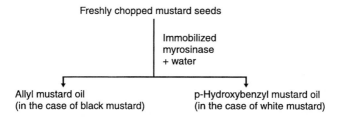

*Fig. 3.7:* Formation of mustard oils from mustard seeds with the aid of immobilized myrosinase *[30, 40]*

*Table 3.14:* Ester production by immobilized lipase *[74]*

| Ester | Conversion (%) |
|---|---|
| Ethyl propionate | 76 |
| Ethyl butyrate | 100 |
| Ethyl caproate | 44 |
| Ethyl heptanoate | 84 |
| Ethyl caprylate | 100 |
| Ethyl laurate | 52 |
| Ethyl isobutyrate | 72 |
| Ethyl isovalerate | 3 |
| Isobutyl acetate | 25 |
| Isoamyl acetate | 24 |
| Isoamyl butyrate | 91 |

In the case of enzymes that need cofactors for their catalytic activity (e.g. alcohol dehydrogenase), the high cost of these cofactors may prevent large-scale use. At present, efforts are being made to regenerate the cofactors by the use of a second enzyme (Fig. 3.8).

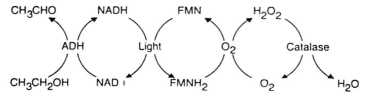

*Fig. 3.8:* Enzymatic oxidation of ethanol to acetaldehyde *[40]*. ADH = alcohol dehydrogenase, $NAD_+$ and FMN = cofactors, catalase = second enzyme to regenerate cofactors

Summarizing the discussion of the production of flavouring substances by biotechnological processes, the following conclusions may be drawn:

a) The biogenetic pathways of many flavouring substances are known *[75-79]*. This knowledge is the prerequisite for current and future biotechnological processes.

b) The potential of plant homogenates and tissue cultures is considered to be useful, although the yields need to be increased in most cases.

c) Microbial transformations and fermentations are in use and appear to be more significant in future than plant homogenates and cell cultures.

d) More than one hundred flavouring substances are already produced with the aid of industrially available enzymes. Technologies have to be developed to use the still unexploited potential of glycosidically bound flavouring substances with the aid of hydrolases *[17, 78]*.

e) Genetic engineering (which was not covered in the present discussion) has generated potentials that will be topics of future research activities.

### 3.2.1.1.3 Examples of Commercially Available Natural Flavouring Substances

A selection of commercially available natural flavouring substances of high purity is given in Tab. 3.15. Note that there are approx. 100 esters.

*Table 3.15: Selection of commercially available natural flavouring substances of high purity*

| | |
|---|---|
| Acetaldehyde | Capsaicin |
| Acetic acid | Carvacrol |
| Acetoin | D-Carvone |
| Acetone | L-Carvone |
| Allyl isothiocyanate | β-Caryophyllene |
| Amarogentin | 1,8-Cineole |
| Anethol | Cinnamic acid |
| Anisaldehyde | Cinnamic alcohol |
| Anisyl acetate | Cinnamic aldehyde |
| Anisyl alcohol | Cinnamyl acetate |
| | Cinnamyl cinnamate |
| Benzaldehyde | Citral |
| Benzyl acetate | Citric acid |
| Benzyl alcohol | Citronellal |
| Benzyl butyrate | Citronellol |
| Benzyl propionate | Citronellyl acetate |
| Butanol | Citronellyl butyrate |
| Butyl acetate | Citronellyl propionate |
| Butyl heptanoate | |
| Butyl isovalerate | γ-Decalactone |
| Butyl 2-methyl butyrate | Decanal |
| Butyric acid | Diacetyl |
| | Diethyl acetal |
| Capric acid | Dihydro cinnamic alcohol |
| Caproic acid | Dihydro cuminyl aldehyde |
| Caprylic acid | 2,5-Dimethyl-3(2H)-furanone |

| | |
|---|---|
| Dimethyl sulfide | Isoamyl acetate |
| Diosphenol | Isoamyl alcohol |
| | Isoamyl butyrate |
| Estragol | Isoamyl caprate |
| Ethyl acetate | Isoamyl caproate |
| Ethyl anisate | Isoamyl caprylate |
| Ethyl benzoate | Isoamyl isobutyrate |
| Ethyl butyrate | Isoamyl isovalerate |
| Ethyl caprate | Isoamyl laurate |
| Ethyl caproate | Isoamyl 2-methyl butyrate |
| Ethyl caprylate | Isoamyl propionate |
| Ethyl cinnamate | Isobutanol |
| Ethyl heptanoate | Isobutyl acetate |
| Ethyl isobutyrate | Isobutyl butyrate |
| Ethyl isovalerate | Isobutyl caprate |
| Ethyl lactate | Isobutyl caproate |
| Ethyl laurate | Isobutyl caprylate |
| Ethyl 2-methyl butyrate | Isobutyl isovalerate |
| Ethyl myristate | Isobutyl laurate |
| Ethyl propionate | Isobutyl 2-methyl butyrate |
| Ethyl pyruvate | Isobutyraldehyde |
| Eugenol | Isobutyric acid |
| | Isopulegol |
| D-Fenchone | Isovaleraldehyde |
| Furfural | Isovaleric acid |
| | |
| Geraniol | Lactic acid |
| Geranyl acetate | Lauric acid |
| Geranyl butyrate | Limonene |
| Geranyl caprate | Linalool |
| Geranyl caproate | Linalyl acetate |
| Geranyl caprylate | Linalyl butyrate |
| Geranyl propionate | Linalyl propionate |
| Glycyrrhizin | |
| | Massoia lactone |
| Heptanal | Menthol |
| Heptanoic acid | Menthone |
| 2-Heptanone | Menthyl acetate |
| Heptyl acetate | Methanol |
| Hexanol | Methyl acetate |
| (E)-2-Hexenal | Methyl benzoate |
| (Z)-3-Hexenol | 2-Methyl butanal |
| (Z)-3-Hexenyl acetate | 2-Methyl butyl acetate |
| (Z)-3-Hexenyl butyrate | Methyl butyrate |
| (Z)-3-Hexenyl caproate | 2-Methyl butyric acid |
| (Z)-3-Hexenyl isovalerate | Methyl caproate |
| (Z)-3-Hexenyl lactate | Methyl cinnamate |
| (Z)-3-Hexenyl 2-methyl butyrate | Methyl heptanoate |
| Hexyl acetate | Methyl isobutyrate |
| Hexyl butyrate | Methyl N-methyl anthranilate |
| Hexyl caproate | Methyl 2-methyl butyrate |
| Hexyl 2-methyl butyrate | Methyl salicylate |
| 4-Hydroxy-2,5-dimethyl-3(2H)-furanone | Monellin |

| | |
|---|---|
| Myrcene | Propyl acetate |
| | Propyl butyrate |
| Naringin | Propyl caprate |
| Nerolidol | Propyl caproate |
| 2-Nonanone | Propyl caprylate |
| Nootkatone | Propyl isovalerate |
| | Propyl laurate |
| β-Ocimene | Propyl propionate |
| Octanal | Pulegol |
| Octanol | Pulegone |
| Octan-3-ol | |
| Octyl acetate | Tartaric acid |
| 3-Octyl acetate | 4-Terpinenol |
| Octyl butyrate | α-Terpineol |
| | α-Terpinyl acetate |
| 2,3-Pentanedione | 2,3,5,6-Tetramethyl pyrazine |
| α-Phellandrene | Thaumatin |
| Phenyl ethanol | 2,3,5-Trimethyl pyrazine |
| α-Pinene | Thymol |
| β-Pinene | |
| Piperine | 2-Undecanone |
| Piperitone | |
| Propanol | Valencene |
| Propionic acid | Valeraldehyde |

A small number of natural flavouring substances is not recognized as natural in the EC, although they are natural in the U.S. The reason for this is that the U.S. have a definition of natural that is slightly different from the EC (EC definition see Tab. 3.3): "The term natural flavour or natural flavouring means the essential oil, oleoresin, essence or extractive, protein hydrolysate, distillate, or any product of roasting, heating or enzymolysis, which contains the flavouring constituents derived from a spice, fruit or fruit juice, vegetable or vegetable juice, edible yeast, herb, bark, bud, root, leaf or similar plant material, meat, seafood, poultry, eggs, dairy products, or fermentation products thereof, whose significant function in food is flavouring rather than nutritional ..." [80-82].

## REFERENCES

[1] Ruttloff, H. and Rothe, M. Ernährung 11, 395-399 (Part 1) and 466-473 (Part 2) (1987)
[2] Hall, R.L. 1968. Food Technol. 22, 162 (1968)
[3] Society of Flavor Chemists. Food Technol. 23, 1360-1362 (1969)
[4] Heath, H.B. Source Book of Flavors, Westport CT, Avi Publishing 1981
[5] International Organization of the Flavor Industry. Code of Practice for the Flavor Industry, Geneva, 1990
[6] Council of the European Communities. Off. J. Eur. Comm. L 184, 61-67 (1988)
[7] Belitz, H.D. and Grosch, W. Lehrbuch der Lebensmittelchemie. Berlin, Heidelberg, New York, Springer 1982, pp. 260-307
[8] Arctander, S. Perfume and Flavor Chemicals (Aroma Chemicals), Montclair, Arctander Publ., 1969
[9] Drawert, F., Kler, A., and Berger, R.G. Lebensm. Wiss. Technol. 19, 426-431 (1986)
[10] Hatanaka, A., Kajiwara, T., and Matsui, K. Reaction Specificity of Lipoxygenase and Hydroperoxide Lyase. In: Progress in Flavour Precursor Studies (Schreier, P. and Winterhalter, P., eds.). Allured Puplishing, pp.151-170 (1993)

[11]  Goers, S.K., Ghossi, P., Patterson, J.T., and Young, C.L. Process for Producing a Green Leaf Essence. US P. 4,806,379 (1989)
[12]  Kanisawa, T., Itoh, H. Method for Preparing Green Aroma Compounds. US P. 4,769,243 (1988)
[13]  Kibler, L.A., Kratzky, Z., and Tandy, J.S. Mushroom Flavor. Eur. P. Appl. 0,288,773 (1988)
[14]  Suga, T. and Hirata, T. 1990. Phytochemistry 29, 2393-2406 (1990)
[15]  Rhodes, M.J.C., Spencer, A., Hamill. J.D. and Robins, R.J. Flavour Improvement Through Plant Cell Culture. In: Bioformation of Flavours (Patterson, R.L.S., Charlwood, B.V., MacLeod, G., and Williams, A.A., eds.). Royal Society of Chemistry, Cambridge, pp. 42-64 (1992)
[16]  Scragg, A.H. and Arias-Castro, C. Biorectors for Industrial Production of Flavours: Use of Plant Cells. In: Bioformation of Flavours (Patterson, R.L.S., Charlwood, B.V., MacLeod, G., and Williams, A.A., eds.). Royal Society of Chemistry, Cambridge, pp. 131-154 (1992)
[17]  Schreier, P. Bioflavours: An Overview. In: Bioformation of Flavours (Patterson, R.L.S. Charlwood, B.V., MacLeod, G., and Williams, A.A., eds.). Royal Society of Chemistry, Cambridge, pp. 1-20 (1992)
[18]  Winterhalter, P. and Schreier, P. Biotechnology: Challenge for the Flavour Industry. In: Flavor Science. Sensible Principles and Techniques (Acree, T.E. and Teranishi, R. eds.). American Chemical Society, Washington, pp. 225-258 (1993)
[19]  Hong, Y. C. and Harlander, S.K. Plant Tissue Culture Systems for Flavor Production. In: Flavor Chemistry of Lipid Foods (Min, D.B. and Smouse, T.H., eds.). The American Oil Chemists Society, Champaign, pp. 348-366 (1989)
[20]  Harlander, S. Biotechnology for the Production of Flavoring Materials. In: Source Book of Flavors, 2nd Edition (Reineccius, G., ed.). Chapman and Hall, New York, London, pp. 155-175 (1994)
[21]  Borejsza-Wysocki, W. and Hradzina, G. Phytochemistry 35, 623-628 (1994)
[22]  Benz, I. and Muheim, A. Biotechnological Production of Vanillin. In: Flavour Science. Recent Developments (Taylor, A.J. and Mottram, D.S., eds.). The Royal Society of Chemistry Information Services, pp. 111-117 (1996)
[23]  Drawert, F. Bioflavor – What Does It Mean? In: Bioflavor '87 (Schreier, P., ed.). De Gruyter, Berlin, pp. 3-34 (1988)
[24]  Westcott, R.J., Cheetham, P.S.J. and Barraclough, A.J. Phytochemistry 35, 135-138 (1994)
[25]  Werrmann, U. and Knorr, D. 1993. J. Agric.Food Chem. 41, 517-520 (1993)
[26]  Dörnenburg, H. and Knorr, D. Food Biotechnol. 10, 75-92 (1996)
[27]  Knorr, D., Caster, C., Dörnenburg, H., Dorn, R., Gräf, S., Havkin-Frenkel, D., Podstolski, A. and Werman, U. Food Technol. 47(12), 57-63 (1993)
[28]  Leete, E., Bjorklund, J.A., Reineccius G.A., and Cheng, T.-P. Biosynthesis of 3-Isopropyl-2-methoxypyrazine and Other Alkylpyrazines: Widely Distributed Flavour Compounds. In: Bioformation of Flavours (Petterson, R.L.S., Charlwood, B.V. McLeod, G., and William, A.A., eds.). Royal Society of Chemistry, Cambridge, pp. 75-95 (1992)
[29]  Romero, D.A. Food Technol 46(11), 122-126 (1992)
[30]  Matheis, G. Dragoco Report. Flavoring Information Service 34, 43-57 (1989)
[31]  Bigelis, R. Food Technol. 46(11), 151-161 (1992)
[32]  Collins, R.P. d Halim A.F. J. Agric Food Chem. 20, 437-438 (1972)
[33]  Yong, L.F.M. and Kok, M.-F. The Effect of Casamino Acids and Glucose Concentrations on Linalool Production by Penicillium italicum. In: Food Flavors, Ingredients and Composition (Charalambous, G., ed.). Amsterdam, Elsevier, pp. 745-751 (1993)
[34]  Latrasse, A., Guichard, E., Piffault, C., Fournier, N. and Dufosse, L. Biosynthesis and Chirality of Some γ-Lactones Formed by Fusarium poae INRA 45. In: Food Flavors, Ingredients and Composition (Charalambous, G., ed.). Amsterdam, Elsevier, pp. 465-470 (1993)
[35]  Matsumoto, M. and Nago, H. Biosci. Biotech. Biochem. 58, 1262-1266 (1994)
[36]  Whitaker, G., Poole, P.R., Cooney, J.M. and Lauren, D.R. J. Agric. Food Chem. 46, 3747-3749 (1998)
[37]  Dulio, A., Fuganti, C. and Zucchi, G. Flav. Fragr. J. 14, 79-81 (1999)
[38]  Schumacher, K., Asche, S., Heil, M., Mosandl, A., Engel-Kristen, K. and Rauhut, D. Z. Lebensm. Unters. Forsch. A207, 74-78 (1998)
[39]  Demyttenaere, J.C.R., and De Pooter, H.L. Flav. Fragr. J. 13, 173-176 (1998)

[40]  Gatfield, I.L. Bioreactors for Industrial Production of Flavours: Use of Enzymes. In: Bioformation of Flavours (Patterson, R.L.S., Charlwood, B.V., MacLeod, G., and Williams, A.A., eds.). Royal Society of Chemistry, Cambridge, pp. 171-185 (1992)

[41]  Cheetham, P.S.J. Novel Specific Pathways for Flavour Production. In: Bioformation of Flavours (Patterson, R.L.S., Charlwood, B.V., MacLeod, G., and Williams, A.A., eds.). Royal Society of Chemistry, Cambridge, pp. 96-108 (1992)

[42]  Cadwallader, K.R., Braddock, R.J., Parish, M.E. and Higgins, D.P.J. Food Sci. 54, 1241-1245 (1989)

[43]  Armstrong, D.W. Selective Production of Ethyl Acetate by Candida utilis. ACS Symp. Ser. 317, 254-265 (1986)

[44]  Armstrong, D.W., Martin, S.M., and Yamazaki, H. Biotechnol. Lett. 6, 183-188 (1984)

[45]  Rabenhorst, J. and Hopp, R. Process for the Preparation of Vanillin, U.S. P. 5,017,388 (1991)

[46]  Larroche, C., Arpah, M., and Gros, J.B. Enzyme Microbiol. Technol. 11, 106-112 (1989)

[47]  Tiefel, P. and Berger, R.G. Volatiles in Precursor Fed Cultures of Basidiomycetes. In: Progress in Flavour Precursor Studies (Schreier, P. and Winterhalter, P., eds.). Allured Publishing, pp. 439-450 (1993)

[48]  Labuda, I.M., Keon, K.A., and Goers, S.K. Microbial Bioconversion Process for the Production of Vanillin. In: Progress in Flavour Precursor Studies (Schreier, P. and Winterhalter, P. eds.). Allured Publishing, pp. 477-482 (1993)

[49]  Kawabe, T. and Morita, H.J. Agric. Food Chem. 42, 2556-2560 (1994)

[50]  Albrecht, W., Heidlas, J., Schwarz, M. and Tressl, R. ACS Symp. Ser. 490, 46-58 (1992)

[51]  Hecquet, L., Sancelme, M., Bolte, J. and Dumuynck, C. J. Agric. Food Chem. 44, 1357-1360 (1996)

[52]  Haffner, T. and Tressl, R. J. Agric. Food Chem. 44, 1218-1223 (1996)

[53]  Gatfield, I.L. H&R Contact 96(1), 3-9 (1996)

[54]  Fronza, G., Fugati, C., Graselli, P. Servi, S., Zucchi, G., Barbeni, M. and Villa, M. J. Chem. Soc., Chem. Commun. pp. 439-440 (1995)

[55]  Corza, G., Revah, S. and Christen, P. Effect of Oxygen on the Ethyl Acetate Production from Continous Ethanol Stream by Candida utilis in Submerged Cultures. In: Food Flavors: Generation, Analysis and Process Influence (Charalambous, G., ed.). Amsterdam, Elsevier, pp. 1141-1154 (1995)

[56]  Stam, H. Technical Implications and Possibilities of Biotechnology/Genetic Engineering Applied to the Production of Savoury Flavours. Presentation on the Third Savoury Conference, Geneva, March 20-21, 1997

[57]  Hagedorn, S. and Kophammer, B. Annu. Rev. Microbiol. 48, 773-800 (1994)

[58]  Duff, S.J.B. and Murray, W.D. Biotechnol. Bioeng. 34, 153-159 (1989)

[59]  Gatfield, I. The World of Ingredients August 1996, pp. 31-35

[60]  Stam, H., Boog, A.L.G.M. and Hoogland, M.: The Production of Natural Flavours by Fermentation. In: Flavour Science. Recent Developments (Taylor, A.J. and Mottram, D.S., eds.). Royal Society of Chemistry Information Services, pp. 122-125 (1996)

[61]  Van der Schaft, P.H., de Goede, H. and ter Burg, N. Baker's Yeast Reduction of 2,3-Pentanedione to Natural 3-Hydroxy-2-pentanone. In: Flavour Science. Recent Developments (Taylor, A.J. and Mottram, D.S., eds.). Royal Society of Chemistry Information Services, pp. 134-137 (1996)

[62]  Thibault, J.F., Asther, M., Ceccaldi, B.N., Couteau, D., Delattre, M., Duarte, J.C., Faulds, C., Heldt-Hansen, H.-P., Kroon, P, Lesage-Meessen, L., Micard, V., Renard, C.M.G.C., Tuohy, M., van Hulle, S. and Williamson, G. Lebensm. Wiss. Technol. 31, 530-536 (1998)

[63]  Schumacher, K., Asche, S., Heil, M., Mosandl, A., Engel-Kristen, K. and Rauhut, D. Z. Lebensm. Unters. Forsch. A 207, 74-76 (1998)

[64]  Krings, U., Gansser, D. and Berger, R.G. Lebensmittelchemie 52, 123 (1998)

[65]  Whitehead, I.M. Food Technol. 52(2), 40-46 (1998)

[66]  Dulio, A., Fuganti, C. and Zucchi, G. Flav. Frag. J. 14, 79-81 (1999)

[67]  Demyttenaere, J.C.R. and De Pooter, H.L. Flav. Fragr. J. 13, 173-176 (1998)

[68]  Matheis, G. Dragoco Report. Flavoring Information Service 34, 115-131 (1989)

[69]  Acilu, M. and Zaror, C. Enzymatic Synthesis of Ethyl Butyrate in Water/Isooctane Systems Using Immobilized Lipases. In: Progress in Flavour Precursor Studies (Schreier, P. and Winterhalter, P., eds.). Allured Publishing, pp. 483-486 (1993)

[70] Lutz, D., Huffer, M., Gerlach, D. and Schreier, P. ACS Symp.Ser. 490, 32-45 (1992)
[71] Engel, K.H. ACS Symp. Ser. 490, 21-31 (1992)
[72] Whitehead, I.M. Food Technol. 52(2), 40-46 (1998)
[73] Gillies, B., Yamazaki, H., and Armstrong, D. Biotechnol. Lett. 9, 709-714 (1987)
[74] Armstrong, D.W., Gillies, B. and Yamazaki, H. Natural Flavors Produced by Biotechnological Processing. In: Flavor Chemistry: Trends and Developments (Buttery, R.G., Shahidi, F. and Teranishi, R., eds.). American Chemical Society, Washington DC, pp. 105-120 (1989)
[75] Matheis, G. Dragoco Report Flavoring Information Service 35, 131-149 (1990)
[76] Matheis, G. Dragoco Report Flavoring Information Service 36, 43-61 (1991)
[77] Matheis, G. Dragoco Report Flavoring Information Service 36, 123-145 (1991)
[78] Matheis, G. Dragoco Report Flavoring Information Service 37, 72-89 (1992)
[79] Matheis, G. Dragoco Report Flavoring Information Service 37, 166-182 (1992)
[80] Code of Federal Regulations. 21 CFR. Ch. 1 (4-1-92 Edition). Subpart B – Specific Food Labeling Requirements. § 101.22 Foods; Labeling of Spices, Flavorings, Colorings and Chemical Preservatives (1992)
[81] Bauer, K. Dragoco Report. Flavoring Information Service 38, 5-19 (1993)
[82] Bauer, K. Labeling Regulations. In: Source Book of Flavors, 2nd Edition (Reineccius, G., ed.). Chapman and Hall, New York, London, pp. 852-875 (1994)

### 3.2.1.2 Nature-Identical and Artificial Flavouring Substances

*Gerhard Krammer*

Since ancient times the delicious taste and aroma of foods, herbs, spices and essential oils have been inspiring human beings in different cultures, geographical locations and ages [1]. Over thousands of years people have developed a wealth of recipes, techniques and technologies for food preparations, mainly driven by flavour, comprising aroma, taste, texture, viscosity, temperature as well as cooling, tingling and pungency [2]. Starting from distillation and extraction in the world of ancient Greece and Rome the medieval era led to an extended use of herbs and spices. In the Renaissance the studies of Lavoiser, Davy, Dalton, Priestly, Scheele and others laid the foundation for modern chemistry [3]. Finally, in the industrial age the curiosity of chemists revealed the chemical nature of numerous flavouring substances. The so-called great cycle of the chemical industry - identification - laboratory synthesis – large-scale synthesis and commercialisation - introduced important aroma chemicals like cinnamic aldehyde, benzaldehyde, methyl salicylate, coumarin, phenyl acetaldehyde and vanillin between 1830 and 1890. Numerous character impact compounds were identified in different segments of our diet such as allyl disulphide in garlic and furfuryl mercaptan in coffee.

The development of modern gas chromatography as well as the combination with olfactometry (GC/O) and in particular the introduction of gas chromatography with mass spectrometry (GC/MS) at a routine level increased the productivity of the great cycle of the flavour industry. Supported by continuously improved nuclear magnetic resonance (NMR) techniques and specific preparative techniques, more than 15,000 chemical compounds with flavouring properties have been reported so far [4]. The use of GC/MS instrumentation in combination with powerful computer systems enabled deep insight into the world of volatile flavour compounds (see 6.2.1). Just recently taste-active molecules as well as trigeminal active compounds have been receiving increased attention while correcting the picture on volatile flavour compounds in modern flavour chemistry [5].

The roots of modern flavour industry grew in local and regional markets in the 19th and early 20th century (see also chapter 1). In Germany Haarmann & Reimer was founded in 1874. Leon and Xavier Givaudan started their company in 1896 in Switzerland. Robert G. Fries and his brother George Fries began selling flavourings in the USA around 1913. In Japan Takasago was founded in 1922. As the business grew and the first restructuring of companies such as the conversion of Naef, Chiut et Cie. to Firmenich et Cie. took place, government authorities started to lay the foundation for the legislation on foodstuffs and in particular flavourings.

In Europe chemically defined substances with flavouring properties, which are obtained by chemical synthesis, are defined in the European Union Flavouring Directive 88/388/EEC [6], article 1 No. 2(b) (ii) and (iii) in two categories (see 7.1 and 7.4.2):

1. **Nature-identical compounds**
   These flavouring substances are characterised by the fact that the flavouring substance obtained by chemical synthesis or isolated by chemical processes is chemically identical to a substance naturally present in material of vegetable or animal origin.
2. **Artificial compounds**
   These flavouring substances are defined by the criterion that the flavouring substance is not chemically identical to a substance naturally present in materials of vegetable or animal origin.

The legislation of some European countries comprises positive lists for artificial flavouring substances, such as ethyl vanillin.

Since this approach is based on the scientific proof of natural occurrence, the International Organisation of the Flavour Industry (IOFI), Brussels (B), established the Working Group on Analytical Methods (WGMA), which examines published data with regard to the validity of chromatographic (e.g. GC retention time) and spectroscopic (e.g. mass spectra, infrared and NMR spectra) data. In this context reference data, artefact formation and the nature of the source material (i.e. food use) are also considered.

This approach is different from the legal regulations relating to use of flavourings in foods, which are described in § 101.22 of Title 21 of the Code of Federal Regulations in the USA. In the USA, the term 'artificial flavour' or 'artificial flavouring' means any substance with flavouring properties, which is not derived from a spice, fruit or fruit juice, vegetable or vegetable juice, edible yeast, herb, bark, bud, root, leaf or similar plant material, meat, fish, poultry, eggs, dairy products, or fermentation products thereof. This definition of artificial flavourings also comprises synthetic flavouring substances and adjuvants, which are listed in §§ 172.515(b) CFR as well as synthetic flavouring substances listed in §§ 182.60 CFR. In practice, this means that any flavouring using synthetic aroma chemicals, with and without natural occurrence, is labelled as artificial flavour (see 7.5.2: USA).

Nature-identical and artificial flavouring substances are produced in order to evoke specific sensorial effects. The potency of the individual compounds is usually described by a set of parameters comprising the odour or taste threshold in the corresponding matrix, i.e. water or oil phase, as well as the dynamics of perception as determined by Steven's law [7]. For many compounds a shift of the specific sensorial properties is observed in different concentrations.

Important physicochemical parameters are volatility, vapour pressure and polarity, which play an important role in the flavour release of solid and paste-like food as well as for beverages [8]. The link between sensorial properties and the molecular structure has always been of highest interest for organic chemists in the past 150 years. Functional groups, stereochemistry, molecular size and heteroatoms have been varied in numerous isomers in order to understand the corresponding sensory profile and related effects. The use of molecular references like (E,E)-2,4-decadienal for the descriptor 'fatty' or (E)-2-hexenal for 'green' has become a standard approach to link sensory with the flavour chemistry of flavourings and finished food products. Analytical techniques like GC/O or the use of mouth model systems [9] are following the same approach. The aroma value concept *[10]* (see 6.2.4) impressively shows that in many foodstuffs only a few flavour compounds play an important role in the authentic flavour profile. The ratio of the concentration of a flavour compound in food and the retronasal odour threshold or taste threshold determine the odour activity value (OAV) or taste activity value (TAV) (see 6.2.4). Together with omission experiments these tools are essential for unmasking the blueprint of a flavour extract. Recently "LC Taste®", a new method for combining instrumental analysis and sensorial analysis, has been developed. Based on a separation via high-temperature liquid chromatography (HTLC) and online tasting of aroma and taste compounds in aqueous matrix, flavour researchers and developers have the possibility for a differentiated retronasal, taste-oriented evaluation of mixtures *[11]*.

Industrial applications of nature-identical and artificial flavourings have to be based on two pillars: a clear understanding of flavour-relevant molecules and at the same time the use of technically feasible, high-purity, high-yield syntheses. In addition, performance criteria like pH, oxidation and chemical stability as well as toxicological safety have to be considered.

Flavour-active compounds cover an enormous dynamic range regarding perceivable concentration, starting with extremely powerful molecules like 1-menthen-8-thiol in the low ppt (ng/kg) range and going up to the few percent level for compounds like acetic acid. Together with the huge variety of chemical compound classes this fact generates additional challenges for the toxicological evaluation of flavouring substances.

In Europe, the EU Commission started to establish an inventory for flavouring substances covering chemically synthesised or chemically isolated flavouring substances chemically identical to flavouring substances naturally present in foodstuffs or in herbs and spices as well as vegetable and animal raw materials normally considered as foods. In addition, chemically synthesised or chemically isolated flavouring substances from other sources and not yet found in nature are also considered. By end of 2004 the list consisted of approx. 2800 compounds as a first inventory comprising natural, nature-identical and artificial substances, based on information received from the flavour industry. All listed compounds are subject to a detailed safety evaluation (see 7.4.2). In 2000, the EU Scientific Committee on Food (SCF), now recognised as the European Food Safety Authority (EFSA) (FSA panel), accepted the evaluation procedure of the World Health Organisation Joint Expert Committee on Food Additives (JECFA) and reviewed the list of flavouring substances already approved by JECFA. These compounds are part of the list of evaluated flavour compounds by the Panel of Expert Scientists (FEXPAN) of the Flavour and Extract Manufacturers' Association (FEMA) in the USA (see 7.5.2: USA). So basically two international expert committees are currently evaluating flavouring substances for their risks: JECFA and EFSA. They perform the 'risk assessment', whilst the responsibility for 'risk management' lies with the Codex Committee on Food Additives and Contaminants (CCFAC) and the EU Commission and its administration (DG Sanco).

For the FEMA approach as well as for the evaluation of flavouring substances of the European register, normal and maximum use levels of aroma-active compounds in different food applications play an important role. The EU approach comprises 16 food categories which cover a broad range of food applications, starting with category 1 (dairy products) and ending with category 16 covering composite foods. Basically, two different methodologies exist, namely MSDI (Maximised Survey-Derived Daily Intake) and TAMDI (Theoretical Added Maximum Daily Intake). The MSDI approach is based on the annual production of substances as stated by the producers divided by the total consumers. The TAMDI system requires a calculation of flavour addition to food and consumption of flavoured foodstuffs per day. For the EFSA concept, the first evaluation is conducted based on the MSDI method following a decision tree, which is published together with the so-called FGE statements [12]. A second interactive step based on a modified TAMDI method helps to identify compounds with a higher use level than the calculated safety level.

**Flavouring Group Evaluation**

All entries of the European inventories are classified in **34 chemical groups**, which represent compounds of consistent chemical, metabolic and biological behaviour. In the case of compounds that are easily degraded to smaller fragments, the most toxic fragment is given higher priority in assigning the appropriate chemical group.

The wealth of powerful aroma chemicals is finally classified based on safety considerations. In order to give more insight, prominent representatives are explained in detail.

Important compounds of **Group 1** are straight-chain primary aldehydes like hexanal (Flavis # 05.008, FEMA 2557) and octanal (Flavis # 05.009, FEMA 2797) with tallowy and fatty notes as well as widely used acids like acetic acid (Flavis # 08.002, FEMA 2006) and the powerful butyric acid (Flavis # 08.005, FEMA 2221). In the same group ubiquitous fruity esters like ethyl butanoate (Flavis # 09.039, FEMA 2427; annual consumption in the USA approx. 10 tons in 1991 [13]) and the apple-like methyl 2-methyl butyrate (Flavis # 09.483, FEMA 2719) are registered.

Chemical **Group 2** comprises cocoa flavour compounds like 3-methyl butanal (Flavis # 05.006, FEMA 2692) and the banana-like 3-methyl butyl acetate (Flavis # 09.024, FEMA 2055).

In chemical **Group 3** important terpene alcohols such as the rose-like geraniol (Flavis # 02.012, FEMA 2507) and aldehydes such as the citrus-like compound citral (Flavis # 05.020, FEMA 2303) are found. The two isomers of citral, namely neral (Flavis # 05.170, FEMA 2303) and geranial (Flavis # 05.188, FEMA 2303) are listed in the same group. The mixture and the two individual isomers are listed under the same FEMA number 2303.

**Group 4** lists compounds with leaf-like notes such as (Z)-3-hexen-1-ol (Flavis # 02.056, FEMA 2563) and (Z)-3-hexenal (Flavis # 05.075, FEMA 2561) as well as esters such as citronellyl acetate (Flavis # 09.012, FEMA 2311) with rose-like, floral profile.

A typical representative for mushroom-like notes, 1-octen-3-one (Flavis # 02.023, FEMA 2805), is found in **Group 5** together with the fatty, fruity tasting 2-undecanone

(Flavis # 07.016, FEMA 3093) and numerous other 2-alkanones, which are responsible for characteristic cheesy notes.

**Group 6** shows additional terpene alcohols like linalool (Flavis # 02.013, FEMA 3045) with flowery profile and α-terpineol (Flavis # 02.014, FEMA 3045) with sweet, fruity aroma. In addition, numerous esters are listed like α-terpinyl acetate (Flavis # 09.065, FEMA 3047), which is characterised by a strong herbaceous odour.

In **Group 7** the α- and β-isomers of santalyl acetate (Flavis # 09.034, FEMA 3007) are mentioned. These compounds have a characteristic sandalwood-like odour and a bitter-sweet taste. Additional compounds in this group are perilla aldehyde (Flavis # 05.117, FEMA 3557) with fatty, spicy notes and safranal (Flavis # 05.104, FEMA 3389) with a characteristic saffron-like odour and taste.

One of the most important compounds in **Group 8** is menthol (Flavis # 02.015, FEMA 2665; world usage 11,800 tons in 1998 *[14]*), well known for its minty aroma accompanied by a strong cooling effect. Furthermore, Group 8 comprises also α-ionone (Flavis # 07.007, FEMA 2594) and β-ionone (Flavis # 07.008, FEMA 2595) with a strong violet-like odour and carvone (Flavis # 07.012, FEMA 2249), which occurs in two different forms. L-carvone exhibits the odour of spearmint, while δ-carvone shows a strong caraway note. Another important compound is nootkatone (Flavis # 07.089, FEMA 3166), which also occurs in two different forms. (+)-Nootkatone has a highly appreciated grapefruit-like profile and (-)-nootkatone is characterised by a terpeny note. β-Damascenone (Flavis # 07.108, FEMA 3420), a very powerful, honey-like, fruity, sweet compound, is also a member of this group together with l-menthyl lactate (Flavis # 09.551, FEMA 3748), a typical cooling compound.

In **Group 9** various organic acids like lactic acid (Flavis # 08.004, FEMA 2611), pyruvic acid (Flavis # 08.019, FEMA 2970) and succinic acid (Flavis # 08.024, registered in Food Chemical Codex) together with various diesters are found. Beside some diols, numerous lactones such as γ-decalactone (Flavis # 10.017, FEMA 2360) with peach-like notes and δ-decalactone (Flavis # 10.007, FEMA 2361) with a sweet coconut-like profile are represented.

**Group 10** is a fairly small group covering secondary aliphatic saturated or unsaturated alcohols, ketones, ketals and esters. Important members are the typical dairy compounds diacetyl (Flavis # 07.052, FEMA 2370) and acetoin (Flavis # 07.051, FEMA 2370).

**Group 11** is the home for mint lactone (Flavis # 10.036, FEMA 3764) and various phthalides, which are well known from the flavour chemistry of celery.

Maltol (Flavis # 07.014, FEMA 2656) and ethyl maltol (Flavis # 07.047) are key compounds in **Group 12**.

**Group 13** sets the frame for the complex world of furanones. Important fruity and caramelic representatives are Furaneol® (Flavis #13.010, FEMA 3174) and the corresponding methoxy derivative mesifuran (Flavis # 13.089, FEMA 3664). In addition the linalool oxides and angelica lactone are in this group.

**Group 14** lists among other compounds various furyl disulphides and in particular the coffee compound furfuryl mercaptan (Flavis # 13.026, FEMA 2493).

In **Group 15** phenyl acetaldehyde (Flavis # 05.030, FEMA 2874) with honey-like, flowery notes has to be mentioned.

**Group 16** contains 1,8-cineol or eucalyptol (Flavis # 03.001, FEMA 2465), a compound with a characteristic camphoraceous odour and a fresh cooling taste.

**Groups 17 and 18** contain among other compounds isoeugenol (Flavis # 04.004, FEMA 2468) and eugenol (Flavis # 04.003, FEMA 2467), respectively.

**Group 19** is dedicated to a few amides like N-ethyl-p-menthane-3-carboxamide (Flavis # 16.013, FEMA 3455).

**Group 20** is a nice collection of wonderful sulphur compounds. Disulphides such as diallyl disulphide (Flavis # 12.008, FEMA 2028) with onion- and garlic-like notes are followed by extremely powerful mercaptans like the grapefruit compound p-1-menthene-thiol (Flavis # 12.085, FEMA 3700). In addition, thiols with tropical sulphury notes complete the picture of interesting molecules.

**Group 21** comprises phenyl derivates like raspberry ketone (Flavis # 07.055, FEMA 2588), which is a compound with sweet, fruity, raspberry-like notes.

In **Group 22** we find cinnamic aldehyde (Flavis # 05.014, FEMA 2286) and cinnamic acid (Flavis # 08.022, FEMA 2288) as well as the corresponding esters.

Vanillin (Flavis # 05.018, FEMA 3107; world usage approx. 12,000 tons in 1990 *[15]*) and benzaldehyde (Flavis # 05.013, FEMA 2127; world usage approx. 13,000 tons in 1994 *[16]*) are major entries in **Group 23**.

**Group 24** represents the world of pyrazines. The sensory profile of the listed compounds ranges from toast to coffee and cocoa notes reminiscent of the wonderful aroma of many culinary pleasures.

In **Group 25** various phenols like thymol (Flavis # 04.006, FEMA 3066) and salicylates are listed. In addition naringin (Flavis # 16.058, FEMA 2769) is also found on the list.

**Group 26** contains a series of aromatic ethers and **Group 27** shows numerous anthranilates.

Pyridine, pyrrols and quinoline derivatives are found in **Group 28**. Thiazole, thiophenes, thiazolines and thienyl derivatives are summarised in **Group 29**. Among miscellaneous compounds **Group 30** lists hydrogen sulphide and ammonia. In **Group 31** we find aliphatic and aromatic hydrocarbons like limonene (Flavis # 01.001, FEMA 2633).

**Group 32** shows various epoxides like β-caryophyllene epoxide (Flavis # 16.043).

In **Group 33** we find amines like trimethyl amine (Flavis # 11.009, FEMA 3241) with a prominent fishy note.

Finally, **Group 34** is dedicated to a collection of amino acids which are important for mouthfeel and taste in the world of savoury, sweet and beverage products.

After the completion of the safety evaluation of the above mentioned flavour compounds, the current register will be transferred into the European positive list of flavouring substances, comprising natural, nature-identical and artificial molecules. The status will not be indicated in the foreseen list. As a consequence, the currently existing positive lists for artificial compounds on the level of country-specific legislation will then be obsolete.

The future use of nature-identical and artificial flavour compounds will be based on a broad set of criteria: sensory properties, safety and use levels and finally performance parameters like stability in application and flavour release.

For the synthesis of flavour-active compounds numerous methodologies have been developed [17]. In many cases natural products served as starting materials such as eugenol from clove oil for the synthesis of vanillin. In case of complex stereochemistry natural materials are still welcome for the synthesis of valuable flavour compounds such as nootkatone, which is obtained by oxidation from valencene.

Certainly economical considerations are an important starting point for the development of a successful aroma chemical. The sensorial profile, impact and the absence of off notes is often underestimated. It is part of the knowledge and the professional skills of a flavourist to understand the relevance of effects like aging, isomerisation and oxidation. Over the years numerous strategies for the synthesis of fairly simple aroma chemicals, like straight-chain esters, and of complex structures, like the different isomers of rose oxide (2S, 4R rose oxide and 2R, 4R rose oxide), have been developed.

A special discipline is represented by the world of sulphur-bearing compounds. In general all representatives of this class are very powerful chemicals, which often leads to their being handled in smaller amounts.

The sensorial quality of sulphur-bearing aroma chemicals is an important criterion, because of the reactivity of, for example, free thiols. For the synthesis of prominent representatives like 8-mercapto-3-menthanone or 1-menthen-8-thiol different synthetic methods have been published.

More than in the past, sensorial delights for millions of consumers will be based on the close interaction of multiple disciplines including flavourists, chemists, toxicologists, technologists and chefs. In this context nature-identical and artificial substances will continue to play a major role in modern flavourings.

## REFERENCES

[1] Buchbauer G., Jäger W., Jirovetz L., Ilmberger I., Dietrich H., Therapeutic Properties of Essential Oils and Fragrances, ACS Symp. Ser., 525, 159-165 (1993)

[2] Corriher S.O., Ccookwise – The Hows and Whys of Successful Cooking, William Morrow, New York (1997).

[3] Rowe D.J., Introduction. In: Chemistry and Technology of Flavors and Fragrances, pp 1-11, Rowe D.J. (Ed.), CRC Press/Blackwell Publishing, Oxford, UK (2005)

[4] Teranishi R. In: Flavor Chemistry – Thirty Years of Progress, Vol. VII, Teranishi R., Wick E.L., Hanstein I. (Eds.), Kluwer Academic/Plenum, New York (1998)

[5] Hofmann T., Schieberle P., pp 388-401, Wiley-VCH, Weinheim (2003)

[6]  Council Directive of 22 June 1988 on the approximation laws of the Member States relating to flavourings for use in foodstuffs and to source materials for their production (88/388/EEC). Official Journal of the European Communities No. L 184/61-15.07.88
[7]  Meilgaard M., Civille G.V., Carr B.T. In: Sensory Evaluation Techniques, 3rd edition, pp. 45-59, CRL Press, Boca Raton (1993)
[8]  Petersen M.A., Ivanova D., Møller P., Bredie W.L.P., Validity of ranking criteria in gas chromatography-olfactometry methods. In: Flavour Research at the Dawn of the Twenty-First Century, Proceedings of the 10th Weurman Flavour Research Symposium, pp 494-500, le Quére J.-L., Etiévant P.X. (Eds.), Editions Tec&doc, London (2001)
[9]  Reinders G., Erfurt H., Boos H., Wellmann R., US Patent 6,547,172 B2
[10] Grosch, W., Review – determination of potent odorants in foods by aroma extract dilution analysis (AEDA) and calculation of odour activity values, Flavour Fragrance J., $\underline{9}$, 147 (1994)
[11] Krammer G.E., Weckerle B., Brennecke S., Weber B., Kindel G., Ley J., Hilmer J.M., Reinders G., Stöckigt D., Hammerschmidt F.J., Ott F., Gatfield I., Schmidt C.O., Bertram H.J., Product Oriented Flavor Research, Mol. Nutrit. Food Res. $\underline{50}$ (4/5), 345-350 (2006)
[12] http://www.efsa.eu.int
[13] Clark G.S., Perfumer & Flavorist, $\underline{16}$, 41-46 (1991)
[14] http://www.roempp.com
[15] Clark G.S., Perfumer & Flavorist, $\underline{15}$, 45-54 (1990)
[16] Clark G.S., Perfumer & Flavorist, $\underline{20}$, 53-60 (1995)
[17] Bauer K., Garbe D., Surburg H., Common Fragrance and Flavor Materials – Preparation, Properties and Uses, 4th edition, Wiley-VCH, Weinheim (2001)

## 3.2.2 Flavouring Preparations and Some Source Materials

### 3.2.2.1 Fruit Juices and Fruit Juice Concentrates

*Martin Simon*

#### 3.2.2.1.1 Introduction

Every fruit has its own characteristic flavour. Over the past 25 years, the industry has learnt to identify and document fruit aroma components by means of gas chromatography.

It was realised that the method of processing fruit juices significantly influences the specific flavour of the fruit and consequently the quality of the juice. When producing concentrate from a high quality fruit juice, the recovery and isolation of the aroma during the concentration process is essential for adding it back to the juice upon reconstitution of the concentrate.

The following factors are detrimental to the aroma as well as the colour and stability of the juice:

- Exposure to high temperatures during pasteurisation and concentration results in deterioration of the aroma and browning of the juice.
- Oxidation of the juice by incorporation of air (change of taste).
- Bacteriological and enzymatical defects due to insufficient pasteurisation and unhygienic processing procedures. Results are insufficient shelf-life (fermentation and mould formation) or enzymatical clarification of naturally cloudy juices (instability).

All juice processing methods have certain techniques in common. The industry constantly evaluates new processing procedures and equipment in order to improve the quality of the product.

#### 3.2.2.1.2 Fruit

Good quality fresh fruit is most important. Only mature and healthy fruit can produce high quality juices. The acknowledgement of this principle leads to the systematic cultivation of the different fruit varieties (e.g. citrus fruit, apples, pineapples, passion fruit, grapes, bananas) in large groves with detailed attention to growing and harvesting conditions.

Such measures assured the success and the highly uniform quality of citrus products from Florida, California and Brazil. It must be emphasised that especially in the United States the cultivation of citrus fruit on an industrial scale has logically initiated a systematic development of the necessary harvesting and processing equipment which in turn has helped to bring the industry to today's high technical level. This for example led to the design and construction of the industry's best citrus juice extractors (BROWN (1) and FMC (2)) and a highly efficient large capacity juice evaporator, the so-called T.A.S.T.E. evaporator (Temperature Accelerated Short Time Evaporator) which can also be equipped with an aroma recovery system (3). This equipment is responsible for the high quality standard of the citrus concentrates produced today in the USA.

Growing conditions in the groves are tightly controlled and the application of pesticides to the plants is strictly regulated by legislation. The products and the entire production facilities and procedures in the juice factories are checked according to standardised methods and supervised by government laboratories. Research laboratories are constantly working on the improvements of horticultural and processing methods.

The application of similar fruit production and juice processing procedures also helped the Brazilians to reach and maintain a high quality level of their citrus products.

There are many citrus producing countries which do not have a well controlled and uniform agricultural fruit production. Juice produced from such fruit cannot possibly be of a similar good quality as from the above mentioned citrus growing areas, even if the processing equipment is the same.

### 3.2.2.1.3 Preparation of Fruit

The fruit delivered to the factory for processing is sampled and checked in the laboratory to determine the degree of maturity, the content of soluble solids, the acid content and other important data. It is then washed and graded to eliminate spoilt fruit not suited for juice production.

### 3.2.2.1.4 Juice Extraction

The fruit is directed to suitable fruit extracting systems for juice recovery. In the case of citrus juices two of the best known juice extractors were mentioned above. While squeezing citrus fruit, undesirable peel oil can easily get into the juice. To avoid this, some extractor systems are eliminating and collecting the peel oil before the juice is extracted. This is achieved by rasping and pricking the peel. The resulting oil raspings are spray washed off the peel and conveyed to a press where the oil/water mixture is separated from the solids to recover as much oil as possible. The oil/water mixture is separated in a two step centrifugation process producing the so-called cold pressed oil. Only after the oil is recovered from the peel the juice is extracted from the fruit in the extractor. This process applies to the BROWN (1) extractor system and those of other extractor manufacturers (e.g. BERTUZZI (4), INDELICATO (5), SPECIALE (6)).

Only the FMC extractor (see Figure 3.9-3.10) and recently, some similar machines from FOMESA (7) and OIC (8) allow a simultaneous collection of rasped peel for oil recovery and juice in separate streams. The juice passes from the extractor to a so-called finisher to remove the coarse pulp particles. Then the juice can be bottled as fresh juice or it can be used for the production of juice concentrate. If necessary, the pulp content of the juice can be further reduced by centrifugation, using so-called desludger type centrifuges (ALFA LAVAL and TETRA PAK (9), WESTFALIA SEPARATOR (10), etc.).

*Fig. 3.9:* The FMC Citrus Juice Extractor

Juice from fruit which is mashed for juice extraction, e.g. apples, pears, pineapples, grapes, cherries, is produced by pressing.

The structure of the mash varies according to the fruit type. A mash with a high fiber content is pressed more easily than mash from very soft fruit. Pineapples for example can be pressed easily with screw presses, whereas special presses have been developed for softer fruit, like apples. For many years, the BUCHER (11) press was commonly used for processing apples. The BUCHER HP press is a hydraulic press working in cycles. The mash is filled into the press and under pressure the juice drains through flexible hose-like drainage filter elements. Once the extraction is completed the pomace is discharged and the process starts over again. Yield and quality of the juice are excellent and the system has proven to operate reliably with many fruits [1].

Another type of press, the continuous beltpress, which was originally used for drying sewage sludge, has been re-designed for fruit processing. Belt presses for the food industry are available from various manufacturers (AMOS (12), BELLMER (13), DIEMME (14), FLOTTWEG (15), KLEIN (16), etc.) [2-5]. The working principle is to press the mash between two sieve belts which continuously pass around cylindrical rollers. At the beginning of the pressing operation the diameter of the rollers is large and the applied pressure is low. As the pressing operation progresses the rollers become smaller and the pressure increases. The tension of the belt and the pressure around the rollers can be adjusted pneumatically and the belt speed can be varied as well. The pressed pomace is removed from the belts by means of plastic scrapers and/ or by rotating brushes and the belts are cleaned with high pressure water on the return

# Flavouring Preparations and Some Source Materials

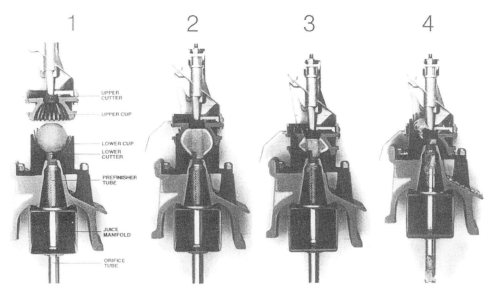

1  **Upper Cutter:** Cuts a plug in the top of the citrus to permit the separation of the peel from the internal portions of the fruit.
   **Upper / Lower Cup:** Supports the exterior of the citrus throughout the squeeze cycle to prevent bursting.
   **Lower Cutter:** Cuts a plug in the bottom of the citrus to allow the internal portions of the fruit access to the prefinisher tube.
   **Prefinisher Tube:** Separates, based on particle size, the internal elements of the citrus.
   **Juice Manifold:** Collects juice and juice sacs.
   **Orifice Tube:** Generates pressure inside the prefinisher tube and collects and discharges membrane and seeds.
2  In this early phase of the extraction cycle, the upper cup moves downward to cause pressure on the citrus so that the top and bottom plugs begin to be cut.
3  As the extractor cycle continues, pressure increases on the citrus causing the internal portions to be forced through the bottom of the fruit and into the prefinisher tube. The peel is now being discharged between the upper cup and cutter.
4  Upon completion of the extraction cycle the internal portions of the citrus are located in the prefinisher tube. At this time, the orifice tube moves upward, placing pressure on the contents of the prefinisher tube. This causes the juice and the juice sacs, due to their small particle size, to flow through the holes of the prefinisher tube and into the juice manifold. Particles larger than the holes in the prefinisher tube are forced through an opening in the orifice tube and discharged out the bottom.

*Fig. 3.10:* The FMC Citrus Juice Extractor Principle

to the mash feed location. The juice coming from the press could contain coarse fruit particles which are removed by an additional simple screening device. When the recovered juice is to be bottled as natural cloudy juice or to be converted into cloudy concentrate it is advisable to remove the insoluble solids by centrifugation. This will avoid formation of bottom pulp in the bottle and it will also aid the concentration process in the evaporator.

The main advantage of this press is the short duration of the entire pressing operation which only lasts a few minutes from mashing the fruit to dejuicing.

The one-sieve belt press is a new development. Also here the belt passes around several cylindrical rollers of different diameters in order to increase the pressure during the pressing operation and to achieve a high yield.

To increase the yield and improve the pressability of the mash it is recommended to treat the mash with enzymes. Specific enzymes, referred to as "press enzymes", have been developed by leading enzyme manufacturers. In order to increase the yield of juice from the mash another specific enzyme for the so-called "mash liquification" was developed and tested.

When producing concentrates instead of fresh juice the juice yield from the mash can be increased by washing the pomace from the first pressing with water (e.g. 1:1) and repeating the pressing operation to recover additional soluble juice solids. Especially designed water extraction systems also result in yield increases (see below). The soluble solids recovered from washing of the pomace can be added to the juice. This procedure is not suitable for fresh juice production but only when processing concentrates due to the dilution with water and the lower soluble solids content in the water extracted juice.

When processing apples, the recovery of juice solids can also be achieved by continuous countercurrent extraction with water. These extraction systems were derived from the extraction of sugar beets in the sugar industry and they work either as hot or as cold extraction.

The extractors work well on freshly harvested, early fruit which has a firm structure. However, as the degree of maturity of the fruit progresses and as the fruit becomes softer, the operation becomes more and more difficult due to the softness of the mash and the decreasing water permeability of the mash layer. Water extraction systems are more and more replaced by the simpler fruit presses.

A similar procedure for the recovery of juice solids is applied in citrus processing. The dejuiced pulp coming from the juice finisher is countercurrently extracted with water in several processing steps. One extraction step consists of a mixing screw and a finisher. In the mixing screw the pulp is mixed with the extraction liquid and in the finisher the liquid is separated from the extracted pulp solids. The liquid is run countercurrently to the pulp from stage to stage. The more extraction steps are built into the process the better the recovery of soluble juice solids from the pulp and the higher the concentration of the recovered liquid. These so-called (soluble) pulp wash solids can be added back to the juice when processing the concentrate.

Another possibility is to produce concentrate from the recovered juice solids as such, so-called "pulp wash concentrate", or "OWP" (Orange Wash Pulp) or "WESO" (Water Extracted Solids) respectively.

The pulp wash liquid from citrus fruit contains a higher amount of pectin than the single strength juice from the juice extractors. In most cases it must be treated enzymatically to achieve a sufficiently high concentration in the evaporator. In order to aid the concentration it is also commonly practised to centrifuge the liquid prior to evaporation to remove some or most of the insoluble solids from the liquid.

The taste of these products is usually clean but flat and less aromatic than normal juice. A result of the higher pectin content, pulp wash products usually have an extremely good turbidity and cloud stability which makes them well suited for the production of drinks with low juice content.

### 3.2.2.1.5 Juices

**(1) Natural Cloudy Juices**

It is common to produce natural cloudy juices and concentrates as standard products from specific fruit varieties such as citrus, exotic fruit and also from apple, pear and berries. Freshly squeezed juices can be bottled as such. The juice is heated (85° – 90°C) and hotfilled into clean bottles. The bottles are closed and then immediately cooled with water as quickly as possible to avoid deterioration of the juice due to the high temperature. The high filling temperature sterilises bottles and product.

In recent years, several new filling systems were developed in order to reduce the heat load necessary for sterilising the juice. For this purpose bottles or soft packages are sterilised before filling. The juice is heated in a closed system and cooled immediately thereafter. The sterilised juice is filled into the sterilised containers at ambient temperature. Glass bottles or cartons and pouches can be used as packaging (Tetra Pak (17), SIG Combibloc (18)).

Apple juice, freshly squeezed and cloudy, sometimes is de-aromatised and then stored after sterilisation in tanks for later filling. This requires tanks which can be safely sterilised by steaming. The juice is sterilised in a similar way to the sterilisation for filling into bottles. The recovered aroma is kept in cold storage and added back to the juice prior to filling. This process is applied by factories which bottle the juice by themselves. If they have sufficient aseptic tank storage they can save the cost for evaporation and concentration of the apple juice. This also has the advantage that the juice can be declared as "Not made from concentrate".

Juices produced for industrial use, especially where transport is involved, are usually concentrated for more economic storage and shipping.

In order to reach the desired concentration it may be necessary to treat the juices enzymatically to avoid jellying at higher concentrations.

**(2) Clarified Juices**

Next to the natural cloudy juices it is also customary to produce clear juices and concentrates, specifically from apple, pear and berries. For this purpose the juice must be clarified which can be achieved by the following three different processing methods:

    (1) Enzymatic treatment and clarification with gelatine, silica sol and bentonite and subsequent filtration.
    (2) Enzymatic treatment and ultrafiltration.
    (3) Enzymatic treatment and clarification by flotation and filtration.

For better aroma it is recommended to recover the aroma prior to the clarification process of the juice and to add it back to the concentrate or later during reconstitution.

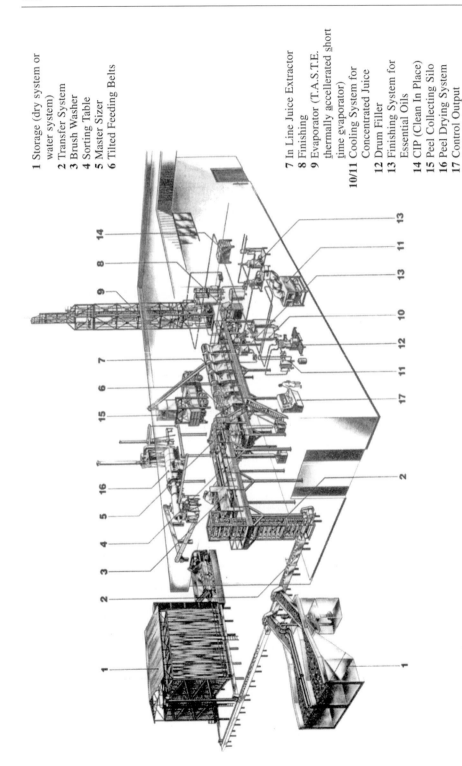

*Fig. 3.11: The FMC Citrus Processing Line*

In order to produce clear juice the cloudy product is treated with pectolytic enzymes to degrade the pectic substances which are responsible for the cloud stability of the cloudy juice. The treatment time depends on the activity of the enzymes and the temperature respectively. Several years ago the industry had no enzymes which were able to work in very acid juice environments (e.g. lemon juice). However, in the meantime new types were developed which are capable of treating acid juices satisfactorily. Some fruit juices contain substances which cannot be removed with pectolytic enzymes and which require other specific enzymatic products to eliminate turbidity, e.g. starch in apple juice requires a suitable amylase.

After the enzymatic treatment the juice can be clarified. Flocculation aids and fining aids (gelatine, silica sol, etc.) help to coagulate the cloudy substances and facilitate their separation by settling and filtration. Bentonites can be used to eliminate proteins and other cloudy substances. Filtration (Kieselgur precoat filtration with plate and frame filters, rotating vacuum filters, and sheet filters as "police filters") is used to produce a crystal clear juice.

A new process uses ultrafiltration to clarify fruit juices (see below). The advantage of this method is that flocculation and fining aids are no longer necessary which results in cost savings. To optimise the throughput it is however recommended to treat the juice enzymatically in order to reduce its viscosity. Properly applied, this process can work very economically.

Several years ago a flotation process for clarification of fruit juices was developed in Canada *[6,6a]*. The juice is enzymatically treated, bentonite is added and the mixture is impregnated with an inert gas (nitrogen, carbon dioxide). Then the liquid is pumped into an especially designed flotation tank in which the solids float to the top from where they can be skimmed off in concentrated form. The liquid is polished by filtration. This process was also tested in Europe.

### 3.2.2.1.6 Concentrates

Juices which are not packed immediately or which are traded as half products are concentrated. Usually juices are made to the following concentrations:

- Orange and Mandarin 65°-66.5° Brix
- Grapefruit 58°-60° Brix
- Lemon to an acid content of 400-500 grams per litre (gpl) corresponding to about 45°-55° Brix.
- Special Citrus concentrates (pulp wash, etc.) 50°-65° Brix. The above mentioned concentrates must be stored deep frozen.
- Apple and Pear 70°-72° Brix. These products can be stored at room temperature, although it is recommended to keep the products at a temperature of approximately 5°C to avoid a temperature related colour change.
- Berries 50°-70° Brix. Storage depends on the degree of concentration and on the demands on quality.
- Pineapple 60° Brix. Storage deep frozen.
- Passion Fruit 30°-50° Brix. Passion Fruit is traded also in unconcentrated form at app. 15° Brix. Storage deep frozen.

- Mango 28°-30° Brix. Also traded in unconcentrated form at app. 12°-15° Brix. Storage deep frozen.
- Banana 28°-32° Brix. Also available as unconcentrated puree. Storage deep frozen.

All concentrates can be also filled aseptically. For this purpose special aseptic filling systems have been developed suitable for filling the product into retail packages as well as into 55 U.S. gallon drums or even larger containers. The advantage of this filling method is that the product can be shipped and stored at normal temperature, although also here storage temperatures of 0°C-5°C are recommended in order to maintain the quality of the product.

**Concentrate Production**

On an industrial scale, today juice concentrates are virtually only produced by evaporation. There are some new developments which are technically highly interesting but of which only a few commercial installations are in operation (see below). Cloudy juices should be depulped before evaporation. This can be achieved by centrifugation using automatically de-sludging centrifuges.

Multi-stage vacuum evaporators are preferably applied for juice concentration. In specific cases, e.g. purees and very sensitive products such as passion fruit, single-stage thin film evaporators offer an advantage. In the citrus industry the most widely used evaporator is the T.A.S.T.E. evaporator (3). It is built under licence in various countries. This is a multi-stage tubular evaporator designed to work with very high product velocities which reduce the retention time of the product in the high temperature zones in the evaporator to a minimum. An important difference to a normal falling film evaporator is the distribution of the liquid in the heads of the tube nests of the individual evaporation stages. The falling film evaporator uses mechanical devices to distribute the liquid into the tubes. In case of plugging or not working properly, this can result in an unequal distribution to the tubes and local overheating of the product. The T.A.S.T.E. evaporator does not work with mechanical devices but uses the principle of "Descending Turbulent Mist". This is caused by thermally expanding the product by flashing it into the top of the stage converting the product into a mixture of vapour and small liquid particles which are equally distributed into the tubes at very high velocities. Inside the tubes the product is heated and additional evaporation is increasing volume and velocity thus it is thermally accelerated to almost sonic velocities. This creates high heat transmission rates and extremely short retention times at high temperatures which diminishes the possibility of product burn-on and damage. This evaporator is manufactured in sizes from 500 kg water evaporation per hour to approx. 90,000 kg water evaporation per hour. It has been successfully used in the citrus industry but also for the concentration of other fruit juices (pineapple, apple, grape, etc.).

Besides the T.A.S.T.E. evaporator also regular falling film evaporators are used successfully (GEA-WIEGAND (19), UNIPEKTIN (20), etc.).

Another evaporator type is the plate evaporator (APV (21), GEA-TUCHENHAGEN (22), SCHMIDT (23)). The advantage of plate evaporators is that the plate sections can be opened for cleaning and inspection. The number of plates can also be adjusted depending on the application. This equipment is used for small to medium capacities (up to 20,000 kg water evaporation/hour) and it can also handle liquids with higher viscosities (tomato puree, mango puree, etc.) *[11-12]*.

All the above mentioned evaporators can be equipped with aroma recovery units which for better heat balance can be integrated directly into the evaporator stages. Apple juice evaporators are designed in such a way that the de-aromatised juice can be taken out of the evaporator with approx. 50°C for clarification. The clarified juice is fed back to the next evaporator stage at the same temperature. The evaporation process is not interrupted and the evaporator can operate smoothly and uniformly.

The Centritherm-Evaporator (27) is very popular for the concentration of passion fruit juice. This is a one-stage rotary thin film evaporator which can be fitted with or without aroma recovery. Originally developed by Alfa Laval for concentrating sensitive products in the pharmaceutical industry, this machine is now sold and serviced by FLAVOURTECH Pty. Ltd (27). For the very efficient recovery of the aroma FLAVOURTECH can also offer a so-called "Spinning Cone Column" which has been successfully applied for the recovery of difficult-to-recover aromas such as coffee aroma.

The concentrate is usually filled into drums for trading. Large citrus factories store their concentrates in deep-frozen tanks and ship it in tank ships or tank trucks. The recovered aroma is stored refrigerated and sold separately. The customer can add it back to his product before bottling.

### 3.2.2.1.7 Aroma Recovery

While producing juices and concentrates it is recommended to recover the aroma of the juice separately in order to add it back at a later time. For this purpose aroma recovery units have been developed which are constantly improved *[7-12]*. Basically there are two different methods:

  (1) Recovery in especially developed separate aroma recovery units.
  (2) Recovery by units integrated into the evaporator.

The working principle of the two types is similar. However, the integrated unit works more efficiently due to a better energy utilisation.

The principle of aroma recovery is to evaporate part of the juice and collect the vapours to recover the aroma. Depending on the type of fruit, 10-40% of the first vapours extracted at the beginning contain the highest amount of aroma and thus are suitable for aroma recovery. The vapour/aroma mixture is separated in a distillation column. The aroma is then concentrated in a rectification column. Usually the degree of concentration is 100-200 fold. To protect the aroma and to keep the temperature as low as possible, such plants are operated under vacuum. Most of the aroma is condensed in the rectification column. However, the gases which are stripped off the rectification column by the vacuum pump still contain highly volatile aroma components which can only be condensed at very low temperatures. For this reason the gases from the rectification column are cooled in a gas deep-freezer (cooling trap) and the highly volatile aroma components are condensed and separated. They can then be added back to the condensed aroma from the rectification column.

The recovered and concentrated aroma must be stored refrigerated if it is stored for any period of time prior to adding it back to the juice for bottling.

The fruit juice aroma is an aqueous liquid with a characteristic individual flavour and a chemical composition, depending on the type and the maturity of the fruit. The quality is judged by gas chromatography and sensorial tests.

The aroma of citrus fruit is a speciality. It consists of the so-called water phase and the oil phase. Citrus juices contain small amounts of volatile oils (0.03-0.06%). During processing there is always a small amount of peel oil that gets into the juice. Also the juice contains a small amount of oil, called juice oil. During evaporation these oils get into the aroma where they create the so-called oil phase. This oil has a very special aroma and is different from the peel oil by analysis (gas chromatography) as well as flavour. Oil and water phase are kept separately and added back to the juice according to individual requirements. Properly applied, the oil phase imparts the special fresh note to the juice which cannot be achieved by adding the water phase only.

The de-aromatised juice is either stored in sterilised condition (e.g. apple and pear juice) or concentrated.

### 3.2.2.1.8 Important Developments in the Past Years

New developments to improve quality and aroma of juice products deserve mentioning. Such developments must not necessarily consist of new working principles but they can also be improvements of traditional processes with newly designed and improved equipment. It usually takes a certain time before new approaches are accepted by industry and market. New approaches can be more expensive and new products need to be introduced into the market and accepted by the consumers.

### (1) Freshly Squeezed Juice

Several years ago a development started in the United States which has also encountered more and more interest in Europe: The production and sale of freshly squeezed juice. The idea is to produce a juice with as little heat treatment as possible. These juices can be divided into two product categories:

– Freshly squeezed juice for immediate consumption with no heat treatment at all. The juice is extracted in small extractors in the presence of the customer in super markets, restaurants or road stands and it is consumed immediately.

– Freshly squeezed juice, produced industrially under conditions as sanitary as possible. The juice, if necessary, can be slightly pasteurised and filled into sterile packaging. It is stored and delivered refrigerated to super markets, restaurants and canteens for sale to the consumer.

In order to achieve the longest shelf life possible, all equipment and packaging which has contact with the product must be absolutely clean in order to reduce bacterial growth to a minimum. The shelf-life of the juice can be increased by pasteurisation at low temperatures which reduces the microbiological count, but does not necessarily produce a sterile product. The flavour of the juice should not be influenced in any way by this treatment.

Filled into sterilised containers (glass bottles, cartons, etc.) and stored at temperatures of 0°C to 2°C, the juice has a shelf-life of approximately 3 weeks. In Florida fresh citrus juices achieve a shelf-life of 6 weeks due to special packaging and absolutely clean processing conditions.

Also in Europe freshly squeezed juices are becoming more and more popular. These can be juices from local fruit varieties, but also from vegetables, citrus fruit and other tropical fruit.

For the production of freshly squeezed citrus juices on a commercial basis FMC (Food Machinery Corporation) (2) has developed an improved extractor, the so-called "Low Oil Extractor", which releases as little peel oil as possible into the juice. With this extractor type an oil content of the juice of below 0.03-0.035% is achieved. Juices containing higher oil levels taste unbalanced and harsh.

De-sludger type centrifugal separators with an internal disc configuration, as used in dairy cream separators (WESTFALIA SEPARATOR (10)), are used to eliminate as much oil as possible from the juice. The newest developments show that oil contents below 0.03% can be reached.

The sale of fresh juice with a short shelf-life requires a well supervised production and distribution system.

## (2) Freeze Concentration

This process was already used by the Florida citrus industry in the 50's. The freshly squeezed citrus juice was deep frozen and the ice crystals were separated from the juice.

At that time aroma recovery units were still unknown. Juices manufactured from concentrate were "flavoured" by adding a small amount of selected peel oil and so-called "cut-back juice". This step could take place at the concentrate stage, but also upon reconstitution of the concentrate to juice strength. "Cut-back juices" were selected juices with excellent aroma which were stored frozen until needed. The freeze concentration process offered the possibility to concentrate "cut-back juices" without loss of or damage to their aroma. This method helped to reduce the storage volume. It also showed an advantage when aromatising concentrates with "cut-back-concentrate" rather than with "cut-back-juice" as the concentration (°Brix) of the final aromatised product is higher.

Large deep-freezing crystallisation units were employed at that time to separate the water from the juice by transforming the water into ice crystals. The next step was to separate the ice crystals from the juice by pusher centrifuges. To reach a high degree of concentration, multi-stage processing steps were necessary. This process was rather costly since it required large amounts of energy for the freezing process. Also, a certain amount of juice substance was lost and could not be recovered from the ice crystals. This was partly due to the difficulties in properly controlling the growth of the ice crystals. After the development of aroma recovery systems for the T.A.S.T.E. evaporator freeze concentration was abandoned.

Apart from Florida, also one Spanish citrus factory produced freeze concentrated juice which was then sold as a premium product.

Recently the freeze concentration process gained new importance when the freeze concentration of GRENCO AG (24) *[14]* was introduced. Large ice crystals can be formed by use of specially designed crystallisation vessels and these crystals can be separated much more easily from the juice and the loss of juice is reduced as well. In addition, the carefully designed process works thermodynamically with the least amount of refrigeration cost. This process is not only applied to juices but also to other products like coffee, beer and wine. The aroma of the concentrated products is practically undamaged and of high quality. Undoubtedly the GRENCO process is an important step towards producing concentrates from fruit juices whilst maintaining the natural and undamaged aroma.

When applying such special processes as to obtain high quality aroma or juice concentrates it makes little sense to subsequently pasteurise or hotfill the reconstituted juice. The heat treatment would damage the aroma and negate the advantage of the previously applied special process. These high quality concentrates are ideally suited for unpreserved, deep frozen juice concentrates which are reconstituted by the final consumer in the home. In the USA similar products are sold in small 6 ounce cans. For many years such concentrates have been produced in T.A.S.T.E. evaporators, re-aromatised and filled straight into retail cans.

This type of product has not yet reached a break through in Europe. However, one hopes that the above described new technology may help to successfully introduce even better aromatised frozen juice concentrates into the European retail markets. After all, they belong to the highest quality of industrially processed juices.

It remains to be mentioned that the same aromatised concentrates are also used for reconstitution and sale in dispensers. These machines keep the concentrate refrigerated, they reconstitute it with water to juice strength and they dispense the ready-to-drink juice into cups for immediate consumption. A large Floridian citrus factory (LYKES PASCO, Inc. (25)) used to distribute a great part of their citrus juice production world-wide in this way (Vitality-Dispenser).

### (3) Membrane Processes

Membrane processes were first used in the fruit juice industry in the middle of the 1970's. Since that time the equipment has been constantly improved and today it has its firm place in fruit juice technology.

Membrane processes are used to filter liquids. Instead of conventional filter materials (e.g. filter cloth, filter candles,) microporous membranes are employed with molecular size pores. First the industry had to learn how to manufacture membranes with controlled pore sizes. To optimise the filtration capacities specific filter structures had to be designed in which the liquid followed well defined flow patterns on one side of the membrane. Many different systems were developed for the varied applications, all having their advantages and also disadvantages, i.e. plate modules, tubular modules, spiral wound membranes, etc. Research and development in this field is far from being exhausted. Today membrane systems are available which are sufficiently resistant to chemical, mechanical and thermal stress. They are produced from plastic materials (polysulphone, polyamid, etc.) as well as from ceramic materials. The pore size of the membrane can be adjusted depending on the application. Ultrafiltration and microfiltration are used for the clarification of fruit juices.

For this application the pore size of the membrane has to be such that the small molecules in the juice e.g. fruit sugars can pass through the membrane (permeate), while large molecules e.g. pectins and cellulose particles are retained (retentate).

In microfiltration (MF) pore sizes are in the range of 0.1 to 12 µm. Viruses, bacteria, colloidal and cloudifying substances with a molecular size of over 1,000,000 MW (molecular weight) are retained.

In ultrafiltration (UF) pore sizes are in the range of 0.001 to 0.1 µm and the cut-off value is 500-500,000 MW. Macromolecules like proteins, enzymes as well as colloids, cloudifying substances, bacteria and viruses are retained.

Since pores are very small the permeate flow, based on the membrane surface area, is relatively small. The flow depends on the pressure applied to the liquid on the membrane surface. The higher the pressure the higher the flow. On the other hand, the membrane surfaces easily plug with slimy substances which decrease the flow until it is so low that the membranes must be periodically cleaned (fouling). The fouling also depends on the pressure on the membranes. To avoid fouling the liquid to be filtered is pumped with high velocity parallel to the membrane surfaces in order to flush fouling substances away from the membrane surface (tangential filtration). Depending on the product, pressure and velocity have to be adjusted for optimal filtration results.

The capacity can be increased further by reducing the viscosity of the liquid to be filtered. This can be achieved by treating fruit juices enzymatically and also by increasing the filtration temperature. The ideal temperature is determined by the temperature resistance of the membrane material and on the other hand by the fact that it must not impair the quality of the product.

Microfiltration and ultrafiltration membranes allow flow capacities of 150 to 500 liters/m$^2$ per hour when operating on water. This is expressed as "water flux" for each membrane type. Naturally the flow capacity for juices is lower. After cleaning of a membrane the water flux should reach its original capacity and it serves as an indication whether the membrane was properly cleaned. It is also an indication for when a membrane needs replacement once it plugs over longer periods.

Ultrafiltration and microfiltration are employed successfully on an industrial scale for the clarification of juices (e.g. apple juice, pear juice, etc.). Ultrafiltration saves filter aids and fining aids which have to be used when operating conventionally with rotating vacuum filters or plate filters and precoat filters. On the other hand, the power consumption of an ultrafiltration system is high and so is the replacement cost of membranes.

Ultrafiltration has also been used for the production of citrus juices with a natural unprocessed taste (BERTUZZI) (4) *[15]*. The fresh juice is split by ultrafiltration into a clarified and a cloudy part. The cloudy retentate contains the pulp, other cloudy substances and microorganisms. This retentate is pasteurised at 85°C – 98°C and, after elimination of the microorganisms, is added back to the clarified permeate which practically does not contain any microorganisms but which contains the natural aroma. The product so treated is packed into sterile packaging and it results in a naturally tasting juice with excellent shelf-life.

In Reverse Osmosis (RO) pore sizes are even smaller than in ultrafiltration. They are in the range of 0.0001 to 0.001 µm and the cut-off value is below 500 MW. The membrane separates not only all substances mentioned under ultrafiltration but also micro-molecules like salts. Practically only pure water is passing through the membrane. In order to achieve economical capacities, high pressure must be applied to the membrane (10 to 80 bar). To prevent fouling and to achieve long operating times between cleaning it is advisable to first clarify the liquid by ultrafiltration before feeding it to the reverse osmosis operation.

This process is used industrially to desalinate sea water for the production of potable water. However, it can also be used for fruit juice concentration. The achievable concentration depends very much on the viscosity of the concentrate. If the viscosity is too high the process is not economical. Since this process is rather expensive due to the high cost equipment it can only be employed for speciality products. In the future new membrane types with larger flow capacities could improve the economy of the RO process.

The water flux of RO membranes is in the range of 40 to 130 liters/m$^2$ per hour.

SEPARA SYSTEMS LP, Santa Clara, Ca. USA (26) *[16]* went one step further with their "FreshNote"(TM) process. First they produced permeate by ultrafiltration from selected juice. The permeate is concentrated by reverse osmosis and then added back to the pasteurised retentate. In this way a concentrate of 42°-51° Brix with natural taste was obtained. The taste of this product was superior compared to the taste of concentrates produced by evaporation.

As previously mentioned, the development of new membranes and further improvements in this field is not yet exhausted. It will be interesting to see further developments of these processes since quite a number of companies are engaged in this technology.

In this regard, a study in the USA is of interest which investigates the loss of the aroma of apple juice when using various filtration methods for clarification (M.F. Sancho u. M.A. Rao, Department of Food Science and Technology, New York State Agricultural Experiment Station, Cornell University Geneva, New York) *[17]*.

### (4) New Juice Extraction Process by Means of Decanter Centrifuges

For many years the leading centrifuge manufacturers have tried to use decanter centrifuges for the extraction of juice from ground fruit mash. Decanting centrifuges are equipped with a long rotating bowl which is conical on one side. Inside the bowl there is a conveyor screw which is turning at a slightly different speed than the bowl. The solid containing liquid is pumped into the rotating bowl where, due to their higher specific gravity, the solids are separated by centrifugal force. The solids travel to the largest diameter in the centrifuge bowl, the bowl wall, where they settle and form a sludge layer. The clarified liquid flows to a pump chamber which is located on the cylindrical side of the bowl with the largest diameter. A built-in centripetal pump (also called paring disk) is conveying the liquid out of the pump chamber and the centrifuge bowl. The sludge layer is conveyed by the conveyor screw (also called scroll) to the conical side of the bowl, moved up the cone and out of the liquid and finally discharged from the bowl through openings at the end of the cone. The amount of sludge which is discharged depends on the differential speed between bowl and conveyor screw.

These machines work quite well when the solids consist of relatively large particles which due to a firm structure convey easily. When, however, the solids are soft and easily destructible as this can be the case with fruit mash (e.g. apple mash), difficulties arise. Partly the solids can already be disintegrated when entering the bowl and being accelerated to bowl speed (homogenisation). The smaller soft particles are more difficult to separate due to their smaller size. They also do not concentrate properly to form a compact sludge. If the differential speed between conveyor screw and bowl is not correctly adjusted to the amount of solids to be conveyed the soft sludge layer on the bowl wall is further homogenised and eventually mixed with the surrounding liquid. Subsequently the solids are leaving the centrifuge bowl in a rather wet, unconcentrated stage with a juice content which is too high. It is therefore of great importance to adjust the speed of the screw conveyor accurately to the amount of solids, which so far was practically impossible with conventional machines also because the amount of solids in the feed liquid is changing constantly. To account for these fluctuations the conveyor screw speed was usually adjusted faster than needed for the amount of sludge coming into the machine to make sure the solids were not piling up inside the bowl, thus eventually filling up the entire bowl volume. However such an adjustment is not suitable for producing a well concentrated sludge, specifically when the solids are soft.

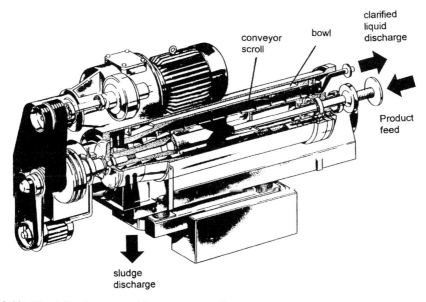

***Fig. 3.12:*** *Westfalia decanter with torque control system*

A new 2-gear-drive was developed which accurately controls the differential speed between bowl and conveyor screw by precise electronic torque control. This new drive instantly accounts for fluctuations in the feed liquid and adjusts the differential speed between bowl and conveyor screw for optimal sludge concentration. This new control device also improves the overall performance of the decanter (WESTFALIA SEPARATOR (10)), *[2, 18-19]*.

As a result of this latest development it is now possible to use decanter centrifuges for the extraction of fruit juices with excellent results (apple, grape, etc.). One specific advantage is the short time needed between crushing of the fruit and juice recovery. Another advantage is the possibility to run the entire process with a minimum exposure to air in a closed system and if needed under a nitrogen blanket to positively avoid oxidation. This has been successfully used to produce very lightly coloured and bright cloudy apple juice concentrates and fruit juices with excellent colour and aroma.

### (5) Treatment of Fruit Juices with Adsorber Resins

Adsorber resin treatment can remove certain undesired substances from juices and can be used to improve colour and taste.

Styrene-divinylbenzene cross-linked copolymer resins constitute one sort of resin with good adsorbing ability; phenolic resins have also been tested. Other types with improved qualities may be developed.

The type of resin has to be specifically determined for the desired application. These adsorber resins must not mistakenly be compared with ion exchange resins. Adsorber resins only adsorb certain molecules but otherwise do not have ion exchange properties. The working principle is similar to that of active charcoal.

Early (not quite mature) citrus juices, lemon, Hamlin and Navel juices can develop the so-called limonin-bitterness. Additionally, there can also be a flavonoid-induced bitterness, like naringin in grapefruit juice. Adsorber resin treatment can successfully eliminate this bitterness below the detectable limit. See also *[20]*.

Apart from taste, also the colour of juices can be improved. This is commercially done with apple juice to obtain a very bright, almost colourless juice which is needed for specific applications and markets.

Adsorber systems are available from BUCHER (11) and UNIPECTIN (20).

Before adsorber installations are planned, it is recommended to check if the food law of the country in which this is to be realised allows their operation.

## Addresses

(1) BROWN Extractor:
Automatic Machinery Corp.
633, North Barranca Avenue
Covina, California 91723-1297
U.S.A.
Phone: +1 626 966 8361
Fax: +1 626 332 7921
Email: info@brown-intl.com

(2) FMC Extractor:
FCM Technologies, Inc.
FMC FoodTech - Citrus Systems
400 Fairway Avenue
Lakeland, Florida 33801
U.S.A.
Phone: +1 863 683 5411
Fax: +1 863 680 3672
Email: citrus.info@fmcti.com
Website: www.fmcfoodtech.com

(3) T.A.S.T.E. Evaporator:
Cook Machinery
P.O. Box 1073
Dunedin, Florida 34697
U.S.A.
Phone: +1 727 796 1367
Fax: +1 727 791 1750
and

FMC Technologies, Italia SpA
Via Mantova 63A
Box 333
1-43100 Parma
Italy
Phone: +39 (0)521 908411
Fax: +39 (0)521 487960
Email: sales.parma@intl.fmcti.com

(4) Alberto Bertuzzi S.p.A.
Via Europa, 11
20047 Brugherio (MI)
Italy
Phone: +39 (0)39 28921
Fax: +39 (0)39 883205
Website: www.bertuzzi.it

(5) Fratelli Indelicato S.r.l.
Via Messina, 86
P.O. Box 57
I-95014 Giarre (Catania)
Italy
Phone: +39 (0)95 932263 - 938266
Fax: +39 (0)95 937864
Email: indelicato@omnia.it

(6) Speciale Francesco
de Speciale Sebastiana & C. snc
Via Torrisi, 18
95104 Giarre (CT)
Italy
Phone: +39 (0)95 931124
Fax: +39 (0)95 930279
Website: www.speciale.it

(7) Fomesa
Food Machinery Española, S.A.
Avda. Jesós Morante Borr«s, 24
46012 Valencia
Spain
Phone: +34 96.3.16.54.00
Fax: +34 96.3.67.79.66
Website: www.fomesa.com

(8) OIC Centenario Ltda.
Av. Major Jos-Levy Sobrinho, 1946
13486-190 Limeira, SP
Brazil
Phone: +55 19 3451.6710
Fax: +55 19 3451.6760
Email: oic@oiccentenario.com.br
Website: www.oicentenario.com.br

(9) Alfa Laval Corporate AB
Rudeboksvægen 1
SE - 22655 Lund
Sweden
Phone: +46 46 36 65 00
Fax: +46 46 32 35 79
Email: info@alfalaval.com
Website: www.alfalaval.com
and
Tetra Pak Dairy & Beverage Systems AB
Bryggaregatan 23
S-221 00 Lund
Sweden
Phone: +46 46 36 10 00
Fax: +46 46 36 54 80

(10) GEA Westfalia Separator
Westfalia Separator Food Tec GmbH
Werner-Habig-Str. 1
59302 Oelde
Germany
Phone: +49 (0)2522 770
Fax: +49 (0)2522 772089
Email: foodtec@gea-westfalia.de
Website: www.westfalia-separator-food-tec.com

(11) Bucher-Guyer AG
Foodtech
CH-8166 Niederweningen
Switzerland
Phone: +41 (0)1 857 22 11
Fax: +41(0)1 857 23 41
Email: foodtech@bucherguyer.ch
Website: www.bucherfoodtech.com

(12) Amos GmbH[*]
Anlagentechnik
Helmuth-Hirth-Str. 8
74081 Heilbronn
Germany

(13) Gebr. Bellmer GmbH
Maschinenfabrik
Hauptstr. 37 - 43
75223 Niefern
Germany
Phone: +49 (0)7233-740
Fax: +49 (0)7233-74215
Email: info@bellmer.de
Website: www.bellmer.de

(14) Diemme S.p.A.
Via Bedazzo, 19
Zona Industriale
48022 Lugo (RA)
Italy
Phone: +039 (0)545 20611 (10 lines)
Fax: +39 (0)545 30358

(15) Flottweg GmbH & Co.KGaA
Industriestr. 6-8
84137 Vilsbiburg
Germany
Phone: +49 (0)8741 301-0
Fax: +49 (0)8741 301-300
Email: mail@Flottweg.com
Website: www.flottweg.com

(16) Alb. Klein GmbH & Co. KG[**]
57572 Niederfischbach/Sieg
Germany

(17) Tetra Pak GmbH & Co.
Frankfurter Str. 79-81
65239 Hochheim/Main
Germany
Phone: +49 (0)6146 590
Email: tphinfo@tetrapak.com
Website: www.tetrapak.de

(18) SIG Combiblock GmbH
Rurstr. 58
52441 Linnich
Germany
Phone: +49 (0)2462 790
Email: info@sig.biz
Website: www.sigcombibloc.com

(19) Gea-Wiegand
Einsteinstr. 9-15
76275 Ettlingen
Germany
Phone: +49 (0)7243 705 0
Fax: +49 (0)7243 705 330
Email: info@gea-wiegand .de
Website: www.gea-wiegand.de

(20) Unipektin AG
Claridenstrasse 25
CH-8022 Zürich
Switzerland
Phone: +41.1.206 54 44
Fax: +41.1.206 5455
Email: mail@unipectin.ch
Website: www.unipectin.com

(21) APV
23 Gatwick Road
Crawley, West Sussex RH10 9JB
Great Britain
Phone: +44 1293 527777
Fax: +44 1293 552640
Email: cbs.apvuk@invensys.com
Website: www.apv.invensys.com

(22) Gea-Process Engineering Division
Tuchenhagen Dairy Systems GmbH
Voss-Str. 11 / 13
31157 Sarstedt
Germany
Phone: +49-(0)5066 990-0
Fax: +49-(0)5066 990-160
Website: www.tuchenhagen.de

(23) API Schmidt-Bretten GmbH & Co. KG
Langenmorgen 4
75015 Bretten
Germany
Phone: +49 (0)7252 53-0
Fax: +49 (0)7252 53-200
Email: info@apischmidt-bretten.de
Website: www.apischmidt-bretten.de

---

[*] This firm is no longer in operation
[**] This firm is no longer in operation

(24) Grenco AG was taken over by:
Niro Process Technology B.V.
P.O. Box 253
5201 AG `s-Hertogenbosch
The Netherlands
Phone: +31 73 6390 390
Fax: +31 73 6312 349
Email: sales@niro-pt.nl
Website: www.niro-pt.nl

(25) Lykes Pasco, Inc.[*]
100, North Highway 301
Dade City, Florida 34297
U.S.A.

(26) Separa Systems LP[**]
317, Brokaw Road, Suite A
Santa Clara, California 95050
U.S.A.

(27) Flavourtech Pty. Ltd.
Phone: +61-2-9418 4022 (Australia)
+44-118-935 7309 (UK)
+1-707-577-7810 (USA)
Email: sales@flavourtech.com.au
Website: www.flavourtech.com.au

## REFERENCES

[1]  Hartmann E., Comparing Apple Juice Yield and Capacity, Flüssiges Obst/Fruit, Processing, $\underline{5}$, 156-162, (1993)

[2]  Colesan F., Jöhrer, P., Einsatz von Bandpressen, Dekantern und Separatoren bei der Frucht- und Gemüseverarbeitung, Flüssiges Obst, $\underline{5}$, 244-246, (1993)

[3]  Bastgen, W., Möslang, H., Bandpressen zur Fruchtsaftgewinnung, Flüssiges Obst, $\underline{10}$, (January 1988)

[4]  Möslang, H., Fruchtsaftgewinnung mit modernen Siebband-pressen, Getränkeindustrie, $\underline{2}$ (1987)

[5]  Reichard, H.J., Dimitriou M., Einbandpressen-Erfahrungsbericht und Weiterentwicklung, Flüssiges Obst, $\underline{5}$, 248, (1993)

[6]  EP 0 090 734 B1

[6a] 'Classic' juice clarification gets better: new method raises yield 2-5% in 2 hours or less, Food Engineering, $\underline{137}$,(March 1984)

[7]  Baier, H., Toran, J., Aromagewinnung und Herstellung von Konzentraten aus Fruchtsäften, Getränkeindustrie, $\underline{2}$ and $\underline{3}$, (1985)

[8]  Baier, H., Toran,J., Wirtschaftliche Aromagewinnungs- und Eindampfanlagen für Fruchtsäfte, Lebensmittel- & Biotechnologie, $\underline{4}$, Fachverlag Vienna, 129-136, (1984)

[9]  Hochberg, U., Progresses in Aroma Recovery, Flüssiges Obst, $\underline{3}$, (1987)

[10] Hochberg, U., Neue Untersuchungen zur Gewinnung von Fruchtsaftaromakonzentrat, Getränkeindustrie, $\underline{19}$, (1986)

[11] Dimitriou, M., Concentration Plants with Aroma Recovery, Operation Experience and New Developments, Flüssiges Obst, $\underline{4}$, (1986)

[12] Dimitriou. M., New Evaporator Type for the Fruit Juice Industry, Flüssiges Obst, $\underline{9}$, (1984)

[13] Florida Department of Citrus, Scientific Research Department, University of Florida -IFAS - CREC, 700 Experiment Station Road, Lake Alfred , Florida 33850-2299, Technical Manual, Florida Orange Juice (Commonly called Fresh-Squeezed Florida Orange Juice), Production/ Packaging/ Distribution. September 1987

[14] Grenco freeze concentration systems...simply a matter of quality, Grenco leaflet (address see above)

[15] Decio, P., Gherardi, S., Freshly Squeezed Orange Juice, Confructa Studien, $\underline{36}$, Nr. V/VI, 162-167, (1992)

[16] Cross, S., Membrane Concentration of Orange Juice, Proc. Fla. State Hort. Soc. $\underline{102}$, 146-152, (1989)

[17] Sancho, M.F., Rao, M.A., Aroma Retention and Recovery during Apple Juice Processing, New York Science Agricultural Experiment Station Special Report No.65, October 1992. (see also.Flüssiges Obst/Fruit Processing, $\underline{5}$, 175-177, (1993) )

---

\* This firm is no longer in operation
\*\* This firm is no longer in operation. It used to be a joint venture of FMC Corp. and Dupont for micro filtration systems

[18] Hamatschek, J., Nagel, B., 100 Years of Centrifuge Engineering and the Use of Decanters for Dejuicing Fruit, Flüssiges Obst/Fruit Processing, 5, 163-167, (1993)
[19] Kern, W., Guldenfels, E. Sieveke, E., The Use of Decanters in Modern Fruit and Vegetable Extraction, Flüssiges, Obst /Fruit Processing 5, 168-171, (1993)
[20] United States Patent 4, 439, 458, March 27, 1984

## 3.2.2.2 Citrus Oils

*Herta Ziegler*
*Wolfgang Feger*

### 3.2.2.2.1 Introduction

Citrus fruits play an important role in modern nutrition. Their pleasant, refreshing taste is always associated with their nutritional value and healthy image. Citrus products constitute an important source of vitamins and bioflavonoids, while offering a fresh, fruity taste.

This accounts for the development of citrus products into a major industrial factor within the last century. The main growing areas are situated in North and South America, southern Europe and northern Africa *[1]*.

Apart from the fresh fruit market, the production of juice, and thus also of citrus by-products, constitutes an important branch of the citrus industry.

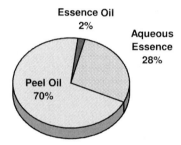

*Fig. 3.13:* Mass balance of orange oil products

The production and concentration of juice lead to downstream products such as ***peel oil***, ***essence oil*** and ***aqueous essence*** (see 3.2.2.1). All three products constitute important raw materials for the flavour industry. The relationship between the three products - depicted in Fig. 3.13 for the example of orange - shows that peel oil constitutes the major product. It is commercially available as 'cold pressed oil', as it is the result of a cold pressing process which is applied during juice production. The ***cold pressed peel oil*** corresponds to the oil content of the oil glands in the citrus fruit's peel (see Fig. 3.14). This type of oil is available for all citrus varieties.

The juice oils that are collected during the concentration of fruit juice have generated increasing interest as a valuable source of raw material. These ***essence oils***, also called recovery oil, oil phase or taste oil, correspond mainly to the oil content of the juice vesicles and, depending on the extractive equipment employed, to the varying amount of peel oil which is present in the juice (see 3.2.2.1).

Apart from the taste oils, the ***aqueous essences*** synonymously called aroma, water-phase, waterphase aroma or essence waterphase are also retained as part of the aroma recovery during juice concentration. Their flavouring potential is mainly used in reconstituted fruit juices. On the one hand, the watery environment exerts a negative impact on many citrus flavourings; on the other hand, the instability and, therefore, the necessity to store the aqueous, low concentrated flavour at freezing temperatures have turned out to be very cost-intensive.

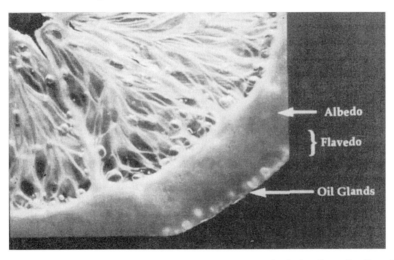

*Fig. 3.14: Cross section of an orange showing the location of oil glands in the flavedo*

Apart from the already described processes, another method of distillative technology has found application, especially for lime fruits. Here entire fruits are submitted to steam distillation which, due to the exposure to thermal and acidic influences, results in the so-called distilled lime oil. The overall flavour profile of these oils differs considerably from that of cold-pressed lime oil.

So far only the oils obtained from citrus fruits have been taken into account. Additionally, essential oils of the leaves, twigs and blossoms of citrus species are also available. Water vapour distillates of leaves and twigs are called *petitgrain oils*, while the distillate of orange blossoms is commercially available as *neroli oil*. Furthermore, also an *orange flower absolute* is produced from orange blossoms [2].

**Citrus Essential Oils – Constituents and Processing**

Citrus oils are widely used for flavouring purposes. Their range of application extends from the food sector to household cleansers and fine perfumery [3].

The flavour properties of the citrus oils are based on compounds like aldehydes, esters and alcohols. Aldehydes play the most important role in citrus flavours. Mainly the saturated C-8, C-10 and C-12 long-chain constituents as well as the terpenoid aldehydes citral (a mixture of neral and geranial) and citronellal form the basic notes of the fresh and pleasant citrus flavour. Also sesquiterpene aldehydes and ketones, such as sinensal or nootkatone, are significant for citrus flavours. The aldehyde content, therefore, is frequently used as a quality criterion for citrus oils. Important citrus alcohols are linalool, octanol, α-terpineol (especially in distilled lime oil) and terpinen-4-ol. Furthermore, esters contribute to the flavour properties of citrus oils. Especially the acetates of geraniol and nerol and, to a lesser extent, citronellol are connected with citrus flavour impressions. Linalyl acetate and methyl N-methylanthranilate are basic esters of bergamot and mandarin oil, respectively. Although flavour quality is closely tied to the composition of these middle polar constituents, they only play a minor role in citrus oils as far as quantity is concerned. The main constituents of citrus oils are non-polar terpenes and sesquiterpenes with limonene contributing the lion's share. Limonene, the main constituent, is present in a range between 60% (i.e. lemon) and 95% (i.e. orange) depending on the citrus variety and geographic origin. The chemical and physical properties of the citrus oils are therefore closely tied to the characteristics of this compound class. In this context, their poor solubility in water (a result of their non-polar structure) as well as their thermal instability and sensitivity to oxidation exert a considerable influence on the properties of these oils. Oxidation, degradation and other transformation products of terpenes, which can form under stress conditions, constitute odour-relevant off-compounds [4-9], which are usually only minor contributors to the citrus flavour. These reactivities are nevertheless able to disturb the citrus flavour associations resulting in a negative impact on the flavoured products.

Apart from the volatile constituents, cold-pressed citrus oils contain a non-volatile fraction in varying amounts (1% in orange oil, up to 10% in cold-pressed lime oil), which is associated with emulsifying and stabilising properties.

While most perfumers employ raw citrus or petitgrain oils for usage in fragrance applications (solubility in alcoholic fragrances is sufficient and no thermal stress is required for fragrance applications), the usage of concentrated citrus oils is often called for in the food industry. Applications in beverages and foods require more sophisticated properties as far as solubility and stability are concerned.

Enhanced stability and improved solubility are features of concentrated citrus oil products as a result of their reduced amounts of terpenes. These properties are of particular importance for flavouring soft drinks, the most important field of application for citrus flavourings in the food sector.

So today more than ever, attempts are made to preclude undesired side-reactions of terpenes by minimising the terpene content of citrus oils.

Basically, different techniques can be employed to optimally concentrate citrus oils: distillative, extractive and chromatographic methods are available; frequently combined multi-step technologies are used. However, cold procedures are apparently the first choice in citrus oil processing (see chapter 2), if the following aspects are considered:

- loss of flavour
- thermal lability of terpenes and flavour contributors
- off-flavour formation
- stabilising effect of non-volatiles
- sensitivity to oxidation
- complexity of the citrus flavour.

The suitability of the various deterpenation methods will be dealt with briefly in the following.

## Distillative Separation of Terpenes (see 2.1.3)

Distillation is often used on an industrial scale *[10, 11]* to produce concentrated citrus oils. Distillation provides a simple method with standard equipment, moderate expenditure and high production capacities.

*Disadvantage.* As any distillative process causes higher temperatures, aspects like thermal instability of terpenes are neglected. This results in radically or thermally induced transformation and crack processes of terpenes, forming precursors which subsequently lead to quick aroma destruction and a negative impact on the aroma profile of the concentrated oils.

## Extractive Methods (see 2.1.1-2.1.2)

This concentration method basically offers a high measure of variability *[10, 12-17]*. The extractive concentration of citrus oils can proceed at low temperatures and, therefore, ensures minimal side reactions. Generally, citrus aroma concentrates of high quality can be produced by extraction. Such products reflect the entire complex aroma pattern: even trace odour components or non-volatiles contributing to the stability of the flavour are selectively extracted. Special product characteristics are adjustable by process parameters and choice of solvent.

*Disadvantages.* Only a limited number of solvents are in accordance with legislative guidelines for food processing, are easy to remove and provide the possibility of reutilisation. Furthermore various processing steps are necessary for recovering the concentrates from the diluted solutions; the method, therefore, is more cost-intensive.

In this context the so-called 'washing' of citrus oils is still employed in the beverage industry. It constitutes a simple, traditional extraction method. Here, the citrus oils are extracted with watery alcoholic solutions and the oil extract is deterpenised depending on water content and oil/solvent ratio *[18]*. At about 50%, the yield for flavour compounds is comparatively low *[19]*. The deterpenised oil is not freed from solvents and, therefore, only encounters a limited range of applications.

## Chromatography

This method allows the fractionation of citrus oils based on the different polarity of terpene and aroma fraction *[20, 21]*. Their different adsorption characteristics on stationary phases can be employed for the selective separation of these complex mixtures of natural substances. The method originates from analytical laboratory techniques and its application on various stationary phases is realised industrially today *[22, 23]*. Adsorption on stationary phases and their subsequent desorption with various solvents is possible for different adsorption materials. Also partition chromatographic methods play an important role in citrus flavour concentration processes, especially for aqueous citrus essences.

Chromatographic processes eliminate high temperatures and even allow the complete deterpenation of citrus oils. Additionally they allow the enrichment of non-volatile compounds. Selectively enhanced fractions of coumarins and tocopherols, natural flavonoids and carotenoids are well suited for usage in special applications which feature excellent stability and capture the desired aroma profile.

*Disadvantage.* Aspects are analogous to extraction methods. Compared to the amount of citrus concentrates produced around the world, these products are only produced on a small scale.

### 3.2.2.2.2 Citrus Oil Varieties

**Bergamot Oil**      *Citrus aurantium* L. ssp. bergamia

Geographical Origin: Italy is the main growing area (especially Calabria), as well as South America, Turkey and China.

Sensory Evaluation: Floral, fruity, greenish, weakly herbaceous, the oil resembles lavender and neroli notes and possesses a weak coriander and nutmeg base note. The oil with its very pleasing odour is mainly used in perfumery.

Production: Pressed by the pellatrice technique from the peel of the fruits from autumn (green oil) to spring (yellow oil).

Fruits and juice, as a result of their high acid content, are not suited for consumption. Cultivation depends entirely on the yield of peel oil which, therefore, is expensive. The confirmation of native origin via suitable detection methods is of special importance.

Physical Data:    $d_{20}$: 0.876 - 0.884      $\alpha_D^{20}$: 8 - 30°

                 $n_D^{20}$: 1.463 - 1.467      $L_{90}^{20}$: 1:1

Chemical Data: Aldehydes (cal. as citral): 0.4 - 1.2%

Depending on production time and growing area, the content of linalool and linalyl acetate varies *[24-30]*. Qualities from autumn production possess higher linalool values (15 - 16%) than spring oils, while, on the other hand, the content of linalyl acetate increases with fruit maturity (up to 33%) *[28, 31]*.

Important constituents of the non-volatile range are bergamottin, 5-geranyloxy-7-methoxycoumarin, citropten and bergapten, which together represent 1.5 - 4% of the oil *[32, 33]*. Due to the high bergapten content, usage in perfumery is limited by law. This is a result of the phototoxicity of bergapten which, when exposed to light, causes severe erythema *[34]*.

*Enantiodifferentiation:* Native qualities contain R(-)-linalool as well as R(-)-linalyl acetate in an enantiomeric purity of over 99% *[35, 36]*. 98% of the limonene content is present as R(+)-limonene.

*Table 3.16: Main constituents (%) of various bergamot oils*

|  | Italy [24] | Calabria [25] | Argentina [26] | China [27] | Turkey [37] |
|---|---|---|---|---|---|
| **Limonene** | 35 | 38 | 42 | 41 | 32 |
| **Linalyl acetate** | 30 | 28 | 28 | 23 | 37 |
| **Linalool** | 12 | 10 | 11 | 13 | 16 |
| **β-Pinene** | 6 | 7 | 5 | 5 | 3 |
| **γ-Terpinene** | 6 | 8 | 0.5 | 5 | 4 |
| **Myrcene** | 1.4 | 1 | 0.9 | 1.5 | 0.8 |
| **α-Pinene** | 1.3 | 1.3 | 0.8 | 1 | 0.8 |
| **Bisabolene** | 0.5 | 0.5 |  | 0.6 | 1.4 |
| **Geranyl acetate** | 0.3 | 0.4 |  | 1.3 | 0.2 |
| **Geranial** | 0.2 | 0.4 | 0.2 | 0.2 |  |
| **α-Terpineol** | 0.1 | 0.1 | 4 | 2 | 0.2 |

**Grapefruit Oil**     *Citrus paradisi* Macfad.

Geographical Origin: Main growing areas are located in North and Central America (Florida, California, Mexico and Cuba) as well as in Israel

Sensory Evaluation: Typical bitter grapefruit note which resembles both bitter and sweet orange. Additionally, the peel oil is characterised as a peel-like, pronounced, watery, juicy, greenish citrus note.

Grapefruit oils are mainly used in flavourings. In perfumery, they are only employed as modifiers for other citrus types (bergamot, bitter orange, lemon etc.).

Production: During juice production from the peel of the fruits.

Physical Data:     $d_{20}$:   0.846 - 0.856     $\alpha_D^{20}$:   90 - 96°

$n_D^{20}$:   1.474 - 1.478     $L_{96}^{20}$:   1:1

Chemical Data: Aldehydes (cal. as decanal): 1.1 - 1.6%

Epoxybergamottin, auraptene, meranzin, epoxyauraptene, bergamottin and isomeranzin are the most important constituents of the non-volatile range. Together with a number of other flavones and coumarins, they form a pattern which is typical for grapefruit oil and make up app. 3-5% of the oil [38, 39]. (+)-Nootkatone is responsible for the typical bitter note.

Apart from the widespread white marsh grapefruit, red and pink grapefruit varieties have encountered increasing popularity. Carotenoids are responsible for the red colour; analytical investigations have shown that the carotenoid pattern depends on the part of the fruit - flavedo, pulp or peel *[40]*. Additionally, a number of hybrids with grapefruit as one parent strain have been developed in recent years with the aim of creating organoleptically new citrus nuances for the fresh fruit market *[41 ,42]*. The variety Sweetie or Oroblanco (hybrid of grapefruit and Pummelo) is already well established on the fresh fruit market and growing harvests now increasingly permit industrial production of juice and oil *[43]*.

*Enantiodifferentiation:*   R(+)-limonene and R(+)-α-pinene are present in enantiomeric purities of over 99% of total limonene and α-pinene *[44, 45]*.

## Grapefruit Essence Oil

For botanical name and origin see above.

Production:   During grapefruit juice concentration (see 3.2.2.1).

Sensory Evaluation:   The intensely fruity, juicy, slightly woody and sulphurous, tropical odour profile of this oil also features a pronounced nootkatone note.

Physical Data:   $d_{20}$: 0.840 - 0.851   $\alpha_D^{20}$: 89 - 100°
                 $n_D^{20}$: 1.469 - 1.477   $L_{96}^{20}$: 1:1

Chemical Data:   Aldehydes (cal. as decanal): 1.5 - 2.5%

*Enantiodifferentiation:*   see grapefruit oil

The especially fresh, juicy grapefruit note is due to the presence of R(+)-para-menthene-8-thiol which can be detected as a trace constituent in the ppb range of grapefruit oils *[46]*. Other important potent odour-active volatiles of grapefruit juice are 1-penten-3-one, 1-hepten-3-one, hexanal, 4-mercapto-4-methylpentan-2-one and 4,5-epoxy-(E)-2-decenal *[47]*.

## Grapefruit Aqueous Essence

For botanical name and origin see grapefruit oil.

Production:   During the concentration of grapefruit juice via TASTE evaporators.

Sensory Evaluation:   The aqueous grapefruit essence is characterised as an aqueous, juicy, fruity, woody, estery, sweaty nootkatone and acetal note.

Physical Data:   $d_{20}$: 0.970 - 0.989   pH: 4.5 - 6.8
                 $n_D^{20}$: 1.335 - 1.343   Vol%: 5 - 15%

Main Components: Ethanol (5-12%), methanol, acetaldehyde

The total aroma content (without the previously listed compounds) amounts to app. 400 ppm and is made up by the following main constituents (the values quoted in parentheses correspond to ppm values) [48]:

cis- and trans-linalool oxide (60-180), α-terpineol (20-40), iso-amylalcohol (15-40), linalool (15-25), C-6-aldehydes (20-60), nootkatone (5-30), t-carveol (10-15), ethyl butyrate (2-15).

*Table 3.17: Main constituents (%) of grapefruit oils*

|  | white Cuba [43] | white Florida [43] | white Israel [49] | Sweetie Israel [43] | Essence Oil USA [48] |
|---|---|---|---|---|---|
| **Limonene** | 94.5 | 93.4 | 93.0 | 94.0 | 88.9 |
| **Myrcene** | 1.87 | 1.92 | 1.97 | 2.00 | 1.62 |
| **Octanal** | 0.51 | 0.50 | 0.29 | 0.62 | 0.33 |
| **Sabinene** | 0.34 | 0.37 | 0.39 | 0.91 | 0.31 |
| **Decanal** | 0.39 | 0.53 | 0.27 | 0.31 | 0.34 |
| **α-Pinene** | 0.51 | 0.48 | 0.59 | 0.58 | 0.29 |
| **β-Caryophyllene** | 0.28 | 0.32 | 0.31 | 0.13 | 2.71 |
| **Geranial** | 0.10 | 0.08 | 0.08 | 0.15 | 0.24 |
| **Neral** | 0.05 | 0.04 | 0.05 | 0.08 | 0.10 |
| **Nootkatone** | tr | 0.37 | 0.30 | tr | 1.00 |
| **Citronellal** | 0.06 | 0.10 | 0.07 | 0.06 | - |
| **Octanol** | 0.05 | 0.04 | 0.02 | 0.02 | 0.02 |
| **Humulene** | 0.04 | 0.05 | 0.04 | 0.03 | 0.29 |
| **Germacrene D** | 0.08 | 0.10 |  | 0.10 |  |

| | |
|---|---|
| **Lemon Oil** | *Citrus limon* (L.) Burm. f. |
| Geographical Origin: | Main growing areas are the USA, Italy, Argentina, Brazil, Uruguay, Greece, Spain and Israel. |
| Sensory Evaluation: | Typical lemon character with a sweet, fresh, sharp, citral-like aroma note, with a weak resemblance to orange. Lemon oil imparts a refreshing touch to many flavour and fragrance blends. |
| Production: | By various pressing processes from the fruit peel (Sfumatrice, Pellatrice and FMC technology). |
| | Depending on time of harvesting, growing and processing technique as well as lemon variety and geographical origin, oils of different qualities are obtained [50-53]. Sicilian winter oils possess the highest priority from a qualitative |

| | point of view, a result of the unique climatic conditions and the fertile volcanic soil. |
|---|---|
| Physical Data: | $d_{20}$: 0.846 - 0.854 $\quad\quad$ $\alpha_D^{20}$: 56 - 66° |
| | $n_D^{20}$: 1.472 - 1.476 $\quad\quad$ $L_{96}^{20}$: 1:1 |
| Chemical Data: | Aldehydes (cal. as citral): 2.7 - 3.9% |
| | A number of coumarin derivatives constitute an important part of the non-volatile fraction. These form a pattern which is characteristic for lemon oil and which can be used for both, quality control and confirmation of native origin [54-57]. The most important constituents of this compound class are: bergamottin, 5-geranyloxy-7-methoxy-coumarin, oxypeucedanin, citropten, 8-geranyloxypsoralen, byakangelicin, isoimperatorin, 5-isopentenyloxy-8-epoxyisopentyloxypsoralen, phellopterin and a few others. |
| *Enantiodifferentiation:* | Native qualities contain R(+)-limonene with an enantiomeric purity of over 98% [58, 59]. |

## Lemon Essence Oil

For botanical name and origin see lemon oil.

| | |
|---|---|
| Sensory Evaluation: | Also in the essence oil, the characteristic lemon odour is closely associated with citral, an isomeric mixture of neral and geranial. Moreover, the essence oil also features a juicy, earthy, aqueous, refreshing note. |
| Production: | During the concentration of lemon juice (see 3.2.2.1). |
| | Depending on production conditions, oils of varying quality are available. As exposure to temperature has a very negative impact on lemon products, only oils obtained by gentle production methods are sensorially valuable. During the distillative production process, all natural stabilizers are removed and the oil, therefore, has to be stored at low temperatures without exposure to air. |
| Physical Data: | $d_{20}$: 0.844 - 0.852 $\quad\quad$ $\alpha_D^{20}$: 66 - 76° |
| | $n_D^{20}$: 1.470 - 1.476 $\quad\quad$ $L_{20}^{90}$: 1:5 |
| Chemical Data: | Aldehydes (cal. as citral): 1.5 - 3.5% |
| *Enantiodifferentiation*: | see lemon oil |

Table 3.18: Main components (%) of lemon oils

|  | Lemon Peel Oils | | | | | Lemon Essence Oil |
|---|---|---|---|---|---|---|
|  | Italy [60] | Spain [61] | USA [62] Calif. (coast) | Uruguay [63] North | Israel [64] | Italy [48] |
| **Limonene** | 65.3 | 66.0 | 59.6 | 67.2 | 71.8 | 66.6 |
| **β-Pinene** | 12.7 | 10.5 | 16.5 | 13.9* | 8.57 | 8.30 |
| **γ-Terpinene** | 9.00 | 10.1 | 9.64 | 8.88 | 7.83 | 10.1 |
| **α-Pinene** | 2.00 | 1.58 | 2.13 | 1.92 | 1.65 | 1.42 |
| **Sabinene** | 2.10 | 1.67 | 2.59 | * | 1.52 | 0.50 |
| **Myrcene** | 1.41 | 1.45 | 1.28 | 1.58 | 1.63 | 1.45 |
| **Geranial** | 1.76 | 1.32 | 1.45 | 1.32 | 1.18 | 2.01 |
| **β-Bisabolene** | 0.58 | 0.84 | 0.59 | 0.51 | 0.66 | 0.41 |
| **Neral** | 1.10 | 0.80 | 0.90 | 0.76 | 0.74 | 1.22 |
| **α-Terpineol** | 0.17 | 0.63 | 0.23 | 0.16 | 0.17 | 1.34 |
| **Geranyl acetate** | 0.32 | 0.58 | 0.65 | 0.31 | 0.20 | 0.56 |
| **Neryl acetate** | 0.37 | 0.57 | 0.56 | 0.42 | 0.32 | 0.60 |
| **t-α-Bergamotene** | 0.37 | 0.50 | 0.38 | 0.34 | 0.39 | 0.24 |
| **Terpinolene** | 0.40 | 0.43 | 0.38 | 0.36 | 0.34 | 0.74 |
| **para-Cymene** | 0.06 | 0.38 | 0.03 | - | 0.20 | 0.53 |
| **α-Thujene** | 0.40 | 0.35 | 0.44 | 0.40 | 0.37 | 0.39 |
| **β-Caryophyllene** | 0.21 | 0.24 | 0.24 | 0.26 | 0.24 | 0.17 |
| **α-Terpinene** | 0.20 | 0.20 | 0.31 | 0.18 | 0.16 | 0.39 |
| **Citronellal** | 0.07 | 0.11 | 0.08 | 0.09 | 0.08 | 0.02 |
| **Nonanal** | 0.11 | 0.11 | 0.23 | 0.10 | 0.11 | ** |
| **Linalool** | 0.29 | 0.10 | 0.25 | 0.12 | 0.11 | 0.3 |
| **Nerol** | 0.03 | 0.09 | 0.03 | 0.03 | 0.04 | 0.07 |
| **Terpinen-4-ol** | 0.02 | 0.08 | 0.11 | 0.04 | 0.03 | 1.02 |
| **Geraniol** | 0.09 | 0.07 | 0.04 | 0.03 | 0.04 | ** |
| **Decanal** | 0.04 | 0.06 | 0.06 | 0.05 | 0.04 | 0.07 |
| **Camphene** | 0.10 | 0.05 | 0.07 | 0.06 | 0.05 | 0.09 |
| **Octanal** | 0.10 | 0.05 | 0.13 | 0.04 | 0.06 | 0.04 |

\* Coelution of β-pinene and sabinene.
\*\* Cannot be characterised unambiguously due to GC peak overlapping.

## Lime Oils

| | |
|---|---|
| Source: | *Citrus aurantifolia* Swingle: Key, Mexican or West Indian lime (small fruit with many seeds) |
| | *Citrus latifolia* Tanaka: Persian or Tahiti Lime (large, seedless fruits) |
| Geographical Origin: | Key lime is cultivated in Mexico, Peru and Haiti, while the Persian lime originates from Florida and Brazil. |
| Production: | |
| 1. Distilled Lime Oils: | After washing the fruits, a screw press is employed to produce a juice-water emulsion. Subsequently, this emulsion is subjected to steam distillation. More than 85% of all commercially available lime oils derive from a distillative process. Sensorially, they differ considerably from cold-pressed oils. |
| 2. Cold-pressed Lime Oils: | Two production procedures are used: |
| | A) After generating the juice-oil-emulsion (see 1), the oil is separated by centrifuging. |
| | B) After washing the fruits, the peel is slightly grated and the oil is washed out with water. The oil then is retained by centrifugation. This method is identical with the Pellatrice production method. |

While both methods are employed for Key limes, cold-pressed Persian lime oils are always produced by method B.

| | |
|---|---|
| Sensory Evaluation: | Distilled lime oils possess a terpeny-like, fresh, sharp citrus note, which characterises the cola flavour. Distilled lime oils are mainly used in soft drinks. Apart from flavouring colas, the combination lemon-lime has also found widespread application. Cold-pressed lime oil is well suited for usage in perfumery, as its fresh, heavy, sweet, earthy and peel-like smooth and balsamic citrus note possesses an exceptionally high sensorial potential and is, therefore, held in high esteem [65]. |

**Physical and Chemical Data:**

|  | Distilled lime | | | Cold-pressed lime | | |
| --- | --- | --- | --- | --- | --- | --- |
|  | Key lime | | Persian lime | Key lime | | Persian lime |
|  | Mexico | Peru | Florida | Type A | Type B |  |
| $d_{20}$ | 0.857-0.867 | 0.854-0.864 | 0.843-0.853 | 0.874-0.884 | 0.873-0.883 | 0.865-0.875 |
| $n_D^{20}$ | 1.472-1.480 | 1.471-1.479 | 1.470-1.478 | 1.482-1.490 | 1.482-1.490 | 1.477-1.485 |
| $\alpha_D^{20}$ | 30-41° | 29-39° | 49-58° | 32-42° | too dark | 44-54° |
| Solubility |  | $L_{90}^{20}$: 1:4 |  |  | $L_{96}^{20}$: 1:1 |  |
| Aldehyde (cal. as decanal): | 0.3-1% | 0.2-1% | 0.5-1.4% | 3.8-5.4% | 4.4-6% | 3.2-4.8% |

As a result of acid-catalysed reactions during steam distillation, a number of terpenes and sesquiterpenes *[66]* of the native cold-pressed oils undergo transformation *[9, 67-69]*. This applies mainly to the pinenes, sabinene, thujene and some sesquiterpenes. While cold-pressed lime oils contain up to 2.3% of the reactive germacrenes, only traces of germacrene B can be found in distilled oils *[70]*. When compared to their educts, the newly formed constituents, mainly alcohols, such as α-terpineol or fenchyl alcohol, possess completely different sensory properties. Additionally, cyclisations and hydratisations of aldehydes result in a reduced presence of these constituents in distilled products. Many compounds which characterise the flavour of distilled lime are formed during production. The extremely different composition of cold-pressed and distilled oils accounts for their completely different flavour profile and they, therefore, have to be considered as a reaction flavouring.

While steam distillates do not contain non-volatile constituents, the cold-pressed oils are rich in coumarin compounds. They contain up to 9% coumarin components, with a profile of bergamottin, 5-geranyloxy-7-methoxycoumarin, 5-geranyloxy-8-methoxypsoralen, citropten, 8-geranyloxy-coumarin, herniarin, bergapten, isopimpinellin, oxypeucedanin and a few other accompanying compounds, typical for native lime oils. As a result of its sensitivity to acid, a considerably lower amount of oxypeucedanin can be found in cold-pressed oil type A than in type B oils *[48, 71, 72]*.

*Enantiodifferentiation:* Cold-pressed lime oils contain R(+)-limonene in an enantiomeric purity of over 96%, while S(-)-limonene reaches values of 7-10% in distilled oils *[48]*.

*Table 3.19: Main components % of lime oils [73]*

|  | Distilled lime | | Cold-pressed lime | | |
|---|---|---|---|---|---|
|  | Key lime | Persian lime | Key lime Type A | Type B | Persian lime |
| **Limonene** | 48 | 58 | 49 | 48 | 52 |
| **γ-Terpinene** | 10.9 | 16.1 | 7 | 8 | 14 |
| **Terpinolene** | 7.7 | 2.7 | 0.4 | 0.4 | 0.6 |
| **α-Terpineol** | 6.3 | 2.1 | 0.3 | 0.3 | 0.2 |
| **α-Terpinene** | 2.5 | 1.1 | 0.2 | 0.2 | 0.2 |
| **β-Pinene** | 2.2 | 6.0 | 20 | 20 | 12 |
| **p-Cymene** | 1.8 | 1.5 | 0.6 | 0.3 | 0.4 |
| **β-Bisabolene** | 1.4 | 0.6 | 2.1 | 2.0 | 1.8 |
| **Myrcene** | 1.3 | 1.6 | 1.1 | 1.2 | 1.3 |
| **α-Pinene** | 1.2 | 2.2 | 2.2 | 2.2 | 2.0 |
| **1.4-Cineole** | 1.0 | 1.0 | 0.01 | 0.01 | 0.01 |
| **1.8-Cineole** | 1.0 | 1.0 | 1.0 | 0.8 | 0.9 |
| **$\Delta^3$-Carene** | 1.0 | 0.2 | 0.01 | 0.01 | 0.01 |
| **γ-Terpineol** | 0.9 | 0.2 | - | - | - |
| **Terpinen-1-ol** | 0.8 | 0.1 | - | - | - |
| **Terpinen-4-ol** | 0.8 | 0.5 | 0.20 | 0.04 | 0.04 |
| **Sabinene** | - | - | 3 | 3 | 2.1 |
| **Geranial** | 0.04 | 0.12 | 2.4 | 3 | 2.8 |
| **Neral** | 0.02 | 0.09 | 1.4 | 1.8 | 0.7 |
| **Geranyl acetate** | 0.08 | 0.09 | 0.3 | 0.3 | 0.3 |
| **Neryl acetate** | 0.08 | - | 0.5 | 0.5 | - |
| **Linalool** | 0.05 | 0.05 | 0.2 | 0.2 | 0.2 |
| **Germacrene A \*** | nf | - | 0.41 | 0.41 | 0.16 |
| **Germacrene B \*** | 0.06 | - | 0.84 | 0.84 | 0.18 |
| **Germacrene C \*** | nf | - | 0.55 | 0.55 | 0.11 |
| **Germacrene D \*** | nf | - | 0.35 | 0.35 | 0.12 |

\* Average germacrenes from [70].

**Mandarin Oil**  *Citrus reticulata, Citrus deliciosa*

**Tangerine Oil**  *Citrus reticulata* Blanco var. Dancy

Geographical Origin: Main growing areas of mandarin are Italy and China. Smaller quantities are produced in the USA, Argentina, Brazil, Spain and Greece. Tangerines are cultivated mainly in Brazil, but also in Spain, Mexico and the USA.

| | | |
|---|---|---|
| Production: | Three different qualities can be distinguished: green, yellow and red mandarin oil, where the latter is the most common. Depending on the degree of ripeness, the fruit is harvested from late autumn to spring in Italy. For yellow and green qualities the harvest is nearly entirely used for oil production, as the juice of the unripe fruits is of little commercial value. These oil specialities are, therefore, of correspondingly high price. The green and yellow oils are produced mainly by the pellatrice technique, while the red mandarin oil is derived from the entire fruit via centrifuging the peel-juice-oil emulsion. | |
| Sensory Evaluation: | Red mandarin is characterised by its amine-like top note. The sweet, heavy flavour with its floral base does, in contrast to all other citrus oils, not possess a refreshing, but a more fatty character. | |
| | On the other hand, both, the green and yellow mandarin as well as tangerine feature a refreshing, fresh-fruity character. | |

Physical and Chemical Data:

| | Italian Mandarin | Tangerine |
|---|---|---|
| **Density $d_{20}$:** | 0.846-0.856 | 0.841 - 0.851 |
| **Refractive Index $n_D^{20}$:** | 1.472-1.478 | 1.471-1.479 |
| **Optical Rotation $\alpha_D^{20}$:** | 65-75° | 85-95° |
| **Solubility $L_{96}^{20}$:** | 1:1 | 1:1 |
| **Aldehydes (cal. as decanal):** | 0.5-1.2% | 0.3-1.1% |

The large number of mandarin varieties [74] results in mandarin and tangerine oils with largely varying composition, also reflected in the oils' sensorial impact (see Tables 3.20 and 3.21) [75].

Italian mandarin oils are characterised by the terpenes limonene (app. 70%) and γ-terpinene (app. 20%). Apart from α-sinensal and long-chain saturated aldehydes there is a number of sensorially important unsaturated aldehydes with citrus-like, aldehydic, fatty and waxy flavour. Also the potent C-11-hydrocarbons (1,3E,5Z)-undecatriene, (1,3E,5E,8Z)-, (1,3E,5Z,8Z)-undecatetraene and methyl N-methylanthranilate characterise the typical taste and odour of red mandarin [76]. The latter compound, which also causes the fluorescent character of mandarin oil, is a main constituent of mandarin leaf oils (see Petitgrain oils).

When compared to Italian mandarin oils, other mandarin [77] or tangerine oils such as Dancy [76] or Murcott tangerine [78] are even more dominated by limonene (app. 95%). Moreover, the composition of the flavour compounds and, therefore, the entire flavour profile, is considerably less complex [76]. This is also evident in the rather different content of methyl N-methylanthranilate and α-sinensal. Apart from the monoterpenes myrcene and α-pinene, octanal, [E,E]-deca-2,4-dienal and wine lactone *[79]*, aroma analysis [see 6.2.4, Grosch] has confirmed linalool as the most important odour constituent of clementine oil.

The non-volatile range of mandarin and tangerine oil is characterised mainly by the methoxyflavones tangeretin, heptamethoxyflavone and nobiletin; this compound class comprises up to 3500 ppm in Italian mandarin oils *[38, 80]*.

| | |
|---|---|
| *Enantiodifferentiation:* | Native qualities of commercial Italian mandarin oils contain R(+)-limonene in an enantiomeric purity of about 98% *[81]*. Mandarin and tangerine varieties, such as Italian clementine *[82]*, nova and satsuma mandarin *[77]*, Murcott tangerine and Chinese tangerine *[78]* even contain >99% R(+)-limonene. |

## Mandarin Essence Oil

For botanical name and origin see Mandarin oil.

| | |
|---|---|
| Production: | During the production of mandarin juice concentrate (see 3.2.2.1). |
| Sensory Evaluation: | Depending on the production conditions, oils of different quality are available. As the mandarin aroma is thermally very sensitive, only those oils that are produced by gentle processes are of sensorial value. Oils of high qualitative value possess a very sweet, fruity, juicy, flowery and aldehydic note. |
| Physical Data: | $d_{20}$: 0.841–0.850 $\quad\quad\quad \alpha_D^{20}$: 65–75° $n_D^{20}$: 1.471–1.479 $\quad\quad L_{96}^{20}$: 1:1 |
| Chemical Data: | Aldehydes (cal. as decanal): 0.4–1.2% |
| *Enantiodifferentiation:* | see mandarin oil |

*Table 3.20:* Main components (%) of mandarin oils

| | Mandarin | | Tangerine | | Mandarin Essence Oil |
|---|---|---|---|---|---|
| | Italy *[83]* | Spain *[84]* | Florida *[85]* | Brazil *[85]* | Italy *[48]* |
| **Limonene** | 70 | 77 | 91 | 94 | 67 |
| **γ-Terpinene** | 19.3 | 13.7 | 3.09 | 0.48 | 20.9 |
| **α-Pinene** | 2.37 | 1.75 | 1.00 | 0.70 | 1.97 |
| **Myrcene** | 1.76 | 1.86 | 2.03 | 1.95 | 1.60 |
| **β-Pinene** | 1.68 | 1.15 | 0.44 | 1.07 | 1.60 |
| **Terpinolene** | 0.88 | 0.62 | 0.13 | 0.04 | 0.98 |
| **Methyl N-methyl-anthranilate** | 0.35 | 0.28 | - | 0.01 | 0.80 |

|  | Mandarin | | Tangerine | | Mandarin Essence Oil |
|---|---|---|---|---|---|
|  | Italy [83] | Spain [84] | Florida [85] | Brazil [85] | Italy [48] |
| α-Terpinene | 0.30 | 0.20 | 0.05 | 0.02 | 0.54 |
| Sinensal | 0.25 | 0.19 | 0.07 | 0.08 | 0.11 |
| Sabinene | 0.25 | 0.21 |  | 0.19 |  |
| Octanal | 0.14 | 0.22 | 0.20 | 0.10 | 0.10 |
| a-Terpineol | 0.15 | 0.20 | 0.02 | 0.02 | 0.44 |
| Linalool | 0.11 | 0.16 | 0.62 | 0.49 | 0.37 |
| Decanal | 0.09 | 0.06 | 0.10 | 0.04 | 0.10 |
| p-Cymene | 0.20 |  | 0.18 | 0.04 | 0.60 |
| Geranial | 0.04 | 0.04 | 0.01 | 0.01 | 0.03 |
| α-Phellandrene | 0.07 | 0.05 | 0.03 | 0.04 | 0.19 |
| Thymol | 0.05 | 0.11 | 0.03 | 0.04 | 0.19 |
| 1.8-Cineole | - | - | 0.63 | 0.69 | 0.02 |
| Geranyl acetate | 0.01 | 0.01 | 0.05 | 0.03 | 0.02 |

*Table 3.21:* Main components (%) of less common mandarin oils

|  | Satsuma Mandarin Uruguay [77] | Nova Mandarin Uruguay [77] | Clementine Italy [86] | Murcott Tangerine Brazil [78] |
|---|---|---|---|---|
| α-Thujene | 0.12 | tr | tr | tr |
| α-Pinene | 0.79 | 0.48 | 0.37 | 0.53 |
| Sabinene | 0.17 | 0.34 | 0.18 | 0.29 |
| β-Pinene | 0.25 | 0.14 | tr | 0.03 |
| Octanal | 0.19 | 0.17 | - | 0.36 |
| Myrcene | 2.01 | 1.93 | 1.83 | 1.85 |
| α-Phellandrene | 0.04 | 0.03 | 0.02 | 0.03 |
| $\Delta^3$-Carene | - | 0.17 | tr | 0.02 |
| Limonene + β-Phellandrene | 91.0 | 93.1 | 94.7 | 94.9 |
| trans-β-Ocimene | 0.03 | 0.44 | 0.02 | 0.04 |
| γ-Terpinene | 3.34 | 0.01 | 0.01 | - |
| Terpinolene | 0.15 | 0.03 | 0.02 | 0.01 |
| Nonanal | 0.05 | 0.05 | - | 0.10 |

# Flavouring Preparations and Some Source Materials

|  | **Satsuma mandarin** Uruguay [77] | **Nova Mandarin** Uruguay [77] | **Clementine** Italy [86] | **Murcott Tangerine** Brazil [78] |
|---|---|---|---|---|
| **Linalool** | 0.20 | 0.88 | 1.15 | 0.37 |
| **Citronellal** | 0.04 | 0.01 | - | 0.12 |
| **Terpinen-4-ol** | 0.02 | 0.01 | 0.02 | tr |
| **α-Terpineol** | 0.10 | 0.08 | 0.09 | 0.04 |
| **Decanal** | 0.13 | 0.32 | 0.27 | 0.36 |
| **Neral** | 0.01 | 0.03 | - | 0.08 |
| **Geranial** | 0.02 | 0.06 | 0.07 | 0.03 |
| **Perillaldehyd** | 0.05 | 0.04 | - | - |
| **Methyl N-methyl-anthranilate** | - | - | 0.02 | - |
| **α-Copaene** | 0.04 | 0.03 | 0.03 | 0.03 |
| **Dodecanal** | 0.02 | 0.07 | 0.06 | 0.07 |
| **Germacrene D** | 0.17 | 0.11 | - | 0.02 |
| **β-Sinensal** | tr | 0.02 | 0.01 | tr |
| **α-Sinensal** | tr | 0.04 | 0.21 | tr |

**Orange Oil**     *Citrus sinensis* (L.) Osbeck

Geographical Origin: Brazil and the USA are main producers of orange oil. As a by-product of orange juice production, orange oil is also available from Israel, Italy, Spain, Argentina, Africa and Australia.

Sensory Evaluation: The sweetish, fresh top note is typical for orange peel oil. Sensorially, this citrus note can be correlated with the attributes fruity, peel-like and waxy, aldehydic, estery, terpene-like.

Orange oil constitutes a highly popular flavouring basis for both fragrances and flavourings. The abundance of the material as well as its moderate price render it attractive for all applications ranging from fine perfumery to household cleaning products.

Production: From the peel of the orange fruit.

Physical Data:    $d_{20}$: 0.840-0.850      $\alpha_D^{20}$: 93-100°

                $n_D^{20}$: 1.469-1.477      $L_{20}^{96}$: 1:1

| | |
|---|---|
| Chemical Data: | Aldehydes (cal. as decanal): 0.8 - 2.0% |

Of all citrus oils, orange oil possesses the highest number of terpene hydrocarbons. It constitutes the best natural source of R(+)-limonene, a highly valued natural solvent. The typical orange flavour is characterised by the presence of aldehydes C8-C12, linalool, α- and β-sinsenal.

A characteristic pattern of various flavones can be detected in the non-volatile part of orange oil: heptamethoxyflavone, nobiletin, tetra-O-methylscutellarein, tangeretin, hexamethoxyflavone and sinensitin.

*Enantiodifferentiation:* Limonene is present in a purity exceeding 99% as the R(+)-enantiomer [45, 58]. Linalool is present in an enantiomeric excess of over 93% S(+)-linalool.

| | |
|---|---|
| **Blood Orange Oil** | *Citrus sinensis* (L.) Osbeck var. Moro<br>*Citrus sinensis* (L.) Osbeck var. Tarocco<br>*Citrus sinensis* (L.) Osbeck var. Sanguinello<br>*Citrus sinensis* (L.) Osbeck var. Sanguigno |
| Geographical Origin: | Sicily |
| Production: | From the fruit peels during juice extraction (FMC). |

Pure qualities of blood orange oil are produced in Italy, since more than 30% of the oranges cultivated in Sicily are blood orange varieties. A blend with Valencia oranges is mostly avoided, as blood oranges are processed during mid-season, from February to April [87, 88]. The commercially available blood orange oils of high quality usually constitute a blend of the varieties listed above: Moro, Sanguinello, Tarocco and Sanguigno.

| | |
|---|---|
| Sensory Evaluation: | When compared to Valencia varieties, blood orange oil is characterised by its intensely fruity, estery, weakly peely, aldehyde note. |
| Physical Data: | $d_{20}$: 0.850-0.860 $\quad$ $n_D^{20}$: 1.471-1.479<br>$\alpha_D^{20}$: 85-95° $\quad\quad\quad$ $L_{96}^{20}$: 1:1 |
| Chemical Data: | Aldehydes (cal. as decanal): 1.0-1.7% |

Both, the composition of the non-volatile sector as well as the enantiomeric ratio of limonene are analogous to that of sweet orange oil.

| | |
|---|---|
| **Bitter Orange Oil** | *Citrus aurantium* L., ssp. amara Engl. |
| Geographical Origin: | The main growing areas of bitter orange are situated in Italy (Sicily), Spain and Brazil. |

| Production: | By Pelatrice-extraction from the fruit peels. |
|---|---|
| | As a result of their sour and bitter taste, the fruits are rarely consumed. Mostly, they are cultivated for the production of peel oil. In Spain the fruit is additionally also used for producing bitter orange marmalade. Just like green mandarin oil and bergamot oil, this oil, therefore, constitutes a rather expensive citrus product, as it is not possible to distribute costs between oil and juice. |
| Sensory Evaluation: | Bitter orange oil possesses an extremely strong fresh note. Its character is less sweet than that of orange oil; its flavour character is floral, aldehydic with a bitter, dry, earthy expression. The oil displays a fresh, aqueous and green aroma note, which allows the association with bergamot. Its fixative effect constitutes an additional positive feature of this pleasing oil. |
| Physical Data: | $d_{20}$: 0.843-0.853     $\alpha_D^{20}$: 86-96° |
| | $n_D^{20}$: 1.471-1.479     $L_{96}^{20}$: 1:1 |
| Chemical Data: | Aldehydes (cal. as decanal): 0.3–1% |
| | While methoxyflavones constitute the non-volatile part of Citrus sinensis oils, bitter orange oil possesses a completely different pattern of coumarins and flavones. Characteristic for this range of bitter orange oil are meranzin, isomeranzin, bergapten, epoxybergamottin, tangeretin, osthol, nobiletin and heptamethoxyflavone [38, 89]. |
| *Enantiodifferentiation:* | While limonene, analogous to Citrus sinensis oils, is present in an enantiomeric purity exceeding 99%, the ratio of the linalool antipodes is inverted. Of its enantiomers, R(-)-linalool is present in values exceeding 70% [48, 80, 90]. |

*Table 3.22: Main components (%) of orange oils*

| | Sweet Orange Oil | | | | | Blood Orange Oil | Bitter Orange Oil | | |
|---|---|---|---|---|---|---|---|---|---|
| | Brazil [48] | USA [48] | Spain [48] | Italy [87] | Israel [91] | Italy [92] | Spain [93] | Brazil [48] | Italy [94] |
| **Limonene** | 94.6 | 93.9 | 93.3 | 94.8 | 94 | 95.2 | 92.3 | 93.2 | 93.5 |
| **Myrcene** | 1.92 | 1.93 | 1.88 | 1.86 | 1.89 | 1.88 | 1.85 | 1.83 | 1.79 |
| **Linalool and Nonanal** | 0.54 | 0.40 | 0.66 | 0.52 | 0.45 | 0.47 | 0.29 | 0.20 | 0.30 |
| **α-Pinene** | 0.50 | 0.53 | 0.52 | 0.51 | 0.52 | 0.50 | 0.48 | 0.50 | 0.56 |
| **Decanal** | 0.30 | 0.37 | 0.34 | 0.23 | 0.32 | 0.17 | 0.14 | 0.14 | 0.12 |
| **Octanal** | 0.26 | 0.30 | 0.37 | 0.27 | 0.28 | 0.15 | 0.02 | 0.07 | 0.14 |

|  | Sweet Orange Oil | | | | | Blood Orange Oil | Bitter Orange Oil | | |
|---|---|---|---|---|---|---|---|---|---|
|  | Brazil [48] | USA [48] | Spain [48] | Italy [87] | Israel [91] | Italy [92] | Spain [93] | Brazil [48] | Italy [94] |
| Sabinene | 0.24 | 0.44 | 0.69 | 0.63 | 0.55 | 0.44 | 0.15 | 0.26 | 0.32 |
| Geranial | 0.09 | 0.08 | 0.09 | 0.10 | 0.09 | 0.07 | 0.02 | 0.07 | 0.07 |
| Dodecanal | 0.09 | 0.10 | 0.09 | 0.04 | 0.06 | 0.03 | 0.03 | 0.05 | 0.02 |
| Neral | 0.07 | 0.08 | 0.07 | 0.06 | 0.05 | 0.05 | 0.01 | 0.05 | 0.05 |
| Citronellal | 0.06 | 0.06 | 0.05 | 0.04 | 0.06 | 0.04 |  |  | 0.01 |
| α-Terpineol | 0.06 | 0.19 | 0.07 | 0.05 | 0.04 | 0.04 |  | 0.10 | 0.04 |
| Octanol | 0.06 | 0.14 | 0.10 | 0.02 |  | 0.01 |  |  | 0.01 |
| Valencene | 0.04 | 0.11 | 0.25 | 0.08 | 0.05 | 0.18 |  | 0.01 |  |
| β-Sinensal | 0.04 | 0.05 | 0.03 | 0.03 | 0.05 | 0.02 |  |  |  |
| Nerol | 0.02 | 0.03 | 0.02 | 0.02 | 0.06 | 0.01 | 0.01 | 0.01 |  |
| β-Pinene | 0.02 | 0.03 | 0.18 | 0.05 | 0.06 | 0.03 | 0.28 | 0.88 | 0.94 |
| β-Caryophyllene | 0.02 | 0.02 | 0.04 | 0.02 | 0.05 | 0.02 | 0.11 | 0.09 | 0.05 |
| α-Sinensal | 0.02 | 0.03 | 0.02 | 0.02 | 0.03 | 0.02 |  |  |  |
| Nootkatone | 0.01 | 0.004 | 0.003 | 0.01 | 0.03 | 0.02 |  |  |  |
| Linalyl acetate |  |  |  |  |  |  | 0.81 | 0.79 | 0.89 |

**Orange Essence Oil**

For botanical name and origin see orange oil.

| | |
|---|---|
| Production: | During the concentration of orange juice (see 3.2.2.1). |
| Sensory Evaluation: | This citrus oil can be categorised as juicy, sweet fruity and weakly peely, green and fatty. |
| Physical Data: | $d_{20}$: 0.840-0.850    $\alpha_D^{20}$: 90-99° |
|  | $n_D^{20}$: 1.468-1.476    $L_{96}^{20}$: 1:1 |
| Chemical Data: | Aldehydes (cal. as decanal): 1.2-1.9% |
| *Enantiodifferentiation:* | Analogous to the ratios of orange oil. |

In recent years, the juice oils of orange have encountered increasing interest. Their sensorial potential, which reflects the sweet and juicy type of orange juice, is highly valued today. In aroma analysis, the highest flavour impact was found for linalool, decanal, octanal, ethyl butyrate, α-pinene, limonene and 6-methyl-octanal [95, 96]. A large number of volatile trace constituents distinguish the juice oil from peel oil. Growing area and production method have considerable impact on oil quality. Brown extractors, which are used in the majority in the USA, influence the juice with peel oil only on a small scale. Therefore, juice oils produced with this method contain components which are typical for peel oils in a lower ratio than oils produced by FMC extraction. The content of ethyl butyrate and valencene is, according to experience, lower in South American oils than in qualities from the USA. Oils from various growing areas differ only quantitatively. Qualitatively all orange essence oils exhibit analogous component patterns.

**Blood Orange Essence Oil**

For botanical name and origin see above.

| | |
|---|---|
| Production: | During the concentration of blood orange juice. |
| Sensory Evaluation: | The character of this oil is juicy, aqueous, estery, fruity, balsamic, honey, only slightly aldehydic and sulphurous with an acid note. The sensorial impact allows the association with the citrus notes of mandarin. The sensory differences between blood orange and orange essence oil reflect the differences of the juices with a number of specific esters, such as ethyl-3-hydroxybutyrate and butyl-2-butenoate, acids such as 3-phenyl-popionic acid and alcohols, e.g. 4-phenyl-2-butanol which are typical for blood oranges. On the other hand, blond orange juices are more dominated by green, fatty, aldehydic constituents [97]. |
| Physical Data: | $d_{20}$: 0.838-0.848    $\alpha_D^{20}$: 92-104° |
| | $n_D^{20}$: 1.469-1.477    $L_{96}^{20}$: 1:1 |
| Chemical Data: | Aldehydes (cal. as decanal): 0.6-1.2% |
| *Enantiodifferentiation:* | As in all other orange oils, R(+)-limonene is present in a purity exceeding 99%. |
| | In contrast to Valencia essence oils, the reduced presence of volatile flavour constituents is significant. |

*Table 3.23:* Main components (%) of orange essence oils

|  | Orange Essence Oils | | Blood Orange Essence Oil |
|---|---|---|---|
|  | USA *[98]* | Brazil *[48]* | Italy *[48]* |
| **Ethanol** | 0.31 | 0.09 | 0.02 |
| **Ethyl butyrate** | 0.13 | 0.05 | 0.02 |
| **Hexanal** | 0.01 | 0.02 | 0.01 |
| **α-Pinene** | 0.46 | 0.52 | 0.45 |
| **Sabinene** | 0.26 | 0.26 | 0.57 |
| **Myrcene** | 1.78 | 1.84 | 1.66 |
| **Octanal** | 0.30 | 0.32 | 0.5 |
| **Limonene** | 92.5 | 94 | 93 |
| **Octanol** | 0.07 | 0.05 | 0.04 |
| **Linalool** | 0.48 | 0.52 | 0.39 |
| **Citronellal** | 0.06 | 0.04 |  |
| **Decanal** | 0.27 | 0.25 | 0.20 |
| **α-Terpineol** | 0.05 | 0.04 | 0.09 |
| **Carvone** | 0.11 | 0.07 | 0.02 |
| **Geranial** | 0.10 | 0.08 | 0.07 |
| **Valencene** | 1.63 | 0.33 | 0.52 |

## Orange Water Phase

For botanical name and origin see orange oil.

Production: During the concentration of orange juice (see 3.2.2.1).

Sensory Evaluation: This matrix features a typical orange pulp note and can be described with the attributes juicy, fruity, estery, green, floral, weakly sulphurous.

Physical Data: $d_{20}$: 0.975 - 0.985    pH: 4.2 - 6.8

$n_D^{20}$: 1.336 - 1.344    Vol%: 10 - 20

Chemical Data: The total aroma content (excluding the components previously listed) usually corresponds to values in the range of 300-500 ppm.

The following compounds make up the aroma profile (the values quoted in parenthesis correspond to ppm values): linalool (45-80); α-terpineol (15-30); (Z)-3-hexenal (10-50); ethyl butyrate (10-30); acetaldehyd diethylacetal (10-

50); ethyl 3-hydroxyhexanoate (10-20); hexanal (8-20); (E)-2-hexenal (5-20); octanol (5-20); octanal (5-10).

Quantitative differences result from different production technologies. Origin from different growing areas is not known to affect the composition of this matrix [99].

The flavouring of reconstituted orange juices is the most important field of application for these aqueous orange essences.

Continuous storage at freezing temperatures is necessary to prevent bacterial destruction and off-flavour formation, processes which easily develop in this highly diluted and acidic medium.

**Petitgrain Oils** [100, 101]

Petitgrain oils are obtained by steam distillation from the leaves and small twigs of citrus plants. As a result of their sweet, flowery and refreshing character, mainly the Petitgrain oils of bitter orange (*Citrus aurantium* ssp. *amara* L.) [100, 102-105] are of commercial interest. The main growing areas are Paraguay, Spain and Italy. The main constituents of bitter orange Petitgrain oil are linalyl acetate (up to 60%) and linalool (17-30%). Terpenes, sesquiterpenes and aldehydes play a minor role.

The Petitgrain oils of the citrus varieties mandarin (*Citrus reticulata* Blanco) and lemon (*Citrus limon* (L.) Burm.) are also commercially available, but to a far lesser extent than for bitter orange. Mandarin Petitgrain oil is characterised by its unique sensory impact, which apart from γ-terpinene (20-25%) results from its high content of methyl N-methylanthranilate (40-50%) [106-108].

On a commercial basis, lemon Petitgrain oil is produced from citrus leaves in Italy. In contrast to other Petitgrain oils, its composition resembles more to the peel oil of the corresponding fruits. The main constituents are limonene (app. 30%), β-pinene (10-20%) and citral (15-28%) [104, 109-114].

The Petitgrain oils of the citrus varieties grapefruit [115], tangerine [116], lime [111] and sweet orange [117, 118] are of minor sensory interest and, therefore, only produced on a limited commercial scale.

*Table 3.24: Main constituents of Petitgrain oils*

|  | **Bitter orange** Spain [102] | **Mandarin** Italy [106] | **Lemon** Italy [109] |
|---|---|---|---|
| **β-Pinene** | 2.5 | 1.9 - 2.5 | 13.6 |
| **Sabinene** | 0.4 | 0.2 - 0.9 | 3.6 |
| **Myrcene** | 2.6 | 0.6 - 0.8 | 1.5 |
| **α-Phellandrene** | tr | tr - 0.1 | 1.2 |
| **p-Cymene** | 0.1 | 3.0 - 5.2 | 0.9 |

|  | **Bitter orange** Spain [102] | **Mandarin** Italy [106] | **Lemon** Italy [109] |
|---|---|---|---|
| **Limonene** | 5.4 | 7.2 - 12.6 | 30.7 |
| **(E)-β-Ocimene** | 2.4 | 0.4 - 0.9 | - |
| **γ-Terpinene** | 0.1 | 23.9 - 28.5 | 2.3 |
| **β-Caryophyllene** | 1.8 | 0.9 - 1.4 | 1.0 |
| **α-Terpineol** | 4.0 | 0.2 - 0.3 | 0.4 |
| **Linalool** | 20.2 | 0.3 -1.1 | 1.8 |
| **Geraniol** | 3.0 | tr | 0.5 |
| **Citronellal** | tr | tr - 0.1 | 1.5 |
| **Geranial** | 0.1 | tr - 0.1 | 10.9 |
| **Neral** | tr | tr - 0.1 | 6.5 |
| **Geranyl acetate** | 3.9 | tr | 2.9 |
| **Neryl acetate** | 2.2 | tr | 7.4 |
| **Linalyl acetate** | 48.9 | tr - 0.1 | 6.5 |
| **Methyl N-methylanthranilate** | tr | 41.6 - 51.9 | 0.8 |

## REFERENCES

[1] Swaine R. L., Perf. & Flav., 13 (6), 1-20 (1988)
[2] Priest N. O., Perf. & Flav., 6 (2), 33-52 (1981)
[3] Mariani M., Imes C. G., Indian Perfumer, 31 (4), 383-398 (1987)
[4] Clark B. C. Jr., Jones B. B., Jacubucci G. A., Tetrahedron Supp., 37(1), 405-409 (1981)
[5] Grosch W., Schieberle P., J. Agric. Food Chem., 36, 797-800 (1988)
[6] Ziegler M., H. Brandauer, E. Ziegler, G. Ziegler, J. Ess. Oil Res., 3, 209-220 (1991)
[7] Balton T. A., Reineccius G. A., Perf. & Flav., 17(2), 1-20 (1992)
[8] von Campe G., Dragoco Bericht, 1, 3-17 (1990)
[9] Clark et al., in Off-Flavours in Foods and Beverages, Charalambous G. (ed.), Elsevier, Amsterdam (1992), pp. 229-285
[10] Moyler D. A., M. A. Stephens, Perf. & Flav., 17 (2), 37-38 (1992)
[11] Tateo F., J. Ess. Oil Res., 2(1), 7-13 (1990)
[12] W. J. D. van Dijck, A. H. Ruys, Perf. Essent. Oil Record, 28, 91-94 (1937)
[13] Ruys A. H., Perf. Essent. Oil Record, 48, 584-588 (1957)
[14] Fleisher A., G. Biza, N. Secord, J. Dono, Perf. & Flav., 12 (2), 57-61 (1987)
[15] Fleisher A., Perf. & Flav., 15 (5), 27-36 (1990)
[16] Fleisher A., Perf. & Flav., 19 (1), 11-15 (1994)
[17] Stahl E., Verdichtete Gase zur Extraktion u. Raffination, Springer Verlag, Berlin 1987
[18] Owusu-Yaw J., Matthews R.F. and West P.F.; J. Food. Sci., 51, 1180-1182 (1986)
[19] Moyler D. in Citrus. Ed. Dugo G. and Di Giacomo A., Francis Taylor, London and New York 2002, pp.391-401
[20] Kirchner J. G., Miller J. M., Ind. Eng. Chem., 44 (2), 318-321 (1952)
[21] Ziegler E., Flavour Ind., 1, 647-653 (1971)
[22] Braverman J. B. S., Solomiansky L., Perfumery Essent. Oil Record, 6, 284 (1957)
[23] Tzamtzis N. E., Liodakis S. E., Parissakis G. K., Flav. & Fragr. J., 5, 57-67 (1990)

| | |
|---|---|
| [24] | Average from: Schenk H.P. et al., Seifen-Öle-Fette-Wachse, 107, 363-370 (1981) and Huet R., Rivista Ital., 63, 310-313 (1981) and: Mazza G., J. Chromatogr., 362, 87-99 (1986) |
| [25] | Average from: Calvarano G. et al., Essenze Deriv. Agrum., 54, 220-227 (1984) and: Dugo G. et al., Flav. Fragr. J., 6, 39-56 (1991) |
| [26] | Average from: Drescher R. W. et al., Essenze Deriv. Agrum., 54, 192-199 (1984) |
| [27] | Average from: Huang Y. Z. et al., Acta Bot. Sinica, 29, 77-83 (1987) |
| [28] | Kirbaslar F.G., Kirbaslar S.I. and Dramur U.; J. Essent. Oil Res., 13, 411-415 (2001) |
| [29] | Dellacassa E., Lorenzo D., Moyna P., Verzera A. and Cavazza A.; J. Essent. Oil Res., 9, 419-426 (1997) |
| [30] | Sawamura M., Poina M., Kawamura A., Itoh T., Song H.S., Ukeda H. and Mincione B., Ital. J. Food Sci., 11, 121-130 (1999) |
| [31] | Raymo V., Gatto S., personal communication |
| [32] | Mondello L., d'Alcontres I.S., Del Duce R., Crispo F., Flav. & Fragr. J., 8, 17-24 (1993) |
| [33] | Gionfriddo F., Postorino E., Bovalo F.; Essenze Deriv. agrum., 67, 342-352 (1997) |
| [34] | Forlot P.P.D., Essenze Deriv. agrum., 70, 147-153 (2000) |
| [35] | Cotroneo A. et al., Flav. Frag. J., 7, 15-17 (1992) |
| [36] | Casabianca H., Graff J.-B., J. High Resol. Chromatogr., 17 (3), 184-186 (1994) |
| [37] | Baser K.H.C., Özek T. and Tutas M.; J. Essent. Oil Res., 7, 341-342 (1995) |
| [38] | McHale D., Sheridan J. B., J. Ess. Oil Res. 1 (4), 139-149 (1989) |
| [39] | Feger W., Brandauer H, Gabris P. and Ziegler H.; J .Agric. Food Chem., 54, 2242-2252 (2006) |
| [40] | Gross J. in Pigments in Fruits; Academic Press, Orlando, Florida 1987, pp 87-186 |
| [41] | Dugo G., Cotroneo A., Verzera A., Dugo G. and Licandro G.; Flav. Fragr., 5, 205-210 (1990) |
| [42] | Ruberto G., Biondi D., Rapisarda P., Renda A. and Starrantino A., J. Agric. Food Chem., 45, 3206-3210 (1997) |
| [43] | Feger W., Brandauer H., Ziegler H., J. Essent. Oil Res., 13, 309-313 (2001) |
| [44] | Mosandl A. et al., Flav. Frag. J., 5, 193-199 (1990) |
| [45] | Casabianca H., Graff J.-B., Jame P., Perrucchietti C., J. High Resol. Chromatogr., 18 (5), 279-285 (1995) |
| [46] | Demole E., Enggist P., Ohloff G., Helv. Chim. Acta, 65 (6), 1785-1794 (1982) |
| [47] | Buettner A., Schieberle P., J. Agric. Food Chem., 47, 5189-5193 (1999) |
| [48] | Own measurements, unpublished |
| [49] | Average from: Boelens, M. H., Perf. & Flav., 16(2), 17-34 (1991) |
| [50] | Vekiari S.A., Protopapadakis E.E., Papadopoulou P., Papanicolaou D., Panou C., Vamvakias M., J. Agric. Food Chem., 50, 147-153 (2002) |
| [51] | Corleone V., Corleone P., La Scala G., Essenze Derivati agrumari, 70, 67-79 (2000) |
| [52] | Verzera A., Dugo P., Mondello L., Trozzi A., Cotroneo A., Ital. J. Food Sci., 11, 361-370 (1999) |
| [53] | Verzera A., Trozzi A., Dugo G., Di Bella G., Cotroneo A., Flav. Fragr. J., 19, 544-548 (2004) |
| [54] | McHale D., Sheridan J. B., Flav. Frag. J., 3, 127-133 (1988) |
| [55] | Ziegler H., Spiteller G., Flav. Frag. J., 7, 129-139 (1992) |
| [56] | Ziegler H., Analytik der Cumarine des Zitronenöls, Doctoral Thesis, University of Bayreuth, Germany, 1992 |
| [57] | Dugo P., Mondello L., Cogliandro E., Cavazza A., Dugo G., Flav. Fragr. J., 13, 329-334 (1998) |
| [58] | Evers P., Krüger A., König W. A., 24th Intern. Symp. on Ess. Oils, Berlin, July 21-24, 1993, p 27 |
| [59] | Mondello L., Catalfamo M., Cotroneo A., Dugo G., Dugo G., McNair H., J. High Resol. Chromatogr., 22, 350-356 (1999) |
| [60] | Average from: Boelens M. H., Jimenez R., J. Ess. Oil Res., 1, 151-159 (1989) and Cotroneo A. et al., Flav. & Frag. J., 1 (2), 69-86 (1986) |
| [61] | Average from: Boelens M. H., Perf. & Flav., 16, (2), 17-34 (1991) |
| [62] | Average from: Staroscik J. A., Wilson A. A., J. Agric. Food Chem., 30, 835-837 (1982) and Cotroneo A. et al., Flav. & Fragr. J., 1 (3), 125-134 (1986) |
| [63] | Dellacassa E., Lorenzo D., Moyna P., Verzera A., Mondello L., Dugo P.; Flav. Fragr. J., 12, 247-255 (1997) |
| [64] | Average from: Boelens M. H., Perf. & Flav., 16, (2), 17-34 (1991) |
| [65] | Chisholm M.G., Wilson M.A., Gaskey G.M., Flav. Fragr. J., 18, 106-115 (2003) |
| [66] | Feger W., Brandauer H., Ziegler H., Flav. Fragr. J., 15, 281-284 (2000) |
| [67] | Chamblee T. S., Clark B. C., Jr., J. Essent. Oil Res., 9, 267-274 (1997) |

[68] Feger W., Brandauer H., Ziegler M., J. Essent. Oil Res., 11, 556-562 (1999)
[69] Dugo P., Cotroneo A., Bonaccorsi I., Mondello L., Flav. Fragr. J., 13, 93-97 (1998)
[70] Feger W., Brandauer H., and Ziegler H., J. Essent. Oil Res., 13, 274-277 (2001)
[71] Dugo P., Mondello L., Lamonica G., Dugo G., J. Agric. Food Chem., 45, 3608-3616 (1997)
[72] Feger W., Brandauer H., Gabris P., Ziegler H., J. Agic. Food Chem., 54 (2006)
[73] Haro L., Faas W. E., Perf & Flav., 10 (5), 67-72 (1985)
[74] Saunt J., Citrus Varieties of the World, Sinclair International, Norwich, UK 2000
[75] Lawrence B., Perf. & Flav., 26 (1), 36-44 (2001)
[76] Näf R., Velluz A., J. Essent. Oil Res., 13, 154-157 (2001)
[77] Verzera A., Trozzi A., Cotroneo A., Lorenzo D., Dellacassa E., J. Agric. Food Chem., 48, 2903-2909 (2000)
[78] Feger W., Brandauer H., Ziegler H., J. Essent. Oil Res., 15, 143-147 (2003)
[79] Buettner A., Mestres M., Fischer A., Guasch J., Schieberle P., Eur. Food Res. Technol., 216, 11-14 (2003)
[80] Dugo G. et al., Flav. & Frag. J., 9 (3), 99-104 (1994)
[81] Dugo G., d'Alcontres I. S., Cotroneo A., Dugo P., J. Ess. Oil Res., 4, 589-594 (1992)
[82] Mondello L., Catalfamo M., Proteggente A.R., Bonaccorsi I., Dugo G., J. Agric. Food Chem., 46, 54-61 (1998)
[83] Average from: Dugo G. et al., Flav. & Frag. J., 3 (4), 161-166 (1988) and: Boelens M. H., Jimenez R., J. Ess. Oil Res., 1 (4), 151-159 (1989) and Dugo G. et al., Flav. & Frag. J., 5, 205-210 (1990)
[84] Boelens M. H., Jimenez R., J. Ess. Oil Res., 1(4), 151-159 (1989)
[85] Hussein M. M., Pidel A. R., Paper No. 261, 36th JFT Meeting, June, Anaheim, California, 1976
[86] Ruberto G., Biondi D., Piatelli M., Rapisarda P., Starrantino A., Flav. Fragr. J., 8, 179-184 (1993)
[87] Dugo G. et al., J. Ess. Oil Res., 6 (2), 101-137 (1994)
[88] Dugo G., Perf & Flav., 19 (6), 29-51 (1994)
[89] Dugo P., Mondello L., Cogliandro E., Verzera A., Dugo G., J. Agric. Food Chem., 44, 544-549 (1996)
[90] Dugo G., Mondello L., Cotroneo A., Bonaccorsi I., Lamonica G., Perf. & Flav., 26 (1), 20-35 (2001)
[91] Average from: Boelens M. H., Perf. & Flav., 16 (2), 17-34 (1991)
[92] Average from: Verzera A., Trozzi A. d'Alcontres I.S., Cotroneo A., J. Ess. Oil Res., 8 (2), 159-170 (1996) and Dugo G. et al., J. Ess. Oil Res., 6 (2), 101-137 (1994)
[93] Average from: Boelens M. H., Flav. & Frag. J., 4, 139-142 (1989) and Boelens M. H., Jimenez R., Int. Congress of Ess. Oils, Elsevier 1988, pp 551-565
[94] Dugo G., Verzera A., d'Alcontres I.S., Cotroneo A., Ficarra R., Flav. & Frag. J., 8, 25-33 (1993)
[95] Widder S., Eggers M., Looft J., Vössing T., Pickenhagen W., in Handbook of Flavor Characterization. Deibler K. D., Delwiche J. (Eds.), Marcel Dekker, New York, Basel 2004, pp 207-216
[96] Högnadottir A., Rouseff R. L., J. Chromatogr. A, 998, 201-211 (2003)
[97] Näf R., Velluz A., Meyer A.P., J. Essent. Oil Res., 8, 587-595 (1996)
[98] Average from: Moshonas M.G., Shaw P.E., J. Agric. Food Chem., 38 (3), 799-801 (1990)
[99] Moshonas M.G., Shaw P.E., J. Agric. Food Chem., 38 (3), 799-801 (1990)
[100] Lawrence B.M., Perf. & Flav., 18 (5), 43-68 (1993)
[101] Dugo G., Bartle K. D., Bonaccorsi I., Catalfamo M., Cotroneo A., Dugo P., Lamonica G., McNair H., Mondello L., Previti P., Stagno d´Alcontres I., Trozzi A., Verzera A., Essenze, Derv. agrum., 69, 79-111 (1999)
[102] Bolens M.A., Oporto A., Perf. & Flav., 16 (6), 1-7 (1991)
[103] Mondello L., Dugo P., Dugo G., Bartle K.D., J. Essent. Oil Res., 8, 597-609 (1996)
[104] Kirbaslar G., Kirbaslar S.I., J. Essent. Oil Res., 16, 105-108 (2004)
[105] Lota M.-L., de Rocca Serra D., Jacquemond C., Tomi F., Casanova J., Flav. Fragr. J., 16, 89-96 (2001)
[106] Mondello L., Basile A., Previti P., Dugo G., J. Essent. Oil Res., 9, 255-266 (1997)
[107] Ekundayo O., Bakare O., Adesomoju A., Stahl-Biskup E., J. Essent. Oil Res., 3, 329-330 (1991)
[108] Fleisher Z., Fleisher A., J. Essent. Oil Res., 2, 331-334 (1990)
[109] Goretti G., Russo M.V., Liberti A., Belmusto G., Essenze Deriv. agrum., 56, 345-358 (1986)
[110] Mondello L., Cotroneo A., Dugo P., Dugo G., J. Essent. Oil Res., 9, 495-508 (1997)

[111] Lota M.-L., de Rocca Serra D., Tomi F., Jacquemond C., Casanova J., J. Agric. Food Chem., 50, 796-805 (2002)
[112] Vekiari S.A., Protopapadakis E.E.; Papadopoulou P., Papanicolaou D., Panou C., Vamvakias M., J. Agric. Food Chem., 50, 147-153 (2002)
[113] Ahmed M., Arpaia M.L., and Scora R.W., J. Essent. Oil Res., 13, 149-159 (2001)
[114] Ekundayo, O. Bakare, A. Adesomoju, E. Stahl-Biskup, J. Essent. Oil Res., 3, 269-270 (1991)
[115] Ekundayo O., Bakare O., Adesomoju A., Stahl-Biskup E., J. Essent. Oil Res., 3, 55-56 (1991)
[116] Blanco Tirado C., Stashenko E.E., Combariza M.Y., Martinez J.R., J. Chromatogr. A, 697, 501-513 (1995)
[117] Mondello L., Cotroneo A., Stagno d´Alcontres I., Dugo G., J. Essent. Oil Res., 9, 379-392 (1997)
[118] Ekundayo, O. Bakare, A. Adesomoju, Stahl-Biskup E., J. Essent. Oil Res., 3, 199-201 (1991)

## 3.2.2.3 Herbs, Spices and Essential Oils

*Maren D. Protzen*
*Jens-Achim Protzen*

Updating the monographs[*] of 60 herbs and spices of major importance for the flavour industry (*Karl, 1998*) special emphasis was placed on the **essential oils** which are obtained from these raw materials as 'natural flavouring extracts'.

The term '*essential oil*' (in French '*huile essentielle*', in Spanish '*aceite esencial*') is linked with the alchemistic word 'quintessence' and its use has been shaped by experience. The German term '*Ätherische Öle*' is related to the Greek word 'Aether/Aither' and in Greek mythology 'Aether' was the description of 'Himmelsduft', i.e. the fragrant atmosphere at Olympus where the gods dwelt. In this context, the often found spelling as 'Etherische Öle' is misleading and questionable – the explanation for the spelling with an 'E' according to IUPAC rules of nomenclature is far-fetched and not convincing at all as, besides the volatility of both products, there is no association with the chemical class of ethers.

Arctander *[1]* defines essential oils as '*volatile materials derived by a physical process from odorous plant material of a single botanical form and species with which it agrees in name and odour and are either distilled (water and/or steam) or expressed*'.

An 'official' definition of essential oils and related products is available in ISO[**] 9235:1997 'Aromatic natural raw materials – Vocabulary'. In this standard not only essential oils but also related solvent extracts like oleoresins, concretes, absolutes and resinoids are defined. To stress the impact of ISO standards for the quality management in the flavour industry, an overview of all relevant standards concerning essential oils is given in Table 3.27 at the end of this chapter[***].

By volume (in tons) the most important essential oils used for flavours are the citrus oils (see 3.2.2.2), the mint oils, eucalyptus oils, clove oils and to a lesser extent some spice oils (Table 3.25).

*Table 3.25: Production figures of essential oils listed by quantity and major use*

|  | Flavour | Fragrance |
|---|---|---|
| Orange | 50 000 | |
| Mint (*arvensis*) crude | 20 000 | |
| Peppermint (*piperita*) | 4 500 | |
| Lemon | 3 500 | |
| Eucalyptus (cineole type) | 3 500 | |

---

[*] Technical remarks: CAS numbers in most cases are given in the title, even if the numbers correspond to the extract, namely the essential oil of the raw material.
[**] International Organisation of Standardisation.
[***] For ISO standards concerning the raw material, e.g. 'Dried Thyme' see search engine at: www.iso.org

|  | Flavour | Fragrance |
|---|---|---|
| Cedarwood |  | 2000 |
| Spearmint | 2 000 |  |
| Clove | 2 000 |  |
| Citronella |  | 1500 |
| Lime | 1 500 |  |
| Citriodora |  | 1000 |
| Lavandin |  | 1000 |
| Patchouli |  | 800 |
| Mandarin/tangerine | 450 |  |
| Nutmeg/mace | 400 |  |
| Grapefruit | 350 |  |
| Cassia/cinnamon bark | 300 |  |
| Tea tree |  | 300 |
| Geranium |  | 250 |

An interesting new aspect regarding the production of essential oils was highlighted by Verlet [2], who associated total production figures with cultivated plants, wild crafting, citrus oils and other tree crops (Table 3.26).

*Table 3.26: Breakdown of production (%) [2]*

|  | Production | |
|---|---|---|
|  | by weight | by value |
| Cultivated plants | 32 | 65 |
| Wild crafting | 1 | 2 |
| Citrus oils | 35 | 15 |
| Other tree crops | 31 | 18 |

In total, there are more than 150 different essential oils which are of commercial interest – estimates on the global production vary widely between 90,000 and 130,000 tons valued between EUR 800 million and EUR 1200 million (EUR 1.2 billion).

The reason for this considerable gap of approximately plus/minus 20% is that no reliable production figures are available from many parts of the world and often the indications obtained from various sources are contradictory. Another reason is that both positive as well as adverse climatic conditions considerably affect crop size (up to ±30%). Also a lack of information regarding the demand for local consumption of the oils in the production areas itself makes statements difficult to ascertain. Often export and import figures do not coincide with one another and therefore the indicated 'guesstimate' allows only an approximation of the dimension of this market.

Very often the quality of essential oils offered in the market is rather dazzling. In many cases, buyers do not have the required experience and knowledge to evaluate offered qualities – and often an attractive price is a big temptation for a compromise concerning the acceptability of a quality.

The combination of a better knowledge of the major, minor and important trace components and their detection by modern, state-of-the-art analytical instruments and methods nowadays enable buyers to assess the quality of a 'seductive' offer analytical wise: but behind all modern and highly sophisticated methods one should also bear in mind that 'figures neither smell nor taste'!

**Angelica, Angelikawurzel, Angélique**
[283-871-1; 84775-41-7]

*Source:* Angelica archangelica L. ssp. archangelica var. sativa (Miller) Rikli (Apiaceae) *[3]*, native to northern Europe and Siberia. Growing areas are located in the Netherlands, Belgium, France and Germany.

*Plant part:* Rhizome with side roots.

*Usage:* In herbal liqueurs and bitter spirits (Boonekamp, Benediktiner, Kartäuser, Chartreuse), in finished flavourings for meat and canned vegetables.

*Main components*: Up to 1% essential oil, coumarins (umbelliferon, osthol, osthenol and others *[4]*), furocoumarins (angelicin, bergaptene, xanthotoxin, etc. *[5]*), oligosaccharides.

Essential Oil:
*Characteristic*: Yellowish to brown colour, bittersweet taste.

*Solubility*: One part oil in four or more parts ethanol (90%) (results often in turbid solutions).

*Composition*: Approximately 90% terpenoids, mainly monoterpene hydrocarbons (pinene, phellandrenes, limonene). Also contains macrocyclic lactones, such as 15-pentadecanolide and 13-tridecanolide, which possess a musky odour and act as important aroma carriers *[6]*. Taskinen and Nykänen *[7]* have performed a detailed GC analysis of the differences between the oil obtained by steam distillation and the products resulting from an alcohol distillate. For latest research results see *[8]*.

*Additional remark*: Even the yellowish essential oil of the seeds is used as an alternative due to its similar composition *[9, 10]*.

*Output:* 1 t (root and seed oil).

**Angostura Bark, Angosturarinde, Ecorce d´Angosture**

*Source:* Angostura trifoliata (Willd.) Elias (Cusparia trifoliata (Willd.) Engl., Galipea officinalis Hancock or Cusparia febrifuga Humb. ex DC. *[11]* (Rutaceae); a tree, 10 to 12 meters high, native to Columbia and Venezuela.

*Plant part:* Bark of branches and young trunks.

*Usage:* In the liqueur industry (Angosturabitter).

*Main components:* 1-2% essential oil, which, according to older literature, contains, among others, cadinene, galipene and galipol, as well as alkaloids (e.g. cuspareine, galipine, angustureine, galipeine *[12]*) and two sensitive glycosidic bitter constituents *[13]*.

**Anise, Anis, Anis vert**
[283-872-7; 84775-42-8]

*Source:* Pimpinella anisum L. (Apiaceae): an annual plant, native to the eastern Mediterranean, which is cultivated mainly in Spain, France, Italy, Greece and other countries of Eastern Europe.

*Plant part:* Fruits. Can be mixed-up and adulterated with the smaller and non-hairy fruits of parsley and with hemlock fruits, which are rounder. For botanical distinction see *[14]*.

*Usage:* For flavouring food, beverages like liqueurs (Pernod, Raki, Ouzo) and oral hygiene products.

*Main components:* 2-6% essential oil, app. 30% vegetable oil, 20% protein, choline and saccharose.

Essential Oil:
*Characteristics:* Colourless to yellowish liquid or crystalline solid (congealing temperature: app. 15 to 20°C, according to its E-anethole content).

*Solubility:* one part oil in three parts ethanol (90%).

*Composition:* The ISO standard for aniseed *[15]* specifies the E-anethole content as 87–94%. Z-anethole (the more toxic isomer and transformation product of E-anethole when exposed to UV light *[16]*) is limited to 0.4%. The typical constituent of aniseed oil, pseudoisoeugenyl 2-methylbutyrate *[17]* can occur up to 2%. For further confirmation of its natural origin by chirality GC, see *[18]*. Further constituents are methyl chavicol (up to 3%), gamma-himachalene (up to 5%), anisic aldehyde (up to 1.4%) and e.g. small amounts of monoterpenes.

Since 2005, the Eur. Pharm. has changed its monograph for Anisi aetheroleum by limiting the source of the oil to Pimpinella anisum L. (formerly P. anisum and Illicium verum Hook). In general, the monograph conforms with the ISO standard, supplemented by the limitation for foeniculin (max. 0.01%, adulteration with star anise oil) and fenchone (max. 0.01%, mixing or adulteration with fennel oil).

*Additional remark:* In the flavour industry the oil has been replaced to a large extend by the less expensive, but also anethole-rich star anise and fennel oils.

*Output:* 10 t (Egypt, Spain).

**Star anise, Sternanis, Badiane**
[283-518-1; 84650-59-9]

*Source:* Illicium verum Hook. F. (Illiciaceae). The plant is native to and cultivated mainly in China. Further producing countries are Vietnam, Japan and the Philippines.

*Plant parts:* Fruits. Can be mixed up with the toxic fruits of Illicium anisatum L. (shikimi fruits), which do, however, not smell like anise but camphorous and bitter.

*Usage:* As a seasoning replacement for anise in East Asia. For the production of essential oil, and anethole (further usage see anise).

*Main components:* 5-8% essential oil.

Essential Oil:
*Characteristics:* Colourless to light yellow liquid or crystalline mass (congealing temperature 15-19°C).

*Solubility:* one part oil in 1.5 to three parts ethanol (90%).

*Composition:* According to ISO *[19]*, the oil contains 86-95% E-anethole, up to 1.0% Z-anethole and up to 6.0% methyl chavicol. Foeniculin can occur up to 3.0%, anisic aldehyde is limited to 0.5%. Further constituents are monoterpenes like α-pinene, α-phellandrene, limonene, linalool and α-terpineol. Besides β-caryophyllene, two further sesquiterpenes are typical trace components for Chinese star anise oil, namely E- and Z-α-bergamotene.

The detection of safrole by GC or TLC hints at adulteration with shikimi fruits, which, however, not always contain safrole. In TLC, safrole can be mistaken, among others, with foeniculin *[20]*. For a review of the latest literature see *[21]*.

*Output:* 500-700 t.

**Sweet Basil, Basilikum, Basilic**
[283-900-8; 84775-71-3]

*Source:* Ocimum basilicum L. (Lamiaceae).

*Plant part:* The fresh or dried leaves of the annual plant, which is probably native to Peninsular India and which is cultivated today in the tropics and temperate zones of the world.

*Usage:* As a seasoning especially for fish, although the herb is very sensitive and quickly loses value. Therefore, the essential oil or an oleoresin is usually used for sauces, canned fish or meat or for flavouring vinegar.

*Main components:* 0.2-0.5% essential oil, proteins, vitamins, saponin, β-sitosterol, carbohydrates.

Essential Oil:
*Solubility:* One part oil in one or two parts (linalool type) or in up to eight parts (methyl chavicol type) ethanol (80%).

*Composition:* As a member of the Lamiaceae family the genus Ocimum is characterised by chemical polymorphism. For O. basilicum several chemotypes are reported in the literature *[22, 23]* but only two types are of commercial interest.

The *linalool type* (European oil) is produced mainly in France and Egypt. The colourless to yellowish liquid has a typical spicy odour and consists of up to 62% linalool, trace to 30% of methyl chavicol and up to 15% eugenol according to AFNOR *[24]*.

The *methyl chavicol type* (exotic oil) is produced mainly in India, Pakistan, Vietnam, Egypt and the Comoros and contains up to 87% methyl chavicol, 1,8-cineole (1-3.5%), linalool (0.5-3%) and methyleugenol up to 2.5% according to ISO *[25]*.

For recent review articles see *[26 , 27]*.

*Additional remark:* The exotic oil is used to isolate pure methyl chavicol, a starting material for the production of synthetic anethole.

Output: 300-350 t (methyl chavicol type); 5-10 t (linalool type).

### Bitter Almond Oil, Bittermandelöl, Huile Essentielle d´Amands Amère
[291-060-9; 90320-35-7]

The essential oil is obtained by steam distillation mainly from degreased apricot and peach kernels, rarely from bitter almonds. The initially clear oil turns yellow during prolonged exposure to air. The oil contains app. 95% benzaldehyde and 2-4% hydrocyanic acid, formed by hydrolysis of the bitter-tasting amygdalin (benzaldehydcyanhydrine gentiobiosid) during distillation. When treated with milk of lime and Fe(II) sulphate, an oil is obtained which is free from hydrocyanic acid and consists entirely of benzaldehyde.

*Solubility:* One part oil in two parts ethanol (70%).

*Usage:* With its intense almond and cherry-like aroma, it is used for flavouring sweets and baked goods, for stone-fruit flavourings.

### Black Currant Buds, Schwarze Johannisbeer-Knospen, Bourgeons de Cassis
[271-749-0; 68606-81-5]

*Source:* Ribes nigrum L. (Saxifragaceae).

*Usage:* The dark green extract (paste) of the leaf buds is mainly produced in France (Absolue Bourgeons de Cassis) and is used for boosting the flavour of black currant preparations.

*Main components:* Up to 0.7% essential oil (70-80% monoterpenes, 15-20% sesquiterpenes) and polyunsaturated fatty acids.

For further information on the essential oil see *[28, 29]*. The sulphur-containing trace constituent 4-methoxy-2-methyl-2-butanthiol has been identified as an important flavour carrier *[30]*.

### Blessed Thistle, Benediktenkraut, Chardon bénit

*Source:* Cnicus benedictus L. (Asteraceae). The plant is native to the Mediterranean region and cultivated in Central Europe.

*Plant part:* Leaves and twig tips.

*Usage:* In bitter liqueurs.

*Main components:* Bitter constituents like the sesquiterpene lactone cnicin (0.2–0.7%) and lactonic lignans *[31]* as well as polyacetylenes and traces of essential oil.

## Borage, Boretsch, Bourrache

*Source:* Borago officinalis L. (Boraginaceae). The plant is native to south-eastern Europe and is cultivated throughout Europe and North America.

*Usage:* The fresh, rarely the dried herb is used for seasoning cucumbers, vinegar, salads and for producing spice extracts.

*Main components*: Resin, mucilage, tannins, soluble silicic acid, ascorbic acid, γ-linolic acid, pyrrolizidine alkaloids and traces of essential oil of unknown composition.

## Buchu Leaf Oils, Buccoblätteröle, Huile Essentielle de Feuilles de Bucco

The essential oil, which is present in a concentration of app. 1.5-2.5% in the leaves of Barosma (syn. Agathosma) species (Rutaceae) from South Africa, has traditionally been used as a flavour booster for black currants. Because both common species, B. betulina and B. crenulata, deliver different types of essential oil, they are treated separately in the following. Nevertheless, hybridisation of both wild growing species is observed.

(a) Buchu betulina oil [283-474-3; 84649-93-4]
*Source:* Barosma betulina (Bergius) Bartl. and Wendl.

*Characteristics:* yellow-brown liquid, sweet mint-like odour with a definite blackcurrant note, bitter taste.

*Solubility:* One part oil in three or more parts ethanol (70%).

*Composition:* (+)-Limonene (about 20%), (-)-menthone (about 10%), isomenthone (20-35%) isopulegone (max. 5%), pulegone (max. 15%) as well as diosphenol and φ-diosphenol (together up to 40%), 8-mercapto-p-menthan-3-on (min 2%) and its S-acetate (up to 0,5%) as odour carriers *[32]*. 8-Mercapto-p-menthan-3-on, which is one of the small number of naturally occurring sulphur-containing terpenoids, was identified also as the carrier of the cassis flavour by Ohloff et al. *[33]*.

(b) Buchu crenulata oil [296-170-0; 92346-85-5]
*Source:* Barosma crenulata (L.) Hook

*Characteristics:* Yellow to dark brown liquid, reminiscent of blackcurrant.

*Composition:* (+)-Limonene (about 20%), (-)-menthone (up to 5%), isomenthone (about 10%), isopulegone (up to 10%), pulegone (about 50%), diosphenols together about 1%, acetylthiole in higher concentration as in B. betulina (increased amount in betulina oil, may be sign of bastardisation). Because the 8-mercapto-p-menthan-3-on content in 'crenulata oil' is much lower than in 'betulina', the cassis aroma is reduced, too.

*Output (together):* 6-7 t (South Africa).

## Capers, Kapern, Câpres

*Source:* Capparis spinosa L. and other Capparis species (Capparaceae).

Capers are the flower buds, usually preserved in vinegar and salt, sometimes in oil, of shrubs native to and cultivated in the Mediterranean area.

*Usage:* For sauces and salads, in fish and meat dishes. The glycoside glucocapparin, which is degraded by the plant enzyme myrosinase to the sharp tasting methyl isothiocyanate, is the most significant constituent.

**Caraway, Kümmel, Carvi**
[288-921-6; 85940-31-4]

*Source:* Carum carvi L. (Apiaceae)

*Usage:* The fruits are used as a seasoning for bread, sausage, potato dishes, cheese and for liqueur production.

*Main components:* 3-7% essential oil, vegetable oil, proteins, hydroxycoumarins.

Essential Oil:
*Characteristics:* Colourless liquid with mild spicy taste which turns yellow when left standing.

*Solubility:* one part oil in max. five parts ethanol (80%).

*Composition:* 50-65% (+)-carvone, up to 45% limonene *[34]* as well as dihydrocarvone, carveol, dihydrocarveol and others. On the chirality GC of carvone see *[35]*.

*Output:* 40-50 t (Netherlands, Poland, Egypt, Hungary)

*Additional remarks:* Roman caraway or cumin (Cuminum cyminum L.) is cultivated in the Mediterranean and is used in the same way as caraway in the local kitchen. Cumin fruit are also a constituent of curry. The taste and odour resembles that of caraway but is less agreeable. The main components of the essential oil according to ISO *[36]* are cuminaldehyde (15-46%), $\gamma$-terpinene (14-32%), $\beta$-pinene (7-20%), p-menth-1,3-dien-7-al (2.8-22%), p-menth-1,4-dien-7-al (1.5-16%) and p-menth-3-en-7-al (0.3-5%). For further constituents see the ISO standard and *[37, 38]*.

In India, caraway can be mixed-up or adulterated with the fruits of Carum bulbocastanum Koch, which is also employed as a seasoning. The fruits are morphologically very similar.

**Cardamom, Kardamom, Cardamome**
[288-922-1; 85940-32-5]

*Source:* Elettaria cardamomun L. Maton (Zingiberaceae).

The plant is mainly cultivated in India, Sri Lanka and Guatemala.

*Plant part:* Fruits as well as seeds. Malabar-cardamoms are held in high esteem, Mysore-cardamoms, which are rounder and possess larger seeds, are less valued. The longer and sharper tasting Ceylon-cardamoms of the species Elettaria major Smith are considered as adulteration.

*Usage:* As a spice for dishes and baked goods (gingerbread), for fruits and marinades, as an ingredient of curry, in the liqueur industry (Angostura, Chartreuse, Goldwasser).

*Main components:* 3-8% essential oil, starch, fat, proteins.

Essential Oil:
*Characteristics*: Almost colourless to pale yellow liquid, pungent and strongly aromatic.

*Solubility:* One part oil in three parts ethanol (70%) (occasionally opalescent solutions).

Composition: According to ISO, the oil contains, depending on origin (Central America/Guatemala or India/Sri Lanka), 1,8-cineole (23-35%), terpinyl acetate (32-45%), as well as linalool (3-7%) and linalyl acetate (4-9%) as main components. Furthermore, α-pinene, sabinene, myrcene, limonene, terpinene-4-ol, α-terpineol and E-nerolidol in mentionable amounts. For further literature see *[39, 40, 41].*

*Output:* 10-15 t.

*Additional remark:* The sensory qualities of extracts produced with ethanol or acetone (oleoresins) exceed those of the essential oil, as apparently some high-boiling constituents are important for the flavour quality *[42].*

**Carrot Seed Oil, Karottensamenöl, Huile Essentielle de Semences de Carotte**
[284-545-1; 84929-61-3]

*Source:* Daucus carota L. (Apiaceae), a species of at least 12 subspecies, within D. carota L. ssp. sativus (Hoffm.) Archang, the root vegetable carrot, is even the source of commercial carrot seed oil *[43].*

*Usage:* Apart from the oleoresin, for spice extracts.

*Characteristics:* Light yellow liquid.

*Solubility:* One part oil in two parts ethanol (90%).

*Composition*: α-Pinene and β-pinene, daucene, E-β-farnesene, caryophyllene, geraniol, geranyl acetate, daucol and as main constituent carotol. For latest literature see *[44].*

*Output:* 5 t. Main producers: France, Hungary, India.

**Cascarilla, Kaskarille, Cascarille**
[284-284-3; 84836-99-7]

*Source:* Croton eluteria (L.) Sw. (syn. Clutia eluteria L.) (Euphorbiaceae).

*Plant part:* Branch and trunk bark of the tree, native to western India.

*Usage:* In the liqueur industry, for flavouring tobacco.

*Main components:* 1.5-4% essential oil, which contains linalool, borneol, terpinene-4-ol, α-terpineol, the methyl ethers of thymol, carvacrol and eugenol, cuparophenol (a phenolic sesquiterpene), sesquiterpene hydrocarbons, cascarilladiene, cascarillone *[45, 46]* and cascarillic acid *[47].* Moreover, cascarilline (bitter constituent), cascarilline A, vanillic aldehyde, resin, tannins, fat and starch are present.

*Output (essential oil):* Less than 2 t (Bahamas, El Salvador).

## Celery, Sellerie, Céleri
[289-668-4; 89997-35-3]

*Source:* Apium graveolens L. (Apiaceae).

*Usage:* The fresh or dried herb is used for seasoning soups, meat broth, vegetables, salads; the ground fruits, or an extract thereof, in celery salt.

*Main components:* Up to 0.8% essential oil in the fresh herb, which is partly lost during drying. Phthalide derivatives as well as diacetyl are aroma carriers *[48]*. The monoterpene fraction represents app. 95% of the herb oil and contains mainly limonene, myrcene and Z-ocimene *[49]*. The essential oil produced from the fruits (they contain up to 2.5% oil) is considered more valuable (see also *[50]* for the corresponding oleoresins).

Essential Oil (seeds):
*Characteristics*: Nearly colourless to brownish-yellow liquid.

*Solubility:* One part oil in maximally up to six parts ethanol (90%).

*Composition:* ISO specifies limonene with 58-79% as main constituent, as well as β-selinene (5-20%) and sedanenolide (3-butyl-4,5-dihydrophthalide) (1.5-11%); β-pinene up to 2%, myrcene up to 1.4% *[51]*. Further phthalides reported for celery oil are e.g. 3-butylnaphthalide, sedanolide (3-butyliden-3a,4,5,6-tetrahydrophthalide), (Z)- and (E)-ligustilide *[52, 53]*. Uhlig et al. describe a separation of the phthalides by HPLC. Their research indicated that among the phthalides, sedanolide, although quantitatively relatively small, contributes most to the characteristic flavour. Also, α-selinene turned out to be important for the aroma *[54]*.

*Output:* 40-50 t in total. Main producers: India, China, Egypt, France, Great Britain and USA.

## Centaury, Tausendgüldenkraut, Petite Centaurée

*Source:* Centaurium erythraea Rafn (Gentianaceae).

*Usage:* The bitter tasting herb is used for producing liqueurs.

*Main components*: The bitter constituents gentiopicrin and swertiamarin.

## Chervil, Kerbel, Cerfeuil

*Source:* Anthriscus cerefolium Hoffm. (Apiaceae).

*Usage:* The fresh herb possesses an anisate aroma (which is lost during drying) and is used as a kitchen seasoning for soups, salads and meat dishes.

*Main components:* Small amounts of essential oil with methyl chavicol and 1-allyl-2,4-dimethoxybenzene as main component *[55]*, bitter principles, flavonoids.

## Cinchona Bark, Chinarinde, Ecorce de Quinquina

*Source:* Cinchona pubescens Vahl (C. succirubra Pav. ex Klotzsch) and other Cinchona species (Rubiaceae).

*Usage:* Ingredient of bitter beverages.

*Main components:* 5-15% bitter tasting alkaloids, mainly chinine, chinidine, cinchonine, cinchonidine, as well as 3-5% tannins, up to 2% chinovin (a bitter tasting triterpene glycoside), resins, starch.

**Cinnamon, Zimt, Canelle**

(a) Sri Lankan (Ceylon) Cinnamon
[283-479-0; 84649-98-9]

*Source:* Cinnamomum verum J.S.Presl (C. zeylanicum Blume) (Lauraceae). The plant is native to Sri Lanka and cultivated in southern India and the Seychelles.

*Plant part:* The inner bark of the shoots of coppiced trees or the leaves.

*Usage:* For sweet dishes, canned fruit and baked goods, sweets, in the liqueur industry and for cola-type beverages.

*Main components:* 0.5-1.4% essential oil, tannins, mucilage.

Bark Oil:
*Characteristics:* yellow liquid.

*Solubility:* one part oil in at least three parts ethanol (70%).

*Composition:* 65-80% (E)-cinnamaldehyde, up to app. 10% eugenol, methyl eugenol, linalool, which, has been reported not to occur in cassia oil *[56]*, (E)-cinnamyl acetate, geranylacetate, benzyl benzoate, caryophyllene, hydrocinnamic aldehyde, benzaldehyde, α-terpineol, pinene, phellandrene, p-cymene and others *[57, 58, 59]*. Note that Cinnamon bark and leaf oil are also specified in the European Pharmacopoeia.

Leaf Oil:
*Characteristics:* reddish brown to dark brown liquid.

*Solubility:* one part oil in at least two parts ethanol (70%).

*Composition:* According to ISO, eugenol (70-83%), benzyl benzoate (2-4%), eugenyl acetate (1.3-3%), (E)-cinnamyl acetate up to 1.8%, (E)-cinnamaldehyde up to 1.3% *[60]*. Further constituents are phellandrene, linalool, β-caryophyllene and others *[61, 62]*.

*Output:* Bark oil: 5 t; leaf oil: 100 t.

(b) Cassia Cinnamon, Cassia-Zimt, Canellier (type Chine)
[284-635-0; 84961-46-6]

*Source:* Cinnamomum aromaticum Nees (C. cassia Blume) (Lauraceae). The evergreen tree or shrub is native to south China; the plant is also cultivated in Japan, Indonesia, Central and South America.

The peeled bark is used in the same manner as Sri Lankan cinnamon.

*Main components:* 1-2% essential oil, sugar, starch, tannins.

Essential Oil:
*Characteristics:* reddish-brown to dark brown mobile liquid.

*Solubility:* One part oil in three parts ethanol (70%).

*Composition:* (E)-cinnamaldehyde (70-88%), (E)-o-methoxycinnamaldehyde (3-15%), coumarin (1.5-4%), (E)-cinnamyl acetate up to 6%, (E)-o-methoxycinnamyl acetate up to 2%, as well as benzaldehyde (0.5-2%), salicylaldehyde, eugenol, cinnamic alcohol and further minor components like styrene, phenylethyl alcohol and aldehyde, (Z)-cinnamaldehyde and acetophenone according to *[63]*.

The commercially available oils are produced by distilling bark, branches and leaves.

*Output:* 300 t. Main producer: China.

**Clary Sage, Muskateller-Salbei, Sauge sclarée**
[283-911-8/ 84775-83-7]

*Source:* Salvia sclarea L. (syn. S. sclarea var. turkestaniana Mottet) (Lamiaceae). Native to the Mediterranean, the plant is cultivated there, in the former Soviet Union and in the USA. The blossoming herb is used.

*Usage:* For herbal liqueurs, for flavouring wine, tobacco and leather goods.

*Main components:* up to 1% essential oil, sclareol (bitter constituent), ursolic and oleanolic acid, rosmarinic acid, resins, mucilage, tannins.

Essential Oil:
*Characteristics:* yellowish to yellow liquid.

*Solubility:* one part oil in less than three parts ethanol (80%).

*Composition:* The European Pharmacopoeia lists linalyl acetate (56-78%), linalool (6.5-24%) and germacrene D (1.0-12%). α-Terpineol is limited to 5.0%, α- and β-thujone to 0.2%. Sclareol (0.4-2.6%) is found mainly in the concrete and is used for the synthesis of ambra fragrance materials. Further constituents are linalool oxide, geraniol, nerol, neryl and terpinyl acetate, citronellol, 1,8-cineole, benzaldehyde and n-nonanal. For further literature see *[64, 65]*.

*Output:* 150 t. Main producers: Russia, China.

**Cloves, Nelken, Clous de Girofle**
[284-638-7; 84961-50-2]

*Source:* Syzygium aromaticum (L.) Merr. et L.M. Perry (Eugenia caryophyllata Thunb.) (Myrtaceae).

*Plant part:* The blossom buds of the tree, native to the Moluccas, are used. Apart from Indonesia, the plant is also cultivated on Madagascar, Reunion, Zanzibar and Pemba (Tanzania).

The fruits (anthophylli) and the clove stems, which both contain essential oil, although considerably less, are considered as adulterations.

*Main components:* Up to 20% (high quality buds) essential oil, resin, fat, oleanolic acid and eugenin, a benzopyrone derivative.

Bud Oil:

*Characteristics:* Yellow to clear brown liquid.

*Solubility:* One part oil in two parts ethanol (70%) (same ratio for the leaf oil).

*Composition:* 75-85% eugenol, 8-15% eugenyl acetate and 2-7% caryophyllene (ISO *[66]*). Further constituents are humulene, α-copaene and α-cubebene. Heptanol-2, nonanol-2 and the corresponding ketones are trace components which are responsible for the fruity note (especially strong in oleoresins). The content of eugenyl acetate is responsible for the quality of the oil when compared to leaf and stem oil. Its content in oleoresins is even higher and these, therefore, represent the sensory qualities of the drug better than the oil *[67]*. For further constituents see *[68, 69, 70]*.

Stem Oil:

*Characteristics:* Yellow to clear brown liquid.

*Composition:* 80-92% eugenol, 0.2-1% eugenyl acetate and 4–17% caryophyllene (ISO *[71]*).

Leaf Oil:

*Characteristics:* Yellow to clear brown liquid.

*Composition:* 83-92% eugenol, 0.5-4% eugenyl acetate and 4–12% caryophyllene (ISO *[72]*).

For further constituents see *[73]*.

The leaf oil is used mainly for the production of eugenol and caryophyllene, because of its harsher note, which does not represent the typical clove flavour.

*Output:* 2000 t (of which 100 t bud oil).

## Cognac Oil, Weinhefen-Öl, Lie de Vin
[297-248-7; 93384-38-4]

The oil is obtained by steam distillation of wine yeast, or as by-product of marc brandy production, which is purified and, if necessary, rectified.

*Characteristics:* The raw oil is a white coloured liquid with wine or cognac aroma. If the yeast gets in contact with copper, a green to greenish-blue product will result.

*Solubility:* One part oil in two to five parts ethanol (80%).

*Composition:* Ethyl and amyl esters of fatty acids, especially capric and caprylic acid.

*Output:* 3 t (France, Germany).

## Coriander, Koriander, Coriandre
[283-880-0; 84775-50-8]

*Source:* Coriandrum sativum L. (Apiaceae). The fruits of the plant, which is cultivated in the Mediterranean, Central Europe, Russia, East Asia and North America, are used.

*Main components:* 0.5-2% essential oil, vegetable oil, proteins, starch, flavonoids, coumarins, isocoumarins *[74]*, such as coriandrine and others.

Essential Oil:
*Characteristics:* Nearly colourless to yellowish, spicy-aromatic and fruity liquid.

*Solubility:* One part oil in two to three parts ethanol (70%).

*Composition*: ISO lists 65-78% linalool, α-pinene (3-7%), γ-terpinene (2-7%), camphor (4-6%), limonene (2-5%), geranyl acetate (1-3.5%), geraniol (0.5-3%), α-terpineol (0.5-1.5%) and myrcene (0.5-1.5%) *[75]* as the main components of coriander fruit oil. Further constituents can be linalyl acetate, borneol, citronellol, nerol, terpinene-4-ol, additionally monoterpene hydrocarbons and aliphatic aldehydes *[76, 77]*. The latter are the main components of the weed (cilantro) oil and their amount is higher in unripe coriander fruits. E-2-tridecanal is held responsible for the bug-like odour of these unripe fruits ('bug dill') *[78]*. Lamparsky and Klimes have performed an exhaustive analysis which also led to the detection of interesting heterocyclic trace constituents *[79]*.

Adulteration of the oil with racemic, synthetic linalool may be detected using chiral gas chromatography, as studies showed the natural enantiomeric ratio of linalool to be as follows: (R)-(-)-linalool (11-15%); (S)-(+)-linalool (85-89%) *[80]*.

*Output:* 150-200 t.

## Davana Oil, Davana Öl
[295-155-6; 91844-86-9]

*Source:* Artemisia pallens Wall. ex DC. (Asteraceae).

*Plant part:* The aerial parts of the plant *[81]*, which is cultivated in India, are used.

*Usage:* For the creation of effects and rounding-off finished flavourings.

Essential Oil:
*Solubility:* One part oil in ten parts ethanol (70%).

*Composition:* Davanones (together app. 40-50%) as main components, nerol, geraniol, isodavanone, davana ether, artemone and davana furan are further characteristic components *[82]*. For additional literature see *[83, 84, 85]*.

*Output:* 2 t (India).

## Dill, Dill, Aneth
[289-790-8; 90028-03-8]

*Source:* Anethum graveolens L. (Apiaceae). The plant is native to the Mediterranean and cultivated all over Europe, the Americas and India.

*Usage:* The fresh herb, before the seeds reach maturity, is used for seasoning salads, fish dishes and for pickling cucumbers, etc. The ripe seeds are employed similarly and are also well suited as a substitute for caraway.

*Main components:* In the fresh herb 0.5-1.5%, in the dried herb 0.05-0.35% essential oil, which contains up to 40% α-phellandrene, as well as limonene, carvone, and dill furan (dill ether, 3,8-epoxy-p-menth-1-en) as main components *[86, 87, 88]*. The latter should be present in a minimum of 5% as an indicator of pure herb oil *[89, 90]*.

The colourless to bluish commercially available oils vary considerably in quantitative composition, depending on whether and in which degree of ripeness, the fruits were present during distillation.

In the ripe fruits 2-4% essential oil is found, which contains the same spectrum of components as dill weed oil, but in different ratios: 30-60% (+)-carvone and up to 40% limonene.

Blank and Grosch [91] have identified carvone as main odour carriers in seed oil, whereas in the leaf oil dill furan, myristicin, methyl 2-methylbutyrate, limonene and -phellandrene are responsible for the typical odour. See also [92].

*Solubility:* One part oil in two or more parts ethanol (80%).

A botanically very similar fruit from Anethum sowa Roxb. ex Flem. is cultivated in Japan and India. The composition of the seed oil is very similar to the one of European oil, the carvone content is lower, and moreover the oil contains caryophyllene, apiol and dill apiol. The latter was reported to lack in European oil [93].

*Output:* 100-150 t (of which < 10 t from seed). Main producers: USA, Hungary, Bulgaria, Russia.

### Eucalyptus Oil, Eucalyptus Öl, Huile Essentielle d'Eucalypte (Cineole-rich)

The oil is produced from the leaves and twigs of Eucalyptus globulus Labillardière [283-406-2; 84625-32-1], Eucalyptus polybractea R. T. Baker [294-961-5; 91771-67-4], Eucalyptus smithii R. T. Baker [294-962-0; 91771-68-5] (Myrtaceae) by steam distillation.

The genuine oil possesses an unpleasant odour, which is removed by subsequent rectification. Simultaneously, the content of 1,8-cineole reaches 70-80%, as required in most pharmacopoeiae. Further constituents, apart from monoterpene hydrocarbons, are α-terpineol, pinocarveol, borneol, α-terpinylacetate, pinocarvone, aromadendrene, globulol and others [94, 95, 96]. The ISO standard [97] defines the three commercially available oils of E. globulus, namely 'raw oil', 'rectified oil 70-75%' and '80-85%', according to the principle of fractionated distillation. With every rectification step, an accumulation of the target compound (1,8-cineole) and of the compounds with a similar boiling point (limonene, p-cymene) is achieved, whereas the components with a considerably lower (α-pinene) and higher boiling point (aromadendrene) decrease continuously.

*Usage:* Sweets, oral hygiene products, beverages.

*Characteristics:* Colourless to light yellow liquid.

*Solubility:* One part oil in five to ten parts ethanol (70%).

*Output:* 3500 t. Main producer: China.

*Additional remark:* The essential oil of E. polybractea mainly originates from Australia, whereas E. smithii is mainly cultivated in South Africa.

## Fennel, Fenchel, Fenouil
(bitter: 283-414-6; 84625-39-8; sweet: 282-892-3; 84455-29-8)

*Source:* Foeniculum vulgare Mill. (Apiaceae). A distinction is made between bitter fennel of the species F. vulg. ssp. vulgare var. vulgare, which is required by some pharmacopoeiae *[98]*, and sweet fennel (var. dulce), which is normally used as a seasoning.

*Usage:* For seasoning baked goods, vegetables, marinades, for fennel honey and candy, for producing liqueurs, for oral hygiene products.

*Main components:* 2-6% essential oil, fatty oil, protein, sugar.

Essential Oil:
*Characteristics:* Colourless to light yellow liquid (congealing temperature: 5-12°C).

*Solubility:* One part oil in 0.5 part ethanol (90%).

*Composition:* E-anethole (bitter fennel 55-75%, sweet fennel 80-95%), fenchone (bitter fennel 12-25%, sweet fennel below 1%), methyl chavicol (2–6%), limonene, α-pinene, α-phellandrene, cis-anethole and anisic aldehyde. For further constituents see *[99, 100]*. Ravid et al. *[101]* employed chirality GC to prove that enantiomerically pure (+)-fenchone is present in fennel oil.

*Output:* 40 t (bitter and sweet fennel oil). Main producers: Russia, Hungary, Bulgaria.

## Fenugreek, Bockshornklee, Fenugrec
[283-415-1; 84625-40-1]

*Source:* Trigonella foenum-graecum L. (Fabaceae). The seeds of the plant, which is cultivated in southern Europe and from North Africa to India, are used.

*Usage:* As an ingredient in blends such as curry powder. Mostly extracts of the roasted seeds are employed in food industry.

*Main components:* 25-45% mucilage, protein, 25-30% fatty oil, traces of an unpleasant smelling essential oil, choline, trigonelline (nicotinic acid-N-methyl betaine), diosgenin and other sterols *[102, 103]*.

Girardon et al. *[104]* have investigated the seeds via headspace GC, the steam distillate and an extract. In further research, they were able to identify 3-hydroxy-4,5-dimethyl-2(5H)-furanone as main aroma carrier *[105]*.

## Galanga, Galgant, Galanga
[291-068-2; 90320-42-6]

*Source:* Alpinia officinarum Hance (Zingiberaceae). Native to southern China, the plant is also cultivated in India and Sri Lanka.

*Usage:* The rhizome is used for producing spirits, bitter beverages and finished flavourings for baked goods.

In India and Indonesia the milder tasting rhizome of Alpinia galanga Willd. is used for seasoning purposes.

*Main components:* Up to 1.5% essential oil, diarylheptanoides, some of which taste sharp, as well as fat, sugar, starch and phlobaphene.

Essential Oil:
*Characteristics:* Greenish yellow liquid with a camphor-like, spicy odour and a cooling, slightly bitter taste.

*Solubility:* One part oil is miscible with 0.2 to 0.5 and more parts ethanol (90%) and soluble in 10 to 25 parts ethanol (80%).

*Composition:* 1,8-Cineole (app. 50%), α-terpineol (5%), camphor (1%), as well as mono- and sesquiterpene hydrocarbons *[106]*. The essential oil of Alpinia galanga has been analysed by Scheffer et al. *[107]*, and 1,8-cineole (47%), terpinene-4-ol (6%), and α-terpineol (6%) were found to be the main constituents. Furthermore, De Pooter et al. identified methyl eugenol, eugenyl acetate, chavicol and, for the first time in nature, chavicyl acetate as important components *[108]*. For the latest literature see *[109]*.

### Garlic, Knoblauch, Ail
[232-371-1; 8008-99-9]

*Source:* Allium sativum L. (Liliaceae).

*Usage:* The bulbs are used fresh, as dried powder, extract or salt for seasoning salads, sauces, meat dishes and sausages.

*Main components:* Alliin, a water soluble and odour-less substance, which is degraded by the plant enzyme alliinase to allicin. The latter constitutes an important aroma carrier. Except for alliin, several homologs are present, such as methyl-, ethyl-, n-butyl- and cycloalliin *[110]*.

Essential Oil:
The essential oil, which is present in the fresh bulb in a concentration of 0.1 to 0.3%, contains diallyl sulphide, diallyl trisulphide and allyl methyl disulphide (derivatives of allicin) as main components. Apart from further allyl polysulphides, the steam distilled oil contains limonene as non-sulphurous component *[111]*. The high pungency of garlic oil makes it necessary to dilute it, commonly in vegetable oil. One gram of oil is equivalent to 900 g of fresh garlic or 200 g of dehydrated garlic powder *[112]*.

*Output:* 20 t. Main producers: China, Mexico, Italy, Egypt, the Balkans.

### Gentian, Enzian, Gentiane

*Source:* Gentiana lutea L., G. punctata L., G. purpurea L., G. pannonica Scop. (Gentiaceae).

*Usage:* The dried roots for bitter liqueurs, gentian schnaps.

*Main components:* The bitter principles, gentiopicrin (up to 3%) and amarogentin (in G. lutea; amaropanin and amaroswerin in the other species), which is only present in small amounts, but possesses a far stronger bitter taste, as well as xanthone derivatives (gentisin and others), tannins, mucilage, considerable amounts of fermentable sugar (gentianose), traces of an essential oil with chroma-4-on-5-carbaldehyde, numerous monoterpenes, sesquiterpenes, diterpenes and fatty acid esters *[113]*.

**Ginger, Ingwer, Gingembre**
[283-634-2; 84696-15-1]

*Source:* Zingiber officinale Roscoe (Zingiberaceae).

*Plant part:* The peeled rhizome of the plant which is native to and cultivated in tropical Asia. Main growing areas are India, China, Nigeria, Indonesia, West Africa, Jamaica, Australia, Malaysia and the Fiji Islands.

Ginger from Jamaica is considered to possess the finest quality. The one from West Africa is mostly unpeeled and very sharp, with the highest essential oil content. As the oil possesses a camphor note, it is considered to be of inferior quality. Malabar ginger is mostly peeled and, like the quality from Australia, possesses a pronounced citrus-like taste and odour.

*Output (2004):* India (235000 t), China (230000 t), Indonesia (151000 t), Nigeria (110000 t), Nepal (90000), Fijis (3300 t), Malaysia (2500 t) *[114]*.

*Usage:* For sweets and baked goods (gingerbread, ginger snaps), candied and for jams, beverages (ginger ale, ginger beer), liqueurs, for sauces and oriental dishes, chutneys and for flavouring oral hygiene products.

*Main components:* 1.5-3% essential oil, 5-8% sharp tasting resin with the homologous gingerols, as well as shogaols, zingerone and others, starch, sugar, fat and organic acids.

Essential Oil:
*Characteristics:* Light yellow to brownish-yellow liquid, not sharp tasting.

*Solubility:* Opalescent solution in ethanol.

*Composition:* Main constituent is zingiberene (>30%) followed by β-sesquiphellandrene. (+)-ar-curcumene, (E,E)-α-farnesene and (-)-β-bisabolene are further important sesquiterpene hydrocarbons. α-Terpineol and citral (up to 9% in Australian ginger) result in the citrus note, whereas nerolidol is responsible for the woody-soapy taste *[115]*. Further constituents are camphene, limonene, phellandrene, borneol, 1,8-cineole, linalool, α-bisabolene, oxygenated sesquiterpene hydrocarbons, etc. For recent analytical data see *[116, 117, 118]*.

*Additional remark:* Often, oleoresins which contain both the essential oil (25-30%) and the pungent constituents are employed in industry.

*Output:* 50-60 t. Main producers: China, India.

**Horseradish, Meerrettich, Raifort**
[283-891-0; 84775-62-2]

*Source:* Armoracia rusticana P. Gärtn. et al. (Cochlearia armoracia L.) (Brassicaceae).

*Plant part and usage:* The root, finely ground to a sharp tasting paste, is used as a seasoning and as a finished side dish with meat dishes and sauces; an ingredient of mustard.

*Main components:* The glucosinolates sinigrin (0.3% in the fresh plant) and gluconasturtiin are hydrolysed by the plant enzyme myrosinase to allyl and α-phenylethyl isothiocyanate, the main constituents of the essential oil, which is present in amounts of 0.1 to 1.4%. For further components of the oil see *[119, 120]*. Furthermore, carbohydrates, ascorbic acid, vitamin B1 and a number of enzymes (lipases, invertases, amylases, proteases, peroxidases) have been found in the root.

*Output of essential oil:* less than 200 kg (France, Belgium, Hungary, India).

### Hyssop, Ysop, Hysope
[283-266-2; 84603-66-7]

*Source:* Hyssopus officinalis L. (Lamiaceae).

*Plant part:* The herb of the plant which is native to and cultivated in the Mediterranean.

*Usage:* Fresh for seasoning sauces, soups and salads, dried for producing herbal liqueurs.

*Main components:* Up to 1% essential oil, tannins, flavonoids (hesperidin, diosmin and others), the bitter constituent marrubiin, ursolic and oleanolic acid, α-sitosterol.

Essential Oil:
*Solubility:* One part oil in 0.5 to eight parts ethanol (80%).

*Composition:* Pinocamphone and iso-pinocamphone (together 40-70%), β-pinene and other monoterpene hydrocarbons, myrtenyl methyl ether, pinocarvone, β-bourbonene, β-caryophyllene, alloaromadendrene, germacrene D, spathulenol, elemol and others. The ISO standard for the essential oil obtained from the leaves is temporarily under revision *[121]*. For latest review articles see *[122, 123, 124]*.

From Spain comes an oil with 1,8-cineole as main constituent, named 'Hyssop decumbens oil', which is used especially in aromatherapy. Its correct botanical source is not known but may be a variety of H. officinalis L.

*Output:* less than 5 t (France, Italy, Russia, Hungary, the Balkans, Bulgaria).

### Juniper Berries, Wacholderbeeren, Baies de Genévrier
[283-268-3; 84603-69-0]

*Source:* Juniperus communis L. (Cupressaceae).

*Plant part:* The false fruits of the shrub or tree, which is common to Europe, North America, northern Africa and the Near East.

*Usage:* As a seasoning for gravies, sauerkraut, canned fish and for producing juniper berry spirits.

*Main components:* 0.8-2% essential oil, 30% invert sugar, tannins.

Essential Oil:
*Characteristics:* Clear, greenish or yellow liquid with an aromatic-bitter taste.

*Solubility:* One part oil in four parts to ten parts ethanol (95 %) (turbid solution).

*Composition:* Up to 80% monoterpene hydrocarbons, mainly α-pinene, myrcene, sabinene, limonene and β-pinene. As oxygenated compounds terpinene-4-ol (up to 10%) *[125]*, borneol, bornyl acetate, terpinyl acetate and a number of other compounds have been detected *[126, 127, 128]*. The composition of the oil strongly depends on the berry's degree of ripeness. The commercially available oils do, in the majority, not constitute pure berry oils, as needles and little twigs are included while harvesting and not removed before distillation. For latest reviews see Lawrence *[129]* and Adams, who published a comprehensive work on the genus in 2004 *[130]*.

*Output:* 10-15 t (the Balkans, Italy).

**Laurel, Lorbeer, Laurier**
[283-272-5; 84603-73-6]

*Source:* Laurus nobilis L. (Lauraceae).

*Plant part:* The leaves of the tree which is native to the Mediterranean.

*Usage:* For seasoning sauces, meat and fish dishes.

*Main components:* 1-3% essential oil, sesquiterpene lactones *[131, 132]*, flavonoids, alkaloids *[133]*.

Essential Oil:
*Characteristics:* Light yellow to yellow liquid.

*Solubility:* One part oil in one to three parts ethanol (80%).

*Composition:* App. 35-60% 1,8-cineole, other important constituents, depending on growing area, are α-terpineol, terpinene-4-ol, α-terpinyl acetate, linalool, eugenol, methyl eugenol, pinenes and others *[134, 135, 136]*.

*Output:* app. 4 t. the Balkans, Turkey, Russia.

**Liquorice, Süßholzholzwurzel, Réglisse**
[283-895-2; 84775-66-6]

*Source:* Glycyrrhiza glabra L. (Fabaceae).

*Plant part:* The peeled or unpeeled root and stolon; the herbaceous plant is mainly grown in Spain, Italy and the former Soviet Union.

*Usage:* Generally a thickened or solid extract (block liquorice), obtained by decoction with water, is employed for producing sweets and for flavouring smoked goods and tobacco products.

*Main components:* 2-15% sweet tasting saponines like glycyrrhizin, a diglucuronid of the triterpene glycyrrhetinic acid, present as potassium or calcium salt. The glycyrrhizin content is reduced by peeling, but the process also removes the bitter tasting resins. Frattini et al. *[137]* and Toulemonde et al. *[138]* report on the volatile aroma constituents. Furthermore, flavonoids (liquiritin, isoliquiritin and others *[139]*), coumarins (umbelliferone, herniarin), sugar and sugar alcohols, starch and asparagin have been detected in the roots.

On the quantitative determination of glycyrrhizic acid in liquorice extracts via HPLC see *[140]*.

**Litsea Cubeba Oil, Litsea Cubeba Öl, Huile Essentielle de Litsea Cubeba**
[290-018-7; 90063-59-5]

*Source:* The essential oil is obtained by steam distillation of the ripe and fresh berries of Litsea cubeba Pers. (Lauraceae) native to the mountain regions of China, Taiwan, Nepal, India and Indochina.

*Characteristics:* Pale to dark yellow liquid.

*Solubility:* One part oil in not more than three parts ethanol (70%).

*Main components:* According to ISO *[141]* geranial (39-45%), neral (25-33%) and limonene (9-15%) are the main components, followed by 6-methyl-5-heptene-2-one (1.8-3%), linalool (1.5-3%), citronellal (up to 1.5%), citronellol, nerol, geraniol and α-pinene.

Further constituents are several monoterpenes hydrocarbons and 1,8-cineole (main component of the leaf oil *[142]*), as well as isoneral, isogeranial and β-caryophyllene *[143, 144]*.

To reach an oil quality of international commercial standard, the crude oil is generally redistilled. The rectified oil contains about 75% carbonyl compounds (chemical analysis) which correspond to 70-72% citral (GC).

A significant part of the oil output is used for the isolation of natural citral, but higher capacities of synthetic citral nowadays compete with the natural product.

*Output:* 500-750 t. Main producer: China (Vietnam).

**Lovage, Liebstöckel, Livèche**
[284-292-7; 84837-06-9]

*Source:* Levisticum officinale Koch (Apiaceae).

*Plant part:* The fresh or dried herb and the root.

*Usage:* The herb for seasoning soups, sauces, meat dishes, the root for producing soup seasonings, finished flavourings for liqueurs and for flavouring tobacco.

*Main components:* 0.1-0.4% essential oil in the fresh herb, 0.3-1.2% in the dried herb; min. 0.3% in the dried roots. Naves *[145]* found up to 70% phthalides as typical flavour constituents: n-dibutyl phthalide, n-butylidene phthalide (ligusticum lactone), its dihydro derivative ligustilid, as well as sedanonic acid and sedanonic acid anhydride. α-Terpinyl acetate (50-70%) constitutes the biggest part of the leaf oil accompanied by β-phellandrene and (Z)-ligustilide in varying percentage, depending on state of maturity *[146]*. The latter compound constitutes the main component in lovage root oil; together with (E)-ligustilide and (Z)-butylidene phthalide, the phthalides account for over 50% of the oil. With n-pentylcyclohexadiene (app. 10%) a somewhat unusual component occurs *[147]*. Its amount in the oil seems to be lower when fresh root material is used for distillation *[148]*. The ISO standard unfortunately gives no information on percentages of oil components *[149]*. For further literature see *[150, 151]*.

*Solubility:* One part (leaf) oil in two to four parts ethanol (80%), one part (root) oil in up to 1.5 parts ethanol (85%), and one part (seed) oil in 0.5 parts ethanol (90%).

*Output:* 1 t. Main producers: the Netherlands, Belgium, France, Hungary, the Balkans, Poland.

**Sweet Marjoram, Majoran, Marjolaine**
[282-004-4; 84082-58-6]

*Source:* Origanum majorana L. (Majorana hortensis Moench) (Lamiaceae).

*Usage:* The dried leaves are used for seasoning sauces, salads and sausages.

*Main components:* 1-3% essential oil, rosemary acid (primarily responsible for the antioxidating effect of sweet marjoram), ursolic and oleanolic acid, phenols like arbutin *[152]*, flavonoids *[153, 154]*.

Essential Oil:
*Characteristics:* Yellow to greenish-yellow liquid.

*Solubility:* One part oil in two parts ethanol (80%).

*Composition:* Main constituents are α- and γ-terpinene, terpinene-4-ol, α-terpineol, linalool, linalyl acetate and cis-sabinene hydrate. The latter, along with its trans-isomer and terpinene-4-ol, is considered as typical aroma carrier of sweet marjoram. The quantitative composition, especially of oils of uncertain botanical origin, varies considerably *[155]*. Oils from India contain eugenol, methyl chavicol and geraniol in marked amounts *[156]*, while these compounds are negligible or absent in European oils. Commercial oils from Turkey contain considerable amounts of carvacrol, some up to 80% *[157]*. Also the production process can influence the oil composition, e.g. cis-sabinene hydrate is partly converted to terpinene-4-ol during steam distillation. The genuine plant additionally contains cis-sabinene hydrate acetate, which is almost totally hydrolysed in the steam-distilled oil *[158]*.

For detailed analysis of the oil see *[159]* and for further information in recent review articles refer to *[160, 161, 162]*.

*Output:* 30 t. Main producers: Morocco, Egypt, Madagascar and Germany.

**Mugwort, Beifuß, Armoise**
[283-874-8; 84775-45-1]

*Source:* Artemisia vulgaris L. (Asteraceae).

*Usage:* The herb is used for seasoning meat and fish and in the liqueur industry for bitters.

*Main components:* 0.03-0.2% essential oil, bitter constituents.

Essential Oil:
*Characteristics:* Light yellow to yellow liquid with a thujone-like taste and odour.

*Composition:* 1,8-Cineole, terpinene-4-ol, borneol, camphor, α- and β-thujone *[163]*. Naef-Müller et al. *[164]* report on unusual monoterpenes.

According to recent investigations, the commercially available oils derive from Artemisia herba-alba Asso (Morocco). Their qualitative composition is similar to the oil of Artemisia vulgaris, greater deviations have been observed quantitatively *[165]*. Nano et al. *[166]* demonstrated varying compositions in essential oils of various plant samples from Piedmont. Vulgarol, a bornane derivative, was detected as a new constituent in some of the oils. Spathulenol and α-cadinol, two sesquiterpene alcohols, as well as davanone were found by Jork and co-workers *[167, 168, 169]*.

Benjilali et al. detected six chemotypes in A. herba-alba (α-thujone, β-thujone, camphor, davanone, cis-chrysanthenyl acetate and chrysanthenone type), which also differed in their geographical origin *[170, 171]*. For the latest literature see *[172]*.

*Output:* 10-15 t (Morocco, Tunisia).

**Mustard, Senf, Moutarde**

(a) White Mustard [284-517-9; 84929-33-9]

*Source:* Sinapis alba L. (Brassicaceae).

*Usage:* The whole seeds for fish marinades, for pickling cucumbers, mixed pickles, ground for producing mustard.

*Main components:* 1.5-2.5% sinalbin, a glucosinolate, which, after the addition of water, is degraded enzymatically by myrosinase into p-hydroxybenzyl isothiocyanate (not water-vapour volatile), glucose and sinapin hydrogensulphate. Sinapin constitutes the choline ester of sinapinic acid. Moreover, the plant contains app. 30% vegetable oil, mainly the glycerides of erucic acid, mucilage, steroids and proteins.

(b) Black Mustard [290-076-3; 90064-15-6]

*Source:* Brassica nigra (L.) W.D.J. Koch (Brassicaceae).

*Usage:* Same as white mustard.

*Main components:* 1-2% sinigrin, which is split enzymatically into allyl isothiocyanate (water-vapour volatile and main constituent of mustard oil), glucose and potassium hydrogensulphate, as well as the bitter-tasting sinapin, phytin, protein, 30-35% vegetable oil and mucilage (less than in white mustard).

The cultivation of Brassica nigra today has been practically replaced by B. juncea (L.) Czern. et Coss. [309-426-4; 100298-73-5], also named 'yellow mustard'. Main exporters of the seeds are the Netherlands, Italy, Eastern Europe and China.

A significant part of commercially available mustard oil is not necessarily of natural origin.

*Output (essential oil):* 1 t.

### Nutmeg/Mace, Muskatnuß/Macis, Muscade
[282-013-3; 84082-68-8]

*Source:* Myristica fragrans Houttuyn (Myristicaceae).

*Plant part:* The kernel (nutmeg) and arillus (mace), which envelops the shell containing the kernel. The tree is native to the Moluccas and cultivated in peninsular India, Sri Lanka, Indonesia and the West Indies (Granada).

*Usage:* As a seasoning for vegetables, baked goods and meat dishes.

*Main components:* Nutmeg: 7-16% essential oil, up to 40% fat (trimyristine), up to 30% starch, as well as sugar, pectines, resins, saponin, lipase. Mace: 4-15% essential oil, up to 25% vegetable oil, up to 30% amylodextrin and lignans *[173, 174]*.

On the detection of treatment with X-rays see *[175]*.

Essential Oil:
*Characteristics:* Colourless to slightly yellow liquid.

*Solubility:* One part oil in five parts ethanol (90%).

*Composition (is assumed to be approximately the same for both, nutmeg and mace):* Mainly monoterpene hydrocarbons (app. 90%), as well as myristicin, elemicin, safrole, eugenol, eugenol methyl ether, terpinene-4-ol, α-terpineol, linalool and others. ISO specifies nutmeg oil (Indonesian type) as follows: α-pinene (15-28%), β-pinene (13-18%), sabinene (14-29%), δ-3-carene (0.5-2%), limonene (2-7%), γ-terpinene (2-6%), terpinene-4-ol (2-6%), safrole (1-2.5%), myristicin (5-12%) *[176]*. For latest review articles see *[177, 178, 179, 180]*.

*Output:* Totally 350-400 t. Main producers: Indonesia, Grenada, Sri Lanka, India.

### Onion, Zwiebel, Oignon
[232-498-2; 8054-39-5]

*Source:* Allium cepa L. (Liliaceae).

*Plant part:* The bulbs (thick, fleshy leaf bases).

*Main components*: A number of sulphur-containing amino acid derivatives, similar to those of garlic, mainly dihydroalliin (trans-$S$-(1-propenyl)-L-(+)-cysteinsulphoxide), methyl- and cycloalliin. The lachrymatory factor 1-propyl sulphide is formed enzymatically from dihydroalliin *[181]*. The essential oil contains methyl-, propyl- dipropyl- and propenyldisulphides, as well as trisulphides and numerous further sulphur-containing compounds *[182, 183, 184, 185]*. Furthermore, sugar, protein, flavonoids, carotinoids, ascorbic acid, vitamin B1, B2 and pantothenic acid have been found in onions.

*Output (essential oil):* Totally 3 t (Europe, Mexico, India).

### Oregano

*Source:* Origanum vulgare L. ssp. hirtum (Link) Ietsw. (or ssp. viride Hayek) (Greek oregano), O. syriacum L., O. onites L. (Turkish oregano) and other Origanum species (Lamiaceae).

Oregano is a comprehensive term for a seasoning which consists of the leaves and blossoms of various Origanum species, which are native to and cultivated in the Mediterranean.

*Usage:* As a seasoning for pizza, for meat and fish dishes.

*Main components:* Up to 3% essential oil with carvacrol, rarely thymol, as main constituent, as well as p-cymene, linalool, borneol, terpinene-4-ol. On the composition of various oregano oils see *[186]*.

Spanish oregano [290-371-7; 90131-59-2] derives from Coridothymus capitatus Rchb. (syn. Thymus or Thymbra capitatus). Again, carvacrol is the main constituent of the essential oil. ISO specifies its content with 60-75%. Further components are p-cymene (5.5-9%), γ-terpinene (3.5-8.5%), thymol (0-5%), β-caryophyllene (2-5%), linalool (0.5-3%), myrcene (1-3%), α-terpinene (0.5-2.5%), terpinene-4-ol (0.5-2%), α-thujene (0.5-2%), α-pinene (0.5-1.5%) *[187]*. For further literature see *[188, 189]*.

Mexican oregano is used in the USA, and derives from various Lippia species, especially Lippia graveolens (Verbenaceae); its taste is reported to be less fresh and pleasant *[190]*. Main constituents are thymol, carvacrol, p-cymene *[191]*.

*Solubility:* One part oil in two parts ethanol (70%), when further diluted, turbid solutions may form.

*Output (essential oil):* 30-40 t (the Balkans, Turkey, Spain).

**Parsley, Petersilie, Persil**
[281-677-1; 84012-33-9]

*Source:* Petroselinum crispum (Mill.) Nyman ex A.W. Hill (var. crispum) (syn. P. sativum Hoffm.) (Apiaceae).

*Usage:* The fresh, sometimes dried, leaves for seasoning soups, salads, meat and fish dishes.

*Main components:* Ascorbic acid, nicotinamide, furocoumarins *[192, 193]*, flavonoids, high amount of chlorophyll and 0.1-0.7% essential oil have been found in the herb.

(a) Leaf Oil

*Characteristics:* Yellow to brownish liquid.

*Solubility:* Soluble in ethanol (95%) (sometimes not completely clear).

*Composition:* p-Mentha-1,3,8-triene is the main constituent, which also constitutes the major aroma carrier *[194, 195]*. Also other monoterpene hydrocarbons, such as β-phellandrene and 1-methyl-4-isopropenylbenzene (α-p-dimethylstyrene), are reported to contribute to the overall flavour profile *[196]*. Further constituents are α- and β-pinene, myrcene, terpinolene, myristicin, apiol and phthalides. The composition varies significantly based on origin and time of harvest *[197, 198]*. In some herb oils the occurrence of Diels-Alder adducts between p-mentha-1,3,8-triene, myrcene or itself (molecular weights of 268 and 270) can be observed in considerable amounts (up to 10%) *[199, 200]*.

(b) Seed Oil

*Characteristics:* Nearly colourless to amber-yellow liquid.

*Solubility:* One part oil in not more than six parts ethanol (85%).

*Composition:* Parsley seeds contain 2-8% essential oil with myristicin, apiol, elemicin and 2,3,4,5-tetramethoxyallylbenzene as main components. ISO specifies the oil as follows: $\alpha$-pinene (10-22%), $\beta$-pinene (7-15%), elemicin (1-12%), myristicin (25-50%), apiol (5-35%), 1,2,3,4-tetramethoxy-5-allylbenzene (1-12%) *[201]*. Also chemotypes with one main component of the latter three constituents do exist.

Further research results for essential parsley oil have been reviewed by Lawrence *[202, 203, 204, 205]*.

*Output:* Leave oil: less than 5 t (Hungary, France, South Africa, USA, Egypt, Australia); seed oil: 10 t (the Netherlands, Poland, France, Hungary, India).

**Pepper, Pfeffer, Poivre**
[284-524-7; 84929-41-9]

*Source:* Piper nigrum L. (Piperaceae).

*Plant part:* Black pepper is based on the mature, partly unripe harvested and dried fruits, white pepper derives from the dried, peeled, ripe fruits. Main growing areas are India, Vietnam, Sri Lanka, Indonesia, Madagascar and Brazil. Green pepper is the green, not fully mature fruit, which is, pickled in salt or vinegar, imported mainly from Madagascar and the Malabar Coast.

*Main components:* Average yield of essential oil is 3.5% in black pepper (up to 2.5% in white pepper). Main constituents are sharp tasting acid amides, especially piperine (all-trans) (up to 10%), the less sharp tasting piperettine in small amounts, piperamine and others *[206, 207]*. Apart from app. 40% starch and 8% vegetable oil, flavonoids, cubebin and other lignan derivatives have been found in traces *[208]*. Furthermore, lipase is present, which has to be taken into account during processing.

Essential Oil:
*Characteristics:* Colourless to greenish liquid without pungency.

*Solubility:* One part oil in three parts ethanol (95%).

*Composition:* App. 95% terpene hydrocarbons, mainly sabinene, limonene, $\alpha$- and $\beta$-pinene, $\delta$-3-carene and $\beta$-caryophyllene *[209]*. Lewis et al. *[210]* reported that a high pinene content results in a turpentine-like, and high $\beta$-caryophyllene content in a flowery odour, while for a strong pepper like top note a generally high monoterpene yield is important. Other authors *[211]* found that sabinene and terpinene-4-ol contribute most to the overall flavour. Further characteristic compounds are germacrene D and $\alpha$-/$\beta$-selinene, the latter occur in higher amounts in oils from Indonesia. Furthermore, linalool, cis- and trans-sabinene hydrate, $\alpha$-terpineol, safrole, eugenol, $\alpha$- and $\beta$-eudesmol, nerolidol, myristicin and other oxygen-containing components have been detected. For further literature see *[212, 213, 214]*.

*Output:* Totally 80-100 t.

In contrast to the essential oil, the oleoresins employed in the industry also contain the pungent constituents. On their production and characterisation see *[215, 216, 217]*, on the determination of piperine via HPLC see *[218, 219]*.

*Additional remark:* The not sharp tasting fruits of another pepper species, cubeb (P. cubeba L.), are also used for seasoning purposes (ingredient of curry). Apart from starch, protein and fatty oil, they contain 6-11% essential oil with δ-cadinene, β-cubebene, α-copaene, cubebol and germacrene D as main components. *Origin and Output:* Indonesia (essential oil: less than 1 t).

**Peppermint Oil and other Mint Oils, Pfefferminz- und andere Minzöle, Huile Essentielle de Menthe Poivrée et autres Huiles Essentielles de Menthe**

(a) Peppermint Oil, Pfefferminzöl, Huile Essentielle de Menthe Poivrée
 [282-015-4; 84082-70-2]

*Source:* Mentha piperita L. (Lamiaceae).

Main growing areas are the western states of the USA (Washington, Oregon, Idaho, Wisconsin, Indiana) and India; smaller areas also exist in Italy, France and England.

*Characteristics:* Colourless, slightly yellow or green liquid.

*Solubility:* One part oil in three to four parts ethanol (70%).

*Composition:* 30-55% (-)-menthol, 14-32% (-)-menthone, 1.5-10% (+)-iso menthone, 2.8-10% menthyl acetate, 3.5-14% 1,8-cineole, 1-9% menthofuran, up to 4% pulegone *[220]*, as well as 3-octanol, trans-sabinene hydrate, piperitone, viridiflorol, mono- and sesquiterpene hydrocarbons. For further constituents see *[221, 222, 223]*. On the differentiation between peppermint oils and the detection of dementholised cornmint oil as a major adulterant see the results of Lawrence et al. *[224]*. The ratio of 1,8-cineole:limonene (min. 2) was even implemented in the European Pharmacopoeia. The latter also limits the isopulegol content for Mentha piperita oil to 0.2%.

On the chirality GC of menthols, menthones and menthol esters see *[225, 226, 227]*.

*Output:* 4500 t. Main producers: USA (3500 t) and India.

(b) Cornmint Oil, Minzöl, Huile Essentielle de Mentha Arvensis
 [290-058-5; 90063-97-1]

*Source:* Mentha arvensis L. var. piperascens Briq.

Main growing areas are India, China and Brazil.

*Characteristics:* Colourless, slightly yellow to greenish-yellow liquid.

*Solubility:* One part oil in four parts ethanol (70%).

*Composition:* 70-80 % (-)-menthol. The commercially available oils are dementholised and again rectified, contain mainly menthol (33-46%), menthone (18-32%), isomenthone (8-14%) and menthyl esters, and are employed for diluting the more expensive piperita oils. The composition of the unrectified oil is similar to that of piperita oil, whereas (E)-sabinene hydrate, viridiflorol and menthofuran are only present in traces. Isopulegol occurs up to 3%.

*Output*: 23,000 t. Main producers: India (20,000 t), China and Brazil.

(c) Spearmint Oil, Krauseminzöl, Huile Essentielle de Menthe Crêpue

*Source:* Mentha viridis L. (syn. *M. spicata* L.) var. crispa Benth. for **Common (Native) Spearmint** [283-656-2; 84696-51-5] and Mentha cardiaca Gerard ex Baker for **Scotch Spearmint** [294-809-8; 91770-24-0].

*Usage:* For flavouring sweets, bubble gums and dental care products.

*Characteristics:* Colourless to yellowish-green liquid.

*Solubility:* One part oil in one part ethanol (80%), may become turbid upon further dilution.

*Composition:* (-)-Carvone, which is present in a concentration of 60-70%, functions as main aroma carrier. Dihydro cumic alcohol and carvyl acetate are also assumed to contribute to the overall aroma profile. Furthermore, limonene, 3-octanol, trans-sabinene hydrate, cis-dihydrocarvone and β-bourbonene occur in marked amounts. Viridiflorol (0.1-0.5%) is only detectable in the native oil type. In its monographs ISO distinguishes between 'Native', 'Chinese (80%, 60%)', 'Indian' and 'Scotch Type' *[228]*. For further literature on the composition of native spearmint and Scotch spearmint see also *[229]*.

*Output:* Totally 1700-2000 t. Main producers: USA, China and India.

**Pimento (Allspice), Piment**
[284-540-4; 84929-57-7]

*Source:* Pimenta dioica (L.) Merr. (P. officinalis Lindl.) (Myrtaceae).

*Plant part:* The fruits of the tree, which is native to Central America and cultivated on Jamaica, are used.

*Usage:* For seasoning marinades, fish products and sausages, baked goods, for producing beverages and spirits, for flavouring tobacco.

*Main components:* 2-5% essential oil, sugar, resin, tannins.

Essential Oil:
*Characteristics:* Light yellow to brown liquid.

*Solubility:* One part oil in two parts ethanol (70%).

*Composition:* Eugenol (65-85%), methyl eugenol, 1,8-cineole and numerous mono- and sesquiterpene hydrocarbons, e.g. myrcene, β-caryophyllene and humulene *[230]*. The commercially available oils also derive from the leaves and possess a similar composition, whereas the content of methyl eugenol in most cases is higher in the berry oil *[231, 232]*. For further research results see *[233]*.

*Output:* Totally 20-30 t (Jamaica).

**Red Pepper, Paprika, Poivre d'Espagne**
[283-403-6; 84625-29-6]

*Source:* Capsicum annuum L. (Solanaceae).

*Plant part:* The ripe fruits, the inner pericarp and seeds are removed entirely or partly when a less sharp taste is desired.

*Main components:* Sharp tasting capsaicinoids (0.3-0.5%); these are vanillyl amides of various acids, whereby capsaicin, the vanillyl amide of an isodecenylic acid, is most important. Furthermore, dihydro-, nordihydro-, homo- and homodihydrocapsaicin and others have been found *[234]*. Sugar (up to 50%), ascorbic acid, traces of essential oil with 2-methoxy-3-isobutylpyrazine as main aroma constituent *[235, 236, 237]*. A steroid saponin mixture, called capsicidin which functions antibacterially against yeast fungi *[238]*, 0.3-0.8% carotenoids, partly esterified with fatty acids (mainly capsanthin, its isomer capsorubin and α-carotene).

Apart from the organoleptic determination of pungency (Scoville heat units *[239]*), the capsaicinoids are determined quantitatively, photometrically or singly best with HPLC (e.g. *[240, 241, 242]*). The quantitative determination of pelargonic acid vanillylamide, which is available synthetically and, therefore, prone to adulteration, is of special importance. This substance is only present in small amounts in genuine mixtures.

As red pepper and the corresponding oleoresins are also used for colouring purposes, techniques allowing their quantitative determination by photometric methods, partly after preseparation by TLC, have been developed *[243, 244]*. Synthetic dyes like sudan I-IV and para red in capsicum preparation are prohibited by European food regulation because of their carcinogenic potential and can be detected sensitively by HPLC-MS/MS methods *[245, 246]*.

### Cayenne Pepper, Chilies (Peperoni)
[288-920-0; 85940-30-3 (C. frutescens)]

*Source:* Capsicum frutescens, C. fastigiatum and other Capsicum species.

As a result of their higher capsicinoid content (0.6-0.9%), the fruits are considerably sharper. Basically, the constituents are the same as those of red pepper. The composition of the essential oil has been investigated by Haymon and Aurand *[247]*, who mainly detected various esters of butyric, valerianic and capronic acid, but no typical aroma constituent.

### Rosemary, Rosmarin, Romarin
[283-291-9; 84604-14-8]

*Source:* Rosmarinus officinalis L. (Lamiaceae).

*Plant part:* Flowering tops and leafy twigs.

*Usage:* Wide range of uses in food processing, because of its properties as antioxidant (though not technically listed as natural preservative) and its capacity to suppress warmed over flavour (WOF) *[248]*.

*Main components:* Up to 2.5% essential oil (app. 1% in fresh leaves), app. 3% rosemarinic acid, which apart from carnosolic acid, a tricyclic diterpenephenol, the bitter constituent carnosol (not genuine but autoxidation product of carnosolic acid) *[249]*, rosmanol and others *[250]* are responsible for its antioxidating effect. Triterpenes *[251, 252]* and flavonoids *[253]* may also contribute to this effect. For this reason oleoresins are used preferably in industry for flavouring purposes.

Essential Oil:
*Characteristics:* Colourless to light yellow and greenish liquid.

*Solubility: Tunisian and Moroccan type*: one part oil in not more than two parts ethanol (80%). *Spanish type*: one part oil in not more than three parts ethanol (90%).

*Composition:* Up to 55% 1,8-cineole, camphor, borneol and its acetate, α-terpineol, terpinene-4-ol, linalool, verbenone, 3-octanone and terpene hydrocarbons. For detailed analyses see e.g. *[254, 255, 256]*. Even if different chemotypes of rosemary exist, only two qualities are of commercial interest (ISO and European Pharmacopoeia give almost identical specifications):

(a) Tunisian and Moroccan Type:

The oil contains α-pinene (9-14%), camphene (2.5-6%), β-pinene (4-9%), myrcene (1-2%), limonene (1.5-4%), 1,8-cineole (38-55%), camphor (5-15%), bornyl acetate (0.1-1.6%), borneol (1-5%), verbenone (up to 0.4%) and others (ISO *[257]*).

(b) Spanish Type:

According to ISO the oil contains α-pinene (18-26%), camphene (8-13%), β-pinene (2-5%), myrcene (2.5-4.5%), limonene (2.5-5.5%), 1,8-cineole (17-25%), camphor (12.5-22%), bornyl acetate (0.4-2.5%), borneol (2-4.5%), verbenone (0.7-2.5%) and others.

*Output:* Totally 120-150 t (Tunisia, Morocco and Spain).

**Saffron, Safran**

*Source:* Crocus sativus L. (Iridaceae).

*Plant part:* Stigmata of the blossom which are still hand-harvested. Main cultivating countries are Spain, Iran, Greece, India, China and Morocco.

*Usage:* For flavouring and colouring rice dishes, sauces, cakes, soups (bouillabaisse), liqueurs, etc.

*Main components:* 0.4-1% essential oil, which contains app. 50% of the odour carrier safranal. For further components of the oil see *[258, 259]*. Safranal is formed by hydrolysis from the glycoside picrocrocin, which causes the bitter taste. Carotinoids, mainly the water-soluble crocin (crocetin digentiobiose ester) and other crocetin derivatives *[260]*, α- and β-carotin, zeaxanthin and lycopin are responsible for the colouring effect.

As saffron is frequently adulterated, e.g. with calendula, arnica, other blossoms such as saflor (Carthamus tinctorius L.), and non-colouring crocus species *[261]*, suitable detection methods have been developed: see *[262, 263, 264]* and its ISO standard *[265]*.

**Sage, Salbei, Sauge**
[282-025-9; 84082-79-1]

*Source:* Salvia officinalis L. (Lamiaceae).

*Plant part:* Leaves. The plant is native to the Mediterranean and cultivated in Europe and North America.

*Usage:* For seasoning meat dishes, sausages, fish and cheese, singly or, most common, in blends.

*Main components:* 1.5-2.5% essential oil, rosemarinic acid (see rosemary), flavonoids *[266, 267]*, tannins, oleanolic and ursolic acid.

Essential Oil:
*Characteristics:* Colourless to yellow liquid.

*Solubility:* One part oil in not more than two parts ethanol (80%) to yield a clear solution.

*Composition:* According to ISO *[268]*, the oil of Dalmatian sage contains, among others, α-pinene (1-6.5%), camphene (1.4-7%), limonene (0.5-3%), 1,8-cineole (5.5-13%), α-thujone (18-43%), β-thujone (3-8.5%), camphor (4.5-24.5%), linalool/linalylacetate (up to 1%), bornyl acetate (up to 2.5%), α-humulene (up to 12%). For the presence of viridiflorol see *[269]*; for a review of recent literature see *[270, 271]*.

*Additional remark:* Under the name 'Spanish sage oil', an essential oil of S. lavandulifolia Vahl [290-272-9; 90106-49-3] is commercially available. As specified in ISO *[272]*, it contains 1,8-cineole (10-30%), camphor (11-36%) and linalool/linalylacetate in higher amounts than the 'Dalmatian oil', various monoterpenes hydrocarbons, borneol, -terpinylacetate and sabinylacetate.

*Output:* Totally 40-60 t. Main producer: the Balkans.

**Savory, Bohnenkraut, Sariette**
[283-922-8; 84775-98-4]

*Source:* Satureja hortensis L. (Lamiaceae).

*Plant part:* The flowering fresh or dried herb. The plant is native to the eastern Mediterranean and cultivated in North America, Europe, western Asia, India and South Africa

*Usage:* For seasoning vegetables, soups, mushrooms, meat and fish dishes, sausages, cucumbers.

*Main components:* Among others, up to 1.9% essential oil, tannins, rosmarinic acid.

Essential Oil:
*Characteristics:* Light yellow to dark brown liquid.

*Solubility:* One part oil in two parts ethanol (80%).

*Composition (essential oil):* Carvacrol is the main constituent. γ-Terpinene, p-cymene, α-thujene, α-pinene, myrcene, α-terpinene and thymol can be found in significant amounts.

*Output:* Totally less than 1 t. Main producers: the Balkans, Hungary, France.

*Additional remark:* Winter savory of the species Satureja montana L. [290-280-2; 90106-57-3] is used similarly in the Mediterranean. The plant is reported to contain slightly more essential oil. On distinguishing both species and on detecting adulterations with Moroccan thyme (Thymus satureoides Cass.) and Senegal savory see *[273]*. For recent literature reviews see *[274, 275]*.

**Tarragon, Estragon**
[290-356-5; 90131-45-6]

*Source:* Artemisia dracunculus L. (Asteraceae).

*Plant part:* The fresh or dried herb. The plant is native to the temperate zones of the Northern Hemisphere and cultivated in Russia, France, Germany and southern Europe. Among the cultivated forms a distinction is made between 'French tarragon', which must be propagated vegetatively, and the less aromatic 'Russian tarragon'.

*Usage:* For seasoning sauces, salads, cucumbers, for producing tarragon vinegar.

*Main components:* 0.1-0.4% essential oil in the fresh, 0.2-0.8% essential oil in the dried herb. Furthermore, flavonoids, isocoumarins *[241]*, coumarins, sterols and others have been found.

Essential Oil:
*Characteristics:* Yellowish to brownish liquid.

*Solubility:* One part oil in at least 0.5-1 part ethanol (90%).

*Composition:* Methyl chavicol (estragol) (up to 80%) is the main constituent of 'French tarragon' and, therefore, of the commercially available oils. Sabinene, limonene, ocimene (both isomers), methyl eugenol and in some oils elemicine are further constituents (besides sabinene, the latter both are main components in Russian and German tarragon leaf oil). For detailed analyses see *[276, 277, 278, 279]*.

*Output:* Totally less than 10 t. Main producers: Italy, France, Hungary.

**Thyme, Thymian, Thym**
[284-535-7; 84929-51-1 (vulg.)]; [284-397-0; 85085-75-2 (zygis)]

*Source:* Thymus vulgaris L., Thymus zygis L. (Lamiaceae).

*Plant part:* The aerial parts of the plant, which is native to the Mediterranean and cultivated in the south of Europe and North America, is used.

*Usage:* For salads, sausages and meat dishes.

*Main components:* 1-2.5% essential oil, rosemary acid, flavonoids *[280]*, tannins, oleanolic and ursolic acid.

Essential Oil:
*Characteristics:* Red to brownish-red liquid ('red thyme oil').

*Solubility:* One part oil in not more than three parts ethanol (80%).

*Composition:* According to Granger and Passet *[281]*, different chemotypes (thymol-, carvacrol-, linalool, geraniol, α-terpineol-, and E-sabinene hydrate-type) exist for T. vulgaris; however, only the thymol type and in special cases the linalool type (e.g. aromatherapy) are of commercial interest. The thymol type of T. vulgaris contains the phenols thymol (30-55%) and carvacrol (1-5%) as well as their precursors in biogenetic pathway, p-cymene (15-20%) and γ-terpinene (5-10%). Additionally linalool (1-5%), methyl ethers of thymol and carvacrol, in minor amounts borneol, camphor, limonene, myrcene, β-pinene, E-sabinene hydrate, α-terpineol, terpinene-4-ol, β-caryophyllen and others are present *[282]*.

The oil of Thymus zygis (main source for thyme oil in Spain), is defined by ISO *[283]* as follows: thymol (37-55%), carvacrol (0.5-5.5%), p-cymene (14-28%), γ-terpinene (4-11%), linalool (3-6.5%), terpinene-4-ol (0.1-2.5%), E-β-caryophyllene (0.5-2%) methyl ether of carvacrol (0.1-1.5), E-sabinene hydrate (up to 0.5%), and further monoterpene hydrocarbons.

In practice, it is nearly impossible to distinguish between the essential oils derived from T. vulgaris and T. zygis.

*Additional remark:* Also the essential oil of wild thyme, T. serpyllum [284-023-3; 84776-98-7], again a polymorphous species bearing several chemotypes, is of commercial interest. A *carvacrol-rich* oil type may contain for example carvacrol (20-40%), thymol (1-5%), p-cymene (10-20%), γ-terpinene (5-15%), borneol, bornyl acetate, 1,8-cineol, citral, geraniol, linalool and others *[284]*.

For the latest literature on thyme see Stahl-Biskup *[285]*, Lawrence *[286, 287]* and references therein.

*Output:* Totally 70-80t.

**Turmeric, Curcuma**
[283-882-1; 84775-52-0]

*Source:* Curcuma longa L. (C. domestica Valeton) (Zingiberaceae).

*Plant part:* The rhizome, which is often scalded when fresh to ensure better drying. The plant is cultivated in India and southern China.

*Usage:* Main ingredient of curry; for seasoning sauces, soups, readymade dishes, colouring ingredient of mustard.

*Main components:* Around 3% curcumin, as well as other curcuminoids like demethoxy curcumin and p-hydroxy cinnamoyl methane as colouring agents *[288]*. 1.5-5% essential oil, which contains up to 50% of the sesquiterpenes turmerone and ar-turmerone. Further constituents, depending on variety and cultivar, are e.g. (+)-ar-curcumene and (-)-β-bisabolene *[289]*, zingiberene, β-sesquiphellandrene, 1,8-cineol, tumerol and several further oxygenated sesquiterpenes *[290]*. The essential oil is responsible for the flavour and aroma of turmeric, even if it is not used as such. The oleoresin is used as spice extract (particularly because of its colouring properties) in the food industry.

## Vanilla, Vanille
[283-521-8; 84650-63-5]

*Source:* Vanilla planifolia Jacks. (syn. V. fragrans auct.) (Orchidaceae).

*Plant part:* The green, unripe fruits (vanilla beans) are subjected to a complex fermentative process for obtaining the aromatic black cured beans. The plant is native to Mexico and cultivated in Madagascar and the Comoros (both origin for the Bourbon quality), Indonesia, Papua New Guinea, India and Uganda. Outside of Central America the blossoms have to be pollinated manually as the humming-birds and butterflies which normally pollinate the blossoms are not indigenous.

*Usage:* In ice cream, soft beverages, for confectionaries, sweets and sweet dishes, cakes and other baked goods, in the liqueur and tobacco industry.

*Main components:* During fermentation, vanillin (up to app. 2.5%) is formed from the odourless glucovanillin by enzymatic hydrolysis. Its aroma is rounded-off and modified by p-hydroxybenzoic acid, p-hydroxybenzaldehyde, p-hydroxybenzyl methyl ether, vanillyl alcohol, vanillic acid, cinnamic acid ester and various other trace constituents *[291, 292]*. Also the resins, gums, amino acids and other organic acids contribute to the typical flavour of the cured beans *[293]*. For further constituents and characterisations of fruits from different growing areas see *[294]*.

Tahiti vanilla (V. tahitensis Moore), which possess a different aroma, is considered as less valuable and can be distinguished from V. planifolia because of its constituents p-anisic acid and p-anisaldehyde *[295]*. The smaller vanillons (Vanilla pompona Schiede) are practically not available anymore.

For the production of vanilla extract ethanol (min. 35%) is used. In the USA 1 l alcohol is used to extract 100 g beans for the 1-fold quality. For a 10-fold quality the same amount of ethanol is used to extract 1 kg of beans. In Europe the ethanol extract frequently is concentrated to yield a dark, viscous extract or oleoresin, wherein, among others, the vanillin content determines the grade of quality.

As vanilla, as a result of its high price, is often adulterated or diluted, a number of publications deal with the detection of piperonal, coumarin, ethyl vanillin, vanitrop and others. Whereas these adulterations can be detected rather easily, the addition of synthetic vanillin, as long as it remains within certain proportions, can only be certified by the expensive isotope analysis (deuterium, $^{13}C$ content) *[296, 297, 298, 299]*.

## Wormwood, Wermut, Absinthe
[284-503-2; 84929-19-1]

*Source:* Artemisia absinthium L. (Asteraceae).

*Plant part:* The herb of the plant, which is native to southern Europe, North Africa and Asia and cultivated also in the Americas, is used.

*Usage:* For producing wormwood wines and bitter liqueurs.

*Main components:* 0.5-1.3% essential oil, bitter sesquiterpene lactones (artabsin, absinthin, anabsinthin), flavonoids; for further constituents see *[300]*.

*Composition of the essential oil:* Mainly α- and β-thujone, thujyl alcohol and thujyl acetate respectively [301]. Some chemotypes contain cis- and trans-epoxyocimene, sabinyl and chrysanthenyl acetate, chrysanthenol as main components [302]. For latest literature see [303].

Wormwood wines or 'absinth' nowadays have seen something of a revival. However, European regulation limits the thujone content in bitter spirits to 35 mg/kg.

*Output:* 5 t.

**Others**

Apart from these typical herbs and spices and the oils and extracts based on them, the flavour industry employs a number of other natural products for rounding-off, modifying and boosting flavour preparations, especially fruit flavours. These often constitute extracts (absolutes, absolues), which usually find application in perfumery. A few examples are the following:

Absolue ciste (leaves and branches of Cistus ladaniferus L.)

Absolue cassie (blossoms of Acacia farnesiana Willd.)

Orris butter (rhizomes of Iris germanica L., I. pallida Lam., I. florentina L.)

Absolue jasmin (blossoms of Jasminum grandiflorum L.)

Massoia oil (bark of Cryptocaria massoia (Lauraceae)). The oil possesses a coconut-peach flavour.

Absolue mimosa (blossoms of Acacia dealbata Lk.)

Absolue violette feuilles (leaves of Viola odorata L.)

Absolue rose (blossoms of Rosa centifolia and R. damascena).

All these raw materials are employed in a very low dosage.

*Table 3.27: ISO standards concerning essential oils (January 2006)*

| No. | Under revision | Title |
|---|---|---|
| 17494: 2001 | | Aromatic extracts, flavouring and perfuming compounds - Determination of ethanol content - Gas chromatographic method on packed and capillary columns |
| 9235: 1997 | | Aromatic natural raw materials - Vocabulary |
| 770: 2002 | | Crude or rectified oils of Eucalyptus globulus (Eucalyptus globulus Labill.) |
| 3794: 1976 | | Essential oils (containing tertiary alcohols) - Estimation of free alcohols content by determination of ester value after acetylation |
| 14714: 1998 | | Essential oils and aromatic extracts - Determination of residual benzene content |
| 7609: 1985 | | Essential oils - Analysis by gas chromatography on capillary columns - General method |
| 22972: 2004 | | Essential oils - Analysis by gas chromatography on chiral capillary columns - General method |

| No. | Under revision | Title |
|---|---|---|
| 7359: 1985 | | Essential oils - Analysis by gas chromatography on packed columns - General method |
| 8432: 1987 | | Essential oils - Analysis by high performance liquid chromatography - General method |
| 21092: 2004 | | Essential oils - Characterisation |
| 1242: 1999 | | Essential oils - Determination of acid value |
| 280: 1998 | | Essential oils - Determination of refractive index |
| 1271: 1983 | | Essential oils - Determination of carbonyl value - Free hydroxylamine method |
| 1279: 1996 | | Essential oils - Determination of carbonyl value - Potentiometric methods using hydroxylammonium chloride |
| 1272: 2000 | | Essential oils - Determination of content of phenols |
| 709: 2001 | | Essential oils - Determination of ester value |
| 7660: 1983 | * | Essential oils - Determination of ester value of oils containing difficult-to-saponify esters |
| 1241: 1996 | | Essential oils - Determination of ester values, before and after acetylation, and evaluation of the contents of free and total alcohols |
| 1041: 1973 | | Essential oils - Determination of freezing point |
| 592: 1998 | | Essential oils - Determination of optical rotation |
| 279: 1998 | | Essential oils - Determination of relative density at 20°C - Reference method |
| 11021: 1999 | | Essential oils - Determination of water content - Karl Fischer method |
| 875: 1999 | | Essential oils - Evaluation of miscibility in ethanol |
| 11024: 1-1998 | | Essential oils - General guidance on chromatographic profiles - Part 1: Preparation of chromatographic profiles for presentation in standards |
| 11024: 2-1998 | | Essential oils - General guidance on chromatographic profiles - Part 2: Utilisation of chromatographic profiles of samples of essential oils |
| 11018: 1997 | | Essential oils - General guidance on the determination of flashpoint |
| 211: 1999 | | Essential oils - General rules for labelling and marking of containers |
| 210: 1999 | | Essential oils - General rules for packaging, conditioning and storage |
| 4720: 2002 | * | Essential oils - Nomenclature |
| 356: 1996 | | Essential oils - Preparation of test samples |
| 3218: 1976 | | Essential oils - Principles of nomenclature |
| 4715: 1978 | | Essential oils - Quantitative evaluation of residue on evaporation |
| 212: 1973 | * | Essential oils-Sampling |
| 11023: 1999 | | Liquorice extracts (Glycyrrhiza glabra L.) - Determination of glycyrrhizic acid content - Method using high-performance liquid chromatography |
| 3525: 1979 | * | Oil of amyris |
| 3475: 2002 | | Oil of aniseed (Pimpinella anisum L.) |
| 3065: 1974 | | Oil of Australian eucalyptus, 80 to 85% cineole content |
| 11043: 1998 | | Oil of basil, methyl chavicol type (Ocimum basilicum L.) |
| 3045: 2004 | | Oil of bay (Pimenta racemosa (Mill.) J.W. Moore) |
| 3520: 1998 | | Oil of bergamot (Citrus aurantium L. ssp. bergamia (Wight et Arnott) Engler), Italian type |

| No. | Under revision | Title |
|---|---|---|
| 8900: 2005 | | Oil of bergamot petitgrain (Citrus bergamia (Risso et Poit.)) |
| 9844: 1991 | * | Oil of bitter orange (Citrus aurantium L.) |
| 8901: 2003 | | Oil of bitter orange petitgrain, cultivated (Citrus aurantium L.) |
| 3061: 1979 | * | Oil of black pepper |
| 590: 1981 | | Oil of brazilian sassafras |
| 7357: 1985 | | Oil of calamus - Determination of cis-beta-asarone content - Gas chromatographic method on packed columns |
| 3523: 2002 | | Oil of cananga (Cananga odorata (Lam.) Hook. f. et Thomson, forma macrophylla) |
| 8896: 1987 | | Oil of caraway (Carum carvi Linnaeus) |
| 4733: 2004 | | Oil of cardamom (Elettaria cardamomum (L.) Maton) |
| 3216: 1997 | | Oil of cassia, Chinese type (Cinnamomum aromaticum Nees, syn. Cinnamomum cassia Nees ex Blume) |
| 11025: 1998 | | Oil of cassia, Chinese type - Determination of trans-cinnamaldehyde content - Gas chromatographic method on capillary columns |
| 9843: 2002 | | Oil of cedarwood, Chinese type (Cupressus funebris Endlicher) |
| 4725: 2004 | | Oil of cedarwood, Texas (Juniperus mexicana Schiede) |
| 4724: 2004 | | Oil of cedarwood, Virginian (Juniperus virginiana L.) |
| 3760: 2002 | | Oil of celery seed (Apium graveolens L.) |
| 3524: 2003 | | Oil of cinnamon leaf, Sri Lanka type (Cinnamomum zeylanicum Blume) |
| 3848: 2001 | | Oil of citronella, Java type |
| 3849: 2003 | | Oil of citronella, Sri Lanka type (Cymbopogon nardus (L.) W. Watson var. lenabatu Stapf.) |
| 3142: 1997 | | Oil of clove buds (Syzygium aromaticum (L.) Merr. et Perry, syn. Eugenia caryophyllus (Sprengel) Bullock et S. Harrison) |
| 3141: 1997 | | Oil of clove leaves (Syzygium aromaticum (L.) Merr. et Perry, syn. Eugenia caryophyllus (Sprengel) Bullock et S. Harrison) |
| 3143: 1997 | | Oil of clove stems (Syzygium aromaticum (L.) Merr. et Perry, syn. Eugenia caryophyllus (Sprengel) Bullock et S. Harrison) |
| 3516: 1997 | | Oil of coriander fruits (Coriandrum sativum L.) |
| 3756: 1976 | | Oil of cubeb |
| 9301: 2003 | | Oil of cumin seed (Cuminum cyminum L.) |
| 9909: 1997 | | Oil of Dalmatian sage (Salvia officinalis L.) |
| 10624:1998 | | Oil of elemi (Canarium luzonicum Miq.) |
| 3044: 1997 | | Oil of eucalyptus citriodora Hook. |
| 14716: 1998 | | Oil of galbanum (Ferula galbaniflua Boiss. et Buhse) |
| 4731: 1978 | * | Oil of geranium (Pelargonium x ssp.) |
| 3053: 2004 | | Oil of grapefruit (Citrus x paradisi Macfad.), obtained by expression |
| 21389: 2004 | | Oil of gum turpentine, Chinese (mainly from Pinus massoniana Lamb.) |
| 9841: 1991 | * | Oil of Hyssop (Hyssopus officinalis Linnaeus) |
| 8897: 1991 | * | Oil of juniper berry (Juniperus communis Linnaeus) |
| 3054: 2001 | | Oil of lavandin abrial (Lavandula angustifolia Miller x Lavandula latifolia Medikus), French type |

| No. | Under revision | Title |
|---|---|---|
| 8902: 1999 | * | Oil of lavandin grosso (Lavandula angustifolia Miller x Lavandula latifolia (L.f.) Medikus), French type |
| 3515: 2002 | | Oil of lavender (Lavandula angustifolia Mill.) |
| 8899: 2003 | | Oil of lemon petitgrain (Citrus limon (L.) Burm. f.) |
| 855: 2003 | | Oil of lemon (Citrus limon (L.) Burm. f.), obtained by expression |
| 3217: 1974 | | Oil of lemongrass West Indian (Cymbopogon citratus (DC) Stapf.) |
| 4718: 2004 | | Oil of lemongrass (Cymbopogon flexuosus (Nees ex Steudel) J.F. Watson) |
| 3809: 2004 | | Oil of lime (cold pressed), Mexican type (Citrus aurantifolia (Christm.) Swingle), obtained by mechanical means |
| 3519: 2005 | | Oil of lime distilled, Mexican type (Citrus aurantifolia (Christm.) Swingle) |
| 3214: 2000 | | Oil of Litsea cubeba (Litsea cubeba Pers.) |
| 8898: 2003 | | Oil of mandarin petitgrain (Citrus reticulata Blanco) |
| 3528: 1997 | | Oil of mandarin, Italian type (Citrus reticulata Blanco) |
| 4730: 2004 | | Oil of Melaleuca, terpinen-4-ol type (Tea Tree oil) |
| 9776: 1999 | | Oil of Mentha arvensis, partially dementholised (Mentha arvensis L. var. piperascens Malinv. and var. glabrata Holmes) |
| 3517: 2002 | | Oil of neroli (Citrus aurantium L. ssp. aurantium, syn. Citrus aurantium L. ssp. amara var. pumilia) |
| 3215: 1998 | | Oil of nutmeg, Indonesian type (Myristica fragrans Houtt.) |
| 14717: 1999 | * | Oil of origanum, Spanish type (Thymbra capitata (L.) Cav.) |
| 4727: 1988 | * | Oil of palmarosa (Cymbopogon martinii (Roxburgh) W. Watson var. motia) |
| 3527: 2000 | | Oil of parsley fruits (Petroselinum sativum Hoffm.) |
| 3757: 2002 | | Oil of patchouli (Pogostemon cablin (Blanco) Benth.) |
| 856: 1981 | * | Oil of peppermint, France, Italy, UK and USA |
| 3064: 2000 | | Oil of petitgrain, Paraguayan type (Citrus aurantium L. ssp. aurantium, syn. Citrus aurantium L. ssp. amara var. pumilia) |
| 4729: 1984 | | Oil of pimento leaf (Pimenta dioica (Linnaeus) Merrill) |
| 11019: 1998 | | Oil of roots of lovage (Levisticum officinale Koch) |
| 9842: 2003 | | Oil of rose (Rosa x damascena Miller) |
| 1342: 2000 | | Oil of rosemary (Rosmarinus officinalis L.) |
| 3761: 2005 | | Oil of rosewood, Brazilian type (Aniba rosaeodora Ducke or Aniba parviflora (Meisn.) Mez.) |
| 3526: 2005 | | Oil of sage, Spanish (Salvia lavandulifolia Vahl) |
| 3518: 2002 | | Oil of sandalwood (Santalum album L.) |
| 4728: 2003 | | Oil of Spanish wild marjoram (Thymus mastichina L.) |
| 3033-1: 2005 | | Oil of spearmint - Part 1: Native type (Mentha spicata L.) |
| 3033-2: 2005 | | Oil of spearmint - Part 2: Chinese type (80 % and 60 %) (Mentha viridis L.var. crispa Benth.), redistilled oil |
| 3033-3: 2005 | | Oil of spearmint - Part 3: Indian type (Mentha spicata L.), redistilled oil |
| 3033-4: 2005 | | Oil of spearmint - Part 4: Scotch variety (Mentha x gracilis Sole) |
| 4719: 1999 | | Oil of spike lavender (Lavandula latifolia (L.f.) Medikus), Spanish type |

| No. | Under revision | Title |
|---|---|---|
| 11016: 1999 | | Oil of star anise, Chinese type (Illicium verum Hook. f.) |
| 3140: 2005 | | Oil of sweet orange (Citrus sinensis (L.) Osbeck), obtained by mechanical treatment |
| 9910: 1991 | | Oil of sweet orange - Determination of the total carotenoids content |
| 10115: 1997 | | Oil of tarragon (Artemisia dracunculus L.), French type |
| 7356: 1985 | | Oils of thujone-containing artemisia and oil of sage (Salvia officinalis Linnaeus) - Determination of alpha- and beta-thujone content - Gas chromatographic method on packed columns |
| 14715: 1999 | * | Oil of thyme containing thymol, Spanish type (Thymus zygis (Loefl.) L.) |
| 11020: 1998 | | Oil of turpentine, Iberian type (Pinus pinaster Sol.) |
| 4716: 2002 | | Oil of vetiver (Vetiveria zizanioides (L.) Nash) |
| 21390: 2005 | | Oil of wintergreen, China (Gaultheria yunnanensis (Franch.) Rehd.), redistilled |
| 3063: 2004 | | Oil of ylang-ylang (Cananga odorata (Lam.) Hook. f. et Thomson forma genuina) |
| 7358: 2002 | | Oils of bergamot, lemon, citron and lime, fully or partially reduced in bergapten - Determination of bergapten content by high-pressure liquid chromatography (HPLC) |
| 4735: 2002 | | Oils of citrus - Determination of CD value by ultraviolet spectrometric analysis |
| 21093: 2003 | | Oil of dwarf pine (Pinus mugo Turra) |
| 7611: 1985 | | Oils of lemon and petitgrain citronnier, and oil of lime obtained by a mechanical process - Determination of citral (neral + geranial) content - Gas chromatographic method on capillary columns |
| 18054: 2004 | | Oils of orris rhizome (Iris pallida Lam. or Iris germanica L.) - Determination of irone content - Method using gas chromatography on a capillary column |
| 7355: 1985 | | Oils of sassafras and nutmeg - Determination of safrole and cis- and trans-isosafrole content - Gas chromatographic method on packed columns |

## REFERENCES

[1] Arctander S., Perfume and Flavor Materials of Natural Origin, Allured Publishing, Reprint (2000)
[2] Verlet N., *Les huiles essentielle: Production mondiale, échanges internationaux et politiques de développement*, PhD Thesis, University of Aix-Marseille (1993)
[3] Lawrence B.M., Perfum. Flav. 24 Nov./Dec., 47 (1999)
[4] Härmälä P. et al., Planta Med. 58, 287 (1992)
[5] Harkar S. et al., Phytochemistry 23, 421 (1984)
[6] Schultz K., Kraft P., J. Essent. Oil Res. 9, 509 (1997)
[7] Taskinen J., Nykänen L., Acta Chem. Scand. B 29, 757 (1975)
[8] Chalchat J.C., Garry R. Ph., J. Essent. Oil Res. 9, 311 (1997)
[9] Formacek V, Kubeczka K.-H., *Essential Oils Analysis by Capillary Gas Chromatography and Carbon-13 NMR Spectroscopy*, 2nd edition, New York, J. Wiley & Sons (2002)
[10] Lawrence B. M., Perfum. Flavor. July/Aug. 14, 4 (1989)
[11] USDA, ARS, National Genetic Resources Program. *Germplasm Resources Information Network - (GRIN)* [Online Database]. National Germplasm Resources Laboratory, Beltsville, Maryland.
[12] Jacquemond-Collet I., Hannedouche S., Fabre N., Fourasté I., Moulis C., Phytochemistry 51, 8 1167 (1999)

[13] Brieskorn C.H., Beck V., Phytochemistry 10, 3205 (1971)
[14] Capeletti E.M., Planta Med. 39, 88 (1980)
[15] ISO 3475:2002 Oil of aniseed (Pimpinella anisum L.), International Organisation of Standardisation, Geneva, Switzerland (2002)
[16] Martin E., Berner Ch., Mitt. Geb. Lebensmitteluntersg. 63, 127 and 132 (1972)
[17] Kubeczka K.H., v. Massow F., Formacek V., Smith M.A.R., Z. Naturforschung 31b, 283 (1975)
[18] Karl V., Schumacher K., Mosandl A., Flavour Fragrance J. 7, 283 (1992)
[19] ISO 11016:1999 Oil of star anis, Chinese Type (*Illicium verum Hook.* f.), International Organisation of Standardisation, Geneva, Switzerland (1999)
[20] Seger V., Miehing H., Hänsel R., Pharm. Ztg. 132, 2747 (1987)
[21] Lawrence B.M., Perfum. Flav. 28 Nov/Dec, 74 (2003)
[22] Lawrence B.M. et al, Flavour Ind. 1, 265 (1970); 2, 173 (1971); 3, 47 and 145 (1972)
[23] Vieira R.F., Simon J. E., Econ. Bot., 52, 207-216 (2000)
[24] AFNOR NF T75-244:1992 Essential oils. Oil of basil, linalol type (*Ocimum basilicium* L.), L'Association française de Normalisation (1992)
[25] ISO 11043:1998 Oil of basil, methylchavicol-type (*Ocimum basilicum* L.), International Organisation of Standardisation, Geneva, Switzerland (1998)
[26] Lawrence B.M., Perfum. Flav. 23 Nov/Dec, 35 (1998)
[27] Lawrence B.M., Perfum. Flav. 29 Sept., 80 (2004)
[28] Latrasse A., Lantin B., Acta Horticult. 60, 183 (1976) and Rivista Ital. 59, 395 (1977)
[29] LeQuere J. L., Latrasse A., Sci. Aliment. 6, 47 (1986)
[30] Rigaud J., Etievant P., R. Henry, A. Latrasse, Sci. Aliment. 6, 213 (1986)
[31] Vanhaelen M., Vanhaelen-Fastr R., Phytochemistry 14, 2709 (1975)
[32] Collins N.F., Graven E.H., v. Beck T.A., Lelyveld G.P., J. Essent. Oil Res. 8, 229 (1996)
[33] Sundt E., Willhalm B., Chappaz R., Ohloff G., Helv. Chim. Acta 54, 1801 (1971)
[34] German Pharmacopoeia, DAB 2005
[35] Ravid U. et alii, Flavour Fragrance J. 7, 289 (1992)
[36] ISO 9301:2003 Oil of cumin seed (Cuminum cyminum L.), International Organisation of Standardisation, Geneva, Switzerland (2003)
[37] Lawrence B.M., Perfum. Flav. 17 Jul./Aug., 42 (1992)
[38] Lawrence B.M., Perfum. Flav. 29 Oct., 88 (2004)
[39] Lawrence B.M., Perfum. Flav. 17 Nov./Dec., 51 (1992)
[40] Lawrence B.M., Perfum. Flav. 23 Mar./Apr., 56 (1998)
[41] Lawrence B.M., Perfum. Flav. 29 Jan./Febr., 56 (2004)
[42] Lewis Y. S., Nambudiri E. S., Krishna Murthy N, 6th Int. Congr. Essent. Oils (Papers) 1974, 65, ref. CA 84, 29258k
[43] Lawrence B.M., Perfum. Flav. 24 Nov./Dec., 52 (1999)
[44] Lawrence B.M., Perfum. Flav. 28 Sept./Oct., 80 (2003)
[45] Claude-Lafontaine A., Rouillard M., Cassan J., Azaro M., Bull. Soc. Chim. Fr. 1973, 2866 and 1976, 88
[46] Hagedorn M. L., Brown S. M., Flav. Fragr. J. 6, 193 (1991)
[47] Motl O., Amin M., Sedmera P., Phytochemistry 11, 407 (1972)
[48] Wilson C.W., J. Food Sci. 35, 766 (1970)
[49] Fehr D., Pharmazie 29, 349 (1974)
[50] Verghese J., Perfum. Flavor. 15 May / June, 55 (1990)
[51] ISO 3760:2002 Oil of celery (Apium graveolens L.), International Organisation of Standardisation, Geneva, Switzerland (2002)
[52] Lawrence B.M., Perfum. Flav. 30 Jan./Febr., 73 (2005)
[53] Lawrence B.M., Perfum. Flav. 20 Jan./Febr., 52 (1995)
[54] Uhlig J. W., Chang A., Jen J. J., J. Food Sci. 52, 658 (1987)
[55] Rigaud R., Sarris J., Sci. Aliment. 2, 163 (1982) ref. C. A. 97, 150553c
[56] Herissét A., Jolivet J., Lavault M., Plant. med. Phytother. 8, 161 (1976)
[57] Lawrence B.M., Perfum. Flav. 27 Jan./Febr., 46 (2002)
[58] Vernin C. et alii, Parfum. Cosmet. Arômes 93, 9 (1990)
[59] Lawrence B.M., Perfum. Flav. 29 Oct., 89 (2004)

[60] ISO 3524:2003 Oil of cinnamon leaf Sri Lanka Type (Cinnamomon zeylanicum Blume), International Organisation of Standardisation, Geneva, Switzerland (2002)
[61] Lawrence B.M., Perfum. Flav. 22 Sept./Oct., 82 (1997)
[62] Lawrence B.M., Perfum. Flav. 29 Nov./Dec., 52 (2004)
[63] ISO 3216:1997 Oil of cassia, Chinese Type (Cinnamomon aromaticum Nees, Cinnamomum cassia Nees ex Blume), International Organisation of Standardisation, Geneva, Switzerland (1997)
[64] Lawrence B.M., Perfum. Flav. 25 Jul./Aug., 62 (2000)
[65] Lawrence B.M., Perfum. Flav. 29 Jan./Febr., 44 (2004)
[66] ISO 3242:1997 Oil of clove buds [*Syzygium aromaticum* (L.) Merr. et Perry, syn. *Eugenia caryophyllus* (Sprengel) Bullock et S. Harrison], International Organisation of Standardisation, Geneva, Switzerland (1997)
[67] Eiserle R. J., Rogers J. A., J. Am. Oil Chem. Soc. 49, 573 (1972)
[68] Lawrence B.M., Perfum. Flav. 20 Sept./Oct., 102 (1995)
[69] Lawrence B.M., Perfum. Flav. 22 May/June, 57 (1997)
[70] Lawrence B.M., Perfum. Flav. 28 Jul./Aug., 88 (2003)
[71] ISO 3243:1997 Oil of clove stems [Syzygium aromaticum (L.) Merr. et Perry, syn. Eugenia caryophyllus (Sprengel) Bullock et S. Harrison], International Organisation of Standardisation, Geneva, Switzerland (1997)
[72] ISO 3241:1997 Oil of clove leaves [Syzygium aromaticum (L.) Merr. et Perry, syn. Eugenia caryophyllus (Sprengel) Bullock et S. Harrison], International Organisation of Standardisation, Geneva, Switzerland (1997)
[73] Vernin G., Vernin E., Metzger, J. et al. Analysis of clove essential oils. In: *Spices, Herbs and Edible Fungi*, Charalambous G., Ed., Elsevier Science, Amsterdam (1994)
[74] Baba K. et alii, Phytochemistry 30, 4143 (1991)
[75] ISO 3516:1997 Oil of coriander fruits (Coriandrum sativum L.), International Organisation of Standardisation, Geneva, Switzerland (1997)
[76] Lawrence, B.M. Perfum. Flav., 19, Jan./Febr., 42 (1994)
[77] Lawrence, B.M. Perfum. Flav., 29, Nov./Dec., 58 (2004)
[78] Reisch J., Schratz E., S. M. J. S. Quadry, Planta Med. 14, 326 (1966)
[79] Lamparsky D., Klimes I., Perfum. Flavor. 13 Oct./Nov., 17 (1988)
[80] Derbesy M.; Uzio R.; Ann. Fals. Chim., (923), 369-378 (1993)
[81] Protzen K.D., personal communication, Paul Kaders GmbH Hamburg
[82] Misra L. N., Chandra A., Thakur R. S., Phytochemistry 30, 549 (1991)
[83] Lamparsky D., Klimes I., *Progress in Essential Oil Research*, Brunke E.-J., Ed., W. de Gruyter, Berlin (1986)
[84] Chandra A., Misra L. N., Thakur R. S., Tetrahedron Lett. 28, 6377 (1987)
[85] Lawrence B.M. Perfum. Flav. 20 Jan./Febr., 54 (1995)
[86] Lawrence B.M. Perfum. Flav. 29 Mar./Apr., 48 (2004)
[87] Lawrence B.M. Perfum. Flav. 25 Nov./Dec., 42 (2000)
[88] Lawrence B.M. Perfum. Flav, 19 Sept./Oct., 90 (1994)
[89] Gupta R, Dill in Handbook of Herbs and Spices, Vol. 1, Peter K.V., Ed., CRC Press (2001)
[90] Straus D.A., Wolstromer R.J. Paper no. 94, Sixth International Essential Oil Congress, San Francisco (1974)
[91] Blank I., Grosch W., J. Food Sci. 56, 63 (1991)
[92] Nitz S., Spraul M. H., Drawert F., Chem. Mikrobiol. technol. Lebensmitt. 13, 183 (1991)
[93] Baslas R. K., Gupta R., Flavour Ind. 2, 363 (1971)
[94] Lawrence B.M. Perfum. Flav. 26 Mar./Apr., 26 (2001)
[95] Lawrence B.M. Perfum. Flav. 22 Jan./Febr., 49 (1997)
[96] Lawrence B.M. Perfum. Flav. 19 Nov../Dec., 61 (1994)
[97] ISO 770:2002 Crude or rectified oils of Eucalyptus globulus (*Eucalyptus globulus* Labill.), International Organisation of Standardisation, Geneva, Switzerland (2002)
[98] European Pharmacopoeia 5.0 'Foeniculi amari fructus aetheroleum' (2005)
[99] Lawrence B.M. Perfum. Flav. 23 Mar./Apr., 52 (1998)
[100] Lawrence B.M. Perfum. Flav. 19 Jan./Febr., 31 (1994)
[101] Ravid U., Putievsky E., Katzir I., Ikan R., Flavour Fragrance J. 7, 169 (1992)

[102] Gupta R., Jain D., Thakur R., Phytochemistry 23, 2605 (1984)
[103] Elujoba A. A., Hardman R., Planta med., 113 (1985)
[104] Girardon P., Bessiere J. M., Baccou J. C., Sauvaire Y., Planta Med., 533 (1985)
[105] Girardon P., Sauvaire Y., Baccou J. C., Bessiere J. M., Lebensm.-Wiss.-Technol. 19, 44 (1986)
[106] Lawrence B. M., Hogg J. W., Terhune S. J., Perfum. Essent. Oil Res. 60, 89 (1969)
[107] Scheffer J. J. C., Gani A., Baerheim-Svendsen A., Sci. Pharm. 49, 337 (1981)
[108] De Poorter H. L., Omar M. N., Coolsaet B.A., Schamp N. M., Phytochemistry 24, 93 (1985)
[109] Lawrence B.M. Perfum. Flav. 30 Jan./Febr., 66 (2005)
[110] Müller BH, Beitrag zur Charakterisierung und Analytik von Alliin und Allicin in Knoblauch und seinen Zubereitungen, Dissertation Universität Erlangen Nürnberg (1991)
[111] Vernin G., Metzger J., GC-MS (EI, PCI, NCI, SIM) SPECMA Bank Analysis of Volatile Sulphur Compounds in Garlic Essential Oils. In *Essential Oils and Waxes*, Linskens HF, Jackson JF, Eds., Springer, Berlin/Heidelberg/New York, pp. 99-130 (1991)
[112] Pandey U.B., Garlic. In *Handbook of Herbs and Spices*, Vol. 1, Peter K.V., Ed., CRC Press (2001)
[113] Glasl H., Wagner H., Tabacchi R., Int. Congr. Med. Plant. Res. München (Papers) 1976
[114] Major food and agricultural commodities and producers. In www.fao.org Food and Agriculture Organisation of the Unites Nations (2004)
[115] Bednarczyk A. A., Kramer A., Chem. Senses Flavor. 1, 377 (1975) and J. Agric. Food Chem. 23, 499 (1975)
[116] Lawrence B.M. Perfum. Flav. 20 Mar./Apr., 54 (1995)
[117] Lawrence B.M. Perfum. Flav. 22 Sept./Oct., 74 (1997)
[118] Lawrence B.M. Perfum. Flav. 30 May, 70 (2005)
[119] Gilbert J., Nursten H. E., J. Sci. Food Agric. 23, 527 (1972)
[120] Grob jr. K., Matile P., Phytochemistry 19, 1789 (1980)
[121] ISO 9841:1991 Oil of Hyssop (*Hyssopus officinalis* Linnaeus), International Organisation of Standardisation, Geneva, Switzerland (1991)
[122] Lawrence B.M. Perfum. Flav. 19 Sept./Oct., 86 (1994)
[123] Lawrence B.M. Perfum. Flav. 20 Sept./Oct., 96 (1995)
[124] Lawrence B.M. Perfum. Flav. 24 May/June, 58 (1999)
[125] European Pharmacopoeia 5.0 'Juniperi aetheroleum' (2005)
[126] Karlsen J., Baerheim Svendsen A., Medd. Nor. Farm. Selsk. 27, 165 (1965) and 29, 13 (1967)
[127] Lamparsky D., Klimes I., Parfüm. Kosmet. 66, 553 and 558 (1985)
[128] Schilcher H., Emmrich D., Koehler C., Pharm. Ztg., Wiss.; 138 (3/4), 85 (1993)
[129] Lawrence, B.M. Perfum. Flav., 21, Jan./Febr., 40 (1996)
[130] Adams R.P., *Junipers of the world: The genus Juniperus*, Trafford Publishing (2004)
[131] Tada H., Takeda K., Chem. Pharm. Bull. 24, 667 (1976)
[132] El-Feraly F. S., Benigni D. A., J. Prod. Nat. (Lloydia) 43, 527 (1980)
[133] Pech B., Bruneton J., J. Nat. Prod. (Lloydia) 45, 560 (1982)
[134] Pino, J, Borges P. and E. Roncal, Nahrung, 37, 592 (1993)
[135] Lawrence B.M. Perfum. Flav. 24 Sept./Oct., 60 (1999)
[136] Lawrence B.M. Perfum. Flav. 18 May/June, 65 (1993)
[137] Frattini C., Bicchi C., Barettini C., Nano G. M., J. Agric. Food Chem. 25, 1238 (1977)
[138] Toulemonde B., Mazza M., Bricout J., Indust. Alim. Agric. 1179 (1977)
[139] Speicher-Brinker A., Ph.D. Thesis, FU Berlin (1987)
[140] ISO 11023:1999 Liquorice extract (*Glycyrrhiza glabra* L.), International Organisation of Standardisation, Geneva, Switzerland (1999)
[141] ISO 3214: 2000 Oil of Litsea cubeba (*Litsea cubeba* Pers.), International Organisation of Standardisation, Geneva, Switzerland (2000)
[142] Cheng M.C., Cheng Y.S., J. Chinese Chem. Soc. (Taipei) 30, 59 (1983)
[143] Lawrence B.M. Perfum. Flav. 31 Jan./Febr., 55 (2006)
[144] Lawrence B.M. Perfum. Flav. 21 Sept./Oct., 62 (1996)
[145] Naves Y. R., Helv. Chim. Acta 26, 1281 (1943)
[146] Bylaite E., Legger A., Roozen J.P., Venskutonis P.R., Spec. Publ. Royal Soc. Chem., 66 (1998)
[147] Stahl-Biskup E., Wichtmann E., Flav. Fragr. J., 6, 249 (1991)

[148]   Venskutonis P.R. In: *Flavours, Fragrances and Essential Oils*, Proceedings of the 13th International Congress of Flavours, Fragrances and Essential Oils, Istanbul, Turkey. Oct. 1995. Baser K.H.C., Ed., Vol. 2, 108, Anadolu Univ. Press, Eskishir, Turkey (1995)
[149]   ISO 11019:1998 Oil of roots of lovage (*Levisticum officinale* Koch), International Organisation of Standardisation, Geneva, Switzerland (1998)
[150]   Lawrence B.M. Perfum. Flav. 24 Mar./Apr., 35 (1999)
[151]   Lawrence B.M. Perfum. Flav. 15 Sept./Oct., 63 (1990)
[152]   Schreck D., Ph.D. Thesis, Marburg (1985)
[153]   Dolci M, Tita S., Riv. Ital. Essenze Profumi 62, 131 (1980)
[154]   Voirin B.et alii, Phytochemistry, 23, 2973 (1984)
[155]   Granger R., Passet J., Lamy J., Riv. Ital. Essenze Profumi 57, 446 (1975)
[156]   Dayal B., Purohit R., Flavour Ind. 2, 477 (1971)
[157]   Baser K.H.C., Özek T., Tümen G., Sezik E., J. Essent. Oil Res., 5, 577 and 619 (1993)
[158]   Surbug H., Güntert M., Harder H. Volatile compounds from flowers. Analytikcal and olfactory aspects. In: *Bioactive Volatile Compounds from Plants*. Teranishi R., Buttery G., Sugisawa H., Eds., 168, ACS Symp. Series 525, Amer. Chem. Soc. Washington, DC (1993)
[159]   Sarer E., Scheffer J. J. C., Baerheim Svendsen A., Planta Med. 46, 236 (1983)
[160]   Lawrence B.M. Perfum. Flav. 25 Sept./Oct., 61 (2000)
[161]   Lawrence B.M. Perfum. Flav. 22 Jan./Febr., 51 (1997)
[162]   Lawrence B.M. Perfum. Flav. 19 July./Aug., 39 (1994)
[163]   Trumpowska M., Olszewski Z., Acta Pol. Pharm. 25, 319 (1968)
[164]   Naef-Müller R. et al., Helv. Chim. Acta 65, 1424 (1981)
[165]   Benjilali B., Richard H., Riv. Ital. Essenze Profumi 62, 69 (1980)
[166]   Nano G. M., Bicchi C., Frattini C., Gallino M., Planta med. 30, 211 (1976)
[167]   Juell S. M.-K., Hansen R., Jork H., Arch. Pharmaz. (Weinheim) 309, 458 (1976)
[168]   Jork H., Juell S. M.-K., ibid. 312, 540 (1979)
[169]   Jork H., Nachtrab M., ibid. 312, 435 (1979)
[170]   Benjilali B., Sarris J., Richard H., Sci. Aliments 2, 515 (1982)
[171]   Benjilali B., Richard H., Liddle P., Suppl.2 Quaderno Agricolo Saint Vincent (Ital.) April 1984
[172]   Lawrence B.M. Perfum. Flav. 19 Sept./Oct., 61 (1994)
[173]   Purushotaman K. K., Sarada A., Indian J. Chem. 19b, 236 (1980), ref. C. A. 93, 3941
[174]   Woo W.S., Wagner H., Lotter H., Phytochemistry 26, 1542 (1987)
[175]   Wilmers K., Gröbel W., Dtsch. Lebensm. Rdsch. 86, 344 (1990)
[176]   ISO 3215:1998 Oil of nutmeg (*Myristicum fragrans* Houtt.), International Organisation of Standardisation, Geneva, Switzerland 1998)
[177]   Lawrence B.M. Perfum. Flav. 30 July/Aug., 59 (2005)
[178]   Lawrence B.M. Perfum. Flav. 25 Sept./Oct., 66 (2000)
[179]   Lawrence B.M. Perfum. Flav. 19 Sept./Oct., 86 (1994)
[180]   Lawrence B.M. Perfum. Flav. 15 Nov./Dec., 62 (1990)
[181]   Brodnitz M.H., Pascale J.V., J. Agric. Food Chem. 19, 269 (1971)
[182]   Brodnitz M.H, Pollock C.L., Vallon P.P., J. Agric. Food Chem. 17, 760 (1969)
[183]   Boelens M., de Valois P.J., Wobben H.J., ven der Gen A., J. Agric. Food Chem. 19, 984 (1971)
[184]   Galetto W.G., Hoffman P.G., J. Agric. Food Chem. 24, 852 (1976)
[185]   Kallio H., Salorinne L., J. Agric. Food Chem. 38, 1560-1564 (1990)
[186]   Lawence, B.M., Perfum. Flav. 9 Oct./Nov., 41 (1984)
[187]   ISO 14717:1999 Oil of origanum, Spanish type [*Thymbra capitata* (L.) Cav, International Organisation of Standardisation, Geneva, Switzerland (1999)
[188]   Lawrence B.M. Perfum. Flav. 18 Jan./Febr., 53 (1993)
[189]   Lawrence B.M. Perfum. Flav. 13 Aug./Sept., 75 (1988)
[190]   Stahl W. H., Skarzynski J. N., Voelker W. A., J. Assoc. Off. Anal. Chem. 52, 1184 (1969)
[191]   Lawrence B.M., Perfum. Flav. 21, May/June, 65 (1996)
[192]   Innocenti G., Dalláqua F., Caporale B., Planta Med. 29. 165 (1976)
[193]   Chaudary S.K. et al., Planta Med. 1986, 462
[194]   Garnero J., Benezet L., Peyron L., Chrétien-Bessière Y., Bull. Soc. Chim. F. 12, 4679 (1967)
[195]   Masanetz C., Grosch W., Flav. Fragr. J. 13, 115 (1998)
[196]   Freeman G. G., Whenham R. J., Self R., Eagles J., J. Sci. Food Agric. 26, 465 (1975)

[197] Kasting R., Andersson J., v. Sydow E., Phytochemistry 11, 2277 (1972)
[198] Simon J.E., Quinn J., J. Agric. Food Chem. 36, 467 (1988)
[199] Shaat N.A., Griffin P., Dedeian S., Paloympis L. The Chemical Composition of Egyptian Parsley Seed, Absolute and Herb Oil. In: *Flavors and Fragrances: A World Perspective,* Lawrence B.M., Mookherjee B.D., Willis B.J., Eds., Elsevier Science, Amsterdam, p 715 (1988)
[200] Unpublished data, these authors (2005)
[201] ISO 3527:2000 Oil of parsley fruits [*Petroselinum sativum* Hoffm.], International Organisation of Standardisation, Geneva, Switzerland (2000)
[202] Lawrence B.M., Perfum. Flav. 30 July/Aug., 57 (2005)
[203] Lawrence B.M., Perfum. Flav. 26 March/April, 34 (2001)
[204] Lawrence B.M., Perfum. Flav. 21 July/Aug., 66 (1996)
[205] Lawrence B.M., Perfum. Flav. 19, March/April, 72 (1994)
[206] Traxler J. T., J. Agric. Food Chem. 19, 1135 (1971)
[207] Grewe R., Freist W., Neumann H., Kersten S., Chem. Ber. 103, 3752 (1970)
[208] Vössgen B., Herrmann K., Z. Lebensm. Unters. Forsch. 170, 204 (1980)
[209] Wrolstadt R. E., Jennings W. G., J. Food Sci. 30, 274 (1965)
[210] Lewis Y. S., Nambudiri E. S., Krishnamurthy N., Perfum. Essenz. Oil Rec. 60, 2659 (1969)
[211] Pino J., Rodriguez-Feo G., Borges P., Rosado A., Nahrung 34, 555 (1990)
[212] Lawrence B.M., Perfum. Flav. 22 March/April, 61 (1997)
[213] Lawrence B.M., Perfum. Flav. 20 March/April, 49 (1995)
[214] Lawrence B.M., Perfum. Flav. 17 May/June, 72 (1992)
[215] Patington J.S., Perfum. Flav. 8 Aug./Sept., 29 (1983)
[216] Verghese J., Perfum. Flav. 14 Nov./Dec., 33 (1989)
[217] Borges P., Pino J., Nahrung 37, 127 (1993)
[218] Rathnawathie M., Buckle K.A., J. Chromatog. 264, 316 (1983)
[219] Weaver K., Neale M.W., Laneville A., J. Assoc. Off. Anal. Chem. 47, 1249 (1983)
[220] European Pharmacopoeia 5.1 'Menthae piperitae aetheroleum' (2005)
[221] Lawrence B.M. Perfum. Flav. 22 July/Aug, 57 (1997)
[222] Lawrence B.M. Perfum. Flav. 18 July/Aug, 59 (1993)
[223] Lawrence B.M. Perfum. Flav. 13 Oct./Nov., 66 (1988)
[224] Lawrence B.M. Chi-Kuen Shu, Perfum Flavor. 14, 21 Nov./Dec. (1989)
[225] Werkhoff P., Hopp R., Perfüm. Kosmet. 66, 731 (1985)
[226] Kreis P., Mosandl A., Schmarr H.G., Dtsch. Apoth. Ztg. 130, 2579 (1990)
[227] Derbesy M., Boyer D., Cozon V., Ann. Fals. Expert. Chim. 84, 205 (1991)
[228] ISO 3033:2005 Oil of spearmint part 1-4, International Organisation of Standardisation, Geneva, Switzerland (2005)
[229] Lawrence, B. M., Perfum. Flavor. 18, 46 Mar./Apr. (1993) and 18, 61 May/June (1993)
[230] Lawrence, B. M., Shu C.-K., Essential Oils as Components of Mixtures. Analysis and Differentiation. In: Flavour Measurement, Ho C.-T., Manley C.H., Eds., 267, Marcel Dekker (1993)
[231] Veek M.E., Russel G.F., J. Food Sci. 38,1028 (1973)
[232] Lawrence B. M., Perfum. Flavor. ,Jan./Febr., 15, 63 (1990)
[233] Lawrence B. M., Perfum. Flavor. , March/Apr., 24, 40 (1999)
[234] Andrews J., In: *Peppers: The Domesticated Capsicums,* Austin, Texas, University of Texas Press (1995)
[235] Buttery R. G., Seifert R. M., Lundin R. E., Guadagni D. G., Ling L. C., Chem. Ind. London 490 (1969) and J. Agric. Food Chem. 17, 1322 (1969)
[236] Luming P. A., De Rijk T., Wichers H. J., Roozen J. P., J. Agric. Food Chem. 42, 977 (1994)
[237] Van Ruth S. M., Roozen J. P., Food Chem. 51, 165 (1994)
[238] Tschesche R., Gutwinski H., Chem. Ber. 108, 265 (1975)
[239] Scoville W., J. Am. Pharm. Assoc. 1, 453 (1912)
[240] Jurenitsch J., Kampelmüller I., J. Chromatogr. 193, 101 (1980)
[241] Cooper T.H., Guzinski J.A., Fisher C., J. Agric. Food Chem. 39, 2253 (1991)
[242] Aczel A., Fresenius Z., Anal. Chem. 343, 154 (1992)
[243] Vinkler M., Kiszel-Richter M., Acta Aliment.Acad. Sci. Hung. 1, 41 (1972)
[244] Buckle K. A., Rahman F. L. M., J. Chromatogr. 170, 385 (1979)
[245] Ma M., Luo X., Chen B., Su Sh., Yao S., J. Chromatogr., in press (Dec. 2005)

[246] Tateo, F., Bononi M., J. Agric. Food Chem. 52, 655 (2004)
[247] Haymon L. W., Aurand L. W., J. Agric. Food Chem. 19, 1131 (1971)
[248] Valenzuela A.B., Nieto S.K., Grasas y Acetias, 47, 186 (1996)
[249] Brieskorn C.H., Dömling H.J., Z. Lebensm. Unters. Forsch. 141, 10 (1969)
[250] Nakatani N., Inatani R., Agric. Biol. Chem. 45, 2385 (1981) and ibid. 46, 1661 (1982), 47, 353 (1983), 48, 2081 (1984)
[251] Brieskorn C.H., Zweyrohn G., Pharmazie 25, 488 (1970)
[252] Ganeva Y. et al., Planta med. 59, 276 (1993)
[253] Brieskorn C.H., Michel H., Biechele W., Dtsch. Lebensm. Rdsch. 69, 245 (1973)
[254] Lamparsky D., Schenk H. P., *Ätherische Öle*, Kubeczka K.-H., Ed., G. Thieme Verlag, Stuttgart (1982)
[255] Lawrence B. M., Perfum. Flavor. , Sept./Oct., 22, 71 (1997)
[256] Lawrence B. M., Perfum. Flavor. , Jan./Febr. 20, 40 (1995)
[257] ISO 1342:2000 Oil of rosemary (*Rosmarinus officinalis* L.), International Organisation of Standardisation, Geneva, Switzerland (2000)
[258] Zarghami N. S., Heinz D. E., Lebensm. Wiss. Technol. 4, 43 (1971) and Phytochemistry 10, 2755 (1971)
[259] Roedel W., Petrzicka M., J. High Resol. Chromatogr. 14, 771 (1991)
[260] Pfänder H., Wittwer F., Helv. Chim. Acta 58, 1608 and 2233 (1975)
[261] Velasco-Negueruela A., Saffron. In: *Handbook of Herbs and Spices*, Vol. 1, Peter K.V., Ed., Woodhead Publishing, Abingdon, p. 277 (2001)
[262] Solinas M., Cichelli A., Ind. Aliment. 27, 7 Jul. / Aug. (1988)
[263] Oberdieck R., Dtsch. Lebens. Rdsch. 87, 246 (1991)
[264] Marini D., Balestrieri F., Ind. Aliment. (Pinerolo) 31, 123 (1993)
[265] ISO 3632-1 and 3632-2:1993 Saffron (*Crocus sativus* L.), International Organisation of Standardisation, Geneva, Switzerland (1993)
[266] Brieskorn C. H., Biechele W., Arch. Pharmaz. (Weinheim) 304, 557 (1971)
[267] Brieskorn C. H., Kapadia Z., Planta Med. 35, 376 (1979)
[268] ISO 9909:1997 Oil of Dalmatian sage (*Salvia officinalis* L.), International Organisation of Standardisation, Geneva, Switzerland (1997)
[269] Karl C., Pedersen P. A., Müller G., Planta Med. 44, 188 (1982)
[270] Lawrence B. M., Perfum. Flavor. 26 May/June, 66 (2001)
[271] Lawrence B. M., *Essential oils 1995-2000*, Allured Publishing (2004)
[272] ISO 3526:2005 Oil of sage, Spanish (*Salvia lavandulifolia* Vahl), International Organisation of Standardisation, Geneva, Switzerland (2005)
[273] Hérisset A., Jolivet J., Zoll A., Chaumont J. P., Plant. Med. Phytother. 7, 121 (1973) and 8, 287 (1974)
[274] Lawrence B. M., Perfum. Flavor. 29 Jan./Febr., 48 (2004)
[275] Lawrence B. M., Perfum. Flavor. 28. Nov./Dec., 48 (2003)
[276] Lawrence B. M., Perfum. Flavor. 27. Jan./Febr., 42 (2002)
[277] Lawrence B. M., Perfum. Flavor. 20 July/Aug., 38 (1995)
[278] Lawrence B. M., Perfum. Flavor. 15 March/Apr., 75 (1990)
[279] Lawrence B. M., Perfum. Flavor. 13 Febr./March, 49 (1988)
[280] Vila R., Flavonoids and further polyphenols in the genus Thymus. In: The Genus Thymus, Stahl-Biskup E., Saez F., Eds., Taylor & Francis, London, p 144 (2002)
[281] Granger R., Passet J., Phytochemistry 12, 1683 (1973)
[282] Stahl-Biskup E., Venskutonis R.P., Thyme. In: *Handbook of Herbs and Spices*, Vol. 2. Peter K.V., Ed., Woodhead Publishing, Abingdon, p. 299 (2001)
[283] ISO 14715:1999 Oil of thyme containing thymol, Spanish type [*Thymus zygis* (Loefl.) L.], International Organisation of Standardisation, Geneva, Switzerland (1999)
[284] Protzen, M.D., unpublished results.
[285] Stahl-Biskup E., Essential oil chemistry of the genus Thmus -a global view. In: *Thyme - The Genus Thymus*, Stahl-Biskup E., Saez F., Eds., Taylor & Francis, London, p. 75 (2002)
[286] Lawrence B. M., Perfum. Flavor. 29 May, 50 (2004)
[287] Lawrence B. M., Perfum Flavor. 28 March/Apr., 52 (2003)
[288] Verghese J., Indian Spices 36(4), 19 (1999)

[289]  König W.A., K-H. Kubeczka et al., J. High Resol. Chrmatogr. 17, 315 (1994)
[290]  Lawrence B.M., Perfum Flavor. 25 Nov./Dec., 32 (2000)
[291]  Klimes I., Lamparsky D., Int. Flavour Food Additives 7, 272 (1976)
[292]  Ramaroson-Raonizafinimanana B., Gaydou E.M. and Bombarda I., J. Agric. Food Chem. 45, 1922 (1997) and 47, 3202 (1999)
[293]  Purseglove J.W., Brown E.G., Green C.L., Robbins S.R.J., Spices, Vol. 2, Longman, New York (1982)
[294]  Adedeji J., Hartman Th. G., Chi-Tang Ho, Perfum. Flavor. 18, 25 Mar. / Apr. (1993)
[295]  Ranadive A.S., J. Agric. Food Chem. 40, 1922 (1992)
[296]  Bricout J., Fontes J.-C., Merlivat L., J. Assoc. Off. Anal. Chem. 57, 713 (1974)
[297]  Hoffman P. G., Salb M., J. Agric. Food Chem. 27, 352 (1979)
[298]  Kaunzinger A., Juchelka D., Mosandl A., J. Agric. Food Chem. 45, 1752 (1997)
[299]  Remaud G.S., Martin Y., Martin G.G., Martin G.J., J. Agric. Food Chem. 45, 859 (1997)
[300]  Rücker G., Manns D., Wilbert S., Phytochemistry 31, 340 (1992)
[301]  Vostrowsky O. et alii, Z. Naturforsch. 36 c, 369 (1981)
[302]  Chialva F., Liddle P. A. P., Doglia G., Z. Lebensm. Unters. Forsch. 176, 363 (1983)
[303]  Lawrence B.M., Perfum. Flavor. 29 May, 63 (2004)

### 3.2.2.4 Flavouring Preparations Based on Biotechnology

*Joachim Tretzel*
*Stefan Marx*

Complex flavours made by biotechnical manufacturing processes are borderline products between natural and nature-identical flavours. The declaration of complex biotechnical flavours is 'Flavour Preparations'. For the manufacturing of those biotechnical steps (enzymatic and microbiological processes) are explicitly approved. There is still some uncertainty, however, with the declaration regarding the origin of the starting materials for the biotechnical reaction and the difficult analytical discernability between physically extracted and biotechnically produced flavours from plant and animal raw materials.

Regardless of this problem biotechnically produced flavour extracts are very attractive. Under properly defined conditions biotechnical reactions allow a significant increase of yield or strength of the flavours compared to purely natural flavours. It is also practised to run biotechnical reactions in such a way that the purity of the produced flavours is increased and undesired side products are suppressed. In comparison to the complicated chemical syntheses of pure isomers of flavours chemicals, the biotechnically manufactured substances are generally homogeneously composed 'by nature'.

These advantages have a positive influence on the economy of biotechnical manufacturing processes. In many cases expensive starting materials can be replaced by cheaper and simpler substrates of the biochemical reaction which results in a favourable cost price for the flavour extract. Compared to commodities which meanwhile include numerous biotechnical products (e.g. technical enzymes), the relatively high prices obtainable for flavour chemicals justify the relatively complicated techniques necessary for biotechnical processes. Production of flavour chemicals is, therefore, an interesting further application of biotechnology in line with e. g. the generation of pharmaceutical products.

Biotechnological manufacturing processes normally are operated under very moderate reaction conditions regarding temperatures and pressures. In addition, they are run in physiological environments without aggressive reagents (acids and bases, solvents etc.). Biotechnically produced flavour extracts, therefore, follow the trend to low intensity processed foods and food ingredients. The taste of a biotechnically manufactured flavour extract very often is more authentical compared to extracts obtained by classical processes like distillation and extraction. The latter are normally operated at elevated temperatures and therefore may exert adverse effects on the fragile flavour components.

With most biotechnologically produced flavour extracts the taste quality of natural fermentation processes can be reached and even surpassed. Normally the taste intensity is significantly higher compared to the usual fermentation process in the raw food material. Even completely new taste types may be accessible biotechnologically . The time necessary for the development of the desired taste in most cases may be significantly shorter. As a consequence the room/time-yield of the process is increased and the costs are decreased, respectively.

In chapter 3.2.1.1.2 biotechnical syntheses leading to pure flavour chemicals are already demonstrated. The metabolic reaction mechanisms for most of these cases are well-known and properly defined. The detailed knowledge available about their biochemistry today even allows the specific modification especially with genetic engineering methodology. The generation of complex mixtures of flavours by biotechnical reactions, however, is not yet very well understood. The application of such processes on an empirical basis, however, is practised successfully and will be described in the following.

The tastes of all natural foods are imparted by complex flavour mixtures consisting of a multitude of different single substances. The total sensoric impression is, however, imparted already by a few significant components, so-called character impact compounds. The biotechnical reaction therefore only has to yield sufficient amounts of these compounds in the correct proportion.

Flavour preparations may be produced by all processes described in chapter 2.2:

- Fermentation with intact micro-organisms
- Enzymatic syntheses or transformation or liberation of flavour imparting substances by isolated enzymes and enzyme complexes
- Cell culture techniques especially with plant cell or tissue cultures.

It is not within the scope of this book to treat the in-situ generation of flavour in finished foods by bacterial or mould fermentation. These methods belong to the traditional biotechnology and are applied on a mostly empirical basis (beer, yoghurt, cheese, bread etc.).

In this chapter we want to focus on 'novel' biotechnical processes. Corresponding to their importance for the flavour industry, reactions with isolated enzymes are described first. Some examples for the flavour generation by fermentation with microorganisms and cell cultures will follow. In these cases results are only available from laboratory and small-scale industrial productions which have not yet been tried on an industrial scale.

### 3.2.2.4.1 Flavouring Preparations by Enzymatic Reactions

The majority of the industrially applied biotechnical processes for the generation of flavour preparations relies on reactions with isolated enzymes and enzyme complexes. The basic principle of enzymatic reactions is mostly hydrolytic decomposition or transformation of the most important biological substance categories like proteins, carbohydrates, lipids and acids.

A large number of standardized enzyme preparations is available today. Table 3.28 gives an overview of these enzymes, their predominate activities, the sources and the main applications.

As a consequence of the high enzymatic activities and the correspondingly low amounts of enzymes necessary to react a certain amount of raw material (significantly less than 1%) the application of enzymes in general is very cost efficient even if the enzyme is lost after completion of the reaction. Enzyme costs are about 0.1-3% of the product value. The use of expensive enzymes is facilitated by the possibility to immobilize the enzyme and to reuse it after the reaction (compare chapter 2.2: Biotechnical Processes).

*Table 3.28:* Commercially available enzyme preparations for production of flavouring materials

| Enzyme | Activity | Source | Application |
| --- | --- | --- | --- |
| Papain | Neutral Proteinase | *Carica papaya* | Meat and Plant Protein Hydrolysis Yields Amino Acids and Peptides: Precursors for Maillard Reactions |
| Bromelain | Neutral Proteinase | *Anana sativa* | |
| Pancreatin Peptidase | Neutral Proteinase Exopeptidase | *Animal Pancreas Aspergillus ssp.* | Bitter-free Protein Hydrolysates |
| Rennet | Acidic Proteinase | *Mucor ssp. and Genetically Engineered Bacteria Calf Stomach* | Coagulation of Milk Proteins in Cheese Production |
| Lipase | Cleavage of Triglycerides | *Aspergillus ssp. Penicillin roquefortii* | Production of Enzyme Modified Cheese (EMC) Chedar flavour |
| Glucanase Lysozyme | Mixed Glucolytic and Proteolytic Activities | *Arthrobacter ssp.* | Yeast Cell Wall Lysis-supporting Autolysis for Production of Yeast Autolysates |
| Pectinase Pectinesterase Pectinlyase Polygalacturonase | Pectolytic Enzymes: Cleavage of Plant (α-1,4) Polysaccharides | *Aspergillus Niger* | Disintegration of Plant Tissue. Reduction of Macerate Viscosity. Improved Extractability of Juice-bound Flavours |
| Cellulase Xylanase | Cleavage of Cellulose (ß-1,4)- and Xylane (ß-1,3)-Polysaccharides | *Aspergillus Niger Aspergillus Oryzae Himicola sp.* | Disintegration of Plant Cell Walls. Improved Extractability of Flavours and Flavour Precursors from Plant Tissue |
| Pectinase | Mixed Pectinase and ß-Glycosidase | *Aspergillus Niger* | Liberation of Glycosidically Bound Flavour Substances from Plant Tissue |
| Lipoxgenase | Oxidation of Unsaturated Fatty Acids | *Glycine max* | Formation of Green Notes |

Enzymatic reactions generally exhibit a high catalytic activity and a high selectivity which even allows to selectively react racemic mixtures of substrates into product consisting of pure isomers. In this way it is possible e. g. to hydrolyse and thereby remove R- or S-isomeric forms of optical active esters by lipases. The non-hydrolyzed other isomer remains unchanged in high purity in the reaction mixture *[1a]*. Enzymes operate under very mild conditions and therefore have a very favourable energy balance. All these properties assign enzymes to the preferred catalysts in biotechnical reactions.

Enzymes are catalysts, i.e. they will not get worn out. It is therefore possible to operate with very low amounts of the enzyme. All these facts point to a very cost-effective possibility for the manufacturing of flavour extracts.

The most important drawbacks of enzymatic reactions are the relative sensitivity of these biocatalysts to elevated temperatures and aggressive media (acid or base). Besides that, most enzymes are stable and active only in aqueous solution. Recent developments show, however, also a certain potential of enzymes in organic non-aqueous solvents and therefore a considerable broadening of the synthetic power of enzyme reactions *[1b]*.

There is a millennium of history of enzyme application in the improvement of food. Cheese production, the production of seasoning sauces, fermentation of raw sausages, tea and tabacco fermentation are only a few examples. In all these cases enzymes have been applied as a part of the living enzyme containing micro-organisms. Today many of those enzymes are characterized and available as pure substances in high concentration. They facilitate a precise operation and safe control of the manufacturing process.

### 3.2.2.4.2 Manufacturing of Natural Meat Flavours

An interesting application of proteolytic enzymes (proteinases) leads to natural meat flavours which impart their authentic taste without additional seasonings or synthetical taste enhancers to a large variety of foods. Figure 3.15 depicts the principle of manufacturing such flavours from natural fresh or deep-frozen meat. Meat of different species (e.g. beef, pork, chicken, fish, seafood) may be used as starting material. The raw material will determine the basic taste of the flavour extract. By proper selection of further processing steps the taste may be differentiated into cooked, boiled, grilled and smoked notes *[2]*.

The enzymatic reaction of the proteins (fruit derived, mould or bacterial proteases) with the meat proteins in the first step leads to exo- and/or endo-cleavage of the peptide bonds. A mixture of free amino acids, low-molecular oligopeptides and high-molecular protein fragments is generated. Unchanged proteins, fats, salts and sugars of the meat remain in the reaction mixture. In the second step the mixture is heated. As a consequence amino acids ($NH_2$ functional groups) and reducing sugars (aldehyde functional groups) react forming Maillard products which enhance the natural basic taste significantly and also intensify the colour. The natural basic tastes which are predominantly coupled to the volatiles of the fat phase remain unchanged and impart the genuine taste within the flavoured end products *[3]*.

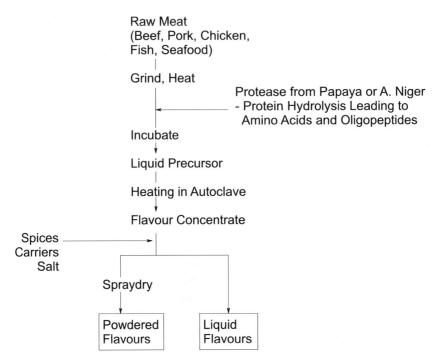

*Fig. 3.15: Enzymatically mediated manufacturing of meat flavour*

### 3.2.2.4.3 Seasonings – HVP

The production of plant derived seasonings from acid hydrolyzed plant proteins is described elsewhere. Recently the acid hydrolysis of plant proteins (wheat gluten, soy protein, etc.) has been questioned especially in the US because of the potential formation of toxic side products. These side products, especially dichloropropanols are formed from plant fats contained in the raw material by the action of concentrated hydrochloric acid which is used in autoclaves at elevated temperatures for hydrolysis *[4]*. The cancerogenic effect of dichloropropanols has been proven in animal experiments.

These seasonings are very important because of their explicit cost-effectiveness and their remarkable taste effects in many foods together with an additional taste enhancement by the liberated glutamic acid and glutamate.

Hydrolysis now has been completed successfully by proteolytic enzymes from bacteria and moulds. In these reactions there are no problematic side products because they operate under mild conditions as mentioned above. The further processing after enzymatic hydrolysis is similar to the classically hydrolyzed products by heating and consecutive Maillard reaction. In the heating step the enzymes are denatured and destroyed *[5]*.

### 3.2.2.4.4 Enzymatically Modified Cheese (EMC)

Lipases (fat-cleaving enzymes) derived from mucor type moulds play an important role in the manufacturing of cheese flavour concentrates (Enzyme Modified Cheese,

EMC). Such 'natural flavour concentrates' as they have to be declared according to the new flavour guideline [6], are used in dosages of 0.5-2.5% e.g. for cheese spreads, cheese pastry, cheese sauces and cheese-containing finished meals [7]. The use of EMC cheese flavours reduces the added amount of cheese by up to 90% and therefore effects a significant price reduction of the end products. In addition, the reduction of cholesterol and calories in the finished menu is important for nutritional considerations.

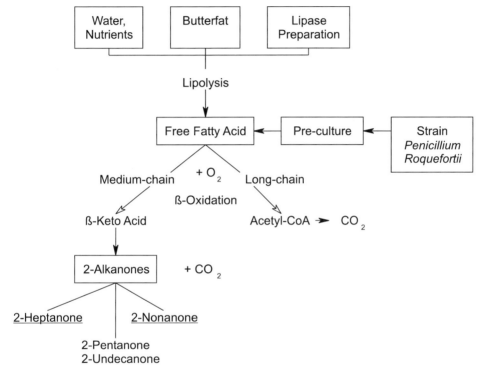

*Fig. 3.16:* Formation of methyl ketones from butterfat (adapted from J.E. Kinsella and D.H. Hwang, [8])

The formation of typical cheese flavours during natural ripening processes are not fully understood yet. The majority of reactions last for an extended time period (months) and comprise oxidative, inter- and intramolecular, enzymatic or microbial (cf. blue cheese) reactions. Substrates are partially very reactive milk-based ingredients which are mainly transformed to volatile flavour intensive compounds like esters, methylketones, aldehydes, lactones and sulphur containing products. The effect of enzymes on the flavour enhancement is also not fully understood. By variation of lipase dosage, reaction time and reaction temperature the production of different flavour notes from milk and butterfat is possible.

Figure 3.16 shows that it is possible to employ an additional fermentation with typical starter cultures following the lipolytic reaction and in this example manufacture a roquefort flavour [8]. The additional application of lipaseesterases for the formation of typical ageing notes in cheese flavour is possible. In addition, amino acid catabolism by lactic acid bacteria leads to the formation of important aroma compounds [9, 10].

### 3.2.2.4.5 Enzymatic Liberation of Flavours from Plant Tissue

Formation of intense flavours during the ripening period of fruit is traditionally well known. In many cases the characteristic flavours are only produced during the autolytic decomposition or during mechanical destruction of plant tissue (apple juice flavour from mash, woodruff in withered clove). The cause of this post-mortem flavour liberation is always a hydrolytic process mediated by plant-borne enzymes.

These enzymes attack preferably the frame substances of a single plant cell or the plant association in the plant tissue. The framing substances consist of polysaccharides, especially cellulose (ß-1,4-glycoside) and in-built pectines (esters of α-1,4-polygalacturonic acid) partially also lignin (complex polymers of coniferylic alcohol) which, however, cannot be cleaved by carbohydrases.

In chapter 3.2.2.1 the manufacturing of juice flavours from fruit is described in more detail.

Fruit flavours are mainly extracted from the essential oils and flavour water phases which are collected during the extraction and concentration of fruit juices. Juice processing itself is today routinely augmented by pectolytic enzymes which are either added to the fruit mash or to the extracted juice.

Already in the mash, polygalacturonases decompose soluble pectins under conservation of protopectin (glue between the individual fruit cells). Pectolytic enzymes with pectinesterase and pectintranseliminase activities effect a significant reduction of juice viscosity especially in berry and stone fruit mashes. The fruit tissue is disintegrated and the colour and flavour yield increased significantly. The activity spectra of the hydrolases reach from a maceration of fruit tissue under conservation of cell structures and soluble pectins to the complete decomposition of all framing substances in the fruit mash and thereby maximum juice and flavour yield. For the latter process enzyme cocktails made from polygalactonurases, pectinesterases and hemicellulases are used. They may also be supplemented by C1-cellulases [11].

The reduction in viscosity of the juice extracted from such mashes facilitates the concentration and significantly increases the juice yield from apple mash by over 10% [12]. Especially important is the enzyme support for juice extraction from fruit containing large amounts of pectins, e.g. black currants, certain grapes and other fruit.

Enzymes have been applied successfully for disintegration of other plant materials to increase flavour yield in the extract. Examples are tea [13], vanilla [14], red grapes [15] and exotic fruits [16]. It has been reported that during vanilla extraction the application of pectinases and cellulases could increase the vanillin content in the extract by a factor of 30 [14]. Recent work has documented the in-situ-liberation of volatile aromatic substances from herbs and other seasoning vegetables directly in the finished food [17]. To this end the cell liquefying-enzymes (proteases, cellulases, pectinases) are administered to the plant material encapsulated together with the seasoning materials in fine distribution. Directly in-situ the taste imparting volatiles are delivered and flavour losses due to industrial processing of fruits are minimized.

# Flavouring Preparations and Some Source Materials

As shown above flavour substances may be mobilized by destruction of cells and framing structures of plants and by extraction of flavour materials dissolved in the plant juices. In addition, flavour precursors may also be liberated from covalent bonds *[18a]*. The majority of these bonds are glycosidic links of flavour chemicals to molecular or particulate plant components. The enzymes used for cleavage of these bonds (β-glycosidases) tend to form equilibria of bound and free forms.

Besides, the liberated flavour chemicals are very reactive and may undergo consecutive reactions in the mixtures forming products which are no longer flavour active. In some cases even off-flavours may be generated. It has been possible to retain the initial products in pure form *[18b]* by quick extraction or distillation processes. The most interesting area for employing this technique is the manufacturing of natural flavour extracts from waste material of fruit processing (hulls, pulps, kernels, etc.).

### 3.2.2.4.6 Flavouring Preparations from Microbial Fermentations

Contrary to microbial flavour generation directly in the food by starter cultures, the technical bioreactions for flavour production with micro-organisms do not use the complete food raw material as substrate. Isolated and purified single components of food are used as substrates for the micro-organisms. Examples are butterfat from butter, proteins from meat, carbohydrates from plant food materials. Microbial material syntheses may lead to chemically defined pure substances (cf. chapter 3.2.1.1.2). It is also possible to obtain complex mixtures of different compounds. Polysaccharides, natural colours and also complex flavour extracts belong to this category. Figure 3.17 outlines the principle of such processes.

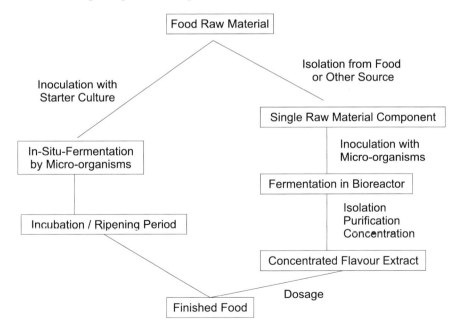

*Fig. 3.17: Microbiological production of flavour preparations*

The utilization of isolated fruit components in bioreactors results in increased yields, higher product purities and better controlled production conditions during fermentation. Additionally, the nutrients for such fermentations can be much simpler. It is possible to use lower alcohols like ethanol and methanol or sugar containing solutions from waste or industrial sugars as easily accessible carbon sources.

In most cases the same micro-organisms are used for the fermentative production of flavour extracts in bioreactors which may also be applied during the natural in-situ food fermentation. In this way a very authentical taste of the concentrated flavour extract may be obtained.

### 3.2.2.4.7 Yeast Autolysates

Yeast extracts and yeast autolysates are very important as taste-delivering and taste-enhancing substances. The production volume of yeast-derived flavour nucleotides comes close to 10,000 t/a with sales per year worth € 350 million. Yeast cells are very robust. Because of their readiness to ferment they may be easily cultivated. The fermentation of yeasts does not call for a complicated technical system in the bioreactor. They can be propagated and may synthesize in non-sterile systems without problems. A certain drawback is that a yeast cell does not release its metabolic products to the surrounding media but retains them within the cell. It is therefore necessary to disrupt the cell walls before the production of the desired substances. This may be effected by mechanical disintegration (high pressure homogenizers and mills), by thermolysis (cooking of biomass in aqueous media), by disintegration in solvents (ethanol) or in highly concentrated salt or sugar solutions.

The most frequently applied technique for cell disintegration of yeast is autolysis. The cell wall is digested by yeast-borne proteases. The process may be supported by externally added proteolytic enzymes from other sources and/or acid or alkali. Consecutively biomass is separated and the aqueous filtrate further processed. In many cases simple concentration by evaporation and drying is sufficient for the production of the yeast extracts or autolysates.

Yeast autolysates may be obtained from waste yeasts from breweries and dairies. If specific flavour types are required also cultured yeasts and yeast lines with special properties are used (species Saccharomyces, Torula).

The finished autolysate may exhibit a multitude of different flavours which may be obtained by heating the extracts and autolysates and by addition of other seasoning components. Today, especially in the area of savoury flavours, a large product range of different tasting yeast autolysates is commercially available. Very often the broth type characteristic for yeast autolysates is dominating. Beside this, a large number of special preparations from yeast autolysates are available with different flavour types, like smoke, vegetables, chicken, cheese, potato, mushroom etc. The different flavour notes may in some cases also be influenced by specially selected substrates (raw materials) for the yeast fermentation.

The taste-delivering principle of a yeast autolysate is mostly derived from the natural contents of the yeasts of special amino acids, especially glutamic acid. Also nucleotides from the naturally occurring ribonucleic acid are important as well as the majority of proteins, peptides and the amino acid patterns of the yeast autolysate. A certain yeast side flavour is always remaining which limits the applicability of yeast autolysates in some ways. Table 3.29 shows a typical composition of a yeast autolysate made by spray drying *[19]*. Yeast extracts and autolysates may even be applied as taste enhancing substances in sweet flavour preparations.

*Table 3.29: Typical composition of a spray-dried yeast autolysate*

| Total Solids | | | 95-97% | |
|---|---|---|---|---|
| Raw Protein | | | 70% | |
| Free Amino Acids | | | 35% | |
| NaCl | | | <1% | |
| Ash | | | 11% | |
| pH | | | 5.0-5.5 | |
| **Amino Acids in % of Total Solids** | | | **Vitamins in mg/100 g Solids** | |
| Asp | 8.0% | Met | 0.7% | Vitamin B1 | 3.0 |
| Thr | 2.3% | Ile | 4.7% | Vitamin B2 | 11.9 |
| Ser | 2.7% | Leu | 3.1% | Vitamin B6 | 2.3 |
| Glu | 11.9% | Tyr | 1.4% | Niacin | 68.0 |
| Pro | 1.8% | Phe | 1.8% | Biotin | 0.17 |
| Gly | 3.1% | Lys | 6.8% | Folic Acid | 3.1 |
| Ala | 5.6% | His | 1.5% | Ca-Pantothenate | 30.0 |
| Cys | 0.4% | Arg | 1.3% | | |
| Val | 3.8% | Ornithine | 0.3% | | |

### 3.2.2.4.8 Cheese Flavours

Enzymatic modification of milk fats with lipolytic enzymes has already been mentioned above. Besides this it is possible to manufacture complex cheese flavours today also by fermentation of raw materials of cheese processing with defined microorganisms. Roquefort and other blue cheese flavours fermented by the mould *Penicillium Roqueforti* are currently in commercial production.

This blue cheese mould is fermented in suspension in the submerged technique. Milk, whey or lipase modified butterfat may serve as substrates. *Penicillium roquefortii* produces a protease and different lipases which also may be applied in isolated and purified form. The fermentation lasts several days and produces mainly different methylketones and secondary alcohols with 5 to 11 C-atoms. Important representatives are 2-heptanone and 2-hexanone, 2-nonanone and the corresponding alcohols. The isolation of the reaction mixture consisting mostly of volatile flavour substances is achieved after fermentation and heat activation of the mould-borne enzymes e.g. by distillation and extraction. Gas chromatograms of the obtained flavour extracts compared with the flavour substances of naturally manufactured Roquefort cheese show a very good correlation in composition and concentration of the taste delivering ingredients *[20]*. The blue mould flavour extracts are used mainly in the flavouring of cheese spreads or cheese containing finished menus.

Some other cheese flavours (Provolone, Emmentaler and other miscellaneous cheesy flavour notes) can be obtained by fermentation of edible fats by means of micro-organisms which are producing butter or propionic acid from these fats.

There is some experience concerning other milk and plant fats which may be fermented by micro-organisms only on a laboratory scale. Examples are butter flavours (generation of diacetyl and acetoin) by suitable bacteria cultures which are derived from milk (*Streptococcus diacetylactis* and *Leuconostoc citrovorum*). Today, buttery flavours are especially important for dietetic foods, like fat reduced butter or for the fortification of plant fats (margarine).

### 3.2.2.4.9 Mushroom Flavours

The characteristic taste of edible mushrooms is mainly derived from the flavour chemicals 3-octanol and 1-octen-3-ol. In order to fermentatively generate mushroom flavours it is possible to cultivate higher mushrooms in a submerged fermenter and propagate them by the formation of mycelia. In the fermentation broth, 1-octen-3-ol is detectable after some time. In the mushroom this substance is formed by a lipoxygenase using unsaturated fatty acids as a substrate *[21]*.

The same substance can also be produced by the fermentation of moulds (aspergillus, penicillium, fusarium).

### 3.2.2.4.10 Fruity Flavours

Micro-organisms also have a potential to form fruity components like alcohols, esters, lactones, ketones and terpenes. These substances are, however, also obtainable by chemical syntheses as well as by relatively simple extraction from the natural fruit. Additionally, fermented fruit flavours in many cases are not very typical for a certain fruit species and therefore may be used only as raw materials for compounded flavours. Presumably this is the reason why the corresponding processes have not yet been transferred onto an industrial scale. This is not true for the fermentative generation of well defined single substances which may serve as natural flavour raw materials for compounded flavours (cf. chapter 3.2.1.1.2).

### 3.2.2.4.11 Beer Flavour

Like many beverage flavours, the flavour of beer is a complex mixture of more than 400 different single substances, most of which have been characterized by gas chromatography/mass spectrometry. The tasting materials primarily are generated during the fermentation of malt sugar amino acids and proteins of the worts by brewer's yeast and in the course of the subsequent ripening period by a number of chemical reactions from certain precursors.

Alcohol-free and alcohol-reduced beers have acquired strong positions in the markets in the last years. There are in principle two different processes for their production:

    (1) Modified fermentation on the basis of worts with reduced extract values with strongly depressed or without any alcohol fermentation.

(2) The production of a normal alcohol containing beer with consecutive alcohol removal by physico-chemical processes (evaporation, membrane extraction).

Both processes result in alcohol-reduced or -free beers with significantly reduced flavour. Therefore, a flavour enhancement of these beers is desirable. In Germany this process has to make use of brewery inherent raw materials because of the German purity law. Besides the production of beer flavours with special extraction processes from normal beer or brewing raw materials *[22]*, flavour has also been obtained by immobilized yeasts *[21]*. This process is controlled in order to suppress the alcohol concentration by short contact times between wort and yeast whereas the taste imparting components are formed very quickly and thus get into the flowing wort. Aldehydes and ketones as well as fatty acids are reduced by immobilized yeast (yeast reduction). For this technique yeast is immobilized on fibers of granular DEAE-cellulose. A fixed bed column reactor with high flow rate is constructed in this way.

Besides the improvement of fermentation performance and simplification of the process, the improvement of product quality is a target for yeast strain development. Genetic modification of yeast strains and the expression of heterologous genes (e.g. acetolactate-decarboxylase) have increased the possibilities of modifying desired or undesired flavour formation.

Another component important for the beer flavour in this context is obtained by the fermentation of cereals with lactic acid bacteria. Flavour substances produced in this way supplement the beer flavour obtained by the yeast fermentation in a very positive way *[24]*.

Another example for the production of natural beer flavour is the extraction of yeast deposits formed during the ripening period in the fermentation tanks. Yeast deposits contain about 10-15% living yeast cells and residual beer with a high content of the desired yeasty flavour substances. A typical natural beer flavour in accordance with the German purity law is obtained *[25]* by distillation and rectification of such yeast suspensions.

### 3.2.2.4.12 Flavour Manufacturing from Cell Cultures

The masterpiece of modern biotechnology is the generation of valuable flavouring substances by means of cell tissue cultures derived from plant and animal organisms. In chapter 2.2 the most important processes for the preparation of and the production with cell cultures are described.

Most technical experiences are based on plant cells. This may be explained by the fact that the majority of all commercial flavour extracts is of plant origin. Compared to intact freely grown plants the in-vitro plant cell culture has advantages because

- it is independent from geographical and climatic variances and harvesting cycles,
- it can be grown without pesticides, fertilizers and other chemical aids and
- it may produce products over extended periods of time.

Plant cells show an extensive repertoire of chemical reaction mechanisms: epoxidation, reduction, oxidation, hydroxylation, isomerisation. It is self-evident that plant cell cultures synthesize as enantioselectively as their mother organisms. Besides the well-known flavour extracts and single substances, also presently unknown naturally flavour chemicals and mixtures of these are in principle obtainable. Therefore the rapid progress in investigating this area is not surprising [26].

Although the main objective of present publications is the synthesis of pure flavour chemicals by plant cells there is a number of experiments with plant cell cultures for the manufacturing of complex flavour extracts. In these cases it is intended to produce flavour extracts using cell cultures of taste-delivering plants similar to the extracts rendered by extraction from the natural mother plant.

For this purpose natural raw materials of the plant metabolism may be used as substrates e. g. mono- and sesquiterpenes with callus cultures of essential oil plants or externally added glycosides with cells of mentha piperita. After adding geraniol to cell cultures of citrus plants geranylic and citronellylic acid could be isolated from the media.

Economically more meaningful would be the ab initio-synthesis of complex natural compounds by plant cells from $CO_2$ and light. For this purpose the chlorophyll containing green parts (chloroplasts) of the plants would be needed which are able to effect photosynthesis. Only laboratory results are available in this area. More advanced results about flavour generation in plant cell cultures are available for:

- vanilla flavour extracts from callus cell cultures of vanilla plants [27]
- truffle flavour by submerged cultures of genuine tuber melanosporum cells [28]
- green flavour complexes from strawberry leave homogenisates [29]. In this case biotransformation of linolenic acid with the leave homogenisate is reported. Among others cis-3-hexenal is formed with a yield of about 10%.

However, all the production processes mentioned here have not yet led to convincing industrial results. Continued development work is necessary, especially regarding increased product yields. Whether e. g. genetic engineering may lead to higher productivity still has to be proven.

The same holds for products from animal cell cultures where even less results are available. With animal cell cultures especially enzymes are accessible which may be used for biotransformations of flavour precursors. Compared to the progress in the production of pharmaceutical substances from animal cell cultures it is only a question of time when processes are developed which allow the economical production of flavour chemicals not yet accessible by other routes. Quick progress is expected, among others, for sweeteners, bitter substances, essential oils, fruity flavours and vegetable flavours from cell cultures.

## REFERENCES

[1a] Klibanov A. M., Advances in Enzymes, in: Biotechnology Challenges for the Flavor and Food Industry, Lindsay R. C., l Willis B. J. (Ed.) London, Elsevier 1989
[1b] Acilu M., Zaror C., Enzymatic Synthesis of Ethyl Butyrate in Water/Isooctane Systems Using Immobilized Lipases, in: Progress in Flavour Precursor Studies, Schreier P., Winterhalter P., Carol Stream, Allured Publishing Corp. 1993
[2] Scheide J., Eigenschaften und lebensmittelrechtliche Probleme thermisch gewonnener Fleischaromen, Nahrung 24, 163-174 (1980)
[3] Mottram D. S., Meat, in: Maarse H. (ed.): Volatile Compounds in Food and Beverages, New York, Dekker 1991
[4] Van Rillaer W., Beernaert H., Determination of Residual 1,3-Dichloro-2-propanol in Protein Hydrolysates by Capillary Gas Chromatography, Z Lebensm Unters Forsch 188, 343-345 (1989)
[5] EP OS 0 429 760 A1 Procédé de préparation d'un agent aromatisant, Société des Produits Nestlé S.A. 1991
[6] EC Flavouring Framework Directive 88-338-EEC
[7] Moskowitz G.J., Noelck S.S., Enzyme Modified Cheese Technology, J. Dairy Sci. 70, 1761-1769 (1987)
[8] Kinsella J.E., Hwang D.A., Crit. Rev. Food Sci. Nutrit. 8, 191 (1976)
[9] Tanous C., Kieronczyk A., Helinck S., Chambellon E., Yvon M. Antonie van Leeuwenhoek 82, 271-278 (2002)
[10] Marilley L., Casey M.G., Int. J. Food Microbiol. 90, 139-159 (2004)
[11] Uhlig H., Enzyme arbeiten für uns, p. 313, München, Hanser 1991
[12] ibid. p. 292
[13] DE OS 34 16 223 A1 Enzymatisches Verfahren zur Herstellung von Instant Tee, Novo Industri A/S
[14] PCT WO 93/25088 Method for Obtaining a Natural Vanilla Aroma by Treatment of Vanilla Beans, and Aroma thus Obtained, v. Mane Fils S.A. 1992
[15] Uhlig H., Enzyme arbeiten für uns, p. 289, München, Hanser 1991
[16] Röhm GmbH, Darmstadt, Firmendruckschrift: Enzyme zur Herstellung von Fruchtsaft und Wein (1993)
[17] DE 43 10 283, Verfahren zur Herstellung von Lebensmitteln, Vösgen W. 1994
[18a] DE/EP 0 416 713 T1 Process for Obtaining Aroma Components and Aromas from their Precursors of a Glycosidic Nature and Aroma Components and Aromas thereby Obtained, Gist Brocades N.V.1991
[18b] Schwab W., Schreier P., Simultaneous Enzyme Catalysis Extraction: a Versatile Technique for the Study of Flavour Precursors, J. Agric. Food Chem. 36, 1238-1242 (1988)
[19] Firmendruckschrift OHLY GmbH, Marl 1992
[20] Kunz B., Trends in der Lebensmittelbiotechnologie, in: Czermak P., (ed.), Lebensmitteltechnologie, Darmstadt, GIT-Verlag 1993
[21] Schindler F., Seipenbusch R., Fungal Flavour by Fermentation, Food Biotechnology 4 (1), 77-85 (1990) oder: DE 40 367 63 Verfahren zur Herstellung von Lebensmittelaromen (1992)
[22] Firmendruckschrift DÖHLER GmbH, Darmstadt (1993)
[23a] Brauwelt 7/8, 302-306 (1993)
[23b] Back W., Verfahren zur Herstellung von alkoholfreiem Bier und Vorrichtung zur Durchführung dieses Verfahrens, Helsinki, Cultor 1994
[24] Lüders J., Technologie mit immobilisierten Hefen, Brauwelt 3, 57-62 (1994)
[25] DE 1 767 040, Verfahren zur Gewinnung von Bieraromastoffen, Aromachemie Erich Ziegler 1971
[26] Berger R.G., Generation of Aroma Compounds by Novel Biotechnologies, in. Aroma Production and Application, Proc. 3rd Wartburg Aroma Symposium, Potsdam, Eigenverlag 1992
[27] Knuth M.E., Flavor Composition and Method, US PS 5068184 (1991)
[28] California Truffle Corp. Woodland, Biotechnology Newswatch 10, 7 (1990)
[29] Goers S.K., Process for Producing a Greenleave Essence, US PS 4,806,379 (1989)

## 3.2.3 Process Flavourings

*Christoph Cerny*

### 3.2.3.1 Introduction

The human being appears to be the only creature that cooks its food before eating. Thermal treatment reduces the microbial count, can destroy toxins and improve digestibility. However, the prime reason to improve and perfect the art of cooking from the fires of the Stone Age to the cooking stoves of famous 3-star chefs like Paul Bocuse and Adrian Ferrà was certainly the better palatability of cooked food.

Unfortunately, the preparation of a square meal from scratch using fresh ingredients from the market requires time, which less and less people are ready to spend. Fortunately, over the years the food industry has developed convenience food, ready-to-eat meals and cooking ingredients, which help to save some time and make cooking easier. The first TV dinner, served in an aluminium tray, was launched in 1954 in the USA. However, industrially manufactured food often lacks aroma because dried ingredients have to be used and aroma compounds are degraded or lost during processing and storage. To compensate for these losses the food industry uses flavourings. In particularly savoury flavourings are very often based on process flavourings.

The development of savoury process flavourings began in the 1950s, and the first pioneering meat flavour patent was published in 1960 *[1]*. The idea of process flavourings is to study and make use of those reactions which occur during cooking of meat, baking of bread or roasting of cocoa or malt, and develop reactions which mimic those processes: 'Process flavourings are produced by heating raw materials which are foodstuffs or constituents of foodstuffs in similarity with the cooking of food' *[2]*.

Process flavourings are typically used at a level of 0.1-2% in culinary products such as bouillons, sauces, but also in snacks, desserts and confectionery. The annual production is estimated at 6750 tons in the USA *[3]* and about 4500 tons in Europe *[4]*. Process flavourings have been reviewed amongst others by van den Ouweland *[5]*, May *[6]*, Salzer *[7]*, Manley *[8, 9]*, Manley and Ahmedi *[3]* and Kerler and Winkel *[10]*.

The fundamental reaction for aroma generation in process flavourings is the Maillard reaction, i.e. the reaction between amino acids or amines and reducing sugars. In meat-like flavourings the Maillard reaction is key to the generation of the basic meat note *[11]*, while the species-specific aroma is due to odorants stemming from lipid degradation *[12]*.

### 3.2.3.2 Definition and General Guidelines

'A thermal process flavouring is a product prepared for its flavouring properties by heating food ingredients and/or ingredients which are permitted for use in foodstuffs or in process flavourings' *[2]*.

The International Organisation of the Flavour Industry has set up guidelines for the production of process flavourings *[2]*. These have been adopted by the Council of Europe *[4]*. Table 3.30 lists the permitted ingredients according to these guidelines. A protein nitrogen source and a carbohydrate source are compulsory in order to produce a process flavouring implying the Maillard reaction as the key mechanism. A fatty acid source and other materials are optional. Other ingredients than those allowed during the reaction, can be added only after the thermal processing is completed.

*Table 3.30: Ingredients for process flavourings [2]*

| Ingredient class | Ingredient |
| --- | --- |
| 1. Protein nitrogen source | Protein nitrogen containing foods, extracts and their hydrolysates:<br>– Meat, egg and dairy products, fish, seafood<br>– Cereals, vegetable products, fruit, yeast<br>Peptides, amino acids including their salts |
| 2. Carbohydrate source | Carbohydrate containing foods, extracts and their hydrolysates:<br>– Cereals, vegetable products, fruit<br>Mono-, di- and polysaccharides |
| 3. Fat or fatty acid source | Fats and oils or foods containing fats and oils<br>Hydrogenated, transesterified, fractionated fats and oils<br>Hydrolysates thereof |
| 4. Other materials | Herbs, spices and extracts<br>Flavouring substances identified in food<br>Water<br>Thiamine and thiamine hydrochloride<br>Ascorbic, citric, lactic, fumaric, succinic and tartaric acid and their salts (Na, K, Ca, Mg, $NH_4$)<br>Guanylic acid, inosinic acid and their salts (Na, K, Ca)<br>Inositol<br>Sodium, potassium and ammonium sulphides, hydrosulphides and polysulphides<br>Lecithine<br>pH regulators: acetic, hydrochloric, phosphoric and sulphuric acid; salts thereof: sodium, potassium, calcium and ammonium<br>Polymethylsiloxane |

The conditions during the flavour reaction are limited as follows:
- The upper temperature limit is set to 180°C.
- The maximum reaction time at 180°C is 15 min with longer reaction times at lower temperatures. For every temperature reduction of 10°C the maximum reaction time is doubled, e.g. 30 min at 170°C, 60 min at 160°C and so on, but not longer than 12 hours *[4]*.
- The maximum pH during processing is 8.0.

### 3.2.3.3 Process Flavour Chemistry

Process flavourings are more related to cooking than to chemical synthesis. The reactions that occur are chemically very complex with hundreds of volatiles and non-volatiles being formed. The final process flavouring is defined by the sum of the sensory effects of all aroma-active volatiles and taste-active compounds.

Model reaction trials and modern analytical methods (gas chromatography/mass spectrometry (GC/MS), gas chromatography/olfactometry (GC/O)) permitted the identification of key mechanisms responsible for flavour generation in process flavourings and some of the most important ones are detailed below. Often chemically complex precursor raw materials (vegetables such as onions, spices, yeast extracts, animal products) are used. Research work on these complex reactions is rare but necessary and allows the discovery of new key odorants and formation pathways. For example, Widder and co-workers [13] discovered a new powerful aroma compound, 3-mercapto-2-methylpentan-1-ol in a complex process flavour based on onion.

The basic reactions contributing to thermal flavour generation are the Maillard reaction, Strecker degradation, lipid oxidation and thiamin degradation. Interaction of products formed by these different mechanisms additionally leads to further flavour components.

#### 3.2.3.3.1 Maillard Reaction and Strecker Degradation

In 1912 Maillard [14] was the first scientist who studied the reaction between glucose and glycine which leads to browning and carbon dioxide formation.

Without any doubt the Maillard reaction and the Strecker reaction play an outstanding role in thermal aroma generation [15-20] and in particular in process flavourings.

One of the first researchers to study the aroma formation during the Maillard reaction was Ruckdeschel [21] in 1914. He described the reaction product of glucose with phenylalanine as rose-like, with leucine as bread-like, with valine as fine roasted and with alanine as similar to coloured malt.

In 1953 Hodge [22] published a scheme giving an excellent overview of the complexity of the Maillard reaction. Fig. 3.18 shows a more simplified scheme of the Maillard reaction.

In the first step an aldose sugar reacts with an amino compound to form the Schiff base **1** which undergoes further reactions and rearranges into an aminoketose **2**, the so-called Amadori compound [23-25]. Fig. 3.19 shows the reaction scheme with glucose as the reducing sugar. Heyns and co-workers [26] found aminoaldoses as first intermediates in a similar mechanism when ketose sugars were used.

The Amadori compounds can react further forming 1-deoxydiketoses (e.g. 1-deoxyglucosone [**3**] from glucose), 3-deoxyaldoketoses (e.g. 3-deoxyglucosone [**4**] from glucose) and 1-amino-4-deoxydiketoses (e.g. 4-deoxyglucosone [**5**] from glucose) [27, 28], also called deoxyglycosones. Figure 3.19 shows the open chain form, but Weenen and Tjan [29] showed that **3** in aqueous solution exists almost exclusively in the form of the cyclic isomers. The formation of 3-deoxyaldoketoses is favoured under acidic conditions while 1-deoxydiketoses are preferably formed at neutral or slightly basic pH values. These dicarbonyl compounds, either in their open-chain or in their cyclic form, are in equilibrium with their enol tautomers, and have a so-called reductone structure (Fig. 3.20). These and other reductones formed in the Maillard reaction are able to reduce other Maillard products, for example 2,3-butanedione *[19]*.

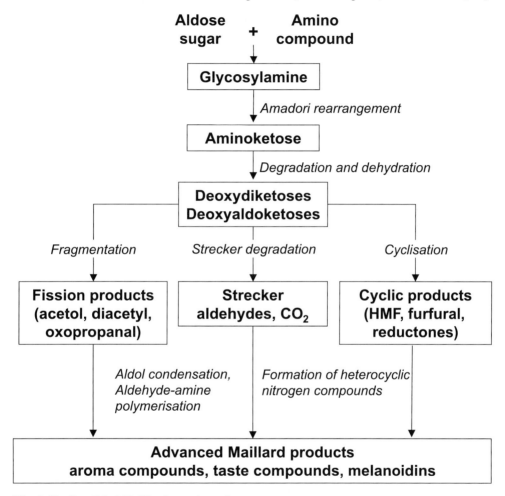

*Fig. 3.18: Simplified Maillard reaction scheme*

The deoxyglycosones can react further in different ways. Cleavage of α-diketones can lead to breakdown products including formic acid (from 3-deoxyaldoketoses) and acetic acid (from the 1-deoxy-2,3-diketoses) *[19, 30]*. Davidek and co-workers *[31]* studied the degradation of the Amadori compound from glucose and glycine, and found an increased formation of formic and acetic acid at higher pH values. In addition, phosphate had a further promoting effect. The acid yield was up to 0.76 mol/mol of Amadori compound. Another cleavage mechanism for deoxyglycosones is retroaldolisation resulting in smaller carbonyl sugar fragments such as glycolaldehyde (**6**) and 2-oxopropanal (**7**) (Fig. 3.21). The formation of small $C_2$ and $C_3$ carbonyls increases with pH during the Maillard reaction *[32, 33]*. These compounds are highly reactive which explains why process flavours with higher pH value mostly show accelerated browning rates during the reaction. Condensation between nucleophilic and electrophilic intermediates of the Maillard reaction is believed to be responsible for the formation of the brown coloured components, the melanoidins.

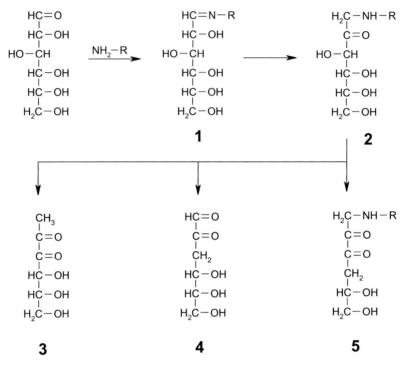

*Fig. 3.19:* Formation of Amadori compounds and deoxyglycosones from glucose

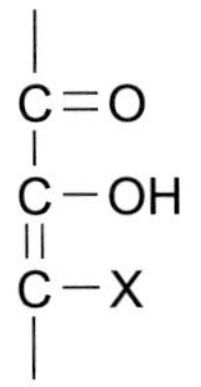

$$\begin{array}{c} | \\ C=O \\ | \\ C-OH \\ || \\ C-X \\ | \end{array} \qquad X = OH, OR, NH_2, NHR, NR_2$$

*Fig. 3.20:* General reductone structure [19]

Strecker [34] discovered that the reaction between amino acids (glycine, alanine, leucine) and the tricarbonyl compound alloxan yields $CO_2$ and aldehydes. The 'Strecker degradation' of amino acids occurs also with dicarbonyl compounds [35] including those that are formed in the course of the Maillard reaction, in particular deoxyglycosones and some of the smaller sugar fragments like **7** and diacetyl. Fig. 3.22 shows the reaction pathway that involves the formation of an imine **8**, followed by decarboxylation and liberation of the resulting aminoketo compound and the Strecker aldehyde from the intermediate **10**. Odour-active Strecker aldehydes which

Process Flavourings 279

contribute frequently to the aroma of process flavours are acetaldehyde, 2-methylpropanal, 2-methylbutanal, 3-methylbutanal, methional and phenylacetaldehyde.

In the presence of oxygen, intermediate **10** can undergo oxidation and give rise to the formation of the aminoketo compound **11** and subsequently an acid, the so-called 'Strecker acid' *[36]*. The formation of Strecker acid, however, is not possible if dicarbonyls such as **4** are present in form of their stable cyclic hemiacetals, as shown by Hofmann and co-workers *[36]*. The same group revealed also that Strecker aldehydes can be formed via an oxidative degradation of Amadori compounds, as shown for the Amadori compound of phenylalanine and glucose *[37]*.

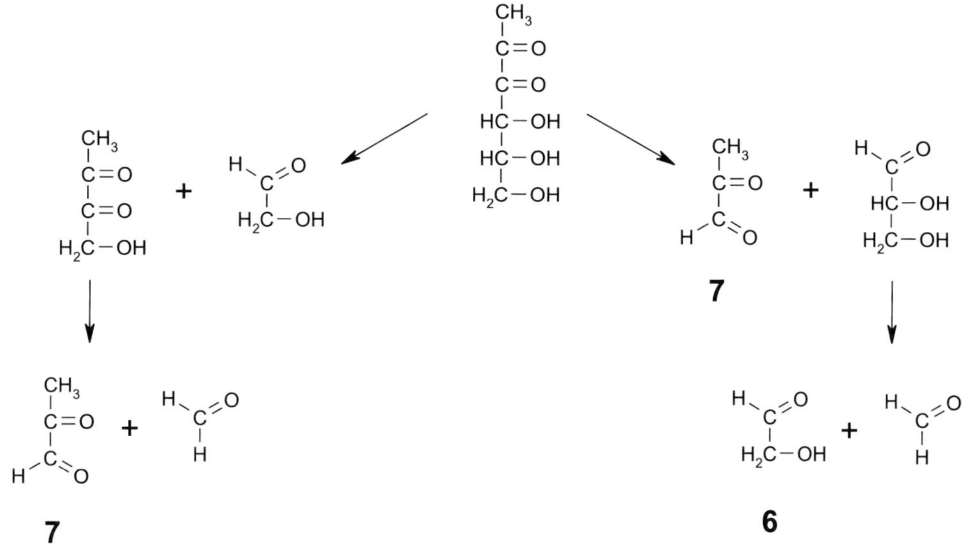

*Fig. 3.21: Fragmentation of 1-deoxyglucosone*

The reaction between reducing sugars and the amino acid cysteine, which serves as a hydrogen sulphide source, especially with pentoses, such as xylose, is of particular importance for meat-like process flavours. The key aroma compounds from this reaction are sulphur molecules. GC/O has revealed 2-furfurylthiol (**12**), 2-methyl-3-furanthiol (**13**), 3-mercapto-2-butanone (**14**), 3-mercapto-2-pentanone (**15**) and 5-acetyl-2,3-dihydro-1,4-thiazine (**20**) as high-impact odorants in a heated cysteine/ribose solution *[38]*. Fig. 3.23 shows the key intermediates for this reaction. The 3-deoxyglucosone (**16**) and furfural (**17**) are intermediates for **12** *[39]*. More than one pathway and intermediate exists for **13** and **15**. 4-Hydroxy-5-methyl-3(2*H*)-furanone (**18**) has been proposed by several authors as an efficient precursor *[40-42]*. A more recent study showed that ribose itself is a more effective precursor for **13** and **15** than **18** *[43]*. These authors proposed 1,4-dideoxypentosone (**19**) as intermediate. For **14** the formation pathway is suggested via 2,3-butanedione *[44]*, a possible further intermediate being 4-hydroxy-2,3-butanedione (**21**) *[45]*.

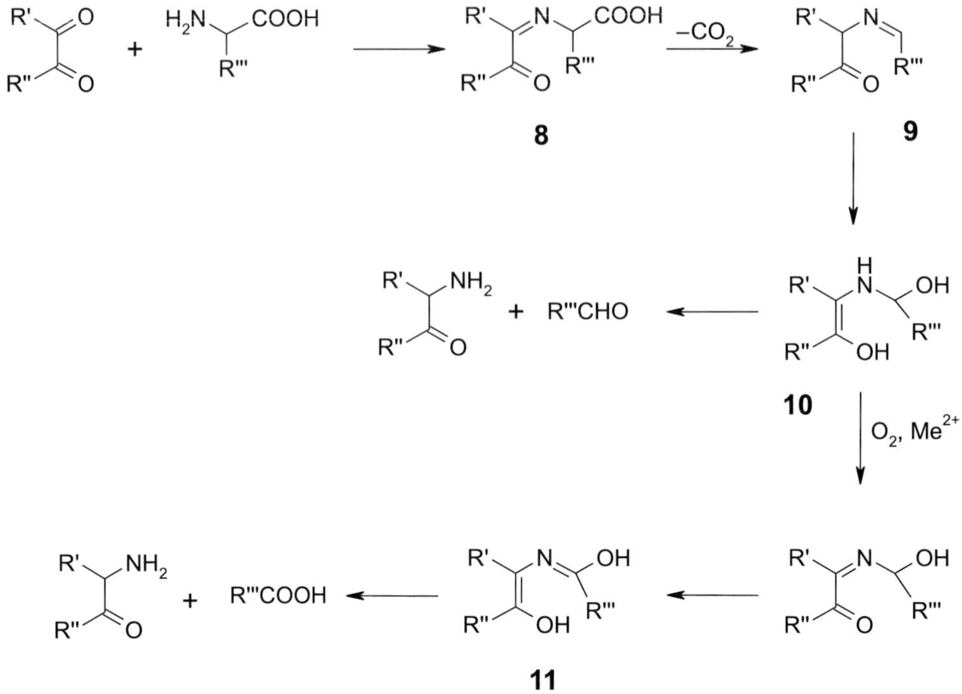

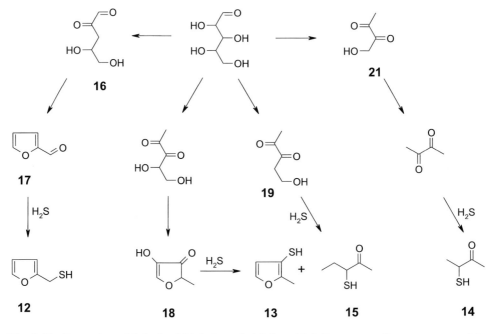

*Fig. 3.22:* Formation of Strecker aldehydes and acids from amino acids and dicarbonyl compounds (adapted from [36])

*Fig. 3.23:* Formation of 2-furfurylthiol, 2-methyl-3-furathiol, 3-mercapto-2-pentanone and 3-mercato-2-butanone from pentose and cysteine

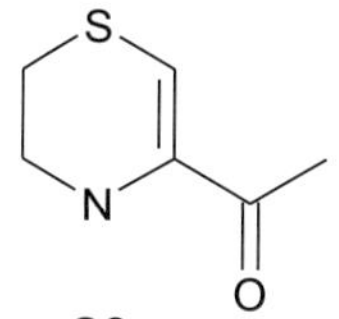

**20**

*Fig. 3.24: Structure of 5-acetyl-2,3-dihydro-1,4-thiazine*

Proline is a particular case in the Maillard reaction because it is, unlike other amino acids, a secondary amine. It represents an important precursor for popcorn-like aroma compounds in the reaction with reducing sugars such as glucose *[46]*. The relevant aroma compounds in this reaction are 2-acetyl-1-pyrroline (**22**), 1-(1,4,5,6-tetrahydro-2-pyridyl)-1-ethanone (**23**) and its isomer 1-(3,4,5,6-tetrahydro-2-pyridyl)-1-ethanone (**24**) *[47]*. The formation pathway (Fig. 3.25) shows the important intermediates 2-oxopropanal (**7**) and 1-pyrroline (**25**) *[48, 49]*. Another amino acid precursor for **22** is ornithine, an amino acid in yeast or yeast extracts *[49]*.

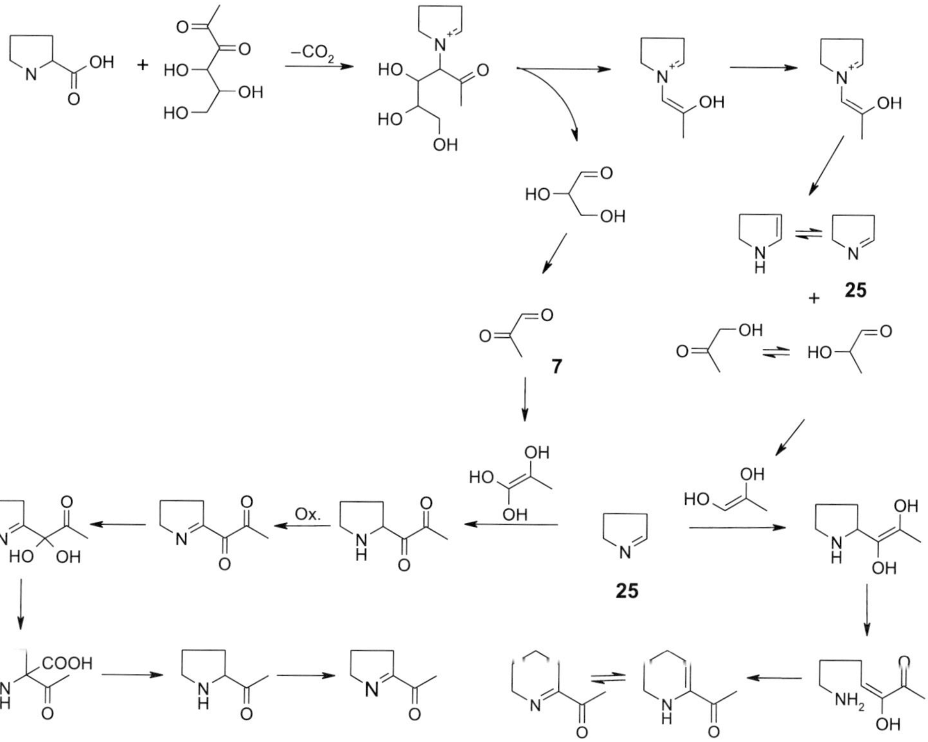

*Fig. 3.25: Formation of 2-acetyl-1-pyrroline and 2-acetyltetrahydropyridine from proline and 1-deoxyglucosone (adapted from [59])*

A particularly important aroma compound formed from reducing sugars during the Maillard reaction is 4-hydroxy-2,5-dimethyl-3(2H)-furanone (**26**, Furaneol). It is a key aroma compound for thermally produced flavours like cooked meat [50], bread [51], coffee [52] and chocolate [53], but also important for fermented food like cheese [54] and soy sauce [55], and even unprocessed fruit like strawberry [56]. Important Furaneol precursors in process flavours comprise rhamnose, glucose and fructose, but it can also be formed from pentoses in the presence of glycine [57], as well as from smaller sugar fragments [58]. Fig. 3.26 illustrates its formation from 1-deoxyglucosone.

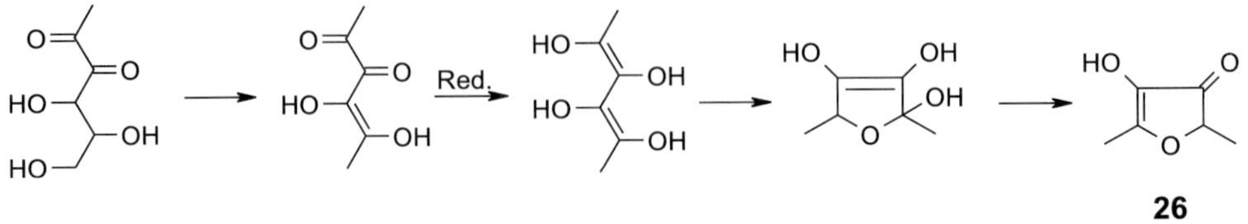

*Fig. 3.26: Formation of 4-hydroxy-2,5-dimethyl-3(2H)-furanone from 1-deoxyhexosone*

### 3.2.3.3.2 Lipid Oxidation

Fats or raw materials that serve as a source for fatty acids are frequently employed in process flavourings. During the flavour reaction, thermal peroxidation of lipids such as triglycerides, fatty acids and phospholipids occurs. This non-enzymatic lipid oxidation, also called autoxidation, leads to a very complex mixture of reaction products, and has to be regarded separately from the enzymatic lipid oxidation which occurs at low temperature and is catalysed by lipoxygenases.

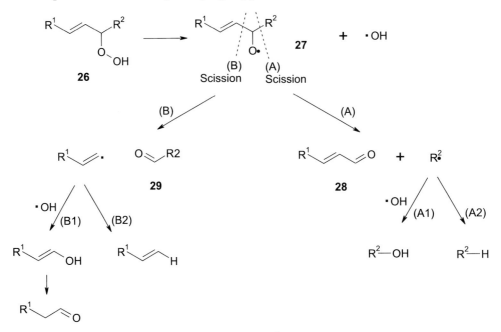

*Fig. 3.27: Homolytic cleavage of monohydroperoxides*

Process Flavourings 283

The most important precursors for lipid oxidation are unsaturated fats and fatty acids like oleic (18:1), linoleic (18:2), linolenic (18:3) and arachidonic acid (20:4). The more unsaturated ones are more prone to oxidation. Lipid peroxidation and the subsequent reactions generate a variety of volatile compounds, many of which are odour-active, especially the aldehydes. That is why lipid oxidation is also a major mechanism for thermal aroma generation and contributes in a great measure to the flavour of fat-containing food. Lipid oxidation also takes place under storage conditions and excessive peroxidation is responsible for negative aroma changes of food like rancidity, warmed-over flavour, cardboard odour and metallic off-notes.

The autoxidation of lipids is a radical reaction and the primary products are monohydroperoxides. Grosch [60] has given an excellent review on the formation of volatiles from these first intermediates. Fig. 3.27 illustrates the homolytic cleavage as a mechanism for lipid peroxide degradation. The unsaturated alkoxy radical **27** formed from the monohydroperoxide **26** can split in two ways. Either the cleavage occurs on the side away from the olefinic bond (scission A), resulting in the formation of an unsaturated carbonyl compound **28**, or scission B between the double bond and the oxygen bearing carbon giving rise to a saturated oxo compound **29**.

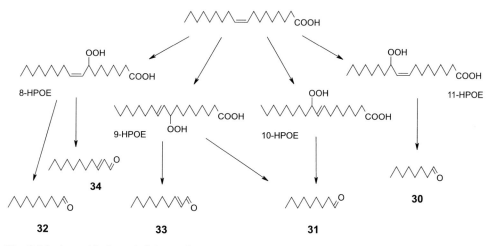

*Fig. 3.28: Autoxidation of oleic acid*

Autoxidation of oleic acid leads to the 4 main hydroperoxy-octadecenoic acids 8-HPOE, 9-HPOE, 10-HPOE and 11-HPOE (cf. Fig. 3.28). Fission generates a multitude of volatiles. Octanal (**30**), nonanal (**31**), decanal (**32**), 2(*E*)-decenal (**33**) and 2(*E*)-undecenal (**34**) belong to the odour-active ones.

Linoleic acid has two non-conjugated double bonds and consequently the theoretical number of hydroperoxides is higher. The main hydroperoxy-octadienoic acids are 9-HPOD, 10-HPOD, 12-HPOD and 13-HPOD (cf. Fig. 3.29). Further reaction produces among others the odorant aldehydes hexanal (**35**), 2(*E*)-heptanal (**36**), 2(*E*)-octenal (**37**), 2(*Z*)-octenal (**38**), 2(*E*)-nonenal (**39**), 3(*Z*)-nonenal (**40**), 2,4(*E*,*E*)-decadienal (**41**) and 2,4(*E*,*Z*)-decadienal (**42**). The potent odorant **41** (odour threshold 0.2 mg/kg in water *[61]*) has a pleasant fried aroma and is a key odorant for meat-like flavours (notably chicken), roasted nuts, butter and heated oils.

The oxidation of arachidonic acid, which is present in low amounts in meat and especially in egg yolk, can in principal generate the same compounds as linoleic acid. Additionally, aldehydes with more than 10 carbons are formed, e.g. 2,6($E,Z$)-dodecadienal and 2,4,7($E,Z,Z$)-tridecatrienal *[60, 62]*. Typical reaction products with low odour thresholds from linolenic acid are 2($E$)-hexenal, 2,6($E,Z$)-nonadienal and 1,5($Z$)-octadien-3-one.

Recent studies by Lin and co-workers *[63]* have shown that not only the fatty acid composition influences the kind and ratio of carbonyls formed during heating of lipids, but also the structure of the reaction medium. The type of self-assembly structure had a significant influence on reaction yields. Egg phosphatidylcholin, which forms a lamellar phase together with water, generated significantly more **41** than phosphatidylethanolamine which adopts a different nanostructure, namely a reversed hexagonal structure.

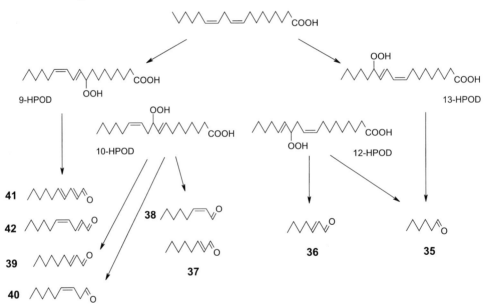

*Fig. 3.29: Autoxidation of linoleic acid*

### 3.2.3.3.3 Thiamine Degradation

Thiamine degradation has a good share in meat aroma formation *[17, 64]*. Neutral and acidic conditions favour the formation of **13** *[65]*, which is a key component in boiled meat *[66, 67]*. It has already earlier been identified in a meat-like process flavouring *[68]*, prepared from cysteine, thiamine, hydrolysed vegetable protein and water *[69]*. Bolton and co-workers *[70]* showed that in model experiments with thiamine, [$^{34}$S]-cysteine, glucose and xylose, only 8% of **13** contained sulphur from cysteine. They concluded that thiamine (**43**) was the primary precursor for the generation of **13** in this system.

Van der Linde and co-workers *[71]* postulated 5-hydroxy-3-mercapto-2-pentanone (**44**) as key intermediate, formed by hydrolysis of **43** (cf. Fig. 3.30). Another postulated intermediate, 3,5-dihydroxypentan-2-one *[70]*, however, generated only trace amounts of **13**, when reacted with cysteine or hydrogen sulphide *[72]*.

Under basic conditions thiamine degrades to 5-(2-hydroxyethyl)-4-methylthiazole (**45**, sulfurol) and the pyrimidine derivative **46** *[71]*. Sulfurol is used in compounded flavourings in the flavour industry. It is almost odourless as such *[73]*; however, it can decompose giving rise to thiazole and its derivatives such as 4-methylthiazole (**47**), 4,5-dimethylthiazole (**48**), 4-methyl-5-ethylthiazole (**49**) and 4-methyl-5-vinylthiazole (**50**) *[71, 74]*, which possess nutty, green notes *[75]*.

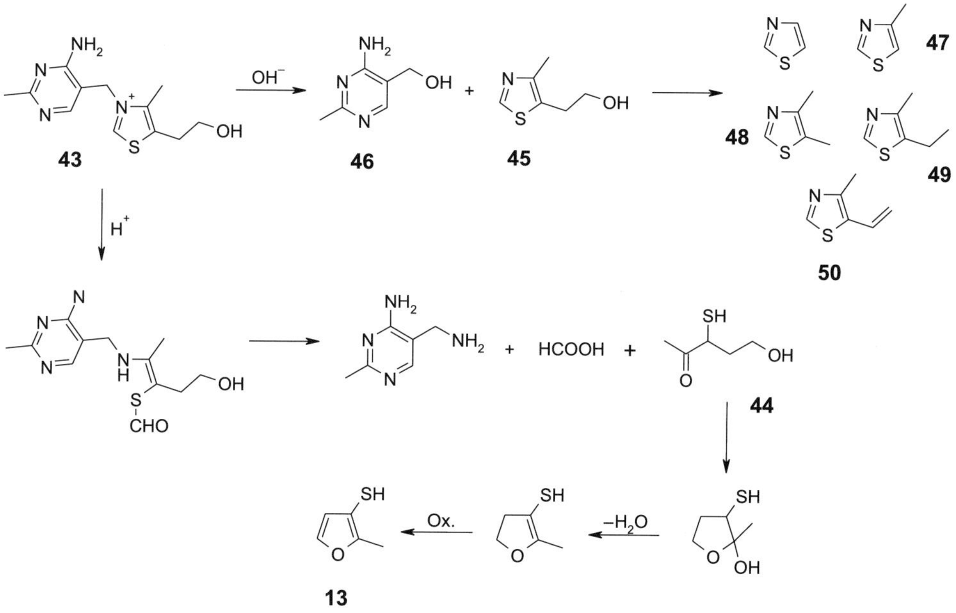

*Fig. 3.30: Thiamine degradation: formation of aroma compounds (adapted from [71])*

### 3.2.3.3.4 Other Reactions

Besides the Maillard reaction, fat oxidation and thiamine degradation, more aroma compounds in process flavours are formed by the interaction of these different reactions, as well as from other precursors (e.g. phenol compounds, terpenes) present in the raw materials used. Three examples are given below but are far from being exhaustive.

A powerful sulphury aroma compound, 3-mercapto-2-methylpentan-1-ol (**51**), has recently been identified *[13]* in a complex process flavouring. It has a low odour threshold of 0.15 mg/L of water. The formation could be traced back to the onions present in the process flavouring and its formation is explained from propanal present in onions via aldol condensation, addition of hydrogen sulphide and enzymatic reduction (Fig. 3.31).

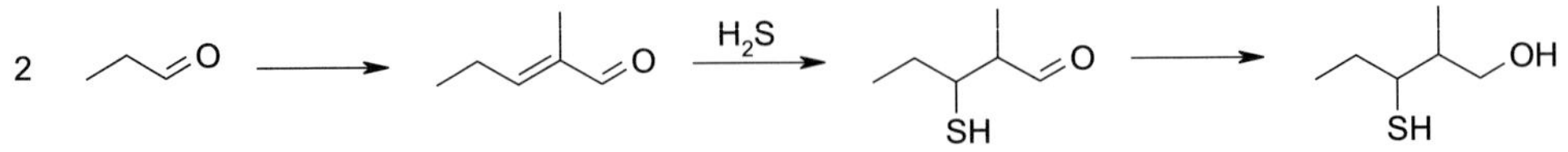

*Fig. 3.31: Hypothetical formation of 3-mercapto-2-methylpentan-1-ol from propanal in onions*

Phenolic aroma compounds can be generated by the thermal radical degradation of phenolic acids such as ferulic acid (**52**), which is a constituent of many vegetable raw materials *[76]*. Fig. 3.32 shows the formation scheme for vinylguaiacol (**53**), vanilline (**54**) and guaiacol (**55**) from **52**.

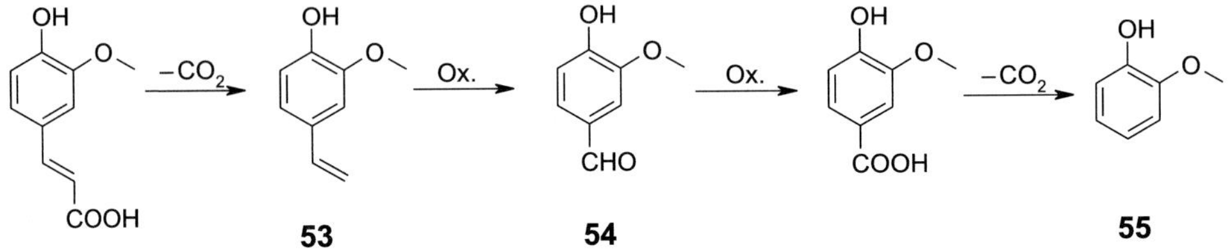

*Fig. 3.32: Degradation of ferulic acid*

The fatty, tallowy-smelling aroma compound 2-pentylpyridin (**56**), which has been identified in roasted sesame seeds *[77]*, is formed from the reaction of a fat oxidation product, 2,4-decadienal, with ammonia from the degradation of amino acids like glutamine and asparagine *[78]* (cf. Fig. 3.33).

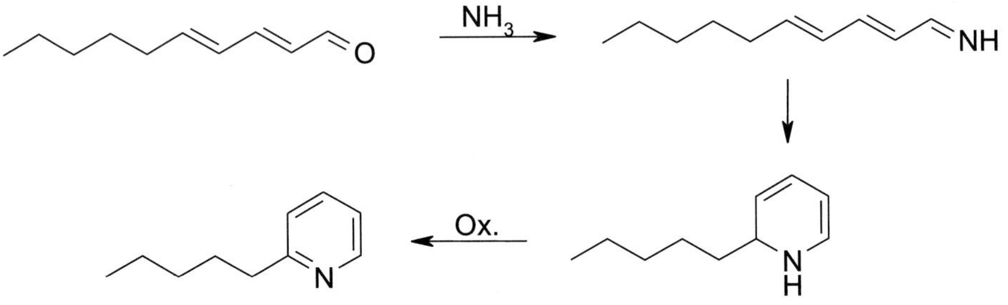

*Fig. 3.33: Formation of 2-pentylpyridine from 2,4-decadienal and ammonia*

### 3.2.3.3.5 Taste Compounds

Taste compounds and flavour enhancers are raw materials used in process flavourings and can also be added to the final product. Current examples are sodium and potassium salts, organic acids, monosodium glutamate (MSG) and the 5'-nucleotides 5'-inosine monophosphate (IMP) and 5'-guanosin monophosphate (GMP).

Moreover, taste compounds can also be generated in the reaction. Maillard observed already in 1911 that cyclodipeptides (diketopiperazines) are formed when amino acids are heated in glycerol at elevated temperatures *[79]*. Certain diketopiperazines have been shown to contribute, together with the purine derivative theobromine, to the bitterness of cocoa *[80]*.

Recently, several other bitter compounds have been found in model reactions. Frank and co-workers evaluated sensorially the HPLC fractions of the reaction products and further analysed the fractions with the lowest taste thresholds by LC/MS *[81]*. This led to the identification of **57** and **58** (Fig. 3.34) in a heated xylose/alanine solution *[82, 83]*.

The very potent taste compound **59** (Fig. 3.34), which has a low taste threshold of only 9 mmol/L, and several other bitter compounds were responsible for the taste of a roasted proline/glucose mixture *[84]*.

A new taste enhancer was isolated from a beef broth and identified as *N*-(1-carboxyethyl)-6-(hydroxymethyl)pyridinium-3-ol inner salt (**60**, alapyridaine; Fig. 3.35) *[85]*. The authors found that it increases the sweetness and umami character when added to a synthetic beef broth. The same compound forms also during the Maillard reaction between glucose and alanine *[86]*. Its formation can be explained via 3-deoxyglucosone and 5-hydroxymethyl-2-furfural as intermediates *[84]*.

*Fig. 3.34: Bitter compounds from the Maillard reaction*

Another taste modifier, *N*-(1-methyl-4-hydroxy-3-imidazolin-2,2-ylidene)-alanine (**61**; Fig. 3.35), was already isolated earlier *[87]*. Its formation can be explained by a thermal reaction between creatinine and lactic acid or alanine, compounds which are also present in process flavourings.

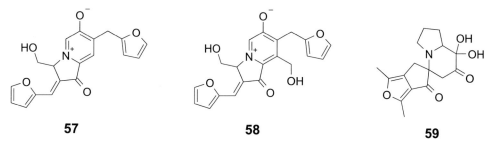

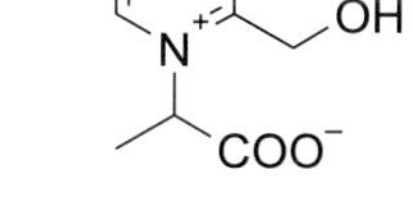

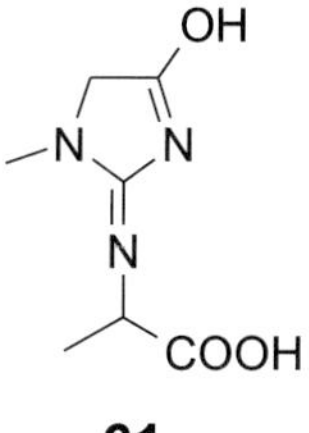

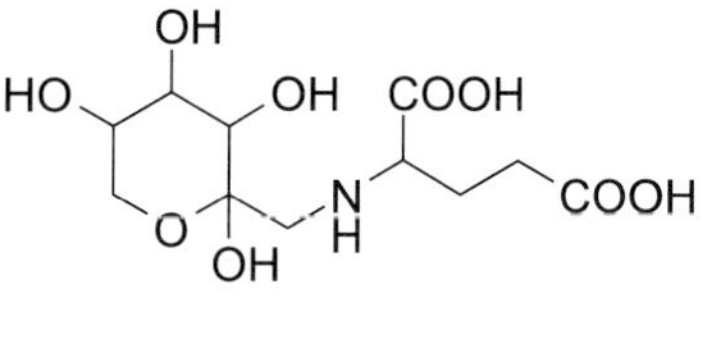

*Fig. 3.35: Structure of new brothy or umami taste compounds*

The Amadori compound between glucose and glutamic acid (**62**; Fig. 3.35) is present in considerable amounts of up to 3% in dried vegetables *[88]*. It was recently isolated from a wheat gluten hydrolysate and found to possess a pronounced umami note [89, 90]. Supposedly it is also present in process flavours made from glucose and glutamic acid.

#### 3.2.3.4 Industrial Process Flavourings

#### 3.2.3.4.1 Savoury Process Flavourings

Savoury process flavourings are widely used in culinary products to boost meat flavour (beef, chicken, pork, lamb, veal) or other notes (fish, seafood, vegetable). The development of meat-like process flavours started in the 1950s and the first patents were published in 1960. To date meat-like and other savoury flavours still represent the largest volumes of all process flavourings produced.

Cysteine is the key sulphur precursor in the centre of the first patents on meat-like process flavourings. Morton and co-workers [1] claim in their pioneering patent application the thermal reaction between a pentose or hexose with cysteine in water under non-alkaline conditions. Other amino acids or amino acid sources like hydrolysed proteins can be added. In a typical example, ribose, cysteine and other amino acids are boiled for 2 hours in water resulting in a flavouring with cooked beef character. Other patents use glyceraldehyde [91] or furan compounds such as furfural [92] instead of reducing sugars in the reaction with cysteine. Another possible carbohydrate source is hydrolysed pectine, which is rich in galacturonic acid. Such a pectine hydrolysate is reacted with cysteine and other amino acids at a pH value between 5 and 7 [93]. Process flavours can also be produced in solvents other than water. For instance, meat flavourings are obtained from ribose, cysteine, methionine and proline when heated in a polyalcohol such as glycerol [94].

Thiamine is another key sulphur precursor used for meat-like process flavours. Bidmead and co-workers [95] used thiamine as precursor in combination with an aliphatic carboxylic acid to obtain roasted meat flavours. The reaction can be carried out in presence or absence of a carbohydrate. For example, refluxing thiamine, cysteine hydrochloride and a carbohydrate-free HVP (hydrolysed vegetable protein) for 4 hours results in a product with beef flavour. Giacino [69] proposes the reaction of thiamine with a cysteine-containing peptide, e.g. glutathione or an amino acid mixture containing cysteine, and adding aldehydes and ketones after the reaction. The reaction is preferably carried out in fat.

Even hydrogen sulphide and its salts can be suitable sulphur precursors for meat-like flavours in the reaction with pentoses, hexoses and other precursors. Gunther [96] claims the reaction between a pentose, such as ribose or xylose, with hydrogen sulphide under acidic or neutral conditions. Meat flavour is also the objective of another patent application, which uses the reaction of inorganic sulphides with pentoses or hexoses, followed by the acidification of the resulting reaction product [97]. The application concentration of the resulting flavouring in food can be as low as 0.5-25 mg/kg. Instead of reducing sugars, more reactive intermediates of the Maillard reaction can be used in the reaction with hydrogen sulphide. Independently, two patent applications assigned to different companies describe the use of heterocyclic ketones (e.g. 4-hydroxy-5-methyl-3(2$H$)-furanone, 4-hydroxy-2,5-dimethyl-3(2$H$)-

furanone and maltol) for the reaction with hydrogen sulphides or inorganic sulphides *[98, 99]*. Heyland *[100]* proposes a somewhat different sulphur source. Hydrolysed onion, garlic and cabbage, which are known to contain sulphur compounds, are used for the reaction with reducing sugars to obtain beef flavourings. Van Pottelsberghe de la Potterie *[101]* describes a process for obtaining a beef flavouring agent by reacting a vegetable protein hydrolysate, which is cysteine-free, together with a reducing sugar such as xylose as well as methionine as the sulphur source. Optionally carboxylic acids can be present.

Animal phospholipids, especially egg phospholipids or egg yolk, are the key precursors for a patent application, which claims the reaction with a sulphur-containing compound like cysteine to produce meaty flavours *[102]*. Phospholipids and monoglycerides can play a fundamental role in the physical structure of a process flavour. Vauthey and co-workers *[103]* use these emulsifiers in water to generate nanostructured mesophases or microemulsions at the reaction temperature. Better aroma yields and different aroma compounds were obtained in these nanostructured reaction media compared to aqueous solutions.

Reaction flavours can be designed to generate the aroma only in the food during heat processing, e.g. in a microwave oven. In this case, the product is a mix of aroma precursors which has to keep associated after the inclusion in the foodstuff, e.g. by encapsulation *[104]*. A typical example is the spray-drying of a solution of cysteine hydrochloride and 4-methoxy-2,5-dimethyl-3(2*H*)-furanone on a carrier, adding the powder to a gravy or hamburger meat and microwaving the product.

### 3.2.3.4.2 Sweet Process Flavourings

Process flavourings for application in sweet and bakery goods (e.g. flavourings with chocolate, malt, caramel, egg, coffee or biscuit tonality) have been known for an even longer time than their savoury counterparts. Sulphur-containing aroma precursors such a cysteine are not the main concern of sweet process flavourings, and in many cases they even have to be avoided in order to prevent the formation of off-flavours.

A patent application from 1918 deals with the formulation of malt flavours *[105]* claiming the reaction of raw materials which are rich in amino acids (e.g. yeast autolysate) with sugars to produce a malt flavour. However, no reference is made to specific amino acids.

Most patent applications for sweet process flavours deal with chocolate flavourings. Russoff *[106]* obtained chocolate flavour from the reaction of sugars with glycine or alanine containing peptides at 90-170°C, depending on the reactivity of the sugar with lower reaction temperatures for pentoses (90-130°C) and higher ones for non-reducing sugars (140-170°C). Instead of pure peptides, partially hydrolysed proteins (e.g. milk proteins) can also be used *[107]*. The reaction of enzymatic gelatine hydrolysates with glucose and fat at 110°C during 40 minutes seems similarly suitable for producing cocoa-like flavour *[108]*. The inventors emphasise the importance of a water level below 30% during the reaction in order to obtain good chocolate-like notes.

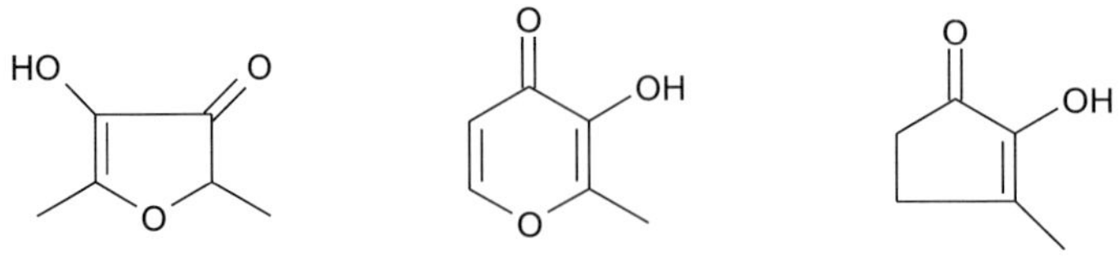

*Fig. 3.36: Cyclic ketones used in process flavours*

Flavourings that are useful for aromatising bakery and chocolate food products can be made from sulphur-free amino acids by the reaction with cyclic ketones (Fig. 3.36) such as 4-hydroxy-2,5-dimethyl-3(2*H*)-furanone (**26**), maltol (**63**) or 2-hydroxy-3-methyl-2-cyclopenten-1-one (**64**, cyclotene) *[109]*. Amino acids of special interest are leucine, valine, proline and hydroxyproline. The reaction is carried out favourably in fat or propylene glycol.

Byrne *[110]* used a combination of the amino acids phenylalanine and leucine in the reaction with a reducing sugar selected from the group consisting of rhamnose, ribose and glucose to make a chocolate flavouring by microwave radiation. The reaction has to be carried out in solvents that can raise the dielectric constant of the mixture, e.g. propylene glycol or glycerol. The same precursors, phenylalanine and leucine, in a specific ratio between 1.5 and 6, together with a reducing sugar, are the subject of another patent *[111]*; however, the reaction is carried out under roasting conditions in a fat matrix. The addition of alanine as a third amino acid is beneficial. A reaction flavour with caramel-like and biscuit-like, cookie-like attributes is obtained from proline and rhamnose by heating them for 10-120 minutes at 100-140°C in a fat-based medium *[112]*.

### 3.2.3.4.3 Industrial Production

The classic approach for producing process flavourings on an industrial scale is the reaction in a batch reactor, very similar to reactors used in the chemical industry. The reactor size can vary from below 100 litres to several cubic metres. Commonly it is made from food-grade stainless steel, equipped with a steam- or oil-heated double jacket, and a stirrer. The reactor may include a high-shear mixer or a colloid mill to allow rapid mixing and dispersion and homogenisation before, during or after the thermal reaction is completed. Most instruments would allow reactions under pressure up to 3-10 bar, and in addition include a vacuum pump. In the majority of the cases the resulting flavourings, after an optional addition of a carrier like maltodextrin, are dried. Common drying techniques include spray-drying as well as vacuum-drying using a vacuum oven (discontinuous) or vacuum belt dryers (continuous), followed by milling and sieving.

Reactors for process flavours can be designed to handle liquids as well as paste products; however, they are not easily suitable for the reaction of solid mixtures with relatively low water content. In the latter case flavour generation by extrusion is a valuable option *[113, 114]*. However, since the residence time of the precursors in the extruder is rather in the range of a few minutes compared to up to several hours in a

# Process Flavourings

batch reactor, the reaction temperatures which are employed are usually higher, typically in the range of 120-180°C, to achieve an acceptable flavour. Both single- or twin-screw extruders are prevalent. After extrusion the product can, if necessary, directly be expanded into a vacuum chamber in order to reduce the moisture content down to a stable level and to interrupt the Maillard reaction instantly [115]. Other alternative technologies to produce process flavours include processing in a drying oven and microwave heating.

### 3.2.3.4.4 Safety

The Council of Europe has set certain limits for the purity of process flavourings [4]. The limits for heavy metals are 10 mg/kg for lead, 3 mg/kg for arsenic and 1 mg/kg for cadmium and mercury. The limit for the carcinogenic benzo(a)pyrene (**65**) is set to 1 µg/kg, which represents the analytical detection limit. Similarly, the level for PhIP (**66**) and the quinoxaline IQ derivative DiMeIQx (**67**), both mutagenic, is restricted to 50 µg/kg flavouring. Flavourings, which are prepared without creatine (**70**) and creatinine (**71**) and without raw materials containing these compounds (e.g. meat ingredients), are not considered to contain significant amounts of IQ mutagens.

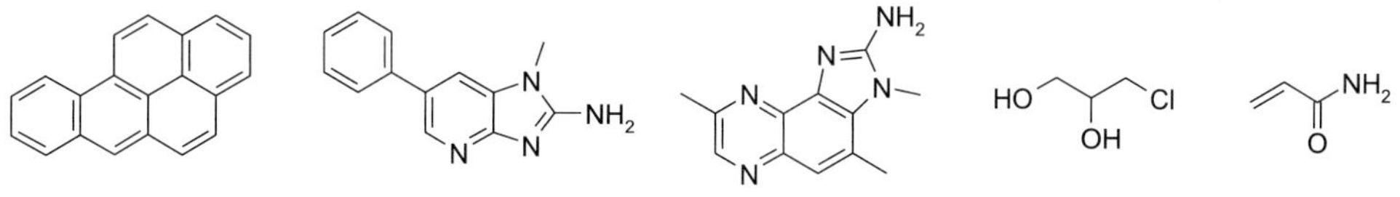

*Fig. 3.37: Structure of toxic compounds in food*

For new developments of reaction flavours, different toxicological studies might be necessary: a gene mutagenicity test in bacteria such as an Ames test, an in vitro gene mutagenicity test in mammalian cells, an assay for chromosome damage in vitro and even a 90 days feeding study in a rodent species [4]. The toxicological tests can be reduced, however, in cases where the new process flavour does not differ substantially in terms of ingredients and processing conditions from one that has already been tested and found acceptable.

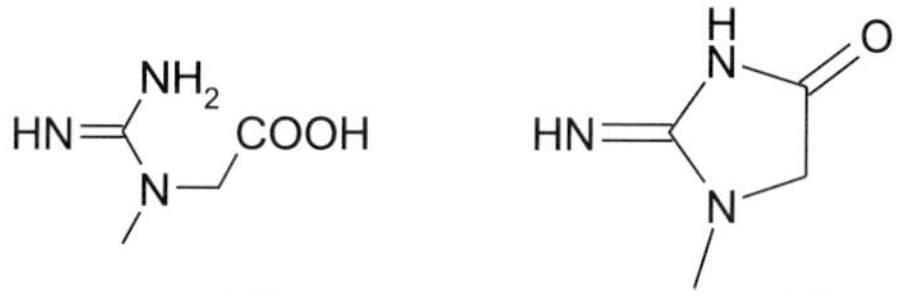

*Fig. 3.38: Structure of creatine and creatinine*

Hydrolysed vegetable proteins that have been obtained from acid hydrolysis (HVP) contain chlorohydrins [116] such as 3-chloro-1,2-propandiol (**68**, 3-MCPD). The European Union limits 3-MCPD in soy sauce and HVP to 20 µg/kg [117]. The level of 3-MCPD in other foods such as toasted bread, grilled cheese and fried batter strongly depends on the cooking conditions, especially the temperature [118]. Process flavours that have been manufactured without HVP and at temperatures lower than 120°C carry a negligible risk. Another potential carcinogen, acrylamide (**69**), has recently been detected in a wide range of foodstuffs. It is formed in the Maillard reaction at higher temperatures from asparagine as the key precursor *[119, 120]*. Again, the risk for acrylamide in process flavours is insignificant if no asparagine is present and reaction temperatures are 120°C or lower.

In the USA process flavourings are generally recognised as safe (GRAS) because the manufacturing process mimics cooking conditions with temperatures mostly below 150°C, which is much milder than home-style cooking like barbecuing, the preparation is similar to cooking a gravy, and the use level of process flavourings is low *[121]*.

### 3.2.3.5 Outlook

The progress in the development of process flavours has never stopped since its beginning, and the following trends are only a glimpse of what research and development of process flavours are addressing:

- Modelling the Maillard reaction to optimise the aroma compound yield and create stronger flavourings.
- Improving the stability of process flavourings in the food application.
- Controlling the release of aroma during the preparation of the final food.
- Using Maillard chemistry to generate taste compounds (sweet, umami).
- Replacing animal products in process flavourings by vegetarian ones.
- Using raw material that meet religious requirements (halal, kosher, Buddhist-conform) and consumer demands (natural, genetically unmodified, organically grown).

## REFERENCES

[1] Morton, I.D., Akroyd, P., May, C.G. Flavoring substances and their preparation. UK Patent, 1960, 836,694, assigned to Unilever.

[2] International Organisation of the Flavour Industry. IOFI Guidelines for the production and labelling of process flavourings. In: Code of practice for the flavour industry, including replacement sheets January 1997, IOFI, Geneva, 1997, F1-F4.

[3] Manley, C.H., Ahmedi, S. The development of process flavours. Trends Food Sci. Technol. 1995, 6, 46-51.

[4] Gry, J. Guidelines for safety evaluation of thermal process flavourings, Council of Europe Publishing, Strasbourg, 1995.

[5] Van den Ouweland, G.A.M., Demole, E.P., Enggist, P. Process meat flavor development and the Maillard reaction. In: Thermal generation of aromas, Parliment, T.H., McGorrin, R.J., Ho, C.T. (Eds.), American Chemical Society, Washington, DC, 1989, 433-441.

[6] May, C.G. Process flavourings. In: Food flavourings, Ashurst, P.R. (Ed.), Blackie, Glasgow, 1991, 255-301.

[7] Salzer, U.J. Reaktionsaromen: Definition, Entwicklung und Herstellung. ZFL 1993, 44, 524-526.

[8] Manley, C. Process flavors. In: Source book of flavors, Reineccius, G. (Ed.), Chapman & Hall, New York, 1994, 139-154.

[9] Manley, C. Process flavors and precursor systems: commercial preparation and use. In: Thermally generated flavors: Maillard, microwave, and extrusion, Parliment, T.H., Morello, M.J., McGorrin, R.J. (Eds.), American Chemical Society, Washington, DC, 1994, 16-25.

[10] Kerler, J., Winkel, C. The basic chemistry and process conditions underpinning reaction flavour production. In: Food Flavour Technology, Taylor, A. (Ed.), Sheffield Academic Press, Sheffield, 2002, 27-59.

[11] Herz, K.O., Chang, S.S. Meat Flavor. Adv. Food Res. 1970, 18, 1-83.

[12] Hornstein I., Crowe, P.F. Flavor studies on beef and pork. J. Agric. Food Chem. 1960, 8, 494-498.

[13] Widder, S., Sabater Lüntzel, C., Dittner, T., Pickenhagen, W. 3-Mercapto-2-methylpentan-1-ol, a new powerful aroma compound. J. Agric. Food Chem. 2000, 48, 418-423.

[14] Maillard, L.C. Action des acides amines sur les sucres: Formation des mélanoidines par voie méthodique. C.R. Hebd. Séances Acad. Sci. 1912, 154, 66-68.

[15] Mabrouk, A.F., Jarboe, J.K., O'Connor, E.M. Water-soluble flavor precursors of beef. Extraction and fractionation. J. Agric. Food Chem. 1969, 17, 5-9.

[16] Hurrell, R.F. Maillard reaction in flavour. In: Food flavours. Part A. Introduction, Morton, I.D., MacLeod, A.J. (Eds.), Science Publishing, Amsterdam, 1982, 399-437.

[17] Baines, D.A., Mlotkiewicz, J.A. The chemistry of meat flavour. In: Recent advances in the chemistry of meat, Bailey, A.J. (Ed.), Royal Society of Chemistry, London, 1984, 119-164.

[18] Mottram, D.S., Salter, L.J. Flavor formation in meat-related Maillard systems containing phospholipids. In: Thermal generation of aromas, Parliment, T.H., McGorrin, R.J., Ho, C.T. (Eds.), American Chemical Society, Washington, DC, 1989, 442-451.

[19] Ledl, F., Schleicher, E. New aspects of the Maillard reaction in foods and in the human body. Angew. Chem. Int. Ed. Engl. 1990, 29, 565-706.

[20] Ames, J.M., Hincelin, O. Novel sulphur compounds from heated thiamine and xylose/thiamine model systems. In: Progress in flavour precursor studies, Schreier, P., Winterhalter, P. (Eds.), Allured Publishing, Carol Stream, 1993, 379-382.

[21] Ruckdeschel, W. Ueber Melanoidine und ihr Vorkommen in Darrmalz. Zeitschr. Ges. Brauw. 1914, 37, 430-440.

[22] Hodge, J.E. Chemistry of browning reactions in model systems. J. Agric. Food Chem. 1953, 1, 928-943.

[23] Amadori, M. Prodotti di condensazione tra glucosio e p-toluidina. Atti R. Acad. Naz. Lincei 1931, 13, 72-77.

[24] Kuhn, R., Weygand, F. Die Amadori-Umlagerung. Ber. Dtsch. Chem. Ges. 1937, 70, 769-772.

[25] Hodge, J.E. The Amadori rearrangement. Adv. Carbohydr. Chem. 1955, 10, 169-205.

[26] Heyns, K., Paulsen, H., Eichstedt, R., Rolle, M. Ueber die Gewinnung von 2-Amino-aldosen durch Umlagerung von Ketosylaminen. Chem. Ber. 1957, 90, 2039-2049.

[27] Beck, J., Ledl, F., Severin, T. Formation of 1-deoxy-D-erythro-2,3-hexodiulose from Amadori compounds. Carbohydr. Res. 1988, 177, 240-243.

[28] Huber, B., Ledl, F. Formation of 1-amino-1,4-dideoxy-2,3-hexodiuloses and 2-aminoacetylfurans in the Maillard reaction. Carbohydr. Res. 1990, 204, 215-220.

[29] Weenen, H., Tjan, S.B. Analysis, structure, and reactivity of 3-deoxyglucosone. In: Flavor precursors: thermal and enzymatic conversions, Teranishi, R., Takeoka, G.R., Guentert, M. (Eds.), American Chemical Society, Washington, DC, 1992, 217-231.

[30] Brands, C.M.J., van Boekel, M.A.J.S. Reaction of monosaccharides during heating of sugar-casein systems: building of a reaction network model. J. Agric. Food Chem. 2001, 49, 4667-4675.

[31] Davídek, T., Clety, N., Aubin, S., Blank, I. Degradation of the Amadori compound N-(1-deoxy-D-fructos-1-yl)glycine in aqueous model systems. J. Agric. Food Chem. 2002, 50, 5472-5479.

[32] Hayashi, T., Namiki, M. Role of sugar fragmentation in an early stage browning of amino-carbonyl reaction of sugar with amino acid. Agric. Biol. Chem. 1986, 50, 1965-1970.

[33] Hayashi, T., Mase, S., Namiki, M. Formation of three-carbon sugar fragment at an early stage of the browning reaction of sugar with amines or amino acids. Agric. Biol. Chem. 1986, 50, 1959-1964.

[34] Strecker, A. Notiz über eine eigenthümliche Oxydation durch Alloxan. Liebigs Ann. Chem. 1862, 123, 363-365.

[35] Schönberg, A., Moubacher, R. The Strecker degradation of α-amino acids. Chem. Rev. 1952, 50, 261-277.

[36] Hofmann, T., Münch, P., Schieberle, P. Quantitative model studies on the formation of aroma-active aldehydes and acids by Strecker-type reactions. J. Agric. Food Chem. 2000, 48, 434-440.

[37] Hofmann T, Schieberle, P. Formation of aroma-active Strecker aldehydes by a direct oxidative degradation of Amadori compounds. J. Agric. Food Chem. 2000, 48, 4301-4305.

[38] Hofmann, T., Schieberle, P. Evaluation of the key odorants in a thermally treated solution of ribose and cysteine by aroma extract dilution techniques. J. Agric. Food Chem. 1995, 43, 2187-2194.

[39] Shibamoto, T. Formation of sulfur- and nitrogen-containing compounds from the reactions of alkanediones with hydrogen sulfide and ammonia. J. Agric. Food Chem. 1977, 25, 206-208.

[40] Van den Ouweland, G.A.M. Components contributing to beef flavor. Volatile compounds produced by the reaction of 4-hydroxy-5-methyl-3(2H)-furanone and its thio analog with hydrogen sulfide. J. Agric. Food Chem. 1975, 23, 501-505.

[41] Whitfield, F.B., Mottram, D.S. Investigation of the reaction between 4-hydroxy-5-methyl-3(2H)-furanone and cysteine or hydrogen sulfide at pH 4.5. J. Agric. Food Chem. 1999, 47, 1626-1634.

[42] Whitfield, F.B., Mottram, D.S. Heterocyclic volatiles formed by heating cysteine or hydrogen sulfide with 4-hydroxy-5-methyl-3(2H)-furanone at pH 6.5. J. Agric. Food Chem. 2001, 49, 816-822.

[43] Cerny, C., Davidek, T. Formation of aroma compounds from ribose and cysteine during the Maillard reaction. J. Agric. Food Chem. 2003, 51, 2714-2721.

[44] Mottram, D.S., Madruga, M.S., Whitfield, F.B. Some novel meat-like aroma compounds from the reactions of alkanediones with hydrogen sulfide and furanthiols. J. Agric. Food Chem. 1995, 43, 189-193.

[45] Cerny, C., Davidek, T. α-Mercaptoketone formation during the Maillard reaction of cysteine and [1-$^{13}$C]ribose. J. Agric. Food Chem. 2004, 52, 958-961.

[46] Tressl, R., Helak, B., Martin, N., Rewicki, D. Formation of flavor compounds from L-proline. In: Topics in flavour research, Berger, R.G., Nitz, S., Schreier, P. (Eds.), Eichhorn, Marzling-Hagenham, 1985, 139-159.

[47] Roberts, D.D., Acree, T.E. Gas chromatography-olfactometry of glucose-proline Maillard reaction. In: Thermally generated flavours: Maillard, microwave, and extrusion processes, Parliment, T.H., Morello, M.J., McGorrin, R.J. (Eds.), American Chemical Society, Washington, DC, 1992, 71-79.

[48] Hodge, J.E., Mills, F.D., Fisher, B.E. Compounds of browned flavor derived from sugar-amine reactions. Cereal Sci. Today 1972, 17, 34-40.

[49] Schieberle, P. The role of free amino acids present in yeast as precursors of the odorants 2-acetyl-1-pyrroline and 2-acetyltetrahydropyridine in wheat bread crust. Z. Lebensm. Unters. Forsch. 1990, 191, 206-209.

[50] Cerny, C., Grosch, W. Evaluation of potent odorants in roasted beef by aroma extract dilution analysis. Z. Lebensm. Unters. Forsch. 1992, 194, 322-325.

[51] Schieberle, P., Grosch, W. Changes in the concentrations of potent crust odorants during storage of white bread. Flavour Fragr. J. 1992, 7, 213-218.

[52] Blank, I., Sen, A., Grosch, W. Potent odorants of the roasted powder and brew of Arabica coffee. Z. Lebensm. Unters. Forsch. 1992, 195, 239-245.

[53] Schnermann, P., Schieberle, P. Evaluation of key odorants in milk chocolate and cocoa mass by aroma extract dilution analysis. J. Agric. Food Chem. 1997, 45, 867-872.

[54] Preininger, M., Grosch, W. Evaluation of the key odorants of the neutral volatiles of Emmentaler cheese by the calculation of odour activity values. Lebensm. Wiss. Technol. 1997, 27, 237-244.

[55] Sasaki, M., Nunomura, N., Matsudo, T. Biosynthesis of 4-hydroxy-2(or 5)-ethyl-5(or 2)-methyl-3(2H)-furanone by yeasts. J. Agric. Food Chem. 1991, 39, 934-938.

[56] Re, L., Maurer, B., Ohloff, G. A simple synthesis of 4-hydroxy-2,5-dimethyl-3(2H)-furanone (Furaneol), an important aroma compound of pineapple and strawberry. Helv. Chim. Acta 1973, 56, 1882-1894.

[57] Blank, I., Fay, L.B. Formation of 4-hydroxy-2,5-dimethyl-3(2H)-furanone and 4-hydroxy-2(or 5)-ethyl-5(or 2)-methyl-3(2H)-furanone through Maillard reaction based on pentose sugars. J. Agric. Food Chem. 1996, 44, 531-536.

[58] Hofmann, T. Charakterisierung intensiver Geruchsstoffe in Kohlenhydrat/Cystein-Modellreaktionen und Klärung von Bildungswegen. Doctoral thesis, Technical University of Munich, 1995, 106-111.

[59]    Hofmann, T., Schieberle, P. 2-Oxopropanal, hydroxy-2-propanone, and 1-pyrroline: important intermediates in the generation of the roast-smelling food flavor compounds 2-acetyl-1-pyrroline and 2-acetyltetrahydropyridine. J. Agric. Food Chem. 1998, 46, 2270-2277.

[60]    Grosch, W. Reactions of hydroperoxides: products of low molecular weight. In: Autoxidation of unsaturated lipids, Chan, H.W. (Ed.), Academic Press, London, 1987, 95-137.

[61]    Guth, H., Grosch, W. Identification of the character impact odorants of stewed beef juice by instrumental analyses and sensory studies. J. Agric. Food Chem. 1994, 42, 2862-2866.

[62]    Blank, I., Lin, J., Arce Vera, F., Welti, D.H., Fay, L.B. Identification of potent odorants formed by autoxidation of arachidonic acid: structure elucidation and synthesis of (E,Z,Z)-2,4,7-tridecatrienal. J. Agric. Food Chem. 2001, 49, 2959-2965.

[63]    Lin, J., Leser, M.E., Löliger, J., Blank, I. A new insight into the formation of odor active carbonyls by thermally induced degradation of phospholipids in self-assembly structures. J. Agric. Food Chem. 2004, 52, 581-586.

[64]    Arnold, R.G., Libbey, L.B., Lindsay, R.C. Volatile flavor compounds produced by heat degradation of thiamine (vitamin $B_1$). J. Agric. Food Chem. 1969, 17, 390-392.

[65]    Reineccius, G.A., Liardon, R. The use of charcoal traps and microwave desorption for the analysis of headspace volatiles above heated thiamine solutions. In: Topics in flavor research, Berger, R.G., Nitz, S., Schreier, P. (Eds.), Eichhorn, Marzling-Hangenham, 1985, 125-136.

[66]    Gasser, U., Grosch, W. Identification of volatile flavour compounds with high aroma values from cooked beef. Z. Lebensm. Unters. Forsch. 1988, 186, 489-494.

[67]    Gasser, U., Grosch, W. Primary odorants of chicken broth. Z. Lebensm. Unters. Forsch. 1990, 190, 3-8.

[68]    Evers, W.J., Heinsohn, H.H. Jr., Mayers, B.J., Sanderson, A. Furans substituted at the three position with sulfur. In: Phenolic, sulfur and nitrogen compounds in food flavours, Charalambous, G., Katz, I. (Eds.), American Chemical Society, Washington, DC, 1976, 184-193.

[69]    Giacino, C. Poultry flavor composition and process. US Patent, 1968, 3,394,017, assigned to International Flavors & Fragrances.

[70]    Bolton, T.A., Reineccius, G.A., Liardon, R. Huyn-Ba, T. Role of cysteine in the formation of 2-methyl-3-furanthiol in a thiamine-cysteine model system. In: Thermally generated flavours: Maillard, microwave, and extrusion, Parliment, T.H., Morello, M.J., McGorrin, R.J. (Eds.), American Chemical Society, Washington, DC, 1994, 270-278.

[71]    Van der Linde, L.M., van Dort, J.M., de Valois, P., Boelens, H., de Rijke, D. Volatile components from thermally degraded thiamine. In: Progress in flavour research, Land, D.G., Nursten, H.E. (Eds.), Applied Sciences, London, 1979, 219-224.

[72]    Zeiler, G. Modellversuche zur Bildung von 2-Methyl-3-furanthiol und 2-Furfurylthiol in gekochtem Fleisch. Doctoral thesis, Technical University of Munich, 1994.

[73]    Stoll, M., Dietrich, P., Sundt, E., Winter, M. Sur les arômes. Sur l'arôme de café. II. Helv. Chim. Acta. 1967, 50, 2065-2067.

[74]    Güntert, M., Brüning, J., Emberger, R., Köpsel, M., Kuhn, W., Thielmann, T., Werkhoff, P. Identification and formation of some selected sulfur-containing flavor compounds in various meat model systems. J. Agric. Food Chem. 1990, 38, 2027-2041.

[75]    Pittet, A.O., Hruza, D.E. Comparative study of flavor properties of thiazole derivatives. J. Agric. Food Chem. 1974, 22, 264-269.

[76]    Belitz, H.-D., Grosch, W., Schieberle, P. Food chemistry, 3rd edition, Springer, Berlin, 2004, 364-366.

[77]    Schieberle P. Studies on the flavour of roasted white sesame seeds. In: Progress in flavour precursor studies, Schreier, P., Winterhalter, P. (Eds.), Allured Publishing, Carol Stream, 1993, 343-360.

[78]    Kim, Y., Hartmann, T.G., Ho, C.-T. Formation of 2-pentylpyridine from the thermal interaction of amino acids and 2,4-decadienal. J. Agric. Food Chem. 1996, 44, 3906-3908.

[79]    Maillard, L.C. Condensation des acides aminés en présence de la glycérine: cycloglycylglycines et polypeptides. C.R. Hebd. Séances Acad. Sci. 1911, 153, 1078-1080.

[80]    Pickenhagen, W., Dietrich, P., Keil, B., Polonsky, J., Nouaille, F., Lederer, E. Identification of the bitter principle of cocoa. Helv. Chim. Acta 1975, 58, 1078-1086.

[81] Frank, O., Ottinger, H., Hofmann, T. Characterization of an intense bitter-tasting 1*H*,4*H*-quinolizinium-7-olate by application of the taste dilution analysis, a novel bioassay for the screening and identification of taste-active compounds in food. J. Agric. Food Chem. 2001, 49, 231-238

[82] Frank, O., Hofmann, T. Reinvestigation of the chemical structure of bitter-tasting quinizolate and homoquinizolate and studies on their Maillard-type formation pathways using suitable $^{13}$C-labeling experiments. J. Agric. Food Chem. 2002, 50, 6027-6036.

[83] Frank, O., Jezussek, M., Hofmann, T. Sensory activity, chemical structure, and synthesis of Maillard generated bitter-tasting 1-oxo-2,3-dihydro-1*H*-indolizinium-6-olates. J. Agric. Food Chem. 2003, 51, 2693-2699.

[84] Ottinger, H. Charakterisierung geschmacksaktiver Verbindungen aus Kohlenhydrat/Aminosäure-Reaktionen mittels instrumentell-analytischer Techniken. PhD Thesis, Technical University of Munich, 2004, 30-117.

[85] Ottinger, H., Soldo, T., Hofmann, T. Discovery and structure determination of a novel Maillard-derived sweetness enhancer by application of the comparative taste dilution analysis (cTDA). J. Agric. Food Chem. 2003, 51, 1035-1041.

[86] Ottinger, H., Hofmann, T. Identification of the taste enhancer alapyridaine in beef broth and evaluation of its sensory impact by taste reconstitution experiments. J. Agric. Food Chem. 2003, 51, 6791-6796.

[87] Shima, K., Yamada, N., Suzuki, E., Harada, T. Novel brothy taste modifier isolated from beef broth. J. Agric. Food Chem. 1998, 46, 1465-1468.

[88] Eichner, K., Reutter, M., Wittmann, R. Detection of Amadori compounds in heated foods. In: Thermally generated flavours: Maillard, microwave, and extrusion, Parliment, T.H., Morello, M.J., McGorrin, R.J. (Eds.), American Chemical Society, Washington, DC, 1994, 42-54.

[89] Schlichtherle-Cerny, H., Amadò, R., Affolter, M. LC-tasting of an acid-deamidated wheat gluten hydrolysate with umami flavour. In: Flavour research at the dawn of the twenty-first century, Le Quéré, J.L., Etiévant, P.X. (Eds.), Intercept, London, 2003, 301-304.

[90] Schlichtherle-Cerny, H., Affolter, M., Cerny, C. Taste-active glycoconjugates of glutamate: new umami compounds. In: Challenges in taste chemistry and biology, Hofmann, T., Ho, C.-T., Pickenhagen, W. (Eds.), American Chemical Society, Washington, DC, 2004, 210-222.

[91] May, C.G., Morton, I.D. Process for the preparation of a meat flavor. US Patent, 1960, 2,934,436, assigned to Unilever.

[92] May, C.G. Flavouring substances and their preparation. UK Patent, 1961, 858,333, assigned to Unilever.

[93] Heyland, S., Philippossian, G. Verfahren zur Herstellung von Aromatisierungsmitteln mit einem Geschmack nach gekochtem Fleisch und deren Verwendung. German Patent (DE), 1982, 2,546,008, assigned to Maggi.

[94] Jaeggi, K. Process for the manufacture of meat flavours. US Patent, 1973, 3,761,287, assigned to Givaudan.

[95] Bidmead, D.S., Giacino, C. Roasted meat flavor and process for producing same. US Patent, 1968, 3,394,016, assigned to International Flavors & Fragrances.

[96] Gunther, R. Flavoring compositions produced by reacting hydrogen sulfide with a pentose. US Patent 3,642,497, assigned to Fritzsche, Dodge & Olcott.

[97] Wiener, C. Meat flavor and its preparation. US Patent, 1972, 3,645,754, assigned to Polak's Frutal Works.

[98] Van den Ouweland, G.A.M., Peer, H.G. Substances aromatisantes destinées à l'aromatisation des aliments. French Patent, 1970, 2,012,073, assigned to Unilever.

[99] Katz, I., Pittet, A.O. Wilson, R.A., Evers, W.J. Foodstuff flavoring methods and compositions. US Patent, 1977, 4,045,587, assigned to International Flavors & Fragrances.

[100] Heyland, S. Verfahren zur Herstellung eines Aromatisierungsmittels mit einem Geschmack nach gekochtem Fleisch und die Verwendung desselben. German Patent (DE), 1977, 2,546,035, assigned to Maggi.

[101] Van Pottelsberghe de la Potterie, P.J. Beef flavor. US Patent, 1973, 3,716,380, assigned to Nestlé.

[102] Lee, E.C., Tandy, J.S. Preparation of flavors. US Patent, 1991, 5,039,543, assigned to Nestlé.

[103] Vauthey, S., Leser, M., Milo, C. An aroma product comprising flavoring compounds. PCT Patent, 2000, 00/33671.

[104]   Mottram, D.S., Ames, J., Mlotkiewicz, J., Copsey, J.P., Anderson, A. Flavouring agents. PCT Patent, 1998, 98/42208, assigned to Dalgety.
[105]   Wallerstein, M., Wallerstein, L. Extract having the flavour and aroma of malt, process of producing the same and beverage therefrom. GB Patent, 1918, 107,367.
[106]   Rusoff, I.I. Process of producing an artificial chocolate flavor and the resulting product. US Patent, 1958, 2,835,590, assigned to General Foods.
[107]   Rusoff, I.I. Flavor. US Patent, 1958, 2,835,592, assigned to General Foods.
[108]   Specht, M., Ruttloff, H., Vetters, M., Frech, I. Verfahren zur Herstellung von Kakao- und Schokoladenaromastoffen. German Patent (DD), 1984, 205,815, assigned to ADW der DDR.
[109]   Pittet, A.O., Seitz, E.W. Würz- und Aromamittel, ihre Herstellung und Verwendung. German Patent (DE), 1972, 2,144,074, assigned to International Flavors & Fragrances.
[110]   Byrne, B. Process for microwave chocolate flavor formulation, product produced thereby and uses thereof in augmenting or enhancing the flavor of foodstuffs, beverages and chewing gums. US Patent 5,041,296, assigned to International Flavors & Fragrances.
[111]   Watterson, J.J., Miller, K.B., Furjanic, J.J., Stuart, D.A. Process of producing cacao flavor by roasting combination of amino acids and reducing sugars. US Patent, 1997, 5,676,993, assigned to Hershey Foods.
[112]   Hansen, C.E., Budwig, C., Kochhar, S., Juillerat, M.A., Armstrong, E., Sievert, D., Spadone, J., Nicolas, P., Redgwell. Chocolate flavour manipulation. GB Patent, 2003, 2,370,213, assigned to Nestlé.
[113]   Baek, H.H., Kim, C.J., Ahn, B.H., Nam, H.S., Cadwallader, K.R. Aroma extract dilution analysis of a beeflike process flavor from extruded enzyme-hydrolyzed soybean protein. J. Agric. Food Chem. 2001, 49, 790-793.
[114]   Villota, R., Hawkes, J.G. Flavoring in Extrusion. In: Thermal generation of aromas, Parliment, T.H., McGorrin, R.J., Ho, C.T. (Eds.), American Chemical Society, Washington, DC, 1989, 280-295.
[115]   Heyland, S., Rolli, K., Röschli, K., Sihver, J.J. Procédé de fabrication d'un agent aromatisant. European Patent, 1988, 286,838, assigned to Nestlé.
[116]   Velisek, J., Davidek, J., Hayslova, J., Kubelka, V., Janicek, G., Mankova, B. Chlorhydrins in protein hydrolysates. Z. Lebensm. Unters. Forsch. 1978, 167, 241-244.
[117]   Commission of the European Communities. Commission regulation No. 466/2001 of 8 March 2001 setting maximum levels for certain contaminants in foodstuffs. Official J. Eur. Communities 2001, 44, L77/1-13.
[118]   Crews, C., Brereton, P., Davies, A. The effect of domestic cooking on the levels of 3-monochlorpropanediol in foods. Food Additiv. Contam. 2001, 18, 271-280.
[119]   Mottram, D.S., Wedzicha, B.L., Dodson, A.T. Acrylamide is formed in the Maillard reaction. Nature 2002, 419, 448-449.
[120]   Stadler, R.H., Blank, I., Varga, N., Robert, F., Hau, J., Guy, P.A., Robert, M., Riediker, S. Acrylamide from Maillard reaction products. Nature 2002, 419, 449-450.
[121]   Lin, L.J. Regulatory status of Maillard reaction flavors. In: Thermally generated flavours: Maillard, microwave, and extrusion, Parliment, T.H., Morello, M.J., McGorrin, R.J. (Eds.), American Chemical Society, Washington, DC, 1994, 7-15.

## 3.2.4 Smoke Flavourings

### 3.2.4.1 Nature, Preparation and Application

*Gary Underwood*
*John Shoop*

#### 3.2.4.1.1 Introduction

'Smoking is, next to drying and salting of food, perhaps the oldest process for preserving and flavouring foods. It has been probably in use since man knew fire for about 90,000 years and possibly longer. The original primary objective to preserve food is achieved by dehydration and by diminishing the number of surface bacteria through the action of certain components of the smoke and thus to enhance its keeping qualities. The secondary objectives of imparting desirable structural and sensory alterations, such as smoke color and smoke flavour, to the final product serve to make the smoked food attractive to the consumer. Today smoking is primarily used for organoleptic purposes [11].

Although this statement may be true for the natural vaporous smoking process, contemporary natural liquid smoke flavourings occupy a unique position in the flavour industry, as flavour is only one aspect of their usage. Depending upon the method of production, smoke flavourings may possess colouring, antimicrobial, antioxidative and protein denaturation as well as flavouring properties. All of these effects are used by the food industry, but it is the meat industry where smoke flavourings are most valued for their non-flavouring properties.

Smoke flavourings, which are produced from the water-soluble constituents from the pyrolysis of wood, contain naturally occurring substances that can be divided into three major classes: acids, carbonyls and phenolics. While the phenolics historically have been thought to be the essential flavour containing class of substances, Maga [1] has shown substantial smoke flavour in neutral, carbonyl and basic fractions as well. It is possible with today's technology to significantly increase the ratio of acids and carbonyls to phenols and still retain a phenolic complement to provide more than enough smoke flavour for most applications.

Application of natural aqueous smoke flavourings to meat is done in large part for the purpose of producing smoked colour. Smoke solutions are applied to the surface of meat where smoked product colour is desired. The carbonyls are known to be the initiators of a brown smoked colour. While aqueous smoke solutions are innately a reddish brown colour, virtually none of this colour is imparted to the product through a staining effect. Rather, carbonyls first react with amines to produce Schiff bases which eventually end up as mealnoidins through complex, multiple-step reactions which are driven by heat and dehydration.

In applications to sausage products which are produced in cellulose casings, the coagulation of surface proteins is an essential functional effect of smoke solutions. While the above mentioned carbonyl-amine reaction is one of two mechanisms of surface modifications, the acidity of natural smoke flavourings is an even more valuable processing aid for meat surface protein coagulation.

The acetic acid and other carboxylic acids present in smoke results in a pH of about 2.5 in aqueous solutions. This low pH treatment of meat surfaces additionally provides an effective means of microbial control which is well documented in the literature [5]. The effect of phenolics in controlling microbial growth is also documented [6].

### 3.2.4.1.2 Manufacturing

As noted above, the water-soluble fraction of wood smoke is the most widely used fraction, particularly where smoke colour is most important. Where only smoke flavour is desired, there are several methods which can be used to produce a natural smoke flavour without forming a significant colour. Through a liquid-liquid extraction process, the less polar smoke constituents are extracted to produce a smoked flavoured vegetable oil [7].

A subsequent extraction from the oil into propylene glycol, polysorbates or other food-grade substances may also be done [8].

A distinction should be made between aqueous smoke flavourings and pyroligneous acid. Pyroligneous acid is the aqueous condensate of the smoke produced in the manufacture of charcoal. On a commercial basis, it is used as a source material for food flavouring compounds. As for its direct use as a flavouring in itself, it has limited application as there is some question with regard to its composition, potential for contaminants and sanitation, since its origin is outside of the food industry. The major features distinguishing pyroligneous acid from aqueous smoke flavourings are flavour profile and the lesser ability of pyroligneous acid to impart smoke colour when applied to food surfaces.

In the USA, the Food and Drug Administration classifies pyroligneous acid as an artificial smoke flavouring. European countries, at present, do not make a distinction between artificial and natural.

Additionally, there are smoke flavourings produced from the water-insoluble tars through a distillation process [9] and on through an alkaline adsorption/organic extraction of whole smoke vapours [10].

Natural phenolic antioxidative activity is an additional benefit in the use of these smoke flavourings. The higher boiling point phenolics, in particular, possess antioxidative functionality similar to that of commercially used synthetic phenolic antioxidants.

Any of the above mentioned flavourings may be added to carriers such as maltodextrin, dextrose, torula yeast, malt flour, salt or non-food, approved carriers such as phosphates or silicas. Individual countries' regulations restrict or limit the use of some carriers.

The one class of substances which is virtually eliminated in all smoke flavourings is that of polycyclic aromatic hydrocarbons (PAH). While there are perhaps as many as ten of these identified as potential problems in smoked foods and smoke flavourings generated, the smoke flavouring industry has traditionally focused upon and tried to eliminate benzo[a]pyrene (b[a]p). Water solubility is the primary mechanism for precluding its occurrence in smoke flavourings generated with vaporous smoke. A total organic concentration of about 30% is sufficient to limit the solubility of this PAH to less than 1 ppb in the aqueous solution with the result that these substances are partitioned into the water-insoluble fraction. The latter compromises primarily nonpolar tars, wood waxes, oils and free phenolics.

Good manufacturing practices are required to achieve complete separation of upper and lower phases to produce low PAH-containing smoke flavouring solutions from crude aqueous products. Various processes such as aging, decanting, filtering, coalescing, centrifugation, etc., can be used to facilitate phase separation.

After producing a b[a]p-free aqueous smoke flavouring concentrate, the oil-based and emulsifier-containing smoke flavourings may be produced.

Although smoke flavourings come in a variety of formulations, aqueous, dry and oil base, all evolve from a base aqueous solution which is produced under the carefully controlled pyrolysis of sawdust prepared from virgin untreated hardwoods like hickory, oak, ash, maple, etc. A more extensive list of wood species used can be found in the report 'Commission of the European Communities' *[11]*. The conditions for pyrolysis are strictly controlled to maximise the colour and flavour components. During the process, it is necessary to elicit phase separation at some point, through the introduction of water, to effect the isolation of soluble smoke flavouring compounds from the heavy insoluble tarry organic phase containing the PAH. The soluble components in the aqueous phase are adjusted to an organic strength of 25-35% weight/volume. A typical analysis for a commercially important aqueous smoke flavouring condensate would be as follows:

| Smoke Component | % Wt./Vol. |
| --- | --- |
| Acids | 6-12 |
| Carbonyls | 10-18 |
| Phenols | 1.0-1.8 |
| Soluble Tar | 0-2 |
| Total Organics | 25-35 |
| Benzo(a)pyrene | < 10 µg/kg condensate *[11]* |
| Benzo(a) anthracene | < 20 pg/kg condensate *[11]* |

Adjustments of the smoke components in the aqueous phase, either through dilution or concentration, can be made to develop a smoke flavour for specific applications. This may include the design of flavourings to be transferred to a functional base material or carrier.

Whole smoke may also be condensed onto a base material such as sugar, salt, maltodextrins, etc., which in turn can be used as a flavouring adjunct or undergo solvent extraction *[11]*.

### 3.2.4.1.3 Application Methods

Depending upon production facilities and processing methods, natural liquid smoke flavourings may be used in a variety of different ways. Also, because of their versatility, they can be standardised in their base formulations and added in control-led quantities to food products to ensure consistency from batch to batch.

The most common methods used for the application of natural liquid smoke flavourings and their derivatives are:

- direct addition
- drench or spray
- atomisation.

Direct addition is the simplest and most accurate means of incorporating smoke flavourings into a food product and the only one that allows provision for use of all types of smoke flavourings; aqueous, oil, dry base and concentrates. It is a method that is applicable to all types of product formulations: meat emulsions, canned meat and fish, barbecue sauce, snack foods, etc.

Although drenching, spraying or atomisation may find limited application in the food industry, they are by far the major methods used for surface applied natural liquid smoke flavourings in the meat and fish industries. Drenching and spraying are widely accepted practices on continuous line operations where large volumes of product are produced in tunnel-type ovens.

## (1) Direct Addition to Meat Products

One of the simplest forms of direct addition is massage or rubs, a process that has been used on meat product for a great number of years and is actually the forerunner of the tumbling process.

Dry smoke flavourings can be added to the dry spice mix to be hand rubbed into the surface of items like pork bellies and beef rounds. This is a rather lengthy process and is also labour intensive. It is generally used by small meat operations that produce speciality products. The massage or rub process has given way to mechanical tumbler and massage-type equipment. The mechanical agitation by action of the tumbling equipment, especially under vacuum, accelerates the diffusion of spices, seasonings, flavourants and curing salts into and salt-soluble proteins out of the muscle components. Surface colour along with flavour can be developed on and in a variety of meat products, depending on the type of natural liquid smoke flavouring used. Products like bacon may be internally injected prior to tumbling to develop a more robust smoke flavour. This is especially the case for value added products like precooked bacon.

Caution must be exercised in the selection of smoke flavourings for use in the process especially if curing salts are added to the tumbler or chopper with the other ingredients. This is especially important if the smoke flavouring is acidic in nature. Normally when sodium nitrite is added to a mildly acidic environment like meat, only small quantities of nitrite are converted to nitrous acid, a chemical intermediate in the curing reaction. Being highly soluble in meat moisture, the nitrous acid converts to nitric acid, nitric oxide and water. The nitric oxide that forms is an important product that enters directly into the curing or colour fixation reaction with myoglobin. In the presence of ascorbates, which are cure accelerators, nitrite is rapidly reduced to nitric oxide. If the pH is lower than 5.5, the conversion of nitrite to nitric oxide may occur so rapidly that the nitric oxide could be converted to nitrogen dioxide, a toxic reddish brown gas. Direct addition to meat products prepared from emulsions is generally accomplished by adding the smoke flavourings at the end of the chopping period prior to incorporation of nitrite. Sufficient chopping time should be allowed to ensure uniform dispersion of the flavouring throughout the emulsion.

Failure to mix thoroughly could result in localisation of the smoke flavouring, which could contribute to nitrite loss.

*Brine addition of smoke flavourings.* Special low-acid smoke flavourings have been developed that are compatible with curing agents such as nitrite and nitrate for use in brine and pickle solutions for cover or direct injection of a variety of meat products including pork bellies. They are very low in acid while retaining high phenol flavouring profiles. As previously mentioned, they are either emulsifier based or sufficiently diluted to ensure complete solubility.

Pork bellies are the raw material from which bacon is produced and the development of contemporary natural liquid smoke flavourings has made it possible to colour and flavour bacon bellies using a variety of methods that are safe, efficient and economical:

- safe from health and environmental concerns that are normally associated with the traditional smoking process which include tars, resins, benzo(a)pyrene and smokehouse emissions
- efficient in that natural liquid smoke flavouring can be applied repetitively to bacon in precise amounts essentially guaranteeing consistency from batch to batch
- economical as a result of shorter processing schedules facilitating oven turnover and through reductions in processing equipment maintenance and cleanup.

Natural liquid smoke flavourings offer the processor the flexibility of developing a smoke flavouring in bacon (1) by adding them directly to a smoke flavoured brine or marinade or by direct injection of the brine into the bacon, (2) by tumbling, (3) by applying the smoke flavouring to the bacon surface after brining by dipping or spraying just prior to heat treating or (4) by atomisation.

Direct addition of acidic smoke flavourings into brines or pickles which offer little buffering capacity could deplete the nitrite before the pickle is pumped into the product. The pH levels in the brines should be maintained at a minimum of 5.5.

## (2) Atomisation of Liquid Smoke on Meat Products

The atomisation of liquid smoke involves impacting a liquid with high-velocity air creating high frictional forces which result in the disintegration of the liquid into fine spray droplets. The intent of the atomisation process is to regenerate the vaporous or gaseous phase of the natural smoke condensate from which the liquid smoke was originally produced. The process is called pneumatic atomisation and is accomplished through the use of internal mix nozzles mounted inside the processing oven. This is a widely accepted method for the application of liquid smoke in batch house-type processing. Schedules for use in the atomisation process in batch house operations invariably include steps for conditioning the product to accept the smoke. Regardless of the product, three principal tenets must be adhered to in the initial stages of processing when using liquid smoke. They are:

- Dry the product to condition the surface to receive the smoke. The surface should be tacky and the product surface temperature should be approximately 50°C.
- Application of smoke. The atomised smoke is generally applied under static conditions. The oven is shut down with the blowers off and exhaust dampers closed.
- Post application of dry heat for approximately 30 minutes at a temperature of about 10°C higher than initial surface conditioning step for the purpose of setting the colour.

## (3) Drenching or Spray Applications on Meat Products

Drenching, spraying or dipping are also commonly used methods for applying liquid smoke to the surface of meat products. Provisions for smoke application are located at the entrance of the oven in the first section or zone of the system. As the stuffed product enters the oven, it is conveyed through a shower of liquid smoke that totally wets the surface of the product. The liquid smoke not only acts to produce a uniform colour but it also contributes to skin formation on the product surface through coagulation of protein. This in turn aids peelability.

General guidelines for smoke application in the drenching process using smoke condensates are:

- 30-40% dilutions of a liquid smoke flavouring (vol./vol.)
- smoke flavouring contact time adjusted to 45-90 seconds
- solution temperatures best if maintained at 40°C
- acidity levels maintained at approximately 3-5%.

In developing optimum concentrations of smoke components needed for colour development on a specific product, it is first necessary to run a series of small-scale tests with the selected smoke flavouring to be used for a specific product. In performing the tests, consideration should be given to length of dip time and smoke component concentration as both can affect colour intensity. Once the concentration of components in millilitres per kilogram (ml/kg) of product is established, they can be scaled up to production levels.

During production runs, it is necessary to maintain consistent smoke component concentrations in the drench solutions. This is generally accomplished through the 'add back' system. In developing the amount of 'add back' smoke to use one should multiply the amount of smoke required for a specific product in ml/kg by the kg/h of the product being processed and divide this by 1000. This will give the approximate number of litres of smoke being used per hour:

$$\frac{(ml/kg) \times (kg/h)}{1000} = \text{litres of smoke used per hour}$$

In preparing the 'add back' solution, one should adjust the smoke flavour component concentration to twice the amount found being used per hour. The 'add back' solution should be used at the rate of approximately twice the normal usage rate found per hour. 'Add back' should be followed up by laboratory analysis for smoke component concentrations. Hourly samples should be taken from the drench solution during a production run for at least the first day or two in order to fine tune the system as to the exact amount of 'add back' needed to maintain a consistent colour.

When using low concentrations of smoke components, it may be necessary to increase acidity levels in the drench solution to ensure skin formation on the surface of the product through protein coagulation and assist in casing peelability. This can be accomplished by adding an organic acid such as acetic or citric acid. Acids are monitored in the drench material by titration against standard base solutions. During production runs, acids should be monitored hourly on a routine basis.

Spray cabinets are also employed in batch house situations in which entire trees or trucks are sprayed or drenched. The same guidelines for smoke preparation and application are used as those for drenching.

There are numerous designs for spray or drench cabinets that can accommodate anything from one stick of product to a complete truck or cage.

Direct dipping of product is also another method of application but not as widely practiced as drenching or spraying. There is increasing interest in this method of application for use in cook-in-bag type products where it is desirable to develop colour after cooking. After processing, the product is stripped from the bag and conveyed through a bath containing the smoke flavouring. Once the smoke flavouring has been applied it is necessary to provide a heat source to develop the colour. Although batch ovens can be used, many processors use continuous convection or impingement-type ovens in tandem with the bath to maintain product flow and minimise handling.

Drenching, spraying and dipping offer the advantages of varying contact times and concentrations to suit a specific product. These are more cost-effective methods because of the opportunity to recycle and reuse the smoke flavouring. Also, since smoke is applied to the product before it goes into the oven, oven cleanup costs are minimised.

### 3.2.4.1.4 Smoke Flavouring Application to Fish and Seafood Products

**Smoke Flavouring Whole Fish and Fillets**

Whole fish and fillets may be smoke flavoured by (1) adding them directly to smoke-flavoured brine solutions, (2) applying the smoke flavouring to the surface of the product after brining by dipping, soaking or spraying just prior to heat treatment or (3) atomisation.

> (1) Brine Addition
> Smoke flavourings may be added directly to the salt solution used for brining fish or fillets. The final level of flavouring usage will depend on the flavour preference of specific market areas.

(2) Surface Application
   The fish or fillets are brined and dried in the usual manner for each type of fish to be smoked. The pre-dried fish are then dipped in or sprayed with a smoke flavouring and heat processed to meet the processing requirements for each type of fish.

(3) Atomisation
   In the atomisation process, the brined fish or fillets are placed on smoke sticks or trays and loaded into the processing oven. The fish are dried until a 'tacky' product surface is obtained before the atomised smoke is applied. This pre-drying step is necessary to remove excess moisture from the surface of the fish and to aid in the solubilisation of protein on the surface to produce a glossy surface. It also aids in the retention of the atomised liquid smoke on the fish surface and development of the desired smoke colour. Heat processing after the smoke application should be sufficient to meet processing requirements set by the applicable regulatory agencies.

## Smoke Flavouring of Fish Sausages

In the preparation of fish sausage products, aqueous, dry or oil base smoke flavourings can easily be added directly to emulsions along with the seasonings. The sausage emulsion can then be stuffed into cellulose, fibrous or collagen casings and then subjected as an external application of liquid smoke flavourings similar to those used for whole fish or fillets. The advantage of external applications of smoke flavourings, aside from colour, is the benefit they offer of skin formation through acid coagulation of surface proteins. Post-application heat processing conditions should be kept as dry as possible (low relative humidity) throughout most of the process to insure proper surface dehydration for optimum smoke colour formation.

## Smoked Flavouring of Canned Fish and Seafood

The appropriate smoke flavouring for canned fish or seafood will be dependent upon the liquid system used in canning the product. With water-packed canned fish, aqueous smoked flavourings can be added directly to the canning solution used to pack the fish. Steaks or fillets can be marinated in an oil-based smoke flavouring for 48 hours and then canned. An alternative method of application would be the addition of an oil-based smoke flavour in the oil pack used for canning fish or seafood products.

## Smoke Flavouring of Minced Fish Products

In most cases, the direct addition of aqueous or oil-based smoke flavourings to fish spreads or pastes is the simplest method of smoke flavouring the product. Aqueous or oil-based flavourings may also be added to the deboned fish in a mincer or silent food cutter prior to emulsification. The recommended rates of addition of the smoke flavourings are based on the final weight of the spread. The finished smoked fish flavour is developed during the pasteurisation of the final product.

### 3.2.4.1.5 Food Product Applications

**Smoke Flavour Batter and Breading Application**

Dry smoke and grilled flavours are used in batter and breading coating mixtures to enhance the roasted, barbecued, grilled or smoky notes to prepared foods. The dry smoke flavouring may be blended with either the pre-dust, batter or breading mix. Flavourings vary in heat stability which should be taken into consideration when selecting a flavouring for a particular type of mix. For example, grill flavours are more heat stable and should be applied in pre-dust mixtures that are enrobed with batter prior to cooking and frying.

**Smoke Flavouring Applications in Snack Foods**

Snack foods include a vast array of food items ranging from cereal-based crackers and corn chips to high-protein snacks such as beef sticks and jerky. In past years, smoking of snack foods was limited primarily to meat snacks, as other snack foods were difficult to handle in the traditional smokehouse. With the development of natural smoke flavourings, snack food manufacturers can now incorporate these flavourings into a variety of products to produce smoked, barbecued, roasted or grilled flavours.

**Peanuts**

Peanuts are normally blanched and then roasted to develop the optimum roasted colour and flavour. If the peanuts are dry roasted, coconut oil is applied to the warm nuts as a dressing. Immediately thereafter, the salt and smoke flavouring mixture is applied while the oil on the surface of the nuts is still molten.

In snack crackers, cookies or candies containing peanut butter as an ingredient, it is recommended that the smoke flavouring be added to the peanut butter to provide an intensified roasted peanut flavour. This allows the snack manufacturer to have a stronger roasted peanut flavour in the product or to reduce the level of peanut butter in the formula.

**Almonds**

Almonds are normally roasted in coconut oil at 120-175°C. Immediately following roasting, oil base flavourings are added to the nuts as a dressing, followed by a seasoning blend composed of fine salt and a smoke-flavoured yeast. If the almonds are dry roasted, the nuts are smoke flavoured by adding a dressing oil composed of an oil base smoke flavouring and coconut oil. Fine salt is then added for seasoning.

**Walnuts**

Walnut halves are dry roasted in coconut oil, an oil base, and then coated with a dressing composed of smoke flavouring and coconut oil. Then fine salt is added to the nuts for seasoning.

**Pecans**

Pecans can be roasted in oil or dry roasted. After roasting, coconut oil is applied as a dressing, followed by application of seasoning blend composed of fine salt, yeast and malt dry base smoke flavourings. Both hickory and mesquite smoke flavourings work well with roasted pecans.

### Soy Nuts, Toasted Corn and Other Seed Snacks

Many of the nut-like snacks can be smoke flavoured similar to peanuts or almonds. Several other smoke flavour seasoning mixtures have also been used successfully on soy nuts and toasted corn products.

### Crackers and Chips

Natural smoke flavourings and grill flavourings can be used in various flavour crackers or in flavour blends for chips. Grill-type flavours have a pleasant fried bacon flavour that is especially well suited for bacon-flavoured crackers. Most of the natural smoke and grill flavours can be used at a rate of 1-5% of the flavour blend.

### Precooked Bacon

Because precooked bacon is a finished ready-to-eat product, it lends itself well to today's fast or convenience food culture in which we live: fast food restaurants, food service, prepackaged food manufacturers, military, etc. The most common practice in processing precooked bacon is to take direct injected pork bellies that have been infused with a natural liquid smoke flavoured brine and process them in a thermal processing chamber according to a cook schedule typically used for bacon. The bacon is then chilled and pressed to square if off for uniform slicing. Bacon at this point is referred to as a 'slab'. The slab is sliced to a processor's specification who, in turn, processes the slices into their ready-to-eat form. It is a common practice to apply a topical spray of a natural liquid smoke flavouring to the bacon slices as they enter the oven. Microwave or high temperature ovens are used in the manufacture of precooked bacon. In order for a product to qualify as prefried bacon, according to the US government, yields are to be no more than 40% bacon.

### Meat Snacks

The three primary types of meat snacks are pork skins, beef sticks and jerky. Smoke flavourings for popped pork skins range from a very light mist of an oil base smoke flavouring applied to the skins right after popping, to incorporation of dry base smoke flavourings in seasonings that are dusted on the skins after processing.

Beef sticks are produced similarly to other dry sausages. Normally, smoke flavourings can be added to the emulsion to provide the optimum smoke flavour intensity. Aqueous base flavourings may also be atomised on the exterior surface during processing to provide a uniform smoke colour.

Beef jerky is traditionally produced by maintaining strips of beef in a brine solution and then drying the strips to reduce the moisture content to a water protein ratio below 0.75 to 1.0. Emulsifier base smoke flavourings are added to brines to produce a smoke flavoured jerky.

Some jerky is presently being produced as a sectioned and formed product in which the meat is chopped or ground to a fine texture, mixed with salt, cure and flavour ingredients and reformed into the proper sized pieces. These flavoured pieces are then dried to produce a shelf-stable product. Normally, smoke flavourings are added to the meat mixture to produce a smoke flavoured jerky product.

## 3.2.4.2 Situation in Europe

*Uwe-Jens Salzer*

Smoke flavourings are officially governed in the European Union by Regulation (EC) No. 2065/2003 of the European Parliament and of the Council of 10 November 2003 on smoke flavourings used or intended for use in or on foods. Regarding smoke flavourings according to the EEC Flavour Directive *[16]* preparations from smoke which are used in traditional processes for smoking food, there are two official publications describing and classifying these products, the first being the "Council of Europe Guidelines concerning the transmission of flavour of smoke to food" *[12]* and the second the "Report on Smoke Flavourings of the EU Scientific Committee for Food" *[11]*. Furthermore the "IOFI Guidelines for the preparation of smoke flavourings" *[13]* are to be mentioned. Some scientific papers have also been published on this matter *[14, 15]*.

The principle of wood pyrolysis is shown in Figure 3.39. The resulting products are smoke and/or charcoal.

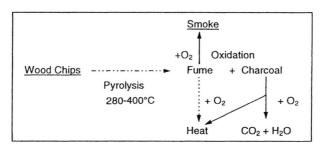

*Fig. 3.39*: *Wood pyrolysis*

The smoke generation from wood can be done by dry distillation, steam distillation, or burning as is shown in Table 3.31.

*Table 3.31*: *Smoke generation*

| System | Dry Distillation | | Steam Distillation | | Burning |
|---|---|---|---|---|---|
| | Without Air | Little Air | Without Pyrolysis | With Pyrolysis | With Air |
| Temperature (°C) | 270-600 | 400-900 | 100-170 | 170-400 | 400-900 |

In general hard wood is used as source material. Identical lists are published in *[12]* and *[13]*.

According to the EC-Regulation on smoke flavourings *[16]*, article 5, the wood shall not have been treated whether intentionally or unintentionally with chemical substances during the six months immediately preceding felling, this must be demonstrated by appropriate certification. Annex I.1 limits the maximum temperature for smoke production to 600°C.

The resulting Regulation (EC) No. 2065/2003 on smoke flavourings *[17]* designates smoke flavourings as "primary products further processed". Primary products have been defined as smoke condensates and fractions thereof. Derived smoke flavourings may be solutions in water or oil or a dry product. For details of the EC-Regulation see chapter 7.4.2.

In summary of these papers and the EC-Regulation, the following can be said:

Smoke flavourings are either smoke condensates or smoke preparations. Smoke flavouring blends produced by mixing chemically defined substances are not on the market because they would be more expensive than the products made from traditional smoke.

Smoke condensates are obtained by condensing smoke in water or another solvent. They may be further fractionated, purified or concentrated. The fractionation steps have two purposes: to obtain products of interesting olfactory properties and to reduce the concentration of undesirable by-products from the smoke. Only the water-soluble fraction is used. The organic phase will be abandoned because a work up of the tar fraction is too expensive. The smoke solution will be filtered in order to remove polycyclic aromatic hydrocarbons (PAHs). According to a Russian patent *[18]* it is also possible to use 2% chitin and 0.5% chitosan for removing PAHs almost quantitatively. Afterwards the components of the smoke solution may be concentrated by distillation. The resulting product will be processed into smoke flavouring preparations.

Smoke flavouring preparations are based on smoke condensates with the addition of other substances. They can be used either to regenerate smoke or they can be used directly in or on food.

The most important flavouring ingredients in smoke are phenols, carbonyl compounds and acids, i.e. acetic acid. In Table 3.32 an example for the composition is shown. This can vary according to the wood type used, moisture content and production process. In total more than 200 components have been identified in smoke, the most important being aldehydes, ketones, diketones, esters, alcohols, acids, furan and pyran derivatives, syringol, guaiacol, phenol and pyrocatechol derivatives and ethers. The amounts may vary considerably depending on wood type and production methods; some components may even be totally absent. Moreover anhydrosugars in higher amounts and lignin dimers and some nitrogenated compounds in lower proportions have been found in oak wood smoke *[19]*. Undesirable ingredients are the PAHs which have shown to be mutagenic and carcinogenic. Therefore it is necessary, besides the moisture content (20-30%) of the wood, to choose a temperature for smoke generation which gives high amounts of desired flavouring ingredients and low amounts of the undesired PAH's. According to Figure 3.40 the temperature has to be <700°C, the optimum being 400-600°C.

# Smoke Flavourings

*Table 3.32: Example for the composition of a smoke condensate*

| Component | Proportion (%) |
|---|---|
| Phenols | 20-35 |
| Carbonyl compounds | 10-15 |
| Acids | 5-12 |
| Esters and other compounds | 10 |
| Not identifiable | 30 |

The manufacturing of smoke flavourings consists in principle of three steps

(1) fixing the smoke

(2) making a smoke extract

(3) final flavouring formulation.

In practice there are many different ways, the criteria being kept secret by the producing companies.

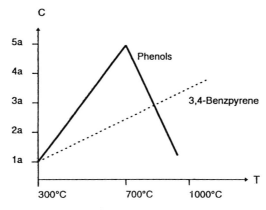

*Fig. 3.40: Development of phenols and 3,4-benzpyrene during wood pyrolysis*

The different possibilities for (1)-(3) are summarized in Table 3.33.

Smoke flavourings are used for meat, fish and fish products either by application on the surface or by direct addition. At the same time they are added to cheese products, soups and sauces, and snacks. Meat products and snacks are the most important applications.

Looking at the composition of a smoke flavouring it has, besides the flavouring properties, also antimicrobial and antioxidative effects which can be used especially in fish products *[20, 21]*.

*Table 3.33: Manufacture of smoke flavourings from smoke*

| Fixation of Smoke | Smoke Extract Production | Smoke Flavouring Formulation |
|---|---|---|
| Condensate | Dilution with water, separation of organic phase | – Adsorption on carrier<br>– Spray drying |
| Collection in:<br>– water<br>– alcohol/water<br>– (oil) | Charcoal filtration, distillation | – Solution in oil, PG etc.<br>– Emulsion |
| Adsorption on<br>– sugar, salt etc.<br>– meat (by-products) etc. | Extraction with alcohol, methanol etc., filtration, distillation | – Solution in alcohol/water |

Regarding quality control it is necessary to look for the level of PAH, expressed as 3,4-benzpyrene. According to Commission Regulation (EC) No. 208/2005 *[22]* the maximum level is 5.0 ppb (g/kg) in smoked meat and fish products, 2.0 ppb in oils and fats for human consumption and 1.0 ppb in foods for infants and young children. Furthermore, CoE and SCF recommend in their guidelines

*benzo(a)anthracene:* giving rise to less than 0,06 µg/kg in the foodstuff as consumed

| | | | |
|---|---|---|---|
| As | max. | 3 | mg/kg |
| Hg | max. | 1 | mg/kg |
| Cd | max. | 1 | mg/kg |
| Pb | max. | 10 | mg/kg |

but without any justification why these contaminants are of concern.

Smoke flavourings have to be authorised before entering the market both in USA and the European Union (EU).

In the EU according to Regulation *[17]* an application shall be sent to the competent authority of a Member State. It shall include information on the type of wood used, production methods of the 'primary product' and the further processing in the production of derived smoke flavourings and on the intended use levels. Moreover the application has to include the qualitative and quantitative chemical composition of the 'primary product' and the characterisation of the portion that has not been identified. Of major importance are the chemical specifications of the primary product and information on the stability and the degree of variability of the chemical composition. The portions which have not been identified, i.e. the amount of substances whose chemical structure is not known, should be as small as possible and should be characterised by appropriate analytical methods, e.g. chromatographic or spectrometric methods. A validated analytical method for sampling, identification and characterisation of the primary product is also mandatory. The Analytical Aspects on Smoke Flavourings committee comprising members of he EU Commission and industry has been working on this matter for some time. The technical work is done in the Joint Research Centre of the Institute for Reference Materials and Methods in Geel (B) under the leadership of R. Simon. A publication in 2004 indicates liquid smoke as an analytical challenge *[23]*.

# REFERENCES

[1] Maga, J.A. & Fapojuwo, O.O., Aroma intensities of various wood smoke fractions, J. Sensory Stud., 1, 9, 1986.
[2] Underwood, et. al, U.S. Pat. No. 4, 959, 232.
[3] Kozlosdki, Z.P., Simon, J., and Wilmomowski M., Investigations on the chemical, toxicological, and biological properties of the polish smoke extract, Proc. Eur. Meet. Meat Res. Workers, 15, 516, 1969.
[4] Donnelly, L.S. Ziegler, G.R., and Action, J.C., Effect of liquid smoke on the growth of lactic acid starter cultures used to manufacture fermented sausage, J., Food Science, 47, 2074, 1982.
[5] Houben, J.H., Bacteriostatic properties of smoke preparations. I Comparative studies of five preparations, Voedingsmiddelen Technol., 7 (41), 8, 1974.
[6] Olsen, C.Z. Smoke flavouring and its bacteriological and antioxidative effects. Acta Aliment. Pol., 3 (3), 313, 1977.
[7] Hollenbeck, U.S. Patent No. 3,480,446.
[8] Underwood, et. al. U.S. Patent No. 4,250,199.
[9] Danuis, et. al. U.S. Patent No. 4,154,866.
[10] Miler, U.S. Patent No. 3,455,248.
[11] Commission of the European Communities – Directorate-General Industry III/E/1 – Scientific Committee for Food: Report on Smoke Flavourings (adopted on 25 June 1993) – CS/FLAV/49 – FINAL July 1993
[12] Council of Europe Guidelines concerning the transmission of flavour of smoke to food – Health protection of consumers (Council of Europe Press, Strasbourg, 1992)
[13] International Organisation of the Flavour Industry (I.O.F.I., 49 Square Marie-Louise, B-1000 Bruxelles): Code of Practice K 1-2/Guidelines for the preparation of smoke flavourings (October 1989)
[14] Aroma geräucherter Fleischerzeugnisse [The aroma of smoked meat products] I-V – T. Gudaszewski – Fleischwirtsch. 67, 1523-1525 (1987); 68, 770-772, 1567-1569 (1988); 69, 271-273, 414-416 (1989)
[15] Neuere Ergebnisse über die Zusammensetzung von Räucherrauch [Recent results on the composition of curing smoke] 1.-3. – K.Potthast + G.Eigner – Fleischwirtsch. 68, 651-655, 991-1000, 1350-1356 (1988)
[16] Council Directive of 22 June 1988 on the approximation of the laws of the Member States relating to flavourings for use in foodstuffs and to source materials for their production (88/388/EEC), Official Journal of the European Communities No. L 184/61 – 15.7.88.
[17] Regulation (EC) No. 2065/2003 of the European Parliament and of the Council of 10 November 2003 on smoke flavourings used or intended for use in or on foods. Official Journal of the European Union No. L 309/1-8 – 26.11.2003.
[18] RU 2.170.021 of 10.07.2001 reported in CA 137:5447.
[19] Guillen M.D. and Manzanos M.J., Food. Chem. 79(3), 283-292, reported in CA 138:88846.
[20] Wang Q. et al. Dahlian Shuinan Xueyuan Xuebao 17(3), 262-266, 2002, reported in CA 139:275959.
[21] Sunen E. et al. Food Res. Int. 36(2), 111-116 (2003), reported in CA 139:35421.
[22] Commission Regulation (EC) No. 208/2005 of 4 February 2005 amending Regulation (EC) No. 466/2001 as regards polycyclic aromatic hydrocarbons. Official Journal of the European Union No L 34/3-5 – 8.2.2005.
[23] Flüssiger Rauch – eine analytische Herausforderung [Liquid smoke – an analytical challenge ] – D. Meier – ForschungsReport 2/2004, 24-27.

## 3.3 Non-flavouring Ingredients

### 3.3.1 Extraction Solvents

*Howard D. Preston*
*updated by: Robert F. van Croonenborgh*


The Annex of the Council Directive on the approximation of the laws of the Member States on extraction solvents used in the production of foodstuffs and food ingredients (88/344/EEC, dated 13 June 1988 as amended (annex I)), is divided into Parts I, II, and III (below) and lists extraction solvents which may be used during the processing of raw materials of foodstuffs, of food components or of food ingredients. Additionally usage of water (e.g. demineralised water, distilled water, potable water) is also permitted. Regulations and permitted extraction solvents, such as n-octyl alcohol and trichloroethylene, are different in the USA and relevant legislation should be consulted.

Directive 88/344/EEC draws up, in its Annex, a list of extraction solvents which may be used during the processing of raw materials for foodstuffs, food components or food ingredients. The conditions for use are specified for a number of such solvents. A safeguard clause allows Member States to provisionally suspend or limit the use of a given extraction solvent – although it conforms to the Directive Cooperation within the Standing Committee for Foodstuffs (see T-3) provides an accelerated procedure for the adoption of specific purity criteria and methods of analysis and sampling.

Amending Directive 92/115/EEC adapts the basic directive as regards, among others, the maximum residue limits and the combined use of certain solvents.

Amending Directive 94/52/EC re-authorises the solvent cyclohexane, used in the preparation of flavourings, under specific conditions of use.

Directive 97/60/EC further amends basic Directive 88/344/EEC by introducing or deleting (because no longer used) or modifying conditions of use of certain solvents and more generally entrusting the Commission under the speedier procedure of the Standing Committee for Foodstuffs (see T-3) adoption of such measures. Complying products are authorised or required in trade as from 27 October 1998 or 27 April 1999, respectively.

Regulation (EC) 1882/2003 specifies the role of the Standing Committee on the Food Chain and Animal Health (see T-3) in the simplified decision-making procedure set out by Directive 88/344/EEC.

# Extraction Solvents

## PART I

Extraction solvents to be used in compliance with good manufacturing practice for all uses*

| Name |
|---|
| Propane |
| Butane |
| Ethyl acetate |
| Ethanol |
| Carbon dioxide |
| Acetone** |
| Nitrous oxide |

## PART II Extraction solvents for which conditions of use are specified

| Name | Conditions of use (summary description of extraction) | Maximum residue limits in the extracted foodstuff of food ingredient |
|---|---|---|
| Hexane*** | Production or fractionation of fats and oils and production of cocoa butter. | 1 mg/kg in the fat or oil or cocoa butter. |
| | Preparation of defatted protein products and flours. | 10 mg/kg in the food containing the defatted protein products and defatted flours. |
| | Preparation of defatted cereal germs. | 5 mg/kg on the defatted cereal germ. |
| | Defatted soya products. | 30 mg/kg in the defatted soya product as sold to the final consumer. |
| Methyl acetate | Decaffeination of, or removal of irritants and bitterings from coffee and tea. | 20 mg/kg in the coffee or tea. |
| | Production of sugar from molasses. | 1 mg/kg in the sugar. |
| Ethylmethylketone**** | Fractionation of fats and oils. | 5 mg/kg in the fat or oil. |
| | Decaffeination of, or removal of irritants and bitterings from coffee and tea. | 20 mg/kg in the coffee or tea. |
| Dichloromethane | Decaffeination of, removal of irritants and bitterings from coffee and tea. | 2 mg/kg in the roasted coffee and 5 mg/kg in the tea. |
| Methanol | All uses | 10 mg/kg |
| Propan-2-ol | All uses | 10 mg/kg |

---

\* An extraction solvent is considered as being used in compliance with good manufacturing practice if its use results only in the presence of residues or derivatives in technically unavoidable quantities presenting no danger to human health.

\*\* The use of acetone in the refining of olive-pomace oil is forbidden.

\*\*\* Hexane means a commercial product consisting essentially of acyclic saturated hydrocarbons containing six carbon atoms and distilling between 64°C and 70°C. The combined use of Hexane and Ethylmethylketone is forbidden.

\*\*\*\* The presence of n-Hexane in this solvent should not exceed 50 mg/kg. This solvent may not be used in combination with Hexane.

PART III Extraction solvents for which conditions of use are specified

| Name | Maximum residue limits in the foodstuff due to the use of extraction solventson the preparation of flavourings from natural flavouring materials |
|---|---|
| Diethyl ether | 2 mg/kg |
| Hexane | 1 mg/kg* |
| Methyl acetate | 1 mg/kg |
| Butan-1-ol | 1 mg/kg |
| Butan-2-ol | 1 mg/kg |
| Ethylmethylketone | 1 mg/kg* |
| Dichloromethane | 0.02 mg/kg |
| Propan-1-ol | 1 mg/kg |
| Cyclohexane | 1 mg/kg |
| 1,1,1,2-tetrafluoroethane | 0.02 mg/kg |

\* The combined use of these two solvents is forbidden.

## 3.3.2 Permitted Carriers and Carrier Solvents

*Howard D. Preston*

The European Parliament and Council Directive no 95/2/EC of 20 February 1995 on food additives other than colours and sweeteners defines 'carriers', including carrier solvents, as substances used to dissolve, dilute, disperse or otherwise physically modify a food additive without altering any technological function (and without exerting any technological effect themselves) in order to facilitate its handling, application and use.

Solvents are prohibited in the following foods in the following countries

| | |
|---|---|
| Baked goods: | Can, Fin, Fr, Italy, Nor, Port, Switz, Ger |
| Cured meats: | Neth. |
| Eggs: | Can, Fin, Spain, US, Ger. |
| Tea, coffee, choc. drinks: | Den, Fin, Ire, Italy, Lux, Neth, Port, Spain, Swed, Ger, UK (tea+choc. drinks) |

Further details are given in Table 3.34 and Annex V of Directive no 95/2/EC (Table 3.35) 'Permitted carriers and carrier solvents'.

*Table 3.34: Carrier solvents*

| Solvent | Permitted | Prohibited |
|---|---|---|
| **Ethanol** | GMP: -biscuits; Bel, Swed.<br>GMP: -soft drinks: Fin, UK<br>US: -GRAS<br>WHO: -GMP | |
| **Isopropanol** | GMP/bread: Ire, Japan<br>GMP/tea, coffee, choc. drinks: Nor<br>GMP/soft drinks: UK<br>US/spice OR, (limited)<br>WHO: GMP with limit | **Bread**: Austria, Bel, Can, Den, Fin, Fr, Is, Italy, Lux, Neth, Nor, Port, Swed, Switz, UK, US, Ger. |
| **Methanol** | Can: GMP/cured meats<br>US/spice/hops (limited)<br>WHO: GMP | |
| **Ethyl Acetate** | Swed: GMP/biscuits<br>UK: GMP/soft drinks<br>US: GMP/ coffee<br>WHO: GMP | |
| **Monoglycerides, Acetylated, Distilled** | UK: GMP/soft drinks | |
| **Glycerol**<br>Alcoholic Beverage | Italy limited<br>GMP: Japan, Nor, Swed, US | Bel, Can, Den, Fin, Fr, Lux, Neth, Switz, Ger. |
| Baked Goods | Can: GMP/ baked goods<br>Den: GMP/flour | **Flour/Bread** Austria, Bel, Can, Fr, Is, Italy, Lux, Neth, Nor, Port, Swed, Switz, US. |

| Solvent | Permitted | Prohibited |
|---|---|---|
| **Glycerol (cont.)**<br>Baked Goods<br>Ger : flour, bread, baked goods | Ire: GMP/ bread, baked goods, cake mixes<br>Japan: GMP/bread and baked goods<br>Spain: special permission baked goods<br>UK : flour + baked goods<br>US : baked goods<br>**5% limit baked goods**<br>Den, Fin, Fr,<br>**5% limit flour**<br>Fin. | **Bread** Den, Fin, UK<br>**Flour**: Ire<br><br>UK/Bread |
| Cured meats | GMP: Can, Fin, Ire, Swed (casings) UK, US. | Bel, Den, Fr, Lux. Port, Switz, Ger. |
| Egg Products | GMP: Aust, Japan | Bel, Den, Fr, Italy, Nor, Port, Swed, Switz. |
| Ices | GMP: Aust, Japan, Port, UK<br>Limits: Bel, Lux, Neth, Den, Swed | Can, Fr, Italy, Nor, Spain, Switz, Ger. |
| Soft Drinks | GMP: Aust, Swed, UK, US, Ger<br>Limits: Den, Ire<br>Unlimited: Japan/ tea, coffee, choc. drinks, soft drinks | Bel, Can, Fr, Italy, Neth, Nor, Port, Spain, Switz. |
| **1,2-Propylene Glycol** | **GMP/Baked Goods**<br>Can, Ire, Swed, Japan<br>**Limits**<br>Fin: baked goods<br>Japan: tea, coffee, choc. drinks<br>Nor: GMP/tea, coffee, choc. drinks<br>Port: GMP/biscuits<br>Spain: baked goods<br>special permission<br>UK: no limit/baked goods<br>GMP/soft drinks<br>US: no limit/baked goods | **Baked Goods**<br>Austria, Bel, Den, Fr, Italy, Lux, Nor, Port, Switz, Ger |
| **Sorbitol** | Bel: GMP/biscuits<br>Den: limit/biscuits<br>Fin: limit/biscuits<br>Fr: limit/biscuits<br>US: GRAS/GMP | |
| **Triacetin** | Bel: limit/egg powder<br>Can: GMP/cake mixes<br>Ire: GMP<br><br>Nor: GMP/tea, coffee, choc. drinks<br>Switz: GMP/cake mixes<br>UK: GMP/soft drinks | **Flour**<br>Austria, Bel, Den, Fin, Fr, Italy, Japan, Lux, Neth, Nor, Port, Swed, US, Ger.<br>**Eggs**<br>Den, Fr, Italy, Japan, Nor, Port, Swed, Switz. |

The fate of 'carbitol', diethylene glycol mono-methyl ether, often used in Italian flavours in the past is uncertain and has been referred to JECFA.

The use of starches, maltodextrin, lactose, glucose syrups, etc. with or without the emulsifying effect of food grade gums allow essentials oils, flavourings or clouding agents to be spray dried so as to maintain maximum retention of the desirable volatiles, or to make them compatible with the end product. Betacyclo dextrin (E 459), which has some technological advantage, is not yet permitted.

***Table 3.35:*** *Permitted carriers and carrier solvents*[*]

Note: Not included in this list are:
    (1) Substances generally considered as foodstuffs;
    (2) Substances referred to in Article 1 (5);
    (3) Substances having primarily an acid or acidity regulator function, such as citric acid and ammonium hydroxide.

| E No | Name | Restricted use |
|---|---|---|
|  | Propan-1,2-diol (propylene glycol) | Colours, emulsifiers, antioxidants and enzymes (maximum 1 g/kg in the foodstuff) |
| E 422<br>E 420<br>E 421<br>E 953<br>E 965<br>E 966<br>E 967 | Glycerol<br>Sorbitol<br>Mannitol<br>Isomalt<br>Maltitol<br>Lactitol<br>Xylitol |  |
| E 400-404 | Alginic acid and its sodium, potassium, calcium and ammonium salts |  |
| E 405<br>E 406<br>E 407<br>E 410<br>E 412<br>E 413<br>E 414<br>E 415<br>E 440 | Propan-1,2-diol alginate<br>Agar<br>Carrageenan<br>Locust bean gum<br>Guar gum<br>Tragacanth<br>Acacia gum (gum arabic)<br>Xanthan gum<br>Pectins |  |
| E 432<br>E 433<br>E 434<br>E 435<br>E 436 | Polyoxyethylene sorbitan monolaurate (polysorbate 20)<br>Polyoxyethylene sorbitan monooleate (polysorbate 80)<br>Polyoxyethylene sorbitan monopalmitate (polysorbate 40)<br>Polyoxyethylene sorbitan monostearate (polysorbate 60)<br>Polyoxyethylene sorbitan tristearate (polysorbate 65) | Antifoaming agents |
| E 442 | Ammonium Phosphatides | Antioxidants |

---

[*] Annex V of directive no 95/2/EC.

| E No | Name | Restricted use |
|---|---|---|
| E 460<br>E 461<br>E 463 | Cellulose (microcrystalline or powdered)<br>Methyl cellulose<br>Hydroxy propyl cellulose | |
| E 464<br>E 465<br>E 466 | Hydroxy propyl methyl cellulose<br>Ethyl methyl cellulose<br>Carboxy methyl cellulose<br>Sodium carboxy methyl cellulose | |
| E 322<br>E 432-436<br>E 470b<br>E 471<br>E 472a<br><br>E 472c<br><br>E 472e<br><br>E 473<br>E 475<br><br>E 491<br>E 492<br>E 493<br>E 494<br>E 495 | Lecithins<br>Polysorbates 20, 40, 60, 65 and 80<br>Magnesium salts of fatty acids<br>Mono- and diglycerides of fatty acids<br>Acetic acids esters of mono- and diglycerides of fatty acids<br>Citric acid esters of mono- and diglycerides of fatty acids<br>Mono- and diacetyl tartaric acid esters of mono- and diglycerides of fatty acids<br>Sucrose esters of fatty acids<br>Polyglycerol esters of fatty acids<br><br>Sorbitan monostearate<br>Sorbitan tristearate<br>Sorbitan monolaurate<br>Sorbitan monooleate<br>Sorbitan monopalmitate | Colours and fat-soluble anti-oxidants<br><br><br><br><br><br><br><br><br><br>Colours and anti-foaming agents |
| E 1404<br>E 1410<br>E 1412<br>E 1413<br>E 1414<br>E 1420<br>E 1422<br>E 1440<br>E 1442<br>E 1450 | Oxidized starch<br>Monostarch phosphate<br>Distarch phosphate<br>Phosphate distarch phosphate<br>Acetylated distarch phosphate<br>Acetylated starch<br>Acetylated distarch adipate<br>Hydroxy propyl starch<br>Hydroxy propyl distarch phosphate<br>Starch sodium octenyl succinate | |
| E 170<br>E 263<br>E 331<br>E 332<br>E 341<br>E 501<br>E 504<br>E 508<br>E 509<br>E 511<br>E 514<br>E 515<br>E 516<br>E 517 | Calcium carbonates<br>Calcium acetate<br>Sodium citrates<br>Potassium citrates<br>Calcium phosphates<br>Potassium carbonates<br>Magnesium carbonates<br>Potassium chloride<br>Calcium chloride<br>Magnesium chloride<br>Sodium sulphate<br>Potassium sulphate<br>Calcium sulphate<br>Ammonium sulphate | |

## Permitted Carriers and Carrier Solvents

| E No | Name | Restricted use |
|---|---|---|
| E 577 | Potassium gluconate | |
| E 640 | Glycine and its sodium salt | |
| E 1505 | Triethyl citrate | |
| E 1518 | Glyceryl triacetate (triacetin) | |
| E 551 | Silicon dioxide | Emulsifiers and colours, max. 5% |
| E 552 | Calcium silicate | |
| E 553b | Talc | Colours, max. 5% |
| E 558 | Bentonite | |
| E 559 | Aluminium silicate (Kaolin) | |
| E 901 | Beeswax | Colours |
| E 1200 | Polydextrose | |
| E 1201 | Polyvinylpyrrolidone | Sweeteners |
| E 1202 | Polyvinylpolypyrrolidone | |

## 3.3.3 Emulsifiers – Stabilisers – Enzymes

*William J. N. Marsden*

### 3.3.3.1 Emulisfiers

**Introduction**

Food production is becoming increasingly industrialised with larger and more centralised production facilities leading to an increase in the requirement for food to travel distance and to undergo prolonged storage.

Consumers require modification of established foods to reflect advances in knowledge about health, to give a wider choice of foods and a higher quality. Convenience is required to compliment more demanding life patterns and products must provide interest and pleasure and not simply basic sustenance.

Moreover, there are now pressures for production methods that make minimum demands on the world's environmental balance: - reduced energy input and more gainful use of raw materials.

Many food products, as a result of these pressures, are now produced using emulsifiers. Although it is not yet true to say that no industrially produced food can be made without emulsifiers, the trend in many product areas is that way.

Processed foods typically exist in the form of a complex, multiphase, multi-component colloidal systems. The key molecules that facilitate the formation and stabilisation of these structures during processing are called emulsifiers.

**Description**

An emulsifier is a surface-active substance with dual functionality which enables them to interact with oil and water to promote a stable emulsion in a food system.

Emulsifiers are a single chemical substance, or mixture of substances, that lower the tension at the oil-water interface (interfacial tension) and have the capacity for promoting emulsion formation and short-term stabilisation.

In molecular terms, emulsifiers have lipophilic (oil soluble) and hydrophilic (water soluble) portions. In some food processes emulsifiers with different ratios of hydrophile to lipophile are required. To meet this need many emulsifiers have been developed but all have included a lipophilic portion derived from fats.

Emulsifiers are derived from natural substances. Emulsifiers are available and are used as semi-purified preparations of natural products or are available as products of synthesis or reaction of natural substances. Table 3.36 indicates a range of common food emulsifiers together with an outline of their origin.

*Table 3.36:* *Common food emulsifier types and origin*

|  | **Common Name** | **Abbreviation** | **Common Source/Origin** |
|---|---|---|---|
| E322 | Lecithins |  | Soy bean, egg |
| E410 | Locust bean gum | LBG | *Cetatonia siligua* |
| E412 | Guar gum |  | *Cyamopsis terragonoloba* |
| E413 | Gum tragacanth |  | *Astragalus spp.* |
| E414 | Gum arabic |  | *Acacia Senegal* |
| E431 | Polyoxyl (40) stearate |  |  |
| E432 | Polysorbate 20 | Tween 20 | Oil, polyhydric alcohol and organic acid |
| E433 | Polysorbate 80 | Tween 80 | Oil, polyhydric alcohol and organic acid |
| E434 | Polysorbate 40 | Tween 40 | Oil, polyhydric alcohol and organic acid |
| E435 | Polysorbate 60 | Tween 60 | Oil, polyhydric alcohol and organic acid |
| E436 | Polysorbate 65 | Tween 65 | Oil, polyhydric alcohol and organic acid |
| E442 | Ammonium phosphatides | EMULSIFIER YN | Mixed ammonium salts of phosphorylated glycerides |
| E461 | Methylcellulose |  | Cellulose |
| E471 | Glycerol monostearate | GMS | Oil and polyhydric alcohol |
| E472 | Esters of monoglycerides | Mono's and Mono-di's | Oil and polyhydric alcohol |
| E472a | Acetic acid esters of mono- and diglycerides of fatty acids | ACETEM | Oil, polyhydric alcohol and organic acid |
| E472b | Glycerol-lacto-palmitic esters of monoglycerides | LACTEM; GLP | Oil, polyhydric alcohol and organic acid |
| E472c | Citric acid esters of mono- and diglycerides of fatty acids | CITREM | Oil, polyhydric alcohol and organic acid |
| E472e | Mono- and diacetyl tartaric acid esters of mono- and diglycerides of fatty acids | DATEM | Oil, polyhydric alcohol and organic acid |
| E473 | Sucrose esters of fatty acids |  | Oil and polyhydric alcohol |
| E475 | Polyglycerol esters of fatty acids | PGE | Oil, polyhydric alcohol and organic acid |

|  | Common Name | Abbreviation | Common Source/Origin |
|---|---|---|---|
| E476 | Poly-Glycerol Poly-Ricinoleate | PGPR | Oil, polyhydric alcohol |
| E477 | Propylene glycol esters of fatty acids | PGME | Oil and polyhydric fatty acid alcohol |
| E481 | Sodium stearoyl-2-lactylate | SSL | Oil polyhydric alcohol and organic acid |
| E482 | Calcium stearoyl-2-lactylate | CSL | Oil polyhydric alcohol and organic acid |
| E491 | Sorbitan monostearate | SMS | Oil, polyhydric alcohol and organic acid |
| E492 | Sorbitan tristearate | STS | Oil, polyhydric alcohol and organic acid |
| E493 | Sorbitan monolaurate | SML | Oil, polyhydric alcohol and organic acid |
| E494 | Sorbitan monooleate | SMO | Oil, polyhydric alcohol and organic acid |
| E495 | Sorbitan monopalmitate | SMP | Oil, polyhydric alcohol and organic acid |
| E496 | Sorbitan trioleate | STO | Oil, polyhydric alcohol and organic acid |

**Hydrophile-Lipophile Balance**

The most widely used system for classifying emulsifiers is the hydrophile-lipophile balance (HLB) concept (Griffin (1949)[1]). A low HLB number means that the emulsifier is lipophilic and a high value means that it is hydrophilic.

Formulae are available for calculating HLB numbers by adding together the weighted contributions of hydrophilicity and lipophilicity from all the different types of chemical groups in emulsifier molecules (Becher (1985)[2]).

The group summation method is useful for the emulsifier technologist since it has long been recognised that an appropriate blend of emulsifiers usually produces a more stable emulsion than one prepared from a single emulsifier of the same HLB value.

Though useful for classifying emulsifiers the HLB concept is of little practical value in food formulations since emulsions are for the most part stabilised by proteins or polysaccharides and there is always an interaction between the emulsion and the other food ingredients.

**Origin of Emulsifiers**

Emulsifiers that are used in the food industry have a number of origins. Modified food proteins and synthesised emulsifiers from an oil or protein base provide for the wide range of functionality now required in food systems.

# Emulsifiers – Stabilisers – Enzymes

Amongst the most common emulsifiers, synthesised emulsifiers are prepared from a reaction between a fat and a polyhydric alcohol which forms the water soluble fraction (Table 3.37).

The lipophilic fraction of the emulsifier is provided from natural oils and fats. Whilst the industry was historically founded on animal or marine fats and oils, a clear trend towards pure vegetable based products has developed. The majority of food emulsifiers are now based upon oils such as soya bean, palm or sunflower.

Furthermore, treatment of monoglycerides with organic acids such as acetic, lactic, citric and tartaric gives further breadth to the range of combinations (Figure 3.41).

*Table 3.37: Common polyhydric alcohols*

| Alcohol | Emulsifier type |
|---|---|
| Glycerol | Partial glycerol esters e.g. monoglycerides |
| Polymerised glycerol | Polyglycerol esters |
| Propylene glycerol | Propylene glycol esters |
| Sorbitol | Sorbitan esters |
| Sucrose | Sucrose esters |

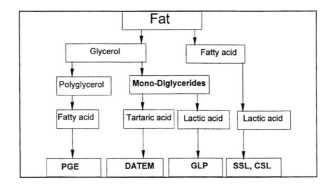

| PGE | Polyglycerol Esters of fatty acids |
| DATEM | Diacetyl Tartaric Esters of Mono-diglycerides |
| GLP | Glycerol Lacto Palmitate (Lactic acid esters) |
| CSL, SSL | Calcium and Sodium Stearoyl lactylates |

*Fig. 3.41: The basis of some of the family of synthesised emulsifiers*

## Applications

The most common type of food products where emulsifiers are used are:

- Margarine, shortenings & baking fats
- Spreads
- Ice cream, whippable fat creams and fillings, desserts

- Bread, fermented goods and flour improvers
- Biscuits and cookies
- Cake
- Pasta
- Noodles
- Extruded snacks
- Soups, sauces, dressings and 'instant' foods
- Cereals.

Emulsifiers have also assumed valuable roles in products such as chocolates (control of fat polymorphism), toffees and caramels, chewing gums, pharmaceutical preparations, soft and liqueur drinks and meat products, in addition to being used as lubricating, release and cutting aids throughout the food industry. In these applications, emulsifiers can be said to be used in roles not directly related to emulsification.

The broad role of emulsifiers in food systems includes therefore the following major functions:

- Stabilisation of Oil in Water and Water in Oil emulsions
  The reduction of interfacial tension which creates an intimate mixture of two liquids that are normally immiscible.
- Starch complexing
  Reaction with starch in bakery products which retard the crystallisation of the starch, thus retarding the firming of the crumb which is associated staling staring.
- Wetting
  Allows a surface to be more attracted to water, such as powders *e.g.* coffee whitener in which the addition of the surfactant aids the dispersion of the powder in the liquid without clumping.
- Crystallisation control
  Control of crystallisation in sugar and fat systems *i.e.* in chocolate where it allows for brighter initial gloss and prevention of solidified fat on the surface.
- Dispersing
  The reduction of the interfacial tension which creates an intimate mixture of two liquids that are normally immiscible *e.g.* oil-water emulsions such as salad dressings.
- Lubrication
  Reduces the tendency of products to stick to containers, transport equipment, cutting knives, wrappers and teeth.

**Advantages**

Emulsifiers give benefits such as product reproducibility, processing tolerance and reliability leading to benefits of yield improvements thus cost savings.

Emulsifiers, as a consequence of their function, have a significant role in texture control. This involved emulsification of fat and water, aeration control and of foam stabilisation in the dairy based industries. In bakery processes, emulsifiers are used as dough strengtheners, as crumb softeners and as aerating agents.

Emulsifier technology is now being used for new product developments such as reduction in total product fat content for low fat margarine, spreads, cream, dressings, biscuits, cookies, cakes and ice cream technologies.

## Development

Emulsifiers are only used at levels sufficient to provide the functionality required, usually only up to about 1% of the food product. They are now often used in combination and are now commonly added directly to the final product mix. The high speed of hydration and aeration properties of many emulsifiers now means that powder or paste forms of the product can be used rather than a pre-hydrated form.

The ease of use of the emulsifier in the production environment has also been addressed. The range of available emulsifiers includes specially developed micro-bead forms which has good free-flowing properties that minimises lumping in warmer climates. This allows the use of automated powder handling systems in production operations. This has been achieved through advancements in spray-chilling technologies (Figure 3.42). In addition, emulsifiers are often available in a paste form to suit handling preferences.

The basis of every type of emulsifier is an oil. In the past animal fats were commonly used but today the emulsifier ranges that are available are based more and more on vegetable fats and are often Kosher certificated.

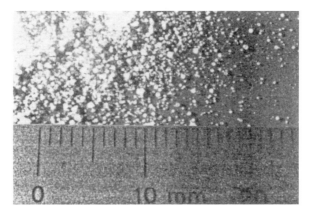

*Fig. 3.42:* Emulsifiers in micro-bead form

## Functionality, Understanding and Advances

*Bread*

Bread production has become increasingly industrialised. Moreover, consumer preference has altered towards loaves of good volume, ready-sliced for convenience and which stay soft for several days. At the same time, increased demand for speciality breads (*e.g.* French baguette type, sour doughs *etc.*) has increased.

Emulsifiers, used either alone or in combination with specially developed enzymes are now established as essential ingredients in large scale bakeries.

The most important emulsifiers are polar lipids which interact with water to form liquid crystalline structures. These are often three dimensional structures which can assume different geometric configurations. The most active is referred to as the layered or lamellar phase. Figure 3.43 shows the lamellar form as a two dimensional structure.

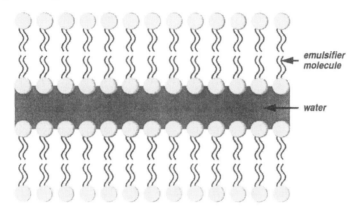

*Fig. 3.43: Two dimensional representation of an emulsifier molecule in the lamellar form*

The three dimensional form can be best visualised as a swiss roll, where jam represents the water and the sponge is the lipid layer with the polar head interacting with the water. These structures are important in the stabilisation of doughs and batters and in particular in promoting foam stabilisation at the air / water interface.

The most important category of polar lipids used in European bread making are DATEM (Diacetyl tartaric acid esters of monoglycerides).

They rely for their performance on their polar character which promotes protein binding and facilitates dispensability in the dough and on their ability to produce the lamella liquid crystalline structures. Their action helps promote air bubble stabilisation.

There is an inverse relationship between volume increase and fermentation tolerance (Figure 3.44).

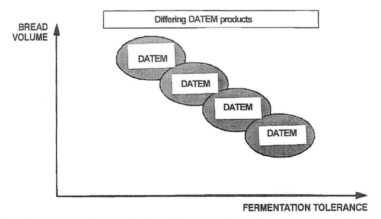

*Fig. 3.44: Performance of DATEM in bread. The numbers refer to differing DATEM products*

These effects correlate with the ratio of polar to non-polar material. As a result of this understanding, work is currently ongoing to develop new DATEM which confer not only high volume but also increased fermentation tolerance.

Other important emulsifiers are monoglycerides and stearoyl-lactylates. These have different functionalities and effects (Table 3.38) and their use depends on legislation, flour types and consumer preference for differing bread types.

*Table: 3.38: Comparison of the effects of mono(di)glyceride and steroyl-lactylate with DATEM in bread*

|  | MONOGLYCERIDES | DATEM | STEAROYL LACTYLATES |
|---|---|---|---|
| TOLERANCE | ✹✹✹✹ | ✹✹ | ✹✹ |
| OVENSPRING | ✹ | ✹✹✹ | ✹ |
| SHAPE | ✹✹ | ✹✹ | ✹✹✹ |
| VOLUME | ✹ | ✹✹✹✹ | ✹✹✹ |
| SOFTNESS | ✹✹ |  | ✹✹✹✹ |
| TEXTURE | ✹ |  | ✹✹ |

*Spreads*

The trend in margarine and spread production is towards reduction in overall fat content of the product. Spreads are predominantly water in oil (w/o) emulsions although some very low fat spreads could be considered as oil in water (o/w) emulsions.

A spread requires an emulsifier in order to be produced and is essentially formed by crystallised fat physically immobilising water droplets. As the amount of fat is decreased then the amount of water increases.

Table 3.39 shows that a decrease from 80% to 40% fat results in a requirement to immobilise 3 times the dispersed (water) phase in half the fat.

*Table 3.39: Reduction in fat content from 80% to 40%*

| Margarine | Low fat spread |
|---|---|
| 80% fat | 40% fat |
| 20% dispersed (water) phase | 60% dispersed phase |

In normal spreads the water droplets are small whereas in the low fat spread the water droplets are much larger and more numerous (Figure 3.45, 3.46).

The amount of fat required in a spread influences the choice of emulsifier. Within the range of emulsifiers, monoglycerides are more polar than triglycerides and therefore have a greater emulsifying power. In addition, the degree of saturation and unsaturation gives the manufacturer the choice of tools to control texture and stability in the full range of spread products.

Powerful w/o emulsifiers such as polyglycerol polyricineolate gives stability in non-crystalline systems, such as warm emulsions, which are important in low fat spread production particularly for very low fat spreads of 20 to 25% fat content.

Other process parameters also play a significant role such as the emulsification regime, emulsification temperature and composition of the water phase.

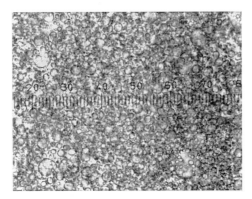

*Fig. 3.45: Immobilised water droplets in a 40% fat spread (x400, each gradation = 2.5 µm).*

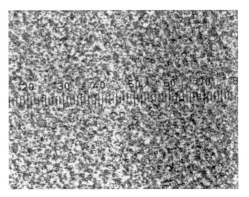

*Fig. 3.46: Immobilised water droplets in a continuous crystallised fat phase in a standard 80% fat margarine (x400, each gradation = 2.5 µm).*

**Future**

The most common emulsifiers used at present (assessed by volume) are food proteins and it is likely that advances in protein understanding will enable production of molecules with advanced emulsification and stabilising properties.

Synthesised emulsifiers also look set to remain key ingredients well into the next century. They are central to low fat food technology and are increasingly being used as mixtures of emulsifiers and in combination with stabilisers in a wider variety of foods.

### 3.3.3.2 Stabilisers

**Introduction**

In the food industry the term 'emulsifier' traditionally refers to a small molecular surfactant. An emulsifier does not necessarily confer long term stability – it simply has the capacity to adsorb rapidly at the fresh interface created during emulsification, thereby protecting newly formed oil-water droplets against immediate re-coalescence.

Long-term stability of 'emulsified systems' is usually provided by proteins or polysaccharides. The role of a good emulsion stabiliser is to keep the droplets apart once they have been formed. This protects the emulsion against processes such as creaming, flocculation and coalescence during long-term storage.

Stabilisers are chemical compounds (usually polymeric) that confer long-term emulsion stability by forming a protective barrier around the surface of droplets or in the liquid between the droplets.

# Emulsifiers – Stabilisers – Enzymes

Some stabilisers can also be considered to have emulsifying properties (such as the hydrocolloid gum arabic which is a genuine emulsifier in its own right). However, stabilisers are used for their ability to control the rheology of the aqueous phase, by either increasing viscosity or conferring gel structure.

## Description

Thickening, gelling and stabilising are important and closely related roles in food systems. Stabilisers are generally very large molecules that form a three-dimensional network in water, thus increasing viscosity. When the concentration is high enough to immobilise the water, a gel is achieved.

Stabilisers can be classified by source according to the following principle groupings:

Plant exudate

– Large polysaccharides of higher plants.

Seaweed extracts

– Macro algal origin (see Figure 3.47)

Plant seed gums

– Extracts from seeds of higher plants.

Plant extracts

– Extracts from higher plants.

Fermentation gums

– Extra-cellular polysaccharides from microbial fermentation

Cellulose derivatives

– Synthetic derivatives of cellulose.

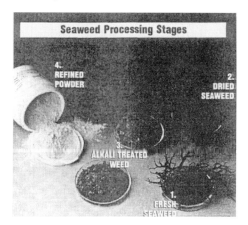

*Fig. 3.47: Seaweed production and extraction facility*

In addition there are also gum derivatives which include propylene glycol alginate and low molecular weight pectin.

Starches, modified starches and gelatin function as water-control agents and are commonly found in many food products. Starch and gelatin are used at relatively high dose levels compared to the inherently more functional gums and hydrocolloids.

A summary of commonly available stabilisers is given in Table 3.40.

*Table 3.40:* *Composition and outline of properties for Emulsified Foods as stipulated in EEC Directive 80/957/EEC*

| E # | Common Name | Effect | Common Source / Origin |
|---|---|---|---|
|  | Gelatin | Water-control Stabiliser Gelling agent | Pork or beef hides, bones and skins |
|  | Starch and rnodified starches | Water-control Stabiliser Gelling agent | Plant carbohydrate |
| E400 to E405 | Alginate (acid and salts) | Stabiliser Thickener Gelling agent | Brown algae: Fucaceae, Laminariales |
| E406 | Agar | Stabiliser Thickener Gelling agent | Red algae: Rhodophyceae |
| E407 | Carrageenan | Stabiliser Gelling agent Thickener | Red algae: Rhodophyceae |
| E408 | Furcelleran | Stabiliser | Red algae: Rhodophyceae |
| E410 | Locust Bean (Carob) gum | Stabiliser Thickener Gelling agent | Endosperm of the seeds of Ceratonia siliqua |
| E412 | Guar gum | Stabiliser Thickener | Endosperm of the seeds of Cyamopsis tetragononolobus |
| E413 | Gum tragacanth | Stabiliser | Gum from the bush of the Astragalus *spp.* |
| E414 | Acacia or Gum arabic | Stabiliser | Gum from the bark of the tress of Acacia *spp.* |
| E415 | Xanthan gum | Stabiliser Thickener | Extracellular carbohydrate obtained from microbial fermentation of *Xanthomonas campestris* |
| E440 | Pectin | Stabiliser Gelling agent Thickener | Apple pomace Citrus fruit peels |
| E460 to E466 | Celluloses | Stabiliser | Cellulose from plants |

# Emulsifiers – Stabilisers – Enzymes

## Application

Stabilisers are used in applications often in conjunction with an emulsifier or at least in an emulsified system. More common food applications for stabilisers include:

- Dressings and sauces
  Pourable dressings; Spoonable dressings; Low and no oil dressings, relishes and pickles, mustard and mint sauces, marinades, canned and frozen sauces, savory dips.
- Bakery
  Cake and premixes, low fat cakes, fruit fillings, glazes, structured glace fruit, meringues, bakery filling creams, cheesecake toppings, jams and jellies.
- Dairy
  Whipped cream, fermented milks, yogfruit preparations, sour cream, cheese spreads, processed cheeses.
- Ice cream and frozen desserts
  Ice creams, sorbets, sherbets, ice lollies.
- Beverages
  Fruit drinks, citrus concentrates, milkshakes, instant drinks, chocolate milks, beers.
- Desserts
  Chilled desserts, instant puddings, instant mousses, topping syrups, instant water jelly.
- Meat and vegetable products
  Structured Fish, meat, vegetables, peppers, tomato products, potato products.
- Other foods
  Coatings, egg products and animal foods.

## Functionality, Understanding and Advances

Stabilising and gelling effects are only obtained after full solubilisation of the stabiliser molecule. When solubilised the molecules reorganise themselves in two basic ways:

- Binding water molecules (thickener effect)
- Network building involving junction zones (gelling effect).

Advances in the industry have led to the understanding on how each hydrocolloid can develop such properties. The development of suitable properties is related to the following:

- Molecular weight
- Functional groups of the molecule
- Temperature
- Molecular interactions.

Although the chemical interactions between molecules clearly differ depending on the molecular composition of the molecule the behaviour of differing stabilisers have some similarities. These are summarised in Figure 3.48.

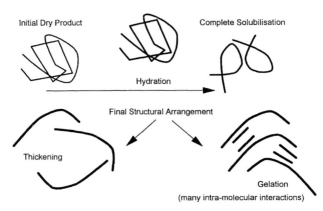

*Fig. 3.48: Dispersion and solubilisation of stabilisers*

## Thickening

When the stabiliser macromolecules cannot be cross linked, the polymer behaves as a thickener. Such colloids lead to viscous solutions, the viscosity of which will depend on the molecular weight.

For a similar concentration, high molecular weight polymers result in solutions with higher viscosities than those from polymers of low molecular weight.

In addition, the form of the macromolecules determine the rheology of the solutions.

## Gelling

Gelation is obtained by the formation of a three-dimensional network. The macro molecules are associated through regions called zones. Free parts remain on the molecule so that one macromolecule can be cross-linked to many others to form a three-dimensional network.

The gel strength and the melting properties (reversible and non-reversible) of a gel are directly related to the quantity of junction zones formed and the nature of the individual bonds involved in the junction zones.

Two steps are generally required for gelation:

– The solubilisation of the macromolecule.
– The association of the macromolecules to form gels.

This usually means heating followed by cooling. For some hydrocolloids (such as alginate) the presence of chelating agents (divalent ions from salts) can generate the gel in cold conditions.

## Application

The selection of a stabiliser for a food system is dependent on the conditions of use:

– pH of the system. Acid systems promote the denaturation of some hydrocolloids.
– Presence of electrolytes. Electrolytes do promote the formation of junctions, therefore increasing the degree of gelation.

- Heat treatments. Most hydrocolloids withstand pasteurisation and sterilisation temperatures at neutral pH.
- Storage stability regime and requirements. Modifications to the structure occur during storage which can lead to syneresis (exudation of water out of the structured system).

**Future**

For the future, stabilisers are set to remain a major ingredient for determining and stabilising the structure of food products.

Enzymic modification of stabiliser molecules is now beginning to produce a new generation of stabilisers which have unique and potentially highly tailored properties in food systems.

### 3.3.3.3 Enzymes

**Introduction**

Enzymes used in food processing are often called "commercial enzymes" and may be derived from plant, animal or microbial sources. These enzymes are produced in large quantities and are not highly refined. In contrast, the highly refined enzymes produced for pharmaceutical or research purposes, are often crystalline enzymes sold in milligram and gram quantities by firms that are specialised in "fine chemicals".

The most widely used animal enzyme is chymosin which is used for milk clotting in the production of cheese. Well known plant enzymes include papain, bromelain and cereal malt. Microbial enzymes have been used in the fruit and cereal processing industries since the 1950's and offer a less expensive source. For example, chymosin (a relatively expensive enzyme found in the stomach of calves) have been replaced by the microbial "rennet" in the production of cheese.

Although enzyme usage was in fact widespread, the first recognition of an enzyme was reported in 1833. It was found that malt extract contained a substance that splits (hydrolyses) starch into glucose. This substance was named "diastase" and is now known as "amylase".

In 1857, Pasteur discovered that the alcoholic fermentation of a sugar solution was caused by microorganisms. However, in the same period Buchner proved that the living cells were not responsible for the fermentation but the substances that were produced by them: the *enzyme's* (Greek for "in yeast"). He ground living yeast cells with sharp sand and squeezed the liberated contents of the cells into a container with antiseptic chemicals in order to kill the remaining living cells. With the thus obtained liquid it was still possible to cause fermentation in a sterile sugar solution.

In 1886 the use of malted barley flour was mentioned for use in breadmaking. The enzymes in the malt converted starch into sugars which could serve as yeast substrate during fermentation. Almost 50 years later, the use of a purified enzyme (lipoxygenase) was described for bleaching the crumb of bread. Since then enzyme usage has become widespread in food applications.

## Description

Today, more than 2300 different enzymes are catalogued by the Enzyme Commission of the International Union of Biochemistry (IUB). The enzyme nomenclature proposed by the IUB uses the name of the substrate (the material that is affected by the enzyme) followed by the suffix "ase". For example, the enzymes affecting amylum (Latin for "starch") are called amylases. Enzymes affecting lipids, cellulose and proteins are named lipases, cellulases and proteases, respectively. However, many enzymes are still named by their original names. Diastase, trypsin, papain, lactase, *etc.* are examples of old names which were used before the official IUB names were proposed. Names of proteases originally had the suffix "in" and are still in use in the traditional industries such as breweries and bread bakeries.

Enzymes are classified nowadays by a four number code. The first number of the code represents the class of the enzyme. There are six different classes (Table 3.41) that are distinguished by the nature of the reaction catalysed by the enzyme. The second number of the systematic cataloguing refers to the group or bonding that is transformed by the enzyme. The third number identifies an acceptor compound or a reaction type or a positional characteristic. The fourth number is for potential further purposes.

*Table 3.41: Enzyme classification*

| Enzyme class ("old" names) | Nature of reaction | Example | EC numbers of examples |
|---|---|---|---|
| 1) Oxidoreductases (oxidases, reductases, dehydrogenases) | Electron transfer | Glucose oxidase | EC 1.1.3.4 |
| 2) Transferases | Group transfer | Glutathione transferase | EC 2.5.1.18 |
| 3) Hydrolases | Hydrolysis | Phospholipase A2 | EC 3.1.1.4 |
| 4) Lyases (aldolases, dehydratases, decarboxylases) | Bond splitting | Pectin lyase | EC 4.2.2.3 |
| 5) Isomerases (racemases, epimerases, isomerases, mutases) | Isomerization | Disulphide isomerase | EC 5.3.4.1 |
| 6) Ligases (synthetases) | Bond formation | Acetyl-CoA ligase | EC 6.2.1.1 |

The EC numbering does not mean that each enzyme is unique in its properties. Several enzymes can act on the same substrate and conversely one single enzyme may act on different substrate.

Hydrolysing enzymes sometimes get an additional prefix to their names, namely 'endo-' or 'exo-'. The prefix 'exo-' indicates that an enzymes is attacking the molecule at one of the ends. When the splitting of the molecule happens somewhere else the prefix 'endo' is used.

## Origin of Enzymes

Sources of microbial enzyme production include:

Bacteria: *Bacillus spp.*
*Streptomyces spp.*
*Klebsiella spp.*

Fungi: *Aspergillus spp.*
*Penicillium spp.*
*Rhizopus spp.*
*Trichoderma spp.*

Yeast: *Saccharomyces spp.*
*Kluyveromyces spp.*

These microorganisms are GRAS (general recognized as safe) and have widespread use in foods. Enzyme production can take place in their submerged fermentation (Figure 3.49) or surface culture (Koji) fermentation processes (Figure 3.50).

Important Classes of Industrial Enzymes

In food (and feed) applications predominantly two classes of enzymes are used. These are hydrolyases (EC class 3), which can be subdivided into carbohydrates, proteases and lipases and oxidoreductases (EC class 1). Enzymes from the other groups are, with only a few exceptions such as glucose isomerase and pectin lyase, not used in bulk quantities for food applications.

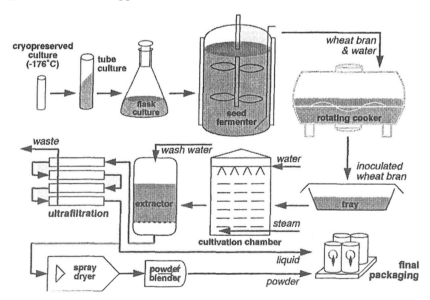

*Fig. 3.49*: *Enzyme production "Surface Culture (Koji) Fermentation"*

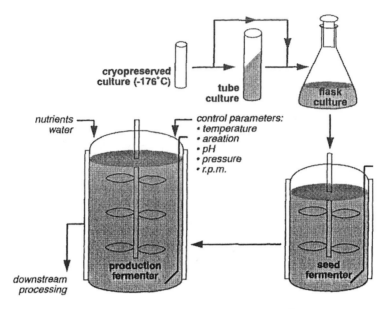

*Fig. 3.50:* Enzyme production "Submerged Fermentation Process"

Carbohydrases

Carbohydrases are the most commonly used enzymes for foods and food processing (Table 3.42).

*Table 3.42:* Carbohydrases used in foods and food processing

| Enzymes | Enzyme | Subclasses Applications |
|---|---|---|
| Amylases | α- and β-amylase, amyloglucosidase, pullulanase | Starch conversion, bread making, brewing, production of HFCS, distilled alcoholic beverages, vinegar |
| Pectinases | polygalacturonase, pectinmethyl-esterase, rhamnogalacturonase | Fruit and juices processing, wine making, coffee and tea fermentation, brewing. |
| Lactases | – | Milk derived product processing. |
| Invertases | – | Conversion of sucrose to fructose and glucose. |
| Cellulases | endo-1,4 β-glucanase, cellobiohydrolase, β-glucosidase | Brewing, cereal processing, fruit and juice processing, food fermentation, wine production, alcohol fermentation, vegetable processing. |
| Hemicellulases | endo- and exoxylanases, pentosanases, mannases, arabinases, galactanases | Bread and biscuit making, brewing, processing of plant derived materials, starch and gluten processing, coffee manufacturing, animal feed |

| Enzymes | Enzyme | Subclasses Applications |
|---|---|---|
| Galactosidases | α-galactosidase, β-galactosidase | Conversion of raffinose (and stachiose) to galactose and glucose and sucrose, animal feed. |
| Dextrinases | – | Sugar beet refining, juice processing. |

Many commercial enzyme preparations are semi-purified and contain not only the specific enzyme activity but other enzyme activties. The side-effects can be positive or negative making enzyme preparations less predictable in food processes [7].

*Proteases*

Production and processing of many food products (*e.g.* cheese, soya sauce and beer) involves enzymatic degradation of the proteins. This has increased the scientific interest in proteases which in turn has led to new applications. Proteases catalyse the degradation of food proteins originate from three sources:

(1) They may be already present in the food material (*e.g.* fish and yeast autolysis products).

(2) They may be excreted by microorganisms growing in the food (*e.g.* proteases excreted by lactic acid bacteria in cheese and yoghurt).

(3) They may be added as isolated enzyme preparations (*e.g.* chymosin in cheese making or papain in meat tenderization).

Proteases can be classified into two major groups: endopeptidases and exopeptidases.

*Endopeptidases*

The endopeptidases can be further subdivided into four groups:

| Main group | Example |
|---|---|
| Serine proteases (EC 3.4.21) | Trypsin, chymotrypsin, subtilisin |
| Cysteine proteases (EC 3.4.22) | Papain, bromelain |
| Aspartic proteases (EC 3.4.23) | Pepsin, chymosin |
| Metalloproteases (EC 3.4.24) | Thermolysin |

As the names already imply, the amino acids serine, cysteine and aspartic acid play an essential role in the catalytic site of the respective enzymes. Modification or blocking of these amino acid residues leads to complete inactivation of these enzymes. The serine proteases have a pH optimum at alkaline pH, the cysteine proteases show maximum activity at more neutral pH, whereas the aspartic acid proteases exhibit optimal activity at acidic pH. The metalloproteases contain an essential metal atom in their active site, usually zinc. They have optimal activity at neutral pH.

*Exopeptidases*

The exopeptidases can be subdivided:

- aminopeptidases
- carboxypeptidases
    serine carboxypeptidases
    metallocarboxypeptidases
    cysteine carboxypeptidases
- dipeptide hydrolases.

Aminopeptidases are widespread in nature but they are hardly commercially available since many of them are intracellular or membrane bound enzymes and thus difficult to produce.

Dipeptide hydrolases are specific for dipeptide and consequently can not be classified as amino- or carboxypeptidases. However, these enzymes are not important for foods.

Exopeptidases are attractive as means of de-bittering protein hydrolysate made with endo-peptidases and a considerable research is invested in these types of enzymes for flavour generation purposes.

*Lipases*

Lipases are fat splitting enzymes that act mainly on triglycerides in addition to mono- and diglycerides and other ester bond containing components in foods. Originally, all lipases used on industrial scale were from animal origin but more recently microbial enzymes have become of interest (Table 3.43).

*Table 3.43:* Lipases of industrial importance

| Industry | Effect | Product |
| --- | --- | --- |
| Dairy | Hydrolysis of milk fat<br>Cheese ripening<br>Modification butter fat | Flavour compounds<br>Cheese, cheese flavour<br>Butter, butter flavour |
| Bakery | Increased shelf life<br>Improved bread flavour | Bakery products |
| Beverage | Improved aroma. | Several beverages |
| Dressings | Quality improvement | Mayonnaise, dressings, toppings |
| Health | Transesterification | Health foods |
| Meat/fish | Fat removal, flavour increase | Meat and fish products |
| Fats/oils | Transesterification<br>hydrolysis | Cocoa butter, margarine<br>fatty acids, mono diglycerides |
| Chemicals | Enantioselective hydrolysis<br>Enantioselective synthesis | Chiral building blocks<br>Specialty chemicals |

Developments in this class of enzymes are so rapidly and promising that they are likely to be a strong influence in the food industry. Examples are the production of entirely new foods and the production of a wide variety of effective flavours or emulsifiers.

Under aqueous conditions lipases catalyse the breakdown (hydrolysis) of fats and other ester bond containing compounds. Moreover, lipases appear to be effective catalysts in organic solvents, *i.e.* under low water conditions. Here lipases do not hydrolyse fats but they are able to reverse their normal reaction and can synthesize fats or ester bonds or they exchange fatty acids between different compounds.

This has opened the way to industrial production of specialty chemicals, flavours, emulsifiers, pharmaceutical and cosmetics. As the molecular structure of several lipases is known, *r*-DNA techniques are being used to optimise lipases.

*Oxidoreductases*

Oxidoredutases are currently not large volume products for enzyme manufacturers but they are increasingly used in specialized applications in food industries such as: colourisation or decolorisation of food products; deoxygenation of foods especially beverages and processing aids particularly in bakery applications.

Table 3.44 gives an overview of the oxidoreductases that are used together with several sources of the enzymes and some specific characteristics.

The wide range of types of oxidation/reduction reactions in foods suggests an enormous potential for oxidoreductases.

*Table 3.44:* Oxidoreductases of industrial importance

| Enzyme | Source | Functions / Characteristics |
|---|---|---|
| Polyphenol oxidase | Mushroom, grape, pear, kiwi, sago, tea, potato, strawberry, olive, beet, cocoa | Enzymatic browning (pigments), ripening of fruits, flavour formation, colour {e.g. strawberries) formation |
| Alcohol oxidase | Mammalian organs | Deoxygenation of foods |
| Peroxidase | Horse radish, apple, bean, grape, barley, potato, cucumber, pea, peanut, tomato | Ripening of foods, several functions in plants, colour formation (off-) flavour formation. |
| Lactoperoxidase | Milk (not human milk), saliva, tears | Anti-microbial effects (viruses, bacteria, fungi, parasites, mycoplasms) |
| Dehydrogenase | Various plants | Co-factor dependent enzymes. Important for metabolism in plants and animals. Only a few food applications known. |
| Sulfhydryl oxidase | Mammalian milk, microbial | Oxidates -SH groups into disulphide bridges, catalyses refolding of denatured proteins |
| Glucose oxidase | Microbial | Anti-oxidant |
| Lipoxygenase | Bean, potato, tomato, maize, strawberry, canola, fish, meat | Oxidation of polyunsaturated fatty acids to hydroperoxides and further into aldehydes, acids, ketones, etc. |
| Dehydrogenase | Various plants | Co-factor dependent enzymes. Important for metabolism in plants and animals. Only a few food applications known. |

**Functionality, Understanding and Advances**

*(1) Enzymes in the Baking Industry*

In many baking processes a combination of enzymes and other baking active ingredients are used to adapt the flour quality to the needs of the individual baker and especially to those of the industrial bakeries.

Flour essentially contains three major components: starch, protein and pentosan. The optimum quality of these flour components is achieved either by the natural enzymes in the grain or by adding extra enzymes. The 'natural' enzymes are influenced by the genotype of the cereal as well as by largely uncontrollable factors including the cereal-growing area, the time of harvesting and the weather conditions during the ripening period. Added enzymes can help compensate for natural deficiencies.

In the past, the enzyme activities present in malt flour were used for flour treatment. Advances in microbial production of enzymes in recent years has lead to the availability and use of a whole range of more targeted enzyme preparations.

Enzymes applied in the baking industry are grouped in three major classes:

- Amylases
- Hemicellulases
- Proteinases.

Fungal amylase has a widespread application in bread baking. As it is inactivated at lower temperatures when compared with malt- and bacterial- amylases, it can only convert damaged starch into fermentable sugars, which are used by the yeast to produce carbon dioxide and alcohol.

In comparison with fungal amylase, fungal hemicellulase is more functional. Hemicellulase is a complex mixture of enzymes. It hydrolyses water-insoluble hemicellulose into water-soluble pentosan and, as a consequence, has a great impact on dough water management.

More aggressive hemicellulase will continue degradation until oligosaccharide are released. This further breakdown is beneficial in applications where low waterbinding is required such as biscuits, crackers and wafers.

Proteinases are not often used for bread baking. They are only applied to strong wheat flours, like those available in North America. Aggressive acting bacterial proteinases are used in the biscuit industry, since a high level of degradation is required and, in addition, the effectiveness of the proteinases is inhibited by the low water content and high fat and sugar ratios in the formulations.

Proteinases help the biscuit manufacturer to become largely independent of the flour quality. For example, the baker can even use flour rich in protein, the flour strength and dough constancy being adapted by proteinases, proteinase / amylase and proteinase / hemicellulase preparations.

A major consideration in the increasingly industrialised bread production and distribution systems is staling. During staling there are changes to many physical properties of bread. The most obvious symptoms of staling are for example: increase in firmness, increase in dryness / crumbliness, loss of taste and aroma and loss of crumb whiteness.

Underlying these visible changes there are a number of changes to the physical properties of dough components, such as a decrease in water absorbtion capacity, decrease in soluble starch and increase in starch crystallinity.

Increase in starch crystallinity is often considered as the most straight forward way of understanding the changes that take place in bread during storage. The consumer associates the softness of bread with freshness. Keeping bread softer for longer will increase the pre- and post-sale shelf-life of the product.

Development in enzyme technology has produced multi-functional enzyme systems that significantly increases the time bread remains soft.

Many major benefits which include:

- Fresh keeping
- Bread volume
- Fine bread texture
- Loaf shape
- Bread browning.

Major benefits in dough processing include:

- Dough development
- Dough volume
- Dough relaxation
- Fermentation tolerance
- Ovenspring.

Interaction with gluten gives a good gluten matrix with fine gas cells, a more flexible dough and a faster dough development. Interaction with starch gives improved dough volume and a better crust colour.

Interaction with other carbohydrate moieties gives a better fermentation tolerance, better shape, increased ovenspring and improved fresh keeping properties.

An overview is given in Table 3.45.

*Table 3.45: Overview of enzyme usage in Bread production*

| Characteristic | Enzymes | Characteristic | Enzymes |
|---|---|---|---|
| Bread volume | Xylanase, Amylase | Proofing time | Xylanase |
| Mixing time | Protease | Tolerance | Xylanase |
| Dough development | Peroxidase, NAGase | Colour/Flavour | Peptidase, Amylase |
| Dough stickiness | NAGase | Fresh-keeping | Amylase |

*Amylases in Baking*

α-Amylase converts starch into dextrin which are in turn further converted by ß-amylases into maltose and by amyloglucosidases into glucose. Raw starch is hardly converted by amylases. Only damaged starch (damaging of the starch granules occurs during wheat milling) and gelatinized starch are real substrates for these enzymes. Amylase reactions constantly supply the yeast with fermentable sugars, leading to a uniform gas production in the dough. The crust colour and flavour are also improved by amylase action. Higher contents of free sugars increase the Maillard browning process. The origin of the amylase added to bread dough is very important. Fungal amylases are inactivated in the oven during baking before the starch is completely gelatinized. These enzymes, therefore, act only on the damaged starch. Bacterial amylases can act also on gelatinized starch because of their higher heat stability. A problem with these enzymes is that their stability is too high and as a consequence these enzymes survive the baking step and keep on hydrolysing the starch. This leads to an unacceptable gummy-like crumb structure.

*Proteases in Baking*

During dough development, the flour proteins form a network of which a simplified model is shown in Figure 3.51.

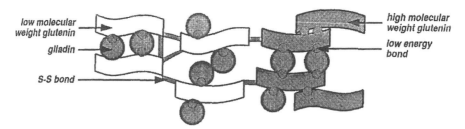

**Fig. 3.51:** *Structure of gluten*

This network is strengthened by the formation of disulphide bridges between certain amino acids (cysteine), the more disulphide bridges the stronger the gluten. Proteases can break down this network by splitting peptide bonds between amino acids. Addition of proteases to a too strong dough leads therefore to improved elasticity and dough handling properties. The gluten structure becomes softer and expands better upon increasing gas pressure. A risk of using proteases for this purpose is over-dosing, leading to decomposition of the dough structure and destruction of the gas holding capacity.

Proteases are especially of use in production of cookies, crackers, wafers and cakes were a weak gluten structure is required. Doughs of weak gluten do not shrink and batters have a low viscosity and do not contain lumps of coagulated gluten.

*Oxidases in Baking*

In contrast to the action of proteases, oxidases can be used to strengthen the dough structure. Although the precise action of proteases is not completely understood, their action most probably leads eventually to an increased number of disulphide bridges in the gluten network. Also coupling of carbohydrate structures to the gluten network may lead to increased strength of the dough.

*Xylanases in Baking*

Xylanases (also named hemicellulases or pentosanases) act on the non-starch polysaccharides (NSP) in the dough. NSP, mainly arabinoxylan structures, are responsible for binding a large part of the water in the dough. Arabinoxylans can bind at least 10 times their own weight of water. The partial breakdown of the NSP structures releases water in to the dough. This water is redivided over the gluten and the starch and is important for optimal dough development and for the additional increase in volume during the baking step (the so-called ovenspring).

For xylanases the source of the enzyme is important, fungal xylanases from *Aspergillus spp.* are generally very useful for bread making whereas for cookies *Trichoderma spp.* xylanases are required because of the differing requirements for water release.

## (2) Brewing

Beer is a beverage made by alcoholic fermentation of a sugar solution derived from grains. Most carbohydrates in grain are present in the form of starch which can not be fermented by brewer's yeast. Before beer can be produced the starch of the grain must be transformed into fermentable sugars by enzymatic digestion.

The most frequently used raw material for beer production is malt. Malt is germinated grain which is rich in enzymes but which is similar to the original grain in external appearance. Barley is the preferred grain for many reasons, one of them is that barley has a good ability to form enzymes during germination.

During this germination hydrolytic enzymes already start degrading the cell wall and the proteins of the grain. In the breakdown of the (1-3) and (1-4)-ß-D-glucans, which are main components of the endosperm cell wall, several ß-glucanases are involved. The result is a 'modified' grain which is less 'rigid' in structure and capable of releasing soluble components during the brewing process.

Starch digestion is rather limited during formation of the malt. The degree of modification of the malt is strongly dependent on the application. Ale malts are heavily modified leading to higher levels of sugars and amino acids which are converted into flavours and coloured products. Traditional ales are therefore dark-coloured beers with strong taste. Pilsner type of beers are very pale beverages which are made from less modified malt. The brewing of these pale products therefore needs more hydrolysis of the malt components during the brewing process itself.

The brewing process starts with making a mixture of hot water and grain, called a mash. The malt is usually the major grain present acting as a source of enzymes. The mash undergoes a program of time and temperature to accomplish the desired hydrolysis and solubilisation of the grain components. At the end of the mash the liquid contains dissolved sugars and other hydrolysed starch and grain components. The liquid, called wort, is separated from the remaining undigested spent grain by filtration using the hulls of the grain as a filter bed. The wort is boiled with flavouring hops, cooled and fermented after the addition of yeast and oxygen.

The enzymes needed for good processing are (α- and β-amylases, ß-(1→ 3)- and ß-(1 → 4) endo glucanases, proteases, peptidase and cellulase. Enzymes similar to those contained in malt are produced industrially and can be used to speed up filtration of the wort and of the finished beer. For optimal filtration *Trichoderma spp.* glucanases which possess cellulase activities are preferred over *Aspergillus spp.* and *Bacillus spp.* glucanases.

*Chill Proofing Enzymes*

Other enzymes are required to obtain good quality beer with good shelf-life. The major use of non-malt enzymes is an enzyme that is added to fermented beer to 'chillproof' the beer. Beer is fermented and aged under chilled conditions. Almost all packed beer is filtered while cold to achieve clarity. In spite of the filtration steps beer becomes cloudy after it is packed, distributed and chilled again for serving. The cloudiness that develops is caused by formation of haze particles called 'chill haze' which are the result of the interaction between peptides and polyphenol compounds.

Modification of the proteins by addition of proteolytic enzymes prevents the precipitation of the peptides upon chilling. Proteases from papaya (papain) were already applied in 1911. It is even more remarkable that the major chill stabilizing enzyme used today is still papain.

*Extension of Shelf Life of Packaged Beer*

Beer is extremely sensitive to oxygen and oxidative staring. With the development of enzymatic chillproofing it in now possible to pack beer and ship it over long distances while maintaining its appearance.

Further development came with new developments in packing technology which made it possible to reduce the air content in the package and with the development of anti-oxidants like sulphite salts. There is, however, a continuing interest in enzymes that react with oxygen to reduce the oxygen level in beer. This is due to the importance of oxidation in beer flavour maintenance. The most readily available enzyme for this purpose is glucose oxidase. A combination of this enzyme from *Aspergillus niger* with catalase from this same organism will eventually remove all oxygen from a solution containing these enzymes, glucose and water. However, in the presence of ethanol, catalase acts as a peroxidase and therefore sulphite addition is still required. The glucose-oxidase-catalase system is still in use to remove the oxygen in the head space of the bottle.

*Fruit Juices*

Pectic substances (pectin) are chain molecules with a rhamogalacturonan backbone. This backbone consists of 'smooth' α-D-1,4-galacturonan regions that are interrupted to a small extent by insertion of 1,2-linked α-L-rhamnosyl residues and highly branched regions with an almost alternating rhamnogalacturonan chain (hairy regions). The side chains of these hairy regions are composed of neutral sugars galactose, arabinose and some xylose. Pectic enzymes are classified according to their mode of attack on the galacturonan part of the pectin molecule.

*Fruit Juice Clarification*

Fruit juice clarification is the oldest and largest use of pectinase. The traditional way to prepare juices is by crushing and pressing fruit pulp. The raw juice is a viscous liquid with a persistent cloud of cell wall fragments and protein complexes. Addition of pectinase lowers the viscosity of the juice and 'modifies' the haze particles. These sediment and can be removed by centrifugation or (ultra)filtration. Pectin lyase or the combination of polygalacturonase with pectin esterase are found optimal for this application.

*Enzymatic Treatment of Fruit Pulp to Increase Juice Yield*

Enzyme treatment of fruit pulp before pressing improves the yield and the colour. Pectin in soft fruits forms a gel-like mass which is very difficult to press. Treatment with pectinase leads to a thin, free running liquid and good pressable pulp. Since some of the pectinase are inhibited by polyphenols, activation of polyphenoloxidase which is present in fruits, is an important aspect of pulp treatment and can be achieved by aeration of the pulp.

*Liquefaction*

Enzymatic liquefaction of the pulp makes pressings redundant. Besides pectinase, the liquefaction process also requires cellulase (endo-cellulase, cellobiohydolase, β-glucosidase). Industrial liquefaction processes became possible when *Trichoderma spp.* cellulase became available. The combination of pectinase and cellulase leads to a very low viscosity liquid. This liquid is treated further by centrifugation and (ultra)filtration to obtain the final clear fruit juice.

*Maceration*

Maceration is the process in which tissue is transformed into a suspension of intact cells. Such pulpy products arc used as base material for pulpy juices, baby foods, dairy products, *etc.*

Maceration requires gentle enzyme action. Besides this, the enzymes which are present in the fruit itself must be inactivated in order to prevent excessive damage to the cells. This is done by heat treatment but has a risk of heat damage. Proper maceration is obtained by mild mechanical treatment followed by treatment with macerating enzymes (a combination of polygalacturonase and pectin lyase). The maceration process is done in a proper way when there is no browning of the pulp, browning being the result of polyphenoloxidase action.

*Starch Conversion*

For the production of starch, sweeteners, syrups and alcohols mainly corn and wheat are used. Corn is often preferred due to its abundant supply and low cost. Its storage stability ensures a continuous source of raw material throughout the year.

*Starch Production*

Starch is produced in so-called refining industries in which aqueous streams are used to facilitate separation of the individual components present in the corn and wheat kernels. The process is called 'wet-milling'.

In 'dry milling' fractions of the kernel are separated by milling, aspiration and screening. After softening by steeping and soaking, the kernel is ground and the germ is separated from the endosperm of the kernel by centrifugation and screening. The remaining material consists of starch and gluten which is also separated by centrifugation. This separation may be facilitated by using appropriate enzymes, like cellulase and hemicellulases. The relatively pure starch is either dried or converted further into sweeteners or syrups.

*Sweetener Production*

Sweeteners are produced by enzyme or acid hydrolysis of starch. The degree of hydrolysis is often expressed as dextrose equivalents (DE). This is the weight percentage of reducing sugars as calculated on dry-solids basis. Starch DE is essentially 0 and complete hydrolysis to dextrose yields a 100 DE product. Complete hydrolysis can be obtained by boiling the starch in dilute sulphuric acid. However, the neutralization after hydrolysis gives many problems and the requirements for acid resistant processing equipment are difficult to meet.

Enzyme hydrolysis avoids these problems and by using special acid temperature stable enzymes the hydrolysis processes have been shown to be very effective. Initial hydrolysis using a thermostable bacterial α-amylase yields a low viscosity hydrolysate of DE 10-15.

Further conversion (saccharification) to dextrose can be done using glucoamylase. Starch consists of polymeric linear α-1,4 linked dextrose units (amylose, ± 25 % of the starch) and polymeric mixed α-1,4/α-1,6 linked dextrose units (amylopectin, ± 75 % of the starch). For enzymatic degradation besides α-1,4 specific amylases also α-1,6 specific amylases (pullulanases) are required. A portion of the dextrose can be converted into fructose with a glucose isomerase yielding a high fructose corn syrup (HFCS).

Many kinds of starch derived products like maltodextrins, oligosaccharide, cyclodextrins, fructose syrups, glucose syrups, crystalline starches, *etc.* can be produced using a variety of enzymes. Examples of enzymes being used are: low pH amylase, low pH glucose isomerase, high- and ultra high temperature stable amylase, raw starch degrading amylase, high temperature stable glucose isomerase, immobilized multi-enzyme reactors, proteases, xylanases, cellulase, *etc.* For the conversion of starch to fuels (bio-ethanol) a whole range of enzymes is required. This may become an interesting new market, since several good and effective enzymes for raw material conversions are available.

*(3) Cheese Ripening*

The most important use of enzymes in the dairy industry is in the coagulation of milk in order to make cheese. Although many proteolytic enzymes can coagulate milk, only a few are suited to make cheese. The general name 'rennet' was introduced to describe all milk-clotting enzymes used for cheese making. Enzymes from plant, animal or microbial origin are allowed for this purpose, of course when other safety criteria (*e.g.* GRAS status of microorganisms) are met.

In milk clotting one of the milk proteins (κ-casein) is partially hydrolysed. This protein is then unable to stabilise milk micelles and leads to a non-enzymatic aggregation of the casein micelles. Both pH and temperature affect milk coagulation. The use of starter cultures lowers the pH and continues doing so even when the curd is already formed.

The milk-clotting enzymes that are added to initiate the coagulation of milk have other proteolytic capabilities that contribute to the ageing of the cheese.

In hard cheese, 25-35% of the insoluble protein of the curd may be converted into soluble protein. In the soft varieties such as Brie and Camembert over 80 % of the insoluble protein can be converted into the water-soluble compounds of amino acids and peptides.

The production of bitter flavours generally is attributed to the formation of bitter peptides. Bitter flavour results when the peptides are formed faster than they can be broken down by the proteases and peptidase of the starter cultures.

Excessive proteolysis, however, leads to excessive loss of fat and cheese yield and adversely affects cheese flavour and texture.

*Lipolysis*

Most lipolytic activity that occurs during cheese ripening is produced by the numerous micro-organisms present in cheese. The micro-flora in cheese is very complex and also changeable because different organisms grow at different stages during the ripening process. Most hard cheese varieties ripen due to the action of bacteria in the cheese. The softer varieties are largely ripened by yeast, mould or bacteria growing on the surface.

Milk lipases may give undesirable rancidity when fresh milk is used for cheese making. The heat sensitivity of these lipases restricts them playing a major role in ripening of cheese made from pasteurized milk. Lipases play a very important role in flavour formation especially in mould cheeses such as Roquefort and Gorgonzola.

*Accelerated Cheese Ripening*

In enzyme modified cheese (EMC) the natural cheese is modified with enzymes or micro-organisms which accelerate the changes that occur normally during cheese ripening. Production of cheese on industrial scale is an expensive process. EMC processes have been developed in order to shorten and thus cheapen the process.

Proteases and lipases or the organisms producing these enzymes are used in manufacturing EMC's. Protein and fat breakdown products play a fundamental role in the flavour development of cheese. The source of the lipases and also the way of addition is important for the final flavour.

An example is the use of pre-gastric lipases for the production of Italian cheese types. In these cheeses the entire stomach of calves is dried and used. This results in a characteristic piquant flavour. The substrate specificity of the lipases, in terms of affinity for certain fatty acid chain lengths, determines the quality and the flavour of the final lipolyses product.

**Future**

For the future, enzymes look set to increase in importance in industrial applications not only in the traditional sectors such as baking, brewing and cheese making, but also in areas as fat modification and sweetner technology. Their role will become more targeted as the exact activity and site of action becomes elucidated. As technology and legislation develops is it likely that the role of $r$-DNA technology will help facilitate the economical production of specific and highly functional enzyme preparations.

**REFERENCES**

[1] Griffin, W.C. J. Soc. Cosmet. Chem. 1, 311-326 (1949).
[2] Becher, P. (1985) in Encyclopedia of Emulsion Technology (Vol. 2) (Becher, P.. ed.), pp 425-512. Marcel Dekker.
[3] Larsson (1986) in "The Lipid Handbook" (F.D. Gunstone, J.L. Hardwood and F.B. Padley, eds.). Chapman and Hall, London and New York, 1986.
[4] Hasenhuetti, G.L. A.I.Chem. E. Symp. Ser. 86, 35-43 (1990).
[5] Randail, R.C., Phillips, G.0. and Williams, P.A. Food Hydrocolloids 2, 131-140, (1988).
[6] Shuster, G. (1981). Manufacture and stabilisation of food emulsions. INSKO, Helsinki, Finland.
[7] Godfrey, T., West, S. Introduction to industrial enzymology. In: Godfrey, T., West, S. (eds.) Industrial Enzymology. 2nd edn. Basingstoke: Macmillan Press and New York: Stockton Press; 1-8 (1996)

## 3.3.4 Flavour Modifiers

*Günter Matheis*

### 3.3.4.1 Definition and Classification

Some food components that have little or no contribution to taste or odour at typical usage levels are capable of enhancing, decreasing, or modifying the taste and odour of foods. These components may be termed "flavour modifiers".

The term flavour modifier is not commonly used and, to my knowledge, there is no generally accepted definition. The term flavour enhancer is commonly known. Various publications emphasize the idea that a flavour enhancer (or flavour potentiator) is a substance added to a food to supplement, enhance, improve, or intensify its original flavour *[1-3]*. Kemp and Beauchamp *[4]* criticized the failure to distinguish flavour enhancer from flavour potentiator. In their view, a flavour potentiator is a substance that increases the perceived intensity of the flavour of another substance, whereas the term flavour enhancer should be used for a substance that increases the pleasantness of the flavour of another substance. Thus, the term flavour enhancer should be restricted to hedonic improvement. Kemp and Beauchamp *[4]* also point out that some substances commonly termed flavour enhancers or flavour potentiators may suppress the flavour of other substances (e.g. monosodium glutamate suppresses sweetness, bitterness and the cooling effect of menthol but not sourness or butteryness). They propose that substances giving such effects be termed "flavour modulators".

The International Organisation of the Flavour Industry (IOFI) defines flavour enhancer as a „substance with little or no odour at the level used, the primary purpose of which is to increase the flavour effect of certain food components well beyond any flavour contributed directly by the substance itself" *[5]*. The EU definition of flavour enhancers is "substances that enhance the taste and/or odour of a food" *[6]*.

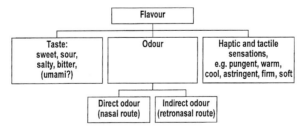

***Fig. 3.52:*** *Flavour of a food*

The flavour of a food consists of odour, taste, and haptic and tactile sensations in the mouth ("mouthfeel") (Fig. 3.52). This means that a flavour modifier may affect odour, taste, and/or mouthfeel of a food. Usually, only taste and/or odour are affected, although maltol and ethyl maltol have been reported to be effective in improving the mouthfeel in low-fat food systems *[7]*. Because of lack of a clear definition, I shall classify flavour modifiers into the five categories shown in Table 3.46. Based on this classification, monosodium glutamate and 5'-ribonucleotides, for example, are taste enhancers (category 1 in Tab. 3.46) and odour suppressors (category 3 in Tab. 3.46) at the same time. The term flavour modifier includes both, taste enhancing and odour suppressing effects, while the term taste or flavour enhancer covers only one effect.

*Table 3.46: Categories of flavour modifiers*

| Category | Examples | Remarks |
|---|---|---|
| (1) Flavour enhancers that show little or no flavour at typical usage level | Monosodium glutamate, purine 5'-ribonucleotides | Enhance taste impressions |
| (2) Flavour enhancers that show flavour at typical usage level | Vanillin*, ethyl vanillin* | Enhance odour impressions, e.g. fruity, chocolate |
|  | Maltol, ethyl maltol, 4-hydroxy-2,5-dimethyl-3(2H)-furanone, 4-hydroxy-5-methyl-3(2H)-furanone | Enhance odour impressions, e.g. fruity, creamy; maltol and ethyl maltol improve the mouthfeel of low fat foods |
|  | 3-Methyl-2-cyclopentene-2-ol-1-one | Enhances odour impressions, e.g. nutty, chocolate |
| (3) Flavour suppressors that show little or no flavour at typical usage level | Monosodium glutamate, purine 5'-ribonucleotides | Mask or suppress odour impressions, e.g. sulfurous and hydrolysate notes |
|  | Gymnemic acids** | Eliminate sweet taste impressions for hours |
| (4) Flavour suppressors that show flavour at typical usage level | Sucrose* | Suppresses unpleasant odour impressions in fruit juices |
| (5) Other flavour modifiers | Miraculin*** | Sour tasting substances are perceived as sweet tasting for approx. two hours |

In the following, the most important flavour modifiers as well as some less important ones will be discussed.

### 3.3.4.2 Monosodium Glutamate, Purine 5'-Ribonucleotides, and Related Substances

#### 3.3.4.2.1 Historical Background

Cooks around the world have known how to prepare good soup using vegetables and meat or bones from time immemorial. Traditionally, the Japanese culture has used

---

\* Although vanillin, ethyl vanillin and sucrose are normally not considered as flavour modifiers, they nevertheless have flavour modifying properties
\*\* Isolated from the leaves of *Gymnema sylvestre,* a tropical plant
\*\*\* Glycoprotein from miracle fruit, the fruit of a West African shrub

dried seaweed, dried fermented bonito (a mackerel-type fish), the dried fungus shiitake, and other natural sources to improve the flavour of various dishes *[2, 8]*.

Although glutamic acid was first isolated from wheat gluten and named after it by the German scientist Ritthausen in 1866, it was only in 1908 that the Japanese scientist Ikeda first attributed the flavour improving effect of seaweed to glutamic acid. He also named this effect "umami". In 1913, investigations of the Japanese scientist Kodama into dried bonito led to the discovery that inosinic acid, known since German scientist Justus von Liebig's mid-19th-century work on beef broth, is another typical umami substance. In 1960, the Japanese scientists Kuninaka recognized the role of 5'-guanylate as another key component of the umami type in shiitake mushroom *[9]*.

Today, the most commonly used umami substances are monosodium glutamate, disodium 5'-inosinate, and disodium 5'-guanylate. They are commercially available world-wide.

### 3.3.4.2.2 Monosodium Glutamate and Glutamic Acid

Monosodium glutamate (MSG) is the monosodium salt of L-(+)-glutamic acid (Fig. 3.53). Only the completely dissociated form of L-(+)-glutamic acid exhibits the umami effect (D-(+)-glutamic acid has no flavour improving properties). At pH 4.0, only 36.0 % of the acid are dissociated. The percentages of dissociation at pH 5.0, 6.0, 7.0, and 8.0 are 84.9, 98.2, 99.8, and 96.9, respectively. This means that only at pH 5.0 to 8.0 glutamic acid shows its optimal effect.

In practise, MSG (which corresponds to the completely dissociated form of the acid) is used almost exclusively to improve the flavour of foods. It is permitted world-wide, although in some countries and some types of foods maximum concentration limits apply. Other salts of glutamic acid that are occasionally used are potassium, calcium, and ammonium glutamates. These are not permitted world-wide.

Glutamic acid and MSG have been originally isolated from natural sources. Glutamic acid is ubiquitous in nature and occurs as building block of all proteins. It is also the most abundant amino acid in almost all proteins. To isolate glutamic acid from natural sources would not cover the quantities currently required by the food industry. Current estimated global demand for MSG is about 500,000 metric tons annually. Presently, the vast majority of MSG is produced through fermentation processes. Among the microorganism strains used for the biosynthesis of glutamic acid, bacteria belonging to the genera of *Corynebacterium* and *Brevibacterium* are widespread. For industrial production, starch, cane and beet molasses, or sugar are employed as the carbon source. An ample supply of a suitable nitrogen source is essential for the fermentation. Ammonium chloride, ammonium sulfate, and urea are assimilable. The ammonium ion is detrimental to both cell growth and product formation. Therefore, its concentration must be kept at a low level. Gaseous ammonia has a great advantage over the ammonium salts in maintaining the pH at 7.0 to 8.0, the optimum for glutamic acid formation. Recent technological innovations such as DNA recombination, cell fusion, and bioreactor development are now being applied for further improvements of glutamic acid formation.

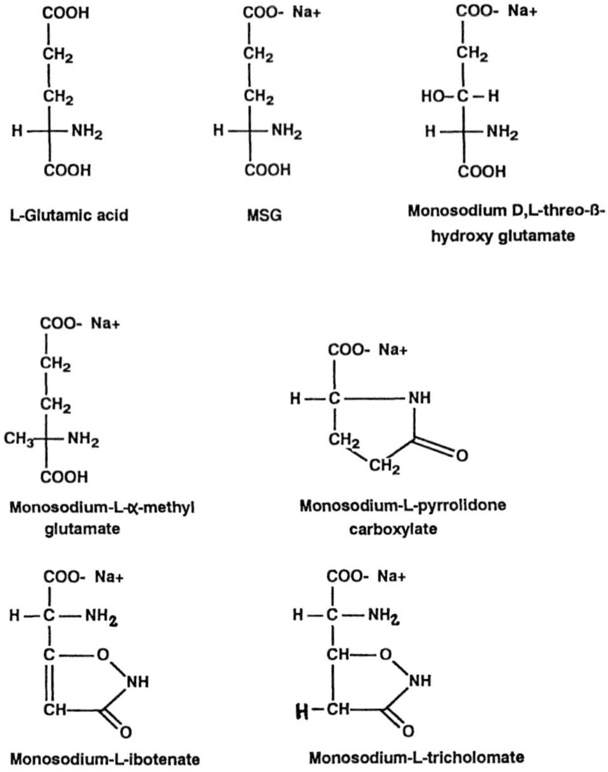

*Fig. 3.53: Chemical structures of glutamic acid, MSG, and selected related substances*

Glutamic acid can also be obtained from acid hydrolysis of proteins (e.g. wheat gluten or corn protein) or by chemical synthesis (e.g. from acrylonitrile) [8].

From glutamic acid, MSG of more than 99 % purity is obtained by neutralization and purification. It consists of odourless white crystals. In aqueous solution, MSG tastes sweet-salty with a detection threshold of 0.01-0.03 % or 100-300 ppm [2, 10].

MSG is not hygroscopic and does not change in appearance or quality during storage. It is not decomposed during normal food processing or in cooking (at pH 5.0 to 8.0). In acidic conditions (pH 2.2 to 4.4) with high temperatures, MSG is partly dehydrated and converted into 5-pyrrolidone-2-carboxylate [2], and the flavour improving effect is lost. At very high temperatures and strong alkaline or acidic conditions, MSG racemizes to D,L-glutamate [2].

Typical applications for MSG are soups, sauces, ready-to-eat meals, meat and fish products, snacks, and vegetable products (with the exception of pickled products because of their low pH of 2 to 3). Typical usage levels of MSG are shown in Tab. 3.47. They range from 0.1 bis 0.6 % in finished food. There appears to be some variability from one person to another as to the preferred optimum level of use. Because MSG is readily soluble in water, recipes often call for dissolving it in the aqueous ingredients of products such as salad dressings before they are added to food.

*Table 3.47:* Usage levels of important umami substances in selected foods

| Food | Usage level (%) | | | |
|---|---|---|---|---|
| | MSG | IMP | GMP | IMP/GMP* |
| Cheese, processed | 0.40-0.50 | | | 0.005-0.010 |
| Dressings | 0.30-0.40 | | | 0.010-0.150 |
| Hamburgers, frozen | 0.10-0.15 | 0.002-0.004 | 0.001-0.002 | 0.001-0.002 |
| Ketchup | 0.15-0.30 | | | 0.010-0.020 |
| Mayonnaise | 0.40-0.60 | | | 0.012-0.018 |
| Meat and fish products, canned | 0.07-0.30 | 0.010-0.015 | 0.004-0.007 | 0.001-0.010 |
| Snacks | 0.10-0.50 | 0.005-0.010 | 0.002-0.004 | 0.003-0.007 |
| Soups and sauces, canned | 0.12-0.18 | 0.004-0.005 | 0.002 | 0.002-0.003 |
| Soups and sauces, dehydrated | 5.00-8.00 | 0.200-0.260 | 0.090-0.110 | 0.100-0.200 |

MSG has no flavour improving effect on a wide range of foods such as confectionary and dairy products, soft drinks, fruit juice drinks, desserts, and others. Moreover, the addition of MSG to such products may even have an adverse effect on their flavour.

It should be noted that many foods naturally contain free glutamic acid. Examples are given in Table 3.48. Protein-bound glutamic acid, which occurs virtually in all proteins, has no umami effect.

*Table 3.48:* Free glutamic acid and 5'-ribonucleotides content of selected foods [2, 8-10]

| Food | Free glutamic acid (%)** | IMP (%) | GMP (%) | AMP (%) |
|---|---|---|---|---|
| Bonito, dehydrated | | 0.630-1.310 | 0 | Traces |
| Cow's milk | 0.002 | 0.115-0.326 | 0.002-0.005 | 0.002-0.013 |
| Cheeses | 0.390-1.200 | | | |
| Fish | | 0.100-0.421 | 0 | 0.001-0.020 |
| Meat | 0.023-0.044 | 0.020-0.200 | 0.002-0.005 | 0.001-0.015 |
| Shiitake | | 0 | 0.045-0.103 | 0.030-0.175 |
| Shiitake, dehydrated | | 0 | 0.126 | 0.100-0.320 |
| Vegetables | 0.047-0.246 | 0.000-0.001 | 0.000-0.001 | 0.000-0.032 |

---

\* Mixture of IMP and GMP (1:1)
\*\* Protein-bound glutamic acid has no umami effect.

In the late 1960s, a condition became apparent through the literature, which was called the "Chinese Restaurant Syndrome" *[11, 12]*. Administration of MSG to "normal" individuals was reported to result in various symptoms such as burning sensation in the back of the neck, facial pressure, chest pain, sweating, nausea, weakness, thirst and headache.

The Joint Expert Committee on Food Additives (JECFA) of the Food and Agricultural Organisation (FAO) of the United Nations and the World Health Organisation (WHO) considered the issue of MSG hypersensitivity ("Chinese Restaurant Syndrome") and concluded in 1987 that "studies have failed to demonstrate that MSG is the causal agent in provoking the full range of symptoms of Chinese Restaurant Syndrome. Properly conducted double-blind studies among individuals who claimed to suffer from the syndrome did not confirm MSG as the causal agent" *[2]*. In 1993, the conclusions of the JECFA were confirmed by the University of Western Sydney, Australia *[13]*. It appears that histamine, allergenic proteins, preservatives, food colourings or high levels of salt could be responsible for provoking the Chinese Restaurant Syndrome *[13, 14]*. In 1995, the Federation of American Societies for Experimental Biology issued a report, which concludes that MSG is safe for the general population at levels normally consumed *[15, 16]*. The panel that prepared the report used the term "MSG Symptom Complex" to describe possible reactions to MSG. The old term Chinese Restaurant Syndrome was deemed misleading and pejorative. The US Food and Drug Administration (FDA) has placed MSG on the list of food additives that are considered "generally recognized as safe" (GRAS). In fact, MSG is cited as an example in the definition of GRAS substances, along with other common food ingredients such as salt, pepper and sugar.

In 1991, the European Union's Scientific Committee for Food (SCF) placed MSG in the safest category for food additives when it determined that it was not necessary to allocate an Acceptable Daily Intake (ADI) for MSG *[17]*.

### 3.3.4.2.3 Purine 5'-Ribonucleotides

5'-Ribonucleotides may contain purine or pyrimidine bases. Only the purine 5'-ribonucleotides have flavour modifying properties. They are building blocks of rinonucleic acid (RNA) and consist of a purine base (e.g. hypoxanthine, guanine, adenine), ribose, and phosphoric acid linked to the 5'-position of ribose (Fig. 3.54). In order to give the flavour improving effect, the purine base within the 5'-ribonucleotide must have a hydroxyl group (or an amino group) in the 6-position (Fig. 3.54), preferably a hydroxyl group (such as in IMP and GMP). An amino group (such as in AMP) decreases the flavour improving effect. Another prerequisite is the phosphate in the 5'-position of the ribose moiety (Fig. 3.54). 2'- and 3'-Phosphates have no flavour modifying effects.

Purine 5'-ribonucleotides were originally isolated from meat and fish extracts and from dehydrated mushrooms. Presently, the commercially relevant members of this group of nucleotides, i.e. inosine monophosphate (IMP) and guanosine monophosphate (GMP), are produced by two procedures *[2, 3, 10]*:

(a) Extraction of RNA from yeast, followed by hydrolytic degradation of RNA with the enzyme 5'-phosphodiesterase from *Penicillium citrinum* or *Streptomyces aureus* (Fig. 3.55) and

(b) fermentation, resulting in ribonucleosides (inosine, guanosine) which are phosphorylated into the corresponding 5'-ribonucleotides (Fig. 3.56).

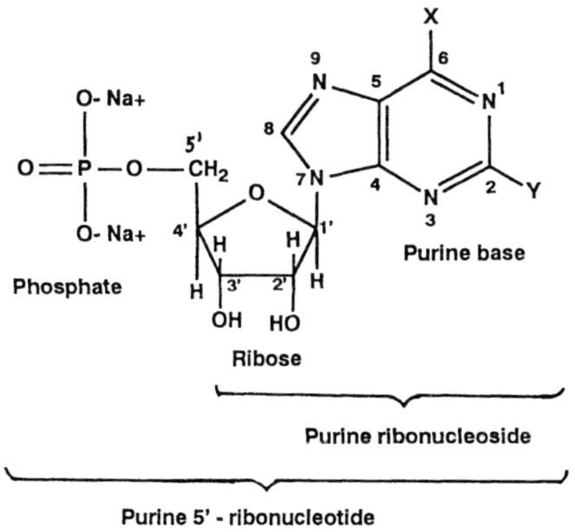

X = OH, Y = H : Inosine monophosphate (IMP) = disodium 5' - inosinate
X = OH, Y = NH2: Guanosine monophosphate (GMP) = disodium 5' - guanylate
X = NH2, Y = H : Adenosine monophosphate (AMP) = disodium 5' - adenylate

*Fig. 3.54:* Examples of disodium salts of purine 5'-ribonucleotides

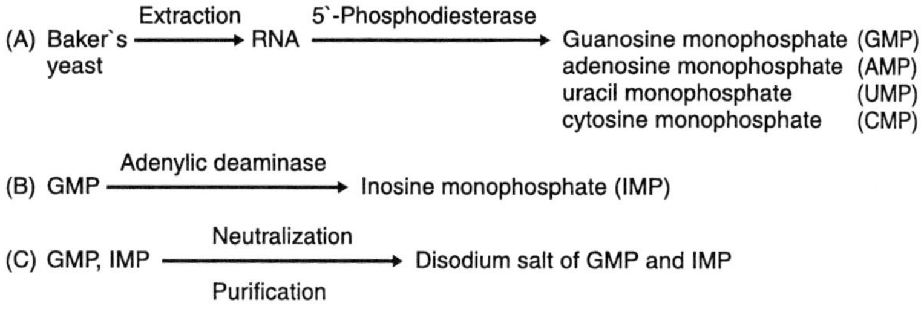

(A) Baker's yeast →(Extraction)→ RNA →(5'-Phosphodiesterase)→ Guanosine monophosphate (GMP), adenosine monophosphate (AMP), uracil monophosphate (UMP), cytosine monophosphate (CMP)

(B) GMP →(Adenylic deaminase)→ Inosine monophosphate (IMP)

(C) GMP, IMP →(Neutralization, Purification)→ Disodium salt of GMP and IMP

*Fig. 3.55:* Production of GMP and IMP from Baker's Yeast's RNA [3]

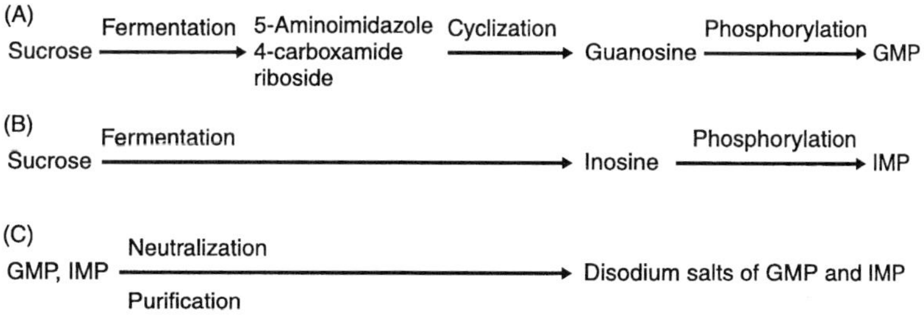

(A) Sucrose →(Fermentation)→ 5-Aminoimidazole 4-carboxamide riboside →(Cyclization)→ Guanosine →(Phosphorylation)→ GMP

(B) Sucrose →(Fermentation)→ Inosine →(Phosphorylation)→ IMP

(C) GMP, IMP →(Neutralization, Purification)→ Disodium salts of GMP and IMP

*Fig. 3.56:* Production of GMP and IMP by fermentation [10]

IMP and GMP are non-hygroscopic salts. They are stable towards heat (up to 120°C) and acidity (optimum pH is 5-7) *[2, 8, 9]*. Taste thresholds in aqueous solution are 25-120 ppm and 12-35 ppm for IMP and GMP, respectively *[2, 8, 10]*. Use of IMP and GMP in liquid foods may present some problems. Many vegetable and animal foods contain phosphomonoesterases which can easily split the phosphomonoester linkage of the ribonucleotides and the flavour improving effect is lost. From a practical standpoint, these enzymes should be inactivated by heating to 85°C before the addition of IMP or GMP.

Typical applications for IMP and GMP are soups, sauces, dressings, ready-to-eat meals, meat and fish products, snacks, and vegetable products. Usage levels range from 0.001 to 0.150 % in finished food (Tab. 3.47). A 1:1 mixture of IMP and GMP is commercially available and extensively used (Tab. 3.47), although there is a current trend to use GMP only because of its stronger umami effect *[8]*. The taste threshold of the IMP/GMP mixture (1:1) is reported to be 60 ppm *[2]*.

IMP and GMP have no flavour improving effects on a range of foods, e.g. confectionary and dairy products, soft drinks, desserts, and others.

Many foods naturally contain purine 5'-ribonucleotides. Examples are given in Tab. 3.46. In animal tissues, IMP is produced from adenosine triphosphate (ATP) *[10]*. Thus, the IMP content of fresh raw meat is usually rather high. In processed meat, however, it is relatively low, since raw meat contains phosphomonoesterase, and the IMP in raw meat is easily lost in processing steps such as thawing, washing, and salting.

### 3.3.4.2.4 Umami Effect

It has been reported that umami substances have no enhancing effects on the four primary taste impressions sweet, sour, salty, and bitter *[2]*. Thus, various authors emphasize the idea that umami is the fifth primary taste impression *[2, 9]*. This is contradictory to the observed facts that umami substances enhance taste impressions and mask odour impressions (Tab. 3.46). MSG, best known for its umami effect, possesses structural features that could elicit all the basic taste qualities (sweet, sour, salty, and bitter) (Fig. 3.57). The basic information as to how umami substances improve palatability of food remains elusive. Nevertheless, notable advances have been made in recent years on the probable mechanism by which MSG, IMP and GMP enhance taste impressions *[18]*. This is illustrated in Fig. 3.58. Fig. 3.58a shows schematically the binding of a flavouring substance A with a taste stimulus A to the surface of a taste receptor located on the tongue. The stimulus A initiates the taste signal A. Fig. 3.58b shows weak binding of MSG to the MSG binding site of the receptor. Because of the weak binding of MSG, stimulus-receptor interaction is still weak. As a result, the taste signal is still weak, although stronger than in the absence of MSG. Fig. 3.58c illustrates the proposed stimulus-receptor interaction when purine 5'-ribonucleotides are bound in addition to MSG. GMP or IMP are thought to allosterically alter the MSG binding site. Both MSG and IMP or GMP are strongly bound and force the flavouring substance to be more strongly bound, as well. This results in a stronger taste signal (or taste enhancement). A similar reaction for a flavouring substance with a taste stimulus B that initiates a taste signal B is depicted in Fig. 3.58d.

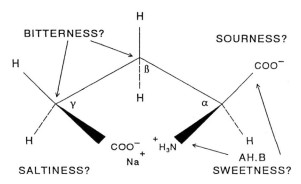

*Fig. 3.57*: *Structure of MSG showing the structural features that could elicit the four primary taste qualitites [3]*

The common structural element of umami substances is the presence of two negative charges in the molecule, spaced 3 to 9 carbon or other atoms apart, preferably 4 to 9 (Fig. 3.59) *[19]*. MSG, IMP, and GMP fit into this concept. AMP, which is less effective, has one negative charge at one end of the molecule and an electronegative atom (i.e. nitrogen) at the other end.

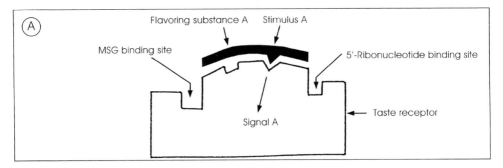
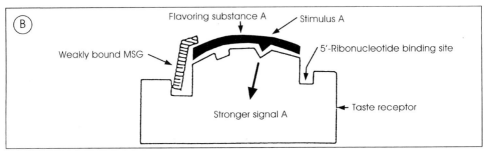
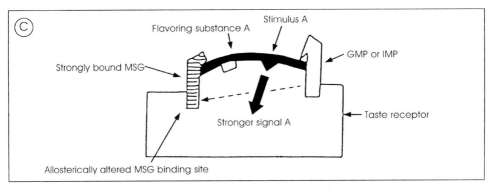
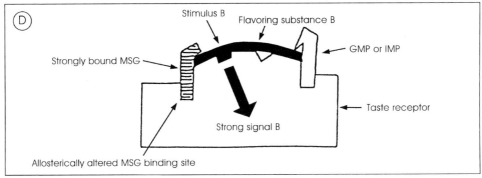

*Fig. 3.58: Proposed interactions of flavouring substances at the taste receptor surface in the absence and presence of MSG, IMP and GMP [18]*

# Flavour Modifiers

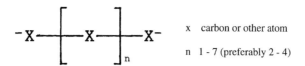

*Fig. 3.59: Common structural element of umami substances*

## 3.3.4.2.5 Other Umami Substances

Substances related to MSG and purine 5'-ribonucleotides include peptides, amino acids (e.g. cysteine, homocysteine, cysteine S-sulfonic acid, aspartic acid, a-amino adipic acid, a-methyl glutamic acid, tricholomic acid, ibotenic acid), pyrrolidone carboxylic acid, 3-methyl thiopropyl amine, and others *[2, 10]*. They are of less commercial interest than MSG, IMP, and GMP. Chemical structures of some of these substances are depicted in Fig. 3.53. Relative umami effects of some are shown in Tab. 3.49. Tricholomic acid and ibotenic acid have been found in the mushrooms *Tricholoma muscarium* and *Amanita stroboliformis*, respectively.

*Table 3.49: Relative umami effect of selected umami substances [2]*

| Umami Substance | Relative effect |
| --- | --- |
| MSG | 1.00 |
| Monosodium D,L-threo-β-hydroxy glutamate | 0.86 |
| Monosodium D,L-homocystate | 0.77 |
| Monosodium L-aspartate | 0.01 |
| Monosodium L-α-amino adipate | 0.01 |
| Monosodium L-tricholomate | 5-30 |
| Monosodium L-ibotenate | 5-30 |

## 3.3.4.2.6 Synergistic Effects

Synergism may be defined as the co-operative action of two components of a mixture whose total effect is greater than the sum of their individual effects. Umami substances show pronounced synergistic effects among each other and with other substances *[1-3, 8-10]*. Table 3.50 illustrates such effects of MSG/IMP and MSG/GMP combinations. Note that a 1:1 mixture of MSG/GMP produces a thirtyfold increase in umami intensity and that intensity goes through a maximum with increasing amounts of MSG (Tab. 3.50). Therefore, it would appear that the most effective flavour potentiator system to use would be the said 1:1 mixture of MSG/GMP. Due to the relatively high cost of 5'-ribonucleotides, however, they are seldom used at 1:1 ratio with MSG. Typically, a 95:5 ratio of MSG/5'ribonucleotides is used in the food industry, the 5'-ribonucleoties being a 1:1 mixture of IMP/GMP. This combination yields approximately a sixfold synergistic effect on MSG.

*Table 3.50:* Umami effect intensities of MSG/IMP and MSG/GMP combinations [1]

| Ratio of | | Relative umami | Ratio of | | Relative umami |
| --- | --- | --- | --- | --- | --- |
| MSG | IMP | intensity | MSG | GMP | intensity |
| 1 | 0 | 1.0 | 1 | 0 | 1.0 |
| 1 | 1 | 7.0 | 1 | 1 | 30.0 |
| 10 | 1 | 5.0 | 10 | 1 | 18.8 |
| 20 | 1 | 3.5 | 20 | 1 | 12.5 |
| 50 | 1 | 2.5 | 50 | 1 | 6.4 |
| 100 | 1 | 2.0 | 100 | 1 | 5.4 |

Synergistic effects between MSG and sodium chloride have also been reported [8]. For maximum palatability of a clear soup, more MSG must be added if only small amounts of salt are used, and vice versa. Optimum amounts of salt and MSG are 0.8 % and 0.4 %, respectively. Thaumatin, an intensely sweet tasting protein has been reported to have a similar synergistic effect to that of 5'-ribonucleotides on MSG, but at significantly lower dosages [8]. Other umami substances with synergistic effects on each other include various amino acids (e.g. L-cysteine, D,L-homocysteine, cysteine S-sulfonate), polypeptides, cycloalliin, and histamine [8].

### 3.3.4.3 Maltol and Ethyl Maltol

#### 3.3.4.3.1 Historical Background

In 1861, maltol (Fig. 3.60) was first isolated from the bark of the larch tree by the British chemist Stenhouse. He also commented on the pleasant odour of maltol and on its slightly bitter and astringent taste. In 1894, Brand isolated maltol from roasted malt.

The first successful synthesis of maltol started from pyromeconic acid, which itself was an expensive material, and gave maltol in low yields. Today, industrial production is carried out by fermentation (kojic acid produced by an *Aspergillus* fungus) combined with chemical synthesis [10,20]. The synthetic routes include oxidation of kojic acid and addition of formaldehyde. By substituting acetaldehyde for formaldehyde, ethyl maltol (Fig. 3.60) is produced [20].

#### 3.3.4.3.2 Maltol

Maltol naturally occurs in many foods, e.g. baked goods, cocoa, chocolate, coffee, caramel, malt, condensed milk, cereals, soy sauce, and beer [10, 20, 21]. It is formed when carbohydrates are heated (Fig. 3.61). It is a white, crystalline powder with a caramel-like odour and a threshold of 35 ppm in water at 20°C [21].

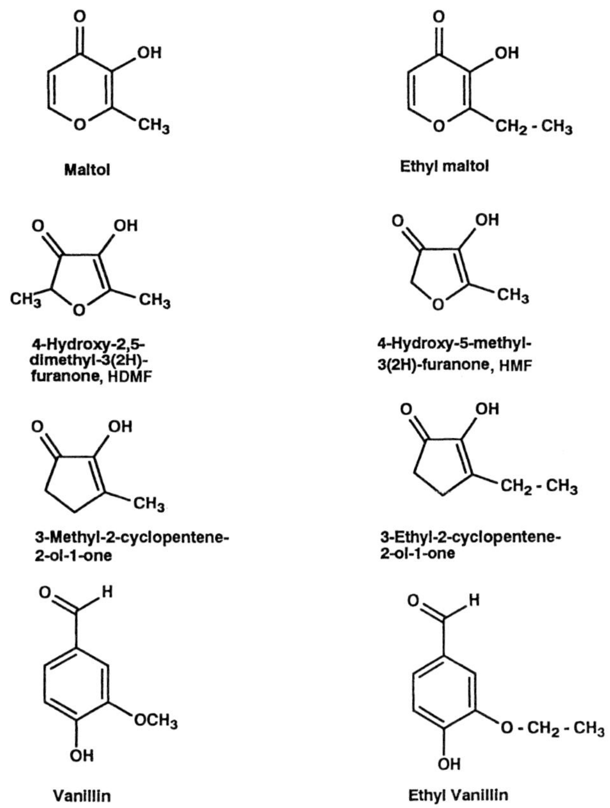

*Fig. 3.60: Chemical structures of some important flavour modifiers*

### 3.3.4.3.3 Ethyl Maltol

Ethyl maltol has not yet been found in nature. It is 4 to 6 times stronger than maltol and can replace maltol in sweet foods. Typical usage levels are shown in Tab. 3.51.

### 3.3.4.3.4 Maltol and Ethyl Maltol

It has been reported recently that maltol and ethyl maltol are also capable of improving the flavour of savoury (or spicy) foods [7]. In salad dressings, they round the spiciness and decrease the "bite" or acid sensation contributed by acetic acid. Both substances have also been reported to be highly effective in improving the perception of low-fat food systems [7]. Low-fat yoghurt, ice cream and salad dressings "taste" richer, fuller and creamier with the addition of ppm levels of maltol or ethyl maltol. In other words, their mouthfeel is improved.

Maltol enhances the flavour of sweet foods. The addition of 5 to 75 ppm maltol may permit a 15 % sugar reduction in various sweet foods [1]. Tab. 3.51 shows some usage levels in selected foods.

*Table 3.51:* Typical usage levels of maltol and ethyl maltol in selected foods [10, 19]

| Food | Maltol (ppm) | Ethyl maltol (ppm) |
| --- | --- | --- |
| Baked goods | 75-250 | 25-150 |
| Beverages | 2-250 | 1-100 |
| Candies | 3-300 | 1-100 |
| Chocolate foods | 30-200 | 5-40 |
| Dairy products | 10-150 | 5-50 |
| Desserts | 30-150 | 10-75 |
| Jams | 40-200 | 15-60 |

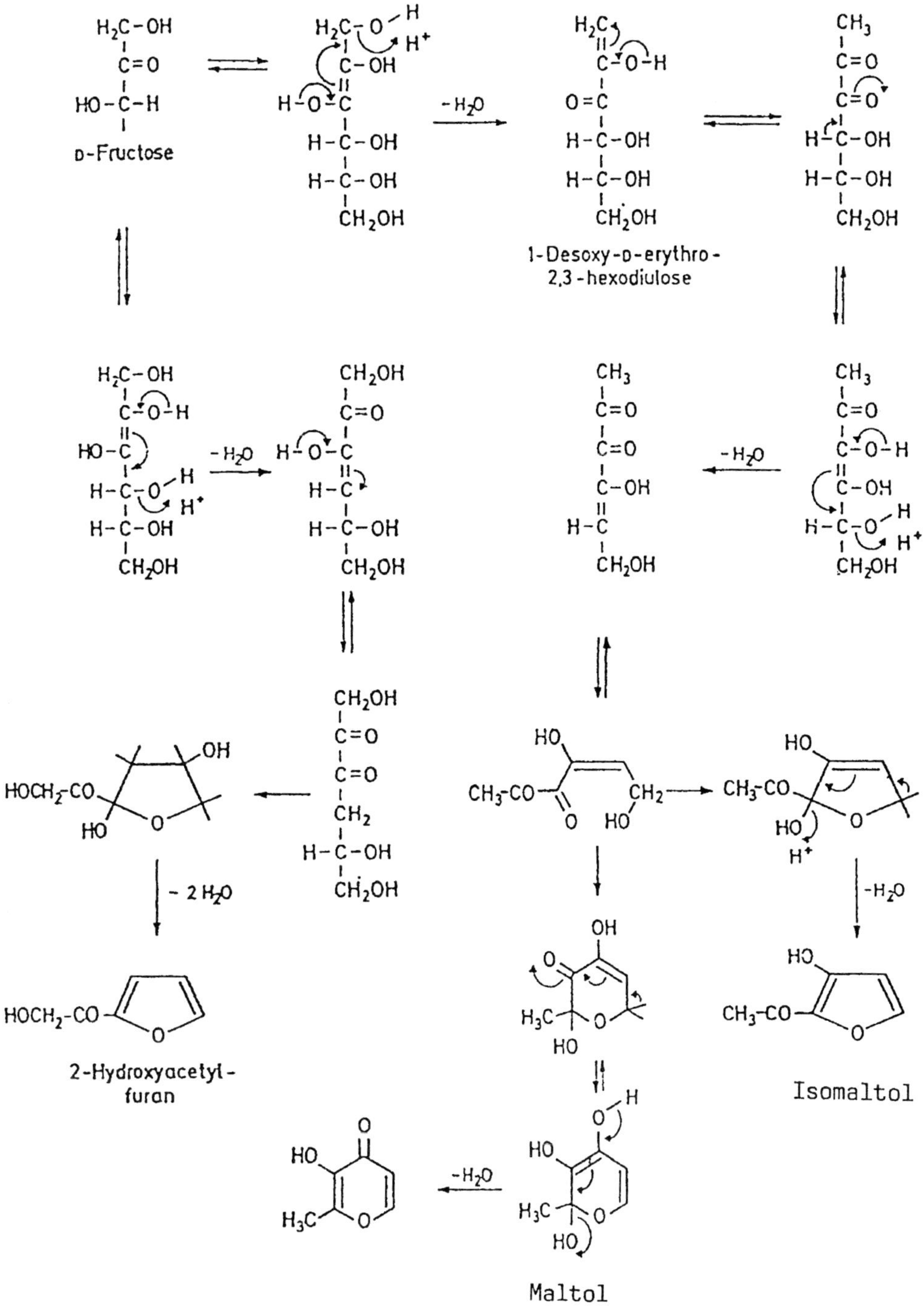

**Fig. 3.61:** *Formation of maltol from carbohydrates [21]*

The JECFA concluded that up to 2 mg/kg/day (120 mg/day for a 60 kg person) was an acceptable level of consumption of ethyl maltol for humans *[7, 20]*. This value is many times greater than the current average consumption levels for both ethyl maltol and maltol [7, 20].

### 3.3.4.4 Furanones and Cyclopentenolones

Both 4-hydroxy-2,5-dimethyl-3(2H)furanone (HDMF; trade name: Furaneol®; Fig. 3.60) and 4-hydroxy-5-methyl-3(2H)-furanone (HMF; Fig. 3.60) contribute to flavour at normal usage level; they are, however, also enhancers of fruity and creamy odour impressions (see Tab. 3.46). Both furanones have caramel-like odours with HDMF possessing an additional burnt pineapple odour [1, 21]. The odour threshold of Furaneol® has been reported to be 0.00004 ppm in water at 20°C [21].

Furaneol® is formed when the sugar rhamnose is heated in the presence of a substance containing an amino group through a Maillard reaction (Fig. 3.62). HMF results from heating of fructose (Fig. 3.63). Both furanones occur in various foods. For example, Furaneol® has been identified in pineapple, strawberries and popcorn; both Furaneol® and HMF have been found in meat broth [21]. Both furanones are applied as flavour modifiers in foods where maltol and ethyl maltol are used [1].

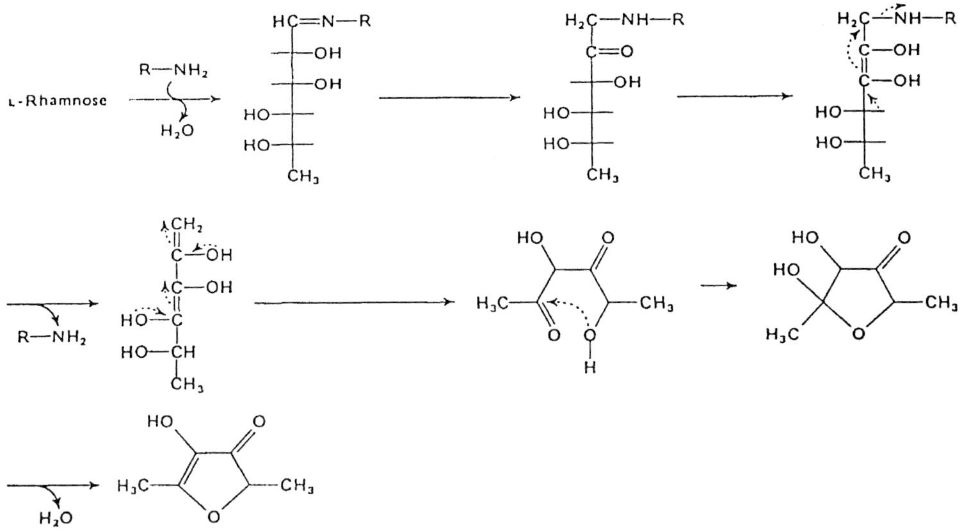

*Fig. 3.62: Formation of Furaneol® from rhamnose [21]*

3-Methyl-2-cyclopentene-2-ol-1-one (MCP; trade name: Cyclotene®; Fig. 3.55) is an enhancer of nutty and chocolate odour impressions (see Tab. 3.46). MCP has been isolated from beech wood tar and identified in various foods (e.g. maple syrup). It is formed when sugars are heated at pH 8-10 (Fig. 3.64) and it has a caramel-like odour. Typical usage levels in selected foods are shown in Tab. 3.52. MCP has also been reported to mask the salty taste impression (10-20 ppm MCP mask 1-2 % of salt) [19]. 3-Ethyl-2-cyclopentene-2-ol-1-one (ECP; Fig. 3.61) may be used to replace MCP.

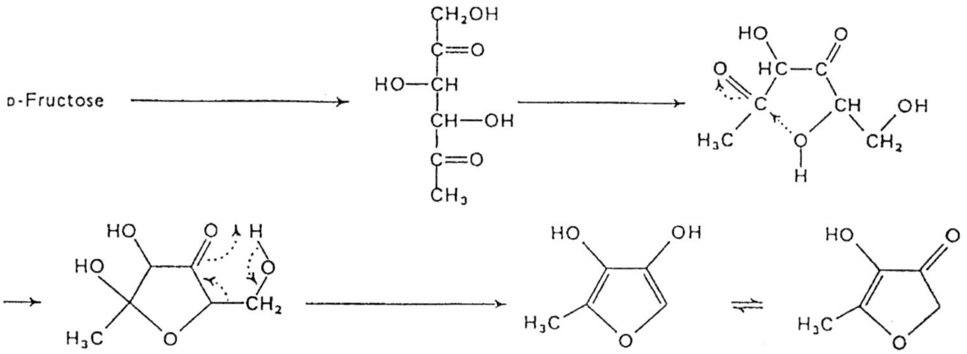

Flavour Modifiers

**Fig. 3.63:** *Formation of 4-hydroxy-5-methyl-3(2H)-furanone (HMF) from fructose [21]*

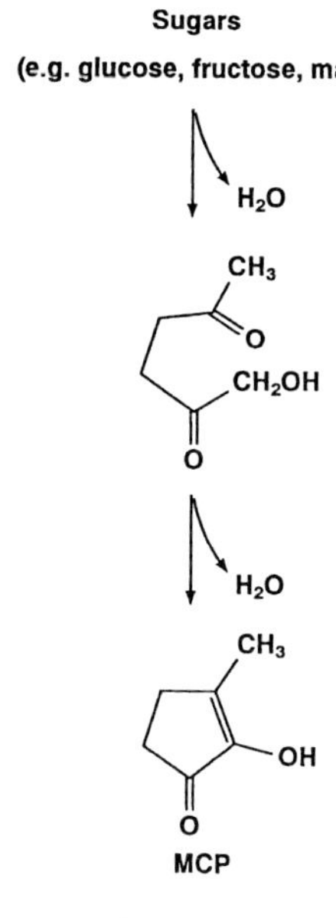

**Fig. 3.64:** *Formation of 3-methyl-2-cyclopentene-2-ol-1-one (MCP) from sugars at pH 8-10 [21]*

**Table 3.52:** *Typical usage levels of MCP in selected foods [19]*

| Food | MCP (ppm) |
|---|---|
| Candies | 15-100 |
| Baked goods | 15-100 |
| Beverages | 10-50 |
| Chewing gum | 5-30 |
| Ice cream | 5-50 |

### 3.3.4.5 Vanillin and Ethyl Vanillin

Vanillin has been known as a flavouring substance since about 1816, and by 1858 the pure chemical had been obtained from ethanolic extracts of vanilla beans. It was not until 1872 that Carles established its correct formulation and in 1874, Tiemann and Haarmann reported it as 3-methoxy-4-hydroxy-benzaldehyde (Fig. 3.60). Finally, Reimer synthesized vanillin from guaiacol and thus proved its chemical structure. For many years, the most important source of vanillin was eugenol, from which it was obtained by oxidation. Today, the major portion of commercial vanillin is obtained by processing waste sulfite liquors, the rest through fully synthetic processes starting from guaiacol [21].

Lignin, present in the sulfite wastes from the cellulose industry, is treated at elevated temperatures and pressures with alkalis in the presence of oxidants (Fig. 3.65). The vanillin formed is separated from by-products by extraction, distillation, and crystallization processes.

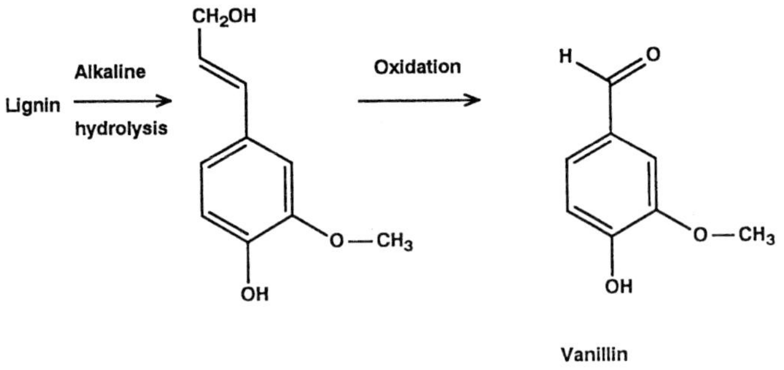

*Fig. 3.65: Synthesis of vanillin from lignin [21]*

Condensation of guaiacol (obtained from catechol) with glyoxylic acid, followed by oxidation and decarboxylation of the intermediates also results in crude vanillin (Fig. 3.66). Various commercial grades of vanillin are obtained by distillation and subsequent recrystallisation.

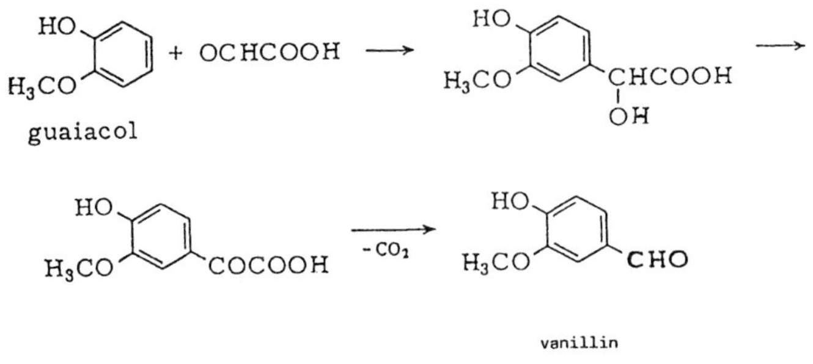

*Fig. 3.66: Synthesis of vanillin from guaiacol [22]*

Vanillin has a vanilla-like odour with a threshold of 0.02 ppm in water at 20°C [21]. Ethyl vanillin (Fig. 3.60), which is 2 to 4 times stronger than vanillin, has not yet been found in nature. It is synthesized from eugenol, isoeugenol, or safrole. Both vanillin

Flavour Modifiers 369

and ethyl vanillin are important flavouring substances. In addition, they enhance fruity and chocolate odour impressions (see Tab. 3.46).

### 3.3.4.6 Other Flavour Modifiers

There are several flavour modifiers which are of minor industrial importance at present. These include natural and synthetic substances. A selection is presented in Tab. 3.53. Studies aimed at using miraculin and curculin as low-calorie sweeteners are in progress [23].

BMP may become a savoury peptide and a flavour modifier of the future [26]. It occurs naturally in beef. One method of production of BMP already initiated by several concerns is biotechnology [28]. The taste threshold of BMP has been reported to be 1.600 ppm [28]. There are also reports in the literature claiming that BMP cannot be considered as a flavour modifier and that its occurrence in beef is highly unlikely [30, 31].

*Table 3.53:* Selection of less important flavour modifiers [2, 21, 23-29]

| Name | Effect | Remarks |
|---|---|---|
| Gymnemic acid(s) | Inhibit(s) sweet taste for several hours (powder sugar tastes like sand and sugar solution tastes like plain tapwater). Salty, bitter, and sour tastes are not affected. | Various triterpene glycosides isolated from the leaves of *Gymnema sylvestre*. |
| Miraculin (trade name: Mirlin®) | Modifies sour into sweet taste. After rinsing the mouth with a solution of miraculin, lemon juice tastes like sweetened lemon juice. Salty, bitter, and sweet tastes are not affected. | Tasteless glycoprotein (molecular weight: 40,000 – 48,000; sugar moiety: 14%) isolated from the miracle fruit, i.e. the fruits of *Richadella dulcifica (Synsepalum dulsificum)*. |
| Curculin | The sweet taste of curculin disappears in a few minutes after holding it in the mouth. Then, water elicits the sweet taste again and black tea tastes like sweetened tea. In addition, curculin modifies sour into sweet taste (like miraculin). | Sweet tasting protein (molecular weight: 28,000) isolated from the fruits of *Curculigo latefolia*. |
| Ziziphin | Inhibits sweet taste. | Triterpene glycoside isolated from the leaves of *Ziziphus jujuba*. |
| Hodulcin | Inhibits sweet taste. | Glycoside isolated from the leaves of *Hovenia dulcis*. |

| Name | Effect | Remarks |
|---|---|---|
| Sodium 2-(4-methoxyphenoxy)propionate (common name: lactisol; trade name: Cypha®) | Decreases sweet taste. | Synthetic compound. Has been identified in roasted coffee. |
| Thaumatin (trade name: Talin®) | Masks bitter taste. | Sweet tasting protein (molecular weight: 22,000) isolated from the fruit of the West African perennial plant *Thaumatoccus danielli*. Although the primary taste property of thaumatin is sweetness, its main use is as a flavour modifier. |
| Lecithin | Masks bitterness and harsh off-notes in mint or menthol flavoured chewing gum. | The primary use of lecithin is as an emulsifier. |
| Neohesperidin dihydrochalcone | Enhances fruity notes and reduces sharp or spicy notes. | Primary use as intense sweetener. |
| Beefy meaty peptide (BMP), also known as „delicious peptide" | Enhances meat flavour. | Octapeptide of the sequence Lys-Gly-Asp-Glu-Glu-Ser-Leu-Ala. |
| Gurmarin | See gymnemic acid(s). | Polypeptide isolated from the leaves of *Gymnema sylvestre*. Has also been synthesized. |

### 3.3.4.7 Final Remarks

Tab. 3.46 classifies flavour modifiers according to their ability to enhance or decrease flavour and whether or not they contribute to flavour at usage level. Another possible classification is based on the types of food they are used in. This is given in Tab. 3.54.

*Table 3.54: Classification of flavour modifiers based on types of food they are used in*

| Type of food | Examples |
|---|---|
| Spicy (or savoury foods) | MSG, GMP, IMP |
| Sweet foods | Maltol, ethyl maltol, Furaneol®, 4-hydroxy-5-methyl-3(2H)-furanone, Cyclotene®, vanillin, ethyl vanillin, Cypha® |
| Sour foods | Miraculin |

It should be mentioned that not only single chemical components but also complex materials are used as umami substances. These include yeast extracts and autolysates (due to high glutamic acid and purine 5'-ribonucleotides content) (Tab. 3.55), as well as hydrolyzed vegetable proteins (due to high glutamic acid content; Tab. 3.55 [26]). Hydrolyzed vegetable proteins (HVP) have been known for a long time in the food industry, but only since the 1930s have these products gained prominence as flavour modifiers. They may be obtained by hydrolysis with mineral acid (e.g. HCl) or by enzymatic hydrolysis (Tab. 3.55). They are used, for example, in soups, gravies, savoury snacks, sauces and ready-to-eat meals. They are stable under varying process conditions, e.g. canning, freezing and heating up to 180°C. HVP preparations now gaining popularity have salt concentrations sometimes less than half of the levels found in traditional HVP. This is achieved by including several crystallization steps after the concentration of the material or by partial replacement of HCl with $H_2SO_4$.

*Table 3.55:* Examples of complex materials used as flavour modifiers

| Material | Remarks |
|---|---|
| Yeast autolysates | Contain up to 6% 5'-ribonucleotides, especially GMP. |
| Hydrolyzed vegetable proteins (hydrolyzed with mineral acid) | Contain up to 17% MSG; 30 to 50 of dry matter is sodium chloride. |
| Hydrolyzed vegetable proteins (enzymatically hydrolyzed) | Contain up to 35% MSG; practically free of sodium chloride. |
| Hydrolyzed vegetable proteins (hydrolyzed with organic acid, e.g. acetic acid) | No commercial importance at present. |

Yeast extracts are concentrates of soluble material derived from yeast following hydrolysis of the cell material, particularly the proteins, carbohydrates and nucleic acids. Hydrolysis is carried out by use of the yeast's own hydrolytic enzymes (autolysis) or by other methods (hydrolysis or plasmolysis). Yeast extracts are commercially available as powders and pastes.

It has been shown that the production of HVP using HCl can lead to the formation of chloropropanols which are known to be carcinogenic [26]. These compounds appear as a result of the reaction between HCl and traces of lipids. Various countries have set maximum levels of chloropropanols in HVP and in food. Some countries are currently reviewing this issue.

Yeast extracts or autolysates have not shown adverse reactions in humans and hence have, thus far, received the "clean label" status of being natural. Products containing glutamic acid (such as HVP and yeast extracts) are considered to be GRAS. In addition to HVP and yeast extracts, various oligopeptides are occasionally used as MSG replacers.

Last but not least, it must be mentioned that sodium chloride (often referred to as "the poor man's flavour enhancer") has a flavour enhancing effect at usage levels below and above its taste threshold (370 to 5000 ppm). Without salt, many foods (both sweet and savoury) have a flat taste. Salt may enhance sweetness and mouthfeel and decrease bitter, sour and metallic sensations [4, 32]. Even in sweet foods such as cakes, candies and toffees, salt has its place. However, its presence is most critical for savoury foods. It is difficult, if not impossible, to make a general statement about the most appropriate salt concentration needed in various food products. This is due to the wide variation in consumer preference and to the individual differences in threshold detection.

## REFERENCES

[1] Heath, H.B. and Reineccius, G. Flavor Chemistry and Technology, Westport, Avi Publishing 1986
[2] Sugita, Y.-H. Flavor Enhancers. In: Food Additives (Branen, A.L., Davidson, P.M. and Salminen, S., eds.). Dekker, New York, pp. 259-296 (1990)
[3] Nagodawithana, T. Food Technol. 46(11), 138-144 (1992)
[4] Kemp, S. E. and Beauchamp, G. K. J. Food Sci. 59, 682-686 (1994)
[5] International Organisation of the Flavour Industry. Code of Practice for the Flavour Industry, Geneva, 1990
[6] European Union. Off. J. Eur. Comm. 38(L61), 1-40 (1995)
[7] Murray, P. R., Webb, M. G. and Stagnitti, G. Food Technol. Int. Eur. 53-55 (1995)
[8] Van Eijk. T. Dragoco Report. Flavoring Information Service 32, 3-17 (1987)
[9] Fuke, S. and Shimizu, T. Tr. Food Sci. Technol. 4, 246-251 (1993)
[10] Oberdiek, R. Alkohol-Industrie 156-164 (1980)
[11] Maga, J. A. CRC Crit. Rev. Food Sci. Nutr. 18, 231-312 (1983)
[12] Taliaferro, P. J. J. Environ. Health 57(10), 8-12 (1995)
[13] Tarasoff, L. and Kelly, M. F. Food Chem. Toxicol. 31, 1019-1035 (1993)
[14] Dayton, L. New Scientist, January 1994, p. 15
[15] Raiten, D. J., Talbot, J. M. and Fisher, K. D. (eds). Analysis of Adverse Reactions to Monosodium Glutamate (MSG). Prepared by the Life Sciences Research Office, Federation of American Societies for Experimental Biology, for the Center for Food Safety and Applied Nutrition, FDA/HHS, 1995
[16] Institute of Food Technologists. Food Technol. 49(10), 28 (1995)
[17] Glutamate Information Service. Food Engin. Int. October 1995, p. 24
[18] Nagodawithana, T. Food Technol. 48(4), 79-85 (1994)
[19] Ney, K. H. Lebensmittelaromen, Hamburg, Behr's Verlag 1987
[20] Le Blanc, D. T. and Akers, H. A. Food Technol. 43, 78-84 (1989)
[21] Belitz, H. D. and Grosch, W. Lehrbuch der Lebensmittelchemie, Berlin, Heidelberg New York, Springer Verlag 1982
[22] Bauer, K. and Garbe, D. Common Fragrance and Flavor Materials, Weinheim, VCH Verlagsgesellschaft 1985
[23] Kurihara, Y. CRCrit. Rev. Food Sci. Nutr. 32, 231-252 (1992)
[24] Lindley, M. G. Sweetness Antagonists. In: Flavor Science. Sensible Principles and Techniques (Acree, T. and Teranishi, R., eds.), Washington DC, American Chemical Society, pp. 117-133 (1993)
[25] Kurihara, Y. and Nirasawa, S. Tr. Food Sci. Technol. 5, 37-42 (1994)
[26] Nagodawithana, T. W. Savory Flavors, Milwaukee, Esteekay Associates 1995
[27] Glass, M., Corsello, V., Orlandi, D. A. and Guzowski, A. US Patent 4 604 288
[28] Spanier, A. M., Bland, J. M., Miller, J. A., Glinka, J., Wasz, W. and Duggins, T. BMP: A Flavor Enhancing Peptide Found Naturally in Beef. Its Chemical Synthesis, Descriptive Sensory Analysis, and Some Factors Affecting its Usefulness. In: Food Flavors: Generation, Analysis and Process Influence (Charalambous, G., ed.), Amsterdam, Elsevier, pp. 1365-1378 (1955)
[29] Ota, M., Tonosaki, K., Miwa, K., Fukuwatari, T. and Ariyoshi, Y. Biopolymers 39, 199-205 (1996)
[30] Van Wassenaar, P. D., van den Oord, A. H. A. and Schaaper, W. M. M. J. Agric. Food Chem. 43, 2828-2832 (1995)
[31] Hau, J., Cazes, D. and Fay, L. B. J. Agric. Food Chem. 45, 1351-1355 (1997)
[32] Gilette, M. Food Technol. 39(6), 47-52, 56 (1985)

## 3.3.5 Antioxidants and Preservatives

### 3.3.5.1 Antioxidants

*Dieter Gerard*
*Karl-Werner Quirin*


Undesired reactions caused by oxygene attack affect food products in numerous ways.

The most noticeable changes which decrease the storage life of food are the following:

- loss of fresh flavour and aroma,
- destruction of oil soluble vitamins,
- decrease of colour, especially of carotenoids and
- subsequent development of rancidity.

These effects are often accelerated by sunlight and heat. Fortunately new modern products and technologies give us the possibility to combat the ravages of oxidation, e.g. improvements of processing practices for flavourings (e.g. $CO_2$-extraction), modern materials and techniques in packaging, such as the use of inert gases to help retard oxidation. However there is still the need to introduce antioxidants directly into flavourings to inhibit the destructive process.

Antioxidants for food have to meet special requirements:

- they have to be effective in low concentrations (100 - 500 ppm),
- they must not change aroma, flavour, colour and structure of the food,
- they must be readily soluble to ensure a homogenous distribution,
- they have to be stable and volatility resistant to certain processes like baking or frying ("carry-through-effect").

Table 3.56 lists the most commonly used antioxidants. The synthetic products BHA, BHT, TBHQ and Propyl Gallate have gained regulatory acceptance (TBHQ not in the EC). They have to be declared in food products with their acceptance number.

*Table 3.56: Regulatory acceptance of food antioxidants [1]*

|  | U.S.A. FDA Regulation | EC Countries E-number |
|---|---|---|
| **Synthetic antioxidants** | | |
| - BHA (Butyl Hydroxy Anisole) | 21 CFR 182.3169 | E-320 |
| - BHT (Butyl Hydroxy Toluene) | 21 CFR 182.3173 | E-321 |
| - TBHQ (Tert. Butyl Hydroxy Quinone) | 21 CFR 172.185 | - |
| - Propyl Gallate | 21 CFR 184.1660 | E-310 |
| **Natural antioxidants** | | |
| - Ascorbic acid | 21 CFR 182.3013 | E-300 |
| - Ascorbyl palmitate | 21 CFR 182.3149 | E-304 |
| - Mixed tocopherols | 21 CFR 182.3890 | E-306 |
| - Sage extract | - | - |
| - Rosemary extract | - | - |

Natural tocopherols, ascorbic acid and ascorbyl palmitate have universal acceptance.

The natural spice extracts from rosemary and sage are not regulated as antioxidants. However, they are GRAS (Generally Recognized As Safe), approved as spice extracts and they have to be declared as such.

Most of the above mentioned antioxidants belong to the substance class of phenolic compounds (see Fig. 3.67-3.69)

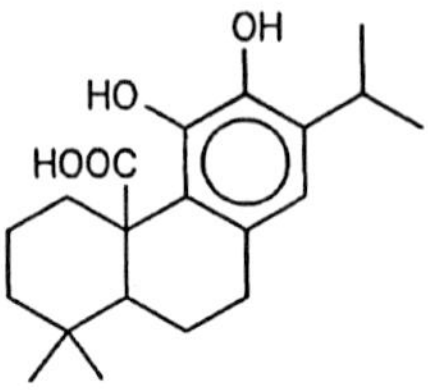

carnosolic acid

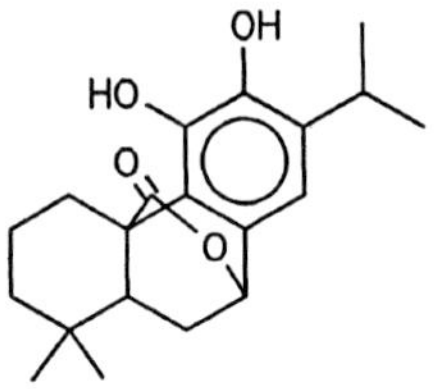

carnosol

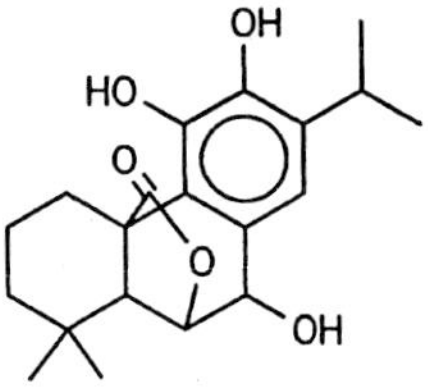

rosmanol

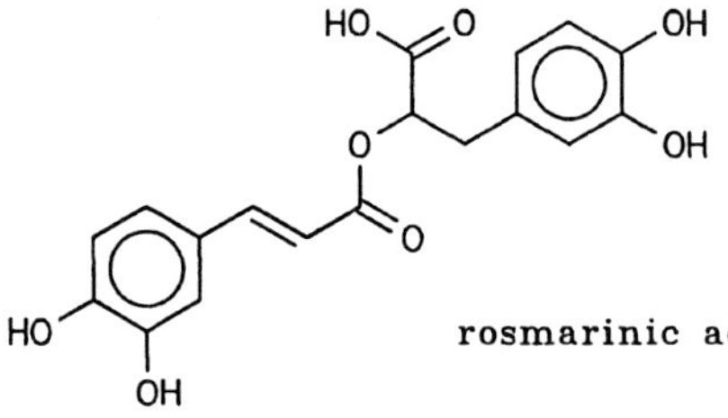

rosmarinic acid

***Fig. 3.67:*** *Natural antioxidant components in rosemary and sage extracts*

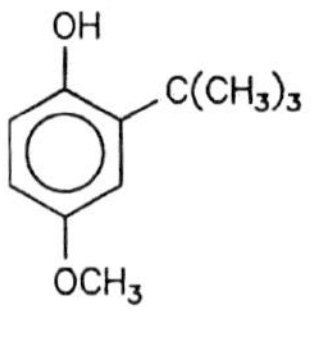

BHA

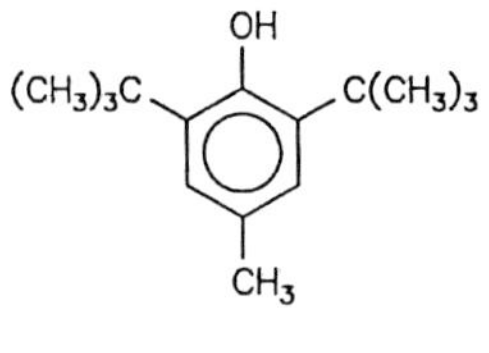

BHT

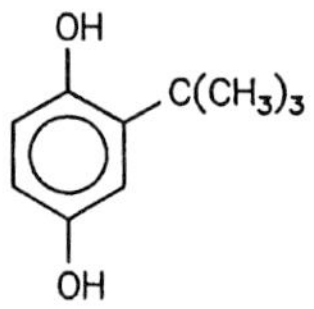

TBHQ

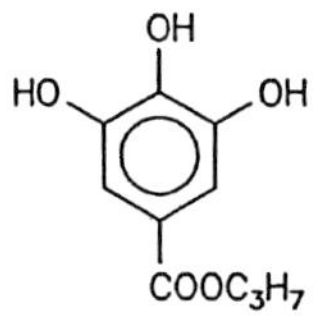

propyl gallate

***Fig. 3.68:*** *Synthetic antioxidants*

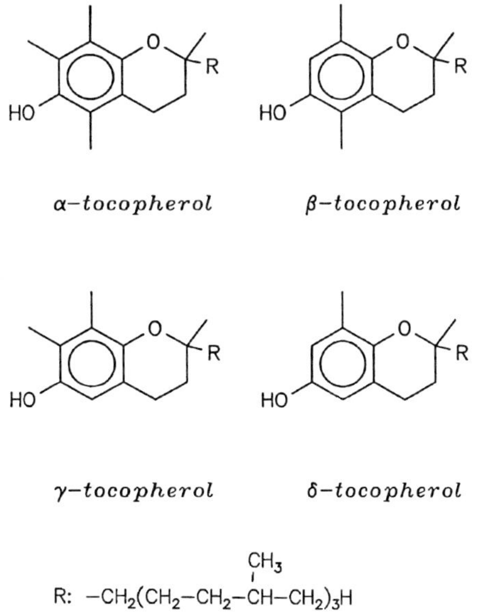

# Antioxidants and Preservatives

α−tocopherol    β−tocopherol

γ−tocopherol    δ−tocopherol

R: $-CH_2(CH_2-CH_2-\underset{\underset{CH_3}{|}}{CH}-CH_2)_3H$

***Fig. 3.69:*** *Natural tocopherols*

The hydroxyl groups are the reaction centres of antioxidant effectiveness. They function as free radical inhibitors by donating a hydrogen ion to a free radical and thus stop the radical reaction.

In the following scheme the effect of the antioxidants as radical inhibitor is illustrated (AH = antioxidant)

(1) RO* + AH   →   ROH + A*
(2) ROO* + AH  →   ROOH + A*
(3) RO* + A*   →   ROA
(4) ROO* + A*  →   ROOA

The effectiveness of the two natural antioxidants rosemary and sage increases with the content of the active phenolic diterpenes *[2-3]*. New and improved extraction techniques based on supercritical $CO_2$-extraction (see chapter 2.1.2) resulted not only in concentrated and very active, but also in well desodourized and almost neutral tasting antioxidants.

Figure 3.70 demonstrates impressively the extraordinary protecting power of rosemary extract *[4]*. A desodourized $CO_2$-extract from rosemary has been tested for the colour stabilization of carotenoids (i.e. paprika oleoresin). Fig. 3.70 shows the colour deterioration of stabilized versus unstabilized paprika oleoresin. For this test the carotenoids have been exposed to energetic radiation of 366 nm at ambient temperature. The colour units were measured as function of the radiation time. It is obvious that the colour reduction of the stabilized product (A) is almost 10 times slower compared to the unstabilized product (B).

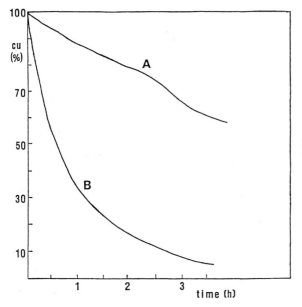

*Fig. 3.70:* Colour decrease of Paprika oleoresin when exposed to radiation of 366 nm at 20°C

The effectiveness of antioxidants can be supported by adding synergists. Synergistic effects occur when the activity of a mixture of two antioxidants is greater than the effect of the sum of the individual components. A synergist for example is citric acid, in its capacity as sequester for traces of metals which can catalyze oxidation processes, in combination with ascorbic acid or ascorbyl palmitate with their properties as reduction compounds. They can regenerate consumed phenolic antioxidants and react with any oxygen present, removing it from the oxidative process.

The "carry-through" effect is the ability of the antioxidant to survive the processing of food. Volatility of antioxidants is a very important consideration in applications where an antioxidant is used to protect baked goods, fried food or other heat treated products. With this the natural antioxidants from rosemary or sage and the tocopherols have advantages versus the synthetic phenolic components which have a distinctly higher volatility. A further advantage of natural antioxidants is the better acceptance by the consumer. This is due to the common view that natural food ingredients are better and safer than synthetic chemicals.

## REFERENCES

[1]  M.E. Dougherty, IFI 27, 1993 (Nr. 3)
[2]  K. Schwarz, W. Ternes, E. Schmauderer, Z. Lebensm. Unters. Forsch. 195, 104 (1992)
[3]  D. Gerard, K.-W. Quirin, A. Cawelius, in prep.
[4]  K.-W. Quirin, D. Gerard, Food Tech Europe, No. 3, 38 (1994)

## 3.3.5.2 Preservatives

*Howard D. Preston*

**Definition**

The European Parliament and Council Directive No 95/2/EC of 20 February 1995 on food additives other than colours and sweeteners states 'Preservatives' are substances which prolong the shelf-life of foodstuffs by protecting them against deterioration caused by micro-organisms.

Man has attempted to preserve food products from spoilage by microorganisms since prehistoric times. Processing such as heating, drying, fermentation, refrigeration and concentration have been used to extend the shelf life of food. Food preservatives such as salt, nitrites and sulphites have been used for many centuries.

Customers nowadays expect food products to be available all year round, to have a reasonable shelf life and to be free from spoilage of pathogenic microorganisms. Chemical preservatives have helped to increase the safety of the food chain, but in some trends of 'free from preservatives', manufacturers must ensure that the customer is aware of the need to refrigerate the product once opened. One must balance the need for food safety against the small risk from artefacts due to the microbial system.

Selection of the appropriate food preservative must consider various factors [1]:

(1) Antimicrobial spectrum of the compounds must be known together with the bioburden of the food product.
(2) Chemical/physical properties of the preservative and the food product - e.g. $pK_a$ and solubility of antimicrobial and pH of product.
(3) Conditions of storage and likely interactions with processing to ensure that preservative remains effective with time.
(4) Food must have highest microbiological quality if preservative is expected to contribute to its shelf life. Preservatives cannot be expected to be able to control a heavily contaminated product.
(5) Safety and legislative acceptance of the preservative system in the end market must be known.

### 3.3.5.2.1 Water Activity, $a_w$

Microorganisms and enzymes need water in order to be active: preservation of foods can be sought by reducing their moisture content to a point where the food-spoilage of pathogenic microorganisms are inhibited. Scott [1,2] called the equilibrium relative humidity (ERH), which is the availability of water in a food medium, **Water Activity** ($a_w$). $a_w$ is defined as the ratio of vapour pressure (p) of solution to the vapour pressure (Po) of the solvent, usually water:

$$a_w = \frac{p}{Po} = \frac{\%ERH}{100}$$

Pure water has an $a_w$ of 1.00 and a saturated NaCl solution an $a_w$ of 0.75. Evaluation of techniques and instrumentation has been investigated by Labuza et al [4]. Table 3.57 illustrates aw and the growth of microorganisms in foods and Table 3.58 the minimal $a_w$ for some microorganisms of significance to public health [5].

*Table 3.57:* Water activity and growth of microorganisms in food [a]

| Range of $a_w$ | Microorganism generally inhibited by lowest $a_w$ in this range | Foods generally within this range |
|---|---|---|
| 1.00-0.95 | *Pseudomonas, Escherichia, Proteus, Shigella, Klebsiella, Bacillus, Clostridium perfringens*, some yeasts | Highly perishable (fresh) foods and canned fruits, vegetables, meat, fish, and milk; cooked sausages and breads; foods containing up to approximately 40% (w/w) sucrose or 70% sodium chloride. |
| 0.95-0.91 | *Salmonella, Vibrio parahaemolyticus, Clostridium botulinum, Serratia, Lactobacillus, Pediococcus*, some molds, yeasts (*Rhodotorula, Pichia*) | Some cheeses (cheddar, swiss, muenster, provolone), cured meat (ham), some fruit juice concentrates; foods containing 55% (w/w) sucrose or 12% sodium chloride. |
| 0.91-0.87 | Many yeasts (*Candida, Torulopsis, Hansenula*), Micrococcus | Fermented sausage (salami), sponge cakes, dry cheeses, margarine; foods containing 65% (w/w) sucrose (saturated) or 15% sodium chloride. |
| 0.87-0.80 | Most molds (mycotoxigenic penicillia), *Staphylococcus aureus*, most *Saccharomyces* spp., *Debaryomyces* | Most fruit juice concentrates, sweetened condensed milk, chocolate syrup, maple and fruit syrups; flour, rice, pulses containing 15-17% moisture; fruit cake; country-style ham, fondants, high-ratio cakes. |
| 0.80-0.75 | Most halophilic bacteria, mycotoxigenic aspergilli | Jam, marmalade, marzipan, glacé fruits, some marshmallows. |
| 0.75-0.65 | Xerophilic molds (*Aspergillus chevalieri, A. candidus, Wallemia sebi*), *Saccharomyces bisporus* | Rolled oats containing approximately 10% moisture, grained nougats, fudge, marshmallows, jelly, molasses, raw cane sugar, some dried fruits, nuts. |
| 0.65-0.60 | Osmophilic yeasts (*Saccharomyces rouxii*) few molds (*Aspergillus echinulatus, Monascus bisporus*) | Dried fruits containing 15-20% moisture; some toffees and caramels; honey. |
| 0.50 | No microbial proliferation | Pasta containing approximately 12% moisture; spices containing approximately 10% moisture. |
| 0.40 | No microbial proliferation | Whole egg powder containing approximately 5% moisture. |

| Range of $a_w$ | Microorganism generally inhibited by lowest $a_w$ in this range | Foods generally within this range |
|---|---|---|
| 0.30 | No microbial proliferation | Cookies, crackers, bread crusts, etc. containing 3-5% moisture |
| 0.20 | No microbial proliferation | Whole milk powder containing 2-3% moisture; dried vegetables containing approximately 5% moisture; corn flakes containing approximately 5% moisture; fruit cake; country-style cookies, crackers. |

a) From Beuchat, [5]

***Table 3.58:*** *Minimal $a_w$ for some microorganisms of significance to public health* [a]

| Microorganism | Minimal $a_w$ | |
|---|---|---|
| | Growth[b] | Toxin production |
| **Bacteria** | | |
| Staphylococcus aureus | 0.86 | |
| | 0.86 | |
| | | < 0.90 (enterotoxin A) |
| | | 0.87 (enterotoxin A) |
| | | 0.97 (enterotoxin B) |
| Salmonella spp. | 0.93 | |
| | 0.94-0.95 | |
| | 0.92 | |
| *Vibrio parahaemolyticus* | 0.94 | |
| | 0.94 | |
| *Clostridium botulinum* | 0.93(A) | |
| | 0.95(A) | |
| | | 0.95(A) |
| | | 0.94(A) |
| | 0.93(B) | |
| | 0.94(B) | |
| | | 0.94(B) |
| | | 0.94(B) |
| | 0.95(E) | |
| | 0.97(E) | |
| | 0.95(E) | 0.97(E) |
| | 0.97(E) | |
| *Clostridium perfringens* | 0.93-0.95 | |
| | 0.95 | |
| *Bacillus cereus* | 0.95 | |
| | 0.93 | |
| | 0.95 | |

|  | Minimal $a_w$ | |
|---|---|---|
| **Microorganism** | **Growth**[b] | **Toxin production** |
| **Molds** | | |
| Aspergillus flavus | 0.78 | |
|  |  | 0.84 (aflatoxin) |
|  | 0.80 | 0.83-0.87 |
| A. parasiticus | 0.82 | 0.87 (aflatoxin) |
| A. ochraceus |  | 0.85 (ochratoxin) |
|  | 0.83 | 0.83-0.87 |
|  | 0.77 | |
| Penicillium cyclopium | 0.81 | 0.87-0.90 (ochratoxin) |
|  | 0.82 | |
|  | 0.83 | |
|  | 0.85 | |
| P. viridicatum | 0.83 | 0.83-0.86 (ochratoxin) |
| Aspergillus ochraceus | 0.81 | 0.88 (penicillic acid) |
|  |  | 0.80 |
|  | 0.76 | 0.81 |
| **Molds** | | |
| Penicillium cyclopium | 0.87 | 0.97 (penicillic acid) |
|  | 0.82 | |
| P. martensii | 0.83 | 0.99 (penicillic acid) |
|  | 0.79 | |
| P. islandicum | 0.83 | |
| P. patulum | 0.83-0.85 | 0.95 (patulin) |
|  | 0.81 | |
|  | 0.83 | |
|  |  | 0.85 |
| P. expansum | 0.83-0.85 | 0.99 (patulin) |
|  | 0.83 | |
|  | 0.83 | |
| Aspergillus clavatus | 0.85 | 0.99 (patulin) |
| Byssochlamys nivea | 0.84 | |
| Trichothecium roseum | 0.90 | |
|  | 0.90 | |
| Stachybotrys atra | 0.94 | 0.94 (stachybotryn) |

[a)] From Beuchat, [5]
[b)] (A), (B), and (E): types of enterotoxins in *Clostridium*.

Scott's work led to a rapid expansion in the use of water activity, $a_w$, which has now become one of the major control variables in the technology of food preservation and its influence on all types of food deterioration. Figure 3.71 shows a food stability map as a function of aw after van den Burg and Bruin, [6], first developed by Labuza [7] and Labuza et al [8].

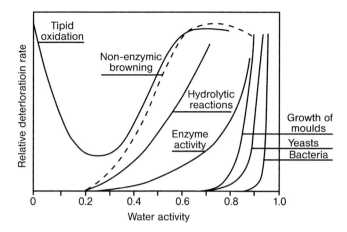

*Fig. 3.71:* Generalized deterioration reaction rates in food systems as a function of water activity (room temperature). {Adapted from Fennema [9] and Heiss and Eichner[10]}

### 3.3.5.2.2 Intermediate Moisture Foods

Intermediate Moisture Foods (IMF's) are characterised by a moisture content of about 15 to 50% and by an aw between 0.60 and 0.85. Traditional IMF's, such as jams, fruit cakes and some ripened cheese are stable at ambient temperatures for various shelf periods (Table 3.59). Water content of IMF's may be lowered to a level which prevents microbial spoilage by the addition of humectants, pH adjustments and anti-microbial agents. Newer IMF's, such as designed for space rations, clinical nutrition and pet foods, can be prepared by adjusting the formulation of the product so that its aw is below 0.86 by use of the following techniques *[11]*:

*Table 3.59:* Traditional intermediate moisture foods

| Food Products | $a_w$ Range |
|---|---|
| Dried fruits | 0.60-0.75 |
| Cake and pastry | 0.60-0.90 |
| Frozen foods | 0.60-0.90 |
| Sugars, syrups | 0.60-0.75 |
| Some candies | 0.60-0.65 |
| Commercial pastry fillings | 0.65-0.71 |
| Cereals (some) | 0.65-0.75 |
| Fruit cake | 0.73-0.83 |
| Honey | 0.75 |
| Fruit juice concentrates | 0.79-0.84 |
| Jams | 0.80-0.91 |
| Sweetened condensed milk | 0.83 |
| Fermented sausages (some) | 0.83-0.87 |
| Maple syrup | 0.90 |
| Ripened cheeses (some) | 0.96 |
| Liverwurst | 0.96 |

*Moist Infusion*

Solid food pieces are soaked and/or cooked in a solution to give the final product the desired aw (desorption).

*Dry Infusion*

Solid food pieces are first dehydrated and then infused by soaking in a solution that contains the desired osmotic agents (adsorption).

*Component Blending*

All IMF components are combined by blending, cooking, or extrusion etc. to achieve a product of the desired aw.

*Osmotic Drying*

Foods are dehydrated by immersion in liquids with an aw lower than the food. By using sugars and salts, water diffuses out of the food into solution while solute diffuses from solution into the food.

These newer techniques *[12,13]* may be achieved by:

- Reduction of aw by dehydration or evaporation.
- Inhibition of enzymatic activity by blanching.
- Microbiological stabilization by thermal/chemical means.
- Control of physical and/or chemical deterioration by addition of antioxidants, stabilisers, emulsifiers and chelating agents.
- Nutrient enrichment.

Labuza *[14]* has reviewed the methods of water binding of humectants.

*S. aureus* is the only bacterium of public health importance that can grow at aw values near 0.86, but many yeasts and moulds are able to grow at aw above 0.70, so antifungal agents may need to be added (propylene glycol, potassium sorbate and sodium benzoate etc.).

### 3.3.5.2.3 Preservatives: Food Additives Directive

The text of the common position on Miscellaneous Additives formally adopted by the Council of the European Union on 10 March 1994 refers to the proposed directive on food additives other than colours and sweeteners.

Annex III, Conditionally Permitted Preservatives and Antioxidants, is divided as follows with details of the maximum permitted level of preservative(s) in the appropriate foodstuff:

Part A: Sorbates, Benzoates and p-Hydroxy Benzoates (Parabens)
Part B: Sulphur Dioxide and Sulphites
Part C: Other Preservatives.

### 3.3.5.2.4 Benzoic Acid and Benzoates

Benzoic acid has been reported in the Weurman list *[15]* in a wide range of berries and fruits such as apple, sour cherry, cranberry, guava, papaya, raspberry, strawberry, passionfruit, cloudberry, plum, mango, cherimoya, loganberry, as well as mushroom, tomato, potato, cassia leaf, nutmeg, pepper, various cheeses, butter, beef, mutton, pork, wines and spirits, and tea.

Sodium benzoate, being more soluble in water (660 g/litre at 20°C) and ethanol, is often preferred to benzoic acid (water 2.9 g/litre at 20°C). Benzoic acid and sodium benzoate are primarily used as antimycotic agents and most yeasts and moulds are inhibited by 0.05-0.1% of the undissociated acid [16]. Food poisoning and spore-forming bacteria are generally inhibited by 0.01-0.02% undissociated acid, but the control of many spoilage bacteria require much higher concentrations. Benzoic acid cannot be relied upon to preserve foods capable of sustaining bacterial growth [16-18]. The antimicrobial activity of benzoic acid resides in the dissociated state (p$K_a$ 4.2) and is related to pH (Table 3.60) and so is of use in acidic foods, soft drinks, cider, tomato ketchup and salad dressings.

*Table 3.60: Summary of some food preservatives*

| Preservative | Organisms affected | Foodstuffs |
|---|---|---|
| E200 Sorbic acid (Sa)<br>E202 Potassium sorbate<br>E203 Calcium sorbate | Moulds | cheese, soft drinks, mead, potato products, olives, egg products, sour breads, salad dressings, dried fruit, fat emulsions, emulsified sauces. |
| E210 Benzoic acid (Ba)<br>E211 Sodium benzoate<br>E212 Potassium benzoate<br>E213 Calcium benzoate | Yeasts and Moulds | soft drinks, alcohol-free beer, low calorie fruit spreads, spirits (less than 15% v/v alcohol) |
| Combined Sa + Ba | | soft drinks, tea concentrates, grape juice, candied/crystalline + glacé, fruit, pickles, semi preserved fish, dairy-based desserts, liquid egg, chewing gum, salads, condiments + seasonings, soups. |
| E214 Ethyl p-hydroxybenzoate<br>E215 Sodium ethyl p-hydroxybenzoate<br>E216 Propyl p-hydroxybenzoate<br>E217 Sodium propyl p-hydroxybenzoate<br>E218 Methyl p-hydroxybenzoate<br>E219 Sodium methyl p-hydroxybenzoate | Yeasts and Moulds | combined use with Sa or Sa + Ba Pâté, jelly coatings of meat products, surface treatment of dried meat products, cereal or potato based snacks, coated nuts, confectionery (not chocolate), liquid dietary food supplements. |

| Preservative | Organisms affected | Foodstuffs |
|---|---|---|
| E220 Sulphur dioxide<br>E221 Sodium sulphite<br>E222 Sodium hydrogen sulphite<br>E223 sodium metabisulphite<br>E224 Potassium metabisulphite<br>E226 Calcium sulphite<br>E227 Calcium hydrogen sulphite<br>E228 Potassium hydrogen sulphite | Yeasts<br>Moulds<br>Bacteria<br>Insects | seafood, starches, cereal and potato snacks, vegetable-dried, pulp or pickled; dried fruits, jams, jellies, citrus and other fruit juices including concentrates, wines, cider, mustard, vinegar, beer, glucose syrup, molasses and other sugars. |
| E230 Biphenyl, diphenyl<br>E231 Orthophenyl phenol<br>E232 sodium/orthophenyl phenol | | surface treatment of citrus fruits |
| E233 Thiabendazole | | surface treatment of:<br>- citrus fruits<br>- bananas |
| E234 Nisin | | semolina and tapioca puddings<br>ripened cheese and processed cheese<br>clotted cream |
| E235 Natamycin | | surface treatment of:<br>- hard, semi-hard and semi-soft cheese<br>- dried cured sausages |
| E239 Hexamethylene tetramine | | provolone cheese |
| E242 Dimethyl dicarbonate | | non-alcoholic flavoured drinks<br>alcohol-free wine<br>liquid- tea concentrate |
| E284 Boric acid<br>E285 Sodium tetraborate (Borax) | | caviar |
| E249 Potassium nitrite | | non-heat treated, cured, dried meat products |
| E250 Sodium nitrite | | other cured meat products<br>canned meat products<br>foie gras<br>cured bacon |
| E251 Sodium nitrate | | cured meat and canned meat products |

| Preservative | Organisms affected | Foodstuffs |
|---|---|---|
| E252 Potassium nitrate | | hard, semi-hard, and semi-soft cheese<br>dairy-based cheese analogue<br>pickled herring and sprat |
| E280 Propionic acid<br>E281 Sodium propionate<br>E282 Calcium propionate<br>E283 Potassium propionate | | pre-packed sliced bread and rye bread<br>energy reduced bread, partially baked pre-packed bread<br>pre-packed fine bakery wares ($a_w > 0.65$)<br>pre-packed rolls, buns, pitta<br>Christmas-pudding<br>pre-packed bread |
| E1105 Lysozyme | | ripened cheese |

### 3.3.5.2.5 Sorbic Acid and Sorbates

Sorbic acid ($pK_a$ 4.8) is more effective in acid foods, best below pH 6.0 and is generally ineffective above pH 6.5. The sorbates work better than sodium benzoate between pH 4.0 - pH 6.0 and are primarily effective against yeasts and moulds. For information on sorbate - nitrite combinations in meat products including bacon see review by Sofos [19].

### 3.3.5.2.6 Acetic Acid and Acetates

Acetic acid ($pK_a$ 4.75) and its salts are used as acidulants and microbials; acetic acid is more effective against yeasts and bacteria than against moulds [21]. Only *Acetobacter* spp., certain lactic acid and butyric acid bacteria are appreciably resistant to acetic acid [16,20,21]. The presence of 1-2% undissociated acid in meat, fish or vegetable products will usually inhibit or kill all microorganisms but, under poor hygiene conditions, more acid-tolerant types may occur. The level of acid may be reduced for foodstuffs being refrigerated or products with high salt or sugar content.

### 3.3.5.2.7 Propionic Acid

Propionic acid ($pK_a$ 4.9) and its salts are highly effective mould inhibitors at concentrations which, when incorporated into bread, for example, are virtually without effect against yeasts. They effectively inhibit many bacterial species at 0.05-0.1% undissociated acid [17] and are used to prevent 'rope' in bread caused by *Bacillus subtilis*. For further information see review by Doores [20].

### 3.3.5.2.8 Citric and Lactic Acids

Citric acid ($pK_a$ 3.1) and lactic acid ($pK_a$ 3.1) have moderate antimicrobial activity and are primarily used for pH control and flavouring.

### 3.3.5.2.9 Sulphur Dioxide

Sulphur dioxide ($SO_2$, $pK_a$ 1.76 and 7.20) and the sodium and potassium salts of sulphite (= $SO_3^{2-}$), bisulphite (= $HSO_3^-$) and metabisulphite (= $S_2O_5^{2-}$) act similarly. Sulphur dioxide has been used since ancient times for fumigation and as an improving agent for wine (Plinius, Naturalis Historia XIV, 129). $SO_2$ possesses antimicrobial activity as well as some antioxidant behaviour. Sulphites react with nucleotides, sugars, aldehydes, disulphide bonds and enzymes with S-H groups: meats and products recognised as containing thiamine should not be treated with $SO_2$. $SO_2$ inhibits yeasts, moulds and bacteria and is used to control undesirable micro-organisms in soft fruits, fruit juices, wines, sausages, fresh shrimp, acid pickles and during the processing of starches. Residues may be reduced by vacuum or heat.

$SO_2$ (50-100 ppm) is added to expressed grape juice to control unwanted moulds, bacteria and yeasts depending upon the condition of the grapes. It is also used in treatment of soft-fruits in order to extend the time available for jam manufacture. Sausage and meat products are treated with $SO_2$ to extend their shelf life under refrigerated conditions. Soft drinks may contain 10 ppm (Continental Europe) -70 ppm (UK), some of which may originate from preserved concentrated fruit juices.

*Further Reading*

Ough, C.S. Sulphur dioxide and sulphites, pp 177-203 in: Antimicrobials in Foods, Branen, A.L. and Davidson, P.M. (Eds.) Marcel Dekker, New York, 1983

### 3.3.5.2.10 Contaminants

**Benzene in Beverages**

In 1990 there were 3 reports of the low level occurrence of benzene in beverages. Bottled mineral water was contaminated with benzene originated from an underground source of carbon dioxide which was inadequately carbon treated [23]. Two manufacturers found benzene in fruit flavoured mineral waters at levels greater than 5.0 µg/kg [24]. The US EPA maximum contaminant level (MAL) for benzene is 0.005 mg/L [25] which is also the Canadian guideline as maximum acceptable concentration (MAC) in drinking water [26]. The manufacturers attributed the benzene to the presence of ascorbic acid and sodium benzoate in combination in their drink formations [27]; omitting either of these additives eliminated the benzene.

A survey for natural benzoic acid was carried out by Nagayama et al [28] and further surveys for benzene in fruits, retail fruit juices, fruit drinks and soft drinks were carried out by the Canadian Health Protection Branch [29] and by the FDA in foods [30]. Decarboxylation of benzoic acid in the presence of ascorbic acid and a transition metal catalyst has been reported to yield benzene [31]. Studies on benzene formation in beverages at the National Laboratory of Food & Drugs in China [32] showed that ascorbic, sodium benzoate and hydrogen peroxide increase benzene formation initially, but when a certain concentration was reached, the effect was reversed: ethanol and $Fe^{3+}$ ions inhibited benzene formation.

Both the Canadian and FDA surveys showed that benzene is formed by benzoates and ascorbic or erythrobic acid in combination, but not when one of the additives is omitted from the formation. The average level of benzene in the Canadian survey of 73 bottled juices and drinks was 0.40 µg/kg which is less than one tenth of the EPA/Canadian Guidelines of 5 µg/kg for drinking water and was not considered to represent a health concern [29].

**Sorbic Acid Derived Off-Notes**

Sorbic acid under certain conditions is known to generate at least two types of off-odours - one geranium-like (wine) and the other a paint, plastic, paraffin-note (cheese).

*Geranium-Aroma*

Burckhardt [33] reported a geranium leaf off-note in German wine preserved with potassium sorbate used to inhibit re-fermentation by yeasts. It was regularly associated with lactic acid formation and Wurdig et al [34] said the geranium odour was caused by 2,4-hexadien-1-ol and its lactate and acetate esters. Crowell and Guymon [35] identified 2,4-hexadien-1-ol and 2-ethoxyhexa-3,5-diene in wines with geranium off-odour, but these compounds were absent in extracts of acceptable wines. It was suggested that the bacterial action on sorbic acid produces sorbyl alcohol (2,4-hexadien-1-ol) which forms the isomer, 3,5-hexadien-2-ol and thence the corresponding ethers, 1-ethoxyhexa-2,4-diene and 2-ethoxyhexa-3,5-diene in equilibrium. Confirmation that 2-ethoxyhexa-3,5-diene was the major compound responsible for the geranium off-odour came from Rymon-Lipinsky [36]. All strains of *Leuconostoc oenos* are capable of reducing sorbic acid to sorbyl alcohol [37].

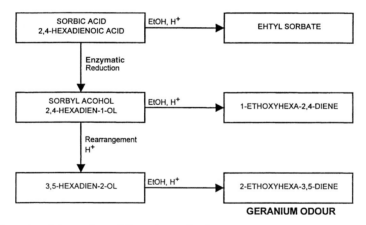

*Fig. 3.72: Postulated formation of Geranium off-oduor*

*Trans-1,3-Pentadiene (Perylene)*

Horwood et al. [38] identified a plastic, paint, paraffin off-note as trans-1,3-pentadiene in products treated with sorbic acid as a mould inhibitor, detectable on taste at 4 ppm in cheese and on odour at 2.5 ppm in brine. Earlier work by Marth et al. [39] indicated degradation of sorbate by *Penicillium* species.

## REFERENCES

[1]  Branen, A.L., Introduction to use of Antimicrobials, p1, in:. Branen, A.L. and Davidson, Antimicrobials in Foods, P. M. (Eds), New York, Marcel Dekker, 1983.
[2]  Scott, W.J., Water relations of staphylococcus aureus at 30°C, Australian J. Bioc. Sci., 6, 549-564, 1953.
[3]  Scott, W.J., Water relations of food spoilage microorganisms, Adv. Food Res., 1, 83-127, 1957.
[4]  Labuza, T.P., Acott, K., Tatini, S.R., Lee, R.Y., Flink, I., and McCall, W., Water Activity Determination: A collaborative study of different methods, J. Food Sci. 41, 910-917, 1976.
[5]  Beuchat, L.R., Microbial stability as affected by water activity, Cereal Foods World, 26, 345-349, 1981.
[6]  Van den Burg, C. and Bruin, S., Water activity and its estimation in food systems theoretical approach, pp 1-61 in: Rockland, L.B. and Stewart, G.F. (Eds), Water Activity: Influences on Food Quality, New York, Academic Press, 1981.
[7]  Labuza, T.P., Proc. 3rd Int. Conf. Food, Sci. Technol., SOS 70, p618, Washington D.C., IFT, 1970
[8]  Labuza, T.P., McNally, L., Gallagher, D., Hawkes, J., and Histado, F., J. Food Sci., 37, 154, 1972.
[9]  Fennema, O. in , Fennema, O. (Ed),'Principles of Food Science', Part I p13, New York, Marcel Dekker, 1976.
[10] Heiss, R., and Eichner, K., Chem. Mikrobiol. Technol. Lebensm., 1, 33, 1971.
[11] Karel, M., Technology and application of new intermediate moisture foods, pp 4-31 in:Davies, R., Birch, G.G. and Parker, K.J., Intermediate Moisture Foods, London, Applied Science, 1976.
[12] Labuza, T.P., Tannenbaum, S.R. and Karel, M., Water content and stability of low-moisture and intermediate-moisture foods, Food Technol., 24, 543,544,546, 548, 550., 1970.
[13] Taoukis, P.S., Breene, W.M. and Labuza, T.P., Intermediate moisture foods, Adv. Cereal Sci. Technol. 1X, 91-128, 1988.
[14] Labuza, T.P., Water binding of humectants, pp 421-445 in: Simatros, D. and Multon, J.L. (Eds), Properties of Water in Foods, Dordrecht, The Netherlands.Martinus Nijhoff, 1985.
[15] Maarse, H., Visscher, C.A., Willemsen, L.C., and Boelens, M.H., Volatile Compounds In Food, Qualitative and Quantitative Data, Supplement 4 and Cumulative Index to the Sixth Edition 1989.Zeist, The Netherlands, TNO Nutrition and Food Research, 1993.
[16] Baird-Parker, A.C., Organic acids, p126 in: Microbial Ecology of Foods, vol. I. Factors affecting life and death of microorganisms: (International Commission on Microbiological Specifications for Foods), New York, Academic Press, 1980.
[17] Chichester, D.F. and Tanner, F.W., Antimicrobial food additives, pp115-184. in: Handbook of Food Additives, 2nd Ed. (Furia, T.E. (Ed).), Cleveland, Ohio, Chem. Rubber Publ. Co., 1972.
[18] Jay, J.M., Modern Food Microbiology, 4th Edition, New York, AVI, 1982.
[19] Sofos, J.N., Sorbate Food Preservatives, Boca Raton, CRC Press, 1989.
[20] Doores, S., Organic acids, pp 75-78, in: Antimicrobials in Foods, Branen, A.L. and Davidson, P.M. (Eds), New York, Marcel Dekker, 1983.
[21] Ingram, M., Ottoway, F.J.H. and Coppock, J.B.M., The preservative acid of acid substances in food, Chem Ind (London) no 42, 1154-1163, 1956.
[22] Dakin, J.C., Microbiological keeping quality of pickles and sauces, Res. Rep. no 79, Leatherhead, England, Food Res. Assoc., 1957.
[23] Anon, Food Chemical News 31, (51), pp 27-28, 1990.
[24] Anon, Food Chemical News 32, (38), pp 22-23, 1990.
[25] Federal Register, 52, 25690-25734, 1987.
[26] Guidelines for Canadian Drinking Water Quality, 4th Edn., Ottawa, Canada, Canadian Publishing Centre, 1989.
[27] Memorandum of meeting, National Association of Soft Drink Manufacturers, Dec. 17, 1990.
[28] Nagayama, T., Nishijima, M., Yasuda, K., Saito, K., Kaminura, H., Ibe, A., Ushiyama, H., and Naoi, Y., J. Food Hygienic Soc. Japan, 24, 416-422, 1983.
[29] Page, B.D., Conacher, H.B.S., Weber, D., and Lacroix, G., A survey of benzene in fruits and retail fruit juices, fruit drinks and soft drinks, J. of AOAC International, 75 (2), 334-340, 1992.
[30] McNeal, T.P., Nyman, P.J., Diachenko G.W., and Hollifield, H.C., Survey of benzene in foods by using headspace concentration techniques and capillary gas chromatography, J. of AOAC International, 76 (6), 1213-1219, 1993.

[31]   Gardner, L.K. and Lawrence, G.D., Benzene production from decarboxylation of benzoic acid in the presence of ascorbic acid and a transition-metal catalyst, J. Agric. Food Chem., 41 (5), 693-695, 1993.
[32]   Chang, Pi-Chiou, and Ku, K., Studies on benzene formation in beverages, Journal of Food and Drug Analysis, 1 (4), 385-393, 1993.
[33]   Burckhardt, R., Lebensm. Chem., 27, 259-260, 1973.
[34]   Wurdig, G., Schlotter, H.A., and Klein, E., Algemeine Deutsche Weinfachzeit, 110, 578-583, 1974.
[35]   Crowell, E.A. and Guymon, J.F., Wine constituents arising from sorbic acid addition, and identification of 2-ethoxyhexa-3,5-diene as a source of geranium like off-odour, Am. J. Enol. Vitic., 26, 97-102, 1975.
[36]   Rymon-Lipinsky von, G.W., Lück, E., Oeser, H., and Lömker, F., Formation and causes of the 'geranium like off-odour', Mitt. Klosterneuss, 25, 387-394, 1975.
[37]   Edinger, W.D., and Splittstoesser, D.F., Production lactic acid bacteria of sorbic alcohol, the precursor of geranium odour compound, Am. J. Enol. Vitic., 37, 34-38, 1986.
[38]   Horwood, J.F., Lloyd, G.T., Ramshaw, .H. and Stark, W., An off-flavour associated which the use of sorbic acid during feta cases maturation., Aust. J. Dairy Technol., 36, 38, 1981.
[39]   Marth, E.H., Capp, C.M., Hazenzahl, L., Jackson, H.W. and Hussong, R.V., Degradation of potassium sorbate by penicillium species, J. Dairy Sci., 49, 1197, 1966.

# 4 Blended Flavourings

*Willi Grab*

# Antioxidants and Preservatives

| | | | |
|---|---|---|---|
| E-2-HEXENAL | Z-3-HEXENAL | E-2-HEXEN-1-OL | Z-3-HEXEN-1-OL |
| E-2-NONENAL | Z-6-NONENAL | E-2,Z-6-NONADIENAL | E-2,E-4-NONADIENAL |
| E-2,E-4-DECADIENAL | Z-6-NONEN-1-OL | E-2,Z-6-NONADIEN-1-OL | CITRAL |
| UNDECA-1,3,5-TRIENE | OCIMENE | alpha-FARNESENE | PRENYL-ETHYL-ETHER |
| ETHYL 2-METHYL BUTYRATE | iso-AMYL ACETATE | METHYL CAPRONATE | ALLYL CAPRONATE |
| ETHYL-E-2-OCTENOATE | ETHYL-E-4-OCTENOATE | ETHYL DECA-2,4-DIENOATE | METHYL JASMONATE |
| METHYL CINNAMATE | gamma-NONALACTONE | gamma-DECALACTONE | delta-DECALACTONE |
| METHYL ANTHRANILATE | BENZALDEHYDE | KETO-ISOPHORONE | NOOTKATONE |

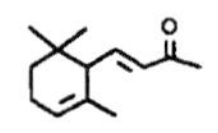

alpha-IONONE

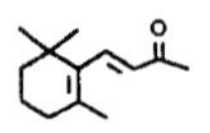

beta-IONONE

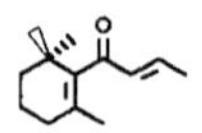

beta-DAMASCONE

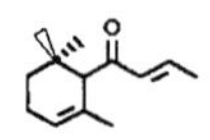

alpha-DAMASCONE

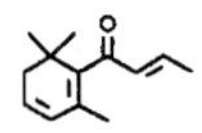

DAMASCENONE

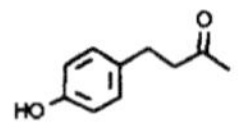

4-(4-HYDROXY-PHENYL) BUTAN-2-ON

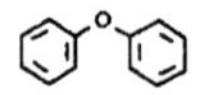

DI-PHENYLE-OXIDE

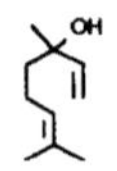

LINALOOL

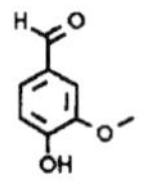

VANILLIN

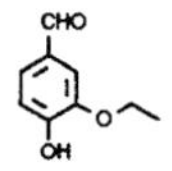

ETHYL-VANILLIN

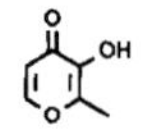

MALTOL

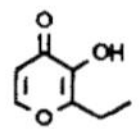

ETHYL-MALTOL

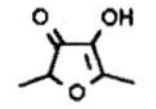

2,5-DIMETHYL-3-HYDROXY-3(2H)-FURANON

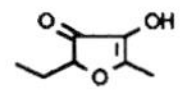

5-METHYL-2-ETHYL-4-HYDROXY-3(2H)-FURANON

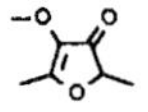

2,5-DIMETHYL-4-METHOXY-3(2H)-FURANON

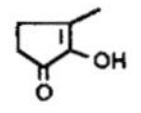

3-METHYL CYCLOPENT-2-EN-2-OL-1-ON

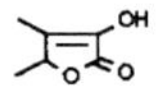

4,5-DIMETHYL-3-HYDROXY-2(5H)-FURANON

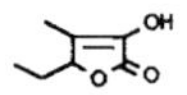

3-HYDROXY-4-METHYL-5-ETHYL 2(5H)-FURANON

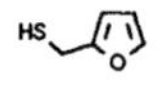

FURFURYL-MERCAPTANE

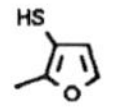

2-METHYL FURAN-3-THIOL

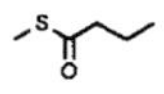

METHYLTHIO BUTYRATE

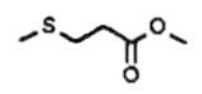

METHYL 3-METHYLTHIO PROPIONATE

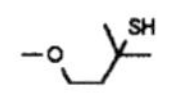

4-METHOXY-2-METHYL BUTANE-2-THIOL

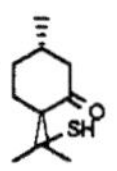

8-MERCAPTO MENTHAN-3-ONE

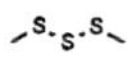

DIMETHYL-TRISULFIDE

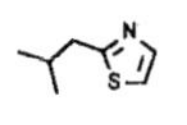

2-iso-BUTYL THIAZOL

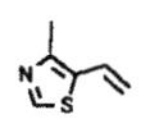

4-METHYL 5-VINYL THIAZOLE

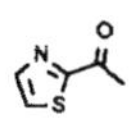

2-ACETYL THIAZOLE

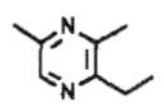

2-ETHYL-3,5-DIMETHYL PYRAZINE

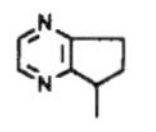

5-METHYL-6,7-DIHYDRO-CYCLOPENTAPYRAZINE

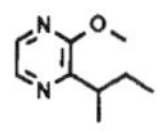

2-METHOXY-3-sec-BUTYL PYRAZINE

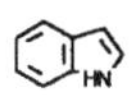

INDOL

# 4.1 Introduction

It is simple to concentrate fruit juices to a fully flavoured concentrate. It is easy to make a concentrated extract of spices with an organic solvent. It is possible to distil off the flavour of apples into a 150 fold concentrated product or to develop a meat flavour by reacting reducing sugars with amino acids. But it is very difficult and very expensive to prepare from strawberries a highly concentrated strawberry flavour with the taste of freshly picked strawberries, which is stable in a baking process for biscuits. The prepared flavouring raw materials discussed in chapter 3 (essential oils, extracts, juice concentrates, process flavourings, biotechnological engineered flavouring products) represent more or less the whole flavour profile. In contrast to them, compounded or blended flavouring products get their final flavour profile only by the well balanced mixture of individual flavouring components, including the above mentioned materials.

The merit of blended flavourings is adaptability and flexibility to adjust the parameters of a flavour to the desired purpose:

flavour profile, solubility, cost, stability, viscosity, colour, density, legislative aspects, health aspects, technological behaviour, heat stability, enzymatic resistance, interactions with food ingredients, various delivery and releasing systems.

The progress of research in understanding flavour development, biosynthesis, composition, perception and technological behaviour allows the flavour industry today to propose an adequate flavouring for almost all requests.

## 4.1.1 Flavour Analysis

The progress in trace analysis (high resolution capillary gas chromatography combined with mass spectrometry, high resolution nuclear magnetic resonance spectroscopy, high performance liquid/liquid chromatography, head space analysis, chiral and isotopic analysis) revealed an enormous mass of information on the formation, composition and interactions of natural flavours. Today we start to understand why and how food smells and tastes, although many questions are not yet answered.

## 4.1.2 Flavour Profiles

The general basic flavours of food can be divided into fruity, meaty, vegetable, spicy and roasted notes. The comparison of the composition of food shows a direct correlation between its composition of tasting materials and the basic tastes (see Table 4.1).

The comparison of the composition of the main, tasting ingredients in food shows the following correlation:

- Fruity notes are related with a balanced sugar/acid ratio, without salt and nucleotides. Some fruits contain bitter or astringent notes.
- Mushroom notes are combined with nucleotides.
- Vegetable notes are based on a balance of sugar, acids and salt.
- Nut notes are mainly based on bitter and astringent notes.
- Meaty notes are based on a balance of salt and nucleotides.

*Table 4.1:* Tasting materials and basic tastes

|  | sugar | salt | acids | nucleotides | bitter | astringent |
|---|---|---|---|---|---|---|
| Strawberry | 50.48 | 0.05 | 9.52 | 0.01 | 0.01 | 0.01 |
| Apple | 2.79 | 0.03 | 4.08 | 0.01 | 0.03 | 0.07 |
| Orange | 55.94 | 0.01 | 8.39 | 0.01 | 0.07 | 0.03 |
| Mushroom | 5.91 | 0.17 | 1.08 | 1.08 | 0.02 | 0.01 |
| Salad | 21.40 | 0.60 | 0.20 | 0.04 | 2.00 | 0.02 |
| Broccoli | 24.27 | 0.39 | 2.91 | 0.10 | 0.49 | 0.10 |
| Carrot | 38.98 | 1.53 | 2.54 | 0.08 | 0.08 | 0.04 |
| Hazelnut | 1.05 | 0.01 | 0.01 | 0.00 | 0.11 | 0.11 |
| Walnut | 1.05 | 0.01 | 0.01 | 0.00 | 0.52 | 1.05 |
| Veal | 0.00 | 0.64 | 0.13 | 0.86 | 0.00 | 0.00 |
| Beef | 0.00 | 0.72 | 0.11 | 0.76 | 0.01 | 0.01 |
| Chicken | 0.00 | 0.59 | 0.11 | 0.73 | 0.01 | 0.01 |
| Salmon | 0.00 | 0.29 | 0.09 | 0.87 | 0.00 | 0.00 |

(percentage in dry, edible matter)

The only way to describe and measure "flavour" is to use descriptor words in analogy to "known standards". The precision of a flavour profile description strongly depends on the clear definition of the standards:

- The descriptor *"floral"* can mean anything of rose, jasmine, hyacinth, lavender, orange flower, lily of the valley smell.
- The descriptor *"floral, rosy"* is more precise as it describes the floral note of a rose. But it is quite obvious, that many different roses with different odours exist: heavy, sultry perfumy, fresh fruity, even minty and lemon notes.
- The descriptor *"floral, rosy, rose oil"* is rather precise because the recognition of the odour of rose oil can be trained in a panel. But as there exist various qualities of rose oil even a more precise description is sometimes necessary.
- Only the descriptor *"floral, rosy, phenyl ethyl alcohol"* identifies a clear, reproducible standard, as phenyl ethyl alcohol is available in any quantities all over the world in a standard quality.

The merit of such a precise description is that its odour is easy to reproduce and it can be shown and explained to other people. The drawback is that no common flavour is just as simple as a pure chemical.

Furthermore unexpected interactions may disturb the perception, for example: 3-methyl cyclopent-2-en-2-ol-one (corylon) has a burnt, roasted, maple and coffee like character.

Allyl hexanoate has a sweet, fruity, candy, pineapple like character.

The combination of 9 parts corylon and 1 part allyl hexanoate is recognized by the majority of people as spicy, celery like.

A professional description of flavours uses as precise descriptors as possible. Chemical analytical results are combined with sensory analysis of the identified components to assess the relative importance and contribution to the flavour profile. Key ingredients or *character impact compounds* (**CIC**) are important components "sine qua non" to impart the typical, product characteristic, flavour, e.g. anethol for anise, eugenol for clove, 3-methyl butyl acetate for banana or ethyl butyrate to improve the juiciness of orange juice.

## 4.1.3 Flavouring Raw Materials

The characterization and description of raw materials is reversed to the description of food flavours and it depends strongly on the individual's experience, expertise, training and vocabulary. The flavouring raw materials are best described in view of their potential use and in analogy to peculiarities of known products of natural or commercial origin.

A complete description of a flavouring raw material has to include:

- *Odour:* from the product; on a blotter, fresh, and throughout the dry-out; intensity character.
- *Taste:* in water, with/without sugar, salt, dosage range, change of character at different concentration levels.
- *Natural occurrence:* analytical results from literature or through own research and a critical evaluation of its importance.
- *Potential use and application:* this is the basic source of creativity for the flavourist.
- *Sensory values:* evaluate the olfactive purity through careful trace analysis using sniffing techniques. Determine threshold value, odour value, anosmia.
- *Physico-chemical parameters:* solubility, vapour pressure, density.
- *Chemical parameters:* chemical structure; reactivity towards functional groups and food ingredients; stability in low pH, heat, enzymatic activity, oxygen, light.
- *Legal aspects:* natural, nature-identical, artificial, GRAS, kosher status, halal, local or company specific limitations.
- *Cost:* put the price into relation to the flavouring cost (= price/dosage). Thus a very expensive material may cause low flavouring costs due to a low dosage level in the finished product.

*Example Benzaldehyde (nature-identical)*

*Odour of the product:* strong, persistent, bitter almond, chemical pungent.

*Odour on the blotter:* bitter almond, sweet, black cherries, evaporates within 24 hours

*Taste in water:* at 0,2-20 ppm: bitter almond, cherry, dry

*Occurrence in food:* mainly fruits, bitter almond, apricot, peach, cherry, cinnamon, plum, black tea

*Potential use:* bitter almond, cherry, grape, red wine, apple pip, apricot kernel, peach kernel, plum, vanilla, pistachio

*Olfactive purity:* 95% benzaldehyde, 4,5% benzoic acid, 0,3% benzyl alcohol, traces of benzoin (1-hydroxy-2-oxo-1,2-diphenyl ethane), benzyl chloride

*Threshold value:* beer: 1,5-3 ppm, wine: 2-3 ppm, 75 ng/l in air

*Odour value:* 81200

*Dosage range (in water):* 0.05-30 ppm

*Solubility in water:* 6580 ppm

*Solubility in organic solvents:* good

*Formula:* $C_6H_5CHO$, molecular weight: 106

*Reactivity:* is easily oxidized to benzoic acid through air oxidation. Reacts with alcohols to form acetals (e.g.: ethanol, 1,2-propanediol); may form Schiff Bases with amines

*Legal aspects:* nature-identical, GRAS, kosher

## 4.1.4 Flavouring Composition

### Flavor Creation - a Vector Addition

$$\vec{Fi}_{tot} = sum(\vec{Fi}_{comp}) + sum(\vec{Fl})_{interactions(Ficompi*compj)}$$

$\vec{Fi}_{tot}$ = total flavor impression

$\vec{Fi}_{comp}$ = flavor impression of individual components

$\vec{Fl}_{interactions(Ficompi*compj)}$ = interaction between two components (synergy, antagony, quality modifier)

- The vector direction is its "flavor quality"
- The vector length is its contribution
- Contribution = dosage for medium intensity/actual dosage
- Medium Dosage: evaluated in water and perceived as „not too strong, not too weak"

*Fig. 4.1: The Flavour Creation as a Theorem of Addition*

# Flavouring Composition

a) contribution ↓ ↓

| strawberry | pineapple | | estery | fruity | fresh | pineapple | jammy | caramel | green | sweet | sulfurous |
|---|---|---|---|---|---|---|---|---|---|---|---|
| 0.1 | 0.2 | methylhexanoate | 60 | 30 | 10 | | | | | | |
| 0.1 | | ethylhexanoate | 50 | 40 | | | | | | 10 | |
| 0.2 | | methythiobutyrate | 40 | 30 | | | 20 | | | | 10 |
| 0.5 | 0.4 | furaneol | | 10 | | | 50 | 20 | | 20 | |
| 0.1 | | cis-3-hexenol | | | 30 | | | | 70 | | |
| | 0.4 | methyl-3-methylthiopropionate | 10 | 30 | | 40 | | | | | 20 |
| 1.0 | | **strawberry** | 19.0 | 18.0 | 4.0 | 0.0 | 29.0 | 10.0 | 7.0 | 11.0 | 2.0 |
| | 1.0 | **pineapple** | 16.0 | 22.0 | 2.0 | 16.0 | 20.0 | 8.0 | 0.0 | 8.0 | 8.0 |

Flavor profile calculations are simple for simple flavor combinations. Results are reasonable. The algorhitm can be used to precalculate the flavor profile of a formula before mixing or to precalculate a formula for a given profile (more speculative)

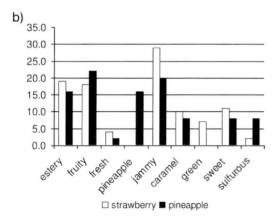

*Fig. 4.2: a Flavour profiles. b Flavour profile, calculated.*

The flavour profile of a composition can be "calculated" or described as a linear combination of the vectors of the individual components using the "flavour quality" as the direction of the vector and the intensity as its length. Components with a similar quality harmonize well and strengthen each other. Components with different quality add a new dimension to the flavouring: a trace of lemon oil in a vanilla flavouring turns it into a fresh cake flavouring.

At an "equal contribution" all ingredients have about the same intensity and therefore the resulting flavour profile is "rounded off" with no dominating component.

Flavourists are trained to detect and describe hundreds of different odours and tastes. They do this in a dynamic way: sniffing a product at intervals shows many aspects of the flavour profile due to changes in the perceived proportions. This is very obvious when sniffing a concentrated flavour on a paper blotter or during food eating and degustation sessions: tea and coffee tasters use a very noisy way to evaporate and separate the flavour in their palate to distinguish minor differences.

David Laing has shown that nobody is able to distinguish more than 3-5 molecules in a single sniff. This can be explained by the detection limits in triangle tests: relative concentration differences below 20% in a mixture are very difficult to differentiate.

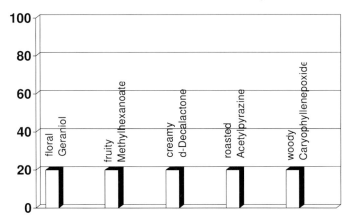

*Fig. 4.3: Sniffing detection limits in compositions*

From this we may conclude that any flavour impression can be generated by only a few character impact molecules. My experience confirms this to a good degree: the selection of the molecules is very critical. However, in order to create a commercial flavour, the flavourist has to mix more than just these few molecules:

- A flavour must be "robust" in the food. A combination of molecules with similar profiles compensates small changes and losses.
- To protect intellectual property, the flavourist will try to "hide" the key molecules.
- A more complex flavour fits to more individual palates and preferences.

Some key impact molecules are very unstable, expensive or just not available on the shelf. They have to be replaced and mimicked by more stable, common raw materials.

A standard compounded flavouring can be divided into various parts:

- the key base contains all the important characterizing trace components.
- general basic materials: esters for fruits, pyrazines for roasted products.
- rounding-off products: vanillin, maltol, 2,5-dimethyl-4-hydroxy-furan-3(2H)-one.
- sweet background, body: 2-methyl-5-ethyl-4-hydroxy-furan-3(2H)-one.

The flavouring concentrate represents the complete flavour profile in the most concentrated form and it also includes general rounding-off materials to impart a basic body for sweet taste (e.g. vanillin, maltol, 2,5-dimethyl-4-hydroxy-furan-3(2H)-one).

The flavouring is diluted with a carrier. This carrier fulfils several purposes:

- it adjusts the technological properties of the flavouring to the need of the users.
- as a diluent it reduces the concentration of the flavouring to a practical, easy to be applied level.
- it reduces the risk of spillages.
- it stabilizes the flavouring.
- it dissolves all flavouring components into a homogenous product.

The basic task in the creation of flavourings is to:

- mix the right components
- in the right proportions.

The example below illustrates this task: a selection of 16 components mixed in the right proportions results in 4 different flavours. The key components (**CIC**) in this example are:

| | |
|---|---|
| Apple: | Hexanal, (E)-2-Hexenal |
| Banana: | Isoamyl acetate, Eugenol, Vanillin |
| Pear: | Heptyl acetate, Citronellyl acetate |
| Pineapple: | Allyl caproate. |

*Table 4.2: Simple flavour combinations*

| | Apple | Banana | Pear | Pineapple |
|---|---|---|---|---|
| 1-Butanol | 30 | 5 | 30 | 1 |
| 2-Methyl butanol | 50 | 5 | 50 | 5 |
| 1-Hexanol | 30 | 5 | 40 | 1 |
| Amyl acetate | 50 | 10 | 20 | 5 |
| Isoamyl acetate | 5 | **150** | 5 | 5 |
| Ethyl butyrate | 5 | 40 | 10 | 10 |
| Amyl butyrate | 5 | 30 | 20 | 20 |
| Heptyl acetate | 5 | 5 | **100** | 5 |
| Ethyl-2-methyl butyrate | 5 | 10 | 5 | 20 |
| Allyl caproate | 5 | 5 | 5 | **120** |
| Citronellyl acetate | 5 | 5 | **40** | 1 |
| Hexanal | **100** | 1 | 5 | 1 |
| (E)-2-Hexenal | **100** | 10 | 30 | 5 |
| Benzaldehyde | 0.1 | 0.2 | 0.2 | 0.1 |
| Vanillin | 1 | **30** | 1 | 30 |
| Eugenol | 0.1 | **2** | 0.2 | 0.1 |
| Ethanol | 693.8 | 686.8 | 638.6 | 770.8 |
| | 1000 | 1000 | 1000 | 1000 |

## 4.1.5 Flavour Release

The profile of an aroma depends on the release of the flavouring molecules during eating or drinking: only molecules that reach the receptors in the nose are perceived as aromatic if the concentration at that moment is above the threshold. From this we may conclude that the composition of a food strongly influences the flavour profile. A given flavour composition, applied in a beverage or in a fatty ice cream, will deliver a different flavour impression: lipophilic components like long-chain fatty aldehydes are dissolved in the fat phase of the ice cream and their vapour pressure above the product is lowered, thus reducing their flavouring impact. In the beverage, they are in a hydrophilic environment, not soluble in water and therefore "pushed" out of the product into the headspace: the higher concentration leads to a high impact of the aldehydes.

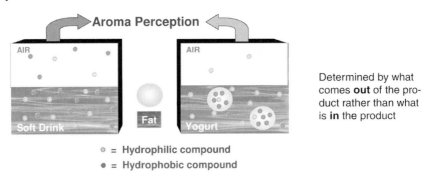

*Fig. 4.4:* Aroma release and perception

Example: A balanced vanilla composition with lemon oil gives a nice, fresh vanilla ice cream. The same flavour applied in carbonated water has a sweet lemon character.

Extensive research in the past 20 years has elucidated the physical and mathematical background of this phenomenon. The main factors influencing the flavour release are the partition coefficients between the aroma molecules and the various phases in food (fat, water, emulsion, solid, air) and the mass transport between the phases. The changes of the food matrix during eating (chewing, salivation, swallowing) further influence the individual perception of a flavour. Reversible and irreversible reactions between the flavouring molecules and the food ingredients modify the flavour release: proteins are strong "binders" of carbonyl and sulphur components, starch and pectin "encapsulate" the molecules.

Example: Flavouring soy milk is extremely difficult, as many important flavour impact molecules are irreversibly bound to the soy protein.

Another effect of "unwanted flavour release" is the loss of flavourings during food processing (baking, extrusion) and storage.

Example: Baking a dough results in an almost complete loss of volatile molecules: the baking process acts like an efficient "water vapour distillation". As a result, the bakery smells fantastic; the product is dull.

The flavour industry has to compensate these effects and balance the flavour in such a way that its expected flavour profile develops after processing, storage and during consumption of the food. The flavourists have different ways to achieve this goal:

***Trial and error*** combined with application trials and stability tests. The flavour industry has a vast experience concerning the performance of their flavourings and "guessing" is in many cases the best approach for a new flavour composition in a known food application.

***Mathematical model calculations*** using verified models and well documented databases. Although these models are very complex, it was possible to develop computer programs and fill databases with a reasonable effort. This program, linked into the daily creation process of the flavourist allows a prediction of the adjustment of a composition for different food applications.

*Table 4.3: Flavour reformulation by computer*

**Original Application:** Drink Yogurt (3.5% dat)
**New Application:** Soft Drink (0% Fat)

| | Concentration for same Perception | | |
|---|---|---|---|
| | | Soft Drink | |
| | Yogurt | Sniff | Mouth |
| *cis*-3-Hexenol | 60 | 48 | 48 |
| Hexylacetate | 220 | 38 | 50 |
| Linalylacetate | 370 | 11 | 50 |
| γ-Undecalactone | 350 | 11 | 50 |
| Total: | 1000 | 108 | 198 |

⇨ **Low fat products have relatively strong smell compared to taste!**

***Encapsulation.*** The flavour industry is using different ways to protect flavourings from losses during storage and processing and to adjust the release in a desired manner. A pure mixture of a liquid flavour with a carrier like starch results in a dry, but still very sensitive adsorbate. Volatile parts are lost and terpenes will be oxidized. Spray drying is often used to transform liquid flavourings into dry powders. Spray drying in combination with coating in double stage dryers gives an additional protection against oxidation. Spray dried powder flavours are extensively used in the food industry. They are easy to handle, relatively stable and possess good solubility in water. Extrusion of a flavour-sugar syrup emulsion, combined with instant chilling and solvent washing gives rather stable products, mainly for the oxygen sensitive citrus range; special treatment is required for tropical environment. The selection of the carrier matrix (mainly gum arabic, maltodextrin and modified starches) is critical for the stability of the end product. Another concept is the encapsulation in a protein or gelatine matrix: these products are not water soluble and quite heat stable. The flavour release depends on mechanical force or higher temperature, an ideal behaviour for heated food products (e.g. baked biscuits), shown in Table 4.5.

*Table 4.4:* Carrier Systems

| | |
|---|---|
| **Adsorbates**<br>Flavour composition is adsobed into porous (carbohydrate) matrix. | • Dry product<br>• No or limited protection of flavour compounds |
| **Spraydries**<br>Flavour emulsion is dried in spraying tower. | • Dry, free-flowing product<br>• Good oxidative stability of encapsulated flavour<br>• Sensitive to humidity<br>• Small particles |
| **Granulates**<br>Flavour emulsion sprayed on nucleus; agglomeration in fluid bed dryer; optionally caoting of particles. | • Dry, free-flowing product<br>• Good oxidative stability of encapsulated flavour<br>• Intermediate particle size<br>• Limited controlled release properties (coatings) |
| **Melt extrudates**<br>Flavour composition is mixed with hot carbohydrate melt, extruded into cold solvent. | • Dry, free-flowing product<br>• Excellent oxidative stability of encapsulated flavour<br>• Moderately sensitive to humidity<br>• Large particles |

*Table 4.5:* Controlled release systems

| | |
|---|---|
| **Fat encapsulation**<br>Flavour composition adsobed by liquid of solid tat; tat dispersed or solidified into small particles | • No significant protection of encapsulated flavour<br>• Particles sensitive to mechanical forces and temperature<br>• Flexible particle size<br>• Limited controlled release properties |
| **Extrudates**<br>Flavour composition encapsulated in high-molecular weight matrix by extrusion process | • Good oxidative stability of encapsulated flavour<br>• Stable particles<br>• (Very) large particle size<br>• Tunable release profiles |
| **Polymer capsules:**<br>• Core-shell<br>Flavour composition entrapped in capsule consisting of solvent core covered by protein shell | • Dry product or liquid dispersion<br>• Limited (oxidative) stability of encapsulated flavour<br>• Flavour can be loaded in 'empty' particles<br>• Flexible particle size<br>• Relatively slow release/burst-like release |
| • Matrix<br>Polymer matrix with dispersed oil or flavour droplets | • Dry product<br>• Limited (oxidative) stability of encapsulated flavour<br>• Flavour can be loaded in 'empty' particles<br>• Tunable release |

## 4.1.6 Application of Flavourings

Flavourings as such are not intended for direct consumption. Only their appropriate inclusion into food systems reveals their value to impart taste and quality. Experience has shown, that flavourings are strongly influenced by the nature of the food matrix, by food processing and through storage conditions:

- heat evaporates the volatile components, induces Maillard reactions.
- fat absorbs liposoluble flavourings, thus reducing their flavour impact.
- proteins react with aldehydes.
- low pH hydrolyses esters, catalyzes chemical reactions.
- oxygen oxidizes lipids, terpenes etc.

Compounded flavourings are well suited to be adapted to the needs of the users:

- an "unbalanced" flavouring reveals a balanced profile in the processed finished food.
- the replacement of unstable aldehydes by equivalent alcohols results in flavourings with improved stability.
- heat stable butter flavourings contain more lactones, less diacetyl.
- precursor systems constantly deliver volatile flavouring molecules.
- encapsulated flavourings are protected against heat, oxygen, acids or enzymes.
- flavour profiles can be adapted to maximum consumer preference.
- original, fantasy and true to nature flavour profiles can be created.

It is important to consider the fact that the dosage of a flavouring in finished food strongly influences its profile: a low dosage just adds some slightly modifying body notes; a regular dosage imparts the characteristic profile: a high dosage may add some chemical off notes. A careful selection of the flavouring as well as an appropriately adjusted dosage are essential for the success of flavoured food products.

## 4.1.7 Flavouring Production

Compounding flavourings on a production scale is in principle simple: just mix the right components in the right proportion. However modern flavouring factories are high tech operations to handle the customer requests in an efficient manner. Hundreds of flavourings are specifically customer made and therefore deliveries are often prepared against customer request only. Approximately 2000 different raw materials of various consistencies (liquid, viscous, solid, crystalline, powder) are used. They are stored under controlled conditions (cold, ambient, warm). They have varying inherent characteristics: they can be sensitive, reactive, volatile, strong smelling, noxious. They are used from very small quantities up to large volumes, stored in tanks. Quality control has to control measurable and non measurable properties like density, refractive index, component concentrations, odour, taste. Flavouring mixtures contain from a few up to dozens of components. They range from parts per million up to 99%. The produced quantities range from 100 g up to several tons batches. Therefore, modern flavouring factories are highly sophisticated: laser scanners are used to identify products and batch numbers, mixing robots automatically mix small to medium quantities. Dispense lines and dispense clusters, supplemented from tank farms mix medium to large quantities of flavourings, all computer controlled. Nevertheless, people are still working in this high-tech, strong smelling environment: there exist no economic computers nor robots for all critical exceptions: mixing small quantities, solid material, viscous material, badly miscible or soluble components into a large number of products still needs the skill of trained people (see Fig. 4.5-4.12).

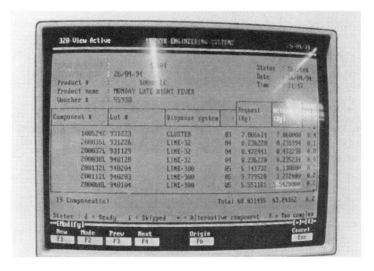

*Fig. 4.5:* Automatic, computer controlled compounding

*Fig. 4.6:* Automatic dosing of flavouring components, mixing tanks for balance

Flavouring Production

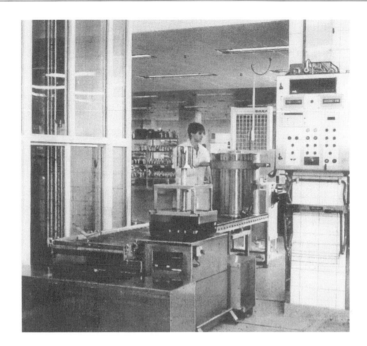

*Fig. 4.7:* Automatic compounding, handling equipment for mixing tanks

*Fig. 4.8:* Raw material storage for automated liquid compounding

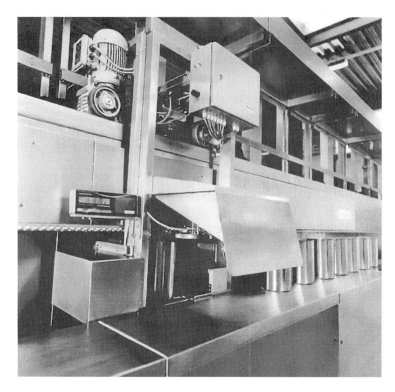

*Fig. 4.9:* Robotic for automated compounding of small batches

*Fig. 4.10:* Cluster of dosing valves for liquid compounding

***Fig. 4.11:*** *Manual compounding of small quantities*

***Fig. 4.12:*** *Flavourist at work*

## 4.1.8 Flavouring Stability

Dealing with flavourings, quality changes in finished products and instability problems with flavourings are often encountered. From our experience we know, that 100% stability is not attainable:

- the intensity is gradually reduced.
- the quality and flavour profile changes before and after application and during the shelf life of the final product.

Whenever we want to understand flavouring stability, we have to know something about:

- the composition of flavourings
- the raw materials

- the characterisation of the flavouring
- how to estimate flavour changes
- the factors responsible for flavour changes
- how to prevent flavour changes.

Traditional flavouring raw materials are produced under rather "harsh" conditions: heat, distillation at high temperatures, concentration, extraction. These result in the destruction of all sensitive substances. Modern flavouring raw materials are produced under controlled and careful conditions: low pressure, low temperature distillation, extraction with low boiling solvents or $CO_2$, careful selection of fresh, high quality raw materials. Therefore sensitive substances survive the production process and influence the quality with all positive (flavour profile) and negative effects (stability).

For an objective description of a flavour we have to identify, evaluate and quantify the sensory properties of the components. The flavour characteristic describes the sensorial behaviour of individual components dependent on the concentration, the threshold value and its odour value.

An objective stability test has to determine whether

- important components disappeared or
- undesirable components with a high odour value developed.

Discussing flavouring stability, we have to distinguish between:

- the stability of the flavouring itself
- its stability in food
- flavour changes during processing
- flavour changes during the storage period.

We distinguish:

*physical stability:*

- evaporation of volatile components
- crystallisation of non soluble material (mainly in liquid flavourings)
- phase separation (in emulsions and washes)
- solubility (in fat-containing food)
- ab- and adsorption effects in complex food systems;

*chemical stability:*

- reactions with food components
- reactions of flavouring components through degradation, rearrangement, oxidation;

*sensory stability:*

- what is the standard sample to compare with (how is it stored?)
- what does the customer expect and remember
- how does the customer evaluate the samples.

The most important factors influencing flavourings are:

- heat treatment (evaporation of volatiles, formation of new flavouring components)
- oxidation (of terpenes, lipids)
- enzyme activity (degradation and formation of flavouring components)
- low pH (acid catalysed reactions)
- fat absorption of liposoluble components
- protein reactions and inclusions.

## 4.2 Fruit Flavours

The flavours of fruits are generally characterized by expressions like:

*juicy:*      high, free water content with balanced, fruit-characteristic, sugar/acid ratio representing the juicy, sweet, sour taste.

*fresh:*      refreshing impression, supported by the acidity and low molecular weight flavouring substances such as acetaldehyde, methyl butyrate, and some green notes like (E)-3-hexenol.

*fruity:*      a general fruity-estery impression of low molecular weight esters like ethyl acetate, ethyl butyrate, butyl acetate, 3-methyl butyl acetate, methyl caproate or ethyl caproate.

*green:*      the fresh, unripe impression of immature fruits represented by (E)-2-hexenal, (E)-2-hexenol, (E)-2-hexenyl acetate, hexanal, often combined with high acidity and low sugar content.

*ripe fruity:*      during the maturation process, the flavour reaches its optimum concentration and balance, often combined with a high sugar/low acid ratio. Ethyl-3-methyl butyrate, 3-methylbutyl butyrate, ethyl caproate, ethyl-3-methyl but-2-enoate, methyl cinnamate are typical examples of this general impression. The ripe, fruity character of fruits is supported by traces of the fruity, caramelic notes of 2,5-dimethyl-4-hydroxy-furan-3(2H)-one, 2,5-dimethyl-4-methoxy furan-3(2H)-one and maltol.

*sweet, creamy:*      a soft sweet impression of vanillin or maltol.

*floral:*      often understood as a non-characteristic general odour note of flowers, even with a slight perfumistic background. Traces of alpha-ionone, beta-ionone, damascenone, geraniol or phenyl ethyl alcohol may be responsible for this character.

*floral, fruity, sweet:*      a general note of flowers with a spicy, food-appealing background represented by linalool, alpha-terpineol

*spicy:*      a general impression of traditional spices like clove, cinnamon, eugenol, cinnamicaldehyde, 3-phenyl propyl acetate,

| | |
|---|---|
| *pip, stone note:* | ethyl cinnamate. But "spicy" is also used in connection with savoury, HVP-like, lovage-like impressions.
the astringent, nutty, woody almond note of the broken seeds may be represented by the bitter almond note of benzaldehyde, the woody character of damascone or the green, nutty note of trimethyl pyrazine. |
| *overripe, fermented:* | the impression of overripe fruits in the first phase of degradation and fermentation. The balance of the flavour is distorted, high concentrations of ethyl esters, mainly ethyl butyrate, ethyl caproate, ethyl octanoate, ethyl decanoate are combined with traces of sulphur components like methyl mercaptane, methyl thioacetate, methyl thiobutyrate. |
| *canned, preserved:* | the canned, preserved note develops during the sterilization process and the following storage time. The most important effects are the loss of the fresh, fruity, typical aldehydic and estery notes and the formation of Maillard reaction products like 5-hydroxy methyl furfural. The sulphur components hydrogen sulphide, methyl mercaptane support the fusty character. A metallic off-taste is often produced by the lipid degradation product (Z)-1,5-octadien-3-one. |
| *cooked:* | during cooking of fruits, most of the volatile esters, aldehydes, alcohols and hydrocarbons are lost by a water vapour distillation, thus losing all the fresh fruity character. The basic cooked note is supported by the acid catalyzed formation of caramel like Maillard reaction products from fruit sugars: caramel notes of 2,5-dimethyl-4-hydroxy-furan-3(2H)-one, 2,5-dimethyl-4-methoxy-furan-3(2H)-one, 5-hydroxy-methyl-furfural, 3-hydroxy-2-methyl-4-pyrone. |

**Character Impact Compounds (CIC):**

All fruit flavourings are fruit-typically characterized by some key ingredients to distinguish them from each other (see the detailed description of the individual fruits).

## 4.2.1 The Flavour of Natural Fruits

Fruits are the main player in the world of flavours: A vast number of fruits grow in nature and are planted throughout the world. They are consumed not only for their nutritional and health value, but mainly for their highly esteemed flavour and taste. More and more fruits are commercialized worldwide from all continents. People experience new tastes and flavours when travelling abroad. Thus processed fruits are used in many industrial food products as main ingredient or as flavour modifier.

## 4.2.2 World of Commercially Important Fruit Flavours

*Table 4.6: World production of fruits*

| Fruits | 360 [in mio tons] |
|---|---|
| Grapes | 60 |
| Oranges | 55 |
| Bananas | 48 |
| Apples | 40 |
| Pineapples | 10 |
| Peaches | 9 |
| Pears | 9 |
| Mandarins, tangerines | 8 |
| Lemons | 7 |
| Plums | 6 |
| Coffee | 6 |
| Grapefruits | 5 |
| Cocoa | 2 |

**Almond** *(Prunus amygdalus, Rosaceae)*
(G: Mandel, F: amande)

The commercial almond is the pit of the peach-like fruit, not consumed as such. The raw almond has a nutty, fatty, slightly green and benzaldehyde character. Almonds are generally roasted to get the typical roasted, nutty, popcorn, slight caramel character with a very low benzaldehyde note. The well known bitter almond oil is isolated from fresh raw bitter almonds and consists mainly of pure benzaldehyde.

*CIC:* Non-enzymatic-browning compounds: 2,5- and 2,6-dimethyl pyrazines, trimethyl pyrazine impart the fresh, green, nutty character, 2-methyl-6,7-dihydro-5H-cyclo-penta-pyrazine is responsible for the roasted-nutty note and 2,5-dimethyl-4-hydroxy-furan-3(2H)-one supports the caramel like, sweet overall impression.

Most commercial almond flavourings contain a high amount of benzaldehyde, which represents more the fresh bitteralmond type.

**Apple** *(Malus sylvestris var. domestica, Rosaceae)*
(G: Apfel, F: pomme, S: manzana, I:pomo, mela)

Due to its high commercial value the flavour of apples belongs to the best known and documented natural flavours. Many hundreds of varieties have been bred and are produced throughout the world. Therefore, each person has its preferred and "typical" apple flavour: a juicy, refreshing, sweet-acid, watery taste is combined with fruity, estery, green notes, supported by species characterizing sweet, floral impressions.

*CIC:* (E)-2-hexenal, (E)-2-hexenol, (E)-2-hexenyl acetate and hexanal are responsible for the fresh, green-fruity basic flavour. Ethyl-2-methyl butyrate, hexyl acetate support the fruity-estery note and additional compounds like 3-methyl butyl acetate, hexyl-2-methyl butyrate, damascenone, linalool impart the specific species character. Benzaldehyde intensifies the pip note.

**Apricot** *(Prunus armeniaca, Rosaceae)*
  (G: Aprikose, Marille, F: abricot, S: albaricoque)

The taste of apricots is sour-sweet with a fresh, fruity, perfumy, sweet, creamy odour. Dried apricots have lost some of the fruity, perfumy odour and are rather sour.

*CIC:* Linalool, phenyl ethanol and traces of damascenone impart the floral, perfumy note, 4-decanolide adds the sweet, heavy lactone character and (E)-2-hexenal, (E)-2-hexenyl acetate impart the fruity, fresh, green topnote.

**Banana** *(Musa sapientum, M.paradisiaca, M. acuminata, M. balbisiana and hybrids thereof, Musaceae)*
  (G: Banane, F: banane, S: platano)

Bananas are the most important single fruit for direct consumption. Over 200 types are produced with different tastes and uses. We can distinguish 2 main uses: direct consumption of ripe, fruity, sweet species and cooking varieties of the more bland, mealy, slightly astringent bananas rich in starch. Bananas have to be artificially ripened near the consumer as ripe bananas are not stable for transportation. The flavour develops from a bland, slightly sour, astringent character over a fruity, fresh, sweet typical estery ripe banana note dominated by 3-methyl butyl acetate to an overripe sweet, creamy, spicy, full-flavoured estery character.

*CIC:* 3-methyl butyl acetate represents the fresh, estery, fruity-sweet character reminiscent of technical solvents. Eugenol is responsible for the spicy part of ripe and overripe bananas. A series of volatile esters (3-methyl butyl butyrate, 3-methyl butyl-3-methyl butyrate, 4-hepten-2-yl acetate) round off the general fruity note and vanillin adds the sweet, creamy part.

**Black Currant** *(Ribes nigrum, Saxifragaceae)*
  (G: Schwarze Johannisbeere, F: Cassis, S: Groselheira negra)

Black currants are of interest mainly in France and the United Kingdom due to their strong, special flavour: a sour, fruity taste with a catty, sulphurous fresh odour.

*CIC:* Ethyl butyrate contribute to the fruity, estery note; linalool, alpha-terpineol, citronellol and damascenone support the floral, fruity ripe character and 1,8-cineol imparts the freshness. 4-methoxy-2-methyl-2-mercapto butane is responsible for typical catty sulphurous black currant note. Extracts of the black currant buds are more green, herbaceous but they also contain the sulphurous CIC. A similar note, 8-mercapto-p-menthan-3-one has been identified in buchu oil and is often used to imitate the catty black currant aspect.

**Cherry** *(Prunus avium, sweet cherry and Prunus cerasus, sour cherry, Rosaceae)*
  (G: Kirsche, Morellen, F: cerises, S: cereza, guinda, I: ciliega)

The basic varieties, sweet and sour cherry, have been crossed to over 600 types. The taste of the stone fruits is acid, sweet and slightly astringent. The balance depends on the ripeness and especially on the species. The flavour is fresh, fruity, green, floral and slightly spicy and develops a strong bitter almond, benzaldehyde character when crushed.

*CIC:* (E)-2-hexenal, hexanal, (Z)-3-hexenol are responsible for the fresh, green impact, (E,Z)-2,6-nonadienal and phenyl acetaldehyde add a full heavy green, fatty character. Linalool, geraniol and damascenone bring along the floral aspect, 3-methyl butyric acid the fruity and eugenol the spicy note. Tannic acids are responsible for the astringent mouthfeel. The very typical black cherry note, benzaldehyde, is overdosed in most cherry flavourings. The art to create a good cherry flavouring is to use the highest possible dosage of benzaldehyde and surround it with other flavouring materials to get the least effect of benzaldehyde.

**Citrus Fruits** *(see also chapter 3.2.2.2)*
(G: Agrumen, F: hesperidees)

Citrus fruits are among the most important fruit crops. Due to the high cross breeding activities of agricultural biologists, a clear identification back to the original species is very difficult or almost impossible.

Their flavour is the most popular flavour in beverages. It is well appreciated due to its refreshing, sour, sweet taste and its refreshing clean flavour. The low cost of the raw material and its high flavour value are other important success factors. The citrus peel is rich in essential oils, which are recovered mainly for flavouring purposes.

**(1) Orange** *(Citrus sinensis, Rutaceae),* see Tab. *3.22 for composition*
(G: Orange, Apfelsine, F: orange, S: naranja, I: arancia)

The taste of orange juice is refreshing, sour, sweet with a fruity, juicy flavour. Some types of oranges may develop a lasting bitterness. It is often overpowered by a contamination by the peel oil. This oil has a strong terpeny, aldehydic, non fruity or juicy character. The taste of blood oranges is more tart with a slight berry, floral note.

*CIC:* Acetaldehyde, although weak in smell and taste, is an important contributor to juiciness and freshness. Ethyl butyrate and (E)-2-hexenol add the fruity, green note and alpha-pinene, octanal and decanal are responsible for the green peely, aldehydic orange note.

**(2) Lemon** *(Citrus limon, Rutaceae),* see Tab. *3.18 for composition*
(G: Zitrone, F: citron, S: limon, I:limone)

We have to distinguish essentially between two different parts: – the lemon juice and the lemon peel. The juice has a very strong acidic taste from the high citric acid content (50 g/l) with a refreshing juicy aroma, far away from the peel oil.

*CIC:* The weak juicy flavour is based on geraniol, geranyl acetate and neryl acetate in combination with linalyl ethyl ether and myrcenyl ethyl ether. The cold pressed peel oil has a high terpene content. Its flavour is dominated by citral, gamma-terpinene and alpha-pinene. Citral is recognized as a very important contributor to the fresh lemon character, although it decomposes within a few weeks in acidic soft drinks.

**(3) Lime** *(Citrus aurantifolia, Rutaceae), see Tab. 3.19 for composition*
(G: Limette, F: citron vert, S: lima, limon)

The flavour of limes is used in 3 different forms:

a) as juice to impart a very special acidic juicy floral character to alcoholic beverages and soft drinks.

b) as cold pressed peel oil to increase freshness and stability in lemon-lime soft drinks.

c) as distilled lime oil, mainly used in soft drinks of the cola or lime type.

*CIC:* The lemon like cold pressed lime oil is characterized by a relatively high citral content, balanced with beta-pinene, gamma-terpinene and neryl acetate. The distilled oil is a mixture of acid catalyzed breakdown products of terpenes: 1,4-cineole, 1,8-cineole and alpha-terpineol are the backbones of this oil, rounded off with gamma-terpinene, other terpene hydrocarbons and alcohols.

**(4) Grapefruit** *(Citrus paradisi, Rutaceae), see Tab. 3.17 for composition*
(G: Pampelmuse, F: pamplemousse, S: toronja)

The taste of grapefruit juice is acid, fruity, bitter sweet. The peel oil has a characteristic terpeny, woody, exotic fruity character.

*CIC:* The bitter taste originates from naringin. Nootkatone imparts the characteristic fresh woody odour. Acetaldehyde and ethyl butyrate improve the juicy note and 1-p-menthane-8-thiol is responsible for the typical exotic grapefruit character.

**(5) Mandarin, Tangerine** *(Citrus reticulata, Rutaceae), see Tab. 3.20 for composition*
(G: Mandarine, F: mandarin, S: naranja-cravo)

The taste of freshly consumed mandarins is very juicy, sweet fresh with the characteristic heavy aromatic aldehydic flavour. The processed juice is not stable. Its flavour turns into an unpleasant phenolic "skunky" off flavour. The peel oil keeps its original aromatic sweet odour and is often used to modify orange flavourings into sweeter, more aromatic products.

*CIC:* The basis of the heavy aromatic flavour is methyl-N-methyl anthranilate in combination with gamma-terpinene. The orange citrus note results from the presence of octanal, decanal and alpha-sinensal.

**Coconut, Macapuno** *(Cocos nucifera)*
(G: Kokosnuss, F: noix de coco)

Fresh coconut milk has a sweet juicy, fruity, slightly creamy, fatty flavour. Roasted coconut meat develops a nice creamy, nutty characteristic odour and taste.

*CIC:* The typical creamy-fatty note is produced by delta-octalactone and delta-decalactone, balanced with free fatty acids C8-C12 and 2-nonanone. The flavouring industry mainly uses gamma-nonalactone (the so called "coconut lactone") to impart the typical coconut flavour. The roasted coconut meat contains in addition pyrazines.

**Grape, Raisin** *(Vitis spez., Vitaceae)*
(G: Traube, F: raisin, S: uve)

Grapes are the most important fruits produced, but 90 % are processed to wine. A wide range of species and hybrids exist, optimized for good wine production. The consumed fruits can be separated into two broad groups:

- the white Muscat type with a very sweet floral fruity character
- the blue concord type with a succulent fruity aromatic aroma, more or less dominated by a typical catty aromatic topnote.

*CIC:* The basic Muscat note is produced from linalool, balanced with further terpene alcohols, geraniol and linalool oxides. The catty odour of the concord grape is based on methyl anthranilate, combined with ethyl-3-mercapto propionate, benzaldehyde adds the touch of a seedy character.

**Hazelnut, Filbert** *(Corylus avellana)*
(G: Haselnuss, F: noisette)

The taste of fresh filberts is somewhat fatty, metallic nutty and is strongly improved by the roasting to become nutty roasted.

*CIC:* The flavour of freshly roasted filberts, is a balanced composition of alkyl pyrazines and aldehydes, derived from the Maillard reaction in combination with 4-methyl-5-hepten-2-one.

**Peach** *(Prunus persica, Rosaceae)*
(G: Pfirsich, F: peche, S: melocoton, I: pesca)

The taste of ripe peaches is dominated by the sweet succulent juice with its very aromatic fruity and fresh aroma. The heavy fruity, fatty, typical peachy note, also found in plums, nectarines, apricots is called "lactony", and it is derived from gamma-lactones.

*CIC:* The typical "lactony" note is produced by 4-decanolide. The flavour industry since years uses mainly the 4-undecanolide, the so called "peach lactone". The lactone has to be balanced with fruity esters like 3-methyl butyl acetate, the floral-fruity note of linalool and some green notes like (E)-2-hexenal.

**Pear** *(Pyrus communis, Rosaceae)*
(G: Birne, F: poire, S: pera, I: pera)

The sweet juicy, fruity taste of pears is accompanied by a typical sweet, fruity, fatty pear note.

*CIC:* The typical fruity, fatty pear note originates from ethyl-(E,Z)-2,4-decadienoate balanced with ethyl-(E)-2-octenoate and ethyl-(Z)-4-decenoate. Farnesene adds a fresh floral terpeny topnote and esters, mainly hexyl acetate, contribute to the fruity, estery character.

**Pineapple** *(Ananas comosus, Bromeliaceae)*
 (G: Ananas, F: ananas)

Ripe, fresh pineapple has a sweet, acid, slightly biting taste with a fragrant, sweet, fruity, fresh aroma combined with a fruity caramel like aftertaste.

*CIC:* Methyl hexanoate represents the fresh, fruity character, ethyl-3-(methylthio)propionate imparts an overripe impression, 1-(E,Z)-3,5-undecatrien adds a fresh green, biting topnote and 2,5-dimethyl-4-hydroxy-furan-3(2H)-one is responsible for the ripe, sugary, fruity, caramelic lasting aftertaste. Allyl hexanoate is often used in compounded pineapple flavourings.

**Raspberry** *(Rubus idaeus, Rosaceae)*
 (G: Himbeere, F: framboise, S: farmbuesa, I: lampone)

Common cultivated raspberries often have nice shape and colour, but their taste is just watery and acidic. Fragrant raspberries develop a delicious fresh, fruity, green, floral, violet like perfume with some seedy, woody background. Ripe raspberries are sweet and very juicy.

*CIC:* The main flavouring components are alpha-and beta-ionone with their floral, violet and typical perfumy, raspberry, woody character. 1-(4-hydroxyphenyl)butan-2-one imparts the fruity, sweet raspberry body. (Z)-3-hexenal is responsible for the fresh, green topnote and 2,5-dimethyl-4 hydroxy-furan-3(2H)-one adds the overripe, almost cooked fruit jammy body.

**Strawberry** *(Fragaria spez., Rosaceae)*
 (G: Erdbeere, F: fraise, S: fresa, I: fragole)

Strawberry varieties are subject to a wide variety of breeding programs due to their commercial importance. Strawberries are world-wide appreciated by their characteristic flavour, but industrial processing and international transports call for colourful, firm, large, stable strawberries. In most cases the delicious flavour suffered. Thus today many consumers know only the dull, acid, green, forced ripened but well shaped, colourful strawberries from the supermarket. They are not aware of the fragrant, sweet, fruity, aromatic, juicy flavour of wild or garden ripened strawberries. Strawberries, like many other fruits, show a very active flavour development: daily changes of flavour concentrations as well as very fast development of new components after cell disruption have been observed.

*CIC:* The basic flavour complex of strawberries is built of 2,5-dimethyl-4-hydroxy-furan-3(2H)-one and 2,5-dimethyl-4-methoxy-furan-3(2H)-one. Both impart the ripe, fruity, caramel, cooked character together with ethyl hexanoate, the fresh fruity, estery note. (E)-2-hexenal and (E)-2-hexenyl acetate are responsible for the fresh, green impression. 2-Methyl butanoic acid leads to a refreshing fruity acidity, S-methyl butanthioate and 4-hydroxy decanoic acid lactone improve the overripe character; linalool characterizes the fruity, floral note of the variety Senga Sengana and methyl anthranilate turns a garden strawberry into a wild strawberry. The flavouring industry uses in addition to these strawberry components other synthetic materials like maltol, ethylmaltol, and ethyl-2-methyl phenyl-glycidate, the so-called "strawberry aldehyde".

## 4.2.3 Sensorially Interesting Fruit Flavours

In the world of fruits many other sensorially interesting fruits exist. Most of them are known only locally, and they are unknown to the consumers in the large markets. Especially many tropical fruits have never been exploited for their commercial value and they run the risk to disappear from the biosphere. Some have found an interest in home made products:

*Blueberry, Billberry (Vaccinium myrtillus), Cranberry (Vaccinium macrocarpon) and Gooseberry (Ribes grossularia)* with their fresh, sour, green, astringent flavour are now produced in larger quantities.

*Naranjilla, Lulo (Solanum quitoense)* with its fantastic sour taste with a touch of apricot, an exotic note of ethyl benzoate. Its juice has a nice green colour when freshly pressed. Unfortunately it has a very delicate skin when ripe and thus it cannot be transported over long distances. Moreover it does not properly ripen after harvesting.

*Jabuticaba (Myrciaria cauliflora)* is a cherry-like fruit from Brazil with a gelatinous flesh and a very sweet, sour taste and a flavour in the direction of concord grape, black currant, raspberry and vanilla.

*Pomegranate, grenadine (Punica granatum)* contains in its kernels a very refreshing juice with a high tannin content (which may cause strong discolouring on cloth). The taste is sour/sweet with a fruity, berry-like aroma, reminiscent of raspberry and red currant with a touch of an earthy, green bell pepper note. The taste of "grenadine syrup" in Europe is a mixture of sweet raspberry with vanillin, quite different from the original fruit and resembles a little bit the "American Cream".

*Sapotilla, Chico (Achras sapota)* is the fruit of the chicle tree, the source of the natural chewing gum base. The fruit has a very sweet taste with a pear-like, soft structure. The flavour is reminiscent of a mixture of pear with hazelnut.

### Blackberry, Brambles *(Rubus fructicosus, Rosaceae)*
(G: Brombeere, F: mure, S: mora, I: mora)

Ripe, wild blackberries have a strong, sweet, heavy fruity, floral musky and sugary flavour.

CIC: The key components are 2,5-dimethyl-4-methoxy-furan-3(2H)-one with its sugary sweet fruity aroma combined with aromatic fruity p-cymen-8-ol. The floral note is supported by dihydroactinidiolide, 14-cyclotetradecanolide and 16-cyclohexadecanolide.

### Cape Gooseberry *(Physalis peruviana, Solanaceae)*
(G: Kapstachelbeere, F: Andenbeere, S: uvilla, membrillo)

The taste of this decorative fruit is very acid, refreshing, fruity and juicy. The flavour is aldehydic fatty, especially on the skin: fine droplets of a strongly flavoured oil develop when ripe.

*CIC:* The fatty aldehydes octanal, nonanal, decanal together with the corresponding fatty acids and their ethyl esters are responsible for the characteristic fatty aldehydic flavour of the skin. (E)-2-hexenal, hexanal and (E)-2-pentenal contribute to the refreshing character, supported by the high acidity of the juice. Methyl-2-methyl butanoate, methyl-3-methyl butanoate and methyl hexanoate impart a fresh, fruity note, rounded off with traces of methyl benzoate, methyl salicylate and the sweet floral note of beta-ionone and damascenone.

**Custard Apple, Cherimoya** *(Annona cherimola, Annonaceae)*
(G: Cherimoya, S: Chirimorrinon)

This delicious fruit started to become available in the northern hemisphere during the past 20 years. Its sweet, creamy, fruity taste is reminiscent of fresh cream with strawberries, pineapple and a floral aromatic exotic background in the direction of ylang-ylang, a flower of the same annona family.

*CIC:* Fruity esters like 3-methylbutyl acetate, 3-methylbutyl-3-methyl butanoate, ethyl-3-methyl butanoate represent the fruity estery character. Linalool and its derivatives, the linalool oxides add a fruity floral note; 2,5-dimethyl-4-methoxy-furan-3(2H)-one, gamma-nonalactone together with the high sugar and low acid content are responsible for the sweet, creamy taste and methyl-2-hexenoate in combination with benzyl esters (acetate, 2-methyl propanoate, butyrate) impart the typical exotic Cherimoya character.

**Soursop** *(Annona muricata, Annonaceae)*
(G: Stachelannone, Sauersack, F: corossol, S: Guanabana)

This tropical fruit, rich in juice and seeds, a native of South America, has a very refreshing sour sweet, juicy taste with a specific green, fruity powerful exotic fresh flavour reminiscent of pineapple. The background is spicy floral perfumistic in the direction of ylang-ylang with cinnamon.

*CIC:* The typical basic exotic green note is produced by methyl-2-hexenoate and methyl hexanoate. Cinnamyl alcohol and methyl cinnamate add the spicy cinnamon note and linalool together with benzyl acetate form the basic floral, perfumistic background.

**Sweetsop, Sugar Apple** *(Annona squamosa, Annonaceae)*
(G:Schuppenannone, F:pomme canelle, S:Anon, P:Fruta do conde)

This tropical fruit from Brazil is difficult to be eaten due to its granular surface. But the creamy pulp has a delicious, very sweet, strawberry, cream, slightly sweet-floral, weak ylang-ylang flavour. The fruity estery character is considerably weaker than in the cherimoyas.

*CIC:* 2,5-Dimethyl-4-methoxy-furan-3(2H)-one and gamma-nonalactone form the sweet, creamy heart of the profile, surrounded with esters like ethyl caproate, ethyl butyrate, ethyl-2-methyl butyrate and floral, fruity notes like linalool and benzyl acetate.

## Cloudberry *(Rubus chaemamorus, Rosaceae)*
(G: Multebeere)

Cloudberries have the same shape as raspberries, but a different, yellow-orange colour. The taste is completely different: no fresh, fruity and violet character, but sour, fruity, juicy, aromatic spicy in the direction of cinnamon, clove.

*CIC:* p-Cymen-8-ol, 2-heptanol and myrthenol impart the basic blackberry note, modified by floral fruity notes of linalool, 2-phenyl ethanol, geraniol and characterized by the aromatic, medicinal note of methyl benzoate embedded in the spicy cinnamyl alcohol, methyl cinnamate and 4-vinyl phenol.

## Durian *(Durio zibethinus, Bombacaceae)*
(G: Stinkfrucht)

Nicknames like "stinkfruit", "the king of fruits" and the separation of people into "durian caters" and "durian haters" clearly shows the polar character of this special fruit. Descriptions like: old cheese, rotten onion, rotten egg, turpentine, creamy, buttery, custard, vanilla cream, brown sherry, strawberry cream, hazelnut cream support its mysterious taste. It is quite obvious that ripe durians produce such a bestial stench that its transport in public transport systems is specially regulated. People unfamiliar with durians avoid the contact with them. Whoever dares to eat a spoon of the yellow, soft, creamy mass surrounding the large seeds, is surprised by its mild, delicious, sweet, creamy taste reminiscent of a combination of strawberries with full fat cream, vanilla cream, a few hazelnuts and a slight topping of fresh, ripe onions.

*CIC:* The main components, responsible for the repulsive odour are 1-propanethiol, ethanthiol and methanethiol, and the corresponding sulfides. Ethyl butyrate and ethyl-2-methyl butyrate impart the fruity, fresh strawberry character. Components like vanillin, 2,5-dimethyl-4-hydroxy-furan-3(2H)-one, 3-methyl-cyclopent-2-en-2-ol-1-one, gamma-decalactone (all not yet identified in Durian) may support the creamy nutty character.

## Guava *(Psidium guajava, Myrtaceae)*(G: Guave, F: goyabe, S: guayaba)
(G: Guave, F: goyabe, S: guayaba)

Ripe guavas develop such a powerful odour that a whole room can be perfumed by their smell. The taste varies according to its degree of ripeness: from sour, green, harsh, over fresh, fruity pineapple and pear like to spicy, cinnamon, creamy, quince-like with some astringent aspects.

*CIC:* The powerful odour is dominated by 2-isobutyl thiazole and 3-pentanethiol. The green notes are lipid degradation products like (E)-2-hexenal, hexanal and higher unsaturated aldehydes, the pineapple-pear like fruity notes are derived from methyl hexanoate, ethyl-2-hexenoate and hexyl acetate. The spicy cinnamon notes are represented by 3-phenyl propyl acetate, cinnamyl acetate, methyl cinnamate, ethyl cinnamate and cinnamaldehyde. Gamma-decalactone and 2,5-dimethyl-4-hydroxy-furan-3(2H)-one and 3-hydroxy-2-butanone add the sweet, creamy body. Beta-farnesene, citronellol, 2-phenylethanol, beta-ionone add the sweet, floral, quincelike part and methyl benzoate and ethyl benzoate impart a characteristic medicinal, exotic topnote.

**Lychee,** *(Litchi chinensis, Nephelium lappaceum, Sapindaceae)*
 (G: Litschi, S: mamoncillo chino)

The fruits has the size of a small plum and a skin like a red speckled egg shell. It contains a very succulent, juicy, watery pulp with a sweet, perfumistic, rosy, fruity taste.

*CIC:* 2-Phenylethanol, citronellol, geraniol and linalool are responsible for the floral rosy background; roseoxide adds an exotic floral rosy topnote. 3-Methyl butyl acetate, 2-methyl-2-buten-1-ol, 1-ethoxy-3-methylbut-2-en and menthol add the fresh, fruity character.

**Kiwi, Chinese Gooseberry** *(Actinida chinensis, Actinidaceae)*
 (G, F, S, I: Kiwi)

The taste of fresh kiwi fruit is refreshing with a very clean acid juice. Its aroma is green like a fresh cut apple with fresh strawberries and gooseberries. Overripe, stale or cooked kiwis have lost their refreshing green, acid character and develop a dull, fatty, estery fruity, non characteristic flavour.

*CIC:* Vitamin C (300 mg/100 g fruit), citric acid and malic acid are responsible for the clean acidic taste of the fresh fruit. A high enzyme activity of a lipoxygenase produces high amounts of (E)-2-hexenal, hexanal and (E)-2-hexenol, all fruity green notes, characteristic of green apples. The cut surfaces develop at prolonged standing 2-heptenal, 2,4-nonadienal and 1-octen-3-one, responsible for stale, fatty odour. Methyl butyrate and ethyl butyrate impart the ripe, fruity, estery, fresh, juicy impression.

**Mango (Mangifera indica, Anacardiaceae)**
 (G: Mango, F: mangue)

To describe the flavour of a mango is a rather difficult task. Since the mango tree is one of the oldest cultivated trees, many different cultivars with different tastes and flavours exist. The taste may vary from very sweet, pulpy like ripe canned peaches to very sour. The consistency of the pulp varies from soft, creamy, buttery to very fibrous. The flavour in general is characteristic sweet, fruity, creamy, floral with a canned peach character, topped with a more or less accentual terpeny, green, resinous note. This terpeny note is a characteristic difference of cultivars: Indian Alphonso mangoes have a weak fresh sweet citrus green terpeny character with a full, sweet, peachy overall taste. The Carabao mangoes of the Philippines are characterized by their fresh, green, herbaceous, parsley-like terpene topnote, combined with a full, sweet exotic fruity apricot like body. Mangoes from Sri Lanka and Malaysia are characterized by their green, resinous, pine needle-like terpeny topnote and a more fibrous pulp. Due to the well accepted fantastic taste of fresh mangoes and their nutritious value as a fresh food, the cultivars are crossed and spread all over the tropical and subtropical zones.

*CIC:* Three basic cultivar types can be distinguished chemically:

- the ocimen type: (Z)-ocimen in combination with beta-caryophyllene impart the sweet, citrus, woody terpene notes like the Alphonso cultivar.

- the carene-type: delta-3-carene with beta-caryophyllene impart the green, woody, herbaceous, parsley-like terpene notes like in the Carabao type.
- the terpinolene type: alpha-terpinolene and beta-selinene impart the green resinous pineneedle-like terpene notes like in the Malaysian mangoes.

These terpene notes are supported by (Z)-3-hexenol, (E)-2-hexenal, (E)-2-hexenol with their fresh, green character. Gamma- and delta-lactones (4-decanolide, 5-decanolide, 4-dodecanolide, 5-dodecanolide-(Z)-7-decen-5-olide) impart the sweet, creamy, buttery, peach and apricot character. 2,5-Dimethyl-4-hydroxy-furan-3(2H)-one and 2,5-dimethyl-4-methoxy-furan-3(2H)-one are responsible for the sweet creamy fruity body. A bouquet of esters imparts the overall fruity character (mainly esters of ethyl-, (Z)-3-hexenyl and butyl alcohol with acetic-, butanoic-, 2-butenoic, 3-hydroxybutanoic- and hexanoic acid).

Alcohols like 3-methyl-2-buten-1-ol, citronellol, nerol, hotrienol, linalool and 2-phenylethanol impart a fresh floral topnote to the whole mango flavour. Artificial mango flavourings may contain diphenyl oxide or styrallyl acetate.

## Melon *(Cucumis melo, Cucurbitaceae)*, Muskmelon, Sugarmelon Watermelon *(Citrullus lanatus, Cucurbitaceae)*
(G: Melone, F: melon, S: melon)

The taste of the muskmelon is very sweet creamy with a strong, sweet, floral fruity-sulphurous and typical melon-like, fatty, green aroma. The taste of watermelon is very watery, juicy, sweet with a weak green, fatty, cucumber-like aroma.

*CIC:* In both melon types the lipid degradation products (Z)-6-nonenol, (Z,Z)-3,6-nonadienol and the corresponding aldehydes are responsible for the typical green, fatty, cucumber melon aspect. Ethyl propionate imparts an overripe character to muskmelon flavour, supported by the sweet, caramelic aspect of 2-methyl-5-ethyl-4-hydroxy-furan-3(2H)-one and the fruity-sulphurous aroma of S-methyl thioacetate and methyl thiobutyrate.

## Papaya, Paw Paw *(Carica papaya, Caricaceae)*
(G: Papaya, F: papaye)

Papayas are used in three different ways:
- unripe, green fruits, as well as the leafs and the stem produce a papain rich latex on scratching. Papain, a proteolytic enzyme, is used in pharmaceuticals, tanning and as meat tenderizer.
- unripe, cooked as vegetable like cucumbers, pumpkins
- small, ripe fruits for fresh consumption.

The small, ripe papayas develop a strong sweet, fruity, floral, spicy aroma. The overripe fruit cannot be stored and develops an unpleasant butyric, sweaty, rotten odour. The numerous seeds have a strong unpleasant radish smell and a bitter burning taste. They are not consumed.

*CIC:* Linalool is responsible for the basic fruity, floral topnote. Methyl-(methylthio)-acetate and butyric acid impart the butyric, rotten character and benzyl isocyanate is the key component for the radish smell of the seeds.

**Passion Fruit** *(Passiflora edulis, Passifloraceae)*
(G: Passionsfrucht, F: fruit de passion, S: maracuja)

From over 400 species of passion fruit, two became commercially important crops: the purple passion fruit, *Passiflora edulis, var. purpurea* and the yellow passion fruit or maracuja, *Passiflora edulis, var. flavicarpa*. Both are well recognized by their delicious, acid, gelatinous, juicy pulp with its characteristic, exotic, sharp, green, fruity, sulphurous and floral, fruity note, whereas the yellow variety is fresher, greener with a stronger sulfury note.

*CIC:* Green, fruity, fatty esters like (Z)-3-hexenyl butyrate, (Z)-3-hexenyl hexanoate, 2-ethyl hexenoate, 2-ethyl-octenoate form the basic fruity body. Nerol oxide and edulan add an ethereal, fresh, green, sharp, floral topnote. The floral note (mainly in the purple variety) is represented by linalool with a complex of ionone derivatives, mainly beta-ionone, dihydro-beta-ionone, theaspirone, damascenone. The key components for the green, exotic, sulfury topnote of the yellow variety are 2-methyl-4-propyl-1,3-oxathiane and 3-methylthio-1-hexanol.

**Plum** *(Prunus domestica, Rosaceae)*
(G: Pflaume, F: prune, S: circuela, I: prugna)

Although there are many different cultivars of plums, we can describe their flavour in general as very sweet to green, acid with a pronounced heavy, sweet, floral, "lactony", spicy, fruity aroma.

*CIC:* The green note is represented by (Z)-3-hexenol, (E)-2-hexenol, (E)-2-hexenal. 4-Decanolide contributes to the heavy, "lactony" character. The heavy floral note is the result of a balanced mixture of linalool, 2-phenylethanol, benzyl acetate. The spicy background, mainly from cooked plums, results from cinnamaldehyde, eugenol, methyl cinnamate with some stone notes of benzaldehyde. The fruity, fatty body is imparted by (Z)-3-hexenyl-2-methyl butyrate and ethyl nonanoate.

**Prickly Pear** *(Opuntia ficus indica, Cactaceae)*
(G: Kaktusfeige, F: figue d inde, S: chumbo, tuna)

The flavour of a prickly pear is reminiscent of a combination of an unripe, granular pear with a floral melon.

*CIC:* (Z)-3-hexenol and (E)-2-hexenal contribute to the unripe, green character; heptyl acetate and ethyl 2,4-decadienoate impart the fruity pear note and 2,6-nonadien-1-ol adds the melon-like topnote.

**Quince** *(Cydonia oblonga, Rosaceae)*
(G: Quitte, F: coing, S: membrillo, I: cotogna)

Fresh, raw quinces cannot be consumed as such due to their very hard, tough texture. But a tree full of ripe quinces produces a powerful, lovely floral, fruity perfume, which is destroyed during cooking.

*CIC:* The floral fruity perfume is the result of a balanced mixture of farnesene, linalool, ethyl-2-methyl-2-butenoate, ethyl (E)-2-(Z)-4-decadienoate, derivatives around beta-ionone (theaspirone, theaspirane (E,E)-6,8-megastigmadien-4-one) and

Processed Flavourings

the quince specific components marmeloxide, quince oxepane: 4-methyl-2-(3-methyl-1,(E)-3-butadienyl) oxepane.

**Starfruit** *(Averrhoa carambola, Oxalidaceae)*
(G: Karambole, F: carambole)

Cut starfruits are very decorative but their taste and flavour is disappointing: a juicy harsh, sour, earthy, medicinal, fruity aroma.

*CIC:* The basic acid taste results from high levels of oxalic and succinic acid. (Z)-3-hexenol and (E)-2-hexenal contribute to the unripe, green aspect and ethyl benzoate with methyl-2-hydroxybenzoate are the typical medicinal notes. Quinoline and ethyl nicotinate add the earthy background.

## 4.3 Other Blended Flavourings

### 4.3.1 Processed Flavourings, *see also chap. 3.2.3*

Processed flavourings play an important role in the flavour industry, as they are an essential part of human food consumption: all heated, cooked, baked, roasted food develop their typical flavour via processing through the Maillard reaction, a reaction path as complex as biochemical pathways. Therefore the flavour industry uses many building blocs of processed flavourings of model systems to create cooked, baked or roasted flavourings. Many of the reaction products of roasted flavourings have a high flavour value, but a short shelf life. Subsequently processed model flavourings are used in their basic form to act as precursors.

**Coffee** *(Coffea arabica, C. canephora, Rubiaceae)*
(G: Kaffee, F: cafe)

Coffee beans develop their characteristic flavour only after a carefully controlled roasting process and this flavour can be appreciated only through the "right" preparation with hot water. The quality of the served coffee is quite different from one place to the next: mild or strongly roasted coffee, prepared by cooking the coffee powder in water or extracted through a filter or in an espresso machine, diluted "brown water" or highly concentrated "Turkish" coffee: the taste, flavour and the preferences vary from cup to cup. The taste may be sweet (depending on the quantity of added sweetener), bitter (depending on the degree of the roasting process) and sour (depending on the preparation method). In the flavour we can distinguish nice, nutty, hazelnut, roasted and caramel or burnt sugar notes. Strongly roasted coffee is characterized by roasted, burnt, tarry notes. In cold coffee we may observe woody, earthy and even chemical, medicinal notes.

*CIC:* The coffee flavour is the result of a very complex Maillard reaction between proteins and carbohydrates (amino acids, reducing sugars), combined with reaction products of lipids, polyphenols and polyamines. Today over 600 components have been identified, contributing to the coffee flavour impression.

Important key components in coffee are:

*Sulphur compounds:* furfuryl mercaptane with its very strong roasted coffee, sulphurous burnt character. This unstable component varies its flavour profile with the concentration. Methanethiol has a strong rotten, disgusting sulphurous odour, but at low concentrations it enhances an interesting, freshly brewed character to the coffee flavour.

*Alpha diketones:* 3-methyl-2-hydroxy-2-cyclo-penten-1-one: a burnt sugar, roasted, lingering aroma.

*Phenols:* guaiacol: smoky burnt, tarry, phenolic or medicinal character.

4-Ethyl guaiacol: smoky, burnt, phenolic, spicy, clove, medicinal.

*Furans:* furfural with its caramel, burnt, ethereal, almond character, is the most prominent sugar degradation product. 2-Methyl-5-ethyl-4-hydroxy-dihydro-furan-3(2H)-one has a strong, sweet, caramel, bread note.

*Pyrrol:* 2-acetyl pyrrol contributes a caramel, burnt, slightly chemical note to the flavour.

*Thiazoles:* 2-acetyl thiazole imparts a roasted, crusty (bread crust, meat crust) note. Trimethyl thiazole has a nutty, green, cocoa note.

*Pyridines:* 2-acetyl pyridine has a roasted, nutty, earthy character.

*Pyrazines:* 5-methyl-6,7-dihydro-(5H)-cyclopenta-(6)-pyrazine imparts a roasted, nutty, coffee character.

**Cocoa, Chocolate** *(Theobroma cacao, Sterculiaceae)*
(G: Kakao, F: cacao)

As with coffee, cocoa gets its characteristic flavour only after the "right" processing: the fermentation of the fresh fruit and the drying and roasting of the seeds. During this roasting process precursor molecules (aromatic acids, lipids, proteins, polysaccharides) react in a complex Maillard reaction to develop the typical cocoa flavour: the bitter astringent, dry harsh taste, combined with a warm roasted, nutty, floral, slightly honey like aroma. The invention of chocolate brought a high tribute to cocoa: a harmonious combination of roasted cocoa powder, cocoa fat, heated milk powder, sugar and vanilla, skilfully processed to eliminate undesirable harsh, acid, green flavour notes, liquefied with lecithin and homogenized to a very fine, melting mass results in a delicious sweet, roasted, nutty, creamy, caramel, floral and honey like flavour with a harmonious bitter, sweet taste, melting on the tongue: the well appreciated chocolate.

*CIC:* The combination of theobromine, diketo piperazines and tannin is responsible for the balanced bitter, astringent taste. Harsh, volatile acids, like acetic acids, are eliminated during the conching process, as well as some green alkyl pyrazines. Strecker degradation aldehydes from leucine, valine, phenyl alanine (2-methyl-propanal, 3-methyl-butanal, phenyl acetaldehyde) and their condensation products (4-methyl-2-phenyl-2-pentenal, 5-methyl-2-phenyl-2-hexenal) form the body of the green, floral, honey like cocoa flavour, supported by linalool and methyl phenyl acetate. The roasted character is imparted by a wide range of heterocyclic nitrogen-

and sulphur components (pyridines, pyrazines, thiazoles, oxazoles). Worth mentioning are:

- 4-methyl-5-vinyl thiazole: nutty, roasted, cocoa
- 2,4,5-trimethyl oxazole: roasted, nutty, cocoa, sweet, green.
- 3-ethyl-2,5-dimethyl pyrazine: green, nutty, roasted.
- Vanillin and maltol (3-hydroxy-2-methyl-4-pyrone) are responsible keys for the sweet, creamy flavour of chocolate.

## Tea *(Camellila sinensis, Theaceae)*
(G: Tee, F: the)

We roughly can distinguish 3 different types of tea:

- non fermented green tea
- partially fermented Oolong tea
- fermented black tea (Darjeeling).

The teas are further heat treated (steamed, roasted). These treatments explain the very complex nature of its flavour: we can detect astringent, bitter taste, green, fruity or fatty fresh notes, the full, sweet, floral, dry, woody, typical tea flavour, some roasted, smoky topnotes.

*CIC:* polyphenols are the precursors for the astringent tannic acids. Lipids are the precursors of the green, fatty notes of green tea: (E)-2-hexenal and (Z)-3-hexenol impart the fresh, green, fruity, grassy topnote. 2-Nonenal, 2,6-nonadienal and 2,4,6-decatrienal add the fatty, cucumber character. Carotenes are the precursors of the main floral, sweet and fruity components. Among them linalool, theaspirone, damascone, damascenone and ionones are the most important ones, rounded off with the woody note of keto isophorone, and methyl jasmonate, which adds a powerful, floral, sweet odour. Alkyl pyrazines and phenols contribute to the roasted, smoky character of black tea supported by the sweaty odour of (Z)-3-hexenoic acid.

## Meat

Meat is an important part of the human diet not only as a protein source, but also, as part of the pleasures of eating due to its wide range of unique tastes, flavours, texture and its variety of preparing it.

The basic taste of meat is a balanced combination of salty, slightly sour and sweet impressions with metallic, typical "meaty" tastes represented by mineral salts, amino acids, oligopeptides and the "umami"-taste of the ribonucleotides (disodium-5-inosinate, disodium-5-guanylate) and MSG (monosodiumglutamate). The flavour can be differentiated by:

- the source of the meat (species, part of the animal)
- the processing (raw, cooked, roasted, fermented, cured)
- the additives (spices, vegetables, fruits) used for its preparation.

All ingredients in meat (proteins, lipids, carbohydrates) react during the normal meat preparation in a very complex chemical reaction to produce the final, highly appreciated flavour. It goes without saying, that each cook finds a new balance of this chemical reaction. An extensive research on meat flavour development in the past 40 years revealed over 1000 individual components contributing more or less to the overall flavour, as well as their possible formation paths. We can differentiate the following basic classes of flavours and their corresponding flavouring components:

- a general, basic, typical but non characteristic "cooked" meat character represented by 2-methyl-3-furanthiol, a reaction product of 5-inosine monophosphate and cysteine.
- fat-derived, species characteristic, carbonyl components like (E,Z)-2,4-decadienal in chicken fat flavour; 4-methyl octanoic acid in mutton fat flavour; (E,Z)-2,4-heptadienal and (E,Z,Z)-2,4,7-decatrienal in fish oil.
- roasted notes from Maillard reactions like alkyl pyrazines, 2-acetyl-3-methyl pyrazine or 2,3,5-trimethyl-6,7-dihydro-5(H)-cyclopenta pyrazine contribute roasted, grilled, burnt, nutty impressions.
- smoked notes of bacon and ham, represented by guajacol with smoked, tarry, medicinal, burnt, phenolic character.
- animal notes in pork from indole and skatole with their fecal, animalic, phenolic impressions.
- ammoniacal, aged, staling fish aroma from trimethyl amine
- green, fresh notes of algae, represented by (E,Z)-1,3,5-undecatrien with an additional galbanum aspect.
- cooked, boiled shrimp note characterized by N,N-dimethyl-2-phenyl ethyl amine.

Meat flavourings are often combinations of the above mentioned chemical classes with reaction products of:

- cysteine, vitamin B1, ribose
- proteolyzed meat with lipolized and oxidized lipids
- hydrolyzed vegetable proteins or yeast autolysates
- nucleotide mixtures.

**Savoury Flavours** *(see also chapter 5.4)*

In order to improve creation efficiency, flavourists use a "building block concept" to create savoury flavours, a combination of base, middle and top notes:

*The Base Notes* impart the basic taste characteristic of savoury flavours like salty, sweet, acidic, umami. A simple combination of sodium chloride, sugar, citric acid, monosodium glutamate and nucleotides may do this job. More sophisticated products are based on the protein degradation products of natural proteins. Historically, acid-degraded plant proteins (HVP) neutralized with sodium hydroxide were a perfect base: they contained the MSG, NaCl and nucleotides in a good balance. Recent results showed the presence of unwanted monochloropropanediol from the nucleophilic substitution of a hydroxy group in glycerine. Therefore, HVP has been replaced by more suitable products like autolysed yeast and fermented plant protein like soy sauce. Due to the fermentation process and the raw materials used, the quality of these products may differ. The base notes are produced in large quantities, but in a limited variability.

***The Middle Notes*** are based on reaction flavourings to impart specific complex roasted, meaty and smoky notes: roasted chicken, boiled beef, smoked ham. They improve the overall taste with a general meaty background and a good mouthfeeling. Meaty middle notes are often based on the reaction of vitamin B1, cystein and reducing sugars. Fat included in the reaction modifies the product into a specific animal direction: chicken, beef, mutton, pork.

The process to produce reaction flavourings is not just simple; therefore the variability of the middle notes is also limited for economic reasons.

***The Top Notes*** are compositions of natural or synthetic flavouring ingredients to impart a very specific desired aroma profile: spicy, peppery, rosemary, lime, buttery, roasted garlic notes for a clear variation. Or specific fatty notes like 2,4-decadienal, 2-nonenal to adjust the chicken flavour to a more natural impression. Top notes are very flexible in production and can be adjusted to customer needs.

The combination of base, middle and top note allows the flavour industry very flexible and fast answers to the demanding food industry. Many savoury applications require dry products; therefore this approach allows a "simple" powder blending operation of prefabricated powder bases.

## 4.3.2 Fermented Flavourings

Fermentation plays an important role in the development of specific, characteristic flavourings. Natural fermentation is used since thousands of years to get wine, beer or cheese. Modern biotechnological processes have been developed to produce flavourings and flavouring building blocks based on natural processes. There are many important fermented products on the markets, produced by traditional and industrial processes:

- the whole range of alcoholic beverages (wine, beer, brandy, whisky, liqueur)
- dairy products (cheese, yoghurt, sour cream butter)
- grain and bean fermentation (bread, shoyu)
- meat fermentation (fish, sausage)
- vegetables (sauerkraut, kimchi).

### 4.3.2.1 Alcoholic Flavourings

Alcoholic beverages are the result of a yeast fermentation of sugar from various sources. Specific treatments and storage leads to a wide variety of products with a balanced flavour: beer, wine, brandy, whisky, rum, sake etc. And when you listen to all the connoisseurs and experts, no two products are the same: plant species, soil, weather, water quality, processing and distillation technology, all influence and change the final quality of the products. Nevertheless it is possible to distinguish some basic flavour notes:

- fusel oil notes: alcoholic, fusel, harsh, burning, fermented, woody notes (2-methyl butanol, 3-methyl butanol).
- wine yeast or cognac oil notes: fermented, soapy, wine yeast (ethyl octanoate, ethyl decanoate, octanoic acid, decanoic acid).

- rum ester notes: estery, rum, nail-varnish-solvent like (ethyl acetate, ethyl propionate).
- pungent sharp topnote: acetaldehyde, propanal, 1,1-diethoxyethane.
- fruity, estery notes: fruity, banana, apple, strawberry, paint, varnish solvent-like (2-methylbutyl acetate, 3-methylbutyl acetate, ethyl hexanoate).

Product specific flavour characters:

- williams pear fruity note: ethyl-(E,Z)-2,4-decadienoate
- cherry, kirsch, bitter almond note: benzaldehyde
- white wine tart taste: tartaric acid
- red wine astringent taste: tannic acid
- beer bitter taste: isomerized cohumulone
- rum floral, dried fruit woody notes: 2-phenylethanol, damascenone
- whisky smoked notes: 4-ethyl phenol, 4-ethyl-2-methoxy phenol
- whisky and beer nutty, malted notes: 2-ethyl-3-methyl pyrazine, 2-ethyl-3,5-dimethyl pyrazine
- white wine sweet, fruity, muscat grape note: linalool
- sherry, sake, walnut, protein hydrolyzate, sugar-like note: 3-hydroxy-4,5-dimethyl-furan-2(5H)-one
- anise brandies sweet, fennel, anise aroma: anethole.

### 4.3.2.2 Dairy Flavourings

The basis of all dairy products is the fresh milk from mammals, mainly from cows. The analysis of the very bland but characteristic, slightly sweet/salty flavour reveals a large number of common flavouring materials at a subthreshold level. Although there is no demand for a fresh milk flavouring, it is a hard challenge to a flavourist to create the well balanced mixture of this flavour. Fresh milk is a delicate nutritional food, sensitive to deterioration by heat, light and microbiological attack. Mankind has used this sensitivity to produce a wide range of dairy products: starting from simple fat separation (to get cream, butter or ghee) to fermentation products like yoghurt, yakult and fresh cheese, right through to the hundreds of different cheeses, based on bacteria or mold fermentation. We should not forget modern industrial dairy products like UHT treated milk, condensed milk or milk powder. All processes include the following basic reactions:

*Heat:* transforms ketoacids to methylketones like 2-heptanone, 2-nonanone with a green fatty metallic blue cheese note. Hydroxyy acids form the corresponding lactones. The creamy, buttery, coconut-like 5-decanolide, 4-dodecanolide, 5-dodecanolide contribute to the sweet creamy buttery flavour in cream and butter. Lactose undergoes a caramelisation reaction to develop sweet, caramelic maltol and 4-hydroxy-2,5-dimethyl-furan-3(2H)-one. Lactose and milk proteins react in a Maillard reaction to roasted, nutty, burnt notes such as 2,5-dimethyl pyrazine.

*Lipases:* release free fatty acids from milk lipids to generate a rancid, butyric, cheesy, fatty, soapy flavour. Most important are the even numbered acids C4 to C20. In cheeses from goats and sheep 4-methyl octanoic acid imparts a strong animal, goaty character.

*Proteases:* degrade the proteins to free amino acids and oligopeptides, thus imparting sweet, sour and mainly a bitter taste to the product. Further reactions lead to the corresponding Strecker aldehydes 2-methylbutanal and 3-methylbutanal with a pungent, malty, cocoa note.

*Sulfurases:* produce free sulphur chemicals like hydrogen sulphide, methylmercapan, S-methyl thioacetate, S-methyl thiobutyrate.

Lactic acid fermentation imparts the clean sour taste of lactic acid to yoghurt. In a side reaction, rather high concentrations of acetaldehyde are formed, which is responsible for the typical yoghurt flavour.

## 4.3.3 Vegetable Flavourings

The taste of vegetables can clearly be distinguished from fruits: the acid/sugar ratio is reduced, the mineral content is higher, the taste is more "salty". In most vegetables, sulphur containing substances impart a more "culinary" flavour. The fruity character is suppressed.

For the composition of vegetable flavourings it is essential to distinguish the flavour from the taste part: this taste is a balanced composition of harsh, slight bitter acids with salt, some taste enhancers (nucleotides) and sugar to impart a slight sweetness.

### Asparagus *(Asparagus officinalis, Liliaceae)*

Asparagus develops the characteristic taste upon cooking: a vegetable-green topnote with a strong sulphurous-sweet characteristic overall flavour.

*CIC:* the typical sulphurous flavour is represented by the high concentration of dimethyl sulphide, combined with traces of 1,2-dithia-cyclo-pentene. The vegetable-green note results from 2-isopropyl-3-methoxy pyrazine, resembling raw potatoes, and 2-sec-butyl-3-methoxy pyrazine, a green bell pepper note.

### Tomato *(Lycopersicon esculentum, Solanaceae)*

Tomatoes are eaten raw, cooked and mainly in the form of canned puree and ketchup. The taste of ripe, flavourful tomatoes (there exist many cultivars without the typical taste and flavour) is well balanced between sweet, sour and salty with an acid green, vegetable like, slight floral aroma.

*CIC:* (Z)-3-hexenal and 2 (2 methylpropyl)thiazole form the fresh, green, vegetable topnote and a trace of damascenone is responsible for the floral background. On cooking, the fresh note from (Z)-3-hexenal is covered by the sweet sulphurous note of dimethyl sulfide. In canned tomatoes, the sulphurous note is supported by the spicy, clove-like aroma of 4-vinyl guaiacol and eugenol.

### Potato *(Solanum tuberosum, Solanaceae)*

This staple food is never eaten in the raw state: the earthy taste is not quite appealing and the starch has to be transformed into an eatable form either by cooking or by baking/frying. Boiled, cooked potatoes develop a characteristic bland balanced creamy, sweet, sulphurous, earthy flavour. Fried potatoes are appreciated for their roasted, crusty, fatty, earthy potato aroma, the crispy structure and the easy, "handy" consumption.

*CIC:* the earthy odour of fresh potatoes is represented by 2-isopropyl-3-methoxy pyrazine. This earthy note is supported by the mushroom character of 1-octen-3-ol. The key component of boiled potatoes is 3-(methylthio)-propanal, balanced with dimethyl sulphide. The high reaction temperatures in baked and fried potatoes start the Maillard reaction to form mainly heterocyclic components: 2-ethyl-3,5-dimethyl pyrazine, 2-ethyl-6-vinyl pyrazine, 5-methyl-6,7-dihydro-(5H)cyclopenta-pyrazine, 2-acetyl-1,4,5,6-tetrahydro-pyridine are responsible for the roasted, nutty cracker-like flavour. The heat-induced degradation of the potato lipids and the frying oil imparts a fatty, tallowy character to the french fried potatoes. (E,E)-2,4-Decadienal, 2-octenal, octanoic acid and decanoic acid are main contributors to this fatty note.

### 4.3.4 Vanilla Flavourings *(Vanilla planifolia, Orchidaceae)*
(G: Vanille, F: Vanille)

The delicious flavour of cured vanilla beans is appreciated since its discovery in Mexico. Its combination with cocoa and milk started the success of chocolate. Its use in soft drinks (Cola) and ice cream and in many other applications makes it one of the most important flavour types. The rich flavour shows many aspects: the basic creamy, sweet odour is surrounded by warm, woody, slight phenolic, smoked notes (the vanilla bean character). Rum notes, combined with dried fruit, slight floral notes round off the whole picture.

*CIC:* Vanillin, the main component in vanilla flavour is the basic key ingredient for the creamy, sweet character. All other volatile flavouring compounds have been identified only in small traces. Among them 2-methoxy phenol and 2-methoxy-4-vinyl phenol are responsible for the phenolic, smoky odour. 4-Methoxy benzaldehyde, 3,4-methylene-dioxy-benzaldehyde, methyl benzoate and methyl cinnamate impart the warm, powdery, aromatic floral character. Vitispirane adds a fruity, floral topnote. Natural vanilla extract blends very well with other flavourings and it has been modified in different directions: ethyl vanillin is used to increase the sweet, creamy vanillin aspect. Tonka beans and coumarin add a full, dried hay, slightly caramel-like custard aspect, supported by the butter notes of diacetyl and 4-hydroxy-decanolide.

### 4.3.5 Fantasy Flavourings

Flavourings are mainly derived from natural prototypes. Many food producers are looking for fantasy flavourings. However the success in the market is rather limited. Consumers are conservative; they do not immediately accept new, unknown tastes. In fragrances, novelties and new ideas are essential for success, but flavourings are compared always to the natural sources.

Nevertheless, there are a few successful fantasy flavourings on the market, and it is the dream of all flavourists to create a new taste not yet known. In most cases modifications and combinations of natural models are the extreme points of acceptance.

## Cola

The cola flavour is an unequalled success story since the invention of the cough syrup by Dr. Pemperton: This fantasy flavour is the dream of all flavourists: to create a new, non-existing taste with a world-wide success which by now lasts over 100 years. Without detracting from the real success and quality of this flavour, we have to set the mystery of the cola flavour into relation to reality. The very well balanced fresh, citrus, lime, lemon topnote with the sweet, spicy, cinnamon, creamy, vanilla heart and the earthy, sweet sour taste can be imitated to a certain level. However, it never will be possible to imitate the success of the beverage: over 100 years of intensive successful marketing and the largest budget for world-wide advertising, combined with a high quality product and good world-wide service are the basis of this success story.

*CIC:* The main key products for a cola flavouring are:

- a well balanced mixture of soluble distilled lime with lemon oil
- a clean cinnamon extract combined with vanilla extract
- a trace of cola nut extract, caffeine and caramel colour
- a balance of the sugar (or sweetener) and citric/phosphoric acid equilibrium.

## Tutti Frutti

Tutti Frutti is another fantasy flavour combination, non existant in nature. It is mainly used in chewing gums. The name suggests a mixture of all fruits. In fact it is mainly a balanced mixture of isoamyl acetate with orange oil, lemon oil and vanillin.

## REFERENCES

Instead of giving a detailed list of literature references, I prefer to list a few informations for a flavourist's library.

Literature on the botany of fruits:

[1]  B. Kranz, Das grosse Buch der Früchte, Südwest Verlag Stuttgart 1988
[2]  A. Fouqué, Espèces Fruitières d'Amerique Tropicale, Institut francais de la recherche fruitières outre mer 1975
[3]  S. Rehm, G. Espig, Die Kulturpflanzen der Tropen und Subtropen, Verlag Eugen Ulmer, Stuttgart 1976
[4]  L. Johns, V. Stevenson, Fruits for the Home and Garden, Angus & Robertson Publishers 1985
[5]  G. Götz, R. Silberstein, Obstsorten-Atlas, Verlag Eugen Ulmer, Stuttgart.
[6]  H.R. Gysin, Tropen Früchte, AT Verlag, Aarau 1984
[7]  F. Bianchini, F. Corbetta, M. Pistoia, Fruits of the earth, Bloomsbury Books 1988
[8]  S. Silva, Frutas Brasil, Empresa das Artes 1991

Literature on flavour composition of food:

[9]   H. Maarse, C.A. Visscher, Volatile Compounds in Food, TNO, Zeist 1989-1994
[10]  H. Maarse, Volatile Compounds in Food and Beverages, Marcel Dekker Inc 1991

General literature on flavours:

[11]  Flavour and Fragrance Materials, Allured Publishing Co. 1993
[12]  H.B. Heath, Source Book of Flavours, AVI Publishing Company Inc 1981
[13]  K.J. Burdach, Geschmack und Geruch, Verlag Huber, Bern 1988

[14]   E. Ziegler, Die natürlichen und künstlichen Aromen, Dr. A. Hüthig Verlag, Heidelberg 1982
[15]   P.R. Ashurst, Food Flavouring, Blackie & Son Ltd., Glasgow 1990
[16]   S. Arctander, Perfume and Flavor Chemicals 2 Volumes, 1969
[17]   S. Arctander, Perfume and Flavor Materials of Natural Origin, 1969.
[18]   E. Guenther, The Essential Oils, 6 Vols., Krieger Publishing 1948
[19]   G.A. Burdock, Fenaroli's Handbook of Flavor Ingredients, CRC Press 2002
[20]   B.M. Lawrence, Essential Oils, 6 Vols., Allured Publishing 1976-2000
[21]   P. Mueller, D. Lamparsky, Perfumes, Art, Science and Technology, Elsevier Applied Science 1991

## Series of Monographs, Textbooks and Reference Books on Food Science by Marcel Dekker Inc.:

[22]   Vol.1: R. Teranishi, I. Hornstein, Flavour Research: Principles and Techniques
[23]   Vol.7: R. Teranishi, R.A. Flat, Flavour Research: Recent Advances.
[24]   Vol.9: H.T. Chan, Handbook of tropical Foods.
[25]   Vol.16: M. O'Mahony, Sensory Evaluation of Food: Statistical Methods and Procedures.
[26]   Vol.18: S.V. Ting, R.L. Rousseff, Citrus Fruits and their Products: Analysis and Technology.
[27]   Vol.20: Y. Kawamura, M.R. Kare, Umami: A Basic Taste.
[28]   Vol.30: S. Nagy, J.A. Attaway, Adulteration of Fruit Juice Beverages.
[29]   Vol. 32: R.H. Matthews, Legumes: Chemistry, Technology and Human Nutrition.

## Series of Monographs, Textbooks and Proceedings of Conferences on the Developments in Food Science by Elsevier:

[30]   Vol. 115: R. Marsili Flavor, Fragrance and Odor Analysis, 2002.
[31]   Vol. 3: I.D. Morton, A.J. MacLeod, Food Flavours.
[32]   Vol. 10, J. Adda, Progress in Flavour Research, Proceedings of the 4th Weurman Flavour Research Symposium, Dourdan, France, 1984
[33]   Vol. 11: J. Hollò, Fat Science, Proceedings of the 16th International Society for Fat Research Congress, Budapest, 1983.
[34]   Vol. 12: G. Charalambous, The Shelf Life of Foods and Beverages, Proceedings of the 4th International Flavour Conference, Rhodes, 1985.
[35]   Vol. 13: M. Fujimaki, M. Namiki, Amino-Carbonyl Reactions in Food and Biological Systems, Proceedings of the 3rd International Symposium on the Maillard Reaction, Susuno, 1985.

## Proceedings of various Conferences:

[36]   R.G. Berger, S. Nitz, Topics in Flavour Research, (Freising-Weihenstephan) H. Eichhorn, D 8051, Marzling 1985
[37]   D.G. Land, H.E. Nursten, Progress in Flavour Research, (2nd. Weurman Symposium, Norwich), Applied Science Publishers LDT, London 1978
[38]   M. Martens, Flavour Science and Technology, (5th. Weurman Symposium, Oslo) J.Wiley & Sons 1987
[39]   G. Charalambous, Frontiers of Flavours, (6th. Weurman Symposium, Geneva, CH) 1990
[40]   R. Hopp, K. Mori, Recent Developments in Flavour and Fragrance Chemistry, (Haarmann & Reimer International Symposium, Kioto) Verlag Chemie, Weinheim 1992
[41]   G. Charalambous, Flavours of Food and Beverages, (Amer.Chem.Soc. Athens, Greece) Academic Press 1978
[42]   G. Charalambous, Analysis of Foods and Beverages, (Amer.Chem.Soc. Chicago) 1977
[43]   P. Schieberle, K.H. Engel, Frontiers in Flavour Science, Deutsche Forschungsanstalt fuer Lebensmittelchemie 2000
[44]   D. Roberts, A. Taylor, Flavour Release, American Chemical Society 2000
[45]   A. Spanier, F. Shahidi, et al. Food Flavors and Chemistry, Advances in the New Millennium, Royal Society of Chemistry 2001

# 5 Application

## 5.1 Introduction

*Günter Matheis*

### 5.1.1 Flavour Binding and Release

The binding and release of flavour [1, 2] are vital criteria in the flavouring of foods. Most foods are actually complex mixtures the ingredients of which interact both with each other and also with the flavouring substances [3]. It is only possible to taste and to smell what has been released in the oral cavity during the consumption of the food. Flavouring substances which cause the impressions sweet, sour, salty, bitter and umami must dissolve in the saliva before they can be perceived. Odour substances must reach the oral and pharyngeal cavity in the gaseous phase before they can be smelt retro-nasally in the olfactory epithelium (Fig. 5.1). Chewing and temperature in the oral cavity encourage the release of the flavouring substances.

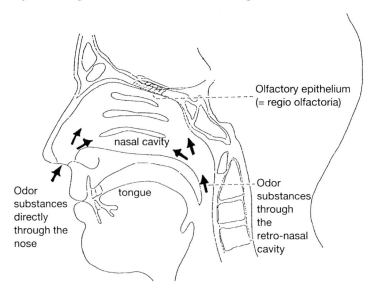

*Fig. 5.1:* Schematic drawing of the human olfactory organ [4,5]

The amount of volatile flavouring substance that is released in the gas phase depends on its vapor pressure. This, in turn, is affected by various factors, including temperature and possible interactions with food ingredients. If an ingredient has a negative effect on the release, we speak of flavour binding. The term binding should not be taken literally here. It is defined as the enrichment of concentration of the volatile substance within or in the proximity of a food ingredient, without specifying precisely the type of binding involved (whether it is physical or chemical, and what type of physical or chemical bond is involved).

For the flavourist, the flavour applications technologist and the food technologist alike, knowledge of the type of such bonds is vital. The elucidation of the type of bond of volatile flavouring substances in the food is extremely difficult. This is due to the extremely complex nature of most foods. For this reason simple (binary and ternary) systems are usually investigated. We shall not go into the various methods here. Suffice it to say that in many cases relatively high concentrations of volatile flavouring substance are used to elucidate the type of bond. In the food itself, however, many of these compounds are only present below the ppm level. This is why model tests do not always reflect the situation in the food.

Furthermore, it is known that biological structures in the food, which are not present in model systems, have an effect on the binding of flavour [6].

Following is a summary of the state of research in the field of interactions between volatile flavouring substances and food ingredients. The investigations have been carried out on many simple systems and a few complex ones.

## 5.1.2 Interactions with Carbohydrates

Looking at the interactions between volatile flavouring substances (e.g. diacetyl, heptanal, heptanone, octanol, menthone, isoamyl acetate and various methyl ketones) and *simple sugars* (mono- and disaccharides) in aqueous systems there are conflicting results. In some cases the volatility of the flavouring substance increased, in others it diminished, and in others again it was not affected to any significant extent [1,7-10]. Nothing is known of the type of interactions, which are presumably physical.

Simple sugars often serve as carrier substances for flavours. Model tests with dry (or virtually dry) sugars in the crystalline state resulted in a number of flavouring substances (e.g. ethyl acetate, butyl amine) only binding weakly to glucose, saccharose and lactose [6,11]. This bond, which presumably involves adsorption on the relatively small surface area of the crystalline sugar, is completely reversible under vacuum at 23°C [6,11]. If the sugars are in the amorphous state (with a larger surface area), the adsorption on saccharose of isopropanyl acetate, phenyl acetate and diethyl ketone, for example, is considerably greater (Fig. 5.2).

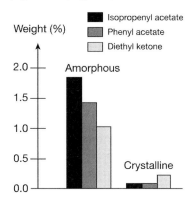

***Fig. 5.2:*** *Adsorption of volatile flavouring substances on amorphous and crystalline saccharose [1]*

# Interactions with Carbohydrates 439

A number of volatile flavouring substances (e.g. acetaldehyde, diacetyl, ethyl acetate, 2-hexanone, butyl amine) bind with varying degrees of strength to the polysaccharides *pectin, guar gum, alginate, agar-agar, cellulose, and methyl cellulose [1,6,9,11,12]*. In most cases, dry, air-equilibrated polysaccharides were studied (these have a 5-10% water content). There is still speculation about the type of bond. Maier *[6,11]* postulates that butyl amine is sometimes bound as a salt to the carboxyl groups of pectin and alginate, but sometimes reacts chemically, to the amide with the same groups (Fig. 5.3). The part of the amine which has reacted chemically, is lost as far as the flavour is concerned. In the case of cellulose, Maier *[6,11]* presumed that hydrogen bridges break out between the cellulose molecules, followed by hydrogen bridge formation between butyl amine and cellulose (Fig. 5.4).

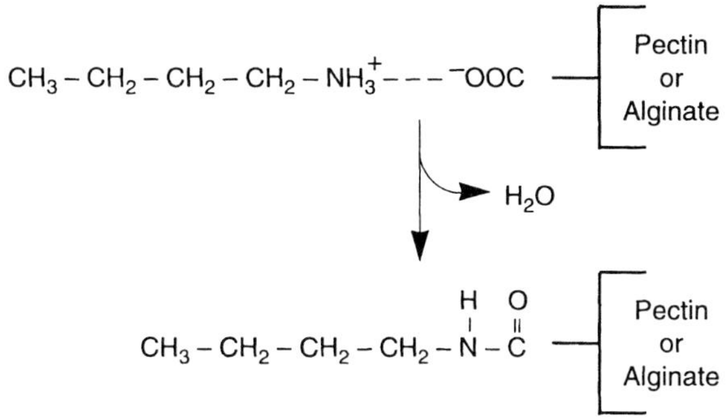

$$CH_3 - CH_2 - CH_2 - CH_2 - NH_3^+ \cdots {}^-OOC \longrightarrow \begin{array}{l} \text{Pectin} \\ \text{or} \\ \text{Alginate} \end{array}$$

$$\downarrow H_2O$$

$$CH_3 - CH_2 - CH_2 - CH_2 - \underset{H}{\overset{}{N}} - \overset{O}{\underset{}{C}} \longrightarrow \begin{array}{l} \text{Pectin} \\ \text{or} \\ \text{Alginate} \end{array}$$

***Fig. 5.3:*** *Possible interactions between butyl amine and pectin or alginate [6,11]*

Considerably more is known about the binding of volatile flavouring substances to *starch*. Most starches consist of two fractions; amylose (unbranched glucose chains) and amylopectin (branched glucose chains) (Fig. 5.5). Starches change their native structure both during and after boiling in water. Hydrogen bridge bonds are weakened in this gelatinization process; the starch grains swell, and part of the amylose and the amylopectin is dissolved. At the same time amylopectin binds large quantities of water, and amylose forms helical structures under water binding. Both amylose and amylopectin are involved in the binding of flavour.

Volatile flavouring substances can be entrapped in the helical structures of amylose (analogous to the well-known starch-iodine inclusion complex). The outside of the helix is hydrophilic because of the hydroxyl groups, the inside is hydrophobic due to hydrogen atoms. In the case of the inclusion complex, the non-polar (hydrophobic) part of a flavouring substance plays the decisive role. Relatively low molecular compounds (e.g. hexanol) are entrapped in 6-fold simple helices (one turn of the helix consists of 6 glucose molecules); higher molecular ones (like ß-pinene) are entrapped in 7-fold simple helices (one turn of the helix consists of 7 glucose molecules) *[13]*.

The amylopectin fraction is also involved in flavour binding. With higher concentration of volatile flavouring substances, the outer branches of the amylopectin form helical structures – as with amylose – in which flavouring substances are entrapped *[14]*.

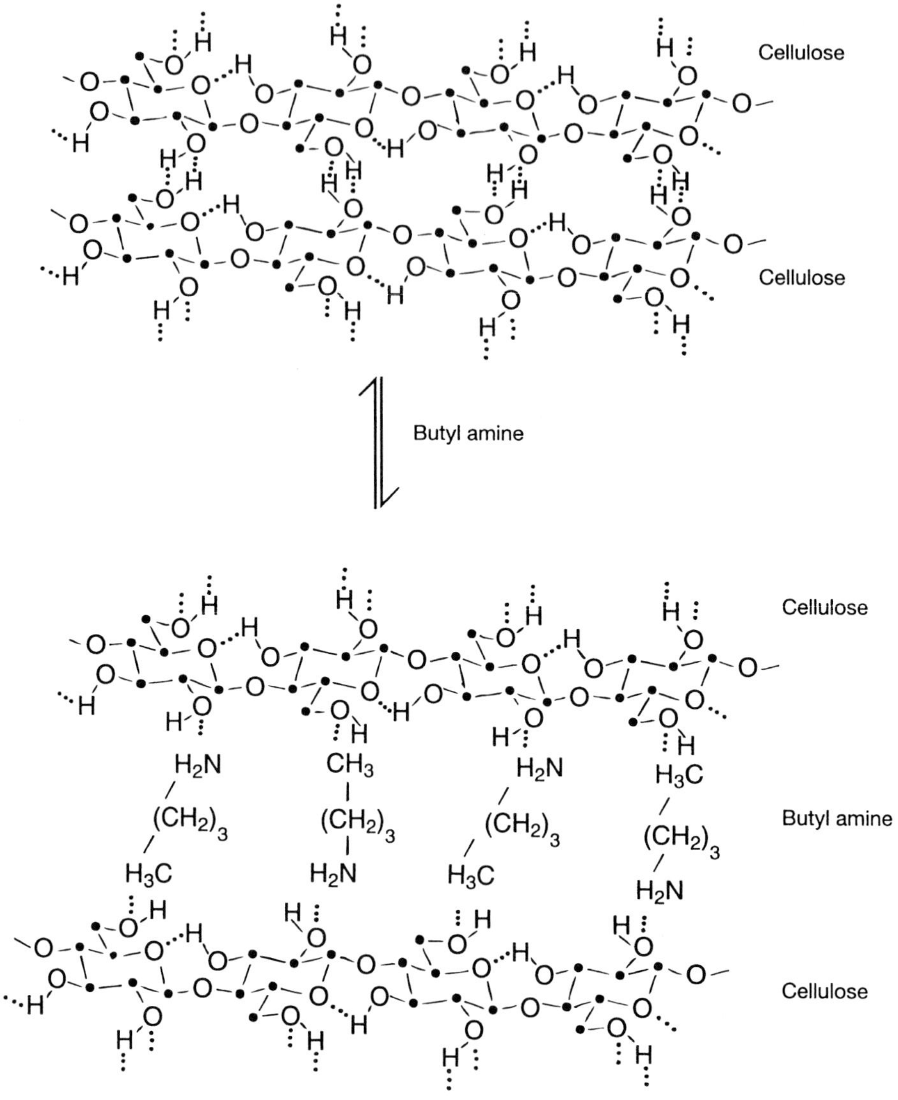

*Fig. 5.4: Formation of hydrogen bridges (...) between butyl amine and cellulose [6,11]*

Different starches show varying flavour binding capacity. Starches with a low amylose content (e.g. tapioca at 17%) and waxy starches consisting only of amylopectin have only a weak binding capability; those with a high amylose content (like potato or maize) have a greater one. Gelatinized potato starch has been researched best.

Interactions with Carbohydrates 441

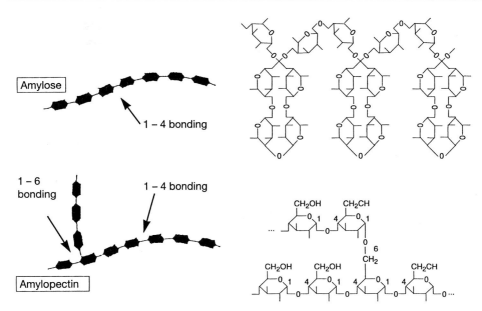

*Fig. 5.5:* Amylose and amylopectin

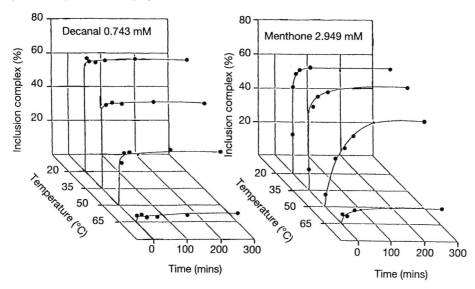

*Fig. 5.6:* The formation of inclusion complexes of potatoe starch with decanal and menthone under isothermal conditions [16]

The model tests were carried out both iso-thermally (i.e. at constant temperature) and in temperature gradients (from warm to cold). Alcohols (e.g. ethanol, hexanol, octanol, decanol), aldehydes (e.g. decanal), ketones (e.g. diacetyl), acids (e.g. caproic acid), esters (e.g. ethyl acetate), terpenes (e.g. menthol, menthone, ß-pinene, limonene), amines (e.g. butyl amine), pyrazines and other classes of substances have been investigated as flavouring substances [6,9,11,13-18]. Fig. 5.6 shows the formation of inclusion complexes of potato starch with decanal and menthone at various temperatures, that is to say, under various isothermal conditions. The process can run relatively quickly (taking less than a minute in the case of decanal) or very slowly (up to 300 minutes with menthone) (Fig. 5.6). In the case of limonene, it takes several days [14-15]. The formation of the inclusion complexes begins at approximately 0.1 mM concentration of a flavouring substance and ends when all helices are "full".

Studies in temperature gradients (e.g., from 80°C down to 20°C) showed that in the case of many volatile flavouring substances the number of inclusion complexes after cooling to 20°C was the same as that formed under isothermal conditions simply at 20°C. In other words, the prior warming does not increase the amount of flavouring substance which is bound after cooling. This is illustrated, using decanal as an example, in Fig. 5.7.

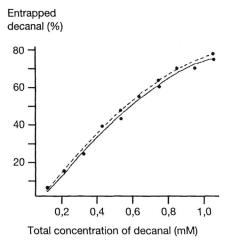

*Fig. 5.7:* The formation of inclusion complexes of potato with decanal under isothermal conditions (20°C; ---) and in temperature gradients from 80°C to 20°C (—) within 14 hours [16]

Volatile flavouring substances in general form relatively weak bonds to gelatinized starches, although the bond is stable (in the case of some flavoring substances up to 65°C) [15]. Moreover, a cooperative binding effect has been observed. This means that in each stage of binding the amount of bound flavouring substance has a positive effect on further binding.

Both, synergistic and antagonistic effects, were observed in the presence of two volatile flavouring substances [1,19-20], depending on their individual initial concentrations. With a high concentration of menthone (menthone is included in a 7-fold helix) there is increased binding of decanal (synergism), whereas with high concentrations of decanal (decanal is included in a 6-fold helix) menthone binding is inhibited (antagonism).

I would like to remind readers that so far we have considered what takes place with gelatinized starches in aqueous media. If such inclusion complexes are dried, they are remarkably stable [1]. The fact that volatile flavouring substances included in starchy foods are only released after a relatively long period of 20 seconds in the oral cavity is of great importance for practical applications [21].

The binding of volatile flavouring substances to dry and air-equilibrated potato starch (which has an 8.2% water content) has also been studied [6,11]. As one would expect, the moist starch binds more flavour than the dry one. Some of the bound substances are not even released in a vacuum at ambient temperatures and above; when water is added, however, the bond is destroyed.

After partial hydrolysis the starches lose a major part of their flavour binding properties. Examples of partially hydrolyzed starch products are *dextrins* (acid or enzymatic hydrolysis) and *maltodextrins* (generally enzymatically hydrolized). Acetaldehyde, ethanol, decanal and limonene only bind weakly to dextrins (presumably by adsorption) [22], while ethylacetate is not adsorbed at all [1]. In the same way, alcohols (such as ethanol, propanol, butanol, pentanol and hexanol) and menthol are only weakly adsorbed on maltodextrins [11, 23].

*Cyclodextrins* are a special type of starch derivative. They come about as a result of the effect of the enzyme cyclodextrin glucosyltransferase on starch. Cyclic products result with 6,7 or 8 glucose units; they are designated as α-, ß-, and γ-cyclodextrin (Fig. 5.8). Although cyclodextrins are not naturally present in food, they will be considered here. In countries where they are permitted for food, they are used, e.g. for the encapsulation of flavourings and colourings and for masking off-flavours [24-29].

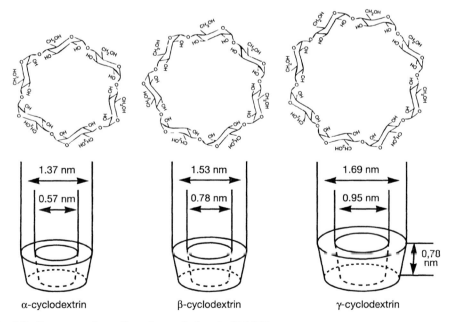

*Fig. 5.8:* Structure of α-, ß- and γ-cyclodextrin [24,28]

β-cyclodextrin is the best suited for encapsulation of flavourings [25-26]. In an aqueous medium it forms inclusion complexes with volatile flavouring substances. The more lipophilic the flavouring substance, the more easily it is entrapped. The interstitial spaces of cyclodextrin are actually non-polar, whereas the surface is polar, determined by the hydroxyl groups. Fig. 5.9 shows schematically the entrapment of benzaldehyde. The lipophilic part (benzene ring) is inserted in the space, the hydrophilic part of the aldehyde (aldehyde group) protrudes. Instable flavouring substances become extraordinarily stable in such inclusion complexes [24,26,30- 31]. Fig. 5.10 shows – as an example – the oxidation stability of benzaldehyde entrapped in ß-cyclodextrin. ß-cyclodextrin can, for instance, entrap 9% benzaldehyde and up to 12% essential oil. There are also a number of cyclodextrin derivatives [24], which shall not be considered here. They are already in use in the pharmaceutical industry [24].

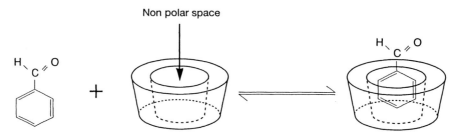

*Fig. 5.9:* Entrapment of benzaldehyde in β-cyclodextrin

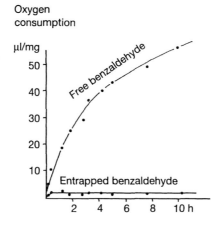

*Fig. 5.10:* Oxidation of free benzaldehyde and that entrapped in β-cyclodextrin to benzoic acid at 37°C (measured by oxygen consumption) [24]

Finally, let us summarize the type of binding of volatile flavouring substances to various carbohydrates – as far as it is known (Table 5.1). In essence, this is a matter of reversible physical and physico-chemical binding (adsorption, inclusion complexes, hydrogen bridges), so that, in principle, flavour release takes place in the oral cavity.

# Interactions with Proteins

*Table 5.1: Type of interactions between volatile flavouring substances and carbohydrates*

| Carbohydrate | System | Type of Bond |
|---|---|---|
| Simple sugars (Mono- and disacharides) | Aqueous | Not known |
| | Dry | Adsorption |
| Pectin and alginate | Dry and moist | Electrovalent bond? Covalent bond? |
| Cellulose | Dry and moist | Hydrogen bridges? |
| Guar gum, agar-agar and methylcellulose | Dry and moist | Not known (with methylcellulose as with cellulose?) |
| Starches and cyclo-dextrins | Aqueous | Inclusion complexes |
| Dextrins and malto-dextrins | Aqueous | Adsorption |

## 5.1.3 Interactions with Proteins

It was proven over 20 years ago that both native and denatured proteins bind volatile flavouring substances *[7,9,12,32-34]*. Studies at the beginning of the 1980's revealed that aldehydes (such as nonanal) and ketones (such as 2-heptanone, 2-octanone, 2-nonanone) formed exceptionally strong bonds to *native proteins* (soy protein, bovine serum albumin, fish actomyosin) in aqueous systems *[35-39]*.

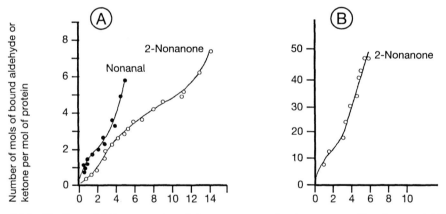

*Fig. 5.11: Binding of (A) nonanal and 2-nonanone to bovine serum albumin, and (B) 2-nonanone to fish actomyosin at 25°C and pH 7.0-7.4 (0.6% protein in aqueous solution) [35,38]*

Fig. 5.11 shows the binding of nonanal and 2-nonanone to bovine serum albumin and that of 2-nonanone to fish actomyosin. The non-linear binding isotherms point to the fact that, in binding, the flavouring substances cause conformational changes in the protein molecules. Native proteins bind aldehydes and ketones through hydrophobic interactions (shown schematically in Fig 5.12) *[35-38]*, whereas alcohols (e.g., buta-nol and hexanol) are bound both, hydrophobically and through hydrogen bridges (shown schematically in Fig. 5.13) *[39]*.

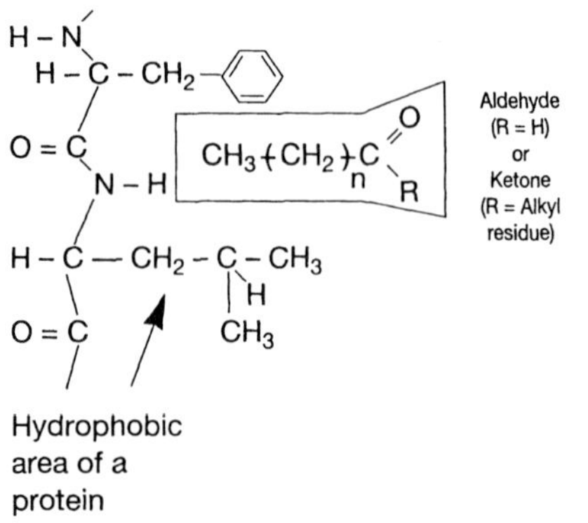

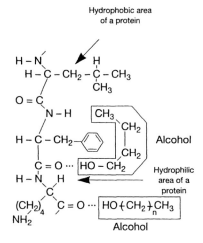

*Fig. 5.12:* Hydrophobic interactions between carbonyl compounds (aldehydes and ketones) and protein

*Fig. 5.13:* Hydrophobic interactions and hydrogen bridge formation (....) between alcohols and protein

Hydrophobic interactions between β-lactoglobulin and 2-alkanones (2-heptanone, 2-octanone, 2-nonanone) have also been reported *[40-42]*. O'Keefe et al. *[43]* stated that binding of carbonyl flavour compounds to soy protein cannot be easily explained as resulting from hydrophobic bonding alone. ß-lactoglobulin was reported to bind β-ionone but not α-ionone, geraniol, nor limonene *[44]*.

In practice, the *heat denatured proteins* are of greater importance than the native ones. Heat treatment is characteristic of many foodstuffs containing protein. Initial investigations into the binding of volatile flavouring substances to heated soy protein in an aqueous medium resulted in increased binding of aldehydes (e.g., hexanal), ketones, (e.g., 2-nonanone) and alcohols (e.g., hexanol) *[33,45]*. These findings – in so far as aldehydes and ketones are concerned – have been confirmed recently; for the alcohols, however, there are contradictory results, namely a reduced binding of butanol and hexanol to heated soy protein *[39]*. As expected, the binding of flavouring substances to protein depends both on temperature and pH *[46]*. Increasing the temperature from 25°C to 50°C improves the binding of heptanal, a pH shift from 6.9 to 4.7 reduces it, but increases the binding of 2-nonanone to milk protein.

Benzaldehyde and limonene have been reported to be bound by whey protein and casein, whereas citral did not bind under the conditions used [47].

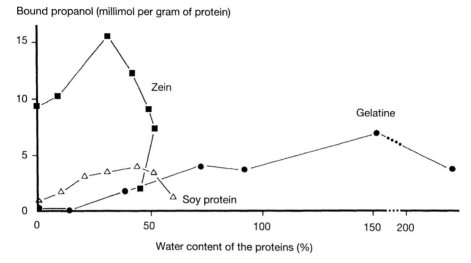

*Fig. 5.14: Binding of propanol to proteins as a function of water content [11]*

*Anhydrous proteins* can also bind volatile substances [6,11]. Fig. 5.14 shows that anhydrous zein – unlike soy protein and gelatine – binds propanol. Fig. 5.14 also shows that for each of the three proteins illustrated there is a maximum water content for propanol binding. As already mentioned, here both hydrophobic interactions (see Fig. 5.12) and hydrogen bridges (see Fig. 5.13) are involved in the binding of the alcohols ethanol and propanol, and of the ketone acetone (Fig. 5.15) to the anhydrous proteins [6].

Regardless of whether the flavouring substances are bound to native and denatured proteins in aqueous systems, or to anhydrous proteins, for practical purposes it is relevant whether the binding is reversible or not. In general, it is true to say that hydrocarbons, alcohols and ketones are reversibly bound (through hydrophobic interactions and/or hydrogen bridge formation; see Figs. 5.12, 5.13, and 5.15).

Some aldehydes are the same (see Fig. 5.13), but a significant number reacts chemically with amino groups in the protein molecules (Fig. 5.16) [1,6,11,16,33,16]. This binding is irreversible due to subsequent reactions, and the part of the aldehyde which has reacted chemically is lost as far as sensory properties are concerned. In this way, for instance, hexanal reacts easily with amino groups of the arginine residues in soy protein and in casein [11].

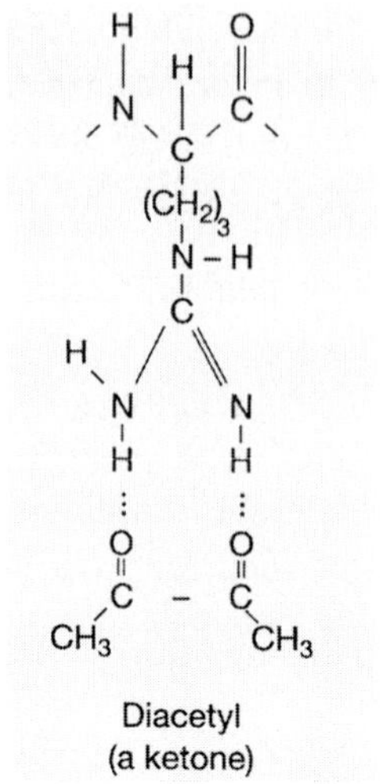

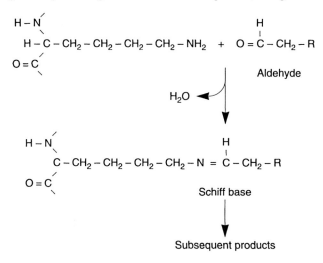

*Fig. 5.15:* Hydrogen bridge binding between ketones and protein, using acetone as an example

*Fig. 5.16:* Reaction of aldehydes with amino groups of protein to form Schiff bases

Esters can also be bound to proteins *[1,6,41,48]*. Possibly, hydrophobic interactions are involved *[48]*. A surprising "salting out" effect was reported for terpenes (limonene, myrcene) in the presence of β-lactoglobulin *[41]*. Nothing is known about the type of physical bond. In general, interactions between volatile flavouring substances and proteins are more complex than those between volatile compounds and carbohydrates. Just as in starches and cyclodextrins, inclusion complexes can also be imagined in the interstices of protein molecules. In particular, lipophilic flavouring substances could be stored in "hydrophobic pockets". In 1988, a detailed report, summarizing the flavour binding capabilities of soy protein, was published *[49]*, and in 1990 more recent papers were presented *[1]* covering binding to bean, milk, and wheat proteins. In conclusion, Table 5.2 gives a summary of the interactions between various classes of volatile compounds and proteins. As with the carbohydrates, these are mainly a matter of reversible physical and physico-chemical binding (hydrophobic interaction, hydrogen bridges), thus in principle ensuring the release of flavour in the oral cavity.

*Table 5.2: Type of interactions between volatile flavouring substances and protein*

| Class of substance | Type of interactions |
|---|---|
| Hydrocarbons | Hydrophobic interaction |
| Alcohols | Hydrophobic interaction |
|  | Hydrogen bridges |
| Aldehydes | Hydrophobic interaction |
|  | Chemical reaction |
| Ketones | Hydrophobic interaction |
|  | Hydrogen bridges |
| Esters | Unexplained |

## 5.1.4 Interactions with Free Amino Acids

Free amino acids can bind with a series of volatile flavouring substances in aqueous media. Ketones and alcohols are reversibly bound by hydrogen bridges to the amino or carboxyl groups of the amino acids (Fig. 5.17), while – as with proteins – some aldehydes react chemically with the amino groups to form Schiff bases (Fig. 5.18). This has been demonstrated for interactions of vanillin with lysine, phenylalanine and cysteine *[50]*.

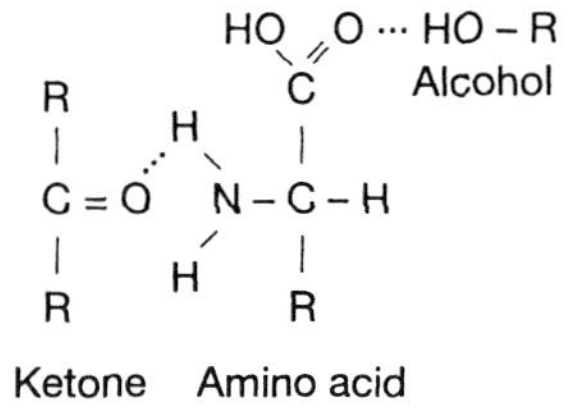

*Fig. 5.17: Hydrogen bridges (...) between amino acid and ketones or alcohols*

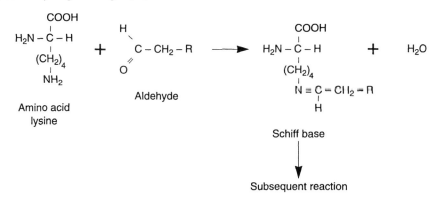

*Fig. 5.18: Reaction of aldehyde and amino acid to Schiff base (using the amino acid lysine as an example)*

The amino acid cysteine reacts in an aqueous medium with aldehydes and ketones to thiazolidine-4-carboxylic acid (Fig. 5.19) [6,50]. This reaction is reversible under heating, particularly when there is an acid pH-value [6].

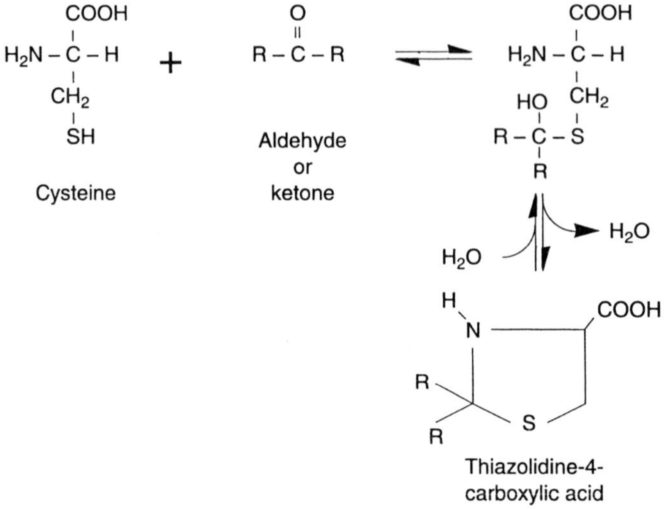

*Fig. 5.19: Reaction between cysteine and aldehyde or ketone [6]*

Dry amino acids adsorb volatile aldehydes (e.g., hexanal), ketones (e.g., acetone, diacetyl), acids and amines; some of the aldehydes and ketones even react chemically with the amino acids (as detailed above). Clearly sufficient residual water is available to enable such reactions to take place.

## 5.1.5 Interactions with Lipids

The most important lipids in foods, quantitatively speaking, are *fats and oils*. They consist predominantly of triglycerides. Triglycerides can bind considerable quantities of lipophilic (i.e. non-polar) and partly lipophilic flavouring substances [1,6,12,20,22,32,47]. The binding capacity of fats (triglycerides which are solid at ambient temperature) is less than that of oils (triglycerides which are liquid at ambient temperature), as is illustrated in Table 5.3, using acetone as an example.

*Table 5.3: Binding of acetone to various media [6]*

| Lipid or other medium | Mols of bound acetone per kg medium |
| --- | --- |
| Tributyrin (liquid) | 93.5 |
| Coffee oil (liquid) | 63.8 |
| Coffee[a] | 4.0 |
| Lecithin (solid) | 4.0 |
| Trilaurin (solid) | 1.3 |
| Cholesterol (solid) | 0.9 |
| Aqueous coffee extract | 0.3 |
| [a] The binding of acetone depends mainly on the oil content of the coffee ||

The quantity of bound flavouring substance also depends on the chain length of the fatty acids in the triglyceride, and on the presence of saturated or unsaturated fatty acids in it. Triglycerides with long chain fatty acids bind less ethanol and ethyl acetate than those with short chain fatty acids *[6]*. Triolein – a triglyceride that only contains the unsaturated oleic acid – binds more flavour than tripalmitin and trilaurin (both of which contain only saturated fatty acids) *[6]*.

In lipid-water mixtures, in oil-in-water emulsions, and in water-in-oil emulsions, the flavouring substances are distributed between the lipid and water phases as a function of the structure of the flavouring substances (more lipophilic or more hydro-philic), the type of the lipid and the temperature, amongst other things. Fig. 5.20 shows the concentration of 2-heptanone in the gas phase above the media: whole milk, skimmed milk, water, edible oil. There is only very little flavouring substance in the head-space above the oil because almost all of it is dissolved in the oil. Because of its fat content, part of the 2-heptanone dissolves in the whole milk (an oil-in-water emulsion), whereas skimmed milk behaves in virtually the same way as water. The greatest concentration of 2-heptanone is to be found in the gas phase above the latter two media. Fig. 5.21 shows the effect of edible oil on the vapour pressure of selected flavouring substances.

Many volatile flavouring substances have a lower vapour pressure in lipids and therefore a higher odour threshold (see Table 5.4) than they do in aqueous systems. In practice, this means that the addition of even small amounts of lipid can significantly reduce the concentration of a flavouring substance above an aqueous system *[32,48]*. Thus, for instance, by adding just 1% oil to an aqueous system, the octanal and heptanal concentration in the gas space is considerably reduced. In the case of the flavouring substances hexanal and pentanal, however, 10% added oil is required to achieve the same effect.

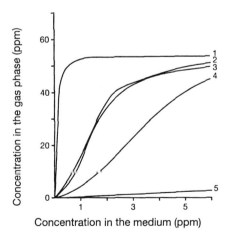

*Fig. 5.20:* Concentration of 2-heptanone in the gas phase above various media (1 = no medium, 2 = water, 3 = skimmed milk, 4 = whole milk, 5 = edible oil) [47]

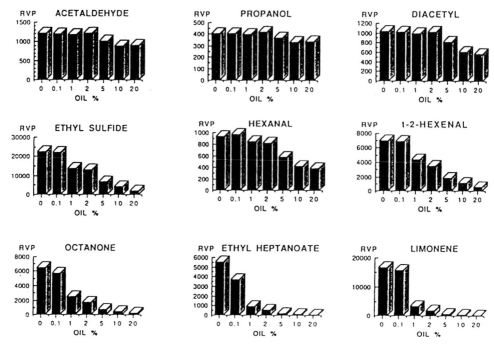

*Fig. 5.21:* Effect of edible oil on the vapour pressure of selected flavouring substances [52]. RVP (= relative vapour pressure) is the headspace concentration of a flavouring substance over a model system containing water, emusifier and oil divided by its headspace concentration above a system containing only water and emulsifier

*Table 5.4:* Odour thresholds of some aldehydes in edible oil and water [1]

| Aldehyde | Odour threshold (ppb) in | |
|---|---|---|
| | edible oil | water |
| Hexanal | 120 | 4.5 |
| Heptanal | 250 | 3.0 |
| Octanal | 430 | 0.7 |
| Nonanal | 1000 | 1.0 |

The way in which an oil is distributed physically in an aqueous phase can affect the headspace concentration of flavouring substances. In the case of dimethyl sulphide, a higher concentration of flavouring substance is needed in the oil-in-water emulsion to achieve the concentration in the gas phase which is measured above the non-emulsified system (Fig. 5.22A). This points to an adsorption of the dimethyl sulphide on the boundary surfaces or to some other type of interaction. In the case of allyl mustard oil, a higher concentration is measured in the headspace of the emulsion than in that of non-emulsified system (Fig. 5.22B). Apparently the mustard oil has little or no affinity with the boundary surfaces.

The amount of flavouring substance bound by fat or oil depends on the chain length of the volatile compound within a homologous series. Thus, for instance, the distribution coefficient in oil-in-water systems increases with the increasing chain length of an alcohol (from ethanol to nonanol) (Fig. 5.23). This means that the concentration in the gas phase decreases as the chain length increases.

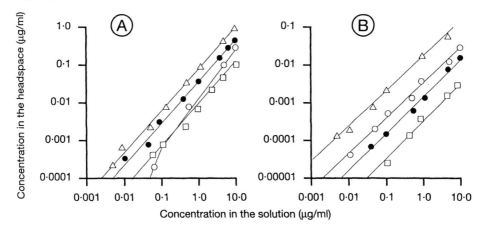

*Fig. 5.22:* Influence of the lipid phase on the headspace concentration of (A) dimethyl sulphide and (B) allyl mustard oil ( $\Delta$ = water, $\square$ = oil, o = emulsion, • = water and oil not emulsified) [32]

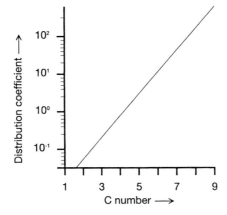

*Fig. 5.23:* Distribution coefficients of aliphatic alcohols in an oil-in-water system [47]

Minimal changes in temperature in the vicinity of the melting point of a triglyceride can change the distribution coefficient of a volatile flavouring substance by up to a factor of five [6]. The fact that the melting point of some food fats is close to the temperature of the oral cavity is of great importance for the flavour release.

Hardly anything has been published on investigations of the interactions between volatile compounds and lipids which are not triglycerides. It is known, however, that *lecithin* and *cholesterol* are capable of binding small quantities of acetone (Table 5.3).

For most volatile flavouring substances the type of binding to lipids can be explained by the distribution laws. A well-known exception is the affinity of dimethyl sulphide to emulsion boundary surfaces, mentioned above. In the case of phenolic compounds, hydrogen bridges could also be involved [6].

## 5.1.6 Interactions with Inorganic Salts, Fruit Acids, Purine Alkaloids, Phenolic Compounds and Ethanol

A well-known phenomenon in *inorganic salts* is the salting-out effect. Adding sodium sulphate, ammonium sulphate or sodium chloride (common salt), for example, in portions to aqueous systems has the effect of driving out some of the volatile compounds into the gaseous phase, or into a solvent which is immiscible with water. Of the salts mentioned above, only common salt has any relevance to food. Additions of 5 to 15% to aqueous systems result in increases of head space concentration of ethyl acetate, isoamyl acetate and menthone up to 25% *[10,32]*. This common salt concentration, however, is way above what is tolerated normally in foodstuffs. In foods with a normal salt content, the salt has virtually no effect on the vapour pressure of volatile compounds *[9,10,32]*. The same is true for calcium chloride *[8]*. The possibility, that the salt content of the saliva has some effect on the vapour pressure cannot be ruled out however *[32]*.

Dry common salt binds acetone, ethanol and ethyl acetate (presumably adsorptively) *[6]*. The adsorption is reversible under vacuum.

The acetone concentration in the headspace above aqueous systems containing *citric acid* is clearly reduced *[8]*, whereas the vapour pressure of diacetyl above aqueous systems is hardly affected by *malic acid*. However, if the system contains both acids (for example 0.7% citric acid plus 0.1% malic acid), the odour threshold of limonene is doubled *[32]*. Nothing is known about the type of interactions. It cannot be ruled out that in foods with a high fruit acid content the release of flavour is negatively influenced (although the dilution of the food by saliva can again reverse the effect).

Aqueous systems containing *purine alkaloids* (e.g., caffeine or theobromine) or phenolic compounds (e.g., chlorogenic acid or naringin) lower the headspace concentration of a number of volatile flavouring substances (e.g., benzaldehyde, benzyl acetate, ethyl benzoate, furfural, various pyrazines and terpenes) *[16,53]*. At the same time, the solubility of the volatile compounds in water increases. There is still some uncertainty about the mechanism of these effects. There is speculation that 1:1 complexes of the alkaloids with one another increase the solubility of the flavouring substances *[16,53]*. In the case of phenolic compounds, chlorogenic acid and naringin are possibly involved in hydrogen bridge bonds *[16]*.

*Ethanol* can – like any alcohol – form acetals with aldehydes (Fig. 5.24). The formation of acetal is reversible. With an acid pH value ethanol and aldehyde are released again. In foods containing ethanol with a pH of 7, part of the aldehyde is bound as acetal.

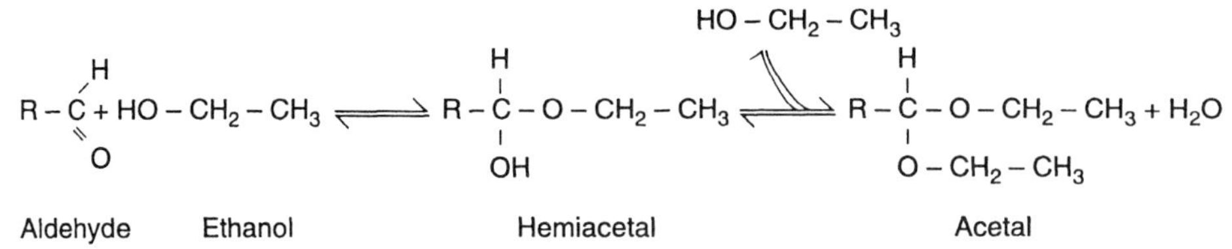

*Fig. 5.24: Hemiacetal and acetal formation from ethanol with aldehydes*

## 5.1.7 Interactions with Fat Replacers

A large body of literature implicates excessive fat consumption in the occurrence of several major chronic diseases. Therefore, governmental and independent health and science organisations have issued recommendations for reducing the quantities of fats consumed. This has generated a strong consumer and industrial interest in fat-reduced or fat-free versions of many foods [52,54-57].

In relation to flavour, fat has three functions in food:

(a) mouthfeel, (b) carrier of flavour, and (c) precursor to flavour. Fatty or oily mouthfeel is a combination of several parameters. These include: viscosity (thickness, body, fullness), lubricity (creaminess, smoothness), adsorption/absorption (physiological effect on taste buds), and others (cohesiveness, adhesiveness, waxiness) [52].

Flavourings consist of a blend of flavouring substances, which can be classified as lipophilic or hydrophilic. Fat or oil serves as carrier of lipophilic, and water as carrier of hydrophilic substances. Tab. 5.4 shows the variation in odour threshold values of selected flavouring substances when placed in water vs. oil. Due to these great variations, reduction in fat levels of foods will affect not only the intensity of the flavour but also its balance, since only little or no carrier system is available for lipophilic flavour components in water. The lipophilic part of the flavour cannot be retained in the food matrix and is released immediately. Fat-reduced or fat-free foods show high flavour impact initially which dissipates quickly, while full fat products gradually build up intensity and dissipate more slowly (schematically depicted in Fig. 5.25).

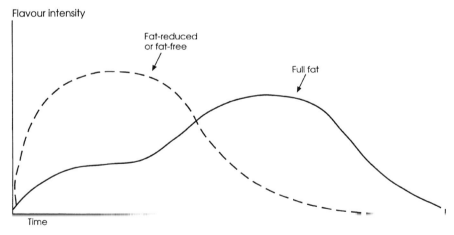

*Fig. 5.25: Schematic representation of time vs. flavour intensity curves of full fat and fat-reduced or fat-free foods when manipulated and warmed in the mouth [53]*

In many conventional food products (e.g. fried foods, cheese products) fat or oil serves as precursor to desirable flavouring substances. Known examples of such substances are the oxidation products of fatty acids. These are not present in fat-free foods. This lack of flavour often cannot be mimicked by simply adding a flavouring to the food matrix, since for many of these flavouring substances fat or oil is the only appropriate flavour carrier.

Currently available ingredients that have, or are claimed to have, fat-mimetic properties may be divided into 5 categories:

(1) Traditionally used emulsifiers, (2) lipid-based fat replacers, (3) carbohydrate-based replacers, (4) protein-based replacers, and (5) mixed-blend replacers. Of these, emulsifiers and lipid-based ingredients are true fat replacers.

The emulsifiers have essentially the same caloric value as fats. Thus, there is no advantage to use them as fat replacers. Lipid-based replacers (also referred to as synthetic fat substitutes) are non-caloric, but the majority of them are not approved for food use at this time (Tab. 5.5). Carbohydrate-based fat replacers include starches and their derivatives (Tab. 5.6) as well as non-starch hydrocolloids (Tab. 5.7). A selection of protein based and mixed-blend replacers is given in Tab. 5.8 and 5.9, respectively.

*Table 5.5:* Examples of lipid-based fat replacers (synthetic fat substitutes) [54, 56]

| Replacer | Trade name | Remarks |
|---|---|---|
| Caprocaprylobehenin | Caprenin[a] | Cocoa butter replacer |
| Dialkyl dihexadecyl malonate | DDM | For high temperature applications |
| Esterified propoxylated glycerol | EPG | For high temperature and cold applications |
| Sucrose polyesters | Olestra | For high temperature and cold applications |
| Trialkoxy citrate | TAC | For margarine and mayonnaise |
| Trialkoxy tricarballylate | TATCA | |

[a] Only Caprenin has FDA approval for use in foods

*Table 5.6:* Examples of carbohydrate-based fat replacers derived from starches [52, 55-59]

| Replacer | Trade name | Remarks |
|---|---|---|
| Starch | Stellar, Remyrise AP | From corn or rice |
| Modified starch | CrystaLean, N-Lite D, Slenderlean, Sta-Slim, Amalean I, Amalean II | From corn, potato, tapioca and other sources |
| Dextrin | N-Oil, Instant N-Oil | From tapioca |
| Maltodextrin | Lycadex, Maltin, Paselli Rice Tnn3 Complete, N-Oil II, Star-Dri, N-Lite B | From corn, potato, rice or tapioca (some are pregelatinized) |
| Blend of hydrolysed wheat, potato, corn and typioca starches | Colesta | |

# Interactions with Fat Replacers

*Table 5.7: Examples of carbohydrate-based fat replacers derived from non-starch hydrocolloids [52, 56-59]*

| Replacer | Trade name | Remarks |
|---|---|---|
| Cellulose | Avicel FD-100 | |
| Gellan gum | Kelcogel BF | |
| β-Glucan | Fibercel, Solka Floc, Oatrim | From yeast, oat and other sources |
| Golden pea fiber | CentuTex | |
| Hemicellulose | Fibrex, Fibrim, AF Fiber | From sugar beets, soybeans and almonds |
| Inulin[a] | Raftiline, Raftincreaming | From chicory roots |
| Konjac flour gel | Nutricol | |
| Pectin | Slendid | From citrus peel |
| Mixture of cellulose and carboxymethyl cellulose | Avicel RC-591 | |
| Mixture of cellulose and guar gum | Avicel RCN-10, Avicel RCN-15 | |
| Mixture of cellulose maltodextrin and xanthan gum | Avicel RCN-30 | |
| Mixture of gum arabic, alginate and modified starch | Pretested Colloid No Fat 102 | |

[a] Also considered a sucrose replacer

*Table 5.8: Examples of protein-based fat replacers [52, 56-59]*

| Replacer | Trade name | Remarks |
|---|---|---|
| Milk-derived solids | Dairylight | Mainly proteins |
| Milk-derived solids | Simplesse 100 | 23% Whey protein, 17% carbohydrates, 2% fat, 2% ash |
| Egg-derived solids | Simplesse 300 | 12% Protein, 10% carbohydrates, 0,1% fat |
| Zein | Lita | Protein from corn |

*Table 5.9: Examples of mixed-blend fat replacers [52, 56]*

| Replacer | Trade name | Remarks |
|---|---|---|
| Sucrose polyester and protein | Prolestra | |
| Egg protein and xanthan gum | Trailblazer | |
| Carrageenan and milk protein | Bindtex | For meat products |
| Oat bran, flavourings and seasonings | LeanMaker | For meat products |
| Pectin and gelatin | Slendid | |
| Polydextrose and soybean oil or milk fat | Ven-Lo | |

| Replacer | Trade name | Remarks |
| --- | --- | --- |
| Mono- and diglycerides, modified pregelatinized starch, guar gum and nonfat dry milk | N-Flate | For cake mixes |
| Modified starch, guar gum, nonfat dry milk and polyglycerol monoester | N-Lite F | For icings and dry frosting mixes |
| Egg and milk proteins corn syrup solids and modified starch | Ultra-Freeze | For frozen desserts |
| Starch, modified vegetable protein and xanthan gum | Ultra-Bake | For baking systems |

Schirle-Keller and co-workers [52,59] studied the interactions of 32 volatile flavouring substances with 9 fat replacers by measuring the equilibrium vapour pressures of the volatiles above systems containing water, emulsifier, fat replacer and volatiles (System 1). A model system containing only water, emulsifier and volatiles (System 2) served as the control for no interaction. Another model system containing water, emulsifier, fat and volatiles (System 3) represented the target interaction. Data are expressed as relative vapour pressure (RVP). RVP is the headspace concentration of a flavouring substance in system 1 or 3 divided by its headspace concentration in system 2. Thus, RVPs are expected to range between 0 and 1. A value of 1 represents no vapour pressure lowering from the system relative to water (or no interaction). Protein-based (Simplesse 100 and 300), carbohydrate-based (Avicel FD-100 and RC-591, N-Oil II, Oatrim, Paselli, and Stellar) and mixed-blend replacers (Slendid) were studied.

Simplesse 100 and 300 exhibited some fat-like interactions with saturated aliphatic aldehydes C6-C10 (hexanal, heptanal, octanal, nonanal, decanal), while carbohydrate-based and mixed-blend replacers showed no interaction (Fig. 5.26 and 5.27). Little or no interaction was noted between any of the fat replacers and the saturated aliphatic methyl ketones (Fig. 5.26 and 5.27). Unsaturated carbonyls showed more interactions with protein-based than with carbohydrate-based and mixed-blend replacers (Fig. 5.26 and 5.27). Of the sulfur components studied, propanethiol substantially interacted with Simplesse 100 and 300 (Fig. 5.26). Of these two protein-based replacers, Simplesse 100 showed some fat-like interaction with limonene (Fig. 5.27) and with the two esters ethyl caproate and ethyl heptanoate (Fig. 5.28).

It appears that protein-based fat replacers exhibit more fat-like flavour interactions than carbohydrate-based and mixed-blend replacers, with Simplesse 100 being the most fat-like in terms of promoting a balanced flavour. Note that Simplesse 100 contains 2% fat (Tab. 5.8). Its fat-like interaction with lipophilic flavouring substances may be due to this small amount of fat.

## 5.1.8 Interactions with Complex Systems and with Foodstuffs

So far, we have considered the interactions of volatile flavouring substances with individual classes of food components (carbohydrates, proteins, amino acids, lipids, inorganic salts, fruit acids, purine alkaloids, phenolic compounds and ethanol), and without the involvement of a second class. In the following section we shall look at more complex mixtures, and at a few foodstuffs [1,6,12,20,22,32,53,60-63].

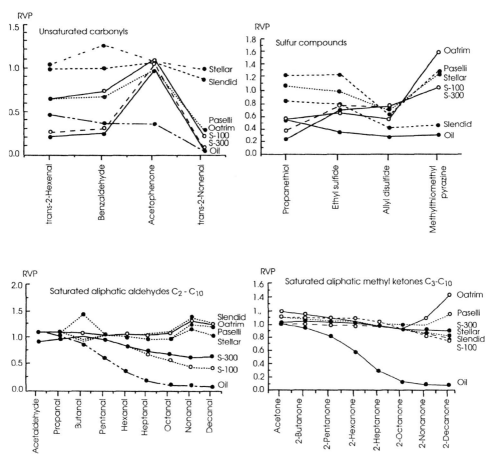

*Fig. 5.26:* Interactions between selected flavouring substances and fat replacers [52]. RVP = relative vapour pressure; S-100 and S-300 = Simplesse 100 and 300, respectively

The amount of hexanal and 2-hexanone which is bound by systems containing either *soy protein plus guar gum* or *guar gum plus an emulsifier* exactly corresponds to the sum of that bound by the individual components. If, however, *all three components* are mixed, there is significantly more binding of flavouring substances. In the system *starch plus glycerin-1 monostearate* the emulsifier forms inclusion complexes with starch. Because of this, less flavouring substance (e.g., decanal, menthone) can be bound to starch, because the inclusion complex is formed by preference with the stearate. The system *protein plus emulsifier* binds less benzyl alcohol than the protein in the absence of the emulsifier.

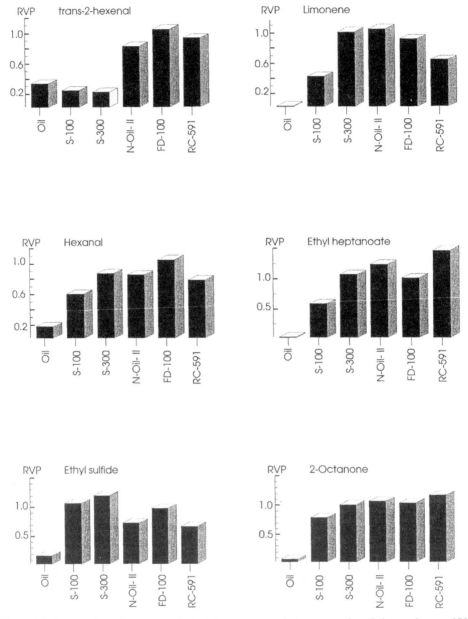

*Fig. 5.27:* Interactions between selected flavouring substances and and fat replacers [52]. RVP = relative vapour pressure; S-100 and S-300 = Simplesse 100 and 300, respectively; FD 100 and and RC-591 = Avicel FD-100 and Avicel RC-591, respectively

Ketones (e.g., acetone) and alcohols (e.g., ethanol) are strongly bound to *whole milk powder* (24.3% lipids, 25.4% protein) and skimmed milk powder (0.6% lipids, 31.6% protein), fairly strongly to *potato flakes* (72.2% polysaccharide, 8% protein), but only weakly to *instant coffee powder* and *strawberry powder*. The interactions of the acetone with *whole milk, skimmed milk and edible oil* have already been shown in Fig.

5.2 and 5.3. Flavoured *black tea* binds considerable quantities of citronellol, citral and phenylethanol. Aggregates of caffeine, theaflavinene, thearubiginene and protein in the tea play a crucial role in this. Such aggregates are referred to as "tea creams".

A detailed study of the flavour binding capacity of a *"model wine"* was published recently [62]. The basis of this model wine is an aqueous solution made of ethanol (10%), fruit acids (pyruvic, malic and acetic acids, totalling 0.71%) and inorganic salts (magnesium sulphate and calcium sulphate, totalling 0.0125%). Even this basic building block binds tiny amounts of volatile flavouring substances (hexanol, isoamyl acetate, ethyl caproate, ß-ionone). The ethanol content presumably is responsible for this [46]. Subsequently various ingredients were added one at a time to this basic building block: protein (sodium caseinate 0.1-1%), amino acid (arginine 0.1-1%), polysaccharide (mannan 0.1-1%) and various phenolic compounds (anthocyanes). This showed that, as far as the macromolecules were concerned, hexanol was bound only weakly, while isoamyl acetate, ethyl caproate an ß-ionone were strongly bound to the protein and the polysaccharide. The low molecular compounds (amino acids, phenolic compounds) virtually had no effect on flavour binding at these concentrations.

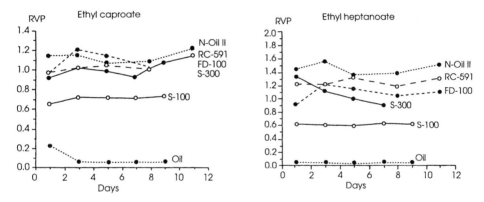

*Fig. 5.28: Interactions between selected esters and fat replacers (during storage at 4°C) [59]. RVP = relative vapour pressure; RC-591 and FD-100 = Avicel RC-591 and Avicel FD-100, respectively; S-300 and S-100 = Simplesse 300 and Simplesse 100, respectively*

Rosin and Tuorila [63] did some comparative research on the odour and flavour intensity of garlic oil and pepper oleoresin in *beef stock* and in *mashed potatoes* with *no fat* and with *10 percent fat* content respectively. Using the same dose the intensity of garlic oil was perceived nearly equal in all three foods, i.e. texture (liquid or solid), carbohydrate content and fat content of the foods had almost no influence. In contrast, pepper oleoresin was perceived far better in the liquid beef stock than in the two types of mashed potatoes. In those containing fat, the oleoresin was perceived weaker than in the fat free potatoes. This allows conclusions as to the flavour binding capacity of the fat.

Harrison and Hills [64] developed a mathematical model to describe flavour release from aqueous solutions containing aroma-binding macromolecules. Seuvre et al. [65] reported that the retention of flavouring substances (benzaldehyde, isoamyl acetate, linalool, 2-nonanone) in a multicomponent medium like the food matrix is influenced by their affinity with the protein when lipid is present at a low level (0.5%).

## 5.1.9 Conclusion

Table 5.10 summarizes the types of interactions of volatile flavouring substances with individual food components. This summary reveals that some reactions which take place at molecular level are known in qualitative terms while others are still unknown. Quantitatively speaking, no exact prognoses can be made about the interactions.

*Table 5.10: Interactions between volatile flavouring substances and individual food constituents*

| Food Component | System | Type of interactions |
|---|---|---|
| Carbohydrates | | |
| – Starches and cyclodextrins | Aqueous | Inclusion complexes |
| – Pectin and alginate | Dry or moist | Electrovalent and covalent bonding? |
| – Guar and agar-agar | Dry or moist | Not known |
| – Cellulose and methyl cellulose | Dry or moist | Hydrogen bridges? |
| – Dextrins and maltodextrins | Aqueous or dry | Adsorption |
| – Simple sugars (mono- and disaccharides) | Aqueous | Not known |
| | Dry | Adsorption |
| Proteins | Aqueous | Hydrophobic inter-actions, hydrogen bridges, chemical reactions |
| Amino acids | Aqueous | Hydrogen bridges, chemical reactions |
| | Dry or moist | Adsorption Chemical reactions |
| Lipids | Lipid/water | Distribution law Adsorption on emulsion boundary surfaces? Hydrogen bridges? |
| Inorganic salts | Aqueous | Salting-out effect |
| Fruit acids | Aqueous | Not known |
| Purine alkaloids | Aqueous | Not known |
| Phenolic compounds | Aqueous | Hydrogen bridges? |
| Ethanol | Aqueous | Acetal formation |

The picture is very confused in most foodstuffs, which are mixtures of several of these individual ingredients. In practical applications this means that – in spite of the basic knowledge of possible interactions outlined here – in flavouring industrially produced foodstuffs, almost every product must be considered individually. For this reason, it is advisable for the food technologist to involve flavour specialists (flavourists, applications technologists, evaluation board) in product development as early as possible. Of course, an essential prerequisite is collaboration between flavour specialists and product developers from the food industry based on mutual trust and confidence. This can extend even as far as providing the flavouring specialists not only with outline ideas but with the exact formulation of the new product. The original ingredients (customer's bases) should be made available. This is the only way to ensure that possible interactions between volatile flavouring substances and the original ingredients can be recognized and taken into consideration. This is the only way to end up with a product which is flavoured optimally, and really marketable (Fig. 5.29).

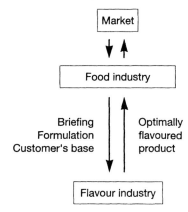

*Fig. 5.29: Collaboration between the food industry and the flavour industry*

## REFERENCES

[1]   Solms, J. and Guggenbuehl, B. Physical Aspects of Flavor Application in Food Systems. In: Flavor Science and Technology (Bessiere, Y. and Thomas, A.F., eds.), New York, John Wiley, pp. 319-335 (1990)
[2]   Overbosch, P., Afterof, W.G.M. and Haring, P.G.M. Food Rev. Intern. 7, 137-184 (1991)
[3]   Birch, G.G. and Lindley, M.G. (eds.) Interactions of Food Components. London, Elsevier 1986
[4]   Thomson, D.M.H. The Meaning of Flavour. In: Developments in Food Flavours (Birch, G.G. and Lindley, M.G., eds.), London, Elsevier, pp. 1-21 (1986)
[5]   Matheis, G. Dragoco Report. Flavouring Information Service 39, 50-65 (1994)
[6]   Maier, H.G. 1975. Proc. Intern. Symp. Aroma Res. Zeist. Pudoc, Wageningen, pp. 143-157 (1975)
[7]   Nawar, W.W. J. Agric. Food Chem. 19, 1057-1059 (1971)
[8]   Voilley, A., Simatos, D. and Loncin, M. Lebensm. Wiss. Technol. 10, 45-49 (1977)
[9]   Land, D.G. and Reynolds, J. 1981. The Influence of Food Components on the Volatility of Diacetyl. In: Flavour '81 (Schreier, P., ed.), Berlin, Walter de Gruyter, pp. 701-705 (1981),
[10]  Ebeler, S.E., Pangborn, R.M. and Jennings, W.G. J. Agric. Food Chem. 36, 791-796 (1988)
[11]  Maier, H.G. Bindung flüchtiger Aromastoffe an Lebensmittelbestandteile. Lecture on the 43rd congress of the Food Industry Research Association (1985)
[12]  Franzen, K.L. and Kinsella, J.E. Chemistry & Industry, pp. 505-509 (1975)
[13]  Solms, J. Aromastoffe als Liganden. In: Geruch und Geschmackstoffe (Drawert, F., ed.), Nürnberg, Hans Carl, pp. 201-210 (1975)
[14]  Rutschmann, M.A. and Solms, J. Lebensm. Wiss. Technol. 23, 70-79 (1990)
[15]  Wyler, R. and Solms, J. Inclusion Complexes of Potato Starch with Flavour Compounds. In: Flavour '81 (Schreier, P., ed.)., Berlin, Walter de Gruyter, pp. 693-699 (1981)
[16]  Solms, J. Interactions of Non-Volatile and Volatile Substances in Food. In: Interactions of Food Components (Birch, G.G. and Lindley, M.G., eds.), London, Elevier, pp. 189-210 (1986)
[17]  Schmidt, E. and Maier, H.G. GIT Supplement 2, 40-45 (1989)
[18]  Rutschmann, M.A. and Solms, J. Lebensm. Wiss. Technol. 23, 80-83 (1990)
[19]  Rutschmann, M.A. and Solms, J. Lebensm. Wiss. Technol. 24, 473-475 (1991)
[20]  Solms, J. Lebensmittelmatrix und Aroma. Paper presented to the forum "Aroma von Lebensmitteln" on 20/21 February 1992 in Bad Honnef, arranged by Behr's Seminare, Hamburg (1992)

[21] van Osnabrugge, W. Food Technol. 43, 74-82 (1989)
[22] King, B., Wyler, R. and Solms, J. Problems of Flavour Application in Food Systems. In: Progress in Flavour Research (Land, D.G. and Nursten, H.E., eds.), London, Applied Science, pp. 327-335 (1979)
[23] Lebert, A. and Richon, D. J. Agric Food Chem. 32, 1156-1161. (1984)
[24] Szejtli, J., Lösungsvermittlung und Stabilisierung durch Cyclodextrine. Paper presented to the seminar "Orale Liquida: Lösungen, Suspensionen und Brausetabletten" 10/12 February 1992 in Bonn, arranged by the Arbeitsgemeinschaft für pharmazeutische Verfahrenstechnik (APV), Mainz (1992)
[25] Szejtli, J., Cyclodextrins and their Inclusion Complexes. Akadeiai Kaido, Budapest 1982
[26] Reineccius, G.A. and Risch, S.J., Perfumer and Flavorist 11, 1-6 (1986)
[27] Reineccius, G.A., Food Rev. Intern. 5, 147-176 (1989)
[28] Asche, W., BioEngineering 7, 27-29 (1991)
[29] Jackson, L.S. and Lee, K., Lebensm. Wiss. Technol. 24, 289-297 (1991)
[30] Szejtli, J. (ed.), Proc. 1st Intern. Symp. Cyclodextrins. Reidel Publishing, Boston 1982
[31] Szente, L. and Szejtli, J., Abh. Akad. Wiss. DDR, Abt. Math. Naturwiss., Techn. pp. 101-106 (1988)
[32] Land, D.G., Some Factors Influencing the Perception of Flavour-Contributing Substances in Food. In: Progress in Flavour Research (Land, D.G. and Nursten, H.E., eds.), London, Applied Science, pp. 53-66 (1979)
[33] Arai, S., Masatochi, N., Yamashita, M., Kato, H. and Fujimaki, M., Agric. Biol. Chem. 34, 1569 – 1573 (1970)
[34] Beyeler, M. and Solms, J., Lebensm. Wiss. Technol. 7, 217-219 (1974)
[35] Damodaran, S. and Kinsella, J.E., J. Agric. Food Chem. 28, 567-571 (1980)
[36] Damodaran, S. and Kinsella, J.E., J. Agric. Food Chem. 29, 1249-1253 (1981)
[37] Damodaran, S. and Kinsella, J.E., J. Agric. Food Chem. 29, 1253-1257 (1981)
[38] Damodaran, S. and Kinsella, J.E. J. Agric Food Chem. 31, 856-859 (1983
[39] Chung, S. and Villota, R., J. Food Sci. 54, 1604 -1606 (1989)
[40] O'Neill, T.E. and Kinsella, J.E., J. Agric. Food Chem. 35, 770-774 (1987)
[41] Jouenne, E. and Crouzet, J. J. Agric. Food Chem. 48, 1273-1277 (2000)
[42] Andriot, I., Harrisson, M., Fournier, N. and Guichard, E. J. Agric. Food Chem. 48, 4246-4251 (2000)
[43] O'Keefe, S.F., Wilson, L.A. Resurreccion, A.P. and Murphy, P.A. J. Agric. Food Chem. 1022-1028 (1991)
[44] Dufour, E. and Haertle, T. J. Agric. Food Chem. 38, 1631-1692 (1990)
[45] Damodaran, S. The Interaction of Carbonyls with Proteins. Dissertation, Ithaca, Cornell University 1981
[46] Mills, O.E. and Solms, J. Lebensm. Wiss. Technol. 17, 331-335 (1984)
[47] Hansen, A.P and Heinis, J.J. J. Dairy Sci. 75, 1211-1215 (1992)
[48] Pelletier, E., Sostmann, K. and Guichard, E. J. Agric. Food Chem. 46, 1506-1509 (1998)
[49] McLeod, G. and Ames, J. Crit. Rev. Food Sci. Nutr. 27, 219-400 (1988)
[50] Chobpattana, W., Jeon, I.J. and Smith, J.C. J. Agric Food Chem. 48, 3885-3889 (2000)
[51] Belitz, H.-D. and Grosch, W. Lehrbuch der Lebensmittelchemie, Berlin, Springer (1982)
[52] Schirle-Keller, J.-P, Reineccius, G.A. and Hatchwell, L.C. The Interaction of Flavors with Fat Replacers: Effect of Oil Level on Flavor Interactions and Data on Homologous Series of Flavor Compounds. Presented at IFT Meeting in New Orleans, June 19-23, 1992
[53] King, B.M. and Solms, J. Interactions of Volatile Flavor Compounds with Caffeine, Chlorogenic Acid and Naringin. In: Flavour '81 (Schreier, P., ed.), Berlin, Walter de Gruyter, pp. 707-716 (1981)
[54] Glicksman, M. Food Technol. 45(10), 94, 96-101, 103 (1991)
[55] Shamil, S., Wyeth, L.J., and Kilcast, D. Food Qual. Pref. 3, 51-60 (1991/1992)
[56] Setser, C.S. and Racette, W.L. CRC Crit. Rev. Food Sci. Nutr. 32, 275-297 (1992)
[57] Bennett, C.J. Cereal Foods World 37, 429-432 (1992)
[58] Plug, H. and Haring, P. Tr. Food Sci. Technol. 4, 150-152 (1993)
[59] Schirle-Keller, J.P., Chang, H.H. and Reineccius, G.A. J. Food Sci. 57, 1448-1451 (1992)
[60] Rutschmann, M.A. and Solms, J. Lebensm. Wiss. Technol. 23, 451-456 (1990)

[61] Rutschmann, M.A. and Solms, J. Lebensm. Wiss. Technol. $\underline{23}$, 457-464 (1990)
[62] Voilley, A., Beghin, V., Charpentier, C. and Peyron, D. Lebensm. Wiss. Technol. $\underline{24}$, 469-472 (1991)
[63] Rosin, S. and Tuorila, H. Lebensm. Wiss. Technol. $\underline{25}$, 139-142 (1992)
[64] Harrison, M. and Hills, B.P.A. J. Agric. Food Chem. $\underline{45}$, 1883-1890 (1997)
[65] Seuvre, A.M., Espinosa Díaz, M.A. and Voilley, A. J. Agric. Food Chem. $\underline{48}$, 4296-4300 (2000)

## 5.2 Flavourings for Beverages

### 5.2.1 Soft Drinks – An Innovative Product Category
*Matthias Saß*

Non-alcoholic beverages, of which soft drinks are a subset, can be divided in three main areas. For the first two categories, legal standards have been established within the European Union; this is not the case for soft drinks (Table 5.11).

Some countries have local legislative directives in place, but there is no harmonised European law for soft drinks. Nevertheless, some horizontal legislation needs to be considered (Table 5.12). Additives, like colours, sweeteners and others, are established within the European Union.

*Table 5.11:* Overview of vertical EU food law related to non-alcoholic beverages

| Legal category | Related vertical food law |
| --- | --- |
| Juices and nectars | Council Directive 2001/112/EC of 20.12.2001 *[1]* |
| Natural mineral water | Council Directive 80/777/EEC of 15.06.1980 *[8]* |
| Soft drinks | No harmonised food law within EU |

*Table 5.12:* Overview of horizontal EU food law, related to non-alcoholic beverages

| Legal category | Related horizontal food law |
| --- | --- |
| Food additives – general | Council Directive 89/107/EEC of 21.12.1988 *[4]* |
| Colours for use in foodstuffs | Council Directive 94/36/EC of 30.06.1994 *[5]* |
| Sweeteners intended for use in foodstuffs | Council Directive 94/35/EC of 30.06.1994 *[6]* |
| Additives for other purposes | Council Directive 95/2/EC of 20.02.1995 *[7]* |
| Nutritional labelling for foodstuffs | Council Directive 90/496/EEC of 24.09.1990 *[2]* |
| Novel foods and food ingredients | Regulation (EC) No 258/97 of the European Parliament and of the Council of 27.01.1997 *[3]* |

Because of no vertical food law applicable to this category, soft drinks are in many cases acknowledged as innovative and flexible.

A technological overview of soft drinks can be visualised as a 'soft drink circle', as shown in Fig. 5.30. This overview shows the broad variability of raw materials used in soft drinks.

*Fig. 5.30: The soft drink circle*

The soft drink circle should be read from inner to outer circle. First of all, the world of soft drinks can be categorised into 'clear' and 'cloudy' products, as further described in 5.2.1.1 and 5.2.1.2. In these chapters, more information can be found on important raw materials, like juices, extracts and emulsions. Within these main categories of soft drinks, it is also important to differentiate between juice specialities, flavours, source of extracts and the principal bases of emulsions.

The three outer closed circles characterise the raw materials that will be used in all types of soft drinks. More information on colours, the important area of health and nutrition and sweeteners can be found in 5.2.1.3. All remarks on single ingredients are focused on their importance in soft drinks.

To return to the two main different types of soft drinks: one is a clear product, with virtually no or only very slight turbidity visible to the consumer; the other type is a turbid product with different degrees of turbidity from slight haze to complete opacity. Figure 5.31 visualises the two categories by measuring the FNU value (Formazin Nephelometric Unit).

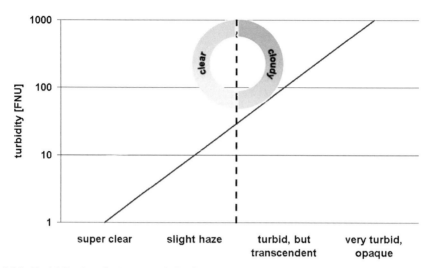

*Fig. 5.31:* Turbidity in relation to soft drink categories

### 5.2.1.1 Clear Soft Drinks

**Soft Drinks Based on Flavours**

A simple way to obtain a clear soft drink is adding clear, water-soluble flavours, so-called essence flavours. The most popular products, usually derived from ethanol-extracted and distilled citrus peel oils, provide a subtle taste of lemon or lime. The dosage of these flavours is around 1 g per litre of final drink. Together with sweetener, acidifier and carbonation, one gets a fizzy clear and colourless drink with citrus taste.

In combination with clear water-soluble dyes, fancy colour shades can be achieved. In conjunction with other fruit flavours, unlimited combinations are possible.

Besides the classic carbonated, clear, sweetened and flavoured soft drink, a new category has appeared on the market during the last few years: 'Near Water Drinks'. Starting in the Japanese market, this soft drink category can now be found around the world. The basic concept is a combination of plain water with low amounts of, preferably natural, sweeteners (see 5.2.1.4), but with no artificial sweeteners added, to underline the natural concept. Thus, the calorie content is in many cases below 20 kcal/100 ml, allowing claims with regard to low calorie content. Functional ingredients, like vitamins and/or minerals, are added to provide extra benefits for the consumer. These drinks are not as sweet as classic soft drinks, but are a refreshing alternative, targeting the adult consumer.

Another important type of soft drink in this category are energy drinks. These are slightly coloured, heavily sweetened, high-caffeine-containing drinks with other functional ingredients, like vitamins and/or amino acids, added. Due to the very high caffeine content, sometimes more than 300 mg/l, three times higher than classic cola drinks, these drinks provide a strong stimulating effect.

### Soft Drinks Based on Plant Extracts

Cola and tea drinks are the most popular examples in this category. The cola flavour is a complex mixture of spice and citrus extracts. Together with colourings from sugar, so-called sugar couleurs, and a special acidifier, phosphoric acid, one gets a deep dark brown coloured, but clear drink with the most popular taste in the world. Caffeine is added as a 'functional' ingredient to provide a stimulating effect and add a hint of bitterness to the taste.

Tea drinks or 'iced teas' are refreshing drinks containing soluble tea solids. Depending on tea extract quality, one can obtain turbid or clear products with a specific astringent tea flavour and red brownish colour. Flavours and colours alone are used for obtaining drinks with an authentic tea flavour and no caffeine to serve the children's market. In this area, 'fruit teas' with bright red hibiscus extracts are also common.

Even if some plant extracts do not lead to a 100% clear beverage with turbidity below 1 FNU, they are nonetheless part of this group, because the usual turbidity that comes from plant extracts in a fresh drink will not exceed 50-100 FNU, the borderline for a cloudy drink.

### Soft Drinks Based on Clear Juices

This is a very demanding category with regard to raw material quality. Because of the complex composition of a juice, it is very difficult to obtain one that is clear and stays clear after reconstitution from concentrate to the final drink. The 'apple spritzer' in Germany is a very popular example. The drinks are sparkling and the juice content is around 50%, with no sugar or other sweetener added. This provides a very refreshing, calorie-reduced drink with low sweetness. To produce this product only from 'super clear' apple juice concentrates without preservatives requires a hygienic plant, strict microbiological control and raw materials from suppliers that guarantee virtually no turbidity, even months after bottling.

### 5.2.1.2 Cloudy Soft Drinks

This is by far the largest category within the world of soft drinks. In many cases, the turbidity provides a criterion for quality to the consumer. Turbidity is a visual proof that there are ingredients in the drink, beside sweeteners and acidifiers.

### Soft Drinks Based on High Juice Content

High juice content is defined by a content of more than 20% by weight in the final drink. In this category, the turbidity derives from a stable colloidal dispersion of fruit cells. Stability and quality of the final drink are mainly based on the quality of fruit juice concentrates formulated into the product. Additional stabilisers, mainly hydrocolloids, might be added.

In this category, citrus juices are by far the most common raw materials. Orange juice is a multidisperse system of suspended fruit particles and emulsified essential oils and provides, in many cases, the backbone of this kind of drink. Acidity is added by using lemon juices. A variety of juices is added to provide extra taste, e.g. passion fruit, pineapple, mango, peach and apricot. Others are added to provide colour as well as taste, e.g. carrot, elderberry, aronia and blackcurrant *[12]*.

These drinks are mainly produced as still drinks. Depending on the other added ingredients, many subcategories have been established. Still drinks are often fortified with healthy functional ingredients. This started in the 1970s by adding vitamins and minerals and continued during the 1980s with calorie reduction using artificial sweeteners. In the 1990s and the beginning of the new century, many other healthy ingredients, e.g. dietary fibre, omega-3 fatty acids, prebiotics, soy extracts and polyphenols, were added (see 5.2.1.5.)

**Soft Drinks Based on Low Juice Content**

Low juice content is defined by a content of 1 to 20% by weight of juice in the final drink. These products are mainly based on citrus juices. As mentioned before, citrus juices provide good turbidity and stability even with low juice content. In combination with special raw materials made from citrus, like cloudy concentrates or peel oils, very stable low-juice drinks can be obtained. Turbidity comes from a complex mixture of dispersions of fruit cells and oil-in-water emulsions, stabilised by fruit pulp particles and hydrocolloid stabilisers, such as locust bean gum *[9]*.

These drinks are marketed as still or carbonated drinks. As with their higher juice-containing counterparts, the replacement of sugar by artificial sweeteners as well as the addition of healthy functional ingredients is common.

**Soft Drinks Based on Emulsions**

One smart and economically efficient way to obtain highly turbid beverages is the use of emulsions *[11]*. In the following, the manner in which emulsions provide optically perfect soft drinks is explained. The turbidity comes from homogeneously dispersed oil droplets in the beverage, which scatter and reflect light. To explain the physical stability a look at Stokes' law will help:

$$v = \frac{2(\rho_K - \rho_M) g r^2}{9\eta}$$

$v$ = velocity of sedimentation or floating of oil droplet
$\rho_K - \rho_M$ = difference in relative density between oil and surrounding medium
$r$ = radius of oil droplet
$\eta$ = dynamic viscosity of the surrounding medium
$g$ = earth gravitational force

The equation shows that the most interesting parameters for the stabilisation of emulsions are the difference in relative density of oil droplet to soft drink medium (Fig. 5.32) and the radius of the individual oil droplets (Fig. 5.33). The dynamic viscosity is of lesser importance as soft drinks are usually of low viscosity and nobody wants to drink products that are thickened like pudding.

The usual relative density of oil is less than that of water. To increase the density, oil-soluble materials, commonly known as 'weighting agents', are used. These are oil-soluble ingredients with a relative density greater than that of water. The main ingredient used in this context is glycerol ester from wood resin (E445). In Europe and the USA, it might be added up to 100 mg/l final drink.

# Soft Drinks – An Innovative Product Category

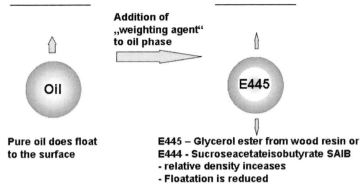

*Fig. 5.32:* Addition of weighting agent to oil phase to reduce buoyancy

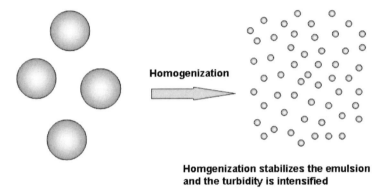

*Fig. 5.33:* Reduction of oil droplet size

It is very important to stabilise the individual oil droplets to prevent coalescence, which will increase droplet size and destabilise the emulsion. This is prevented by using a stabiliser, as shown in Fig. 5.34. For beverage emulsions the most important stabilisers are gum arabic (E414) and octenyl succinate starch, a modified food starch (E1450).

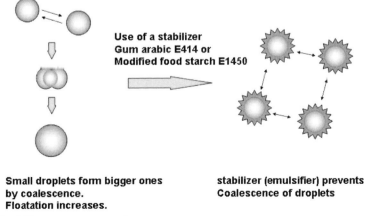

*Fig. 5.34:* Stabiliser to prevent coalescence

All emulsions in this area are oil-in-water (o/w) emulsions. Raw materials for this kind of emulsions can be derived from several sources. The main categories are listed in Table 5.13.

*Table 5.13: Types of emulsions*

| Raw material | Stabilisers | Usual dosage | Turbidity range | Taste |
|---|---|---|---|---|
| Weighted essential oils (from citrus peels) | Hydrocolloids, like gum arabic or modified starch | 1-2 g per litre | 100-500 FNU | Citrus-like |
| Unweighted vegetable oils | Hydrocolloids, like gum arabic or modified starch | 10-50 g per litre | 1000-2000 FNU | Neutral |
| Cows milk (yoghurt) | Proteins, pectin | 100-200 g per litre | 500-2000 FNU | Depending on source, animal |
| Soy bean extract | Proteins, pectin | 100-200 g per litre | 500-2000 FNU | Neutral, vegetable |

In some cases, emulsions are combined with juice-containing drinks to provide additional turbidity and/or taste to the final drink *[10, 11]*.

### 5.2.1.3 Colours for Soft Drinks

Generally, the visual appearance, and especially the colouring, of foods currently have a significant effect on purchasing decisions. Colours that the consumer associates with the product's raw materials also tend to be linked with freshness, quality and good flavour in the finished product. Despite synthetic colours, natural colours and colouring foodstuffs are increasingly being used.

The following parameters can affect the stability and character of the colours used and must, therefore, be taken into account when applying colours *[13]*:

- pH value of the matrix: the colour can take on different nuances, depending on the pH value of the matrix in which the colour is to be used.
- Light: some colours may fade when exposed to light. In this case, opaque packaging, such as cartons or cans, is recommended.
- Oxygen: carotenoids may fade when exposed to oxygen. Colour changes caused by oxidation can be delayed by adding antioxidants like ascorbic acid.

**Colour Shade: Yellow to Orange**

For colours ranging from yellow to orange, beverage manufacturers have a choice of various natural colorants, depending on the product to be coloured.

Riboflavin (also known as vitamin B2) is often used, for example, in energy drinks, as it is valued not only for its yellow colouring effect, but also for its positive properties as a functional ingredient.

Lutein provides a brilliant yellow colouring within a wide range of pH levels. There is also a positive side effect attributed to this natural colour extract from the genus Tagetes: lutein is supposed to have positive effects on the long-term health of the human eye.

Citrus oil is a natural flavouring derived from citrus peel that has a colouring effect with a light yellow shade.

Mixed carotenes, such as those derived from palm oil, also serve as a yellow colorant. They are made up of a mixture of various carotenes, including alpha-, beta-, and several other carotenes.

Beta-carotene is the colorant most frequently used in the beverage industry. Depending on the particular beverage, it produces various colour hues from yellow to red-orange.

Extracts from red pepper are also used as a colouring foodstuff in order to colour beverages without the use of additives.

**Colour Shade: Red**

Carmine, betanin and anthocyanins are the ingredients most frequently used to give beverages a red-coloured hue.

Carmine is the most stable, natural red, water-soluble pigment. It is extracted from the insect Dactylopius coccus costa and for this reason is available only as a colorant. The beverage industry uses special acid-stable qualities that achieve colour hues ranging from orange to red-violet, depending on the production method used.

Beets contain betanin as a colouring pigment and there remains some shade of red at almost any pH level. Because of their low heat stability, beets are mainly used in cold produced, aseptically filled beverages such as milk drinks.

The colouring pigment anthocyanin can be extracted from fruit and vegetable concentrates, including elderberries, chokeberries, purple carrots and red grapes. Anthocyanins are available as a colouring foodstuff, juice concentrate or colour extract.

Anthocyanins have a much better light and heat stability than, for example, beets. This makes them especially suitable for use in fruit juice drinks, spritzers, etc. They are characterised by rich, brilliant red colour hues. Due to their chemical structure, anthocyanins from vegetables such as purple carrots or red cabbage are much more stable than fruit anthocyanins.

The colour hue of the different anthocyanins varies as a function of the anthocyanin mixture typical of the particular fruit, ranging from strawberry red (red-orange) to blueberry red (red-violet). The colouring of anthocyanins is also highly dependent on pH. The best pH value for a red colour is from 2 to 3.8. A higher pH results in colours from blue to green (the colour pigment destabilises and precipitates). This is why it is absolutely essential that anthocyanins are used within the correct pH range and that the pH value is correctly adjusted before they are added to the beverage.

**Colour Shade: Brown**

Caramel colour is the most frequently used colorant for brown beverages. It is available in four different types: Type I, II, III, IV (E150a, b, c, d). Each type is used for specific applications, e.g.:

- Type I for whiskey
- Type II for brandy and non-alcoholic beverages
- Type III for beer and protein containing drinks
- Type IV for non-alcoholic beverages such as colas.

Because of their molecular weight and different charges, caramel colours react in many different ways when combined with other beverage constituents. This means it is extremely important to know exactly which type can be used for which application.

Dark malt can be used as a colour alternative for some caramel colours and listed as a colouring foodstuff. However, malt also has a flavouring effect that can be disturbing in some applications. The dosage is around five times that for Type IV caramel colour, which makes it almost impossible to use malt without significantly affecting the flavour. For this reason, malt extract is often used in typical malt beverages.

**Colour Shade: Green**

Spinach and stinging nettle extracts are used as green colouring foodstuffs in many preparations. These extracts are not acid-stable because their colour pigment is magnesium chlorophyll, which is unstable at low pH and changes to olive-green or grey. Copper salts can be added to the chlorophyll to prevent this degradation. The resulting copper chlorophyll retains its stable green colouring even at a low pH.

### 5.2.1.4 Sweeteners for Soft Drinks

An important characteristic of soft drinks is their sweet taste, which is derived from natural sweetening concentrates as described above. However, other raw materials are also used. Table 5.14 provides an overview of the various sweeteners.

*Table 5.14:* Overview of sweeteners

| Sugar | Other sugars | Sugar substitutes | High-intensity sweeteners | |
|---|---|---|---|---|
| | | | Artificial | Natural origin |
| Sucrose | Glucose<br>Inverted sugar<br>Corn syrup<br><br>High-fructose corn syrup (isoglucose)<br><br>Honey | Fructose<br><br>Maltitol<br>Sorbitol<br>Isomalt<br>Xylitol<br>Mannitol<br>Trehalose<br><br>Tagatose[a]<br>Isomaltulose[a] | Acesulfame-K<br>Aspartame<br>Cyclamate<br>Saccharine<br>Sucralose | Neohesperidin DC<br>Thaumatine<br><br>Stevia extract[b]<br>Lo-Han extract[b] |

[a] Novel food
[b] Not permitted in the European Union

## Sugar Types

The various types of sugar have traditionally been the most common sweeteners employed in soft drinks. Sucrose, invert sugar syrup, high-fructose corn syrup and corn syrup perform three functions in beverages:

- sweet taste
- contribute mouthfeel
- intensify overall flavour.

The sensory appearance of sucrose is used as the reference value for the desired sweetness impression. Comparable sweetening can be achieved with invert sugar syrup or high-fructose corn syrups.

Sucrose from cane or beets is a disaccharide composed of fructose and glucose. It is highly soluble in water. For this reason, liquid sucrose syrups with a sucrose content of up to 65% are available for industrial use. In the presence of the low pH values typical of soft drinks, sucrose is gradually 'inverted', meaning it is split into its two components, glucose and fructose. The process occurs automatically as a function of storage temperature, time and pH value. It can also be referred to as 'maturing', since it rounds off the flavour of the beverage. Consequently, sucrose inversion has a desirable effect on flavour and is an important factor, for example, in the production of cola beverages. Invert sugar syrup is produced by partially inverting sucrose to fructose and glucose. The inversion process improves the sugar's solubility, which makes it possible to produce syrups with up to 73% dry matter. The 'maturing' of the beverage through sucrose inversion described above can already be accomplished by using invert sugar, which has a positive effect on flavour.

Corn syrups are made from raw materials containing starch (corn, wheat). Starch is a polymer of glucose. When corn syrups are produced, this polymer is hydrolysed into individual glucose molecules. Depending on how thoroughly the process is applied, it can result in maltodextrin, which still contains polymer groups, or very strong glucose syrups. The sweetening power of these syrups is much lower than the sweetening power of the same amount of sucrose or invert sugar, making them unsuitable for usage as sweeteners in beverages. However, these products are extremely useful when combined with other sweeteners to improve mouthfeel. High-fructose corn syrup (HFCS) is produced from corn syrup in which part of the glucose has been converted to fructose. Another name for HFCS is 'isoglucose'. Dependent on the conversion ratio the sweetening power is increased up to 95% (HFCS 42) or even 104% of the level of an equal concentration of sucrose syrup. HFCS is employed as a soft drink sweetener, especially in North America where large quantities of glucose syrup can be extracted from corn under economically favourable conditions. All the sugar types described here contain approximately four kilocalories per gram of dry matter.

## Sugar Substitutes

Sugar substitutes have similar sweetening power and technological properties as sucrose but with a different physiological effect. The sugar substitute most commonly used in beverages is fructose (also called fruit sugar). Fructose is absorbed relatively slowly and raises the blood sugar level only slightly. For this reason, it is often employed in diet soft drinks in combination with artificial sweeteners. Fructose is also suitable for use in sport drinks, where sustained carbohydrate availability for prolonged exercise is desired.

Trehalose is a relatively new sugar substitute. In 2001, the EU approved this substance as a novel food. Like fructose, it is metabolised very slowly, making it ideal for diabetic products and sport drinks. However, its sweetening power is only half that of sucrose. A similar product is isomaltulose, approved as a novel food ingredient in 2005, which provides long energy release (low glycemic) with the same caloric content of sucrose but only 40% of sweetness level, making it suitable for sport drinks. It has no cariogenic effect. This is only important for confectionary, because in acid soft drinks, the main effect on teeth health is provided by the low pH.

Tagatose is a novel food ingredient approved in 2006, which provides less sweetness than sucrose and is low glycemic. In the European Union, it is only allowed in soft drinks up to usage levels of 1% to avoid laxative effects.

**Polyhydric Alcohols**

Polyhydric alcohols include xylitol, mannitol, sorbitol, maltitol, lactitol and isomalt. Although these products generally have a lower sweetening power than sucrose, they also have fewer calories, only 2.4 kilocalories per gram of dry matter. They also do not cause tooth decay, and for this reason are mainly used in anti-cariogenic chewing gum and confectionaries. However, polyhydric alcohols also have an undesirable laxative effect that becomes apparent with excess consumption and as a function of individual constitution. For this reason, the European Union does not allow the beverage industry to use polyhydric alcohols as a sweetener.

**Artificial Sweeteners**

Artificial sweeteners are substances that are many times sweeter than sucrose. This means that smaller amounts can be employed to obtain the same level of sweetness. However, sweeteners provide no mouthfeel, with the result that their sweetness profile is usually perceived differently: products taste 'empty'. As the calorie content of artificial sweeteners is negligible due to their low usage level, they are used in calorie-reducing concepts. By combining different sweeteners, it is possible to round off the flavour profile until it is very similar to sugar. The combination of sweeteners results in 'synergies', meaning the intensification of the sweetening power of the individual substances. This technique is used in sweetener mixtures, in many cases available as flavour systems with highly sophisticated formulations. Several flavour system suppliers offer sweetening concepts, which use sugars or sweetening concentrates that improve mouthfeel as well as the flavour-enhancing and masking technology.

Table 5.15 provides information on the basic properties of artificial sweeteners. Note that legal limits are subject to change (status as of January 2004).

*Table 5.15: Overview of high-intensity sweeteners*

| Name | Sweetening power | ADI value | US limit | EU limit | Solubility at 20°C |
|---|---|---|---|---|---|
| **Cyclamate** | 35 times sweeter | 7 mg/kg bw/day | Prohibited | 250 mg/l | 20% |
| **Saccharin** | 450 times sweeter | 5 mg/kg bw/day | 12 mg/serving | 80 mg/l | 66% |
| **Aspartame**[*] | 200 times sweeter | 40 mg/kg bw/day | Quantum satis | 600 mg/l | 1% (hot) |
| **Acesulfame K** | 200 times sweeter | 9 mg/kg bw/day | Quantum satis | 350 mg/l | 25% |
| **Thaumatin** | 1000 times sweeter | Undetermined | 0.1-0.5 mg/l as a flavour enhancer | 50 mg/l | (very soluble) |
| **Neohesperidin** | 2000 times sweeter | 5 mg/kg bw/day | Flavour enhancer | 30 mg/l | 0.05% |
| **Sucralose** | 400-600 times sweeter | 15 mg/kg bw/day | Quantum satis | 300 mg/l | 25% |

Sweetening power: as compared to a 10% sucrose solution in water; ADI value: acceptable daily intake in mg per kg of body weight (bw) per day; US limit: maximum amount permitted in soft drinks in the USA; EU limit: maximum amount permitted in soft drinks in the EU; solubility: maximum solubility of the raw material in water.

[*] For the sake of persons with the metabolic disorder phenylketonuria, aspartame must carry the health warning 'Phenylketonurics: Contains Phenylalanine'.

## Natural Sweetening Concentrates

Natural sweetening concentrates are fruit concentrates that are specially selected and, in some cases, further treated, for usage as an alternative to sugar in beverages. In specific cases, fruit juice concentrates are also used as sweeteners. Such sweeteners must be derived from juices with especially low acidity, such as apple, grape or pear from specific growing regions. Gentle production and fining processes are applied to these juices to produce very light-coloured juice concentrates that are ideal for use as natural sweeteners.

Other fruit juices can also be used as sweeteners after undergoing physical treatment methods to neutralise and decolourise them. Such fruit concentrates are used as alternatives to natural fruit sugar.

The manufacturing of such products requires extensive knowledge of treatment methods and raw material procurement.

## 5.2.1.5 Healthy Nutrition – Functional Drinks

For years, the popularity of functional ingredients in foods and beverages has been growing. Such ingredients add value to the product. Consumers use these products because they want to ingest health-promoting substances and thereby contribute to a healthy diet. In the past years nutritional science has shown that some specific diets lead to a better health status and lower risk of getting diseases that reduce lifetime, especially cardiovascular problems, diabetes and cancer. The most recognised diets in this area are the 'Mediterranean' and the 'Traditional Asian' diet. Many concepts of healthy drinks relate to these diets or their main ingredients. Examples are soy, omega-3-rich fish oil, more servings of fruits and vegetables and dietary fibre, as shown in Fig. 5.35.

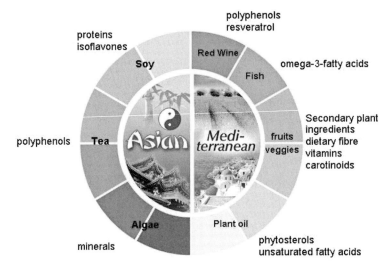

*Fig. 5.35: Asian and Mediterranean diets*

Beside these 'philosophical' approaches to a healthy diet, many other concepts are applied to functional or healthy foods. The fortification with single ingredients provides a certain effect on the body. Removal of less healthy ingredients is realised in 'low-fat', 'light' or 'sugar-free/reduced' or 'low-sodium' products. Also, new technologies could provide healthy nutrition, such as low-temperature processing to avoid loss of ingredients or non-thermal sterilisation technologies [14].

The 'functional food' concept constitutes a worldwide trend. That it is viewed very differently in the USA, Europe and Asia is clearly evident from the fact that there is no universally accepted definition of the term 'functional food'. The following are some examples of the international definitions of 'functional food'.

**Japanese Definition**

Japan is the only country with a definition of 'functional food' that is legally binding.

The Japanese Ministry of Health defines functional food as a processed food with ingredients that, in addition to their nutritional-physiological properties, serve to promote specific body functions. They must be 'real' foods, i.e. not pills, capsules or powder. In Japan, functional ingredients must be derived from naturally occurring ingredients. In Asian cultures, people are traditionally familiar with the idea that certain foods have a positive effect on health and, in some cases, can even be used therapeutically.

## US Definition

The Institute of Medicine for the National Academy of Sciences defines functional foods as those in which the concentrations of one or more ingredients have been modified to enhance their contribution to a healthful diet. This includes nutritionally modified foods but excludes foods with a naturally high nutritional content.

## European Working Definition

This working definition was developed within the framework of an EU project (FU-FOSE, Functional Food Science in Europe) [14]:

> A food can be regarded as 'functional' if it is satisfactorily demonstrated to have a beneficial effect on one or more target functions in the body in a way that is relevant to an improved state of health and well-being and/or to a reduction of risk of disease. Functional foods must be foods, and they must demonstrate their effects in amounts that can be expected to be consumed in a normal diet.

Functional foods are foods, not pills, capsules or any other dietary supplement. It must be possible to demonstrate their effects scientifically. Their beneficial effects exceed normal nutritional effects and are especially relevant for an improved state of health and/or a reduction of risk of disease. Functional foods are consumed as part of a normal food pattern.

As part of the FUFOSE project, a list of target functions that can be positively affected by functional ingredients was developed [14]:

- early development and growth
- regulation of basic metabolic processes
- defence against oxidative stress
- physical performance and fitness
- cardiovascular health
- intestinal health
- cognitive and mental performance.

All the worldwide initiatives relating to functional additives revolve around the scientific basis for the assertion that consuming these products has positive effects on the health of the consumer. A scientific basis is very useful for advertising products. In this aspect for which the regulations in the USA, Japan and Europe are very different.

## Health Claim Concepts – The FOSHU Concept in Japan

FOSHU products are foods expected to have a specific health benefit as a result of certain constituents they contain, which have been added or removed. The benefit of such an addition or removal must have been scientifically proven, and permission must be granted to make claims regarding these benefits. FOSHU products must be consumed as part of a normal diet. They must be in the form of ordinary foods (i.e. not pills, capsules, etc.). In 1993, the first FOSHU product to be approved in Japan was hypoallergenic rice in which the allergenic protein was destroyed by means of proteases.

**Health Claim Concepts – The US – FDA Approval of Health Claims**

The US Food and Drug Administration (FDA) permits certain health claims on food labels. The FDA determines the wording to be used in the claims and also specifies the composition of the food. Approval is general. As soon as a food meets the defined specifications, the manufacturer can use the health claim.

**Health Claim Concepts – The European Approach**

There is currently no legal basis for health-related information on foods. Another EU project began as an offshoot of the FUFOSE project with the object of establishing a basis for permitting health claims: PASSCLAIM (process for assessment and scientific substantiation of health claims). This project was finalised in 2005. The basis for permitting health claims is solid scientific evidence that a functional food 'enhances function' or 'reduces disease risk'. For this purpose, biomarkers should be defined for the above-mentioned target functions that can be used as scientifically substantiated, measurable variables *[15]*.

**Topics to be Taken into Account for Health Food Development**

All developments start from the desired claim! Prior to the formulation of products, it is very important to understand the motivation of the consumer. Does the consumer really understand the benefits that the food promises? Is it likely that the consumer can understand that the food has an effect on his or her individual health? Is the health effect measurable? It is very important to take these factors into account. The number of failures in the area of development and marketing of functional foods is high. Table 5.16 gives a checklist for the development of healthy drinks *[16]*.

*Table 5.16:* Important topics for food development with health ingredients

| Topic | Parameter | Result |
| --- | --- | --- |
| Physical property of the ingredient | Solubility | Oil-soluble; water-soluble; insoluble |
| Recommended dosage | Recommended daily allowance OR<br>Effective amount | Per serving<br><br>Per daily consumed portion |
| Content<br>  in formulation<br>  in final product | Analytical determination | Possible<br>OR<br>impossible |
| Stability | Loss during processing<br>Loss during storage | Amount of active substance |

| Topic | Parameter | Result |
|---|---|---|
| Standardisation | Content of functional ingredient? Combination with other raw materials possible | How much? Possible/impossible |
| Sensory properties | Consumer acceptance | |
| Carrier substances | Made from/with GMO? Allergenic potential | Yes/no |
| Chemical properties | pH-value aw-value acidity etc. | According to method of measurement |
| Storage of final product | Temperature | Ambient Refrigerated |
| Packaging | Material | Glass, PET, can, flexible pouch, carton |
| Packaging | Properties of material | Light protection Oxygen permeability Fragility |

## Examples for Healthy Food Ingredients and Concepts

There is a steadily increasing number of healthy ingredients that are used for healthy soft drink concepts. Table 5.17 provides an overview of some ingredients and their health effects.

*Table 5.17: Examples for healthy food ingredients*

| Food | Ingredient | Nutritional effect |
|---|---|---|
| Green tea | Catechins | Antioxidative |
| Fish oil | Omega-3 fatty acids | Heart and cardiovascular health |
| Vegetable oil | Phytosterols | Lowering cholesterol |
| Fruits: general | Pectin | Soluble dietary fibre |
| Fruit: apples | Flavonoids, e.g. quercetine | Antioxidative |
| Fruit: red grapes | Anthocyanins | Antioxidative |
| Fruit: acerola | Vitamin C | Antioxidative |
| Milk | Specific peptides | Lowering blood pressure |
| Coffee beans | Caffeine | Stimulant |
| Cacao beans | Flavonoids | Antioxidative |
| Soy beans | Protein | Lowering cholesterol |
| Vegetable: carrots | Beta-carotene | Antioxidative |

| Food | Ingredient | Nutritional effect |
|---|---|---|
| Vegetable: onions | Quercetine | Antioxidative |
| Vegetable: chicory | Inulin | Prebiotic, dietary fibre |
| Cereals | Cellulose | Insoluble dietary fibre |
| Exudate from acacia trees | Gum arabic | Soluble dietary fibre |
| Carbohydrates | Fruit sweetener | Low glycemic index |

In many cases, the described health effect is antioxidative. Antioxidative properties of a food ingredient can influence different health-related effects in the body. The energy metabolism in our body cells is based on oxidation, which can also lead to the formation of unwanted substances, e.g. free radicals. Antioxidants protect the cells from such damage. The health effect of antioxidative substances varies, among others, from cardiovascular protection to cancer prevention and strengthening of the immune system.

As regards measuring and assessing health effects of food in our body, more information can be found in *[15]*.

### 5.2.1.6 Flavour Systems for Convenient and Efficient Soft Drink Production

A flavour system is a semi-finished product used as a base in the production of non-alcoholic and light-alcoholic beverages as well as confectionery and ice cream. Normally, a flavour system contains all the ingredients necessary for producing a standardised finished product minus water, sugar and acid. It is tailored to customer requirements and developed in collaboration with the customer. Flavour systems are available either with or without preservatives and in differently sized packaging, including canisters, tote packaging and tank trucks.

Despite significant fluctuations in the quality and availability of raw materials, a flavour systems manufacturer offers beverage producers a selection of suitable raw materials and processing technologies, thus providing them with a high-quality product standardised in all its major parameters and tailored to their particular application. An example of a typical specification is shown in the Fig. 5.36.

Raw materials selection and processing technologies are the main factors determining the flavour system's shelf life and areas of application, as well as its other properties. Because product properties are expected to meet increasingly high standards and are growing more international in character, raw material procurement is of primary importance. The inspection, testing and purchasing of raw materials must be performed onsite worldwide, which places huge demands on raw materials buyers. Only globally active organisations and companies can manage such transactions effectively.

## Product specification

| | |
|---|---|
| Product number | : 35035000380000 |
| Product description | : Orange compound |
| Applications | : Beverages |
| Standard usage | : 14 kg : 1000 l |

## Composition

| | |
|---|---|
| Ingredients (in descending order) | : Orange juice concentrate, water, acid citric acid E 330, vitamin C, natural flavours, stabiliser locust bean gum E 410, preservative sodium benzoate E 21, colour beta-carotene E 160a |
| Limited ingredients | : Sodium benzoate: 1.18 g/kg (=benzoic acid: 1 g/kg) |

## Physical/chemical data

| | | Target | Min | Max | Unit | Remarks |
|---|---|---|---|---|---|---|
| Brix | : | 46.0 | 45.0 | 47.0 | °Bx | Refractometric (20°C) |
| Total acidity (calculated as citric acid) | : | 18.18 | 17.86 | 18.50 | g/100g | Titrimetric (pH 8,1) |
| Density | : | 1.229 | 1.222 | 1.236 | g/ml | Calculated (20°C/20°C) |

## Microbiological data

| | | | |
|---|---|---|---|
| Aerobic, mesophilic total viable count | : | <150/g | IFU No. 2 |
| Aerobic, mesophilic sporeforming bacteria | : | <100/g | IFU No. 6 |
| Coliforms and E. coli | : | neg/10g | |
| Lactic acid and acetic acid bacteria | : | neg/10g | IFU No. 5 |
| Yeasts | : | neg/10g | IFU No. 3 |
| Moulds | : | neg/10g | IFU No. 4 |

## Best before

| | |
|---|---|
| Refrigerated (+5 to +8°C) | : 180 days in sealed original packaging |
| Ambient (+20°C) | : 63 days in sealed original packaging |
| Recommendations | : Stir contents before dispensing! Re-opening containers should be avoided |

*Fig. 5.36:* Specification of an orange flavour system

### 5.2.1.7 Use of Flavour Systems at the Consumer Level

The ingredients in a flavour system are present in exactly the right amounts for producing a particular product. The use of flavour systems provides beverage manufacturers with more security and efficiency in the production process. The most important flavour system quality parameters (Brix, acidity, sensory impression and stability) are standardised. The stability of the flavour system is already tested for the particular beverage during the development phase.

Because consumers have become more concerned with food quality and safety, it is more important than ever that raw materials be traceable. Flavour systems suppliers are responding to this trend with the highest degree of transparency and corresponding certification.

Flavour systems are available year-round, always at the same consistently high, standardised level of quality. By reducing the number of components, beverage manufacturers can also minimise their storage capacities and keep their purchasing contacts at a reasonable number. They can also eliminate fluctuations in the quality of raw materials and, consequently, of their beverages, while reducing the number of raw materials used in production to a minimum.

### Processing instructions "finished drink batched flow production"

especially prepared for:
Ziegler - Flavourings

Date printed: 27.03.2006
Date of issue: 16.02.2006

| Recipe: Preparation of orange soft drink | Kilogram | Litre |
|---|---|---|
| Pre-Water | 10.000,000 | 10.000,000 |
| Orange compound 3503500058 | 350,000 | 284,784 |
| Reverse sugar syrup 72,7% | 3.450,000 | 2.540,875 |
| Citric acid monohydric 50% | 57,500 | 46,958 |
| Post-Water | 12.127,383 | 12.127,383 |
| Finished drink | 25.984,883 | 25.000,000 |

| | | | |
|---|---|---|---|
| Relative density | : | 1,0394 | g / ml |
| Beverage Brix | : | 10,30 | °Bx (refr. 20 °C) incl. 0,1°Water-Brix |
| Beverage acidity | : | 9,92 | ml 0,5 N KOH / 100 ml |
| Beverage acid | : | 3,17 | g / l ( calculated as citric acid anhydrid, pH 8,1) |

| | | | |
|---|---|---|---|
| Packaging size: | : | 1,000 | 0,500 |
| Dosage per package: | : | 1.000,00 | 500,00 |
| Number of packages: | : | 25.000 | 50.000 |

| | | | Ingredient: °Bx/refr./20°C | Ingredient: Acid g/100g/CA/ pH 8,1 | Ingredient: Density g/ml/20°C |
|---|---|---|---|---|---|
| Ingredients per litre beverage: (The brix, acidity and density figures for dilution refer to the solutions) | : | 14,000 gr. Orange compound 3503500058 | 46,0 | 18,18 | 1,2290 |
| | | 100,326 gr. Sugar | 71,6 | 0,00 | 1,3578 |
| | | 1,150 gr. Citric acid Monohydrate | 40,0 | 45,80 | 1,2245 |

The processing instructions are based on a water brix value of 0,1 °Bx.
The acidity calculation is based on a water carbonate hardness of 21(dH) degrees.

Brix variation due to inversion when using sucrose are not taken into account.
The calaulation of an inverted sugar syrup quantity is based on its dry substance.

CO2-Content:    6,0 g/ltr.

Issued and checked by:    Matthias Sass

These processing instructions were written conscientiously. The applicant has to check the data for correctness and completeness. For various reasons it is not possible to take the scale of variations of the used raw materials into account. The above recipe is based on rated value of the raw material specification.

***Fig. 5.37**: Example for a processing instruction*

For all these reasons, ready-made ingredients are a profitable choice for small and medium-sized companies with only limited resources in the areas of logistics, product development or procurement. Ready-made flavour systems are an important part of the supply chain and can contribute to a reduction in processing costs. This is a benefit that broad-based multinational corporations can also appreciate.

Suppliers of flavour systems offer customers more than just semi-finished products. They also provide qualified support in the development of concepts and products as well as on-site technical and microbiological consulting and support. One example is the preparation of detailed processing instructions for the use of the flavour system (see Fig. 5.37).

The timely transfer of international trends helps beverage manufacturers to stand out from the competition and prepare for coming trends well in advance.

## REFERENCES

**European Union law (see also http://europa.eu.int/eur-lex/lex/en/index.htm)**

[1] Council Directive 2001/112/EC of 20 December 2001 relating to fruit juices and certain similar products intended for human consumption http://europa.eu.int/eur-lex/lex/LexUriServ/LexUriServ.do?uri=CELEX:32001L0112:EN:HTML)

[2] Council Directive 90/496/EEC of 24 September 1990 on nutrition labelling for foodstuffs (http://europa.eu.int/eur-lex/lex/LexUriServ/LexUriServ.do?uri=CELEX:31990L0496:EN:HTML)

[3] Regulation (EC) No. 258/97 of the European Parliament and of the Council of 27 January 1997 concerning novel foods and novel food ingredients (http://europa.eu.int/eur-lex/lex/LexUriServ/LexUriServ.do?uri=CELEX:31997R0258:EN:HTML)

[4] Council Directive 89/107/EEC of 21 December 1988 on the approximation of the laws of the Member States concerning food additives for use in foodstuffs intended for human consumption (http://europa.eu.int/eur-lex/lex/LexUriServ/LexUriServ.do?uri=CELEX:31989L0107:EN:HTML)

[5] European Parliament and Council Directive 94/36/EC of 30 June 1994 on colours for use in foodstuffs (http://europa.eu.int/eur-lex/lex/LexUriServ/LexUriServ.do?uri=CELEX:31994L0036:EN:HTML)

[6] European Parliament and Council Directive 94/35/EC of 30 June 1994 on sweeteners intended for use in foodstuffs (http://europa.eu.int/eur-lex/lex/LexUriServ/LexUriServ.do?uri=CELEX:31994L0035:EN:HTML)

[7] European Parliament and Council Directive 95/2/EC of 20 February 1995 on food additives other than colouring and sweeteners (http://europa.eu.int/eur-lex/lex/LexUriServ/LexUriServ.do?uri=CELEX:31995L0002:EN:HTML)

[8] Council Directive 80/777/EEC of 15 July 1980 on the approximation of the laws of the Member States relating to the exploitation and marketing of natural mineral waters (http://europa.eu.int/eur-lex/lex/LexUriServ/LexUriServ.do?uri=CELEX:31980L0777:EN:HTML)

**Technology**

[9] Carle, R., Cloud stability of fruit-rich tropical fruit nectars. Flüssiges Obst $\underline{65}$(11), 692-698 (1998)

[10] Shachman, M., Flavour emulsions for carbonated soft drinks. Food Review $\underline{27}$(12) 17, 19 (2000)

[11] Buffo, R., Reineccius, G., Beverage emulsions and the utilization of gum acacia as emulsifier/stabilizer. Perfumer and Flavorist $\underline{25}$(4), 24-44 (2000)

[12] Bates, Morris, Crandall, Principles and practices of small- and medium-scale fruit juice processing. FAO Agricultural Service Bulletin 146 (2001)

[13] More information on the application of colours in food: http://www.wild.de/wild/opencms/en/index.html (click on 'Color Finder')

**Healthy nutrition – functional food**

[14] Margaret, A., Concepts of Functional Foods, ILSI Europe Concise Monograph (2002)

[15]  Aggett, P.J., Asp, N.G., Contor, L., Cummings, J.H., Howlett, J., Mensink, R.P., Prentice, A., Rafter, J., Riccardi, G., Richardson, D.P., Saris, W.H.M., Westenhoefer, J., PASSClaim: Process for the Assessment of Scientific Support for Claims on Foods, ILSI (2005)

[16]  Erbersdobler, H.F., Meyer, A.H., Praxishandbuch Functional Food, Behr's Verlag (1999); continuously updated collection on topics of functional food (Loseblattsammlung)

## 5.2.2 Alcoholic Beverages

*Günther Zach*

### 5.2.2.1 Introduction

Alcoholic beverages are ethyl alcohol-containing liquids intended for human consumption with specific organoleptic properties. The alcohol contained must be derived through fermentation, or fermentation and subsequent distillation, from agricultural raw materials. The relevant components are fermentable sugars and/or starches, which based on enzymatic activities, are converted into fermentable sugars. Worldwide, alcoholic beverages can be categorized based on variables like fermentation ingredients, processing techniques, treatment, blending components, sweetening, alcohol levels and consumption habits as follows:

- Generic distilled spirits
- Liqueurs/cordials
- Wine/wine-based alcoholic beverages
- Beer/beer- & malt-based alcoholic beverages
- Cocktails/spirit-based alcoholic beverages

### 5.2.2.2 Generic Distilled Spirits

Distilled spirits are alcohol-containing beverages with low or no extractive content with or without flavour addition.

#### 5.2.2.2.1 Production of Distilled Spirits

Generally speaking there are two basic processing techniques applied in the production of distilled spirits:

    a) The fermentation distillation process
    b) The compounding process

Raw materials can be mainly described with the following criteria:

- Sugar-containing materials
  (sugar cane, sugar beet, molasses, fruits, fruit derivatives, residues, fermented milk, palm juice and sugar-containing tropical plant material, etc.)
- Starch and starch-containing raw materials
  (grain types, potatoes, topinambur, manioc, tapioca, chicory, etc.)

The conversion of starch material into fermentable sugars is mainly handled through the addition of malt. Based on the starting materials as well as the desired finished products, various distillation techniques are being applied. In the case of brandy, rum, fruit brandies and grain spirits/whisky, rather simply constructed stills are being used; consequently, together with ethyl alcohol, also fermentation by-products or raw material specific flavouring components, such as fusel oils, esters, aldehydes, volatile acids and essential oils become part of the distillate. In order to increase the alcohol concentration in the distillate several distillations/redistillations, so-called rectifications, are necessary.

In the case of neutral spirits, the intention is to optimally purify the distillate and therefore separate aromatic components, such as fermentation by-products (fusel oils), as much as possible and to ideally obtain absolutely neutral ethyl alcohol. Today's most common raw materials for the production of beverage alcohol are potatoes, grain and molasses. For the fermentation, so-called distillers' yeasts, in particular top-fermenting yeast types are used. Since the measures are not sterile and increased temperatures accelerate the fermentation process, the yeast strains must provide good fermentation capabilities such as high resistency against alcohol, acidity and increased temperatures.

The conversion of starch into fermentable sugars is mainly based on enzymes, naturally occurring in malt. After the starch conversion has been achieved, the temperature of the mash is decreased to approximately 30°C and the yeast (often bakery yeast) is added.

After approx. 48 h of fermentation, the fermented mash shows an alcohol content of approx. 6-10% by vol. When distilled in continuous stills with 3 or more columns, the alcohol level in the distillate can be enriched up to 96.6% by vol. The purification of the raw distillate is normally carried out in continuous rectification equipment, for which the raw alcohol is usually diluted to approx. 15% by vol. before rectification in order to better separate fusel oil components. The first fractions of the rectified distillate – the so-called head fraction – contain significant quantities of acetaldehyde, methyl alcohol and low-boiling esters; the middle part – heart fraction – represents so-called neutral spirits with an alcohol content of approx. 96.6% by vol. The tail fractions contain higher alcohols and higher esters etc. Since head and tail fractions contain organoleptic properties undesirable for neutral alcohol, they usually are employed as technical alcohol or have lately also been isolated and used as starting raw materials for the production of natural aroma components.

### 5.2.2.2.2 Grape Brandy & Fruit Spirits

The outstanding characteristic of the above mentioned brandies is their distinguishing taste and odour; they contain approx. 38% by vol. alcohol. They are denominated as natural or original spirits. In a single distillation process a distillate with low alcohol content is produced (raw distillate), which often contains the specific organoleptic components of the raw materials (example: juniper brandy).

The most important factor during the production process is the best possible retention of the distinguishing, specific odorous and flavouring components (esters, essential oils) or the enhancement of these compounds (fermentation products, cognac oil). In order to mature the taste, the storage of the finished product is the decisive factor.

**Brandy (from Grape Wine)**

Brandy is the liquid which is produced on the basis of wine distillate and possesses an alcohol content of at least 36.5-40% by vol. The denomination Cognac is reserved for brandies originating from specific areas of France (Charente etc.). Armagnac which is produced in Southern France is qualitatively comparable to Cognac. Brandy was originally produced in France from fermented grape juice which was distilled in stills of the simplest kind (copper alambiques) on open fire and often without prior removal of the yeast. In the first step of this process a raw distillate is gained which, after redistillation, represents the final brandy. The production of this spirit soon spread to other countries (Germany, Russia, Spain, Hungary, California and Australia) and nowadays is also carried out in continuously operating production plants. In Germany, imported distilling wines are used as a base and are enhanced by means of brandy distillate addition.

The resulting brandy distillate (52-86% by vol. ethanol) can be called an intermediate material for the production of brandy. The final product "brandy" made from the distillate is not finished until it is treated according to the "Cognac method", i.e. after an improving maturation period in wooden casks (min. 2 years for Cognac). Oak wood is the preferred material for these casks. During the maturation process the wine distillate absorbs aromatic compounds of the wood as well as its colouring components which lead to the typical more or less golden yellow colour of the brandy. Oxidation and esterification processes refine odour and taste. A common method for improving quality is the addition of – often brandy-based – extracts from e.g. oak wood, prunes, green walnuts, dried almond shells or vanilla extracts; however, this addition must meet the legal requirements of the producing countries ("Harmless blending and colouring agents" / example: bonificateurs or blending sherry). Another common method is the treatment by means of filters and purifying agents. The required alcohol concentration (drinking strength) is achieved by dilution of the distillates. In order to avoid expensive and time-consuming cask maturation, there have been numerous attempts of developing artificial aging methods for brandy as well as for other spirits.

## Fruit Spirits

Products of this kind are denominated as Williams Pear Spirit, Kirschwasser, Plum Spirit, Blueberry Spirit, Raspberry Spirit etc. The production of Kirschwasser and Plum Spirit is described as an example for these fruit brandies.

Kirschwasser is mainly produced in Southern Germany (Black Forest Kirschwasser), France and Switzerland by crushing the different kinds of sweet cherries, either as a whole fruit or by crushing part of the kernels, and leaving the mashed mass to fermentation for several weeks (in most cases together with pure culture yeast). The fermented mash is then distilled in copper stills on open fire or with vapour while removing the first runnings and the tailings. The resulting distillate has an alcohol content of approx. 60% by vol. and more and is marketed as clear, colourless fruit spirit with an alcohol content of 40-50% by vol. Kirschwasser is also used as an additive for different liqueurs (e.g. Curacao, Cherry Brandy, Maraschino etc.).

For the production of Plum Spirit, the fully ripe plums are treated similarly, however, without crushing the kernels. The result is a product which is often sold as Slivovitz. Apart from Germany, main producing countries are Switzerland, Yugoslavia, Romania, the Czech Republic, Hungary and France. In addition to the common plum, the highly aromatic cherry plum is also processed. Cherry Plum Spirit is used as valuable component for flavouring purposes in fruit liqueurs.

Other fruit spirits are distilled from fresh or frozen fruits or their juices (blueberries, raspberries, strawberries, currants, apricots, peaches and others) while adding alcohol. Pomaceous fruit spirits are made from fresh fermented apples or other pomaceous fruits, from the whole fruit or its juices without the addition of sugar-containing substances, sugar or alcohol of another kind with a minimum alcohol content of approx. 38% by vol.

### 5.2.2.2.3 Juniper-based Spirits

Juniper is a spirit which is often produced from grain distillate while adding juniper distillate and/or fermented juniper raw distillate. Juniper spirit is produced exclusively by using juniper distillate derived from the whole berry or from its fermented, watery infusions. The berries of the juniper plant (*Juniperus communis*) are processed in Germany, Hungary, France and Switzerland. "Juniper brandy" (Steinhäger) is exclusively produced by means of distillation while using juniper raw distillate from fermented juniper mash with a minimum alcohol content of 38% by vol. Pure juniper spirits are also used as intermediate materials for the production of other spirits with juniper flavour such as Genever (Dutch), which is distilled from a mash of grain and caramelized malt.

Gin is another spirit produced from juniper distillate and spices and has an alcohol content of approx. 38% by vol. Dry Gin usually has a higher alcohol content of approx. 40% by vol.

*Example Formula for Gin "English Style" (40% v/v)*

Step 1: Gin Distillate, 72% v/v

| | | |
|---:|---|---|
| 6000 | g | Juniper Berries |
| 1600 | g | Coriander Seeds |
| 400 | g | Lemon Peel, fresh |
| 3200 | g | Angelica Root |
| 200 | g | Malt, roasted |
| 160 | g | Lavender Flowers |
| 120 | g | Orange Flowers |
| 80 | g | Hop Flowers |
| 40 | g | Cinnamon Bark |
| 40 | g | Ginger Root |
| 24 | g | Allspice Seeds |
| 30 | l | Neutral Spirits, 96% v/v |
| 20 | l | Water, demineralised |

*Yield:*   36   l   *Gin Distillate, 72% v/v*

Process:
Macerate for 1-2 days without adding water, distil slowly under vacuum. While the distillate is still running, add water and de-alcoholise to zero. Catch tails separately. Set in cool storage for 24 hours and filter. Rectify distillate to 72% v/v.

Step 2: Formula for "English Style" Gin, 40% v/v

| | | |
|---|---|---|
| 386.70 | l | Grain Neutral Spirit, 96% v/v |
| 40.00 | l | Gin Distillate, 72% v/v |
| QS | l | Water, demineralised |
| 1000.00 | l | Total Volume, 40% v/v |

**5.2.2.2.4 Rum**

The main production countries for rum are Jamaica, Cuba, Barbados, Puerto Rico, British and Dutch Guyana, Brazil, Mauritius and Martinique. The rum in those sugar cane growing countries is produced from cane sugar syrup and freshly squeezed juice; often side-products such as skimmings, molasses, press residues, their extracts and dunder from prior distillations are also employed. The sugar-containing, diluted solutions are left to a spontaneous fermentation at max. 36°C and are subsequently distilled in simple pot stills. In order to increase and stabilize the aroma, flavouring plant parts are sometimes added to the mash and thus rums with very different flavours like pineapple, Russian leather or rather neutral flavour can be found on the market. In general, rum is classified as drinking rum and blending rum. The rum destined for exportation shows an alcohol content of approx. 76-80% by vol. (original rum). For the use as drinking rum it is usually diluted to 50% by vol. or blended with neutral spirits and water. Among all consumed spirits, rum has the most intensive flavour, which is influenced by maturation periods in wooden casks with atmospheric oxygen while absorbing extraction substances from the oak wood and through the formation of ester as well as other flavouring substances. Original rum contains approx. 80–150 mg/100 ml total acids of which a large percentage is formic acid and acetic acid. The combination and content of esters is of utmost importance for the evaluation of rum.

Over the last couple of years, the rum segment has grown as a result of flavoured line-extensions of brands, such as spiced rums as well as other fruit-flavoured rums (similar to the vodka segment).

Additionally, Cachaca, a Brazilian sugar cane molasses based distilled spirit, has spread internationally as a member of the rum family, due to the popularity of Caipirinha cocktails.

**5.2.2.2.5 Arrack**

Arrack is produced by fermentation and distillation using rice, sugar cane, molasses or sugar-containing plant juices, especially the juice of the palm trees. Producing countries are mainly Java, Sri Lanka, the Malabar Coast and Thailand. Well-known brand names are Batavia and Goa. In comparison to rum, there are fewer different types of arrack, which are imported as original arrack with an alcohol content of 56-60% by vol. Due to its strong flavouring properties arrack is an especially important blending material for different industries, e.g. the liqueur, confectionery and baking industry.

### 5.2.2.2.6 Gentian Spirits

Gentian or gentian spirits represent a product which is produced through simple distillation, gained from the mash of fermented gentian roots or from gentian distillate obtained from non-fermented gentian roots. The raw material for the product are the roots of different gentiana species which, in fresh condition, contain considerable amounts of fermentable sugar (6-13%) and different glycosides as carriers of bitter substances. Main production areas are the countries of the Alps, Germany, Tyrol, France and Switzerland.

### 5.2.2.2.7  Grain Spirits – Korn & Whisky

Typical representatives of this class of spirits, which are based on grain distillates, are whiskies and Korn. Different kinds of grain, e.g. rye, wheat, oat, barley, corn etc., which are ground and gelatinized with sulphuric acid-containing water are used for the production. Saccharification is achieved after stirring approx. 15% malt into the pre-mash tub at a temperature of approx. 56°C and proceeds rather fast at a temperature range of 55-60°C under the influence of malt amylase. Subsequently, the product is heated to 62°C for inactivation and then immediately cooled down to 19-23°C. Grain spirits from rye, wheat or barley are always produced through a mashing process which means the lauterization of grains does not occur. The production of malt spirits on the other hand can include lauterization. The sweet mash is fermented by means of a special yeast and subsequently submitted to a distillation process. In smaller production plants simple stills are used for distillation, continuous high performance distilling columns are employed for distillation and rectification on an industrial scale. Depending on the applied processing techniques, yields of 30-35 l alcohol from 100 kilos grain (e.g. rye) are achieved. Different production processes lead to grain spirits of widely differing character and varying quality. The simple still appliances result – after unsophisticated separation of head and tail – in very characteristic products which, however, contain a lot of grain fusel oils. High-tech distillation processes allow the almost complete separation of fusel oils and produce grain spirits of high alcohol content from which mild-tasting grain spirits with pure taste and less pungent flavour are produced. A well-adapted maturation in barrels is very important for the final taste of all these products.

Depending on the desired type of whisky, different processing methods are employed. The raw material for Scottish malt whisky is malt, which was dried over peat or coal smoke. The peated malt is mashed with water at a temperature of 60°C and the lauterized mash subsequently fermented at a temperature of 20-32°C. The distillation is conducted in two steps, often in simple pot stills. During the second distillation step undesired components are separated from the raw distillate with the head and tails. During the production of the Scottish grain whisky – after saccharification of the starch and fermentation in mostly continuous equipment – a distillate is produced which is more neutral and has less flavour than malt whisky. In both cases, the distillate with an alcohol content of approx. 63% by vol. is submitted to a maturation process, preferably in used Sherry casks or charred oak casks, in order to fully develop its flavour. Subsequently, the product is diluted to an alcohol content of approx. 43% by vol., suitable for consumption. Depending on the desired taste, the malt whisky is blended with grain whisky.

American whisky is produced from corn, rye and wheat by means of saccharification with malt fermentation of the lauterized mash and through double distillation in column distilling units. The product is then stored in charred casks made from oak wood. Bourbon whisky contains at least 51% corn distillate and grain whisky at least 80%. Rye whisky has a minimum content of 51% rye and wheat whisky contains mostly wheat distillate.

#### 5.2.2.2.8 Vodka

Vodka is a spirit produced from neutral alcohol, mainly grain distillate, following a special process with few additives. During the production process the characteristics, especially the softness of the taste, must be achieved. Multiple charcoal filtration steps followed by corresponding rectification are typical procedures employed in this context. Additives must be hardly noticeable taste-wise, the content of extractives for example is approx. 0.3 g /100 ml and the alcohol content approx. 40% v/v.

Flavoured vodka with tastes such as lemon, herbs, honey, pepper etc. has been gaining more and more popularity, a process already discernible in the countries of origin (Russia, Poland and Finland) for decades.

Today, one can find a wide range of fruit-flavoured or fruit-infused vodka varieties in the market due to their mixing properties in the cocktail segment (e.g. Martinis).

*Example Formula for Cranberry Flavoured Vodka (35% v/v)*

| | | |
|---:|---|---|
| 866.30 | l | Vodka (triple distilled), 40% v/v |
| 10.00 | l | Cranberry Flavour, nat., 35% v/v |
| QS | l | Water, demineralised |
| 68.80 | l | Sugar Syrup 72.7% w/w |
| 1.20 | l | Citric Acid Solution 50% i. Water |
| 1000.00 | l | Total Volume, 35% v/v |

#### 5.2.2.2.9 Spirits Specialities

Many spirits are produced "cold" by simply blending purified spirits of different origin with water; they are given local, geographically determined names. They often contain spices/flavourings such as grain raw distillates, caraway, anise, fennel or other distillates as well as sugar, flavours, essential oils and other flavouring materials and are then classified as flavoured spirits. Some examples will be described below.

#### Aquavit

Aquavit is a spirit which is predominantly flavoured with caraway. It is produced from a distillate of herbs, spices or botanicals and has an alcohol content of at least 35% by vol.

*Example Formula for Aquavit (40% v/v)*

Step 1: Aquavit Distillate, 65% v/v

| | | |
|---:|---|---|
| 2000 | g | Caraway Seeds |
| 1800 | g | Lemon Peels |
| 600 | g | Coriander Seeds |
| 400 | g | Dill Seeds |
| 150 | g | Orris Root |
| 150 | g | Cassia Bark |
| 100 | g | Anise Seeds |
| 80 | g | Clove Buds |
| 40 | g | Chamomile Flowers, Roman |
| 10 | g | Cognac Oil |
| 60 | l | Alcohol-Water Mixture 45% v/v |
| Yield: 40 | l | Aquavit Distillate, 65% v/v |

Process: Macerate botanical blend in alcohol-water mixture for approx. 2 days; after first distillation filter through diatomaceous earth and rectify to desired strength.

Step 2: Formula for Aquavit, 40% vol

| | | |
|---:|---|---|
| 372.90 | l | Grain Neutral Spirits, 96% v/v |
| 20.00 | l | Grain Spirits Concentrate, 80% v/v |
| 40.00 | l | Aquavit Distillate, 65% v/v |
| QS | l | Water, demineralised |
| 30.00 | l | Sugar Syrup 72.7% w/w |
| 1000.00 | l | Total Volume, 40% v/v |

## Anise Spirits

Within the spice-flavoured spirits segment, anise today still plays an important role. Mediterranean countries like Greece produce Ouzo, Turkey bottles Raki, but also France and Spain have their traditional anise-based spirit specialities. Anethol is the key flavouring component contained in Anise seed essential oil.

*Example Formula for Anise Spirit (35% by vol)*

"Spice Essence made from Essential Oil 1:100"

| | | |
|---:|---|---|
| 2.00 | g | Anise Oil nat. |
| 3.00 | g | Star Anise Oil nat., rectified |
| 0.20 | g | Juniper Berry Oil, rectified |
| 0.20 | g | Lemon Oil, single-fold |
| 680.00 | g | Alcohol, 96% v/v |
| 314.60 | g | Water, demineralised |
| Yield: 1000.00 | g | Spice Essence, 65% v/v |

Process:

Draw terpenes in separation funnel, filter solution over diatomaceous earth. Optionally, above formula can also be distilled under vacuum conditions.

Formula for Anise Spirit (35% v/v)

| | | |
|---|---|---|
| 10.00 | l | Spice Essence nat., 65% v/v |
| 357.80 | l | Grain Neutral Spirits, 96% v/v |
| QS | l | Water, demineralised |
| 1000.00 | l | Total Volume, 35% v/v |

**Bitters**

Bitter spirits are produced using bitter and aromatic plants and fruit extracts and/or distillates, fruit juices, essential oils, with or without the addition of sugar or glucose syrup. Products of this type are for example Boonekamp, English Bitter, Spanish Bitter or Angostura Bitter (named after a town in Venezuela (today Ciudad Bolivar), it is the most popular cocktail spice made from bitter herbs and other aromatic botanicals):

*Example Formula for Angostura Bitter (45% v/v)*

Step 1: Angostura Bitter Extract, 50% v/v

| | | |
|---|---|---|
| 4500 | g | Angostura Bark |
| 3000 | g | Carobe |
| 3000 | g | Sandalwood, red |
| 3000 | g | Raisins |
| 2000 | g | Sweet Orange peel |
| 1500 | g | Bitter Orange Peel |
| 1500 | g | Bitter Orange Fruits, unripe |
| 1500 | g | Galanga Root |
| 1500 | g | Gentian Root |
| 1500 | g | Cinchona Bark |
| 1500 | g | Vanilla Beans |
| 1000 | g | Cinnamon Bark, Ceylon |
| 1000 | g | Lemon Peel |
| 1000 | g | Massoia Bark |
| 750 | g | Ginger Root |
| 750 | g | Cardamom |
| 700 | g | Mace |
| 700 | g | Clove Buds |
| 500 | g | Cedoary Root |
| 60 | l | Alcohol 96% v/v |
| 40 | l | Water, demineralised |

Yield:   100  l   Extract, 50% v/v (filtered)

Process:

Macerate for approx. 5 days; expression of soaked botanicals required. Alcohol recovery by dry distillation of botanicals is recommended. Chill-filtration of macerate is important to avoid sedimentation of insoluble extractives.

Step 2: Formula for Angostura Bitter, 45% v/v

| | | |
|---:|---|---|
| 389.30 | l | Grain Neutral Spirits, 96% v/v |
| 40.00 | l | Brandy, 40% v/v |
| 13.00 | l | Dark Rum, 40% v/v |
| QS | l | Water, demineralised |
| 100.00 | l | Angostura Bitter Extract, 50% v/v |
| 30.00 | l | Malaga Wine, 17% v/v |
| 70.00 | l | Sugar Syrup 72.7% w/w |
| 1000.00 | l | Total Volume, 45% v/v |

Colour: Red-brown

### 5.2.2.2.10 Other Spirits

Special spirits of regional and partly increasing international importance are Mexican Tequila and Mescal from Mexico derived from the agave plant, as well as spirits from the Middle East made from raisins, figs or dates and Asian specialties, such as Shochu from fermented rice mash.

### 5.2.2.3 Liqueurs

Liqueurs are alcoholic beverages with an alcohol content of approx. 15-20% by vol. and a sugar content of at least 10 g/100 ml and which can be flavoured with fruits, spices, extracts and other flavouring materials.

### 5.2.2.3.1 Fruit Juice Liqueurs

The essential, taste-determining component of fruit juice liqueurs is the juice of the fruits after which the liqueur is named. Depending on the legal requirements of the different countries, the minimum content of fruit juice can range from to 1-12 up to 40 l per 100 l in the finished product. Other additives, such as fruit essences, natural and/or artificial (nature-identical) flavouring components as well as colours, sugar acidulants and others can be incorporated depending on legal requirements. The most popular products of this kind are pineapple, strawberry, cherry or raspberry liqueurs.

*Example Formula for Premium Cherry Juice Liqueur (25% v/v) – Juice Content 40%*

| | | |
|---:|---|---|
| 242.40 | l | Grain Neutral Spirit, 96% v/v |
| 25.00 | l | Cherry Distillate (Kirschwasser), 50% v/v |
| QS | l | Water, demineralised |
| 2.00 | l | Vanilla Extract single-fold, 60% v/v |
| 30.00 | l | Cherry Extract, 18% v/v |
| 220.00 | l | Sugar Syrup 72.7% w/w |
| 50.00 | l | Glucose Syrup 42DE |
| 60.00 | l | Cherry Juice Concentrate, depectinised 65% w/w |
| 1000.00 | l | Total Volume, 25% v/v |

Natural red colour

A special type of cherry liqueur is cherry brandy which is essentially produced from cherry juice, Kirschwasser, sugar or glucose syrup, neutral spirit and water.

*Example Formula for Cherry Brandy (35% v/v) – Juice Content 40%*

|  |  |
|---|---|
| 266.50 l | Grain Neutral Spirits, 96% v/v |
| 30.00 l | Cognac Distillate, 65% v/v |
| 80.00 l | Cherry Distillate (Kirschwasser), 50% v/v |
| 30.00 l | Cherry Kernel Distillate, 50% v/v |
| 5.00 l | Rum Distillate Jamaica, 75% v/v |
| QS l | Water, demineralised |
| 2.00 l | Vanilla Extract single-fold, 60% v/v |
| 2.00 l | Cocoa Extract, 45% v/v |
| 100.00 l | Cherry Marc Infusion, 18% v/v |
| 220.00 l | Sugar Syrup 72.7% w/w |
| 100.00 l | Glucose Syrup 42DE |
| 60.00 l | Cherry Juice Concentrate, depectinised, 65% w/w |
| 1000.00 l | Total Volume, 35% v/v |

Natural red colour

### 5.2.2.3.2 Fruit-flavoured Liqueurs

Fruit-flavoured liqueurs are spirits which are produced using natural and/or nature-identical or maybe even artificial flavour components, distillates and extracts. Liqueurs of that kind are made from apricots, peaches, citrus fruits etc. Denominations such as "Triple" or "Triple Sec" are quite common for citrus liqueurs.

*Example Formula for Curacao Triple Sec (38% v/v)*

Step 1: Curacao Triple Sec Distillate:

|  |  |  |
|---|---|---|
| 3500 | g | fresh Curacao Peels |
| 850 | g | fresh Bitter Orange Peels |
| 250 | g | Lemon Peels |
| 25 | g | Cardamom |
| 40 | g | Mace Flowers |
| 15 | g | Cloves |
| 7 | l | Alcohol, 80% by vol |
| 14 | l | Water |

Yield: 15 l Distillate, 52% v/v (double distilled)

Process:

Extraction for 1-2 days with alcohol then water is added, distilled in black-flow distillate, 15 l distillate are retained, filtrated and redistilled to 10 l total yield

Step 2: Triple Sec Liqueur Formula, 38% v/v, colourless

| | | |
|---:|---|---|
| 300.00 | l | Alcohol, 96% v/v |
| 100.00 | l | Curacao Distillate, 52% v/v |
| 60.00 | l | Wine Distillate, 70% v/v |
| 7.00 | l | Arrack, 58% v/v |
| QS | l | Water, demineralised |
| 250.00 | l | Sugar Syrup, 72.7 w/w |
| 50.00 | l | Glucose Syrup 42DE |
| 1000.00 | l | Total Volume, 38% v/v |

Note:

The addition of flavourings like vanillin (1-2 g), Rose Water (1/4-1/2 l) or Orange Flower Water (1/2-1 l) is quite common.

### 5.2.2.3.3 Coffee, Cocoa and Tea Liqueurs

Coffee, cocoa and tea liqueurs are produced with extracts made from the corresponding raw materials. Emulsion liqueurs are e.g. chocolate, cream and milk liqueurs, mocha-with-cream-liqueurs, egg liqueur (egg cream Advokat) and liqueurs with the addition of eggs. The widely consumed egg liqueur is a preparation made from alcohol, sugar and egg yolk.

*Example Formula for Coffee Liqueur (28% v/v)*

Step 1: Coffee Extract, 30% v/v

| | | |
|---:|---|---|
| 3000.00 | g | Freshly roasted Coffee Beans, ground |
| 10.00 | g | Cinnamon Bark, Ceylon |
| QS | l | Alcohol-Water Mixture, 30% v/v |

Yield: 32.00 l Coffee Percolate, 30% v/v (filtered)

Step 2: Formula for Coffee Liqueur, 28% v/v

| | | |
|---:|---|---|
| 164.80 | l | Grain Neutral Spirit, 96% v/v |
| 20.00 | l | Arrack Distillate, 58% v/v |
| 20.00 | l | Brandy Distillate, 65% v/v |
| QS | l | Water, demineralised |
| 320.00 | l | Coffee Percolate, 30% v/v |
| 2.00 | l | Vanilla Extract single-fold, 60% v/v |
| 270.00 | l | Sugar Syrup 72.7% w/w |
| 80.00 | l | Glucose Syrup 42DE |
| 0.10 | kg | Common Salt, diluted in water |
| 1000.00 | l | Total Volume, 28% v/v |

Colour: Dark brown

### 5.2.2.3.4 Herbs, Spice and Bitter Liqueurs

These liqueurs are produced with fruit juices and/or botanicals, natural essential oils, natural essences and sugar (anise, caraway, curacao respectively bitter orange, peppermint, ginger and numerous others). Many of these herb, spice and bitter products were first produced from the original formulas developed in the monasteries all over Europe. The monks' formulas – handed down through centuries – are still partly employed for botanicals and processing techniques (extractions).

### Herbal Spirits – Raw Materials
Classification according to Active Agents

**Essentiao Oil Botanicals**
(e.g. Citrus Fruits)

**Bitter Agent Botanicals**
(e.g. Dandelion Root)

**Saponine Botanicals**
(e.g. Quillaja Bark)

**Spice Botanicals**
(e.g. Cloves)

**Tannin Agent Botanicals**
(e.g. Oakwood Bark)

**Mucilaginous Botanicals**
(e.g. Fenugreek Seeds)

**Others**
(e.g. Coffee Beans)

**Glycoside Botanicals**
(e.g. Limetree Flowers)

**Botanicals with Fatty Oils**
(e.g. Almonds)

**Silicic Acid Botanicals**
(e.g. Corn Horsetail)

*Fig. 5.38: Herbal spirits – raw materials*

*Example Formula for Italian Bitter Liqueur (38% v/v)*

Step 1: Italian Bitter Herbs Extract, 40% v/v

| | | |
|---:|---|---|
| 3000.00 | g | Bitter Thistle Herb |
| 2500.00 | g | Gentian Root |
| 2500.00 | g | Lemon Peel |
| 1500.00 | g | Cinnamon Bark |
| 1500.00 | g | Cascarilla Bark |
| 1000.00 | g | Mace |
| 1000.00 | g | Woodruff Herb |
| 1000.00 | g | Chamomile Flowers |
| 1000.00 | g | Spearmint Leaves |
| 65.00 | l | Neutral Spirits, 96% v/v |
| QS | l | Water |

Yield: 150.00 l   Bitter Herbs Extract, 40% v/v

Process:

Extraction for approx. 5 days, express herbs and store in cool area for sedimentation; chill-filter before use.

Formula for Italian Bitter Liqueur, 38% v/v

| | | |
|---:|---|---|
| 335.00 | l | Grain Neutral Spirits, 96% v/v |
| QS | l | Water, demineralised |
| 150.00 | l | Bitter Herbs Extract, 40% v/v |
| 240.00 | l | Sugar Syrup 72.7% w/w |
| 1000.00 | l | Total Volume, 38% v/v |

Colour: Ruby red

*Example Formula for Monastery Herb Liqueur (42% v/v)*

Step 1: Monastery Herb Distillate, 65% v/v

| | | |
|---:|---|---|
| 300 | g | Lemon Balm Leaves |
| 300 | g | Coriander Seeds |
| 150 | g | Hyssop Herb |
| 80 | g | Peppermint Leaves |
| 120 | g | Artemisia Herb |
| 100 | g | Angelica Root |
| 80 | g | Angelica Seeds |
| 80 | g | Ambrette Seeds |
| 50 | g | Cardamom |
| 25 | g | Spearmint Leaves |
| 25 | g | Cinnamon Bark, Ceylon |
| 25 | g | Clove Buds |
| 25 | g | Mace |
| 25 | g | Thyme Herb |
| 25 | g | Sage Herb |
| 25 | g | Celery Seeds |
| 20 | l | Water |
| 17 | l | Neutral Spirits, 96% v/v |

Yield:  25  l  Herbs Distillate, 65% v/v

Process:

Approx. 3 days extraction and subsequent copper distillation

Step 2: Formula for Monastery Herb Liqueur, 42% v/v

| | | |
|---:|---|---|
| 162.30 | l | Grain Neutral Spirit, 96% v/v |
| 250.00 | l | Herbs Distillate, 65% v/v |
| 100.00 | l | Brandy Distillate, 65% v/v |
| 2.00 | l | Rum Distillate, 75% v/v |
| QS | l | Water, demineralised |
| 2.00 | l | Saffron Flower Extract, 60% v/v |
| 250.00 | l | Sugar Syrup, 72.7% w/w |
| 200.00 | l | Honey Solution, dewaxed 17% v/v |
| 1000.00 | l | Total Volume, 42% v/v |

Colour: Bright yellow

*Example Formula for Citrus Bitter Liqueur (36% v/v)*

Step 1: Citrus Bitter Extract, 40% v/v

| | | |
|---:|---|---|
| 7000.00 | g | Bitter Orange Fruits, unripe |
| 2000.00 | g | Bitter Orange Peel |
| 2000.00 | g | Curacao Peel |
| 1500.00 | g | Buckbean Leaves |
| 1500.00 | g | Gentian Roots |
| 1500.00 | g | Cinnamon Bark |
| 900.00 | g | Cloves Buds |
| 500.00 | g | Ginger Root |
| 250.00 | g | Elecampane Root |
| 200.00 | g | Poplar Buds |
| 200.00 | g | Cardamom |

Yield: 150.00 l   Citrus Bitter Extract, 40% v/v

Process:

Macerate for approx. 5-8 days; express botanicals and chill-filter.

Formula for Citrus Bitter Liqueur, 36% v/v

| | | |
|---:|---|---|
| 276.00 | l | Grain Neutral Spirit, 96% v/v |
| 50.00 | l | Brandy Distillate, 65% v/v |
| QS | l | Water, demineralised |
| 150.00 | l | Bitter Herbs Extract, 40% v/v |
| 50.00 | l | Malaga dark, 18% v/v |
| 260.00 | l | Sugar Solution, 72.7% w/w |
| 50.00 | l | Glucose Syrup 42DE |
| 1000.00 | l | Total Volume, 36% v/v |

Colour: Dark brown

*Example Formula for French Herbs Liqueur (38% by vol)*

Step 1: French Herbs Extract & Distillate, 40/65% v/v

|         |   |                             |
|--------:|---|-----------------------------|
| 2000.00 | g | Walnut Shells, green        |
| 1200.00 | g | Hyssop Herb                 |
| 1000.00 | g | Angelica Root               |
|  500.00 | g | Angelica Seeds              |
|  500.00 | g | Artemisia Herb              |
|  500.00 | g | Ambrette Seeds              |
|  450.00 | g | Orris Root                  |
|  400.00 | g | Balm Leaves                 |
|  350.00 | g | Basil Herb                  |
|  150.00 | g | Thyme Herb                  |
|  150.00 | g | Cinnamon Bark               |
|  100.00 | g | Tumeric Root                |
|  150.00 | l | Alcohol-Water Mixture, 40% v/v |
|  (40.00 | l | Water)                      |

Yield: 30.00 l   Herbs Extract, 40% v/v
       70.00 l   Herbs Distillate, 65% v/v

Process:

Extraction for 2 days; 30 litres of extract are removed, the rest is distilled after adding 40 litres of water to obtain 70 litres distillate.

Formula for French Herbs Liqueur, 38% v/v

|          |   |                                 |
|---------:|---|---------------------------------|
|  278.00  | l | Grain Neutral Spirits, 96% v/v  |
|  100.00  | l | Brandy Distillate, 60% v/v      |
|      QS  | l | Water, demineralised            |
|   30.00  | l | Herbs Extract, 40% v/v          |
|   70.00  | l | Herbs Distillate, 65% v/v       |
|  300.00  | l | Sugar Syrup 72.7% w/w           |
|   50.00  | l | Honey (diluted in warm water)   |
|    1.00  | l | Elecampane Root Extract, 60% v/v |
| 1000.00  | l | Total Volume, 38% v/v           |

Colour: Yellow or green

*Example Formula for Half and Half Liqueur (35% by vol)*

| | | |
|---:|---|---|
| 286.20 | l | Grain Neutral Spirits, 96% v/v |
| 0.30 | l | Orange Oil, bitter |
| 0.01 | l | Angelica Root Oil |
| 30.00 | l | Orange Flower Water |
| 5.00 | l | Rum Distillate, 75% v/v |
| 60.00 | l | Brandy Distillate, 65% v/v |
| QS | l | Water, demineralised |
| 35.00 | l | Angostura Bitter Macerate, 50% v/v |
| 100.00 | l | Port, 15% v/v |
| 275.00 | l | Sugar Syrup 72.7% w/w |
| 20.00 | l | Caramel Syrup, 50% Solution i. Water |
| 1000.00 | l | Total Volume, 35% v/v |

### 5.2.2.3.5 Other Liqueurs

Other liqueurs are crystal liqueurs containing sugar crystals (e.g. crystal caraway, an especially aromatic caraway liqueur with high sugar content and an alcohol content of up to 40% by vol.). Liqueurs with other additives besides the above mentioned sugar crystals, such as gold leaves, are still very popular. Another product of this category is vanilla liqueur where the flavour must be – depending on the respective legislation – derived exclusively from vanilla beans, and finally honey liqueur (Bärenfang) with an original honey content of approx. 25 kg per 100 l finished product.

Cream-based emulsion liqueurs play an important role; dairy cream, eggs and fat-containing ingredients, such as nuts, are emulsified together with alcohol by means of homogenisation to form a stable liquid without or with little separation.

Typical flavours are chocolate, noisette, caramel, butterscotch, coffee etc.

*Example Formula for Honey Liqueur (38% v/v)*

| | | |
|---:|---|---|
| 250.00 | kg | Honey |
| 30.00 | l | Sugar Syrup 72% w/w |
| 50.00 | l | Glucose Syrup 42DE |
| QS | l | Water, demineralised |
| 37.90 | l | Grain Neutral Spirits, 96% v/v |
| 30.00 | l | Brandy Distillate, 65% v/v |
| 22.00 | l | Lemon Peel Distillate, 70% v/v |
| 2.00 | l | Elecampane Root Extract, 60% v/v |
| 1000.00 | l | Total Volume, 38% v/v |

### 5.2.2.4 Wine

Wine is a beverage produced by means of complete or partial alcoholic fermentation of fresh or mashed grapes or fermentation of grape juice. The cultivation of vines has played an important role in the Mediterranean countries for centuries and even nowadays Italy, France and Spain hold a 50% share of the world production of wine, even though the significance of North and South America as wine-producing countries is growing continuously.

Depending on geographical and botanical origin, preparation and processing technique, numerous different types of wine can be produced from the raw material grape which basically occurs in 2 forms, red and white grapes. Basically, the grapes or the grape juice are exposed to wine fermentation which happens either spontaneously through different yeasts on the fruit surface, mostly however through the addition of pure-culture yeasts, if necessary after pasteurization of the grape must. For the production of sparkling wine, special yeasts are used which produce a dense and grainy sediment that can be easily separated from the fermented product. Yeast preparations of this kind are also added to the grape must in the fermentation containers (oak wood casks, stainless steel tanks etc.).

The grape must has to ferment slowly (up to approx. 20 days) and under cool conditions (12-14°C for white and 20-24°C for red wine). The main fermentation for low-content products starts already after 1 day and reaches its most active point after 3-4 days. In order to protect the product against air intake (oxidation/browning effect), extraneous infection (acidic acid bacteria) and to preserve the carbonation, the loss of liquid in the fermentation container is compensated by adding back the same kind of wine. At the end of the main fermentation period (5-7 days), the sugar has been almost completely fermented into alcohol. Protein, pectin and tanning components, tartrates and cell fragments settle – together with the yeast – at the bottom of the container. The sedimentation of a part of the tartaric acid as the corresponding potassium salt depends on temperature, alcohol content and pH-value.

After the appropriate time, the young wine is drawn from the top of the fermentation containers into large, sulphurised casks. The purpose of the storage is the further enhancement of the odorous and taste-related components. In general, the wine is bottled after 3-9 months while maturation continues. Duration of the maturation and shelf-life vary and depend on the quality of the wine. Only premium wines can be stored for 10-12 years without a loss of quality, the average storage period is anywhere between 4 to 8 years. An important process during maturation is the degradation of malic acid into lactic acid, a process which is related to various microorganisms. As a result of this process the wine looses acidity and gains a milder taste.

Still existing turbidities will be eliminated by appropriate methods of clarification and stabilisation of the wine. Turbidity-causing matters are mostly proteins as well as oxidized and condensated polyphenols. Moreover, metal ions may lead to discolouring and sedimentation. Common methods of wine clarification are precipitation (fining), filtration or centrifugation. Clarification by means of filtration is achieved with cellulose or diatomaceous earth.

The chemical composition of the different wines varies widely and is influenced by climatic conditions, soil, location, type and treatment of grapes, grape must and wine. Important factors for the analysis of wine are extractives, alcohol content, sugar, acid, ash, colour, nitrogen compounds and bouquet components. Value and quality of the wine are furthermore determined by the contents of ethanol, extractives, sugar, glycerol, acidity and bouquet components. Due to the large number of value-determining components an evaluation and classification of wine is only possible by means of a combination of chemical analysis and sensory tests.

In addition to the many different types of wine, there are specialty wines such as dessert wines which show a relatively high extract content; this is achieved through the addition of concentrated grape juice or by means of alcohol addition to stop fermentation and preserve the residual sugar. Another kind of specialty wines are sparkling wines that are produced on the basis of different techniques (e.g. en tirage) which lead to the characteristic high level of carbonation in the finished product.

### 5.2.2.4.1 Wine-like Beverages

**Fruit Wines**

The pressed juices of apples, pears, cherries, prunes, peaches, red and black currants, gooseberries, blueberries, cranberries, raspberries, rose-hips and rhubarb are used for the production of fruit wines.

In a first step, the apple and pear mash is separated from the solid matters and then fermented as pressed must. The mash of berries on the other hand, is in most cases fermented directly in order to extract the colourants.

Pure-culture yeasts are used in order to avoid extraneous fermentation. The degree of fermentation in low-nitrogen berry musts can be increased by adding small quantities of ammonium salts. Up to 3 g/l of lactic acid will be added to low-acidic musts, mainly musts derived from pears. In order to achieve a pure fermentation and to improve the often acidic character of berry and stone-fruit wines, it is common practice to add sugar water. Juices from pomaceous fruits are improved through blending in approx. 10% water and sugar.

**Malt Wines / Mead**

Malt wines are produced by fermentation of malt extracts (extracts from ground malt). The flavour is acidified with bacteria from lactic acid, the formation of acid is adjusted by means of heating to 78°C and subsequently fermented to an alcohol content of approx. 10-13.5% vol. with pure-culture yeast. The resulting products show the characteristics of dessert wines; they can be distinguished from authentic dessert wines by their high content of lactic acid and the taste of malt extract.

Mead is produced by fermentation of a honey/water mixture (max. 2 l water for 1 kg of honey). Mead has already been a widespread beverage since early historic times and is still consumed in Eastern and Northern Europe.

Other wine-like products are palm wine and agave wine (pulke), maple and tamarind wines as well as rice wine (sake), a sherry-like beverage which is consumed mainly warm.

**Vermouth**

Vermouth was first produced in Italy during the last four decades of the 18$^{th}$ century, later also in Hungary, France, Slovenia and Germany. For the production process, either Woodruff herb (Artemisia absynthium) is extracted with wine, or an extract of the plant is added. The addition of other botanicals such as thyme, gentian or others is common practice.

**Wine Coolers / Wine-based Drinks**

Stagnating volumes in wine consumption, the increasing demand for profile variability and the consumers' willingness to experiment with taste led to the establishment of a flavoured, wine-based drinks segment in many countries.

Additionally, the individual, country-specific alcohol taxations have increased the attractiveness of this segment also for drinks producers.

Today, the range of products available to the consumer extends from classical wine-based products like Sangria or Mulled Wine to fruit-flavoured, juice-based wine cocktail concepts and includes even so-called flavoured wines.

### 5.2.2.5 Beer

Beer is mainly produced from barley malt, hops, yeast and water. Apart from barley malt, other starch- and/or sugar-containing raw materials are important, e.g. other types of malt (wheat malt), non-malted grains, also called raw fruit (barley, wheat, corn, rice), starch flour, products from starch degradation and sugar. For the application of these additional raw materials the use of enzymes is necessary. The stimulating and intoxicating properties of beer are due to ethanol; its aromatic properties are a result of hops, the malt products and the numerous flavouring components produced during fermentation. Its nutritional properties are due to the considerable content of non-fermented extract (carbon hydrates, proteins) and finally the refreshing effect is due to carbonation which is an important, value-adding component.

**Barley**

Among the raw materials, barley is the most important one which can not be substituted by any other grain. Summer barley is preferably used as brewing barley since it has the most suitable properties. Barley of high brewing quality produces large amounts of malt extractives, shows high starch content but only low protein content (9-10%), good germinating properties and high germinating energy.

Wheat malt in combination with barley malt is used for the production of top-fermenting beers. Besides barley malt, non-malted grains – called cereal – are added at a percentage of 15-20%, e.g. barley, wheat, corn and rice. Since the cereal grain shows only low enzymatic activities, the use of enzymatic preparations with alpha-amylase and proteinase activity is required. The processing of cereal may lead to disruptions and therefore sometimes extracts are used which were gained by enzymatic hydrolysis or by acidic hydrolysis from barley, wheat or corn. These can be purchased on the market in the form of syrup or powder. Barley syrup for example may be used in an amount up to 45% of the total mash. For the production of non-hopped malt extracts and hopped flavour concentrates, traditional flavourings are concentrated by means of vaporization under vacuum or by freeze-drying. For the production process, they are rediluted to the desired concentration. The content of bitter matters and the probability of turbidity are in general lower as a result of flocculation of tannin substances and proteins during the concentration process.

## Hops

Hop is an important and indispensable raw material for beer production. Acting as a clarifying agent, it causes a precipitation of protein matters in the wort, changes the character of the brew towards a specific flavour and bitter taste, increases the storage properties of beer due to its content of antibiotic active substances besides alcohol and carbonation, and supports the formation of foam through its pectin content. The bitter matters are the most critical components of hop. In fresh hop, these exist as alpha-bitter acids and beta-bitter acids; these are highly reactive compounds which produce a large number of subsequent products during drying, storage and processing by means of isomerisation, oxidation and polymerization. The quality and intensity of the bitter taste of these subsequent products vary considerably. Consequently, the evaluation of hop is determined by the composition of the fraction of bitter matters which can differ widely. During the cooking of the flavourings, the humulones are converted into iso-humulones which are more soluble and bitter than the original compounds.

## Water

The specific properties of many beers and types of beer can be traced back to the available brewing waters in which the residual alkalinity plays an important role. Waters with a low residual alkalinity are suitable for stronger, hopped light beers and waters with a high residual alkalinity are more suitable for dark beers. The treatment of brewing water consists mainly in the removal of carbonates. However, nowadays it is possible to treat any water according to the desired type of beer.

## Yeast

The yeasts used for the production of beer are exclusively of the saccharomyces species; among these, there are so-called top-fermenting yeasts which are active at temperatures above 10°C and so-called bottom-fermenting yeasts which can be used for temperatures as low as 0°C. The top-fermenting yeasts (Saccharomyces cerevisiae) settle at the surface of the beer during the fermenting process whereas bottom-fermenting yeasts (Saccharomyces carlsbergensis) form a sediment at the bottom of the fermenting kettle.

The germination process of the grain is initiated by soaking it with water and the resulting green malt is subsequently transformed into more or less dark and flavourful kiln-dried malt by means of drying and roasting. The kiln-dried malt is stored for approx. 4-6 weeks until it is used. The water which is added to the grain starts the germination and enzyme building process. The ideal germination temperature is 15-20°C. The process takes approx. 7 days for the production of light malt and approx. 9 days for the production of dark malt.

The kiln-drying process transforms the germinated grain – so-called green malt – with a water content of approx. 42-45% into storable kiln-dried malt with a water content of approx. 2-3%. At the same time the roasted flavor and colour are created through Maillard reactions. In a first step, the grain is dried for approx. 12 hours at a temperature of 35-40°C during which the water content for the production of light malt is reduced rapidly to approx. 10%, whereas for the production of dark malt it is kept at more than 20% over a longer period in order to intensify hydrolytic processes. In a second step, the grain is heated to roasting temperature in approx. 2 hours and roasted for 4-5 hours at 80°C (light malt) or 105°C (dark malt). Subsequently, the malt is cleaned and purified.

Special malts are produced for all kind of special purposes. Dark caramel malt is kept at a temperature of 60-80°C for a short period of time for the saccharification of starch and then roasted to the desired shade of colour at a temperature of 150-180°C.

The malt, which is rich in colour, is free of amylase; it is a good foam builder and is used mainly for the flavouring of malt and bock beers. Light caramel malt is produced through a similar process; however, it is dried at a low temperature after starch conversion. It is still enzymatically active, lightly coloured and increases the mouth-feel and foaming properties of beers produced from light malt.

Coloured malt is produced by roasting of kiln-dried malt without prior saccharification at a temperature of 190-220°C. It is used to increase the colour of dark beers. The ground malt is dispersed in water, a process which also causes a hydrolysis of starch and other malt components through the malt enzymes. By means of filtration, these flavours are gained in a fermentable, clear solution which is then cooked for the flavourisation with hop.

**Mash**

For the mash, the ground malt is combined with brewing water and partially degraded by the malt enzymes and solubilised. Approx. 8 hectolitres are needed for 100 kg of malt. Important for the composition of the wort, and thus for type and quality of the beer, are pH-value and temperature. The alpha-amylases of the malts show optimum efficiency at a temperature of 72-76°C and a pH-value of 5.3-5.8. The beta-amylases have their optimum between 60-65°C and a pH-value of 4.6, and the proteinases between 55-65°C and a pH-value of 4.6. Since wort has a neutral pH-value of approx. 6.0, there are no optimum conditions without the correction of pH-value.

The separation of the wort from the insoluble residues (leuters), also called treber, is achieved by the traditional process in a cylindrical container with double bottom; the upper one is a sieve on which the treber forms a natural, approx. 35 cm thick filter layer. The resulting filtrate (flavouring) is collected in the brew kettle (hop kettle) and combined with hop or hop products. The quantity of added hop depends on the type and quality of beer.

The quantity (hop per hectolitre) for different beers is:

| | |
|---|---|
| Light Lager: | 130-150 g |
| Dortmunder beer | 180-220 g |
| Pilsener | 250-400 g |
| Dark Munich beer | 130-170 g |
| Malt beer and dark Bock | 50-90 g |

The content of bitter substances is closely related to the dosage of hop. The flavouring is steamed down to the desired concentration by cooking it for 70-120 min. The protein coagulates, different valuable hop components are diluted and the enzymes inactivated.

# Alcoholic Beverages

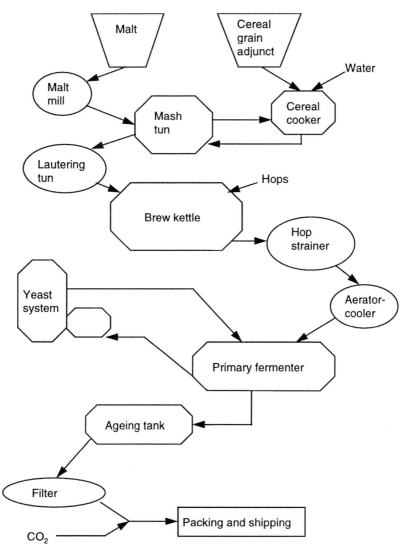

*Fig. 5.39: Brewing process*

## Fermentation

The bottom fermentation is divided into main and secondary fermentation. For the main fermentation, the cooled-down flavouring, which contains approx. 6.5-18% malt extract, is combined in fermentation containers with yeast (approx. 1 litre per hectolitre) and fermented at a temperature of 6-10°C for 6-10 days; within this period approx. 90% of the fermentable extract is transformed.

For the secondary fermentation (maturation), the young beer is stored in the storage cellar for 1-2 months at a temperature of 1-2°C. The top fermentation proceeds at higher temperatures of 18-25°C. Therefore, the fermentation period is shortened to 2-7 days. A secondary fermentation may take place in both, tanks or bottles. Top fermentation methods are mainly employed in England and Belgium and, for wheat beer production, in Germany.

Nowadays, precoated filters with diatomaceous earth are mainly used for the separation of yeast. For overseas exportation, the product is pasteurized after prior adsorptive elimination of on-heat-resisting colloids through treatment with bentonite, silica gel etc. Essential for the quality of the beer is the elimination of temperature fluctuations during storage and transport.

*Table 5.18:* Different types of beer: According to the used yeasts the beers are classified in top-fermenting and bottom-fermenting beers

|  | **Lagers** | **Ales** |
|---|---|---|
| Yeast Type | *Saccharomyces uvarum (carlsbergensis)* | *Saccharomyces cervisiae* |
| Yeast Behaviour | Settles to the bottom of fermenter | Rises to the top of the fermenter |
| Fermenting Temperature | 7 – 15 °C | 18 – 22 °C |

*Table 5.19:* Classical beer types

| **Type** | **Character** | **Origin** | **Alcohol** (% w/v) | **Flavour** features |
|---|---|---|---|---|
| **Bottom fermented** | | | | |
| Bock | Lager | Bavaria, US, Canada | 6 | Full bodied |
| Doppel bock | Lager/ale | Bavaria | 7-13 | Full bodied, estery |
| Dortmunder | Lager | Dortmund | 5+ | Light hops, dry, estery |
| Light beers | Lager | US | 4-5 | Light hops, light body |
| Munchner | Lager/ale | Munich | 4-4.5 | Malty, dry, mod. Bitter |
| Pilsener | Lager | Pilsen | 4.5-5 | Full bodied, hoppy |
| Vienna (marzen) | Lager | Vienna | 5.5 | Full bodied, hoppy |
| **Top fermented** | | | | |
| Ales | Ales | UK, US, Canada, Australia | 2.5-5 | Hoppy, estery, bitter |
| Alt | Ale | Dusseldorf | 4 | Estery, bitter |
| Barley wine | Ale/wine | UK | 8-10 | Rich, full, estery |
| Kolsch | Ale | Cologne | 4.4 | Light, estery, hoppy |
| Porter | Stout | London, US, Canada | 5-5.7 | Very malty, rich |
| Provisie | Ale | Belgium | 6 | Sweet, alelike |

| Type | Character | Origin | Alcohol (% w/v) | Flavour features |
|---|---|---|---|---|
| Salsons | Ale | Belgium, France | 5 | Light, hoppy, estery |
| Stout (bitter) | Stout | Ireland | 4-7 | Dry, bitter |
| Stout (Mackeson) | Stout | UK | 3.7-4 | Sweet, mild, lactic sour |
| Strong/old ale | Ale | UK | 6-8.4 | Estery, heavy, hoppy |
| Trappiste | Ale | Belgium, Dutch abbeys | 6-8 | Full bodied, estery |
| **Wheat-malt beer** | | | | |
| Gueuze-lambic | Acidic ale | Brussels | 5+ | Acidic, estery |
| Hoegaards wit | Ale | East of Brussels | 5 | Full bodied, bitter |
| Berliner weissen | Lager | Berlin | 2.5-3 | Lightly flavoured, mild |
| S. German weizenbier | Lager/ale | Bavaria | 5-6 | Full bodied, low hops |

### 5.2.2.6 Malt/Beer-based Alcoholic Beverages

A rather new segment currently developing in Europe is based on very common home use habits: the bottled ready–to-drink mix of beer and lemonade or cola.

Breweries have implemented innovations such as citrus-flavoured beer, slightly sweetened and acidified to offer trendy and fashionable new products for the younger generation of consumers as well as attract female customers.

Today, one can find coloured beer-based mixes in even exotic flavours and coloured appearance.

Many breweries have managed to fill their brewing capacities this way.

*Example Formula for Lemon Brew (2.5% v/v)*

```
   40.00  l    Sugar Syrup 72.7% w/w
    2.90  l    Citric Acid Solution 50% i. Water
    0.70  l    Trisodium Citrate Solution 25% i. Water
    2.00  l    Ascorbic Acid Solution 10% i. Water
    2.50  l    Lemon Extract nat,, 55% v/v
  451,90  l    Water, carbonated (6 g/l)
  500.00  l    Lemonade
  500.00  l    Beer, 4.9% v/v
 1000.00  l    Lemon Brew, 2.5% v/v
```

Note: If colour is required it is recommended to use natural colour

### 5.2.2.7 Cocktails/Spirit-based Alcoholic Beverages

This partly quite young category of alcoholic beverages is distinguished by the relatively low alcohol content (1.5 – approx. 10% by vol). The original concept for these beverages derives from bar and cocktail formulas for alcoholic mixed drinks.

The classical way of producing these alcoholic beverages with a low alcohol strength is based on the simple mixing of beer, wine or spirits with soft drinks of all kinds (e.g. whisky & cola, rum & cola, beer & lemonade, wine & fruit juice etc.).

Today, the decision for a suitable source of alcohol for low-alcoholic beverages is often based more on tax savings and cost efficiency of the source of alcohol than on the raw material nature of the alcohol base.

The production of coolers is handled in general by preparation of a throw syrup with simple further dilution with water (plus carbonation) in a second step.

*Example Formula for Cola Whisky (8% v/v)*

*Step 1: Cola Throw Syrup 1 + 5:*

| | | |
|---:|---|---|
| 606.00 | l | Sugar Syrup 72.7% w/w |
| 3.60 | l | Sodium Benzoate Solution 23.5% i. Water |
| 4.60 | kg | Phosphoric Acid 85% |
| 11.00 | l | Cola Flavour Emulsion nat. |
| 0.60 | kg | Caffeine, cryst. |
| QS | l | Water, purified |
| 1000.00 | l | Cola Throw Syrup 1 + 5 |

Step 2: Formula for Cola Whisky, 8% v/v

| | | |
|---:|---|---|
| 186.10 | l | Whisky Original, 43% v/v |
| 136.00 | l | Cola Syrup 1 + 5 |
| QS | l | Water, purified |
| 1000.00 | l | Cola Whisky, 8% v/v |

Note: 6–8 g/l Carbonation as required; colour adjustment with caramel colour as desired.

### 5.2.2.8 Production of Flavours for Alcoholic Beverages

In addition to the raw materials commonly used for the production of flavours, such as synthetic flavouring substances, uniform natural flavouring substances (gained by fermentation respectively biotechnological methods), essential oils, absolutes, essences etc., aqueous alcoholic distillation and extraction are processes which are specifically used today for the production of raw materials for the sector of alcoholic beverages, just as they have been for many years in the spirits industry.

# Alcoholic Beverages

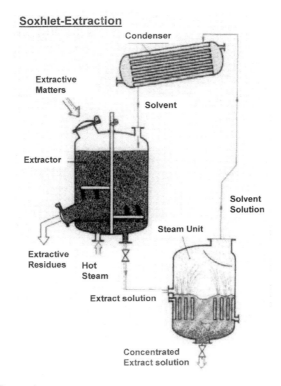

*Fig. 5.40:* Soxhlet-Extraction

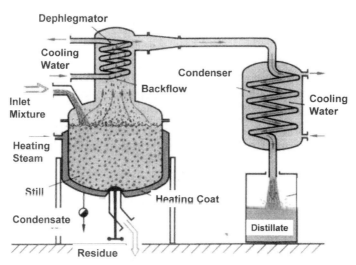

*Fig. 5.41:* Batch Distillation

## REFERENCES

Haeseler G., Wüstenfeld H., Trinkbranntweine und Liköre, Berlin, Paul Parey Verlag 1964

Belitz H.-D., Grosch W., Lehrbuch der Lebensmittelchemie, Berlin, Springer Verlag 1987, ISBN-3-540-16962-8

Rose A.H., Economic Microbiology, Volume 1 Alcoholic Beverages, London, Academic Press Inc. Ltd. 1977, ISBN-0-12-596550-8

Burdock, George A. Ph.D., Fenaroli's Handbook of Flavor Ingredients Volume II, Boca Raton Fla, CRC Press Inc. 1994, ISBN-0-8493-2711-3

Hardwick, William A. Ph.D., Handbook of Brewing, New York, Marcel Dekker Inc. 1995, ISBN-0-8247-8908-3

Kolb E., Fauth R., Frank W., Simson I., Stroehmer G., Spirituosentechnologie, Behr's Verlag 2002, ISBN-3-86022-997-4

# 5.3 Flavourings for Confectioneries, Baked Goods, Ice-cream and Dairy Products

## 5.3.1 Flavourings for Confectioneries

*Jan-Pieter Miedema*
*Josef Brazsák*


### 5.3.1.1 Introduction

The variety of confectioneries with respect to both composition and processing is very high compared with other groups of processed food. The common property of all confectioneries is, however, that the ingredients representing the majority of their mass, such as the principal ingredient sugar and others such as corn syrup, invert sugar, vegetable fat and water, do not have characteristic flavours (except chocolate products). Adding flavourings and ingredients having characteristic flavour (taste) is necessary in order to achieve the required taste sensation. It is evident therefore that flavour is the quality determining factor for most confectionery products. The difficulty is in selecting the most suitable flavouring for a given product and recommending its dosage. When selecting a flavouring and determining its dosage the following aspects of application need to be considered:

### (1) Product property

The added flavouring must not alter the required properties of the product (e.g. a powder flavouring shall be added to a powder product; oil-based, non water-soluble flavourings must not be added to clear jellies because the product becomes cloudy, etc.).

### (2) Shelf life of the product

Sweets generally carry a relatively long shelf life (4 to 24 months). However, generally they do not include any preservative or antioxidant. Only some of the packaging is air-tight and as such aroma-tight.

### (3) Physical-chemical properties of the ingredients

In certain cases chemical reactions may occur between the flavouring and some of the ingredients. The rate of aroma adsorbing and the proportion of the lipophilic-hydrophilic components may influence the so-called 'mouth-feel' to a considerable extent. Another important factor is the inherent flavour of the ingredients which should be enhanced or suppressed. Certain flavourings are reactive with basic or acidic agents (role of pH).

### (4) Characteristics of the manufacturing process

In many cases the flavouring is added at high temperatures (e.g. high boilings can have temperatures from 100°C to 130°C). Due to the high temperature part of the flavouring may evaporate. The speed of other types of chemical reactions is also dependent on the temperature. Foaming of the product increases the possibility of oxidation.

Before a decision is made which flavouring to use it is necessary to carry out trials in the laboratory. These experiments also provide important information for the development of flavourings.

The principal types of confectioneries may be distinguished according to identity or similarity of the manufacturing characteristics and / or in the raw materials used. Accordingly the principal types can be as follows:

- High boilings:
  - formed by pressing:
    - without filling
    - with filling
  - formed by moulding
- Fondants
- Jellies and gums
- Caramels and toffees
- Chewing gum
- Compressed tablets
- Dragee (sugar coated or chocolate coated)
- Chocolate and cocoa products.

### 5.3.1.2 High Boilings

High boilings (hard candy) is a type of confectionery which has a glass-like texture and a hard brittle structure. Its basic ingredients are saccharose and corn-syrup with a water content of 1 – 3% in the final product, which is achieved by cooking. Depending on the required principal flavour, ingredients such as flavours citric or tartaric acid, concentrated fruit juice, vegetable extract, honey, malt extract, milk components, colourings can be added.

The typical composition of a high boiling is as follows [1]:

| | |
|---|---|
| Sugar | 50-70% |
| Corn syrup (42DE) | 30-50% |
| Citric acid | 0-1% |
| Flavouring | 0.1% |
| Colouring | 0.01% |

The manufacturing process is shown in Figure 5.42.

The first step is the preparation of the basic solution. The most important aspect, when preparing the basic solution, is the complete solution of the sugar. The corn syrup added during solution and the invert sugar developing during manufacturing impede recrystallization of the saccharose later on. The manufacturing method and the type of equipment used determine the suitable proportion of sugar to corn syrup. The longer the cooking and the greater the concentration, the less corn syrup has to be added.

The solid content of the basic solution produced through dissolving and pre-cooking is about 70-75%, its boiling point ranges from 105-107°C.

Continuous cookers are used almost exclusively for this operation.

The basic solution with 70-75% solid content must be concentrated to a sugar mass of 98-99% solid content. In order to achieve such sugar concentration under atmospheric pressure the cooking temperature should be 145-150°C. At this point a negative phenomenon occurs, namely part of the sugar caramelizes. In order to avoid this normally vacuum or microfilm cookers are used. The vacuum cookers operate at reduced pressure ranging from 50 to 100 millibar, therefore the cooking temperature can be between 130-135°C.

According to the traditional manufacturing method the sugar mass at 130-135°C is transferred to the 'cooling table'. As a result the temperature of the sugar mass falls to 100-110°C. When this temperature range is achieved flavouring, acid and colouring are added and folded in.

An even distribution of the flavouring and colouring can be achieved by folding in during cooling, i. e. as the temperature of the sugar mass drops to 70-80°C over 5-10 minutes.

Sizing rollers or drop rollers at a controlled temperature give the final shape to the product at a temperature of 70-80°C. In the case of manufacturing high boilings with centres, the filling is added by a dosing pump through a pipe which is positioned in the centre line of the sugar ribbon at the same temperature. Flavourings are added to the filling at 70-80°C. The last operation is cooling. The temperature of the product falls to 30-35°C and it becomes solid while being transported on an air cooled conveyor belt.

In modern continuously operating closed systems the flavourings, colourings and acids are added to the sugar mass at elevated temperatures of about 130-135°C. It should be noted that cooling is more intensive than with the traditional manufacturing method.

The manufacturing of moulded high boilings is very difficult, particularly from the point of view of adding the flavourings. In this method the sugar mass is not cooled, therefore the flavouring must be added to the sugar mass at 135-140°C.

To manufacture the so-called pulled high boilings with air bubbles included during the cooling operation, the temperature of the sugar mass is reduced to 70-80°C. It is followed by the so-called 'pulling', foaming operation.

The high temperatures (more than 100°C) at which the flavouring is added, will result in a certain loss of flavouring substances during the cooling time for the candy.

To get a well-flavoured hard candy, the following considerations have to be taken into account:

- to compensate for the loss of flavour a higher dosage of the flavouring is necessary (2-3 times compared to jellies and gums)
- the flavouring should not contain too many volatile compounds or the volatile compounds might have to be overdosed

– the solvent used in the flavouring should have a high boiling point and it should dissolve well in the sugar matrix (e.g. propylene glycol).

The taste options for hard candy are generally fruit flavours and also cough drops types (herbals, mint, eucalyptus etc.).

In the past only citrus and cough drops tastes could be based on natural flavourings. Thanks to biotechnology it is now also possible to create natural flavourings in other tonalities for hard candy, although the price is rather high.

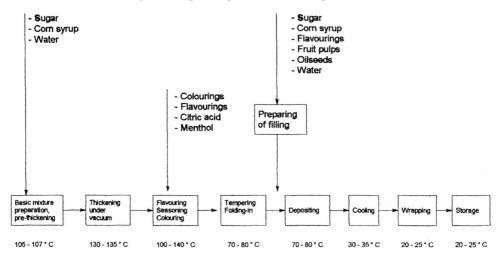

*Fig. 5.42: Flow Diagram of high boilings manufacture*

### 5.3.1.3 Fondants

The fondant mass is more or less a plastic suspension having saccharose crystals of very small size (about 0.01 mm) in a saturated saccharose, corn-syrup or saccharose-invert sugar solution [1].

The final product is made from the fondant mass directly through moulding or extrusion. Fondants are generally used as semi-finished products for manufacturing other types of confectionery (e.g. chocolate bars with soft centre, chocolate coated products, etc.).

A typical composition of fondant is as follows:

| | |
|---|---|
| Sugar | 60-80% |
| Corn syrup (42DE) | 20-40% |

The manufacturing process is shown in Figure 5.43.

The first step is preparing the basic solution. It includes dissolving granulated sugar in water, adding corn syrup and heating the solution to 106-108°C. Quality properties of the finished product (e.g. crystal size, invert sugar content) can be influenced by varying the proportion of corn syrup. The basic solution is made in the same equipment used for manufacturing high boilings.

The pre-cooked basic solution is fed by a pump to the continuous fondant cooker. Boiling temperature range is between 116-118°C when water is evaporated intensively. The solid content of the concentrated solution ranges from 86% to 90%.

Crystallization of the fondant is achieved by intensive cooling and beating. During this operation the solid phase of the fondant develops; the water content is reduced further and air is absorbed to a small extent. The temperature of the fondant discharged from the crystallizing cooler is about 50-60°C.

If colouring and flavouring are required, depending on the application, they can be added at the last phase of the crystallizing cooling. Generally these ingredients are added in separate pans equipped with beaters and cooling jackets.

The temperature of the fondant is between 50-60°C when flavourings, citric acid and thickeners are added.

Fondants are formed generally in moulds. The fondant is reheated to 65-70°C and poured into moulds made of starch powder in high capacity continuous machines. Silicon rubber moulds are used in certain types of these machines. Generally, the finished product can be taken out of the mould after 8 hours of cooling.

The fondant products must be coated with a thin layer of saccharose crystals or with chocolate in order to prevent drying out. The form can be created by crystallization and drying of the coating solution onto the surface of the fondant. The latter is achieved by coating the fondant with chocolate.

Fondant is rather neutral in flavour.

Since the production of fondants does not involve high temperatures, normal water soluble flavourings based on propylene glycol and alcohol as preferable carriers can be used.

Besides peppermint, all types of fruit flavours enhanced with citric acid are suitable. In addition, almond, vanilla and floral notes are used in flavouring fondants [1].

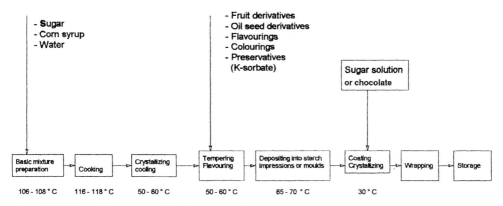

*Fig. 5.43: Flow Diagramm of fondant manufacture*

## 5.3.1.4 Jellies and Gums

Jellies and gums are elastic gels. A number of factors influence the development of their characteristic consistency. The ratio of sugar to corn syrup, the quantity and quality of other additives (e.g. jellifying agents, consistency stabilizers, acid and buffer ingredients) and the water content of the finished product play an important role. According to the water content, two types can be distinguished. The water content of the jellies is generally 20-24%, while that of the gums is 12-20%. In most cases pectin or agar is used for jellies while gelatine, modified starch or gum arabic are added to gums.

The typical composition of such products is as follows [1]:

| | |
|---|---|
| Sugar | 40-60% |
| Corn syrup (42DE) | 40-60% |
| Citric acid | 0-1.5% |
| Gelling agents | |
| – Pectin (150 US SAG) or | 1.3-2.0% |
| – Agar or | 1.2-1.5% |
| – Gelatine (200BLOOM) or | 8-12% |
| – Modified starch | 5-10% |
| Flavouring | 0.15-0.20% |

The manufacturing process is very similar to fondant processing. The difference is that the crystallizing cooling is not applied. Open boiling pans and continuous cookers (JET COOKER) are used for boiling. Gelling agents, dissolved in water, are added to the basic solution while it is being boiled. The gelling effect of the pectin depends on the pH, and can only be influenced by adding buffer ingredients (i. e. sodium citrate). The typical pH-value of such jellies is 3.3-3.5.

The final boiling point for concentration depends on the water content required in the finished product. 106-110°C is characteristic for jellies (solid content 76-78%). When gums are produced the basic solution is heated to 120-130°C. As it cools to below 100°C, the gelatine, previously dissolved at 60-65°C, will be added.

The solution which already contains gelling agents is then further cooled prior to flavouring. For pectin jellies and gums, 90°C is the appropriate temperature for flavouring, while for agar jellies this temperature is 60°C. According to traditional manufacturing methods, flavouring is added in mixing pans. When modern continuous equipment is used, acid and flavouring are injected at the final phase of manufacturing just before moulding.

Moulding is similar to that for the fondants, i. e. the mixture is deposited into starch moulds. The deposited jelly or gum achieves its final consistency and water content within 4-24 hours. When modified starch is used as the gelling agent, the products are set 24-48 hours in the so-called drying chamber at 30-50°C before demoulding.

The demoulded products need protective coating which can be made of sugar, oil or wax and also chocolate.

# Flavourings for Confectioneries

Products which are whipped to incorporate air into the mass, i. e. Turkish delight, montelimar, marshmallow, can be included in this type of confectionery. Air is incorporated into the basic solution before moulding or extrusion. The flavouring has to be increased.

Most jelly and (wine) gum confectionery products are made with fruit flavourings in combination with citric acid. The most suitable flavourings are water soluble. Because lower temperatures are involved in the production methods, the use levels of flavourings are lower (50% of the hard candy levels).

Starch jellies normally need more flavouring because starch has an inherent taste and inhibits the release of flavouring substances into the mouth. Because of the typical taste of the starch, the applied flavouring has to harmonize with the starch character.

## 5.3.1.5 Caramel and Toffee

Caramels and toffees contain basic ingredients like saccharose, corn syrup and fat. Milk (mostly condensed milk), gelatine, fondant mass, emulsifier, sorbitol, flavouring, fruit pulp, cocoa, coffee etc. are additional ingredients. Caramels compared to high boilings have higher water content (4-8%) which results in a plastic consistency in the mouth.

The caramels, depending on the type and ratio of the ingredients, may have amorphous or crystalline texture.

The typical composition of the caramels is as follows [1]:

| | |
|---|---|
| Sugar | 40-70% |
| Sweetened condensed milk | 5-30% |
| Corn syrup | 20-30% |
| Vegetable fat and/or butter | 5-15% |
| Salt | 0.5% |
| Lecithin | 0.3% |
| Flavouring | 0.2% |

The manufacturing process is shown in Figure 5.44.

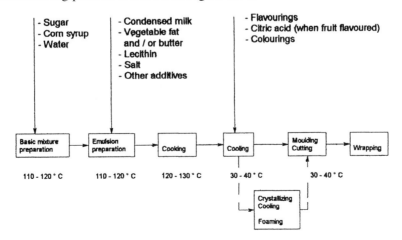

*Fig. 5.44:* Flow diagram of caramel manufacture

The basic solution in these cases is prepared by boiling the super-saturated sugar solution to which corn syrup is added. The boiling temperature is 120°C.

When the emulsion is made the pre-heated condensed milk, fat and emulsifying agent (lecithin) are added to the hot sugar solution. A relatively stable water-fat emulsion can be produced by intensive mixing.

During boiling at the final temperature of 120-130°C the mixture, now containing milk, is caramelized (Maillard reaction). Intensive mixing is needed during boiling. When manufacturing toffees with fruit flavourings and citric acid caramelizing is undesirable. Such toffees usually do not contain milk solids. Their typical consistency is derived from vegetable fats.

Boiling is followed by cooling. When the temperature of the mixture has been reduced to 90 -100°C, flavourings, citric acid, and colourings are added.

In traditional processing this cooling process happens on a so-called 'cooling table'. Modern plants feature continuous coolers. In certain cases the flavouring is added to the hot mix which results in considerable aroma losses.

The mass is cooled to 31-35°C and it is then passed through forming rollers which are in most cases combined with the wrapping machine.

Caramels of crystallized texture have a relatively high water content (6-10%) together with a low reducing sugar content of 10-12%. Under such conditions the saccharose in the caramel of amorphous texture crystallizes a few days after manufacturing. The crystallization can be sped up by adding fondant (10-15%), which is mixed in during cooling.

Caramel with a foam texture can be obtained by adding ready-made stable foam (Frappé) to the mixture. Alternatively the so-called 'pulling' operation is applied.

In toffees and caramels the flavouring dosage is high (even higher than in high boilings), not because of the high temperatures, but mostly because of the fat in which the flavouring is dissolved and which causes a slow flavour release in the mouth.

When the caramels are pulled and air is incorporated into the candy mass, flavourings resistant to oxidation are required. The most popular taste directions are caramel, cream, butter, coffee, cocoa, vanilla. In the case of products where dairy ingredients and sugars are used, a browning reaction (caramelization) takes place. The flavour type has to harmonize with the inherent flavour of the candy. In caramels with a fruity taste, oil-soluble fruit flavourings can be used. Flavour types such as orange, lemon, raspberry, cherry, strawberry, blackcurrant and tropical fruits are typical.

### 5.3.1.6 Chewing Gum

The plastic consistency of this product is achieved by non water-soluble natural gums and/or synthetic thermoplastic ingredients (gum base). Other components, such as sugar, corn syrup, acids, flavourings, etc. are dispersed by kneading into the structure of the giant molecules of the gumbase. Two basic types of chewing gums are known, the 'chewing gum' of harder consistency and the 'bubble gum' of softer consistency which can be blown.

Flavourings for Confectioneries

The typical composition of chewing gums is as follows:

| | |
|---|---|
| Ground sugar | 50-60% |
| Corn syrup (85% solid content) | 20-30% |
| Gum base | 15-25% |
| Glycerin | 0.5% |
| Flavouring | 0.5-1.0% |

In sugar-free gums, sorbitol and xylitol replace the ground sugar and the corn syrup is replaced by sorbitol syrup.

The manufacturing process is shown in Figure 5.45.

During the preparatory phase of manufacturing, the gum base and the corn syrup must be heated to at least 60°C. Certain authors [2] recommend 90-100°C as appropriate temperatures.

At 60-100°C the gum mixture and the corn syrup are fed into a jacketed kneader which features a Z-shaped. After 5-10 minutes mixing (homogenization), 50% of the ground sugar is added to the mixture. Mixing continues for about 10 minutes, when the rest of the ground sugar and other ingredients are added. Flavourings are metered in at a temperature of 60-70°C. Part of the flavouring is absorbed onto the surface of the ground sugar while the rest is integrated into the structure of the gum base. This is one of the reasons why the quantity of flavourings used in chewing gums is larger than for other types of confectionery. The flavouring integrated into the gum base modifies its structure. The consistency becomes softer or harder and thus it is important to establish the right composition of the flavouring and its proper dosage in pre-production scale-up trials. In order to reduce unfavourable effects and to achieve prolonged taste sensation it is usual to substitute one part of the liquid flavouring with spray-dried (encapsulated) versions.

After kneading is completed the gum base is compressed at 60°C by extruders which drive residual air-bubbles out of the base.

The chewing gum is formed into ropes or ribbons at a controlled temperature. The final shape (ball, sheet, fruit shape) of the finished product is produced from the air cooled rope or ribbon.

The shaped chewing gum should be cooled to 15-18°C by chilled air on a conveyor belt.

Chewing gums which are not pancoated with sugar will be wrapped immediately after cooling.

A large proportion of chewing gums is coated with sugar which acts as protection against drying. At the same time it offers the possibility to add further flavouring and it also improves the appearance of the product. The panning process will be described in the section on panned work (see 5.3.1.8).

The most suitable flavouring compounds for chewing gum are oil-based or oil-soluble for the following reasons:

- as the flavouring is dissolved in the lipophilic gum base, the flavour will have longer lasting qualities.
- water-soluble carriers like alcohol or propylene glycol will affect the structure of the gum base.

Although flavourings are added at rather moderate temperatures, high dosages are required. This is not because of flavour loss, but because of the slow flavour release from the gum base in the mouth. Increasing the longevity of a flavour in chewing gum is a problem which chewing gum producers and also flavour houses have tried to improve in many ways. There are various patents dealing with this problem, e.g. flavouring substances which develop during chewing or combinations of powdered and liquid flavourings or specialized solvent systems, etc.

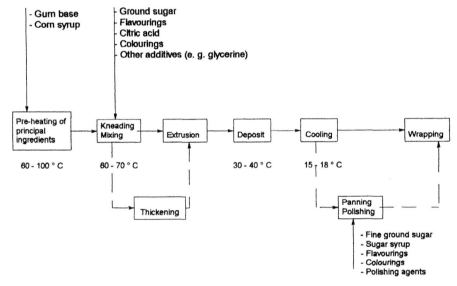

*Fig. 5.45: Flow Diagram of chewing gum manufacture*

Apart from mint, most flavourings used in chewing gum are nature-identical because of the high use level and the related high flavouring costs for natural flavourings. With flavourings produced nowadays by biotechnology it is possible to create natural flavours which are strong enough to flavour the gum satisfactorily.

**5.3.1.7 Compressed Tablets**

The principal ingredients of compressed tablets are ground sugar, dextrose or sorbitol. Typically in this type of product the cohesive force is created by high compression which keeps the grains together. Cohesion can be increased by adding binders such as gelatine, modified starch, gum arabic. Magnesium stearate is usually used to ease shaping.

The typical composition of compressed tablets is as follows:

| | |
|---|---|
| Ground sugar | 95-98% |
| Binder (e.g. gelatine) | 0.7-0.8% |

| | |
|---|---|
| Water | 4-5% |
| Magnesium stearate | 0.4-0.5% |
| Citric acid | 0-1% |
| Flavouring, powdered | 0.2-0.4% |

With ground sugar being the principal material, all other ingredients must be granulated first. Granulation of dextrose and sorbitol is not needed.

The first step when making a granulate is mixing of the powdered ingredients (e.g. citric acid, colourings, fruit powder) with the ground sugar. Surface moistening is the next operation generally using the aqueous solution of 10-20% of the binder (gelatine, gum arabic, etc.). Liquid flavourings are also added at this point. Before further processing the mixture of the granulated materials is dried for 6-8 hours. With fluid bed dryers this operation requires 20-30 minutes. Temperature of the drying air is 50-60°C.

The dried mixture is ground to a standard grain size of 2-3 mm which then is fed again into a powder mixer. Separating agents and flavourings are added to the mixture.

The last operation is forming by tabletting.

Effervescent tablets are a special group in this category. Processing is identical to compressed tablets except for the addition of sodium carbonate during and citric acid after granulation.

Sorbitol is the principal ingredient of sugar-free compressed tablets. To enhance the sweet taste usually artificial sweeteners (aspartame, cyclamate, saccharin) are used.

Flavourings used in compressed tablets are mainly encapsulated powder flavourings. Taste directions are mainly mint and citrus. They either have to be stabilized against oxidation or they have to be compounded without ingredients liable to oxidize or these ingredients have to be removed (e.g. terpenes).

### 5.3.1.8 Panned Work

These products can be categorized partly according to coating types and partly to the basic ingredients of their centres.

Grouping according to the coating can be as follows:

- Soft coated panned goods
- Hard coated panned goods
- Chocolate coated panned goods

If categorized according to the ingredients of the centre, the group of panned work is extremely large. All kinds of ingredients which have sufficient stability and can be formed into spheres, can be coated. A few examples for these are: moulded jelly, fondant, high boiling, chewing gum of ball and pillow shape, roasted nuts, raisin, etc.

The processing method used is almost identical for all types of panned goods. First an appropriate layer, if applied in revolving pans, coats the centre of the product. Later the coated product is polished.

Two methods are used to coat the centres with sugar: 'soft panning' and 'hard panning'.

In soft panning the centres are placed into the revolving pan and their surfaces are moistened with a coating syrup based on coloured and flavoured sugar syrup with corn-syrup content. Later fine caster sugar is dispersed onto the surface of the goods. Layers are created around the centres as required to obtain the desired size.

In the case of hard panning, the concentration of the sugar syrup is higher (66-82%) and the surface of the product is not coated with fine caster sugar. The product is dried by blowing hot air onto it. Coating is continued until the required product size is achieved. When flavourings are added, the basic flavour of the semi-finished product must be considered. As liquid flavourings should be added to the sugar syrup it is usual to use water-soluble compounds.

Due to the hot air used for drying, the flavour loss is considerable. In consideration of this loss and because the coating is rather thin, the flavouring has to be either very concentrated or it has to be used at a high level to create a sufficient taste sensation.

The typical flavour directions are vanilla and mint, but all kinds of fruit flavourings can be used as well.

Liquid chocolate is sprayed onto the centres when chocolate coated panned goods are produced. When the chocolate mass is evenly spread over the surface it is solidified by cool air.

Polishing again takes place in revolving pans. The polishing material is generally some kind of wax derivative. The polishing material is applied directly to hard panned and soft panned goods, while chocolate coated products need a sealing layer prior to polishing based on gum arabic.

### 5.3.1.9 Chocolate and Cocoa Products

The principal ingredients of both categories are the cocoa mass and cocoa butter made from cocoa bean.

The processing of cocoa beans and the manufacturing processes for chocolate and cocoa are shown in Figure 5.46.

Cleaning and sizing of the fermented cocoa bean with a water content of 6-8% are the first step. Any extraneous matter and broken beans are removed and selected sizing of whole beans is performed by specialized equipment. Sizing of the beans is important in order to later achieve uniform roasting.

Cocoa beans are roasted between 140 and 180°C in roasters heated by steam or gas. The internal temperature of the beans must not be higher than 130°C. Roasting time is from 30 to 50 minutes. The water content of the bean is reduced to 1.5-2.5% during this process, its colour and flavour change, the pH-value is reduced and the husk is loosened.

The husk is removed because it decreases the quality of the finished product. In order to ease separation of the husk from the bean, the roasted beans are ground to a particle

# Flavourings for Confectioneries

size of 3-5 mm. The husk is then removed from the nibs by screening and by using air stream separation.

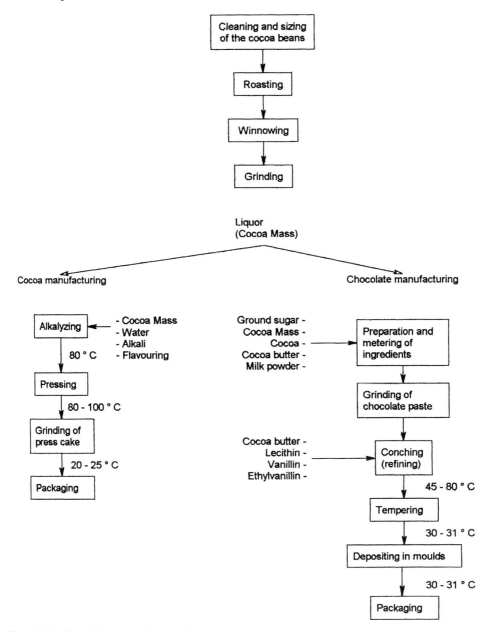

*Fig. 5.46: Flow Diagram of chocolate and cocoa manufacture*

The left over nibs are ground to be low 20 micron and then heated. This result in the liquid cocoa butter separating from the crushed cells and the cocoa nibs turning into a cocoa mass. This cocoa mass is the semi-finished product which is used in both, chocolate products and cocoa.

The quality of the cocoa mass is determined by the properties of the cocoa bean, the roasting conditions and the grinding rate. At least three varieties of cocoa bean should be mixed to obtain good quality cocoa mass. It is useful if one of them is 'spice bean'.

The grinding equipment used are the differential and rod mill and the ball mill.

### 5.3.1.9.1 Massive Chocolate Products

The principal ingredients of the massive chocolate product are the cocoa mass, cocoa butter, ground sugar and milk powder. Typical compositions of massive chocolate products are shown in Tab. 5.20.

*Table 5.20:* Composition of chocolate products

|  | Dark chocolate | Milk chocolate |
|---|---|---|
| Cocoa mass | 35-50% | 10-40% |
| Ground sugar | 32-48% | 20-40% |
| Cocoa butter | 5-15% | 5-20% |
| Salt | 0.2-0.5% | 0.2-0.5% |
| Flavouring (vanillin or ethylvanillin) | 0.01-0.02% | 0.01-0.02% |
| Lecithin | 0.3-0.6% | 0.3-0.6% |
| Milk powder | - | 10-25% |

The cocoa butter can be partly or wholly substituted by other fats. If more than 20% of the cocoa butter is substituted by another fat then the product is regarded as chocolate imitation. In addition to vanillin and/or ethylvanillin or other flavourings are added to the so-called chocolate imitations (e.g. chocolate, coffee, hazelnut, orange).

Cocoa butter may be substituted only by special fats (such as CBE, CBS and CBR fats) which have similar physical properties.

The first operation of chocolate processing is to homogenize the cocoa mass, ground sugar, milk powder and part of the cocoa butter in a mixer which can be a kneading machine with a Z-shape mixing arm or a continuous screw mixer.

The chocolate paste, having 28% fat content, is further worked through five-roll refiners or machines which work on different principles. The objective of this operation is to reduce the size of the majority of the solid particles to under 20 micron.

The fine ground chocolate powder or paste is conveyed into refining equipment where the rest of the cocoa butter, emulsifying agent (lecithin) and the flavourings are added.

The heated liquefied chocolate paste is conched under intensive mixing for 24-48 hours. During conching the dark chocolate is kept at 60-80°C, while the temperature range for milk chocolate is 45-50°C. The addition of flavouring materials (vanillin, ethylvanillin) usually is just before finishing this operation. During conching important changes to the physical and chemical properties of the chocolate paste occur (e.g. water and volatile acids evaporate, complex chemical compounds develop etc.). The result of these changes is the development of the final aroma characteristic of chocolate. Conching of chocolate imitations does not exceed 6 hours. The finished chocolate paste is pumped into the storage tanks where its temperature is maintained at 45-50°C.

The chocolate paste needs tempering before further operations take place. Tempering is required because of the polymorph property of the cocoa butter. Polymorphy is a phenomenon when a given material is able to produce crystal-modifications of different physical properties. Out of four possible modifications only one, the β-crystal modification is stable. Through tempering it is possible to obtain β'-crystal modification which is transformed into stable β-modification in the finished product.

Tempering is not required with certain chocolate imitations because the fat does not directly produce stable crystals (non-temper fats).

Massive chocolate products are deposited at 30-31°C into pre-heated (30-32°C) metal or plastic moulds. The moulds are then conveyed through a cooling tunnel where the temperature of the product is reduced to 8-9°C. The volume of the cocoa butter and chocolate shrinks during cooling and a contraction takes place which eases demoulding.

In the case of enriched massive chocolate products the enriching ingredients (e.g. roasted hazelnuts, raisins, etc.) are added to the tempered chocolate paste or metered into the moulds before chocolate is deposited.

### 5.3.1.9.2 Filled Chocolate Bars

A common property of the filled chocolate bars is that they have centres of different composition and consistency.

The basic operations of depositing chocolate mass are as follows:

- pre-heating of the metal or plastic moulds
- depositing of the chocolate paste into the moulds and spreading evenly by means of vibration
- development of outer layer on the chocolate bar by turning the mould to 180°C
- solidification of the outer layer in cooling tunnel
- tempering of the filling mass at 32-35°C, and filling
- cooling of the centres
- covering of the centres with a layer of chocolate mass
- final cooling
- demoulding.

Among the fillings which give character to the finished product are the well-known liquid liqueur fillings, fillings with jelly consistency, fondant and fat based fillings. The liqueur fillings are invert sugar solutions containing 2-30% alcohol. Natural flavourings complementing the liqueurs are added to the sugar solution.

Pectin of special properties is used in the manufacture of jelly centres. Fruit flavour types are most commonly used. The cream centres based on fondant, in addition to the basic fondant recipe, contain sugar solution, enriching materials such as coffee, milk powder, fruit pulp, etc. and also a preserving agent (potassium sorbate). Citric acid is added at 0.5-0.6% to fruit flavoured fillings. The temperature of the fondant cream should not exceed 60°C at any stage of the filling processing.

Fat based cream fillings, in addition to sugar, also contain 25-35% so-called cream-fat. The cream-fats typically have a low melting point of 30-32°C and a plastic, cream-like consistency within the temperature range of 20-25°C. Natural enriching agents such as coffee, cocoa, milk powder, coconut shreds etc. and flavourings are used to flavour the cream. Processing cream fillings is similar to processing chocolate paste, with the exception that conching is not required.

Naturally, water containing fillings require water-soluble flavourings and fat based fillings oil-soluble flavourings. The flavour type has to go together with the favour of the chocolate.

### 5.3.1.9.3 Extruded Bars

The manufacturing process is almost reverse to the method described above. The centres which are coated with chocolate or any other type of coating paste must be processed first. They should have a composition which will result in the right consistency after demoulding. The centres can be based on fondant, caramel, cream fat or even on bakery products. The enriching agents such as coconut shreds, cocoa, coffee, ground roasted hazelnut, ground croquant, raisin, etc. and flavourings together with colourings are added to the basic mix in temperature controlled kneading machines. During kneading the temperature of the mix must not exceed 40°C. Foamy consistency is produced either by adding foaming agents, or simply by intensive mixing. When selecting flavourings it should be considered that the fillings can be based on either fondant or hydrogenated vegetable fat, so that it is possible to use water or fat soluble flavourings.

Once the centres are cooled to 30-35°C they are deposited or extruded in a number of layers. Extrusion is made by screw or roll type extruders. The rope or ribbon is cut after cooling to the required size and coated generally with a tempered chocolate paste.

### REFERENCES

[1] Ashurst P. R., Food Flavourings, Glasgow, London, Blackie and Son 1991
[2] Meiners A. Kreiten K., Joike H., Silesia Confiserie Manual No. 3, Neuss, Silesia Gerhard Hanke KG 1983

## 5.3.2 Flavourings for Baked Goods

*Jan-Pieter Miedema*
*Bernhard Ludwig*


### 5.3.2.1 Introduction

Baked goods are made from flour, the main ingredient, as well as sugar, eggs, milk, fat, aerating agents etc. as minor ingredients. When added together they form the dough mix which is then heated in an oven to achieve the finished baked product.

For this chapter we will divide the baked goods into two groups:

- *baked goods*, which are considered as basic foodstuffs like bread and rolls
- *fine bakery products*, such as pastries, cakes, biscuits, etc.

The products of the first group will be discussed only briefly since they are generally not flavoured and the addition of flavourings is prohibited in most countries. Breads and rolls develop their typical taste through complex microbiological and chemical processes. In the first step where yeast ferments the flour, a multitude of flavour substances is generated. All these substances undergo a second process when they are heated in the oven. Degradation reactions and others take place resulting in a typical bread taste caused by a complex mixture of flavouring substances [1].

Some special breads are flavoured by adding spices (e.g. caraway, coriander, fennel, anise, etc.) to the dough.

As breads and rolls are produced more and more on an industrial or semi-industrial scale, bakers can no longer spend several hours preparing a dough. Baking aids are used to speed up this process. A problem that arises from these new production methods is the insufficient development of flavouring substances in the dough, resulting in a bread with a poor flavour profile. The obvious solution would be the use of bread or sour dough flavourings. However, as mentioned above, in most countries the flavouring of bread is not permitted.

The second group of bakery products that will be discussed in more detail are fine bakery products, which are allowed to be flavoured.

### 5.3.2.2 What are Fine Bakery Products and what is a Dough?

Fine bakery products are made from specific raw materials which are baked, roasted, dried, hot extruded or otherwise processed. They contain ingredients which determine the nutritional value as well as contribute to the sensory properties.

Fine bakery products, including pastries, are quite different from bread in that the recipe requires additional fats and/or sugars in a ratio of at least 10 parts to 90 parts of ground grains and/or starch.

Instead of grinded grain products, other grain products from wheat or rye and/or products of other types of grains could be used; in this case the appropriate minimum quantities apply. Also whole grains can be processed [2].

With reference to the base-recipe, the dough predominantly consists of flour-like ingredients such as flour and added starch. In addition, other determining characteristics of dough are its consistency, how it kneads and bakes and the type of processing to different pastries.

With reference to the working process, kneading and mixing are predominant in processing dough.

The way of making a dough smooth is often determined by the amount of non-flour ingredients used in the recipe. Since high proportions of fat and sugar or insufficient water impair the yeast fermentation, rich recipes use physical or chemical relaxants or a combination of the two instead of yeast.

The main factors to bind liquids and/or to influence the consistency of the dough are wheat glutens, pentosanes and mechanically damaged starch (through the grinding process of the corn) or hydrocolloids (for example thickeners).

The consistency of a dough varies from being elastic to being plastic which causes the dough to be more or less kneadable or mouldable *[3]*.

### 5.3.2.3 Flavourings for Fine Bakery Products

Flavourings used in bakery products are exposed to high temperatures. Temperatures reached during the baking process can vary from 100°C in the centre of a dough to 250°C on the outside. Parallel to the high temperature, another effect is caused by the heat: the evaporation of water from the dough. A considerable amount of flavour molecules is removed with the disappearing water. This effect is a kind of steam distillation. A process that helps bigger flavour molecules to stay in the baked product is the rapid development of a crust around the dough, which functions as a kind of filter.

Flavourings can be tailored for a successful application in bakery products in several ways:

- The use of flavourings at higher dosages or the use of more concentrated flavourings.
- The use of a solvent system for the flavouring substances with a high boiling point and which is lipophilic (against steam distillation): oils and fats.
- The use of special encapsulation methods for the flavouring substances.
  There are encapsulation methods which ensure that the flavouring system stays intact during baking and will be released only by the chewing action in the mouth.
- The use of so-called precursor flavouring systems.
  These systems contain substances that react with each other during baking, forming new substances which give the bakery product its taste.
- The addition of the flavouring after the baking process, e.g. by injection or by adding it to the coating or glaze, to prevent exposure to high temperatures.

**Three main possibilities to flavour a bakery product can be differentiated:**

- flavouring of the dough
- flavouring of the (fruit, cream) filling that is used in the bakery product

– flavouring of the coating or icing.

**Depending on the user, different flavourings for fine bakery products are on the market:**

*Flavourings for Use in Households*

These are flavourings packed in small bottles or bags to be used at home. Generally, one unit is enough to flavour one cake or pie. They are sold in the supermarket in liquid or powder form.

*Flavourings for Use in Small Handicraft-Type Bakeries*

Small and medium sized bakeries cannot work with highly concentrated industrial flavours which are only available in bigger quantities. They use so-called handicraft flavourings in the form of a paste which can be added easily to the dough and which are sold in standardized containers. Suppliers of these products make a lot of efforts in demonstrating these products in all kinds of bakery products to give the bakers new ideas. The suppliers need an extensive distribution system.

*Flavourings for Industrial Use*

These flavourings are often tailor-made for the final product. They are highly concentrated, mostly liquid and very economical due to their low usage rate. Powdered industrial flavourings are generally used in baking powder mixes for household or handicraft use.

**Types of Flavourings**

The most common flavourings used for fine bakery products are vanilla, lemon, butter, rum and almond. Local specialities such as Panettone (Italy), Stollen (Germany) or Bunspice (U.K.) are made with special combinations that give the bakery product its particular taste.

During the winter time spices or spice flavourings such as cinnamon, coriander, cardamom etc. are popular. In most European countries, the use of flavourings for suggestive purposes, e.g. to achieve a butter flavour without the use of butter, is not allowed.

In Table 5.21. some typical fine bakery products with their respective flavouring types are listed.

*Table 5.21: Fine bakery products and their flavourings*

| Type of Bakery Products | Flavouring/Flavouring combination |
|---|---|
| yeast bakery products general | butter-vanilla, lemon, vanilla combination of these, possibly with cardamom |
| – braided loafs | – butter, butter-vanilla |
| – raisin based bakery products | – rum, lemon |
| – panettone | – butter, vanilla, lemon, citrus combination |
| biscuits, sponge mixes general | vanilla, lemon |
| – cookies | – fruit flavours: mostly citrus |

| Type of Bakery Products | Flavouring/Flavouring combination |
|---|---|
| filling mixes<br>– poppy-seed mixes<br>– nut mixes<br>– marchpane mixes<br>Mixes, general<br>– as enhancer<br>– sacher mix | – lemon, rum<br>– nut, lemon, rum<br>– lemon, almond<br>lemon, almond, vanilla<br>– nut, chocolate<br>– cardamom, vanilla, cinnamon |

## REFERENCES

[1] Maarse H., Volatile compounds in foods and beverages, New York, Marcel Dekker inc. 1991, 41-71
[2] Brot und feine Backwaren 2, Frankfurt am Main, DLG Verlag 1985
[3] Brot und feine Backwaren 2, Frankfurt am Main, DLG Verlag 1985
[4] Seibel W., Feine Backwaren, Berlin und Hamburg, Verlag Paul Parey 1991

## 5.3.3 Flavourings for Ice-Cream

*Updated by:*
Alan D. Ellison
John E. Jackson

*Based on the manuscript of:*
Jan-Pieter Miedema
Jürgen Brand

### 5.3.3.1 Introduction

One of the highest growth areas in the food industry is the ice-cream sector. Because of an increasingly better organisation of refrigerated transport systems, the long distance supply of many households with freezing possibilities as well as the increasing popularity of ice-cream the market is expanding at an increasing rate.

Whereas 100 years ago ice-cream was a rare dish, today it is an industrially produced mass market product.

It can reach the consumers mainly in three different ways:

- handcrafted ice-cream
- novelty ice-cream
- bulk ice-cream.

Since the beginning of the nineties new developments such as ice-cream with ripple sauce, frozen yoghurt or ice-cream with vegetable fats have been available in the market alongside the traditional ice-creams such as milk-ice, soft-ice, sherbet and fruit- and water-ice. Due to the great number of modern, open minded consumers and different tastes in different countries the ice-cream market is continuously changing. The product life cycle of ice-cream brands is very short.

Variations on traditional ice cream can most often be classified as the subtraction of ingredients the consumers deems less desirable (such as fat or sugar) or the addition of ingredients the consumer deems as desirable (fibre, vitamin and mineral enrichment). This yields yet another consideration on behalf of the product developer in selecting an appropriate flavouring system as discussed later in this chapter [9].

### 5.3.3.2 Classification

Ice-cream has become a food which is kept under strict legal controls. Unfortunately, international classifications as well as constant changes in hygiene regulations only allow a rough classification which strongly depends on the respective countries. In some regions of the world, the use of milkfat is not feasible and alternative sources of vegetable fat are permitted in the formulation of frozen confections [9].

The most important criteria are:

- milkfat or vegetable fat content
- milk protein content

– solid (non fat solids)
– kind of flavouring.

Minor criteria are the content of egg and the classification by taste or consistence.

### 5.3.3.3 Ingredients

Table 5.22 shows typical ice-cream compositions which comply with the German standard for ice-cream. The most important ingredients groups are:

- milk products:    cream, milk, yoghurt, buttermilk, butter and butterfat
- sweeteners:       sucrose, dextrose, corn syrup, invert sugar, artificial sweetener
- fruit components: fresh fruit, fruit preparations, fruit concentrates
- food preparations: nuts, vanilla, cocoa, coffee, liqueurs, cheese
- flavourings:      natural, nature-identical or artificial
- colours:          synthetic colour or food preparations with colour gining effect
- stabilizers:      macro-molecules such as: agar-agar, gelatine, locust been gum, alginate, carrageenan, guar gum, sodium carboxymethyl cellulose, pectin
- emulsifiers:      mainly mono- and diglycerides of fatty acids
- air, nitrogen:    as whipping agent
- water:            food grade.

*Table 5.22:* Typical ice-cream compositions: (from Kessler [2])

| Milkfat | milk solids non fat | sweetener sucrose | stabiliser & emulsifier | water | whipping volume | start of freezing | German name of ice-cream |
|---|---|---|---|---|---|---|---|
| % | % | % | % | % | % | °C | |
| 8 | 6 - 7 | 13 - 16 | 0.25 - 0.3 | 59 - 60 | 70 - 100 | -2.5 | Eiskrem |
| 12 | 10 | 15 | 0.35 | 63 | 70 - 100 | -2.5 | Eiskrem |
| 10 | 10 - 11 | 13 - 15 | 0.3 - 0.5 | 63 - 65 | 70 - 100 | -2.5 | Eiskrem |
| 6 | 11.5 | 14 | 0.45 | 69 | 70 - 100 | -2.5 | Einfacheiskrem |
| 3 | 13 | 14 | 0.5 | 70 | 70 - 100 | -2.5 | Einfacheiskrem |
| 2.1 | 6.5 | 14 | 0.55 | 77 | 70 - 100 | -2.5 | Milchspeiseeis |
| 1 - 3 | 1 - 3 | 28 - 30 | 0.4 - 0.5 | 64 - 70 | 25 - 40 | -3.5 | Fruchteis (Sherbet) |
| - | - | 30 - 32 | 0.4 - 0.5 | 68 - 70 | 15 - 25 | -3.5 | Fruchteis (Sorbet) |
| - | - | 15 - 20 | 0.5 | 80 - 85 | 0 | -3.5 | Kunstspeiseeis |
| 3 - 6 | 11 - 15 | 12 - 15 | 0.4 - 0.5 | 60 - 65 | 30 - 60 | -2.5 | Softeis |

*Typical form as delivered to the consumer [9]:*

| | |
|---|---|
| Ice-cream | bulk, extruded in shape |
| Milk ice | bulk or moulded |
| Water ice | usually moulded |
| Sherbet | bulk or moulded |
| Sorbet | bulk |
| Soft-serve ice-cream | cup/cone |
| Yoghurt ice | bulk |

### 5.3.3.4 Flavourings

The flavourings for ice-cream should be:

- water soluble to have an immediate effect in the mouth. The flavour release of fat soluble flavourings is too slow.
- sterile or heat stable for microbiological reasons. Ice-cream is a very sensitive product from the hygienic point of view.
- not reactive with lipids.
- containing little to no terpenes [9]
- shelf stable for one year at -18°C.

Locally there are certain legal restrictions in the use of nature-identical and artificial flavourings (e.g. the German Speiseeis-VO 1933 with amendments and German „Leitsätze für Speiseeis und Speiseeiserzeugnisse vom 27. April 1995)

In the USA a standard of identity exists for ice-cream vanilla flavours. This standard of identity contains three categories. Category 1 vanilla-flavoured ice-cream can contain only all natural vanilla flavour. Category 2 can contain a blend of natural vanilla flavour plus an equal fold of artificial vanillin per gallon. A 'fold' of vanillin is equal to 1 oz of vanillin per gallon of natural vanilla flavour. Category 3 is the general realm of artificial vanilla flavours.

Industrial producers of bulk ice-cream or novelty ice-cream either use concentrated flavourings with a use level between 0.5-2 : 1000 or they add flavour as part of fruit or other food preparations.

So-called handicraft ice, made in ice shops, hotel kitchens etc. use less concentrated standardized flavourings for smaller batches with or without fruit and/or colouring agents. The basic flavouring substances for the different categories of ice-cream are the same.

The volatility of the flavourings is decreased due to an ice consumption temperature between -9 and -12°C.

The most popular taste directions are:

- vanilla
- strawberry
- chocolate.

### 5.3.3.5 The Influence of Other Ingredients on the Ice-cream

*Milkfat*

Milkfat is an ingredient of major importance in ice-cream in order to balance the mix properly. The main sources are fresh cream, butter and butterfat. In regions where the use of milkfat is not feasible, coconut or other vegetable fats with similar melting points have been substituted. In many cases a butter or cream flavour is added to these products to compensate for the absence of milkfat. The use of condensed milk can result in a cooked note. The milkfat has no influence on the freezing point. A high content limits the consumption, a low content reduces the rate of whipping.

While a higher level of fat gives a rich impression of mouthfeel, the fat also diminishes the perception of an added flavour. Therefore, we may see a quicker expression of the added flavour in a reduced-fat ice-cream product, sherbet, sorbet or water ice. The flavouring dose level may be adjusted down in these products *[9, 11]*.

*Milk Solids Non Fat (MSNF)*

MSNF improve the palatability, the nutritional value as well as the economic aspects. The main components are:

>   protein, lactose and minerals

A high content results in a sandy body because the high lactose content may crystallize. The protein makes the ice more compact and smooth. MSNF have no relevant flavour effect except a salty, slightly sweet note.

*Sweeteners*

The most important sweeteners are sucrose, dextrose, fructose, corn syrup, sugar alcohols and invert sugar. The average content is between 12% and 20%. A part of the sucrose should be replaced by corn syrup in order to maintain product properties and sweetness. The main function of sweeteners is to enhance the flavour and give it the desired sweetness. The sugar influences viscosity and the content of solids.

Sugar and the total solids of the mix depress the freezing point. Proper freezing point and subsequent draw temperature of the product from the barrel freezer influences many attributes of the ice-cream. Departure from this freezing point can slow throughput of the ice-cream in production at the plant. Additionally, the product may suffer from shrinkage in the package over its shelf life or prematurely develop oversized ice crystals and iciness *[9, 10]*.

With the number of reduced sugar ice cream products introduced into the market over these past few years, manufacturers have added sugar alcohols or lower levels of monosaccharide simple sugars to efficiently move the freezing point of the ice-cream mix. For example, dextrose is almost twice as effective as sucrose in freezing point depression. These ingredients often have a cooling effect in the mouth and thus alter the expression of added flavouring systems.

Artificial sweeteners are used for diabetic or calorie-reduced ice-cream. These may include sweeteners used singly or in combination for a synergistic effect. The impression of sweetness on the tongue may last longer or fall off when compared to a traditional sugar-sweetened product and flavouring systems may also be introduced to diminish this effect *[9]*.

*Stabilizers*

Stabilizers are macro-molecules which prevent the formation of large ice crystals. The average ice crystal size should be between 40-50 µm. Stabilizers increase viscosity, improve the smoothness and resistance to melting. The average content depends on the kind of stabilizer, but it is typically between 0.01% and 0.2%. The use of stabilizers corresponds to the content of fat and solids. Stabilizers are important when the fat content is below 12% or the total content of solids is below 40%.

Typically, stabilisers are added to control water in the system, thus influencing the rheological properties of the aqueous phase of the ice-cream. This affects the flavouring system to varying degrees based upon the blend of stabilisers used. Sometimes flavours are muted by over-stabilisation.

A product developer may experiment with different blends of stabilisers or ultimately increase dosage to compensate [9].

*Emulsifiers*

The emulsifiers influence the whipping quality of the mix. They produce a dry, smooth and stiff finished product. The dryness is often related to the impression of textural creaminess and warm eating characteristics in the mouth. Flavours such as caramel or vanilla may be chosen to complement this warm eating characteristic. This is as opposed to fruit flavours which may be better showcased in a crisp, refreshing and somewhat colder eating system [9, 10].

### 5.3.3.6 Sensory Aspect of Ice-cream

The most important criteria of ice-cream are:

| criteria | *factor |
|---|---|
| appearance | 3 |
| mouth feel | 5 |
| taste/flavour | 8 |
| melting characteristics | 1 |
| whipping volume | 2 |
| non fat solids | 2 |
| microbiology | 7 |

* factor of the German Food Association in the official sensory test for milk ice-cream [1].

As the factors of the sensory test methods show, flavour is the most important aspect of ice-cream. Added flavourings improve many details which contribute to an overall positive perception.

### 5.3.3.7 The Production of Ice-cream

The production of industrial ice-cream is shown in (Fig. 5.47).

The flavouring can be added before, during or after the ageing. The flavouring is added either directly to the mix after the pasteurisation or together with the fruit or food preparation. The preparation should be sterile or pasteurized.

Generally, the flavour should be added at the lowest possible temperature in the process to prevent evaporative losses of volatile flavour nuances [9].

The procedure of ageing and freezing is a very important process. During the ageing process the macro-molecules need time for their hydration. This time has to be balanced between economy, hydration of the stabilizers, flavour building and the danger of microbiological infection.

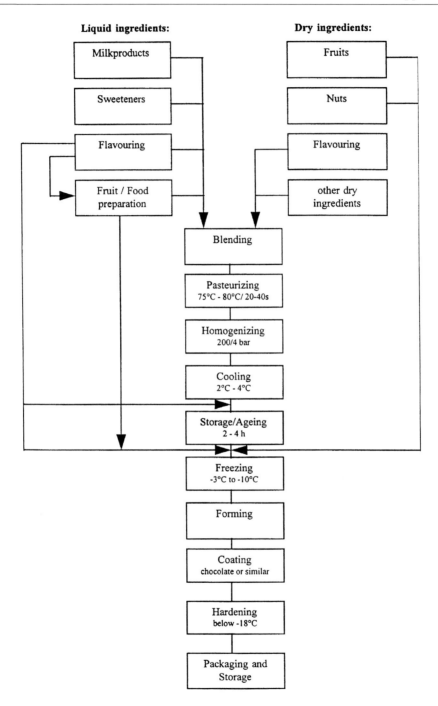

**Fig. 5.47:** Flow chart "production of industrial ice-cream"

The freezing process influences the building of the phases of ice, air and fat. The maximum size of ice crystals may not exceed 60 µm, otherwise the crystals could be detected by the consumer.

## REFERENCES

[1] DLG Qualitätsprüfung DLG-Prüfbestimmungen für Milch und Milchprodukte einschließlich Speiseeis, 1989/90, Hrsg. Deutsche Landwirtschafts-Gesellschaft Frankfurt am Main
[2] H.G. Kessler, Lebensmittel- und Bioverfahrenstechnik - Molkereitechnologie, Verlag A.Kessler, Freising 1988 S. 427-434
[3] Ice-cream Stabilisation, Iain C.M. Dea, Leatherhead Food Research Association (UK), IFI Nr. 1-1991, 9-13
[4] Funktionelle Eigenschaften von Emulgatoren in Eiskrem, hergestellt unter Verwendung verschiedener Fett-Typen, Sören Olsen, Brabrand/Dänemark, ZSW 3/1993, 124-130
[5] Arbuckle, Wendell Sherwood , 1977, Ice cream.
[6] Deutsche Lebensmittelbuch-Kommission: Leitsätze für Speiseeis und Speiseeishalberzeugnisse, Bundesanzeiger Nr. 101 S.5990, 31.Mai 1995
[7] Verordnung über Speiseeis vom 5.Juli 1927, zuletzt geändert: Vierte Verordnung zur Änderung der Verordnung über Speiseeis vom 24. April 1995
[8] Timm, F.(1985). Speiseeis. Verlag Paul Parey. Berlin, Hamburg
[9] Peters, S.W.P.G., Heinsman, N.W.J.T. (eds.), Dairy Manual: Ice Cream, Quest International, Global Flavour Training Department, 2005
[10] Halford, B., Ice cream, Chemical and Engineering News $\underline{82}$(45), 51 (2004)
[11] Ohmes et al., Sensory and physical properties of ice creams containing milk fat or fat replacers, Journal of Dairy Science $\underline{81}$, 5 (1998)

## 5.3.4 Flavourings for Dairy Products and Desserts

*Jan-Pieter Miedema*
*Eckhard Schildknecht*


### 5.3.4.1 Introduction

Until around 1975 dairy products in Europe were very much specific to individual countries or even regions. As Europe becomes more unified dairy products also tend to grow across the borders, however, increasing the variety within the regions. It is a great opportunity for the dairy industry and for food ingredient manufacturers to play their part in the unification of Europe, since this process involves high levels of technology, cooperation (teamwork) and marketing skills.

In the beginning only a limited range of unflavoured dairy products was on the market. Then, after manufactures started adding (fruit) preparations or flavourings, there was a significant increase in the number and type of dairy products on the market.

Also flavourings were adapted more and more to an 'European taste'.

This chapter describes manufacturing of the different flavoured dairy products by direct addition of flavourings or (fruit) preparations, the technology involved in their manufacturing as well as their further processing.

The type of flavourings applicable for particular dairy products will be also discussed.

### 5.3.4.2. Production of Flavoured Dairy Products

The following flow chart (Figure 5.48) gives an overview over the most common flavoured dairy products. It shows the production technology of yoghurt, fresh cheese products, milk rice and directly flavoured dairy products.

#### 5.3.4.2.1 Directly Flavoured Dairy Products

Products like puddings, custards, milk drinks belong to this category and they can be divided into two main groups:

    (1) Desserts
    (2) Milk Drinks

The production technology for desserts is largely determined by the shelf life, the consistency of the product and by the equipment available. The flavourings can be either liquids or powders. They are added to the milk or the water at the beginning of the process together with starches, hydrocolloids and colours (either individually or as a premix). A new dosage version is the aseptic injection of coloured flavouring after the UHT treatment before the aseptic filling. This has the advantage of fresh colour and flavour.

Depending on market requirements dessert and milk drinks are heated under UHT, sterilisation or fresh product conditions. The filling temperatures vary from hot fill to cold aseptic filling. Heating and cooling process with sterile products are carried out in tins, cups, PEHD and PET bottles, cartons or glass jars, using either continuous or batch systems. The flow chart (Fig. 5.48) shows the technology involved in the manufacturing of UHT puddings and UHT milk drinks.

### 5.3.4.2.2 Sour Milk Products

Sour milk products are always cultured dairy products with lactic acid bacteria (depending on the food legislation of the respective country). After increasing the dry matter, pasteurisation and incubation of culture, they are processed into yoghurts of set, stirred or drinking consistency, with or without a final heat treatment.

Yoghurt products are manufactured by using special micro-organisms, normally lactobacillus bulgaricus and streptococcus thermophilus. Lactobacillus bulgaricus converts proteins into amino acids. The bacteria streptococcus thermophilus produces the typical yoghurt flavour by using these amino acids as nutritional substrates.

In the case of direct flavouring of yoghurt during its manufacturing the flavouring can be added to the milk at the beginning. More often a (fruit) preparation is added to the yoghurt during the production process using either dosing or mixing equipment.

### 5.3.4.2.3 Fresh Cheese Products with (Fruit) Preparations

This description covers all kind of fresh cheese products with ingredients such as (fruit) preparations, flavourings, spices and herbs. Using stabilisers, fresh cheese products, can also be pasteurised. Country specific regulations determine the nutritional claims.

In the manufacture of fresh cheeses, e. g. quark, the pasteurised skim milk is inoculated with micro-organisms (Sc. lactis, Sc. cremoris). To accelerate the thickening, the enzym chymosin is added. After ripening – with a pH value of about 4.6 – the coagulated milk must be pumped through a separator, possibly including an ultrafiltration system, in order to separate the sour whey. The ultrafiltration would separate the whey into 2 phases: the permeate (water soluble) and the retentate (protein phase). Finally the quark, retentate, cream, (fruit) preparations, flavourings or spices and herbs are added.

Further quark production technology includes full concentration of the skim milk, ultrafiltration without separator as well as the thermoheating method.

### 5.3.4.2.4 Milk Rice with (Fruit) Preparations

For the manufacturing of rice with (fruit) preparations the production technology is shown in Figure 5.48.

Rice, starch, milk and water are added into the mixing tank and are heated at a preset temperature to about 60°C for pre-soaking. If required, flavourings and sugar are added at that point. The mix is then transferred to a scrape heat exchanger and the mixture is heated to approx. 115-121°C with a holding time of 15 to 20 minutes.

The product is then cooled to 15-20°C, the (fruit) preparation is added and the product is filled cold aseptically.

Production is also possible with an batch rice cooker system.

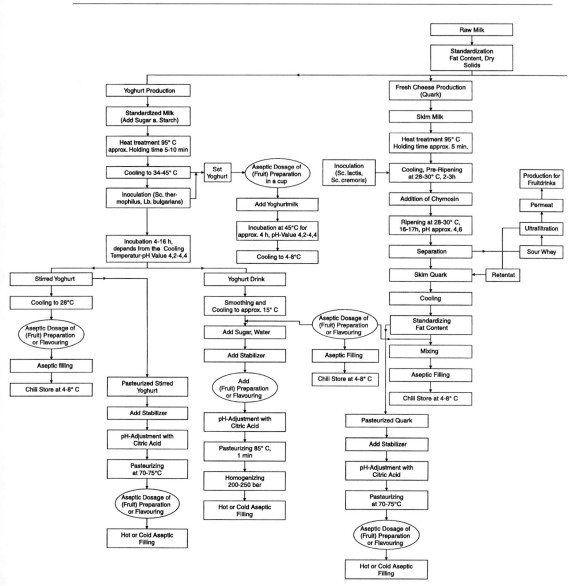

**Fig. 5.48:** Flow Chart Milk Products

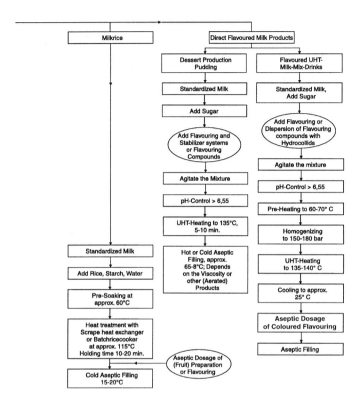

It should also be mentioned that for microbiological reasons only spore-free (fruit) preparations should be used for industrial dessert products in the neutral pH-area.

### 5.3.4.3 Flavouring Components for Dairy Products

#### 5.3.4.3.1 General

Milk in its natural state does not have a pronounced odour and flavour. Through various kinds of processing milk can be changed into other products which taste differently, e. g. butter, cheese, yoghurt, quark etc. Certain processes as well as storage, transportation, packaging material can cause so-called 'milk defects'. From a sensory point of view these are described as cow shed smell, light taste, cooked taste and oxidation taste, only to mention a few.

In order to offer the consumer a greater variety of dairy products other then milk, UHT milk, sour milk products, cream, butter or cheese, the dairy industry has developed a variety of flavoured dairy products and later in the sixties, dairy products containing (fruit) preparations. The latter category originated from a close cooperation between the dairy industry, fruit manufacturers and flavour houses. Fruit juice processing companies, marmalade manufacturers and also flavour houses established departments for (fruit) preparations which resulted in a great variety of preparations.

These (fruit) preparations have multifunctional purposes in dairy products. They provide fruit, structure, colour, stability and, mostly through added flavourings, a variety of tastes.

### 5.3.4.3.2 Flavourings for Dairy Products

Dairy products can be flavoured with either natural, nature identical, artificial or even process flavourings. The type depends on what the producer wants to declare on the ingredient list: 'natural flavourings' or only 'flavourings'.

Today natural flavourings can be used without technical restrictions since in the nineties biotechnology and generally improved technologies provide a much wider, better and more economical range of natural flavouring substances.

In most cases the use of natural flavourings as well as their advertisement on labels is decided by marketing. The dairy producer has to take into account the higher prices of natural flavourings in order to achieve a more sophisticated product or a better image which in turn justifies higher sales prices and results in increased sales.

In Europe the trend regarding natural flavourings in dairy products changes every few years, depending on social, economical and also public opinion factors.

Technically the most suitable flavourings for dairy products are water soluble without solvent or, with propylene glycol or alcohol respectively.

In the case of citrus flavourings it is important that they do not contain terpenes, because of their interference with polymer based packaging materials.

If the flavouring is added as part of a premix with powder stabilisers and colours, powder flavourings are used. Except for vanilla, most powder flavourings are encapsulated spray-dried flavourings with maltodextrin and/or gum arabic as carriers.

The suitability of flavour types for the different dairy products depends on many factors. In the case of direct flavouring it is advisable to test different flavourings in a product prototype which is made up under the same conditions and with the same ingredients as the new dairy product to be developed. If the flavour is introduced indirectly via a (fruit) preparation it is important to test as if it was prepared under the conditions the (fruit) preparation is to be produced. Too much heating, cooling and evaporating may cause flavour losses and changes. It is also important that the flavour harmonises with that of the genuine dairy product.

The most widely used flavour types for neutral dairy products are vanilla and chocolate. Caramel, butterscotch, coffee etc. are also being used.

The most important taste for dairy products with a lower pH is strawberry. Other taste directions are raspberry, blueberry, peach, passionfruit, cherry, several citrus types, banana, kiwi, etc. Flavour combinations such as peach – passion fruit, kiwi – gooseberry etc. play also an important role.

In calorie-reduced dairy products the lack of richness from the fat is often compensated with cream-, butter- or fat flavourings.

### 5.3.4.3.3 Manufacture of (Fruit) Preparations

(Fruit) preparations are always applied when a typical taste, smell, fruit content - (smooth or pulp structure, based on concentrate, puree or fruit pieces) are looked for in the end product. Sometimes (fruit) preparations are added in order to obtain a particular viscosity or consistency in the finished product. This carry-over effect is not at least due to the hydrocolloids and modified starches contained in fruit preparations.

Depending on how the desired fruit has to show up in the end product, with or without fruit pieces, flavour addition, colours content (natural, synthetic), and viscosity, different technologies are used in manufacturing (fruit) preparations.

For so-called low acid products there is special equipment to obtain an end product with neutral pH.

The selection and quality of the raw materials such as fruit, sugar, flavouring, colour, hydrocolloids, starches and water is of critical importance.

In general, individual quickly frozen (IQF) fruit is used. However, in special cases (e. g. pine-apple) tinned fruit and aseptic packed fruit can also be used. Hydrocolloids like starches or modified starches are also important components of the formulation. In order to give the end product an appealing appearance food colours such as beet root concentrate, red and yellow juice concentrates or synthetic colourings are added.

Sucrose, isoglucose, fructose, dextrose and sweeteners are responsible for the solid content and sweetness of the preparation. Water quality also plays an important role.

Every ingredient has to be strictly controlled before it is used in production; it must not be used unless the required standard is obtained.

(Fruit) preparations are manufactured by either continuous systems (e.g. Ohmic heating, tube or plate heat exchanger, scraped surface heat exchanger) batch or microwave systems. The plant environment, the plant itself and the C.I.P. (Cleaning in Place) equipment are subjected to an intensive quality management.

After an initial heating step the pH and ° Brix values are measured. The finished product is very much influenced by temperature treatment. For (fruit) preparations to be used in finished products with a pH of less than 4.2 it is normal to heat the fruit preparation to 85-98°C with an appropriate holding time (so-called high acid products). For (fruit) preparations to be used in finished products with approx. pH 4.2 to 6.8, a temperature of approx. 138°C with adequate holding time must be obtained in order to have sporefree results.

The final cooling stage is identical to the heating stage in the sense that it depends on the size of fruit pieces we want, which colour and viscosity. After production there is a detailed microbiological and physio-chemical control.

### 5.3.4.3.4 Use of Flavourings in (Fruit) Preparations

Flavourings are responsible for the finishing touches of a fruit containing dairy product.

Flavourings are important to compensate for the loss of flavour and smell due to the heating of the fruit preparation and pasteurisation of the milk product, and, of course, they provide the desired taste to the end product.

Trials need to be carried out in the development laboratory and the pilot plant in order to optimise parameters for the addition of flavourings, e. g. the required quantity.

Flavourings and colourings can be added at different stages of the production. Which and how much flavouring and colour is added, mostly depends on how long the flavourings are exposed to which temperatures. This, of course, depends on the equipment the preparation is manufactured with.

**REFERENCES**

[1] J. Klupsch Saure Milcherzeugnisse - Milchmischgetränke und Desserts, Gelsenkirchen-Buer, Verlag Th. Mann 1984
[2] Dr.-Ing. Spreer Edger, Technologie der Milchverarbeitung, Leipzig, Fachbuchverlag 1988
[3] Richtlinie für Fruchtzubereitungen zur Herstellung von Milchprodukten, Teil I, II, Hamburg, Behr's Verlag 1971
[4] Prof. Dr. H. O. Gravert, Die Milch: Erzeugung, Gewinnung, Qualität, Eugen Ulmer Verlag, 1983
[5] Prof. Dr. A. Fricker, Berlin, Heidelberg, New York, Tokyo; Springer Verlag, 1984 'Lebensmittel mit allen Sinnen prüfen'
[6] Alfa-Laval Food Engineering AB, P.O. Box 64, S-22100 Lund, Sweden 'Dairy Handbook', TETRA LAVAL

# 5.4 Flavouring of Dehydrated Convenience Food and Kitchen Aids

*Chahan Yeretzian*
*Imre Blank*
*Stefan Palzer*

## 5.4.1 Introduction

Dehydrated convenience food and kitchen aids are important industrial segments within the larger category of culinary products. Their widespread usage in modern households is mainly due to the convenience they bring to everyday cooking, while guaranteeing good results in relatively short time (quick preparation), with little effort and at an affordable price. To describe the role of dehydrated convenience food and kitchen aids as culinary products, we will first outline the different ways commonly used for categorising culinary products.

Culinary products cover a range of distinct categories, all sharing the characteristic of having a savoury flavour. This has to be seen in contrast to the other main flavour category: sweet. As culinary products encompass such a large range of products, it is common to divide it into sub-categories. This may be achieved in a variety of ways, depending on the perspective and purpose of such a classification.

*Division according to product format*

Culinary products can be divided into three different textures. These are liquid (e.g. liquid seasoning), paste (mayonnaise) and dehydrated (bouillon cubes).

*Division according to distribution logistics*

Depending on the shelf-life characteristics of the products, three different types of distribution logistics have to be used. These are ambient (room temperature), chilled (4°C) and frozen (-20°C) distribution.

*Division according to usage (solutions)*

Depending on the role of the product in relation to the overall eating and cooking pattern of consumers and the consumption situation, we differentiate between kitchen aids, snacks and main dishes.

The first two classifications are mostly technical and related to internal processes and technologies of the food industry (manufacturer, distributor, retailer perspective). Differentiation according to *texture* is mainly related to technologies at the stage of product manufacturing and linked to shelf-life (aseptic filling, multi-hurdle technologies). Differentiation according to *distribution logistics* refers again to shelf-life, but is linked to distribution systems and storage. The third classification differs from the first two in the sense that it introduces a consumer's perspective.

While all three classifications have their merits, a product-, process- and technology-oriented view of the business is not sufficient any more. Today, and more so in the future, innovation and growth are more and more linked to a better understanding of the customer's needs and wants. Increasingly, success rests on a customer-based view. To succeed, we first have to think of the business in terms of the problem it is solving for its customers (solution provider), and the consumer benefit(s) associated with the solution, rather than the product it is selling to them. First we want to know who our customers (current and potential) are. We need to understand not just their habits and practices (observational insight), but their needs and wants (motivational insight), including those they cannot articulate. Then we have to delight them with our brands and products. Only in a second step are technologies becoming relevant, as they are a means for offering solutions to important costumer needs.

The third scheme of classification (kitchen aids, snacks, main dishes) integrates some consumer perspective. It refers to different modes of usage, irrespective of the production technologies and distribution logistics involved.

While the third type of classification is in line with a global trend in the food industry (and well beyond the food industry), it still falls short of a real consumer-centric approach. In order to spot real business opportunities, food companies are today in a process to better understand the (unsatisfied) needs and motivations of their actual and potential costumers, to whom they want to offer profitable solutions. For which needs a company decides to offer solutions depends on what the company considers as its 'core' – markets, brands, categories, (proprietary) technologies, know-how – as well as on metrics related to short and long-term expectations on return on invested capital (ROIC), so as to focus the efforts entirely on it.

In this transformational process, we are observing a trend away from product-category thinking to a consumer-share view [1]. While traditionally the metrics of success have been market share, measured one product category at a time, the success metrics are increasingly moving towards share of customer and return-on-customer (ROC). Without customers there is no project, no products, no brand, no business. While many companies still think and do business in terms of product-portfolio and brand-portfolio management, a company, in its roots, is a portfolio of customers. This will increasingly affect the way we valuate a company. If you have a customer for a business, you can almost certainly get the capital you need [2].

Referring to the third scheme of classification, the majority of products belonging to kitchen aids are 'flavour solutions' and 'recipe solutions', products that are added at different stages of cooking and food preparation, in different formats and dosages to provide flavour to the dish. Typical examples of flavour solutions are bouillons or seasonings, while fixes and sauces belong to recipe solutions. Flavour solutions and recipe solutions together form the 'Flavour World', all types of kitchen aids that are solutions to flavour dishes.

The title of the chapter 'Dehydrated convenience food and kitchen aids' refers to a segment within culinary convenience food where flavours are of utmost importance and a key attribute driving consumer preferences.

Considering the importance of flavour with regard to the quality and consumer acceptance of culinary products, several food companies were in the past very active in the field of flavours and had large in-house divisions for flavour research and produc-

# Introduction

tion. More recently, the major players in the food industry have steadily focused their efforts on core business(es) and divested activities outside this core. This led some food companies to divest their flavour activities. The two most resounding examples were the divestments of Quest by Unilever (in 1997) and of FIS by Nestlé (in 2002). On the one hand, the flavour industry has gone through a massive consolidation, with the five global big players – Givaudan, IFF, Firmenich, Symrise, Quest – holding now a global market share of the flavour business of approximately 50%. This consolidation is probably going to continue, leading to an increasing concentration of the business on a few big players. On the other hand, it is not expected that new major entrants to the flavour industry will appear soon on the horizon to counterbalance this consolidation (barriers to entry from technologies, know-how and customer relationship).

At this point, it is important to clarify a few terms that are often used in the context of this chapter. Flavour is usually divided into the subsets taste and smell, which are perceived in the mouth and the nose, respectively [3]. The terms 'aroma' and 'odour' are not well defined and often used as synonyms. Odour is best reserved for the smell of food before it is put into the mouth (nasal perception) and aroma for the retronasal smell of food in the mouth. In this paper, we mainly use the terms 'aroma' and 'taste', as well as 'flavour' comprising sensory notes imparted by both volatile and non-volatile compounds (odorants and tastants). In the public domain, however, 'taste' is often used as a synonym for 'flavour'. Proper definitions of these terms are:

*Aroma* is the sensation that results when olfactory receptors in the nose are stimulated by particular chemicals in gaseous form, and is synonymous with odour and scent. Around 10,000 different olfactive receptors have been detected in humans, while a well-trained 'nose' can differentiate up to about 1000 aroma compounds.

*Taste* is the sensation that results when taste buds in the tongue and throat convey information about the chemical composition of a soluble stimulus.

*Flavour* is the sensory impression of a food while eating. It is determined by the three chemical senses of taste, smell, and the so-called trigeminal senses, which detect chemical irritants in the mouth and throat.

## 5.4.1.1 Flavour World

When consumers are asked which criteria they consider 'very important' for choosing food, 'taste' systematically ranks highest. Consumers rarely compromise on taste. Here, kitchen aids provide convenient flavour solutions for enhancing the flavour of dishes, and hence assist consumers to succeed in what is very important to them.

Figure 5.49 represents a possible scheme for illustrating the roles of different kitchen aids with regard to the overall flavour of a dish. Bouillons are added, often early during the preparation of a dish, to define its basic taste direction, its taste foundation. While they deliver a taste direction like chicken, meat or vegetable, they are applied to a large range of dishes.

Sauces and fixes are much more specific to a particular dish and are used to enhance the personality (the characteristic note) of the dish. They are also termed 'recipe solutions', delivering the flavour (plus other attributes such as texture, colour, binding) to one particular recipe. Finally, seasonings are most often added at the end, as a final and individual flavour touch, to complement the taste of a dish, to round it off, stress a particular 'taste-kick' or enhance a taste note particularly appreciated by the consumer. Since seasonings are often on the table, they help to fit the final flavour of a dish to the palates of the individuals, and are often individually dosed.

The combination of bouillons, fixes, sauces and seasonings enhances the flavour of complete dishes and allows for a personalised flavouring. It should yet be stressed that the illustration of the functions and usages of the different kitchen aids, as shown in Fig. 5.49, is a schematic one; e.g. when observing people preparing a dish, occasionally seasonings are used during food preparation very much like bouillons. Nevertheless, Fig. 5.49 illustrates the primary usage of the different kitchen aids, and shows how various kitchen aids may be combined to deliver good and personalised flavours to culinary dishes.

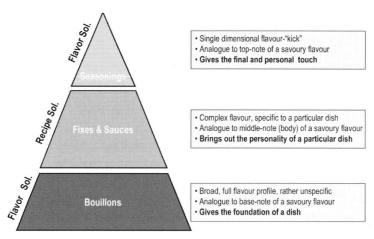

*Fig. 5.49*: *The flavour-world pyramid of a culinary dish*

### 5.4.1.2 Savoury Taste: Umami and Saltiness

It is generally recognised that taste is a combination of five taste modalities for all of which the corresponding receptors have been identified. These are sweet, sour, bitter, salty and umami. While all five are important, two are closely associated with the flavour of a savoury dish: umami and salty. Saltiness and umami are key taste attributes of culinary dishes. We know today a range of compounds that elicit an umami taste sensation (or close to umami). However, the one compound most typical for umami taste is monosodium glutamate (MSG). It is widespread in nature and part of our natural diet through products such as cheese, tomatoes or meat. The compound typical for saltiness is the sodium ion ($Na^+$). While MSG and $Na^+$ are naturally occurring in most common foods, consumers are increasingly concerned about their level of salt ($Na^+$) consumption and about the usage of MSG for enhancing the umami taste in culinary dishes. In some countries, reduction of $Na^+$ and of added MSG in their daily intake is today a growing and important consumer demand and large efforts and resources are employed by the food and flavour industry as well as in academic research, to find solutions for Na$^+$ and MSG reduction, while keeping a good and full savoury taste experience in culinary dishes. Considering the role Na$^+$ and MSG play in the flavour of culinary dishes, we will shortly outline the main consumer concerns associated with these ingredients and summarise ongoing efforts to address these.

#### 5.4.1.2.1 Salt

Sodium is a chemical element, and as such, cannot be synthesised by the body. It is, therefore, an essential nutrient that must be provided by the diet. Sodium is required by the body for several important biological functions including regulation of osmotic pressure within individual compartments of the body, active transport of many essential nutrients into cells, as well as in neurotransmission processes.

Common salt, known chemically as sodium chloride (NaCl), is, in its purest form, a crystalline compound comprising 39.34% sodium and 60.66% chlorine. On a weight basis, therefore, a factor of 2.54 is used for converting an amount of sodium into the corresponding amount of sodium chloride. Most dietary sodium comes from common table salt (sodium chloride), but sodium also occurs naturally in food as sodium salts of several compositional acids, such as sodium citrate and MSG.

The main function of salt is generally considered as giving saltiness to food. Yet, salt is also used in processing food for reasons other than taste. One of the longest-established and most important reasons for using salt in food is as a preservative. In a variety of products, the reduction of water activity in a product is a very important way of controlling the growth of food spoilage and pathogenic organisms. Increasing the salt concentration in the product can effectively reduce the water activity. Improvement of texture and appearance are also an important reason for adding salt during food processing. Salt is used in bread, for example, for improving the texture of the dough by increasing the water retention of gluten, the colour of bread crust and the keeping quality of the bread itself. Salt is also used in cured meat and fish to enhance the binding of proteins with water to make the meat more tender.

The relationship between the intake of dietary sodium (Na$^+$) and the risk of developing hypertension (high blood pressure) has been the focus of considerable research and has become an increasingly important matter for public health nutrition policy over the last few decades *[4, 5]*. The disease for which there is the most scientific evidence for a causative link with sodium intake is hypertension (high blood pressure) and the resultant associated increase in the risk of cardiovascular and cerebrovascular morbidity. Although some degree of controversy still exists as to the extent to which the health of a population as a whole is affected by high sodium intakes, it is now generally accepted that high levels of dietary sodium negatively affects the health of approximately 30% of individuals of a population.

Today, there are four well-known approaches to reducing the sodium content of food products *[6]*.

- **Reduction** of the actual amount of sodium or salt – simply put less in. The simplest way of reducing the sodium content in a food product is to take some sodium out without replacement. In several products, 5% reduction will not lead to a significant change in saltiness and hence can be applied. Yet,

when going beyond a 5% salt reduction, this strategy has to be complemented with careful consumer preference studies, to ensure that the product will not be rejected by consumers for lack of taste.
- **Substitution** of sodium with other cations, particularly potassium and calcium. Today, a large range of 'salt replacers' are offered on the market place. While these may compensate for some of the loss of saltiness from $Na^+$ reduction, they often do not fully match the taste of sodium and may introduce some (unpleasant) side tastes.
- **Redistribution** of sodium in a food product (especially between liquid and solid compartments) so as to increase the amount of sodium that reaches the receptors on the tongue. In order to elicit a saltiness sensation, the sodium ions must reach the taste buds. Hence what matters for the saltiness impression is the fraction of 'accessible' sodium ions in contrast to the sodium that one swallows without ever being perceived. This can be influenced by the way the sodium is distributed over the different constituents of the food products.
- **Optimisation** of ingredient interactions (such as the use of acid and spices) to enhance the perception of a salty taste. It is well known that the different taste modalities like acidity and sweetness are affecting the overall saltiness impression. Hence, reducing sweetness or increasing acidity may increase the saltiness impression without changing the actual salt level. Yet, again this strategy is applicable as long as the final sensory profile is appreciated by the consumer.

Besides these well-known approaches, several companies are actively looking for new and innovative solutions for sodium reduction. Very active in this field are also all major flavour companies, all offering their own proprietary solutions. It is expected that research will soon yield interesting and new opportunities.

### 5.4.1.2.2 Umami

What exactly is the umami taste? The Japanese word 'umami' can be translated into English as savoury, essence, pungent, deliciousness and meaty. Umami is associated with an experience of perfect quality in a taste. It is also said to involve all the senses, not just that of taste.

Over 1200 years ago, Asian cooks began adding a type of seaweed found in the Pacific Ocean to their soup stocks. They had discovered that foods cooked in this seaweed broth simply tasted better. What these chefs did not know was that the broth's unique flavour enhancement quality was due to the high levels of naturally occurring glutamate in the seaweed. Finally, in 1908, the link between glutamate and the seaweed was discovered. Professor Kikunae Ikeda, at Tokyo Imperial University, isolated MSG from the seaweed and unlocked the secret of the plant's flavour-enhancing properties.

MSG has repeatedly come under attack and it is claimed to provoke various symptoms collectively referred to as 'Chinese Restaurant Syndrome' However, scientific studies carried out over the last 30 years have failed to confirm such an association. Several scientific symposia and conferences were held on this issue. The first significant international symposia on glutamic acid were held in Milan in 1978 [7] and in Hawaii in 1979 [8]. Since these early symposia, very intense research on MSG has been conducted. In October 1998, the scientific community met again in Italy (Bergamo) to review the current knowledge in the field, issuing comprehensive conference proceedings [9]. In addition, in 1998 a special issue of Food Review International was entirely devoted to Umami [10]. The scientific community is clear about the fact that in spite of intense research, no scientific proof for the negative effects could be reported.

In addition, several renowned international bodies have issued comprehensive reports:

- **FAO/WHO:** In 1987, the Joint FAO/WHO Expert Committee on Food Additives (JECFA) evaluated the bulk of scientific material [11]. Their conclusion was 'ADI not specified' (ADI = acceptable daily intake), which means that they do not put any upper limit on the MSG consumption. There was also no special note for infants. The evaluation of the JECFA means that the total dietary intake of glutamates arising from their use at the level necessary to achieve the desired sensory effect as food additives and from their normal naturally occurring levels in food does not present any hazard to health.
- **EU:** The European Communities' Scientific Committee for Food (EC/SCF) independently confirmed the safety of MSG [12]. At the meeting of the Scientific Committee for Food on the First Series of Food Additives of Various Technological Functions, the Committee allocated an ADI of MSG as 'not specified.'
- **FDA:** MSG is considered GRAS (generally recognised as safe) by the FDA [13].

Irrespective of this scientific evidence, there remains the fact that some consumers (in some countries) are concerned about consumption of added MSG and ask for products without added MSG. Therefore, food companies are actively looking for solutions to reduce or eliminate added MSG in their food recipes, without sacrificing the sensory quality of the products.

*'Taste' versus Nutrition*

Not only does food nourish the body, it provides nourishment for the soul. Why do you choose to eat certain foods? Most likely the reason is taste. When we eat food, a multitude of emotions is triggered. Food brings pleasure and conviviality to our lives. We like to eat what tastes good to us. Survey after survey has found that what people choose to purchase, prepare and eat is based on taste. We all know that nutrition, convenience and cost are important factors in food selection, but taste rules and it always will. Even the most nutritious foods will not be willingly accepted and regularly consumed if they have poor sensory quality.

Well aware of the pivotal importance of taste for consumer preference, several major food companies acknowledge and stress the link between taste and nutrition. The difference between liking and disliking is important because personal dietary choices are largely based on sensory preferences. Hence, for a global food company, taste is not an end in itself. The challenge for the food industry is to fit what consumer want (food that tastes good) with what they need (balanced diet), delivering nutritional value and health through products they enjoy [14, 15].

## 5.4.2 Savoury Flavours

Imparting a particular flavour to culinary products is part of culinary art that is required in both home-style cooking and foodservice to deliver tasty food to consumers. Kitchen chefs and food developers use home-style preparations as the reference to be achieved, considering known constraints such as freshness upon storage (shelf-life), time of preparation (convenience), consistent ingredients quality (cost, availability), need for flexibility to cope with consumer trends and regulatory issues, etc.

The most frequently used principles for flavouring culinary food products are:

- adding finished flavours as food additives to the food to be consumed
- developing flavour upon food processing using appropriate raw materials, ingredients and processes
- using pre-reacted flavours that develop in situ the final flavour note just prior to consumption.

The first approach is the domain of flavour houses which offer various types of customised flavour solutions to food companies. While this service usually leads to rapid solutions, it is in most cases of short-term nature and requires improvement on regular basis to maintain the competitive advantage. However, in some cases it requires sharing of proprietary technology and direct involvement in the food concept development phase that may limit collaboration.

On the other side, food companies rely on the second approach ensuring a strong competitive edge by using proprietary technologies and recipes that can hardly be copied. In-depth knowledge of ingredients and processes allow the development of products with a unique flavour and also modulating the flavour profile depending on consumer trends. This approach may be hampered by the existing technology, limiting the flexibility to develop rapid solutions.

Pre-reacted flavouring systems offer the advantage of delivering freshly developed flavours to increase 'freshness' and compensate for possible flavour losses upon storage. Successful application of this approach requires a well-defined procedure for preparing the food that is to be followed by the consumer, e.g. heating time and temperature, amount of water or other ingredients, etc.

Finally, the way of imparting flavour to the culinary product depends very much on the type of product (dry, chilled, frozen), processing parameters (heat load), shelf-life requirements (from a few weeks to 24 months), degree of convenience (home, foodservice), etc. In all cases, the aim is to deliver a 'tasty' food appreciated by the consumer at the point of purchasing and consumption.

The savoury flavour pyramid, consisting of the three major elements base, middle and top notes (Fig. 5.50), is a convenient way of summarising the various technical approaches to developing an overall 'tasty' culinary food.

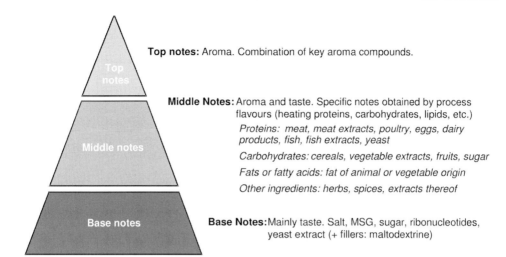

*Fig. 5.50: The savoury flavour pyramid, a mixture of base, middle and top notes*

Many variations exist in terms of segmentation of the savoury pyramid. While the role of the three major blocks is rather well defined, they can hardly be separated from each other. As an example, the middle notes may comprise both volatile aroma and non-volatile taste compounds, which may have the function of top notes and base notes, respectively. Also base notes such as yeast extracts may impart aroma compounds contributing to top notes. The key aspect here is the fact that various ingredients may have different functionalities. These building blocks have to be used in a way adapted to the target food to achieve the desired flavour complexity. This modular approach allows higher flexibility in product development and is a means for delivering defined functional properties, such as aroma, taste, texture, colour, etc., thus fulfilling the requirements of the flavour-world pyramid of culinary dishes (Fig. 5.49).

Depending on the culinary dish, these building blocks may be suitable on their own or in combination to complement the flavour-world pyramid with the aim of achieving a desirable flavour sensation that is characteristic for the food preparation. Other considerations are regulatory hurdles and the tendency of consumers to prefer natural, organic food. This, of course, depends very much on the cost required for developing natural flavouring systems or manufacturing processes maintaining the authentic and fresh flavour of the food product upon storage. Following, we will elaborate more on the various building blocks (flavour notes) and their application using examples.

### 5.4.2.1 Base Notes: Flavour Body and Taste Enhancement

Base notes contribute to the intrinsic taste of culinary products. This can be achieved by using basic savoury ingredients such as meat extract, bone-stock, yeast extract, fermented soy sauce, wheat gluten sauce, vegetable powders, herbs and spices. They provide a complex mixture of taste-active and taste-modifying compounds, some of them still unknown, in a typical and balanced composition. The basic taste can be fortified by defined tastants (salt, sugar, organic acids) and taste enhancers such as MSG and 5′-ribonucleotides (Fig. 5.51) to boost certain flavour characteristics.

As previously discussed, there is a tendency (e.g. in Europe) to reduce the amount of certain compounds (e.g. MSG, salt). Current research aims at identifying alternatives for these molecules without compromising quality. By applying state-of-the-art screening tools, such as taste dilution analysis (TDA) *[16]*, receptor-based assays *[17]* and molecular modelling *[18]*, a number of MSG alternatives have recently been reported (Fig. 5.51).

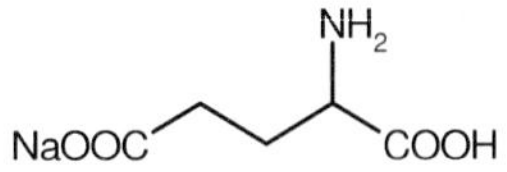


Monosodium glutamate (MSG)

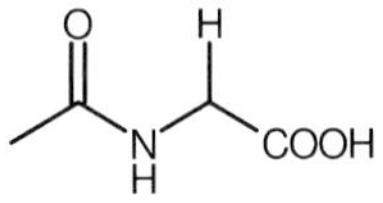


*N*-Acetylglycine (NAG)

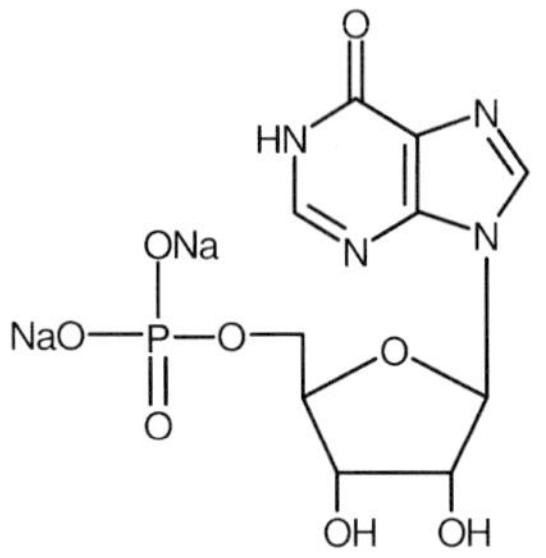


Inosine monophosphate (IMP)

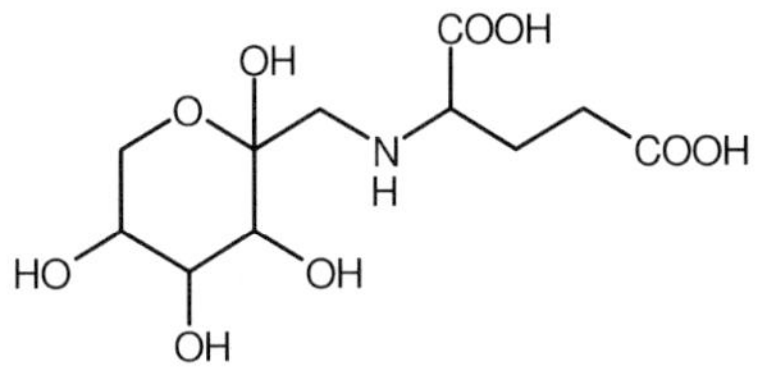


*N*-(1-Deoxy-D-fructos-1-yl)-
L-glutamate (DFG)

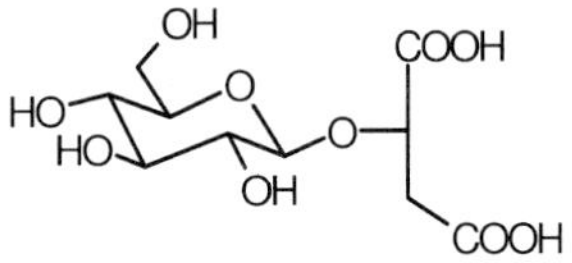


(*S*)-Malic acid 1-*O*-D-
glucopyranoside (Morelid)

***Fig. 5.51:*** *Molecules imparting or enhancing the umami-like taste in culinary products*

Molecular modelling has suggested N-acetylglycine (NAG) as umami-like, which was confirmed by sensory tests *[18]*. The Amadori compound N-(1-deoxy-D-fructos-1-yl)glycine (DFG) has been described as umami-like, or bouillon- and seasoning-like with a mouthfeel effect, similar to MSG *[19]*. These attributes were reinforced in the presence of salt in a synergistic manner. DFG is found in nature, i.e. high amounts of up to 3.6 g of DFG per 100 g of dry matter have been reported in dried tomatoes *[20]*. The potential of this glycol-conjugate was also demonstrated in food-type applications. As indicated in Table 5.23, DFG impacted the taste profile of a model bouillon, as MSG does. The model bouillon base was essentially free of any taste enhancer such as MSG or IMP. The pure bouillon base was evaluated with an overall intensity of 1 and was perceived as bland, very salty with only a weak glutamate-like taste. The bouillon spiked with DFG, however, elicited a pronounced bouillon-like umami taste with an intensity of 2-3, which was close to that with added MSG.

**Table 5.23:** *Influence of monosodium glutamate (MSG) and N-(1-deoxy-D-fructos-1-yl)-L-glutamate (DFG) on the taste profile of a model bouillon base[a] [19]*

| Additive | Quality attributes | Intensity |
|---|---|---|
| No additive | Salty, bland, fatty, weak umami taste | 1 |
| MSG | Umami, bouillon-like, salty, slightly sweet | 3 |
| DFG | Umami, bouillon-like, salty, fresh | 2-3 |

[a] The bouillon base (19 g/L; pH 5.8) was evaluated either in the absence or presence of additives (10 mol/L) at 65°C by five sensory panellists on a scale from 0 (not perceived) to 3 (intensely perceived).

Another recent example is the identification of (S)-malic acid 1-O-D-glucopyranoside, also referred to as Morelid, by the TDA approach *[21]*. It has been isolated from morel mushrooms and identified as a new umami-like tasting compound with a taste threshold of 6 mol/L. A comparative taste profile analysis of a 20 mM aqueous solution of Morelid and a 4 mM aqueous solution of MSG showed similar intensities for the umami-like taste *[21]*. However, the oral evaluation of the glycoside did not show any salty note as compared to MSG.

Taste research has evolved considerably over the last years what will certainly result in new compounds and lead structures. The three approaches mentioned, i.e. TDA, receptor-based assays, and molecular modelling, are complementary methods (or tools) that may reveal new taste-active and taste-modifying compounds. The TDA method may especially help discovering taste modulators, because the corresponding receptors and processes are largely unknown. However, many other parameters must be checked, apart from technical taste testing, to evaluate the commercial potential of these new molecules, i.e. stability, safety, cost, availability, range of application, etc. Therefore, just a few molecules may succeed to be widely applied to culinary products.

### 5.4.2.2 Middle Notes: Process and Reaction Flavours

More characteristic flavour attributes are derived from specific flavourings such as beef, chicken and fish flavours. Very often, they are used as process or reaction flavours, which are based on hydrolysed proteins or yeast extracts obtained by enzymatic hydrolysis or fermentation. They may contain specific aroma and taste molecules to impart the desired note, including taste enhancers and top notes. Specific animal notes can be generated by using animal fat (beef, chicken) or fish-derived ingredients. Process flavours are used as dry powders obtained by spray-drying, extrusion, and other drying processes. Water activity is an important parameter in this context to ensure storage stability. It should be in the range of 0.1 to 0.3 to ensure product-keeping quality. Process flavours are combined with other food ingredients by dry-mixing or they are added to industrial food preparations that are heat-processed. Again, resistance against flavour deterioration by heat-induced preservation processes (e.g. pasteurisation, sterilisation, concentration, drying) is a major issue in product development and application.

Pre-reacted flavours can be used to develop the final flavour upon cooking. The general idea is to enrich the flavouring composition with suitable intermediates that readily develop the flavour upon final heat treatment. As an example, 2-(1-hydroxyethyl)-4,5-dihydrothiazole (HDT) has been reported in model systems as the direct precursor of the roasty smelling odorant 2-acetyl-2-thialzoline (2-AT) [22]. The intermediate HDT can be formed by fermentation with baker's yeast using suitable precursors (Fig. 5.52A) [23]. Thermal treatment may then lead to 2-AT as key odorant imparting roasty notes, e.g. in a microwave oven (Fig. 5.52B).

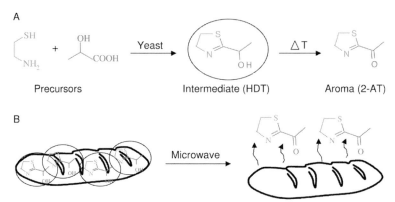

*Fig. 5.52:* Concept of combining bio-approaches and thermal generation of roasty flavours based on 2-(1-hydroxyethyl)-4,5-dihydrothiazole (HDT) as key intermediate releasing 2-acetyl-2-thiazoline (2-AT) upon heating

The impact of HDF as a precursor for improving the roasty notes of baked goods was evaluated using two types of pizzas, frozen and chilled [23]. An aqueous solution of HDT (1.6 mg/mL) obtained by bioconversion was mixed with the classical ingredients of the pizza recipe to reach 5 mg per 50 g of raw dough. Thirty assessors were asked to describe the aroma quality of the freshly prepared samples by smelling the headspace above the sample (Table 5.24). The addition of HDT resulted in an improvement of the roasted, toasted and popcorn-like notes as compared to the reference with 99.9% confidence level in triangle tests.

*Table 5.24:* Sensory evaluation of samples containing HDT

| Sample | Frozen pizza | Refrigerated pizza |
|---|---|---|
| HDT mixed into dough | No difference ($p^* = 0.14$) | Significantly different ($p = 0.001$) |
| HDT as surface coating | Significantly different ($p = 0.01$) | Significantly different ($p = 0.002$) |

* The p value represents the significance level of the experimental data obtained.

Flavouring preparations obtained by bio-approaches using enzymes (bio-transformation) or micro-organisms (fermentation) have the advantage of being considered natural. In addition, they may round off the taste by fermentation and due to the contribution of yeast as source of flavour impact compounds. In general, the combination of bio-approaches with thermal treatment is an elegant way of improving the overall flavour of culinary products.

### 5.4.2.3 Top Notes: Key Compounds of the Culinary Aroma

One of the most delicate steps in product development is to impart the characteristic flavour note. This concerns both aroma and taste. While the proper taste quality can be achieved by a careful selection of ingredients representing the base note (Fig. 5.50), delivering the typical aroma of a culinary product is in many cases a real challenge. This may be due to losses of odorants caused by high volatility and chemical reactions, thus leading to a misbalanced aroma profile, perceived as weak or even as an off-flavour that is no longer characteristic for the food product.

Apart from generating aroma in situ, this means just prior to consumption, the use of top notes is another well-known approach. This aroma creation is usually performed by flavourists. One critical step is to adapt the aroma formulation to the food environment in order to impart the desired food aroma. Specific impact molecules are used to achieve aroma intensity and specificity. They are summarised in the FEMA/GRAS list consisting of more than 4000 items which are generally recognised as safe by the Flavor and Extract Manufacturers Association. Currently, there is a general trend of (i) using new high-impact aroma chemicals [24] and (ii) moving from the 'trial-and-error' approach to more systematic sensory evaluation methods [25]. The advantages of creating flavours using qualitative sensory data and quantitative intensity ratings have recently been summarised: i.e. shorter time of development phase, greater accuracy in ingredient selection and use level, higher quality, reduction of inventory, etc.

Recent research resulted in new and potent impact odorants found in foods or Maillard reaction samples [26-28]. Systematic studies revealed common structural elements of sulphur-containing compounds required for developing a basic meat flavour (Fig. 5.53) [29]. The 1,2-oxygen, sulphur grouping has been suggested as a common element of the savoury olfactophore (Fig. 5.54).

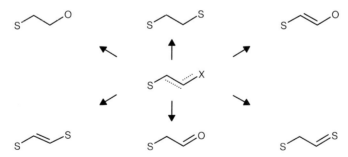

*Fig. 5.53: Subclasses of sulphur-containing compounds with meat flavour*

The systematic use of gas chromatography coupled with olfactometry [27, 28] in the last 20 years has resulted in a number of new high-impact aroma chemicals found in natural extracts, food products and reaction flavours. In general, sulphur-containing odorants play a particularly important role in food products and savoury flavours [30]. Some of them are shown in Fig. 5.54. Usually, the odour threshold is one key attribute showing the potential impact of the odorant. This may be as low as 0.00002 µg/L water reported for bis-(2-methyl-3-furyl)disulphide (BMFD) (Fig. 5.55) found in cooked meat with a typical meaty, sulphury note.

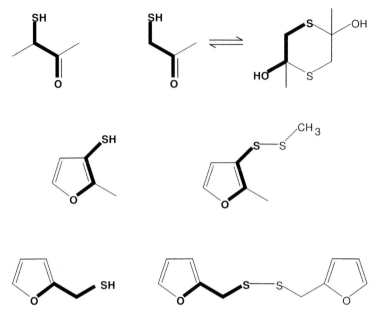

*Fig. 5.54:* The 1,2-oxygen, sulphur grouping as common element of the savoury olfactophore (according to D. Rowe, Oxford Chemicals)

However, not only the odour threshold is an important parameter to consider, but also the specific note that the odorant may impart. 2-Isobutylthiazole is a well-known example that is essential in tomato top notes, but has a relatively high threshold of 3 µg/L water. Another recent example is the identification of 2-heptanethiol (Fig. 5.55) in bell pepper extracts having a sulphury, onion-like and vegetable-like note, reminiscent of bell pepper at lower concentrations [31]. This odorant can impart the characteristic bell pepper note to complex top notes despite its moderate threshold value of 10 µg/L water.

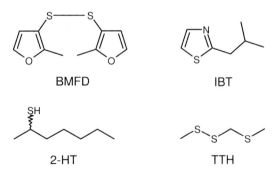

*Fig. 5.55:* Sulphur-containing impact odorants used in top notes to impart specific notes, e.g. bis-(2-methyl-3-furyl)disulphide (BMFD, meaty), 2-isobutylthiazole (IBT, cooked tomato), 2-heptanethiol (2-HT, bell pepper), 2,3,5-trithiahexane (TTH, cooked broccoli)

Similarly, 2,3,5-trithiahexane (TTH) was found to contribute to the typical cooked broccoli note when boiled in water (Fig. 5.56A) compared to vapour-cooked broccoli

(Fig. 5.56B), which was described as more cabbage-like, sulphury and pungent [32]. The formation of TTH in broccoli required the combination of an enzymatic process to generate precursors followed by heat-induced reactions to form this unusual aroma-impact compound. The enzymatic part of the formation pathway was associated with cutting of broccoli florets into small parts, which activated endogenous enzymes. Both perception thresholds and olfactive descriptors of TTH confirmed its potential role in imparting the specific character of freshly cooked broccoli. Therefore, TTH may be a key component in top notes for reinforcing the overall cooked broccoli character and achieving desirable flavour notes in culinary products known from home-style preparations.

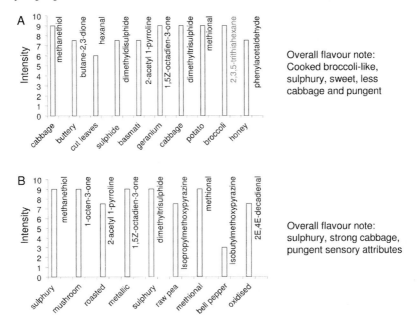

*Fig. 5.56:* Impact of 2,3,5-trithiahexane (TTH) on the overall aroma of (A) cut, frozen, and water-boiled broccoli as compared to (B) vapour-cooked broccoli

### 5.4.3 Manufacturing of Dehydrated Convenience Foods

#### 5.4.3.1 Legal Aspects

The Codex Alimentarius for bouillons and consommés (revised codex standard for bouillons and consommés Codex STAN 117-181, Rev. 2-2001) includes legal requirements for the formulation of dehydrated bouillons. For dehydrated soups, sauces, seasonings and prepared dishes no Codex standard exists. According to the Codex standard bouillons and consommés should contain hydrolysates and extracts. In addition, they might also contain seasonings and/or flavouring substances. As flavouring substances natural flavours and flavouring substances or nature-identical flavouring substances are permitted for bouillons and consommés. Their addition is limited by good manufacturing practice (GMP).

However, in some countries, like Italy, there are local laws, which have to be respected. Dehydrated convenience foods manufactured or imported and sold in Italy by Italian companies have to comply with the Italian law for such products. For manufacturing bouillons and seasonings based on MSG, a special law exists in Italy (legal requirements for the 'preparato per brodo e condimento a base di glutammato' DPR 30-05-1953 n. 567). Within this law, the addition of flavours is restricted to a maximum of 1% of the dehydrated mass. Nevertheless, companies outside of such countries with specific regulations might produce such products according to their country-specific laws or according to the Codex standards. In case they sell these products in their country the authorities of the country with the specific regulations have to accept that such a product is imported and sold in their country (principle of free trade).

If flavour products are added to dehydrated convenience foods, they have to be mentioned in the ingredient list at a position according to the added quantity. Ingredients representing large quantities within the formulation are added first in the ingredients list. The manufacturer of food has the choice of labelling flavours composed of different substances just as flavour or of listing the single components separately within the ingredients list. It is also possible to mention only some components of the flavour separately in the ingredient list and to summarise the remaining substances under the expression 'flavour'.

### 5.4.3.2 Flavour Application during Manufacturing of Dehydrated Convenience Foods

Dehydrated convenience foods are powder mixes which are sometimes agglomerated in fluidised beds or extruded as wet powder masses to form granules. Frequently such powder mixes are also compacted to bouillon or seasoning tablets and cubes. Beside this, pasty and partially dehydrated culinary products are offered for seasoning applications. Table 5.25 shows different dehydrated convenience foods and their typical flavour content.

*Table 5.25: Dehydrated convenience foods and their flavour content*

| Product | Tabletted and powdered bouillons | Seasoning powders | Sauce powders | Dehydrated soups | Recipe mixes | Dehydrated prepared dishes |
|---|---|---|---|---|---|---|
| Flavour content of the dehydrated form (%) | 1-20 | 10-25 | 3-15 | 1-10 | 10-25 | 1-8 |

A wide range of flavouring substances is used in powdery or pasty convenience foods. Reaction flavours based on hydrolysed plant proteins, natural flavours, artificial flavours and meat, vegetable, yeast and spice extracts are added to the products to generate the desired flavour profile. The most common savoury flavour types used for dehydrated convenience foods are chicken, beef and vegetable flavours. However various other flavour types are used while formulating dehydrated convenience foods. Amongst them are mutton, crawfish, fish, wine and various spice and herb flavours.

These flavours are added as powders, pastes or in liquid form to the powder mass. Sometimes encapsulated flavours are used for providing a sufficient shelf-life (12-24 months) or reducing flavour losses during cooking. Encapsulation of savoury flavours used in dehydrated culinary products can be achieved by applying different encapsulation processes. One possibility is to extrude an emulsion containing the flavour component together with an amorphous carbohydrate mix (containing e.g. maltodextrines). During this melt extrusion glassy carbohydrate capsules, in which the liquid flavour/oil mix is embedded in form of small droplets, are generated. Depending on the chosen carbohydrate composition, moisture-resistant flavour granules are obtained. Another possibility is to spray-dry an emulsion composed of the flavour containing oil and a carbohydrate solution. Spray drying provides a limited protection of sensitive top notes and losses of volatile flavour components during storage are sometimes reduced. Such spray-dried flavour powders can be coated with hardened or fractionated palm fat to improve their moisture sensitivity. Especially for dehydrated convenience foods that are not hermetically packed (e.g. bouillon tablets), such encapsulated flavours are useful for extending their shelf-life and avoiding liquefying within the product while storing at high relative humidity. An additional advantage of encapsulation can be a more or less controlled release of the flavour component upon re-hydration.

Liquid flavours are used because they are cheaper than powdered products or because they generate a quick and intensive flavour release during dehydration. In addition, liquid flavours might provide the dehydrated product with a strong smell observed while opening the packaging. In products packed non-hermetically they can generate a distinguishing smell around the shelves at the point of sale by a constant release of flavour through the packaging.

Prior to the addition during the manufacturing of the final product, the flavours sometimes require a pre-treatment. Flavour powders that are caked during storage due to sintering processes require a milling or force sieving to avoid flavour lumps in the final product, which would result in a local overdosing of the flavour component. Since small amounts of powdered flavours are difficult to dose, they are often pre-mixed in larger batch sizes and then divided in several portions, each required for a single batch of finished product. Especially difficult to dose are liquid flavours or extracts which are highly concentrated and, thus, often very viscous. Such liquid flavour components are often pre-mixed with a powder carrier (maltodextrine, dextrose syrup or salt) or other powdered ingredients included in the final product. These powder/liquid mixes are then added during mixing of the dehydrated convenience food.

During processing, flavour components are mainly added while mixing the powder mass. Sometimes liquid flavour is also added after an agglomeration step to avoid losses during the drying of the agglomerated product. However, there are also several potential sources for significant flavour losses during manufacturing, storage and preparation of dehydrated convenience foods. Figure 5.57 includes a flow chart illustrating the manufacturing process of dehydrated convenience foods.

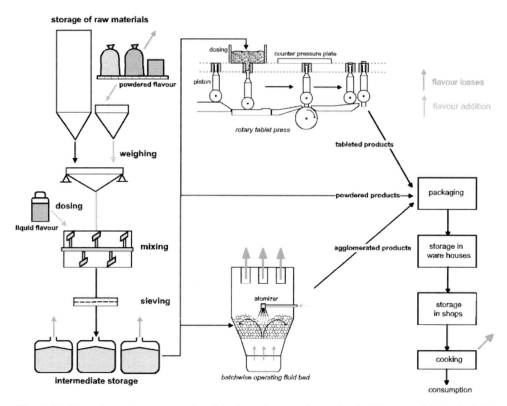

*Fig. 5.57: Manufacturing process for dehydrated convenience foods (flavour addition is indicated in green; flavour losses are marked red).*

Powdered dehydrated convenience foods are mixed in ribbon, paddle or ploughshare mixers. Since the volumes of single recipes are limited and because some components of the mix are added in fairly small quantities, mixing is performed batchwise with a batch load of 350-2000 kg. Another advantage of a batchwise mixing is the traceability provided by single batches. In case of a product defect, it is relatively easy to identify the non-conforming batches while manufacturing batchwise.

Generally, the mixing process for dehydrated convenience foods is divided into several mixing steps. Adding flavours to simple powder mixes like powdered seasonings or soup and sauce powders can be done in any mixing phase.

Mixing powder masses used to form tablets, the flavours are added in a later mixing phase. In the first phase, the crystalline components are dosed and mixed with each other. In the second mixing phase, fat is added to the crystalline substances to allow a proper distribution of the fat due to high friction between the moving coarse crystals. After these two phases the flavour components and finally the garnishes are added to the mix. Masses with a fat content of up to 20% have to rest for 1-2 days before they are compacted into tablets in single-punch rotary tablet presses. Pasty masses with a fat content between 20 and 30% are used to form tablets by dosing and forming. In case of high-fat products, the fat provides a moisture protection of the flavouring components.

# Manufacturing of Dehydrated Convenience Foods

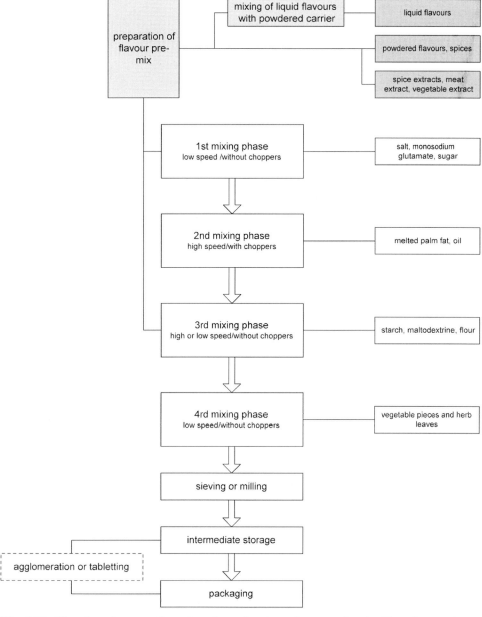

*Fig. 5.58: Flowchart for manufacturing of powdered, agglomerated and tabletted convenience foods (flavour components are highlighted in green; process steps for flavour preparation are highlighted in orange)*

Some dehydrated convenience foods are extruded as wet powder mass or they are agglomerated in a pneumatically or mechanically fluidised bed. Due to the addition of water during the process they require a drying step. Drying, which is done in pneumatically fluidised beds or vacuum ovens, can cause significant losses of volatile components. To reduce such losses, flavours which are added before drying have to be encapsulated. Another possibility is to use liquid flavours which are added after drying and cooling by spraying them onto the agglomerated powder mass.

Figure 5.58 shows the different mixing steps while manufacturing dehydrated convenience foods.

### 5.4.3.3 Physical and Chemical Flavour Changes during Processing and Shelf-Life

During storage of the pure flavour as well as during manufacturing and storage of the finished product, the flavour might undergo physical or chemical changes.

During storage often a loss of volatile components is observed. Especially flavours in dehydrated convenience foods are sensitive to such losses because such products are highly porous systems with a large surface area available for mass transfer between the solid and the gaseous phase. In addition often oxidation of single components is observed, as dehydrated convenience foods are in most cases not packed under a protective nitrogen atmosphere or by applying vacuum. Flavour losses and flavour degradation are also observed during drying after an agglomeration or extrusion step. Losses of volatile components and oxidation of sensitive substances both result in a changing flavour profile. Another well-known process changing the flavour profile and the colour of the product is the Maillard reaction. Flavours rich in proteins or amino acids, such as reaction flavours that are not fully reacted or flavouring compounds which contain especially yeast extract, are sensitive to the Maillard reaction. Due to the Maillard reaction an unpleasant roast note might be developed within the product.

In addition to such chemical changes, often physical changes of the flavour powder are observed. Flavour products manufactured by spray or vacuum oven drying are often amorphous solids or they contain amorphous particles. The viscosity of such amorphous solids strongly depends on temperature and their moisture content. At a specific temperature called the glass transition temperature, the viscosity drops by 3-4 orders of magnitude. The solids texture changes from a glassy into a rubbery state. In this rubbery state, the powder particles can sinter together depending on the viscosity and the time available for sintering *[33, 34]*. Such sinter processes might happen very fast during mixing processes in which moisture is added to the powder. Powdered flavours, intermediate powder masses stored before packaging and packed dehydrated convenience foods can show caking and lumping during storage.

Increasing the moisture content or the temperature further, the amorphous solid can liquefy. If the amount of amorphous flavour components is high enough, liquefying can even result in significant texture changes of the final product. Bouillon tablets might undergo a post-hardening and later they can even be transformed into a pasty mass. Figure 5.59 shows the glass transition temperature of different flavour powders or ingredients used for flavour formulation.

# Manufacturing of Dehydrated Convenience Foods

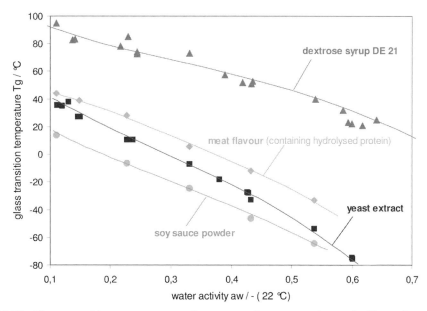

*Fig. 5.59: Glass transition temperature of a savoury flavour powder and of ingredients used for formulating savoury flavours (measured by differential scanning calorimetry DSC; heating rate: 5 °C/min, second scan)*

Exceeding the glass transition temperature by more than 5°C, a caking of the powder mass can be expected within some days. If the difference between process and glass transition temperature is larger than 20-40°C, an instantaneous stickiness and a reduced flowability of the powder is observed [33]. Another problem related to the water sensitivity of flavour substances is the undesired agglomeration of culinary powders during mixing. To some powder masses a small amount of moisture is added during mixing. Due to the increasing humidity within the mixer, the flavour particles become sticky and, thus, they adhere to each other or to the equipment surface forming lumps or a crust.

According to Fig. 5.59, soy sauce powder is very moisture-sensitive because its glass transition temperature is very low at a given water activity. Consequently, the process temperature easily exceeds its glass transition temperature significantly. If this is the case soy powder particles start to soften and sintering will occur. As a result of the sinter bridges built, the soy powder will start to cake and form lumps. As a consequence, flavour powders containing a significant amount of such ingredients which are sensitive to moisture like soy sauce powder or yeast extract are also moisture sensitive. In the final product they might lead to caking or textural changes if their amount which is in the recipe is high enough. Adding glucose syrup or even low DE maltodextrines to savoury flavour powders can improve their moisture stability. This might be especially useful for stabilising flavour powders in non-hermetically packed dehydrated culinary products like bouillon tablets and cubes.

## 5.4.4 Conclusions

Dehydrated convenience food and kitchen aids represent a major segment within the whole savoury category. Here we have discussed several issues related in particular to the flavouring of this segment. In chapter 5.4.1 we set the framework and introduced concepts and classifications relevant in this context. In chapter 5.4.2, the savour flavour pyramid was discussed based on specific examples. Finally chapter 5.4.3 elaborated on technological aspects of the subject. All these are critical elements towards delivering good-tasting dehydrated convenience foods and kitchen aids that reflect consumer needs and preferences.

The flavour pyramid of culinary dishes, as introduced in chapter 5.4.1, is a simple scheme for visualising and organising the role and contribution of various commercial products to the taste of a culinary dish. We conclude our excursion into the world of the flavours of dehydrated convenience food and kitchen aids by illustrating the flavour pyramid of culinary dishes with some typical examples of commercial products from the three categories bouillon, fixes and sauces and seasoning in Fig. 5.60.

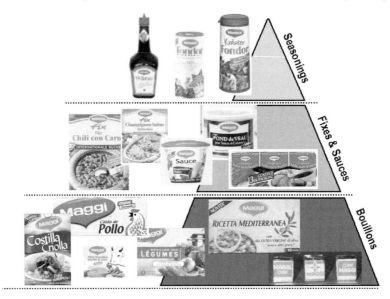

*Fig. 5.60:* *The flavour pyramid of culinary dishes, illustrated with commercial products*

**REFERENCES**

[1] Peppers, D., The One To One Manager: Real-world Lessons in Customer Relationship Management, Currency Doubleday, New York, 1999
[2] Peppers, D., Rogers, M., Return On Customer: Creating Maximum Value From Scarest Resources, Marshall Cavendish Business, Singapore, 2005
[3] Acree, T. E., Bioassays for flavor. In Flavor Science. Sensible principles and techniques (Acree, T.E., Teranishi, R., eds.), ACS Professional Reference Book, American Chemical Society, Washington, DC, pp. 1-20, 1993
[4] Kaplan, N.M., Clinical Hypertension, Williams & Wilkins, Baltimore, MD, 1990
[5] Antonios, T.F., MacGregor, G.A., Salt intake: potential deleterious effects excluding blood pressure. J. Hum. Hypertens. 9, 511-515 (1995).

| | |
|---|---|
| [6] | Angus, F., Phelps, T., Clegg, S., Narain, C., den Ridder, C., Kilcast, D., Salt in Processed Food: Collaborative Research Project, Leatherhead Food International, report no. 00193, March 2005. |
| [7] | Filer, L.J. Jr., Garattini, S., Kare, M.R., Reynolds, W.A., Wurtman, R.J., Glutamic Acid: Advances in Biochemistry and Physiology, Raven Press, New York, 1979 |
| [8] | Boudreu, J.C., Food Taste Chemistry, ACS Symposium Series 115, American Chemical Society, Washington, DC, 1979 |
| [9] | J. Nutr. 130 (2000) (entire volume) |
| [10] | Food Rev. Int. 14(2,3) (1998) (entire volume) |
| [11] | Joint FAO/WHO Expert Committee on Food Additives. In Toxicological Evaluation of Certain Food Additives and Contaminants, WHO Food Additives Series No. 759, Cambridge University Press, New York, p. 97, 1987 |
| [12] | Commission of the European Communities. In Food Science and Technical Reports of the Scientific Committee for Food, 25th series, EUR 13416 EN, Office for Official Publications of the European Communities, Luxembourg, 1991 |
| [13] | Analysis of Adverse Reactions to Monosodium Glutamate, prepared for Center of Food Safety and Applied Nutrition, Food and Drug Administration, Department of Health and Human Services, Washington, DC, 20204, under FDA contract no. 223-92-2185, by Life Sciences Research Office, Federation of American Societies for Experimental Biology, 9650 Rockville Pike, Bethesda, MD 20814-3998, July 1995. |
| [14] | German, J.B., Yeretzian, C., Watzke H.J., Personalizing foods for health and delight, Food Technol. 58(12), 26-31 (2004) |
| [15] | Yeretzian, C., Pollien P., Lindinger, C., Ali, S., Individualization of flavor preferences: toward a consumer-centric and individualized aroma science. Comprehen. Rev. Food Sci. Food Safety (CRFSFS) 3(24), 1-8 (2004) |
| [16] | Frank, O., Ottinger, H., Hofmann, T., Characterization of an intense bitter-tasting 1H,4H-quinolizinium-7-olate by application of the taste dilution analysis, a novel bioassay for the screening and identification of taste-active compounds in foods. J. Agric. Food Chem. 49, 231-238 (2001) |
| [17] | Li, X., Staszewski, L., Xu, H., Durick, K., Zoller, M., Adler, E., Human receptors for sweet and umami taste. Proc. Nat. Acad. Sci. 99, 4692-4696 (2002) |
| [18] | Grigorov, M.G., Schlichtherle-Cerny, H., Affolter, M., Kochhar, S., Design of virtual libraries of umami-tasting molecules. J. Chem. Inform. Comp. Sci. 43, 1248-1258 (2003) |
| [19] | Beksan, E., Schieberle, P., Robert, F., Blank, I., Fay, L.B., Schlichtherle-Cerny, H., Hofmann, T., Synthesis and sensory characterisation of novel umami-tasting glutamate glycoconjugates. J. Agric. Food Chem. 51, 5428-5436 (2003) |
| [20] | Eichner, K., Reutter, M., Wittmann, R., In The Maillard Reaction in Food Processing, Human Nutrition and Physiology (Finot, P.A., Aeschbacher, H.U., Hurrel, R.F., Liardon, R., eds.), Birkhäuser Verlag, Basel, pp 63-77, 1990. |
| [21] | Rotzoll, N., Dunkel, A., Hofmann, T., Activity-guided identification of (S)-malic acid 1-O-D-glucopyranoside (Morelid) and g-aminobutyric acid as contributors to umami taste and mouth-drying oral sensation of morel mushrooms (Morchella deliciosa Fr.). J. Agric. Food Chem. 53, 149-4156 (2005) |
| [22] | Hofmann, T., Schieberle, P., Studies on the formation and stability of the roast-flavor compound 2-acetyl-2-thiazoline. J. Agric. Food Chem. 43, 2946-2950 (1995) |
| [23] | Bel Rhlid, B., Fleury, Y., Devaud, S., Fay, L.B., Blank, I., Juillerat, M.A., Biogeneration of roasted notes based on 2-acetyl-2-thiazoline and its precursor 2-(1-hydroxyethyl)-4,5-dihydrothiazole, in Heteroatomic Aroma Compounds, ACS Symposium Series 826 (Reineccius, G.A., Reineccius, T.A., eds.), American Chemical Society, Washington, DC, pp. 179-190, 2002. |
| [24] | Rowe, D., More fizz for your buck: high-impact aroma chemicals. Perf. Flav. 25(5), 1-19 (2000) |
| [25] | Heinze, R., Focused flavor creation. Perf. Flav. 28(4), 40-47 (2003) |
| [26] | Schieberle, P., Hofmann, T., Identification of the key odorants in processed ribose-cysteine Maillard mixtures by instrumental analysis and sensory studies. In Flavour Science. Recent Developments, 8th Weurman Flavour Research Symposium (Taylor, A.J., Mottram, D.S., eds.), Royal Society of Chemistry, Cambridge, pp 175-181, 1996 |
| [27] | Blank, I., Gas chromatography-olfactometry in food aroma analysis. In Techniques for Analyzing Food Aroma (Marsili, R., ed.), Marcel Dekker, New York, pp. 293-329, 1997. |

[28] Grosch, W., Evaluation of the key odorants of foods by dilution experiments, aroma models and omission. Chem. Senses 26, 533-545 (2001)

[29] Sun, B., Tian, H., Zheng, F., Liu, Y., Xie, J., Meaty aromas: characteristic structural unit of sulfur-containing compounds with a basic meat flavor. Perf. Flav. 30(1), 36-45 (2003)

[30] Blank, I., Sensory relevance of volatile organic sulfur compounds in food. In Heteroatomic Aroma Compounds, ACS Symposium Series 826 (Reineccius, G.A., Reineccius, T.A., eds.), American Chemical Society, Washington, DC, pp. 25-53, 2002

[31] Simian, H., Robert, F., Blank, I., Identification and synthesis of 2-heptanthiol, a new powerful flavor compound found in bell peppers. J. Agric. Food Chem. 52, 306-310 (2004)

[32] Spadone, J.-C., Matthey-Doret, W., Blank, I., Formation of methyl (methylthio)methyl disulphide in broccoli (Brassica oleracea (L.) var. italica), 11th Weurman Flavour Research Symposium, Roskilde, Denmark, 21-25 June 2005.

[33] Palzer, S., Zürcher, U., Kinetik unerwünschter Agglomerationsprozesse bei der Lagerung und Verarbeitung amorpher Lebensmittelpulver. Chemie Ingenieur Technik 76(10), 1594-1599 (2004)

[34] Palzer, S., Desired and undesired agglomeration of amorphous powders. Proceedings of the 8th International Symposium on Agglomeration, Bangkok, March 2005, pp. 251-264

# 6 Quality Control

# 6.1 Sensory Analysis in Quality Control

*Helmut Grüb*

## 6.1.1 Introduction

There are a number of critical aspects in Quality Control (QC) of flavourings:

- Flavourings contain between a few hundred and over a thousand ingredients.
- Essential ingredients are often far below the detection limits of analytical measuring instruments.
- Normal analyses usually take a long time (too long).
- Normal analyses provide little information on the product itself.

In contrast, under appropriate conditions, sensory analysis or sensory quality testing of flavourings is:

- fast
- effective
- informative.

## 6.1.2 "Sensory"

"Sensory" is the science of measurement with the human senses. Panelists measure subjective and individually to produce an only one objective final result. Mention of the human senses normally brings to mind the usual 5 senses, smell, taste, vision, hearing and touch. In both the sensory and the physiological sense we better speak of the modalities sense of smell, sense of taste, sense of seeing, sense of hearing and sense of touch (tactile or haptic sense).

These modalities include other senses which are also very important for Sensory, such as sense of temperature and sense of pain. All these senses resp. modalities are important for quality testing of flavourings. Before we give a brief outline of them, various terms need to be defined, i.e.

*Organoleptic (ISO 5492 [1])*

Relating to an attribute or a property of a sample perceived by the sense organs.

A degustation is an act(ion) of gustation organoleptically. (In sensory nowadays 'organoleptic' is normally an obsolete term because of confusion with sensory)

*Sensory (ISO 5492 [1])*

Relating to the use of sense organs.

Within the meaning of 'Sensory' sensory in combination with other attributes additionally means performing actions with sensory panels. Therefore Sensory Testing means testing with a sensory panel.

*Organoleptic Quality Control (OQC) / Organoleptic Quality Testing (OQT)*

Quality testing normally is an organoleptic procedure as only certain characteristics of recurring products are being ascertained.

Although the testers are examined by an initial sensory selection and also receive a period of training, subsequently they are normally not subject to sensory control and sensory monitoring in regular intervals. They automatically train themselves in their own domain (see also further below).

*Sensory Quality Control (SQC) / Sensory Quality Testing (SQT)*

SQC is carried out by sensory assessors (panelists). In this case, the characteristics (mostly of the whole product) are measured with sensory test methods in order to provide a solution to a given problem.

Sensory panelists are selected, trained and monitored in regular intervals. Their qualification and performance are documented completely.

*Aroma*

Aroma is defined as 'the overall impression obtained from smell and taste perceived retronasal when tasting' (see below).

*Flavour*

Flavour is made up of the aroma and trigeminal perceptions when tasting (such as haptic aspects and/or hotness, strong spiciness as an aspect of the sense of pain).

## 6.1.3 The Senses

*Sense of Smell*

As far as olfaction is concerned, it is essential for the purposes of quality testing that a strict distinction is made between two types of smelling:

– Nasal Smelling

Nasal smelling is normal smelling through the nose. The active substances producing the smell pass through the nostrils on their way to the olfactory epithelium in the roof of the nasal cavity and from there through the nasopharynx to the oral / pulmonary cavity from where they are then exhaled. The process corresponds to breathing with the mouth closed.

The olfactory perception can be reinforced by sniffing i.e. by swirling the air intermittently over the ethmoid bone below the olfactory epithelium. The swirling effect causes the active components producing the smell to flow past the olfactory epithelium several times (instead of just once), thereby reinforcing the effect (particularly important in quality testing in the case of weak samples or samples with only very minor differences between them).

# The Senses

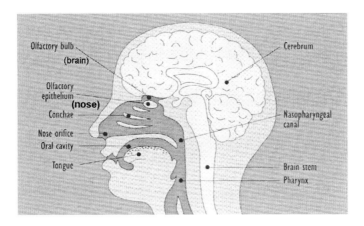

*Fig. 6.1:* Nasal and retronasal smelling

– Retronasal Smelling

Retronasal smelling occurs when an aromatic product is placed in the mouth, the mouth is then closed and the product is "eaten", causing the active components producing the smell to rise through he nasopharynx to the olfactory epithelium (this can be demonstrated quite simply by holding the nose closed, thus creating a counter-pressure which prevents the active components from rising from the mouth to the nose).

This effect can also be reinforced by smacking one's lips or slurping. This causes additional air to be sucked into the mouth, so that the active parts of the substances producing the smell are carried and swirled past and "round the back of" (= retro) the ethmoid bone.

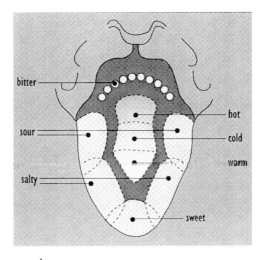

*Fig. 6.2:* Primary tastes on the tongue

*Sense of Taste*

Apart from the five normal basic tastes – sweet, sour, salty, bitter and umami– a great many other actions occur in the mouth, although their importance here is limited. The basic taste sensations are experienced in different areas of the tongue. For example, sweetness is tasted more at the tip of the tongue, sour sensation move on the sides towards the back of the tongue and saltiness on both sides towards the front (Fig. 6.2). Bitter sensations are perceived predominantly towards the back of the tongue, although the precise location varies between individuals, in some cases almost as far back as the throat. This is why it is often difficult to ascertain bitterness. It is best to lick a small sample, so that if possible more papillae are reached.

It is also often difficult because there are several "types" of bitterness. Saltiness is quite different, the only salty taste being that of common salt (sodium chloride).

Umami (glutamate, ribotides, etc.) is perceived in the mouth to be „harmonized".

*Smell and Taste is "Aroma" or*

'Why an aroma has to be tested 'through' the mouth and not only 'through' the nose!'

Quality testing of "aromas" is often only done by smell (olfaction) alone, i.e. through the nose. This is a serious error. Because it can be seen clearly from the above that an "aroma" needs to be taken into the mouth, simply because of the different, retronasal olfactory effect. For example, the different temperature in the mouth means that the active components producing the smell are carried past the olfactory epithelium retronasally and therefore differently.

*Sense of Seeing*

Colour is vitally important as a part of vision (in sensory terms: the colour component of the characteristic vision). There are physiological as well as psychological reasons for this. But it would be too complicated to embark on an explanation here. At all events, it is permissible to allow colour to be an influence when an aroma is being tested, unless the colour itself is being tested.

Often natural products in particular are subject to colour fluctuations, which are not necessarily related to taste.

*Sense of Hearing*

Sense of hearing plays virtually no part in testing flavourings, although it does when testing normal products.

*Sense of Touch*

There is also little stimulation of the sense of touch when flavourings are being tested.

*Trigeminal Stimuli*

Trigeminal stimuli are triggered by the trigeminal nerve (Nervus trigeminus). As with taste and smell, we do not know precisely how this works. Trigeminal stimuli are, for example, basic, metallic, astringent, etc. In principle, the sense of temperature and pain also belong here. The sense of temperature is difficult to describe. For this reason, only menthol will be mentioned here. Menthol causes a "chemical" stimulation – sensorically so-called chemaesthetic stimulation – and is not strictly related to the normal sensation of cold. This cold sensation is produced over a period of time and is therefore not easy to test. It is important that this effect have to be neutralised between samples. Soft white bread or warm drinks are effective in achieving this.

The sense of pain is also very complex. It can be stimulated in many different ways. In the case of flavourings, sharpness is of particular importance.

### 6.1.4 Man as a Measuring Instrument

So what could be more obvious than to use man himself as a measuring instrument so that these sensations, which are of such importance for a flavouring, can be measured. Because the ultimate "tester" of the product, the consumer, is also a human being. Furthermore the human senses are far superior in this respect to normal measuring instruments. As the complex pattern of a product is composed of many "sensory pictures" from many 'sensory mosaics', a normal measuring instrument cannot record these. Quite apart from this, psychological influences also pay an important part. We therefore also try to objectify this subjective human characteristic by means of "Sensory".

Technically speaking, as mentioned earlier, quality testing in both ways, sensory or organoleptic, using man more or less as a measuring instrument, has considerable further advantages, such as

*Speed*

Good assessors can certainly check more samples in a short time than a measuring instrument.

*Effectiveness and Information Value*

Sensory or organoleptic quality testing is very effective as high quality information is obtained on the quality of the product. For example, the assessor can immediately describe the sensory difference and appropriate measures can be implemented in the productive process. As the consumer, i.e. the customer for whom the product is manufactured, is in principle an untrained assessor. From the sensory point of view to a great extent he is a layman, therefore sensory control must be relevant to the consumer, as it is carried out by trained assessors.

### 6.1.5 Sensory Quality Measuring Methods

*The Assessors*

The problem in quality testing is the number of assessors. In Sensory Analysis, which is closely allied with statistics here, between 7 and 11 panelists are normally required for this type of testing. But no company carries out initial quality testing with over 7 people. So a compromise has to be found.

An initial prerequisite is that assessors and the test system are organised in such a way that the individual assessor is not overburdened *[2]*. As described above, the senses adapt very quickly. So the sensory organs (primarily the nose) need sufficient time to recover between individual tests. This is easily achieved by organisational measures, such as selecting the standard, entering data into the computer, etc. This gives a time factor of approximately 10-15 minutes per test, although the actual test requires only a fraction of this time and lies somewhere in the middle of the time frame. This gives the assessors 10-15 minutes recovery time between one sample and the next.

The second precondition is that in practice the assessor is an 'expert'. He will have become an 'expert' by appropriate (sensory) training and monitoring (subject to sensory preselection) and familiarisation with the sample material and specific information on the products.

Another precondition is filled by the work itself, namely on the job training.

So if there are at least 2 assessors testing the samples, this is acceptable, although not statistically assured.

*Statistical Confidance and Significance Level*

All the comparison tests mentioned below are analysed according to Bernoulli's binomial distribution *[3]*. For this, however, 3-5 assessors are required mathematically and in practical terms at least 5-7. These figures can, if necessary, be achieved with a sensory panel. Experience has shown, however, that sufficient confidence can be achieved with a statistical significance level of 5% (significant [*]). This applies in particular to quality testing or the difference between "specialist/expert" and consumer/layman. Of course, there is nothing to prevent the confidence or significance level being raised to 1% (highly significant [**]) or even 0.1% (very highly significant [***]). These significance levels naturally vary with different products. For example, the significance level should be higher in the case of a product which is used by the consumer on a daily basis. In this case the "aroma" pattern can to some extent be memorised by the consumer, who will notice minor differences. A typical example of this could be toothpaste which a normal consumer uses every morning and every evening. The picture is very different in the case of products bought occasionally by consumers, such as high quality confectionery. The significance level must therefore be carefully considered and balanced. Because there is also the possibility that the second case (high quality confectionery) will be the more common, but the first (toothpaste) can also arise. So a housewife will not purchase a stew every day and will certainly not compare two cans with each other, but empty the two out and cook them together. But, of course, there are also large kitchens which supply stews every day and here the cook, not the consumer, is the assessor, as the consumer would rarely eat the same stew on two consecutive occasions. This section, i.e. setting significance levels, is highly complex, but also very important (see above). Some experience is required.

## Methods

*General remarks on the test methods*

If a product which has a standard in the market is manufactured or if there is some other standard, it must be assumed that when it reaches the consumer it will be the same as usual. Sensory or other quality testing methods are therefore almost exclusively those which detect minor differences [4].

*Arrangement of samples*

The way, i.e. the sequence, in which samples are arranged is important. There is again a psychological and associated sensory motive for this.

One of the psychological motives is that assessors begin with the sample on the left (at least in countries which write from left to right) so that the first sample, the one on the left, normally leaves the strongest (psychological) impression. A great deal of sensory practice is required to eliminate or "neutralise" this first impression. So from the sensory point of view the aim is to try and prevent this effect as far as possible or at least to "neutralise" it.

In the triangle test (TRIO test), for example, this is achieved by the way in which the samples are arranged. Formerly, the 3 samples were arranged in the form of an (equilateral) triangle in order to compensate for the above effect by different distribution and equal distance between samples. This is intended to give assessors the feeling that it does not matter which sample they take.

Nowadays, the samples are arranged in a row. There are various possibilities for achieving a "neutral" arrangement:

- Each assessor is given a trio of samples, randomly distributed in a different sequence (assuming a difference between the samples): this gives six possible different arrangements: AAB, BAA, ABA, BAB, ABB, BBA.
- From these arrangement results two groups of test samples each with three TRIOs randomly distributed. One group starts on the left with the A sample, sequence is AAB, ABA, ABB, and the other on the left with the B sample, i.e. BAA, BAB, BBA.
- Each assessor is given 2 Trios of samples, randomly distributed, one with an A sample on the left and the other with a B sample on the left, i.e. for example AAB and BAB or
- the same but the samples randomly mixed etc.

*Questioning and questionnaire*

Questionnaires should be designed in such a way that on the one hand the assessor understands all the questions correctly and equally and on the other hand that he or she is unable to draw any conclusions from the questions as to the problem/result.

Samples can simply be compared in this way, for example, but questions can also be asked about the difference between them. But this way then at least psychologically suggests a difference to the assessor. This can have a decisive influence on the test.

The counterpart to this is that, depending on the actual problem, a direct question is asked, e.g. "Which of the two products is the most bitter?" This direct method of questioning involves sensory and psychological problems for the test, be they advantageous or disadvantageous, i.e. a difference is already put directly to the assessor, who will then pay particular or exclusive attention to bitter. The assessor will also evaluate bitterness considerably more precisely that if he was unaware of this. If the problem involves a quality test, for example, this type of questioning is preferable to a "more neutral" question such as "compare the samples".

Questioning and questionnaires must therefore also be approached very cautiously.

*The Test Methods*

### Testing for minor differences: **DUO tests**

DUO tests are particularly suitable for this, i.e. tests in which pairs of samples are compared. These are tests such as the "Paired Comparison Test" (DIN Standard 10 954 *[5]*, ISO Standard 5495 *[6]*) and the "Comparison Test" which does not involve a direct question about the difference. [Although it is called a Comparison Test in English, it actually deals with differences]

- The Paired Comparison Test (ISO 5495 *[6]*)

Two test samples or several pairs of samples are tested for minor differences. At least 7 assessors are normally required for this. The pairs of samples are coded with (2 or 3 digit) random numbers. The questions asked in quality testing would then be as follows:

(1a) Is a difference between the two samples detectable?
(1b) Of the two samples which is the more ........
(2) Questioning only No.1b.
(3) Compare and describe the two samples.

For questioning and the arrangement of pairs of samples, see above in the relevant sections. Analysis is carried out using the relevant significance table S1 (ISO 5495) (see Table 6.1): If, for example, out of 8 replies (assessors) 7 are able to distinguish one sample, there is a significant [*] difference between them.

### Testing for very minor differences: **TRIO tests**

In Trio testing, 3 samples are compared with each other, 2 of which are the same, so that in effect a comparison of two samples is being conducted.

- The Triangle Test (ISO 4120 *[7]*)

This is called a triangle test because in the past the samples were arranged in a triangle (for problems with this see "Arrangement of samples" above).

The question here is as follows:

– Examine the three test samples in the order specified.
– Please circle the number of the test sample which You decide is different..
– (It is essential that You make a choice).

# Sensory Quality Measuring Methods

The samples are coded as described above. They are arranged in taking the appropriate precautions mentioned above. The analysis is also carried out here using the relevant significance table S2 (ISO 4120) (see Table 6.2).

### Testing for differences: **DUO-TRIO tests**

As the name suggests, this DUO-TRIO test (ISO Standard 10399 *[8]*) lies between the two types of test mentioned above. As a rule, it is a DUO test in which 2 samples are tested against a THIRD sample (standard, control, etc.) thus making it a TRIO. The difference to the triangle test, however, is that in principle only 2 samples are being "processed". As with the Duo test, the statistical probability that it will be done CORRECTLY is assumed to be 50% (p= 1/2), whereas with the triangle test it is only 33 1/3% (p = 1/3).

The test can be varied, e.g. several samples can be tested using several pairs (DUOs) against one control.

- The DUO-TRIO test (ISO 10399)

The samples are arranged in such a way that the individual marked controls stand separately and the assigned pair(s) of samples stand next to them.

The question is as follows:

- You are provided with three samples:
- The left hand sample marked C (C = Control] or S (S = Standard) or for example the red beaker marked C or the sample marked ### is the standard (control, reference, etc.).
- Which of the other two samples is the same as the standard (control, reference, etc.) ?
- Examine the samples from left to right and write down the numeral of the sample which is same as the standard (control, reference, etc.) !

Everything else applies by analogy with the other tests. Statistical analysis is with the use of the relevant significance table as with DUO test.

*Table 6.1:* Significance Table for DUO Test and DUO-TRIO Test [9]

| Number of replies | Minimum number of correct replies for a significance level of | | | Number of replies | | | |
|---|---|---|---|---|---|---|---|
| | 5% | 1% | 0,1% | | 5% | 1% | 0,1% |
| 7 | 7 | 7 | | 54 | 34 | 36 | 39 |
| 8 | 7 | 8 | | 55 | 35 | 37 | 40 |
| 9 | 8 | 9 | | 56 | 35 | 38 | 40 |
| 10 | 9 | 10 | 10 | 57 | 36 | 38 | 41 |
| 11 | 9 | 10 | 11 | 58 | 36 | 39 | 42 |
| 12 | 10 | 11 | 12 | 59 | 37 | 39 | 42 |
| 13 | 10 | 12 | 13 | 60 | 37 | 40 | 43 |
| 14 | 11 | 12 | 13 | 61 | 38 | 41 | 43 |
| 15 | 12 | 13 | 14 | 62 | 38 | 41 | 44 |
| 16 | 12 | 14 | 15 | 63 | 39 | 42 | 45 |
| 17 | 13 | 14 | 16 | 64 | 40 | 42 | 45 |
| 18 | 13 | 15 | 16 | 65 | 40 | 43 | 46 |
| 19 | 14 | 15 | 17 | 66 | 41 | 43 | 46 |
| 20 | 15 | 16 | 18 | 67 | 41 | 44 | 47 |
| 21 | 15 | 17 | 18 | 68 | 42 | 45 | 48 |
| 22 | 16 | 17 | 19 | 69 | 42 | 45 | 48 |
| 23 | 16 | 18 | 20 | 70 | 43 | 46 | 49 |
| 24 | 17 | 19 | 20 | 71 | 43 | 46 | 49 |
| 25 | 18 | 19 | 21 | 72 | 44 | 47 | 50 |
| 26 | 18 | 20 | 22 | 73 | 45 | 47 | 51 |
| 27 | 19 | 20 | 22 | 74 | 45 | 48 | 51 |
| 28 | 19 | 21 | 23 | 75 | 46 | 49 | 52 |
| 29 | 20 | 22 | 24 | 76 | 46 | 49 | 52 |
| 30 | 20 | 22 | 24 | 77 | 47 | 50 | 53 |
| 31 | 21 | 23 | 25 | 78 | 47 | 50 | 54 |
| 32 | 22 | 24 | 26 | 79 | 48 | 51 | 54 |
| 33 | 22 | 24 | 26 | 80 | 48 | 51 | 55 |
| 34 | 23 | 25 | 27 | 81 | 49 | 52 | 55 |
| 35 | 23 | 25 | 27 | 82 | 49 | 52 | 56 |
| 36 | 24 | 26 | 28 | 83 | 50 | 53 | 56 |
| 37 | 24 | 27 | 29 | 84 | 51 | 54 | 57 |
| 38 | 25 | 27 | 29 | 85 | 51 | 54 | 58 |
| 39 | 26 | 28 | 30 | 86 | 52 | 55 | 58 |
| 40 | 26 | 28 | 31 | 87 | 52 | 55 | 59 |
| 41 | 27 | 29 | 31 | 88 | 53 | 56 | 59 |
| 42 | 27 | 29 | 32 | 89 | 53 | 56 | 60 |
| 43 | 28 | 30 | 32 | 90 | 54 | 57 | 61 |
| 44 | 28 | 31 | 33 | 91 | 54 | 58 | 61 |
| 45 | 29 | 31 | 34 | 92 | 55 | 58 | 62 |
| 46 | 30 | 32 | 34 | 93 | 55 | 59 | 62 |
| 47 | 30 | 32 | 35 | 94 | 56 | 59 | 63 |
| 48 | 31 | 33 | 36 | 95 | 57 | 60 | 63 |
| 49 | 31 | 34 | 36 | 96 | 57 | 60 | 64 |
| 50 | 32 | 34 | 37 | 97 | 58 | 61 | 65 |
| 51 | 32 | 35 | 37 | 98 | 58 | 61 | 65 |
| 52 | 33 | 35 | 38 | 99 | 59 | 62 | 66 |
| 53 | 33 | 36 | 39 | 100 | 59 | 63 | 66 |

*Table 6.2:* Significance Table for TRIO Test [10]

| Number of replies | Minimum number of correct replies for a significance level of ||| Number of replies | ||| 
|---|---|---|---|---|---|---|---|
| | 5% | 1% | 0,1% | | 5% | 1% | 0,1% |
| 5 | 4 | 5 | | 53 | 24 | 27 | 30 |
| 6 | 5 | 6 | | 54 | 25 | 27 | 30 |
| 7 | 5 | 6 | 7 | 55 | 25 | 28 | 30 |
| 8 | 6 | 7 | 8 | 56 | 26 | 28 | 31 |
| 9 | 6 | 7 | 8 | 57 | 26 | 28 | 31 |
| 10 | 7 | 8 | 9 | 58 | 26 | 29 | 32 |
| 11 | 7 | 8 | 10 | 59 | 27 | 29 | 32 |
| 12 | 8 | 9 | 10 | 60 | 27 | 30 | 33 |
| 13 | 8 | 9 | 11 | 61 | 27 | 30 | 33 |
| 14 | 9 | 10 | 11 | 62 | 28 | 30 | 33 |
| 15 | 9 | 10 | 12 | 63 | 28 | 31 | 34 |
| 16 | 9 | 11 | 12 | 64 | 29 | 31 | 34 |
| 17 | 10 | 11 | 13 | 65 | 29 | 32 | 35 |
| 18 | 10 | 12 | 13 | 66 | 29 | 32 | 35 |
| 19 | 11 | 12 | 14 | 67 | 30 | 33 | 36 |
| 20 | 11 | 13 | 14 | 68 | 30 | 33 | 36 |
| 21 | 12 | 13 | 15 | 69 | 31 | 33 | 36 |
| 22 | 12 | 14 | 15 | 70 | 31 | 34 | 37 |
| 23 | 12 | 14 | 16 | 71 | 31 | 34 | 37 |
| 24 | 13 | 15 | 16 | 72 | 32 | 34 | 38 |
| 25 | 13 | 15 | 17 | 73 | 32 | 35 | 38 |
| 26 | 14 | 15 | 17 | 74 | 32 | 35 | 39 |
| 27 | 14 | 16 | 18 | 75 | 33 | 36 | 39 |
| 28 | 15 | 16 | 18 | 76 | 33 | 36 | 39 |
| 29 | 15 | 17 | 19 | 77 | 34 | 36 | 40 |
| 30 | 15 | 17 | 19 | 78 | 34 | 37 | 40 |
| 31 | 16 | 18 | 20 | 79 | 34 | 37 | 41 |
| 32 | 16 | 18 | 20 | 80 | 35 | 38 | 41 |
| 33 | 17 | 18 | 21 | 81 | 35 | 38 | 41 |
| 34 | 17 | 19 | 21 | 82 | 35 | 38 | 42 |
| 35 | 17 | 19 | 22 | 83 | 36 | 39 | 42 |
| 36 | 18 | 20 | 22 | 84 | 36 | 39 | 43 |
| 37 | 18 | 20 | 22 | 85 | 37 | 40 | 43 |
| 38 | 19 | 21 | 23 | 86 | 37 | 40 | 44 |
| 39 | 19 | 21 | 23 | 87 | 37 | 40 | 44 |
| 40 | 19 | 21 | 24 | 88 | 38 | 41 | 44 |
| 41 | 20 | 22 | 24 | 89 | 38 | 41 | 45 |
| 42 | 20 | 22 | 25 | 90 | 38 | 42 | 45 |
| 43 | 20 | 23 | 25 | 91 | 39 | 42 | 46 |
| 44 | 21 | 23 | 26 | 92 | 39 | 42 | 46 |
| 45 | 21 | 24 | 26 | 93 | 40 | 43 | 46 |
| 46 | 22 | 24 | 27 | 94 | 40 | 43 | 47 |
| 47 | 22 | 24 | 27 | 95 | 40 | 44 | 47 |
| 48 | 22 | 25 | 27 | 96 | 41 | 44 | 48 |
| 49 | 23 | 25 | 28 | 97 | 41 | 44 | 48 |
| 50 | 23 | 26 | 28 | 98 | 41 | 45 | 48 |
| 51 | 24 | 26 | 29 | 99 | 42 | 45 | 49 |
| 52 | 24 | 26 | 29 | 100 | 42 | 46 | 49 |

## REFERENCES

[1] ISO 5492: Sensory analysis – Vocabulary (Rev.1992)
[2] Jellinek, G., Sensory Evaluation of Food, Weinheim, VCH Verlagsgesellschaft 1985
[3] O'Mahony, M., Sensory Evaluation of Food – Statistical Methods and Procedures, Marcel Dekker, Inc., N.Y. USA 1986
[4] Gacula Jr., M.C., Design amd Analysis of Sensory Optimization, Food and Nutrition Press, Inc., Trumbull, Connecticut USA 1993
[5] DIN 10954: Sensorische Prüfverfahren – Paarweise Vergleichsprüfung (1997)
[6] ISO 5495: Sensory Analysis – Methodology – Paired Comparison Test (1983)
[7] ISO 4120: Sensory Analysis – Methodology – Triangle Test (2004)
[7a] DIN ISO 4120: Sensorische Prüfverfahren – Dreiecksprüfung (2004)
[8] ISO 10399: Sensory Analysis – Methodology – Duo-trio Test (1991)
[8a] DIN 10971: Sensorische Prüfverfahren – Duo-Trio-Prüfung (2003)
[9] Table 6.1/1: Significance Table from [6] or [8]
[10] Table 6.1/2: Significance Table from [7]

## 6.2 Analytical Methods

### 6.2.1 General Methods

*Dirk Achim Müller*

#### 6.2.1.1 Introduction

An effective quality control for the manufacture of flavouring products and their application in the food industry, places the knowledge regarding varying analytical procedures and their application to the fore.

Flavouring products, as they are presently prepared in the flavour industry, are mainly very complex mixtures of chemically derived flavouring materials and flavour extracts of varying compositions as well as various carriers and other components (e.g. additives).

It is important for the quality control of individual flavouring components, as well as whole flavouring mixes, to have analytical methods which allow testing for identity as well as specific screening for components limited by law.

Particularly in view of food regulations in the various countries and differing customs regulations, it is necessary to carry out appropriate controls. In fact, procedures are required which give even more accurate evidence than is expected by law.

#### 6.2.1.2 Determination of Colouring Principles by UV/VIS-Spectroscopy

UV/VIS-spectroscopy is commonly used in the quality control laboratories of the flavour industry particularly for those products where colour characteristics are important. The possible measurement of the absorption or transmission of the samples in wavelength maxima as well as the intake of a spectrum over the entire wavelength area between 250-800 nm is important for the routine check of colour identity. By constant measuring parameters (cuvette, dilution, etc.) exact evidence of the colour intensity can be established. This analysis is very important for coloured flavouring preparations, fruit- and plant extracts or essential oils with colouring properties, as well as for testing for the presence of non permitted colouring materials *[1]*.

#### 6.2.1.3 Determination of Heavy Metal Contamination

An important analysis regarding toxicological and legal requirements of flavourings is the control of heavy metal contaminations. Most of the heavy metals show toxic effects in humans, even in trace quantities. Their determination can only be accomplished using trace analysis techniques. In practice, the different analytical techniques "Atomic Absorption Spectrometry" (AAS) and "Inductively Coupled Plasma-Atomic Emission Spectrometry" (ICP-AES) have been employed successfully. Both methods require complete dissolution of the sample by decomposition.

### 6.2.1.3.1 Atomic Absorption Spectrometry (AAS)

The most common method is the AAS. This method offers a high selectivity, low detection limits and the possibility to detect a wide range of heavy metals. For different elements this method requires the use of some technical variants:

*Flame – AAS*

This technique is useful for determination of e.g. Ag, Ni, Cd, Cr, Zn, Cu, Pb, Co, Fe, Mn, at concentration ranges of 50 µg/l. The sample solution is sprayed as aerosol in the approx. 2300°C hot acetylene flame. The measurable absorption at the elements specific fixed wavelength is proportional to the requested quantity [2].

*Graphite-Furnace – AAS*

With this method heavy metals are atomized electrothermally. The detection limits are about 100-times lower than those for the Flame–AAS. Problems with interference signals at this very low measuring range (below 1µ/l) can be overcome by using the Zeemann correction technique [2].

*Cold-Vapour – AAS*

The Cold-Vapour technique is required for determination of mercury because mercury is insensitive to the other AAS techniques. This method utilizes the unique characteristic of mercury to exist in the atomic form at room temperature. This method can only be used for determination of mercury because of unspecific interferences for other elements [2].

*Hydride – AAS*

Elements which can form volatile hydrides can be determined by using the Hydride-AAS technique, because only this method reaches the necessary sensitivity for detection. This method is required for As, Se, Sb, for example. By using this technique it is necessary to transform the required elements completely to their hydride form. The volatile hydrides are evaporated into the flame, atomized and their specific signal is taken for determination [2].

### 6.2.1.3.2 Inductively Coupled Plasma-Atomic Emission Spectrometry (ICP-AES)

This new technique which shows several advantages has been introduced for heavy metal detection over the recent years. The Inductively Coupled Plasma-Atomic Emission Spectrometry allows the simultaneous determination of almost all elements in all fields of elementary analysis. This technique is not very widespread in the laboratories due to the high cost and the complicated operation.

The nebulized sample is exitated in an argon-plasma at approximately 10000 Kelvin. A spectrometer with a polychromator system can determine several elements by measuring their specific emission wavelength with photomultipliers as sensitive detectors. The system requires one photomultiplier for each detectable element. The detection limits of this modern technique are usually between those of the Graphite-Furnace- and the Flame–AAS. The advantages of this system are the very high measuring speed, the parallel detection of approx. 30-40 different elements and highly significant signals without interferences due to the high temperatures of the plasma [3].

### 6.2.1.4 Modern Chromatographic Techniques

Generally quality control for most flavourings requires quantitative details of the individual components or of the entire composition. In such cases "gas chromatography" (GC) should be chosen for volatile compounds and "high performance liquid chromatography" (HPLC) for non-volatile compounds.

The current methods require a preceding clean up and concentration of the sample.

### 6.2.1.4.1 Sample Preparation Methods

For varying requirements, the following systems have proved to be efficient:

*Solid/liquid-extraction*

The commonly used method is the Soxhlet-Extraction by using diethylether and/or pentane as solvent. The sample is extracted continuously by using an Soxhlet-extractor over a period of several hours. The extract is dried and concentrated by evaporation of the solvent. The residual extract can be applied directly to GC-analysis. This method is used for solid or pasty flavourings as well as for whole fruits or plants and their parts where the composition of the volatile compounds is required [4].

*Liquid/liquid-extraction*

The sample material is diluted with demineralized water and extracted with organic solvent like pentane/dichloromethane by using the funnel-separator or rotation-perforator method over a longer period. The extract is dried and concentrated by solvent evaporation. The concentrated extract can be applied to GC-analysis. This method is applicable for liquid extracts and flavouring compounds [4].

*Simultaneous distillation/extraction*

The sample is diluted with the 10-fold quantity of distilled water and extracted for several hours with dichloromethane or diethylether using a "Simultaneous Distillation-Extraction Apparatus" (SDE). The organic phase is dried and concentrated by solvent evaporation using a Vigreux column. The concentrated extract can be applied to GC-analysis. This method is applicable for solid, viscous or pasty extracts and flavouring compounds [4].

*Liquid/solid-extraction*

The sample is diluted with an appropriate solvent. The solution is applied on to a cartridge containing an appropriate solid phase material (silica gels, different RP materials, ion exchange materials). The required compounds are dissolved subsequently from the cartridge by using different solvents. The eluent can be used directly for chromatographic analysis. The advantages of this clean-up method are: a small amount of solvent yields a high concentration of the sample; speed; no interference from carbohydrates [5-7].

*Solid-phase microextraction (SPME)*

A coated fused silica fibre is directly introduced in the liquid sample or in the headspace above the sample. Respective ingredients from the sample material are absorbed onto the fibre material until equilibrium is reached. The fibre is removed from the sample and directly applied into the injection system of the GC. The absorbed compounds are thermally desorbed into the GC column for analysis. The method is solvent-free and requires no special additional equipment. It is used for the analysis of special groups of compounds, depending on the enrichment based on the type of fibre used *[32-36]*.

*Stir bar sorptive extraction (SBSE)*

A bar magnet encapsulated in glass which is coated with a special polymer is used for stirring the aqueous sample. During this stirring period the respective analytes are adsorbed on/in the coating material. After equilibration has been reached, the bar is removed, washed with ultrapure water and dried. Then the bar is inserted in the thermal desorption system which desorbs the compounds from the coating and transfers them to the GC column. The method is similar to the SPME but offers a wider range of capacity for organic compounds *[37, 38]*.

### 6.2.1.4.2 High Performance Liquid Chromatography (HPLC)

### (1) Principles

Today HPLC is an integral part of routine analysis of low- or non volatile flavouring components as well as of thermally instable or decomposing flavouring substances.

Figure 6.3 shows the systematic construction of HPLC-systems:

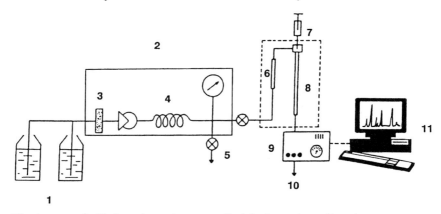

1  mobile phase vessel with degassing system
2  pumping system with gradient control unit
3  prefilter unit
4  pulse suppressor
5  pressure relief valve
7  injection system with syringe
8  column
9  detection system
10 ancillary detection systems or fraction collector
11 workstation

***Fig. 6.3:*** *Schematic diagram of the HPLC System*

# General Methods

According to the column material mainly used for separation, HPLC procedures can be divided into the following categories:

*Adsorption Chromatography*

- Normal Phase Chromatography

    In this case a polar material (e.g. silica gel or aluminium oxide) is used as stationary phase for the column. As for the mobile phase an apolar solvent is used (e.g. hexane, tetrahydrofurane). The separation follows through varying adsorption of the sample molecules at the stationary phase.

- Reversed Phase Chromatography (RP)

    The introduction of the chemically bonded apolar reversed phase materials as stationary phases is responsible for the common use of HPLC in routine laboratory analysis, because the sample can be applied directly from the aqueous solution into the HPLC-system. For the mobile phase, polar solvents such as water or aqueous buffer solutions or gradient mixes of aqueous solvents with methanol or acetonitrile are used mainly.

*Ion Exchange Chromatography*

The stationary phase contains anionic or cationic groups which interact with the ionic groups of the sample ingredients to reach separation.

*Affinity Chromatography*

This highly selective separation procedure is used mainly for isolation and analysis of bioactive components (e.g. lipids, proteins).

The stationary phase consists of a special chemically modified carrier which, with the sample molecules, results in a reversible interaction. Generally water is used for mobile phase *[8-10]*.

## (2) Common Detection-Systems

Detection systems mainly used in routine analysis are universal UV/VIS-detectors, selective fluorescence-detectors, electrochemical-detectors and, in recent years, the Diode-Array-detector.

*UV/VIS-Detection*

These detectors are commonly used in HPLC analysis. The absorption of the substances in the eluate by a specific wavelength normally between 190 and 700 nm is proportional to their concentration. This technique is widely used for quality control of flavourings, e.g. for the quantification of the vanillin/p-hydroxybenzaldehyde ratio for natural vanilla products, for the determination of caffeine in extracts and drinks, or for the detection and quantification of restricted ingredients such as quassine, coumarin or β-asarone in natural extracts *[11-12]*.

*Fluorescence-Detection*

The fluorescence detector is a selective detection system with a high sensitivity for fluorescing substances. Molecules with fluorescing properties show fluorescence-emission by excitation with UV-light at specific wavelength. The measurable emission is proportional to their concentration. This method requires fluorescing functional groups at the molecule or chemical derivatization with a fluorescing functional group. The derivatization can be performed before HPLC-separation, so called pre-column derivatization, or directly after HPLC-separation (postcolumn derivatization).

Examples for the fluorescence detection system are the determination of quinine in soft drinks and the important quantification of benzo[a]pyrene and other polycyclic hydrocarbons (PAH) in smoke flavourings or smoke preparations [13-14]

*Electrochemical-Detection*

The electrochemical detection system requires reducible or oxidable substances. The eluted substance is reduced or oxidized by an electrode. This not very common detection system is a useful tool for determination of amino-compounds and other organic nitrogen-compounds, phenols and quinones [8].

*Diode-Array-Detection (DAD)*

Diode-Array detection systems become more and more standard detectors in HPLC-analysis. Based on the principle of UV/VIS-spectrophotometers, DAD facilitates the simultaneous detection and registration of UV-chromatograms at different wavelength and the spectra of the single substances. This additional spectrum information is often required for the definite identification of the single substance peaks in UV/VIS-chromatograms [15].

### (3) HPLC-MS-Coupling Technique

HPLC-MS coupling is the most useful technique for the analysis of thermally instable or non-volatile compounds. Due to the technical difficulties in the past, HPLC-MS is not very common in laboratories. The main problems that may occur in HPLC-MS analysis are the high dilution (low concentration) of the compounds in the HPLC eluate. Secondly, often non-volatile inorganic substances may be present in the eluate, especially by running RP-chromatography using buffer solvents. If gradient elution is used, the composition of the solvent changes during analysis. These problems prevent the widespread use of the technique in routine analysis of flavourings. Nevertheless, this technique offers a very useful tool for the identification and determination of some very important compounds or contaminants, such as coumarines, aflatoxines, heterocyclic aromatic amines, pesticides or pharmaceutical residues.

The introduction of the thermospray interface in the mid to late 1980s provided the first efficient LC-MS connecting technique. With the relatively new interface techniques of electrospray interface and the complementary atmospheric pressure chemical ionization interface (APCI), the full potential of the LC-MS system can now be achieved.

In both electrospray and APCI, the ionization takes place at atmospheric pressure.

Using the electrospray interface, the eluate is electrically charged at a potential of typically 3-4 kV and vaporized by the Coulomb energy. During vaporization the charged droplets emit particle ions of the dissolved appropriate analytes. After passing a series of focusing lenses these are detected by the MS-system. This electrospray interface system allows one to obtain mass spectra from highly polar and ionic compounds, whereas non-polar or low-polarity compounds cannot be properly analysed.

The APCI interface uses a combination of heat and nebulizing gas or pneumatic nebulizer to disperse the eluent into small droplets. The ionization takes place using the corona discharge. As a result of the ability to deal with flow rates of 0.5 to 2 ml/min and a higher tolerance for buffers, it can also be performed with 4.6 mm columns. The APCI system enables the recording of mass spectra from non-polar or only slightly polar compounds. Due to the soft ionization technique the determination of the molecular weight of the analyte is often possible.

These interface techniques for connecting the HPLC with the MS-system are very sensitive for most of the substances of interest to the flavour industry. Therefore, HPLC-MS coupling techniques have become an increasingly powerful tool for quality control of flavourings, especially for the analysis of complex mixtures like process flavourings or contaminants present in such complex mixtures. New developments in the area of mass detection systems, such as time-of-flight (ToF) mass analysers and tandem mass spectrometry systems or the features of matrix-assisted laser desorption ionization (MALDI) techniques, may enhance the analytical capabilities of these systems in the near future *[16, 17, 28-31]*.

### 6.2.1.4.3 Gas Chromatography (GC)

**(1) Principles**

This technique is the most commonly used chromatographic method for analysis of volatile flavouring compounds.

The systematic structure of the GC-system is shown in Figure 6.4:

Apart from the sample material the following capillary columns (stationary phases) are suitable for flavour analysis:

      polar phases :    Carbowax-20M, FFAP, DB-Wax
      midpolar phases : SPB-35, DB-1701, HP 225
      nonpolar phases : DB-1, HP-5

Usually helium is used as carrier gas. The temperature gradient used for the separation follows from the known or supposed composition of the volatile components (normally between 25° and 250° Celsius) and it also depends on the stability of the chosen column material.

The range of information gained after successful separation will depend on the type of detector used *[18-19, 39]*.

**(2) Common Detection Systems**

*Flame Ionization Detector (FID)*

This detector system allows quantitative evidence of all occurrences of isolated components (peaks), however no direct evidence for the identification of individual substances. It is routinely used for testing the identity of flavourings by comparing the chromatographic results with the standard quality of the product in question [19].

*Electron Capturing Detector (ECD), Nitrogen-Phosphorus Detector (NPD)*

These selective detector systems allow the specific, quantitative determination of substances with definite molecule groups. The ECD is suited especially for the determination of halogenated materials, whereas the NPD is specific for the determination of nitrogen and phosphorus containing molecules respectively. These systems are used for detection of halogenated solvents or pesticides with halogen-, phosphorus- or nitrogen-groups, which are covered by world-wide differing legal requirements [19].

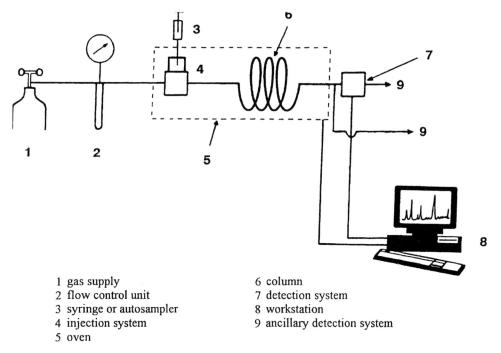

1 gas supply
2 flow control unit
3 syringe or autosampler
4 injection system
5 oven
6 column
7 detection system
8 workstation
9 ancillary detection system

*Fig. 6.4:* Schematic diagram of the GC system

### (3) Multidimensional Gas Chromatography

Element specific detectors and GC coupling techniques are helpful in analyzing the composition of complex natural extracts or flavourings. But even in high resolution capillary gas chromatography, complete resolution often cannot be achieved because of insufficient separation of certain compounds by overlapping of interesting substances present in the mixture. These natural chromatographic limitations can be largely overcome by using multidimensional gas chromatography (MDGC). The principle is that compounds showing insufficient separation on one column are transferred by flow switching to a second column of different polarity where complete resolution is achieved. The direct coupling of capillary columns coated with different stationary phases leads to a much higher efficiency of separation. The main goal for applying MDGC is to enhance the resolution of the single component in complex mixtures for identification even by using coupling techniques as MDGC-MS or MDGC-FTIR [20-21, 40-41].

### (4) Modern Techniques

*Gas Chromatography-Mass Spectrometry – Coupling (GC-MS)*

The most common coupling technique used in quality control of flavouring compounds and other flavouring materials is the gas chromatography-mass spectrometry coupling system. This combination of two gas-phase techniques can be applied to a wide range of problems. With a good interface, any substance which elutes from the GC can be examined by the GC-MS. The Mass-Selective-Detection system (MSD) used in routine analysis of flavourings, enables qualitative and quantitative statements of the structure of single components. After GC-separation the single components are transferred to the MSD using a specific interface. In the MSD-ion source the molecules are ionized by "electron impact" conditions (EI) at conventional
70 eV. This leads to the fragmentation of these molecules by formation of different ions. The separation of the ions according to their different mass-to-charge ratios is achieved with magnetic or electric fields or both combined. The separated ions impinge on the electron multiplier detector system which collects and amplifies these signals. The individually recorded signals are specific for each fragment; the whole mass spectrum of all fragments is characteristic for the compound. Identification of the components is possible by comparing the achieved mass spectrum with spectra of well-known substances. Ionization sensitive molecules of unknown substances for which standard spectra are not available, are difficult to elucidate from EI-mass spectra because they contain only low abundances of the informative high-mass ions.

When the ion source is changed to "chemical ionization" conditions (CI), the resulting mass spectrum almost always contains abundant ions at one atomic-mass-unit greater than the molecular weight, owing to proton-transfer from the reactant gas (methane, iso-butane, ammonia), This gives dedicated information regarding the exact molecular weight of the molecule.

The structure of unknown components is far easier to elucidate by examination of both, EI- and CI- mass spectra. This is typical for many components and implies that no one ionization technique is always superior to another. In order to achieve quantitative data, the abundance of ions has to be measured, since in an ionized compound, the absolute abundance of any of its ions is related to the quantity of that substance. During the residence of the compound in the ion source, repetitive scanning with a mass spectrometer can provide several complete mass spectra. For a typical total ion chromatogram (TIC), in which mass spectra may be recorded every second, any single $m/e$ would be sampled 5-30 times, being focused on the electron multiplier detection system each time for a few microseconds only.

Another possibility of measuring a mass peak is the selected ion monitoring method (SIM). By using this method a thousand-fold increase in sensitivity is obtained compared with the TIC, because for the entire time that a compound resides in the ion source, the detector records only the selected ions. Selected ion monitoring is capable of measuring amounts of flavouring substances in the pico- or femtogram range. SIM is especially useful in the field of quality control for detection, identification and quantification of limited or prohibited flavouring substances in complex mixtures or preparations [22-24].

Beside the most commonly used ion separation systems, such as the quadrupole mass analyser, the ion-trap mass analyser and the double-focusing and tri-sector mass analyser, a new system has encountered increasing interest. This system, the time-of-flight (TOF) mass analyser offers fast scanning opportunities coupled with the possibility of detecting ions with high m/z ratios. This provides a useful tool for rapid quality control of sensorial profiles as well as the detection and identification of toxins and proteins or protein-bound substances [42].

The following example shows a strawberry flavour, analysed by the GC/MSD coupling technique using the TIC-method (Fig. 6.5). The identification of peak No.: 5 (ethyl isovalerate), peak No.: 14 (alpha-ionone) and peak No.: 21 (ethyl 3-phenylglycidate) by the mass-spectrum shows the usefulness of this modern analysis technique.

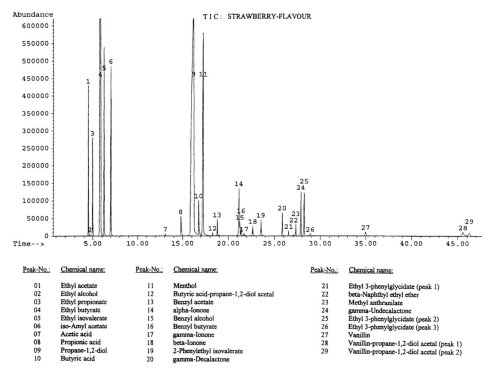

*Fig. 6.5:* Total Ion Chromatogram (TIC) of strawberry flavour

*Gas Chromatography-Fourier Transform Infrared Spectroscopy*
*– Coupling (GC-FTIR)*

A relatively new and modern analytical method in flavour analysis is the GC-FTIR-coupling technique. This method together with GC-MS-coupling represents the most sophisticated and flexible analytical technique with respect to the quality and the quantity of information.

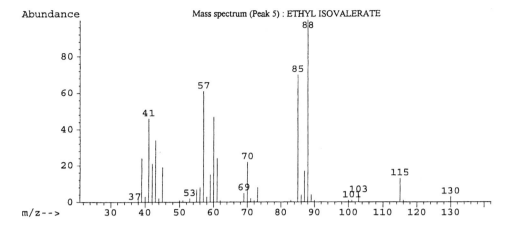

*Fig. 6.6:* Mass spectrum of peak No. 5, identified as ethyl isovalerate

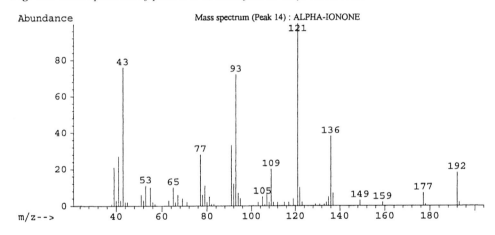

*Fig. 6.7:* Mass spectrum of peak No. 14, identified as alpha-ionone

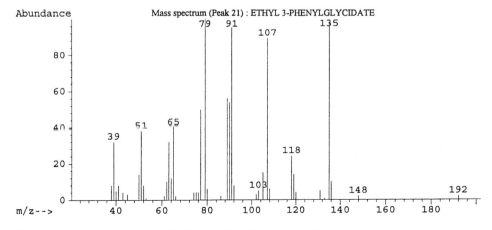

*Fig. 6.8:* Mass spectrum of peak No. 21, identified as ethyl 3-phenylglycidate

After separation, the out-flow from the GC-column passes continuously through an interface, the "light pipe". The inner surface of the interface is gold coated for higher reflection of IR-light and chemical inertness. The light pipe is set to high temperatures (between 280 and 350°C) to prevent trapping of substances. Infrared radiation from the IR-source passes into the interferometer where the beam is split, one half to a fixed mirror and the other half to a moving mirror. After reflection from the mirrors the light beams are recombined. Depending to the moving mirror, interference will occur for every wavelength of IR-light. The combination of all interference patterns of all IR-wavelengths results in an interferogram. Parallely a laser beam controls the change in optical path difference due to the movement of the mirror. The recombined interfering IR-beam is sent through the light pipe containing the volatile substances to a dedicated IR-detection system. The acquired raw data from the specific absorption of varied wavelength of the IR-beam by the single substances are converted by several mathematical steps (Fourier transformation) to obtain the infrared spectrum which is characteristic for each component. Characteristic group frequencies are absorption bands that appear with a relatively high intensity in the wavelength range, characteristic for a certain functional group. For structural information of unknown compounds the spectrum can be interpreted through the use of known structure-absorption correlations by checking which absorption bands are present and which bands are absent. Identification can be achieved by comparison of the IR-spectrum with available spectra from common data bases.

A further advantage of this modern technique is the non-destructive nature of the IR-detection. This makes the out-flow available for additional detection techniques such as FID or MSD, or fraction collection of interesting substances [25-27].

The following FTIR spectra show the usefulness of this modern analysis technique by the ensuing verification of peak No.: 5 (ethyl isovalerate), peak No.: 14 (alpha-ionone) and peak No.: 21 (ethyl-3-phenylglycidate) detected by GC-FTIR-spectrometry in a strawberry flavour.

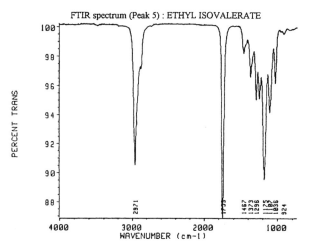

*Fig. 6.9:* FTIR spectrum of peak No. 5, identified as ethyl isovalerate

General Methods                                                                 599

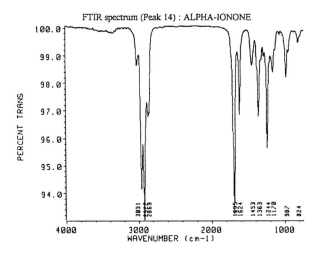

*Fig. 6.10:* FTIR spectrum of peak No. 14, identified as alpha-ionone

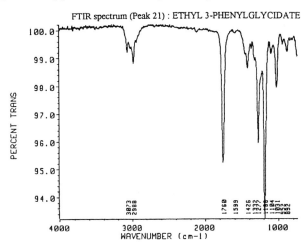

*Fig. 6.11:* FTIR spectrum of peak No. 21, identified as ethyl 3-phenylglycidate

These systems illustrate the current standard of analytical procedures employed by the flavour industry for the quality control of raw materials and finished products.

## REFERENCES

[1]  Hesse, M., Meier, H., Zeeh, B., Spektroskopische Methoden in der organischen Chemie, Stuttgart-New York, Georg Thieme Verlag 1987
[2]  Welz, B., Atomabsorptionsspektroskopie, Weinheim, Verlag Chemie 1983
[3]  Hoffmann H.-J., Rohl R, Plasma-Emissions Spektrometrie, Band 5, Berlin-Heidelberg-New York-Tokyo, Springer Verlag 1985
[4]  Sandra, P., Bicci, C. (Eds.), Capillary Gas Chromatography in Essential Oil Analysis, Heidelberg-Basel-New York, Dr. Alfred Huethig Verlag 1987
[5]  ICT (Ed.), Cyclobond Handbook, A guide to using cyclodextrin bonded phases, Vol. 4 Frankfurt, ICT Handelsgesellschaft mbH 1988

[6] Wörner, M., Gensler, M., Bahn, B., Schreier, P., Anwendung der Festphasenextraktion zur raschen Probenvorbereitung bei der Bestimmung von Lebensmittelinhaltsstoffen, *Lebensm. Unters. Forsch.*, 189, 422-425 (1989)
[7] Analytichem (Ed.), Proceedings of the 2nd International Symposium of Sample Preparation and Isolation using Bonded Silicas, Harbour City, Analytichem 1985
[8] Unger, K. K.(Ed.), Handbuch der HPLC, Teil 1, Darmstadt, GIT-Verlag 1989
[9] Engelhardt, H. (Ed.), Practice of High Performance Liquid Chromatography, Berlin-Heidelberg-New York, Springer Verlag 1986
[10] Dean, P. D. G., Johnson, W. S., Middle, F. A. (Eds.), Affinity Chromatography, Oxford, IRL Press 1985
[11] L.M.L. Nollet (Ed.), Food Analysis by HPLC, Marcel Dekker, Inc. New York, Basel, Hongkong, 1992
[12] Jürgens, U., Zur hochdruckflüssigchromatographischen Analyse von Aromen: Untersuchung von Lebensmitteln mit Vanillegeschmack, *Deutsche Lebensmittel-Rundschau*, 77 (3), 93-96 (1981)
[13] Wehry, E. L. (Ed.), Modern Fluorescence Spectroscopy, Vol. 1 – 2, New York, Plenum Press 1976
[14] Stijve,T., Hischenhuber, C., Simplified Determination of Benz(a)pyrene and Other Polycyclic Aromatic Hydrocarbons in Various Food Materials by HPLC and TLC, *Deutsche Lebensmittel-Rundschau*, 83 (9), 276-282 (1987)
[15] Owen, A. J. (Ed.), Diodenarray-Technologie in der UV/VIS- Spektroskopie, Waldbronn, Hewlett-Packard GmbH 1988
[16] Yergey, A. L., Edmonds, C. G., Lewis, I. A. S., Vestal, M. L., Liquid Chromatography/Mass Spectrometry – Techniques and Applications, New York, London, Plenum Press 1990
[17] Engelhardt, U. H., Wagner-Redeker, W., HPLC-MS-Kopplungen – Technik und Anwendungen, *GIT Spezial Chromatographie*, No. 1, 5-9 (1994)
[18] Jennings, W., Analytical Gas Chromatographie, Orlando, Academic Press 1987
[19] Wittkowski, R., Mattissek, R. (Eds.), Capillary Gas Chromatography in Food Control and Resarch, Lancaster-Basel, Technomic Publishing Company 1993
[20] Krammer, G., Bernreuther, A., Schreier, P., Multidimensionale Gas-Chromatographie, *GIT Fachz. Lab*, 3, 306-312 (1990)
[21] Nitz, S., Multidimensional Gas-Chromatographie in Aroma Research, in : Topics in Flavour Research, Proc. of the Int. Conference Freising-Weihenstephan, Berger, R. G., Nitz, S., Schreier, P. (Eds.), Marzling, Eichhorn-Verlag 1985
[22] Chapman, J. R., Practical Organic Mass Spectrometry Chicester-New York, J. Wiley 1985
[23] Reineccius, G. A.,Gas Chromatography and Mass Spectrometry in Quality Control and Research, *Food Sci. Technol,*. 45, 67-81 (1991)
[24] Ramaswani, S. K., Briscese, P., Gargiullo, R. J., von Geldern, T., Sesquiterpene Hydrocarbons : From Mass Confusion to Orderly Line-Up, in : Proceedings of the 10th International Congress of Essential Oils, Fragrances and Flavours, 1986, Lawrence, B. M., Mookherjee, B. D., Willis, B. J. (Eds.) Amsterdam-Oxford-New York-Tokyo, Elsevier Science Publishers 1988
[25] Herres, W., HRGC-FTIR : Capillary Gas Chromatography – Fourier Transform Infrared Spectroscopy, Therory and Application, Heidelberg-Basel-New York, Dr. Alfred Huethig Verlag 1987
[26] Schreier, P., Mosandl, A., Aromaforschung heute, *Chemie in unserer Zeit*, 19 (1), 22-30 (1985)
[27] Poucert, C. J. (Ed.), The Aldrich Library of FTIR-Spectra, Ed. I, Vol. I + II, Aldrich Chemical Company Inc. 1985
[28] Abian, J., The coupling of gas and liquid chromatography with mass spectrometry, *J. Mass Spectrom.*, 34, 157-168 (1999)
[29] Niessen, W.M.A., Liquid Chromatography-Mass Spectrometry, 2nd edn., Chromatographic Science Series, No. 79, New York, Marcel Dekker 1999
[30] Abian, J., Oosterkamp, A.J., Gelpí, E., Comparison of conventional, narrow-bore and capillary liquid chromatography/mass spectrometry for electrospray ionization mass spectrometry: practical consideration, *J. Mass Spectrom.*, 34, 244-254 (1999)
[31] Cotter, R.J., The new time-of-flight mass spectrometer, *Anal. Chem.*, 71, 445a-451a (1999)
[32] Pawliszyn, J., Theory of solid-phase microextraction, *J. Chromatogr. Sci.*, 38, 270-278 (2000)
[33] Shirey, R.E., Optimization of extraction conditions and fibre selection for semivolatile analytes using solid-phase microextraction, *J. Chromatogr. Sci.*, 38, 279-288 (2000)

[34]  Grimm, C.C., Lloyd, S.W., Batista, R., Zimba, P.V., Using microwave distillation solid-phase microextraction gas chromatography-mass spectrometry for analysing fish tissue, J. Chromatogr. Sci., 38, 289-296 (2000)
[35]  Demyttenaere, J.C.R., Dagher, C., Verhé, R., Sandra, P., Flavour analysis of Greek white wine using solid phase microextraction-capillary GC/MS, in Flavour Research at the Dawn of the Twenty-first Century, Proceedings of the 10th Weurman Flavour Research Symposium Beaune, Le Quéré, J.L., Étiévant, P.X. (Eds.), London/Paris/New York, Lavoisier and Intercept 2003
[36]  Demyttenaere, J.C.R., Sánchez Martínez, J.I., Téllez Valdés, M.J., Verhé, R., Sandra, P., Analysis of volatile esters of malt whisky using solid phase micro-extraction-capillary GC/MS, in Flavour Research at the Dawn of the Twenty-first Century, Proceedings of the 10th Weurman Flavour Research Symposium Beaune, Le Quéré, J.L., Étiévant, P.X. (Eds.), London/Paris/New York, Lavoisier and Intercept 2003
[37]  Sandra, P., David, F., Vercammen, J., New developments in sorptive extraction for the analysis of flavours and fragrances, in Advances in Flavours and Fragrances, Swift, K.A.D. (Ed.), Cambridge, Royal Society of Chemistry 2002
[38]  David, F., Tienport, B., Sandra, P., Stir-Bar Sorptive Extraction of Trace Organic Compounds from Aqueous Matrices, LC-GC Europe, July 2003
[39]  Jensen, S.M., Christensen, L.P., Edelenbos, M., Kjeldsen, F., Quantitative analysis of volatile compounds in vegetables by gas chromatography using large-volume injection technique, in Flavour Research at the Dawn of the Twenty-first Century, Proceedings of the 10th Weurman Flavour Research Symposium Beaune, Le Quéré, J.L., Étiévant, P.X. (Eds.), London/Paris/New York, Lavoisier and Intercept 2003
[40]  Mondello, L., Lewis, A.C., Bartle, K.D., Multidimensional Chromatography, Chichester, John Wiley 2002
[41]  Marriott, P.J., Shellie, R. Fergeus, J., Ong, R., Morrison, P., High resolution essential oil analysis by using comprehensive gas chromatography methodology, Flav. Fragr. J., 15, 225-239 (2000)
[42]  Sies, A., Hirsch, R., Löscher, R., Tablack, P., Guth, H., Direct thermal desorption and FAST-GC-TOF-MS for a rapid quality control of hazelnuts, in Flavour Research at the Dawn of the Twenty-first Century, Proceedings of the 10th Weurman Flavour Research Symposium Beaune, Le Quéré, J.L., Étiévant, P.X. (Eds.), London/Paris/New York, Lavoisier and Intercept 2003

## 6.2.2 Stable Isotope Ratio Analysis in Quality Control of Flavourings

H.-L. Schmidt
R. A. Werner
A. Roßmann
A. Mosandl
P. Schreier


Flavouring compounds are, like any other organic matter, composed of atoms of the bioelements carbon, hydrogen, oxygen and in some cases also of nitrogen and sulphur. These elements occur in nature as mixtures of stable isotopes, among which the "light" ones are by far dominant, whereas the abundance of the "heavy" ones is of the order of 1 atom-% or less (Table 6.3). Only in the case of carbon also a radioactive isotope ($^{14}C$) is of importance in the present context: This isotope is continuously produced by a natural (since 1954 also artificial) nuclear process; it is present at a constant level (in 1950 before anthropogenic activities: 13.56 ± 0.07 disintegrations per g C and min [1, 2]) in any recent biomass, but due to its half life time of about 5730 ± 40 y, it is completely decayed in fossil material (which is usually millions of years old). On this basis organic compounds synthesised from fossil carbon (petrol descendants) can hence be distinguished from natural products (e.g. [3, 4]). Recently also radioactive hydrogen ($^{3}H$, tritium) has been used in the authentication process, e.g. of benzaldehyde [5]. This approach has e.g. also been tested in 1978 for the distinction between nature-identical and biogenic vinegar [6].

The natural abundances of the stable isotopes given in Table 6.3 are global average values, whereas the actual exact values are subjected to small local and temporal variations, due to the slightly different behaviour of isotopologue molecules (by IUPAC definition a molecular entity that differs only in isotope composition, means number of isotopic substitutions; in contrast, isotopomer ["isotopic isomer"] molecules are isomers, having the same number of each isotopic atom but in different positions) in the course of chemical reactions or physical processes ([8], kinetic and thermodynamic isotope effects, respectively, see below). The corresponding shifts are so small that they cannot be indicated in the atom-% scale; therefore they are expressed in δ-values[*], differences of the isotope ratio R, e.g., ($[^{13}C]/[^{12}C]$) of the sample and an international standard relative to this international standard:

---

[*] Most isotopic ratio data in this chapter are reported in ‰ (per mill [32]) deviation from the international isotope standards V-SMOW (Vienna Standard Mean Ocean Water) for oxygen and hydrogen; V-PDB (Vienna PeeDee Belemnite) for carbon; AIR-$N_2$ (atmospheric nitrogen gas) for nitrogen; and V-CDT (Vienna Cañon Diablo Troilite) for sulphur. The virtual non-existing standard V-PDB is defined by adopting a $δ^{13}C$-value of +1.95‰ for NBS 19 (NBS: former National Bureau of Standards, now NIST: National Institute for Standard and Technology, USA) carbonate relative to V-PDB [17] because the supply of the original standard PDB [16] (the carbonate skeleton of a fossil cephalopod, *Belemnitella americana*, from the Cretaceous formation in South Carolina) has been exhausted for more than 3 decades. *continued on page 604.*

**Table 6.3:** Characteristics and standards of the stable isotopes of the bioelements. After references [7-14]. Abbreviations: V-SMOW = Vienna Standard Mean Ocean Water [15]; V-PDB = Vienna PeeDee Belemnite (Carbonate) [16, 17]; AIR-$N_2$ = atmospheric $N_2$ [18-20]; V-CDT = Vienna Cañon Diablo Troilite [21-23]

| Element | Stable isotopes | Rel. natural abundance [atom-%][a] | Atomic mass [AMU][b] | Nuclear magnetic resonance | | | International standard | | Gas for analysis |
|---|---|---|---|---|---|---|---|---|---|
| | | | | spin I | frequency[c] [MHz at 9.4 Tesla] | rel. sensitivity[d] | name | isotope ratio R = [isotope]/[main isotope] [× 10^6] | |
| Hydrogen | $^1$H | 99.985(1) | 1.00782522 | 1/2 | 400.000 | 1.0 | V-SMOW [e] | 155.76 | $H_2$ |
| | $^2$H | 0.015(1) | 2.0141022 | 1 | 61.402 | $1.45 \times 10^{-6}$ | | | |
| Carbon | $^{12}$C | 98.90(3) | 12.0000000 | 0 | - | - | V-PDB [f, g] | 11180.2 | $CO_2$ |
| | $^{13}$C | 1.10(3) | 13.0033543 | 1/2 | 100.577 | $1.76 \times 10^{-4}$ | | | |
| Nitrogen | $^{14}$N | 99.634(9) | 14.0030744 | 1 | 28.894 | $1.01 \times 10^{-3}$ | AIR-$N_2$ | 3678.2 | $N_2$ |
| | $^{15}$N | 0.366(9) | 15.0001081 | 1/2 | 40.531 | $3.85 \times 10^{-6}$ | | | |
| Oxygen | $^{16}$O | 99.762(15) | 15.9949149 | 0 | - | - | V-SMOW [e, g] | 379.9 | $CO_2$, CO |
| | $^{17}$O | 0.038(3) | 16.9991334 | 5/2 | 54.227 | $1.08 \times 10^{-5}$ | | 2005.2 | |
| | $^{18}$O | 0.200(12) | 17.9991598 | 0 | - | - | | | |
| Sulphur | $^{32}$S | 95.02(9) | 31.972074 | 0 | - | - | V-CDT [h] | 44150.9 | $SO_2$, $SF_6$ |
| | $^{33}$S | 0.75(4) | 32.97146 | 3/2 | 30.678 | $1.72 \times 10^{-5}$ | | | |
| | $^{34}$S | 4.21(8) | 33.967864 | 0 | - | - | | | |
| | $^{36}$S | 0.02(1) | 35.96709 | 0 | - | - | | | |

[a] Natural abundance values are followed by uncertainties in the last digit(s) of the stated values [14 and literature cited therein]; [b] AMU: atomic mass unit, relative to $^{12}$C = 12.000000; [c] the strength of earth's magnetic field is $6 \times 10^{-5}$ T; [d] for natural isotope abundance spectra; [e] additional water "standards" are: SLAP (Standard Light Antarctic Precipitation) with $\delta^2$H: -428‰, $\delta^{18}$O: -55.5‰ [15, 24] and GISP (Greenland Ice Sheet Precipitation) with $\delta^2$H: -189.88‰, $\delta^{18}$O: -24.72‰ [25], $\delta^2$H-values are reported vs. V-SMOW; [f] Russe et al. [26] give a new value for the $^{13}$C/$^{12}$C ratio of V-PDB of (11101 ± 16) × $10^{-6}$; [g] $\delta^{18}$O-values of carbonate material can also be expressed in per mill relative to V-PDB (with a $^{17}$O/$^{16}$O ratio of 381.9 × $10^{-6}$ and a $^{18}$O/$^{16}$O ratio of 2067.2 × $10^{-6}$, calculated from the above given oxygen isotope ratios for V-SMOW using a $\delta^{18}$O-value of +30.92‰ for V-PDB vs. V-SMOW and a λ of 0.5281 [27]) by assigning a $\delta^{18}$O-value of -2.2‰ exactly to NBS 19 calcite, $\delta^{18}$O-values of all other oxygen-bearing substances should be expressed relative to V-SMOW; [h] The V-CDT- scale is defined by assigning a $\delta^{34}$S-value of -0.3‰ exactly (vs. V-CDT) to the international available silver sulphide reference material IAEA-S-1 [28, 29]. The $^{34}$S/$^{32}$S ratio of the virtual non-existing standard V-CDT can then be calculated to 0.0441509 [30, 31]. The "heavy" hydrogen $^2$H is also called deuterium (D).

$$\delta^{13}C\ [‰]_{V\text{-}PDB} = \left( \frac{[^{13}C_{Sample}]/[^{12}C_{Sample}]}{[^{13}C_{Standard}]/[^{12}C_{Standard}]} - 1 \right) \times 1000 \tag{1}$$

The δ-values of a whole organic compound or of a given position within a molecule are determined by that of its starting material, and by isotope effects in the course of its synthesis, hence report and preserve the origin and the conditions of its formation, and this is the basis of origin and history assignment of an organic compound [33].

### 6.2.2.1 Correlations between (Bio-)Synthesis and Isotope Content or Pattern of Organic Compounds

The fundamental basis for the isotopic composition of an organic compound is that of the primary source of the elements in question. As already mentioned, fossil carbon compounds do not contain any $^{14}C$. The primary sources of any carbon in recent biological matter are the carbon dioxide of the atmosphere and the hydrogen carbonate of the hydrosphere. The $\delta^{13}C$-values of these two compounds (-7 to -8‰, and ± 0‰, respectively, relative to the international standard V-PDB, Table 6.3) differ due to a temperature-depending thermodynamic isotope effect [16, 34-36] on the isotope distribution in the system: $CO_2 + H_2O \leftrightarrow HCO_3^- + H^+$. Hence any organic matter synthesised in the same way by organisms in an aquatic system is by about 7‰ "heavier" than the corresponding one produced by land plants [37]. Aquatic plant $\delta^{13}C$-values cover the same range as those of the terrestrial plants. In marine environments, there is a trophic level-depending decrease in $\delta^{13}C$-values from algae to plankton [38]. Influences on the global $\delta^{13}C$-value of plant biomass are connected to light intensity and quality [39, 40], transpiration rate (temperature, relative humidity) [41], yields of crops [42], salinity [43], physiological state of the plant's organ [44, 45], the relative altitude of location [46, 47], the grade of air pollution [48-50] and to local isotope abundance variations of atmospheric $CO_2$, e.g. by industrial production of "fossil" $CO_2$ [51, 52] or by contributions of soil- [53] and plant-respired $CO_2$ [54] and finally to actual seasonal conditions [55].

Sources of oxygen in organic material produced by plants are $CO_2$, $H_2O$ and $O_2$ [56] [$\delta^{18}O$-values around +41‰ for air $CO_2$ above that of water after equilibration [57], (the $\delta^{18}O$-value of $CO_2$ entering the photosynthetic reduction is influenced by various factors), -15 to ±0‰ for $H_2O$ in non-extreme climates, and +23.8‰ for $O_2$ [58-60])]; depending on the reaction involved to the introduction of a given oxygen atom, wide ranges for $\delta^{18}O$-values in the organic material are possible (Table 6.4).

---

\* The $\delta^{18}O$- and $\delta^2H$-values reported are normalized on a scale that the $\delta^{18}O$- and $\delta^2H$-values of SLAP (Standard Light Antarctic Precipitation) vs. V-SMOW are -55.5‰ and -428‰. For the exact description of this procedure see [24].
CDT is not isotopically homogenous, and has also been exhausted for a long time [28]. Therefore it is proposed to adopt for the IAEA-S-1 sulphur standard a $\delta^{34}S$-value of -0.30‰ vs. a defined, hypothetical V-CDT (Vienna Cañon Diablo Troilite) standard [28]. Laboratories should specify that the reference V-CDT is used, indicating a measuring calibration through IAEA-S-1 (IAEA: International Atomic Energy Agency, Austria). Positional $^2H$ (deuterium)-concentrations are often given in ppm values (see 6.2.2.3.2).

***Table 6.4:*** *Range of δ-values of natural products from land plants. Values in [] for $C_4$-plants. $δ^2H$-values of carbohydrates from CAM-plants can be up to +50‰. Most $δ^{18}O$-values are correlated to those of the water present at the (bio)synthesis of the compounds. The $δ^{13}C$- and $δ^2H$-values of isoprenoids depend on the biosynthetic pathway; products from the mevalonate pathway are generally more depleted in $^{13}C$ and less depleted in $^2H$ as compared to those from the deoxyxylulose 5-phosphate pathway. Monoterpenes from higher plants and bacteria of interest in the present context are normally synthesised via the deoxyxylulose 5-phosphate pathway, sesquiterpenes can originate from the mevalonate pathway*

| Compound group | $δ^2H[‰]_{V-SMOW}$ | $δ^{13}C[‰]_{V-PDB}$ | $δ^{15}N[‰]_{AIR-N2}$ | $δ^{18}O[‰]_{V-SMOW}$ |
|---|---|---|---|---|
| Carbohydrates and direct descendants | -170...-30 [-100...±0] | -30...-21 [-15...-9] | - | +20...+40 |
| Fatty acids and descendants, isoprenoids | -400...-150 | -36...-28 [-16...-12] | - | +5...+10 |
| Amino acids and descendants | -200...-100 (estimated) | -29...-25 [-14...-11] | -10...+10 | -5...+10 |
| Phenylpropanoids and other aromatic compounds | -150...-70 | -32...-26 [-21...-14] | - | ±0...+5 |

The situation is quite complicated with water as the primary source of hydrogen (and partially oxygen) in organic material. Its evaporation from the sea is accompanied by isotope discriminations, leading to a relative enrichment of light isotopes ($^1H$, $^{16}O$) in the vapour phase, and later on to a relative enrichment of the corresponding heavy isotopes ($^2H$, $^{17}O$, $^{18}O$) in the condensates originating from this vapour *[61, 62]*. This has for consequence an isotope abundance gradient of meteoric water depending on the distance of a given area from the ocean (continental effect) and its local conditions (temperature of the site, latitude, altitude, amount of precipitation) *[61-66]*. The $δ^2H$- and $δ^{18}O$-values of the precipitations are correlated to each other ("meteoric water line") *[67]*. Moreover, the local climate has an influence on the transpiration of terrestrial plants. While there is a negligible isotope fractionation during uptake of groundwater by roots or during transport up to shoots *[68, 69]*, the preferred evaporation of "light" water molecules through leaf stomata results in an enrichment of the "heavy" isotopes in the remaining leaf tissue water *[70-72]*. Again a relationship between $δ^2H$- and $δ^{18}O$-values has been found for leaf water *[73]*, and therefore, the $^2H$- and $^{18}O$-content of this matter can be indicative for latitude and climatic conditions of its origin *[74-76]*. This water is also the source of hydrogen (and partially oxygen) in the organic matter produced by the plant during photosynthesis *[77, 78]*. Shifts of δ-values occur due to isotope effects implied in reactions of the biosynthesis, but secondary exchanges of oxygen and hydrogen atoms bound to certain functional groups with that of the surrounding water may also occur *[79-82]*. Therefore only hydrogen atoms bound to carbon are indicative for the original correlation between leaf water and organic matter *[74, 83, 84]*. For oxygen as indicator of the origin of a substance, hydroxyl- and, with some restrictions, carbonyl-oxygen atoms can be used.

$δ^{15}N$- and $δ^{34}S$-value measurements are important parts of the multielemental average isotope analysis for origin and authenticity checks of foodstuffs *[85-88]*; in context with flavour isotopic analysis they are an exception because there are only very few nitrogen- or sulphur-containing compounds of relevance. Systematics of the relative $^{15}$N-characteristics of nitrogen-containing substances have recently been compiled *[89]*, and $^{15}$N-patterns of some amino acids with more than one nitrogen atom have been described *[90]*.

Apart from the isotope abundance of the primary material, kinetic isotope effects of many reactions in the biosynthesis of organic compounds contribute to their isotope abundances and patterns. A kinetic isotope effect is the ratio of the rate constants for a given reaction for the "light" and the "heavy" isotopologue molecules. As its value depends on the relative mass of the atoms in question, it is generally very large for hydrogen, ($k_{1H}/k_{2H}$ can attain values up to 20), while it is for carbon, nitrogen and oxygen between 1.0 and 1.1 *[91]*. Kinetic isotope effects preferably accompany a formation or fission of a given bond; however, in the biosynthesis of secondary plant material they will only become manifest in case of a partial turnover in connection with pool sizes and relative turnover rates after metabolic branching *[33]*. Kinetic isotope effects will generally result in an enrichment of the "heavy" isotope in the remaining substrate and a depletion in the product, because the "lighter" isotopologues react faster as compared to the "heavy" ones. For a more detailed introduction to the theoretical background of isotope effects see references *[92-94]*.

The most well-known and important carbon isotope discriminations of practical significance are those accompanying the primary $CO_2$-fixation by the photosynthesis in terrestrial plants. The material from $C_3$-plants (most trees, bushes, herbs and grasses; the name is derived from the number of carbon atoms of the first photosynthetic product, phosphoglyceric acid) is, due to the larger isotope effect on the ribulosebisphosphate-carboxylase reaction *[95, 96]*, relatively "light" ($\delta^{13}$C-values from -30 to -24‰) *[97, 98]*, whereas that from $C_4$-plants (sugar cane, corn, sorghum and millet; the name comes from the four C-atoms containing primary product of the phosphoenolpyruvate-carboxylase reaction, oxaloacetate) is relatively "heavier" (smaller isotope effect *[99, 100]*; $\delta^{13}$C-values between -16 and -10‰) *[101, 102]*; products from CAM-plants (crassulacean acid metabolism, occurring in succulents, orchids and some tropical grasses; both $CO_2$-fixing reactions) lie in between these areas *[103-105]* (Table 6.4). Kinetic carbon isotope effects on further reactions in the biosynthesis have for consequence additional but small $\delta^{13}$C-value shifts *[54, 106]*, because the reactions proceed in a *quasi*-closed system. However, the corresponding $\delta^{13}$C-values are typical for given groups of compounds from the same origin, and indicative for their metabolic correlation in between different groups (see intermolecular standardisation, 6.2.2.3.1. and 6.2.2.4.4.). Correspondingly, even the relative isotope abundances in different positions (isotopic pattern) of a molecule are typical *[33, 107-109]*.

Analogously, the primary reduction processes in the photosynthesis *[110]* and any further hydrogen transfer in the course of the synthesis of a given compound are accompanied by hydrogen isotope effects and, in addition to the content of the primary source, the leaf water, these isotope effects contribute to the mean $\delta^2$H-value and $^2$H-pattern of organic compounds *[111]*. In general, most organic substances are depleted in $^2$H as compared to the plant tissue water, from which the hydrogen atoms originate. According to our present knowledge these $^2$H-depletions accompany the reduction of the primary photosynthetic products, the formation of aromatic compounds, the synthesis of acetyl-CoA and isopentenyl-pyrophosphate, as well as further reactions in the synthesis of isoprenoids. As a consequence of different metabolic directions, four main classes of compounds with typical different relative deuterium contents result: the carbohydrates, representing the highest deuterium content *[112]*, followed by the phenylpropane derivatives (aromatic substances) *[113]*, then ordinary lipids (aliphatic substances *[114, 115]*) and finally, with the lowest deuterium content, the isoprenoids *[116, 117]*. Products of CAM-plants have to be treated as an exceptional case of hydrogen isotope content. They show a remarkable relative enrichment of deuterium in carbohydrates *[115]* and also in some substances of their secondary metabolism (aromatic substances *[118]*), but not in their lipids *[115]*.

Most natural compounds besides carbohydrates contain only one or a few oxygen atoms. As already mentioned, these can originate from three primary sources. Oxygen atoms bound in carboxyl and carbonyl groups can equilibrate with water. Therefore their $\delta^{18}$O-value can be indicative for the geographical origin of natural compounds *[56, 82, 119, 120]*. Some OH-groups, mainly of phenols and terpenes, have $O_2$ for oxygen source, introduced by a monooxygenase reaction implying a kinetic isotope effect. Their $\delta^{18}$O-values can be indicative for the distinction between natural and adulterated compounds *[56, 121]*. Quite different is the situation with nitrogen: most dominant is the $\delta^{15}$N-value of the primary source, and isotope effects have a secondary importance for isotope signatures.

It is evident that all reflections concerning the influence of the primary source of an element, and isotope effects in the course of a synthesis of a molecule, are analogously true for synthetic compounds. Hence the determination of average $\delta$-values and of isotopic patterns will often already permit the assignment of a substance to its origin*.

Recently, enormous progress has been made in understanding the systematics of in vivo isotope discrimination and in understanding and even predicting isotopic patterns of natural compounds *[33, 56, 89, 111]*. As has been demonstrated, these fingerprints of origin and biosynthesis of natural compounds are created by the overlap of influences from equilibrium isotope effects under conditions of metabolic steady state, from kinetic isotope effects in connection with branching of synthesis chains and from mechanisms of enzyme catalysed reactions.

On this basis it is possible to supplement the average $\delta$-value ranges of natural compounds given in Table 6.4 by some of their isotopic pattern characteristics in order to differentiate them from synthetic or biotechnological analogues. Due to the complex biosynthetic correlations in the formation and the extremely complicated

---

* In the present context, according to EU (European Union) food laws, flavourings are called "natural", when they have been isolated from a biological source by a physical process or have been synthesized from natural precursors by microorganisms or enzyme catalysed reactions (biotechnological procedure), they are called "nature-identical" when they have been (partially) synthezised from plant or fossil carbon material by chemical methods, but are, as to their structure, identical to the corresponding natural compounds, they are "synthetic" or "artificial" when a natural counterpart does not exist. In the USA and in some countries of Asia the situation is quite different. For the exact definition see corresponding food laws.

analytical demands of the analysis of $^{13}$C-patterns preferably $^2$H- and $^{18}$O-patterns are of practical importance. While the analysis of the latter ones demands sometimes degradations or indirect analyses by calculations from δ-value differences, the former are routinely determined by NMR measurements. The following general rules may be helpful:

**1. $^2$H-patterns:** Natural *aromatic compounds* descending from the shikimic acid pathway show relative positional $^2$H-abundances in the sequence p > o ≥ m; when the p-position is hydroxylated, the sequence is inverted into m > o.

*Isoprenoids* from the mevalonate pathway show in general a $^2$H-depletion of the methylene groups relative to the methyl and vinyl positions, while those from the deoxyxylulose-phosphate pathway (in the present context preferably monoterpenes) are characterised by a distinct relative $^2$H-depletion of the vinyl groups. Chains of *fatty acids* and or descendants of acetogenins show an alternating $^2$H-abundance with relative $^2$H-depletions in the odd (C=O-deriving) positions.

**2. $^{18}$O-patterns:** In natural compounds the $δ^{18}$O-values of *hydroxyl groups originating from monooxygenase reactions* (most phenolic OH-groups, some OH-groups in isoprenoids) range between +3 and +7‰ vs. V-SMOW. Oxygen functions introduced from water by a lyase reaction must be $^{18}$O-depleted relative to this water, those originating from carbonyl or carboxyl groups are, after isotopical equilibration with water, relatively $^{18}$O-enriched by ~ +28 and ~ +19‰, respectively, above this water. Natural but also certain nature-identical esters are characterised by an extreme $^{18}$O-enrichment (up to +40 to +60‰) of the carbonyl-O and a modest $^{18}$O-content of the ether-O (< +10‰).

### 6.2.2.2. Methods for the Determination of Average Isotope Abundances in Organic Compounds

The most common isotope analysis in practice of flavouring authenticity assignment is a average δ-value determination. This implies the combustion (C) or pyrolysis (P, in the present context mostly performed in the presence of carbon) of the (defined) organic samples in order to convert them into gas molecules like $CO_2$, CO and $H_2$ for the isotope ratio analysis of carbon, oxygen and hydrogen, respectively. As these gases are formed from any organic matter, a characteristic value for a given compound will only be obtained when it is rigorously purified and separated from any foreign source of the element prior to combustion. Classic methods for this purpose (e.g. distillation, crystallisation) demand relatively large amounts of starting material (gram amounts). Chromatographic methods can be applied to much smaller samples but gas chromatography is limited to volatile compounds (yet in the case of flavouring compounds the most preferred method), while HPLC implies the separation of the fractions from solvent and buffer. Most common and advantageous in the present context is the on-line coupling of gas chromatography with combustion or reductive pyrolysis (pyrolysis in the presence of carbon) to isotope ratio mass spectrometry (GC-IRMS); a special chapter is dedicated to this method in this contribution (6.2.2.2.2.). Very recently a system for the on-line determination of the $δ^{13}$C-values of water-soluble analytes after HPLC separation has been proposed *[122]*; it applies inorganic (carbon-free) buffers and wet oxidation. Any of these multiple separation processes can imply small but dangerous isotope effects leading to unpredictable separations within chromatographic peaks *[123-126]*. It is therefore indispensable to test the method with standard substances, and especially to collect the complete peak of the eluted compound from the chromatographic separation preceding the elemental analytic procedure *[127, 128]*.

Generally, it will be advantageous and even indispensable in flavour isotope analysis to obtain as much information as possible from each sample, therefore to aim at a multi-element isotope ratio analysis. The combinations of carbon, oxygen and hydrogen isotope data would not only provide information on the botanical and genetic origin of the compound, but also on climatologic and biochemical conditions of its synthesis, in particular, as interpretations and even predictions about average and positional $\delta^{18}$O- and $\delta^{2}$H-values of natural compounds on the basis of their biogenesis are increasingly available *[56, 111]* (see 6.2.2.1); this will certainly facilitate the discrimination between natural and synthetic products. Furthermore, as natural flavours and fragrances are mostly composed of several compounds with biogenetic and from here isotopic correlations, also a multi-component isotopic analysis would be advantageous. All these opportunities can only be realised by the combination of a chromatographic separation process with a suitable elemental analytical conversion and IRMS (isotope ratio mass spectrometry).

### 6.2.2.2.1 Classic Elemental Analysis Procedures and Isotope Ratio Mass Spectrometry

The precise determination of the tiny differences in isotope abundances of organic material can only be performed by means of a highly developed quantitative mass spectrometric method, isotope ratio mass spectrometry (IRMS), which is only applicable to pure simple gases. These gases have to be produced from the organic material by methods of elemental analysis. The principles of the procedures in use are summarised in Fig. 6.12. Details are subject to the procedures recommended by the corresponding equipment producers (e.g. Elementar Analysensysteme GmbH, Hanau, Germany, and CE Instruments, former Carlo Erba, Rodano (Milano), Italy). The gases obtained are separated by GC, cryogenic methods or reversible adsorption and then transferred (either on-line or in individual vessels) to the mass spectrometer.

The common instruments use electron-impact ionisation, an ion separation by a (permanent or an electro-) magnet, high voltage ion acceleration and a double or triple Faraday cup set for ion current measurement. The ion current for the compound with the lower abundance isotopologue is amplified 100 to 1000 times in correspondence to that for the main isotopologue. For more details about the technical side of mass spectrometers and the measuring procedures, see e.g. references *[134, 135]*.

The molecules analysed in stable IRMS measurements are hydrogen gas ($^{1}$H$^{1}$H and $^{2}$H$^{1}$H, masses 2 and 3) for the hydrogen isotopes, CO$_2$ ($^{12}$C$^{16}$O$_2$, $^{13}$C$^{16}$O$_2$, $^{12}$C$^{16}$O$^{18}$O, masses 44, 45, 46) for carbon isotopes, CO ($^{12}$C$^{16}$O and $^{12}$C$^{18}$O, masses 28 and 30) for oxygen isotopes, N$_2$ ($^{14}$N$_2$ and $^{14}$N$^{15}$N, masses 28 and 29) for the nitrogen isotopes and SO$_2$ ($^{32}$S$^{16}$O$^{16}$O and $^{34}$S$^{16}$O$^{16}$O, masses 64 and 66) or SF$_6$ *[136]* for sulphur isotopes. The measurement of $\delta^{13}$C-values on m/z 45/44 of CO$_2$ needs a correction of the contribution of $^{17}$O in $^{12}$C$^{16}$O$^{17}$O to mass 45. The correction is done on the basis of the constant ratio between $^{18}O$ and $^{17}O$ in terrestrial systems [16, 137]. The electron impact ionisation spectrum of $SO_2$ gas also shows interfering ion currents on mass 66 ($^{18}O$-correction for $\delta^{34}S$-determination; see e.g. [138, 139]). Very important is the $H_3^+$-correction [140, 141] for the measurement of $\delta^2H$-values; the determination of the corresponding correction factor in on-line system is described in [142, 143].

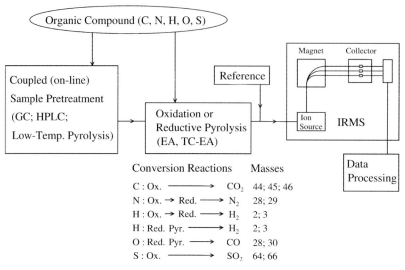

*Fig. 6.12: Sample pretreatment and conversion of organic compounds into gases for isotope ratio mass spectrometry (IRMS). The reductive pyrolysis is normally performed on glassy carbon (for a summary see e.g. [129]). In case of coupled HPLC-IRMS for $\delta^{13}C$ determination wet oxidation is used [122]. In the routine isotope ratio analysis of water often isotope equilibration with gases are used: for $\delta^{18}O$-analysis $CO_2$ [130], for $\delta^2H$-analysis $H_2$-gas in the presence of Pt [131]. Recently an on-line method for $\delta^{18}O$ and $\delta^2H$ in water has been described showing dual-inlet like performance [132]. Low-temperature pyrolysis in connection with GC is used for compound fragmentation with the aim of partial isotope pattern analysis [133]. TC-EA = thermo conversion elemental analyser*

The conversion of organic compounds into the measurement gases can be performed either by the classic off-line or increasingly by the more modern on-line procedures (see below). The classic methods often depend on vacuum systems and cryogenic gas separations and demand high amounts of isolated analytes and are time- and labour-intensive. They might still be necessary for special applications, where on-line methods are not yet adapted for routine purposes (e.g. isotope analyses of molecule parts after specific degradation reactions [144, 145]), but even here on-line degradation processes offer new possibilities [133, 146, 147]. Further information on classic $\delta^{13}C$-value determinations can be found in [148-150], for $\delta^{15}N$ in [151, 152], for $\delta^2H$ in [153-158], for $\delta^{18}O$ in [159-164] and for $\delta^{34}S$ in [165]. Schematic designs of classic devices for off-line $\delta^{13}C$- and $\delta^2H$-determinations can be found in [166] and literature cited therein. For the classic off-line batch methods the mass spectrometer has to be equipped with a dual-inlet system for an automated alternating inlet of sample and a corresponding reference gas [167]. The internal reproducibility of measurements by IRMS instruments with double inlet systems in δ-units is 0.01-0.02‰ for $^{13}C$, $^{15}N$ and

$^{18}$O, for $^2$H it is about 2‰ *[134]* using 1 ml gas. Today, modern equipment consists of couplings of elemental analyser (for bulk samples *[168-172]*) or GC *[167, 173, 174]* and HPLC devices *[122]* via conversion units (for compound-specific measurements) to an IRMS. Recently also couplings for measurement of $\delta^{18}$O- and $\delta^2$H-values were available. For more information see e.g. *[167, 175]* and chapter 6.2.2.2.2.

The reference principle for on-line methods has to be different as compared to classic dual-inlet isotope ratio analysis, as it is not possible to introduce sample and reference gas in exactly the same manner into the mass spectrometer. Reference gas pulses have to be set at different points during one run, but also laboratory standards with known δ-values have to be analysed periodically and exactly in the same manner and under the same conditions as the samples under investigation (identical treatment principle *[13]*). A general problem of on-line HPLC and GC methods in isotope ratio analysis is the lack of international reference materials suitable to fulfil the above mentioned requirements. This implies the necessity to establish suitable laboratory standards by EA (elemental analysis) or classic measurements. Meanwhile, also efforts in supplying suitable international organic standards have been overtaken by the IAEA (e.g. caffeine, glutamic acid *[176, 177]*). This programme (benzoic acid reference materials with different $\delta^{18}$O-values) will also support the on-line EA and GC $\delta^{18}$O-measurement by reductive pyrolysis (carbon reduction) methods (standardisation problems are compiled in *[178]*).

### 6.2.2.2.2 On-line Coupling of Gas Chromatography to Stable Isotope Ratio Mass Spectrometry

The isotopic characterisation of organic compounds by classical elemental analysis with isotope ratio mass spectrometry (EA-IRMS) demands milligram amounts of pure compounds, which can not always easily be provided. Most analytes in flavour characterisation are volatile, and therefore, after their extraction, coupling of (capillary) gas chromatography (cGC) with IRMS would be an ideal tool for the isotope ratio analysis of the individual substances. For $^{13}$C- *[127, 128]* and $^{15}$N-analysis *[179, 180]* this has been realised for a long time by combining GC to IRMS via a combustion (C) unit (GC-C-IRMS), even for polar substances after their derivatisation *[181, 182]*.

Several procedures and systems for the on-line determination of the $\delta^2$H- and $\delta^{18}$O-values of water and organic compounds have been investigated in the last few years *[183-194]* and problems arising from isotope effects on the GC-separations of analytes and their pyrolysis (P) products have been discussed *[195-202]*. Today most of the complications have been overcome and the technical prerequisites for the on-line GC-C-IRMS analysis of carbon and the GC-P-IRMS analysis of hydrogen and oxygen isotopes in volatile organic compounds have been optimised and become available for routine application *[203-205]*.

The actual elemental amounts needed for the on-line isotope analysis are, depending from the instrumental equipment, in the range of 1 nmol C, 15 nmol H and 5 nmol O; the measurement precisions are 0.2, 3.0 and 0.8 [‰]$_{Standard}$, respectively *[206]*. A corresponding experimental device is described in Fig. 6.13.

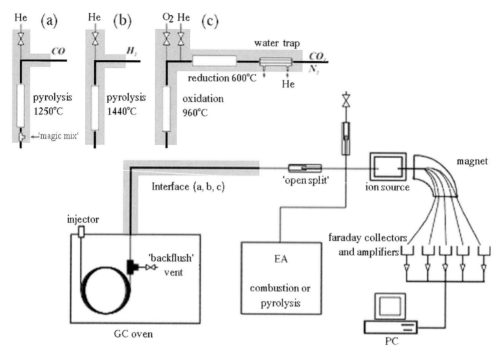

*Fig. 6.13:* Schematic diagram of a multifunctional GC-IRMS system. The device consists of a gas chromatograph coupled to an isotope ratio mass spectrometer via a pyrolysis or a combustion interface a, b and c for oxygen, hydrogen and carbon isotope analysis, respectively, and is additionally equipped with an elemental analyser (EA). After [205] with kind permission of the American Chemical Society

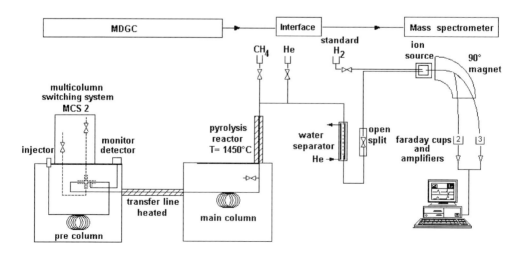

*Fig. 6.14:* Schematic diagram of a multi-dimensional gas chromatographic-pyrolysis-isotope ratio mass spectrometry (MDGC-P-IRMS) device. After [210] with kind permission of the American Chemical Society. For application see Fig. 6.20

Instead by solvent extraction *[207]*, aroma compounds from aqueous media, e.g. fruit juices, can even be separated and enriched by techniques of solid phase micro extraction (SPME), preferably from the headspace *[208]*; corresponding devices can often be directly connected to GC systems. These techniques provide the complete spectrum of the individual compounds of an aroma. As it will normally not be possible and even not necessary to analyse all components of the complex mixture, the separation of its main compounds may demand a multi-dimensional (MD) gas chromatographic system *[209]* as displayed in Fig. 6.14 *[210]*. Examples for the multi-element/multi-compound isotope analysis by such systems will be given later (6.2.2.4.4, *[211]*); they can even integrate the identification of the compounds by molecular mass spectrometry and a simultaneous determination of the enantiomer ratios of isomers *[210, 211]*. The importance of enantiomer analysis as a tool for authenticity assessment is extensively treated in chapter 6.2.3.

### 6.2.2.3 Intermolecular, Site-specific and Positional Isotope Ratio Analysis

As outlined before, natural aromas and fragrances are complex mixtures. The relative concentrations of their individual components are not only characteristic for origin and actual conditions of plant growth and processing, but also of great importance for quality and price of the products. The preferred methods for their corresponding analytical characterisation are chromatographic processes (see above) but, in some cases, chemical or biochemical fractionations of individual components in connection with the isotopic analysis of their molecular moieties can provide additional valuable information about their origin and synthesis. For example, natural esters have isotopic correlations between their molecular parts, while nature-identical materials may not have. It has recently been shown that, as expected, hydrolysis does not interfere with the carbon and oxygen isotope abundance of the alcoholic moiety of esters, and shifts only modestly its $\delta^2H$-value *[212]*. Hence, a hydrolysis of such molecules and an IRMS of their parts will give information on their origin. Finally the measurement of isotopic patterns of the molecules themselves will provide a fingerprint of their biological or chemical synthesis, respectively *[109, 213]*. Various methods for the determination of inter- and intramolecular isotope ratio analysis are available and of great significance in aroma authenticity analysis. Recently the deuterium-pattern analysis by $^2$H-NMR (see 6.2.2.3.2.) has made significant progress and has been developed to the most powerful tool of stable isotope analysis, e.g. of aromatic compounds and terpenes, for origin assignments.

### 6.2.2.3.1 Mass Spectrometric Determination of Intermolecular and Intramolecular Isotopic Correlations

The "intrinsic" or intermolecular standardisation (evaluation of isotopic correlations in between molecules of the same origin) demands the separation, purification and isotopic analysis of at least two components of the same source, which is, in practice of routine analysis of flavourings, normally only realisable by GC-C-IRMS or by multi-dimensional (MD)GC-C-IRMS. Instrumentation and practical performance have already been outlined in 6.2.2.2.2.

Very informative examples for this principle are GC-C-IRMS analyses of monoterpenes from essential oils. Assuming that all terpenes from a given natural source are originating from the same precursor, their relative amounts, derived from the peak heights (integrals) of a GC detector, and their $\delta^{13}C$-values (from GC-C-IRMS) must correlate within a balance between the different branches of their metabolic sequences. A deviation from this balance will help to detect the addition of a compound from a foreign origin. The practical performance is outlined in 6.2.2.2.2; examples are found in 6.2.2.4.4.

Esters are very common components of fruit aromas, and glycosides may occur as precursors of spices like mustard oils. The hydrolysis products *[214]* of esters can be extracted and analysed by GC-C-IRMS; aglycons obtained by enzymatic hydrolysis of the glycosides *[108]* can be analysed in the same way. Other molecules can be split into defined fragments by suitable methods, and some can even be degraded by pyrolysis *[133]*. This is possible for acetic acid in the form of sodium acetate in the presence of sodium hydroxide at 600°C; ethanol is oxidised to acetic acid prior to this degradation. The reaction product methane is analysed by GC-C-IRMS *[184]* in modification after *[106]*. A partial degradation of unsaturated compounds can be performed by ozonolysis *[215, 216]*. A more sophisticated and detailed combined enzymatic and chemical degradation is restricted to primary compounds like carbohydrates *[107]*, organic acids *[106, 217-219]*, amino acids *[220]* and their descendants; their pattern analysis is, however, more of importance in the quality control of wine and juices *[123, 221]*.

### 6.2.2.3.2 Stable Isotope Pattern Analysis by Nuclear Magnetic Resonance Measurements

Nuclear magnetic resonance (NMR) is a property of "non-symmetric" atom nuclei with odd numbers of protons and/or neutrons, hence a resulting nuclear spin and magnetic moment (see Table 6.3) capable of orientating in a few defined directions relative to a strong external magnetic field under resonance conditions. These are modulated under the influence of an additional alternating magnetic field as produced by electromagnetic radiation in the radio frequency range (Fig. 6.15), provided the energy transferred corresponds to the quantum difference of two orientation states. For a proton in a field of 9.4 T, the resonance frequency is of the order of 400 MHz *[11]*. If this proton is part of an organic molecule, which is normally the case, the magnetic field in its surroundings is additionally modulated by the overlapping magnetic fields of adjacent nuclei, the field of which can even overlap the external fields; also coupling of several neighboured elemental magnets can occur. This results in a slight shift of the resonance absorption (chemical shift) and the splitting of the signals (coupling constants). In reverse, the exact position of a resonance signal provides information about the surroundings of a given magnetic nucleus within a molecule, e.g. the kind, number and relative orientation of its neighbouring nuclei. As the induced shifts are very small in comparison to the absolute resonance frequency, the chemical shift is expressed in ppm relative to a signal of an intrinsic standard such as tetramethylurea (TMU) generally used also as quantitative reference in $^2_1H$- and $^{13}_6C$-NMR.

# Stable Isotope Ratio Analysis in Quality Control of Flavourings

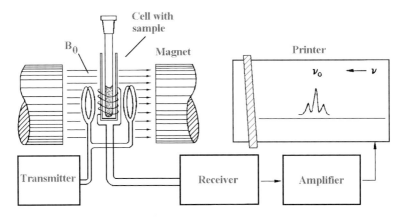

*Fig. 6.15:* Scheme of an NMR device. The field strength influences resolution and sensitivity of the measurement. $B_0$ = permanent magnetic field. Transmitter and receiver for superimposed radio frequency field. Adapted with kind permission from [222]

Due to the low abundance of the rare isotopes of the bioelements in natural products it is, for statistical reasons, improbable that an individual molecule out of a given population contains (if at all) more than one deuterium ($^2_1H$) nucleus or two adjacent $^{13}_6C$-nuclei. Therefore coupling between two equivalent nuclei is practically not occurring in the present context, and the corresponding $^2H$- or $^{13}C$-NMR spectra become very simple, yielding just one resonance signal for each individual type of nucleus with a relative signal intensity proportional to the number of identical nuclei. These signals reflect simultaneously the relative number of independent isotopomers within the total molecule population (Fig. 6.16).

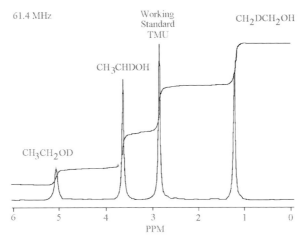

*Fig. 6.16:* Typical $^2H$-NMR spectrum (61.4 MHz) of ethanol with TMU (tetramethylurea) as working standard. Adapted with kind permission from [223]. Signals of the isotopomers

As a prerequisite for a quantitative site-specific NMR analysis, the assignment of the signals to a defined molecular position or the label position of the isotopomer must be known. In a deuterium spectrum this will normally be the case, because it is qualitatively identical to the corresponding $^1$H-NMR spectrum. The resulting isotope ratio D/H is expressed in ppm relative to the standard (not to be mistaken for the ppm for the chemical shift!). The correlation of this notation to the $\delta^2$H-scale is displayed in Fig. 6.17 [223-225]. A quantitative determination of D/H ratios is only possible through IRMS calibration, and therefore normally the signal area of an intrinsic "working standard", tetramethylurea (TMU), is measured as a reference together with the signals of the analyte (Fig. 6.16) [223].

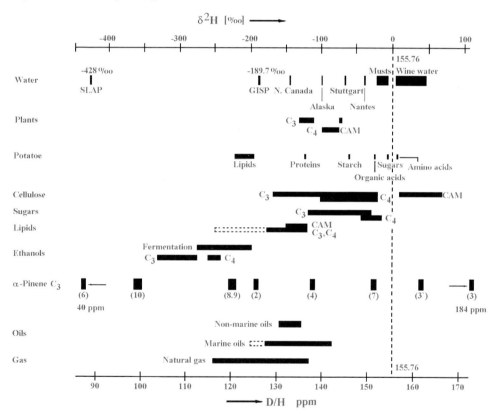

*Fig. 6.17:* Natural abundance areas of deuterium in various kinds of products and correlation of the ppm- and δ-scale (equ. 6, p. 618) for deuterium (D = $^2$H). Adapted with kind permission from [223]. Numbers under the signals for α-pinene refer to positions in the molecule

An NMR spectrometer for SNIF-NMR® (Fig. 6.17) (site-specific natural isotope fractionation NMR) measurements [226] must be specifically equipped and adapted, e.g. for deuterium analysis by a high field magnet (e.g. 9.4 T, corresponding to 400 MHz ($_1^1$H) and 61.4 MHz ($_1^2$H) resonance frequency, or 11.4 T, corresponding to 500 MHz ($_1^1$H) and 76.8 MHz ($_1^2$H) resonance frequency), a specifically adapted $_1^2$H-NMR probe with fluorine lock and proton decoupling, highly stable electronics and software for spectra acquisition and data processing/treatment. Instrumental details and sophistication have to be adapted to the tasks in question and are subject of information by the corresponding producers.

The signals of a deuterium NMR spectrum are relatively broad, but for their quantitative evaluation it is of advantage that the quadrupole relaxation mechanism of the excited state, and the lack of an Overhauser effect permit relatively fast signal accumulations. The required signal-to-noise ratio is obtained in a 9.4 T magnet with a sample of 1 g after 700 scans, equivalent to 5 hours of measurement time [227, 228]; smaller samples demand more powerful (and much more expensive) instrumentation or longer measuring times; however, in practical terms of flavouring analysis, 200 mg of substance and several thousand scans (corresponding to several hours of measurement) are the lower limit at present. However, very recently a principle for the reduction of the measuring time, and the replacement of an intrinsic standard by an electronic calibration signal, called the ERETIC (electronic reference to access *in vivo* concentration) method, has been proposed [229].

In many cases it is sufficient to establish the relative quantitative D/H ratios between the different positions of a molecule, the reference being the mean global ratio ($\overline{D/H}$) of the compound measured by IRMS (provided the applied preparation methods do not alter the $^2H/^1H$ ratio of the exchangeable, O- or N-bound hydrogen atoms). In this case, the site-specific ratio $(D/H)_i$ or the relative molar fraction $f_i$ of the corresponding isotopomer results from [230]:

$$(D/H_i) = f_i \times (\overline{D/H}) \qquad \text{or} \qquad (D/H)_i = F_i \times (\overline{D/H}) \qquad (2)$$

The experimental molar fraction $f_i$ is directly accessible from the corresponding signal area $S_i$ as integrated by the instrument, $F_i$ is the statistical molar fraction, as calculated from $P_i$, the number of equivalent deuterium positions of type i [231].

$$f_i = S_i / \Sigma_n S_i \qquad F_i = P_i / \Sigma_n P_i \qquad (3)$$

In reverse ($\overline{D/H}$) can be calculated from the sum of the positional ratios:

$$(\overline{D/H}) = \Sigma_n F_i (D/H)_i \qquad (4)$$

In case of a comparison or relation between several samples, as for the check of an adulteration of alcoholic beverages [223, 232], the use of an intrinsic working standard WS, mixed in an exactly known (isotopic) ratio with the analyte A, is generally necessary. If M are the molar masses, m the actual masses of the two components, P the number of equivalent deuterium atoms, $t_A$ the purity of A (in %) and $T_{Ai}$ the ratio of the signal area for the site i of A and the working standard WS, $(D/H)_i$ is calculated according to:

$$(D/H)_i = \frac{P_{WS}}{P_{Ai}} \times \frac{M_A}{M_{WS}} \times \frac{m_{WS}}{m_A} \times \frac{T_{Ai}}{t_A} \times \overline{(D/H)}_{WS} \qquad (5)$$

As the signals of different compounds are independent, theoretically even mixtures could be investigated. The masses $m_{WS}$ and $m_A$ have to be weighed exactly for the sample preparation, also $t_A$ has to be determined very accurately. Recently, a publication [233] showed the application of quantitative $^1H$-NMR for the on-line determination of the required mass ratio ($m_{WS}/m_A \times 1/t_A$), which facilitates the sample preparation for the site-specific $^2H$-NMR. However, the limitations would be imposed by the lower total amount of each component and the overlap of signals. In some cases even the signals of the working standard may interfere with the signals of the analyte. The normally used working standard is tetramethylurea (TMU; Tetramethylurea (STA 003)), available from the Institute of Reference Materials and Measurements (IRMM) in Geel, Belgium. The $(D/H)_i$-values can be converted to positional $\delta^2 H_i$-values on the $[‰]_{V-SMOW}$ scale with

$$\delta^2 H_i \ [‰]_{V-SMOW} = 1000 \times [(D/H)_i - 155.76] / 155.76 \qquad (6)$$

For the evaluation and interpretation of the large amount of individual data (and taking into account their standard deviations) statistical multivariant methods can be used [230]. By these methods $(D/H)_i$-ratios of several positions can be determined and connected to other properties or information on the analyte ($\delta^{13}C$-value, enantiomer purity) and lead to clusters of data from which interpretations on specific factors like the production area of a flavouring may be drawn. By multivariante and discriminant analysis it is possible today not only to discriminate between natural and nature-identical compounds but also to assign natural compounds to different geographical and botanical origins. Very recently it has even become possible to determine the (D/H)-ratios of prochiral hydrogen positions of the same methylene group of a fatty acid chain by embedding the sample into a polypeptidic chiral liquid crystalline solvent [234].

In spite of the fact that patterns can increasingly be understood and interpreted, and that the relative sensitivity of NMR for $^{13}C$ is theoretically higher by two orders of magnitude as compared to deuterium (Table 6.3), so far quantitative evaluations for this nucleus are very rare. While the principle feasibility of the site-specific quantitative $^{13}C$-NMR has been shown and patented [235], and a non-statistical $^{13}C$-distribution has already been measured with artificially enriched material [236], the application to material with natural $^{13}C$ abundance provides serious problems. Due to the fact that carbon isotope effects are much smaller than hydrogen isotope effects, the relative differences between given positions to be expected are lower. Furthermore, in contrast to $^2H$-NMR, $^{13}C$-NMR is implied with long relaxation times and large nuclear Overhauser effects, which results in long measuring times. A final problem is the difficulty of an absolute identical excitation over the whole field which may demand individual calibrations of several parts of the spectrum. Nevertheless, first characteristic $^{13}C$-distributions in natural and synthetic acetic acid and in vanillin of different origins have been obtained [237]. The work on vanillin has very recently been taken up and successfully promoted to practical applicability [238, 239].

Summarising the advantages and problems of NMR as a tool for authenticity control in the flavouring and aroma field one can state:

- NMR is the only method which yields complete isotopic patterns, an absolutely non-adulterable characteristic of a natural product. Due to the fact that these patterns can more and more be interpreted by biochemical correlations [33], they are actually the most potent base for the assignment of a compound to its climatic origin and authenticity.

- The instrumentation is very expensive (two or three times as much in comparison to IRMS), the amounts of analyte (0.2 to 1 g pure compound) are rather high, and a standardisation by IRMS is needed. Promising recent developments contribute to reducing the measuring time.

- The isotope abundance of structurally identical positions (e.g. symmetrical molecule parts, prochiral centers) can routinely so far not be differentiated but work to overcome this problem is in progress.

- The method is so far practically limited to deuterium; recent results on $^{13}C$ are promising.

### 6.2.2.4 Results of Stable Isotope Ratio Measurements in Authenticity Checks and in the Verification of the Natural Origin of Flavouring Compounds

We believe that it is, in respect to practical demands, most useful to report the results of isotope analyses according to classes of interesting compounds rather than to individual isotopes or methods. The corresponding order is given by nature and its main pathways for the biological formation of the materials. As already explained, isotopic fractionations in the course of the corresponding biochemical reaction sequences lead to characteristic isotope contents and patterns in natural and synthetic flavourings, respectively. The δ-values displayed in the first three chapters are compiled from different laboratories, demonstrating different botanical and climatic characteristics, but also individual variations in the measurements. In the last chapter results on mixtures and correlations of the same compounds are given.

#### 6.2.2.4.1 δ-Values of Aromatic Substances

Aromatic compounds in the chemical meaning of the word are generally synthesised in plants through the shikimic acid pathway from erythrose 4-phosphate and phosphoenolpyruvate. Shikimic acid by itself is the direct precursor of some phenols, benzoic acid derivatives and of quinones. By the integration of a second phosphoenolpyruvate unit, phenylpropanes from the type of cinnamic acid are produced. From results of isotope abundance measurements it is well established that the stepwise formation of the products of this pathway is accompanied by a corresponding depletion of $^{13}C$ relative to the primary plant products, the carbohydrates *[240-242]*. This depletion by about 3 to 6‰ is very well demonstrated through the difference of $δ^{13}C$-values of sugars and aglycons of phenolic glycosides (see intramolecular standardisation). Independently, methyl groups, often found in the form of ethers in aromates, are generally depleted by another few ‰ relative to the aromatic nuclei *[237, 243, 244]*.

The $δ^{13}C$-values found for natural aromatic substances from $C_3$-plants are usually within the range of -26 to -32‰, while the deuterium content of these products ($δ^2H$ -50 to -150‰) is relatively close to that of the carbohydrates of the same origin ($δ^2H$ -30 to -170‰), even though in special cases biochemical reduction steps in the course of the biosynthesis of these products may be accompanied by remarkable deuterium depletions *[245, 246]*. Secondary modifications (oxidation, methylation) usually cause only small additional fractionations of the hydrogen isotopes.

As already mentioned, the oxygen atoms in phenylpropane derivatives originate from different sources ($CO_2$, $O_2$, $H_2O$) with different $δ^{18}O$-values. Nevertheless, the average range found for natural aromates of $δ^{18}O$-values varies from +10 to +15‰ (Table 6.5).

**Table 6.5:** *Range of $\delta^2H$- and $\delta^{13}C$-values of aromatic compounds occuring in flavourings. Data include results from IRMS and from GC-C-IRMS measurements. Other $\delta$-values of some compounds are:*

a. $\delta^{15}N$-value [‰]$_{AIR-N2}$:
   Dimethylanthranilate: n. +1.8, n.i. -2.4 [247]
   Methyl N-methylanthranilate: n. +2.0 to +4.7, n.i. -5.0 [248]
   Methylanthranilate: n. +2.5, n.i. -2.7 [248]

b. $\delta^{18}O$-value [‰]$_{V-SMOW}$:
   Benzaldehyde: bitter almond +19.3, benzalchloride +6 to +7.5, toluene +14.1[249, 250]
   Estragole: n. +2.7 to +15.1, n.i. +13.3 to +24.8 [251]
   Methyleugenol: n. +2.7 to +8.9, n.i. +5.5 to +6.6 [251]
   Cinnamic aldehyde: cinnamon bark +14.9, n.i. +20.2 [249, 250]
   Vanillin: n. +11.8 and +12.5, n.i. +0.3 and +9.8 [252]; n. +8.3 to +10.6, ex lignin +6.5, ex guaiacol -2.8 [191]; ex beans average +11.6, -OH +5.8, -OCH$_3$ +2.7, >CO +26.3, n.i. average +11.1, -OH +9.1, -OCH$_3$ -0.9, >CO +25.1 [ 253]; ex beans average +13.1, mean –OH/-OCH$_3$ +9.5, n.i. average +14.2, mean –OH/-OCH$_3$ +10.2 [254]
   Raspberry ketone: n. average +13.3, n.i. average +29.9 and +9.1 [255]
   Coumarin: n. average +20.3 to +23.6, -O- –0.5 to + 2.3, n.i. average +12.6, -O- +5.6 [265]

| Substance | Status* | Origin/Source | $\delta^2H$ [‰]$_{V-SMOW}$ | $\delta^{13}C$ [‰]$_{V-PDB}$ | References |
|---|---|---|---|---|---|
| Vanillin | n. | vanilla beans | -54 to -96 | -19.8 to -20.8 | [256] |
| | n. | vanilla beans | n.d. | -16.8 to -20.7 | [257, 258] |
| | n. | vanilla beans | -52 to- 115 | -18.5 to -21.5 | [3] |
| | n. | *Vanilla planifolia* | n.d. | -18.7 to -20.3 | [259] |
| | n. | *Vanilla tahitensis* | n.d. | -16.8 | [259] |
| | n. | various sources | n.d. | -18.4 to -20.8 | [260] |
| | n. | ferulic acid | ??? | | |
| | n.i. | wood lignin | -170 to -190 | -26.9 to -28.7 | [256] |
| | n.i. | wood lignin | n.d. | -26.8 to -27.4 | [257, 258] |
| | n.i. | wood lignin | -170 to -204 | -26.5 to -27.5 | [3] |
| | n.i. | eugenol | n.d. | -29.9 to -31.7 | [257] |
| | n.i. | guajacol | n.d. | -29.0 to -36.2 | [257] |
| | n.i. | guajacol | -17 to -23 | -24.9 to -26.1 | [3] |
| | n.i. | | n.d. | -27.8 to -30.5 | [260] |
| Cinnamic aldehyde | n. | cinnamon bark | -85 to -138 | -26.5 to -28.8 | [242] |
| | n. | cinnamon bark | n.d. | -26.0 to -27.9 | [4] |
| | n. | various sources | -75 to -148 | -26.0 to -30.1 | [3, 261] |
| | n. | *Cassia vera* | -101 to -150 | -23.7 to -27.1 | [204] |
| | comm. | ? | +523 to -110 | -23.5 to -25.8 | [204] |
| | n.i. | | -94 to +238 | -28.0 to -30.7 | [242] |
| | n.i. | various syntheses | -11 to -43 | -28.8 to -29.2 | [3, 261] |
| | n.i. | toluene oxidation | +464 to +569 | -24.6 to -26.7 | [3, 261] |
| Benzaldehyde | n. | bitter almond apricot kernel | -100 to -146 | -28.8 to -30.4 | [113] |
| | n. | apricot kernel | -84 to -86 | -27.5 to -28.0 | [3, 261] |
| | n. | bitter almond | -82 to -152 | -27.1 to -30.0 | [3, 261] |
| | n. | bitter almond, fruit, fruit kernels | -113 to -148 -111 to -188 | n.d. | [262, 263 264] |
| | n.i. | benzalchloride | -11 to -68 | -26.4 to -30.4 | [113] |
| | n.i. | benzalchloride | -47 to -67 | -28.5 to -30.0 | [261] |
| | n.i. | toluene oxidation | +753 to +802 | -26.3 to -27.3 | [113] |
| | n.i. | toluene oxidation | +526 to +720 | -24.6 to -28.6 | [3, 261] |
| | n.i. | benzalchloride | -78 to -85 | n.d. | [262, 263, 264] |
| | n.i. | toluene | +420 to +668 | | |

* n.d. = not determined, n. = natural, n.i. = nature-identical, syn. = synthetic, comm. = commercial

| Substance | Status[*] | Origin/Source | $\delta^2H$ [‰]$_{V-SMOW}$ | $\delta^{13}C$ [‰]$_{V-PDB}$ | References |
|---|---|---|---|---|---|
| Anethole | n. | star anise | -67 to -102 | -25.9 to -33.0 | [3] |
| | n. | star anise | -72 to -108 | -28.4 to -29.6 | [265] |
| | n.i. | | -39 to -110 | -30.7 to -32.1 | [3] |
| | n.i. | | -46 to -65 | -27.7 to -32.8 | [265] |
| trans-Anethole | n. | *Foeniculum vulgare* | -86 to -91 | n.d. | [266] |
| | n. | *Foeniculum vulgare* | -56 to -98 | -25.2 to -29.7 | [267, 268] |
| | n. | *Illicium verum* | -84 to -96 | n.d. | [266] |
| | n. | *Illicium verum* | -74 to -106 | -26.6 to -30.5 | [267, 268] |
| | | *Pimpinella anisum* | -51 to -81 | -25.1 to -27.3 | [267, 268] |
| | n. | anise, fennel oil | -46 to -99 | -24.2 to -29.6 | [269] |
| | n.i. | | -45 | n.d. | [266] |
| | n.i. | | -48 to -145 | -27.5 to -30.0 | [268] |
| | n.i. | | -20 to -79 | -24.8 to -32.1 | [269] |
| Estragole | n. | essent. oils | -105 to -193 | -32-2 to -36.4 | [251] |
| | n.i. | | -3 to -155 | -29.9 to -32.8 | [251] |
| Eugenol | n. | | -89 to -116 | -31.6 to -32.9 | [265] |
| Isoeugenol | n. | | -99 | -31.8 | [265] |
| Methyleugenol | n. | essent. oils | -107 to -217 | -32.2 to -41.1 | [251] |
| | n.i. | | -126 to -155 | -35.0 to -37.4 | [251] |
| Methyl salicylate | n. | *Gaultheria procumbens* | -146 | -32.7 | [270] |
| | n. | birch bark | -56 | -33.2 to -33.7 | [3] |
| | n. | | -130 to -163 | -32.7 to -33.1 | [3] |
| | n.i. | | -111 | -27.7 | [270] |
| | n.i. | | -50 to -97 | -28.3 to -30.8 | [3] |
| Benzyl acetate | n. | | -113 | -29.1 | [265] |
| | n.i. | | +352 | -26.5 | [265] |
| Methyl cinnamate | n. | | -147 | -29.1 | [265] |
| | n. | basil oil | -126 to -133 | -28.9 to -29.0 | [271] |
| | comm. | ? | -85 to -191 | -25.7 to -28.5 | [271] |
| | n.i. | nat. educts | -162 to -169 | -25.6 to -30.1 | [271] |
| | syn. | synth. educts | -126 to +287 | -27.3 to -30.6 | [271] |
| Raspberry ketone | n. | raspberry | n.d. | -29 to -36 | [272] |
| | n.i. | | n.d. | -24 to -27 | [272] |
| Dimethylanthranilate | n. | | -94 | -35.6 | [158] |
| | n.i. | | -132 | -38.5 | [158] |
| Methyl N-methyl-anthranilate | n. | orange | n.d. | -29.5 to -34.0 | [248] |
| | n.i. | | n.d. | -37.1 | [248] |
| Methylanthranilate | n. | orange | n.d. | -31.5 | [248] |
| | n.i. | | n.d. | -31.5 | [248] |

[*] n.d. = not determined, n. – natural, n.i. – nature identical, syn. = synthetic, comm. – commercial

Natural *vanillin* is an exceptional case among plant aromates, because its sources *Vanilla planifolia* or *Vanilla tahitensis* belong to the CAM-plants; correspondingly the range of the $\delta^{13}C$-values for this compound reported in the literature is from -16.8 (*V. tahitensis*) to -21.5‰ (*V. planifolia*) (Table 6.5). This is remarkably different from that of nature-identical vanillin ($\delta^{13}C$ = -24.9 to -36.2‰) which is produced from lignin (by oxidative degradation), cinnamic aldehyde (by retro aldol condensation), eugenol (by isomerisation and degradation) or guaiacol (by chemical synthesis) *[3, 256, 259]*. Therefore, the $\delta^{13}C$-value by itself would normally be a reliable proof for the authenticity of vanillin. However, adulterators have meanwhile learned to imitate the $\delta^{13}C$-value of natural vanillin by blending "nature-identical" vanillin with products artificially enriched in $^{13}C$; as this enrichment is usually performed in the methoxy or carbonyl group, the proof of this adulteration is possible by positional isotope analysis *[244, 273]* (see 6.2.2.4.4). However, meanwhile this kind of adulteration is surpassed, because "natural" vanillin can be obtained by fermentation from the natural precursors isoeugenol (source essential oils) or ferulic acid (source rice bran or sugar beet pulp). The discrimination of this natural $C_3$-plant vanillin from nature-identical products is only possible by the combination of multielement IRMS with the NMR analysis of the $^2H$-patterns of the samples (see 6.2.2.4.4, Table 6.14).

Back to natural vanillin "ex beans": depending on the origin from CAM-plants (higher $\delta^2H$-values *[274, 275]*), also the deuterium content of authentic vanillin is more positive as compared to that of aromatic substances from $C_3$-plants. Natural vanillin usually has a $\delta^2H$-value from -50 to -90‰, whereas the nature-identical products from wood lignin show values ranging from -170 to -200‰. Vanillin from guajacol may have $\delta^2H$-values close to -20‰ (and often unusual $\delta^{18}O$-values) due to a deuterium enrichment in the course of its synthesis. As a matter of fact, a mixture of lignin vanillin and guajacol vanillin could simulate the $\delta^2H$-value of natural vanillin; however, even then its $^{13}C$-content would differ from that of the authentic compound, and the $^2H$-pattern would be different from that of natural products.

Recently published results obtained by very detailed studies of vanillin by $^2H$-NMR and $^{13}C$-GC-C-IRMS including the application of minor components as intrinsic standards (e.g. p-hydroxybenzaldehyde) together with compositional analyses should enable analysts to even detect sophisticated adulterations of vanilla flavour *[276-278]*.

The $\delta^{13}C$-value of natural *cinnamic aldehyde* as reported so far ranges from -26 to -30‰. Nature-identical cinnamic aldehyde, which is normally produced by condensation of benzaldehyde and acetaldehyde *[279]*, has been found to show generally a more positive $\delta^{13}C$-value, but this difference would not be sufficient for a definite discrimination between the two origins *[3, 261]*. Therefore the partially positional $^{13}C$-analysis on the aromatic core after oxidative elimination of the side chain should be the basis for a more reliable determination of the origin of this substance (see 6.2.2.4.4). Independently, the hydrogen isotope analysis will allow a reliable differentiation between natural and nature-identical cinnamic aldehyde in most cases, as the latter products are remarkably enriched in deuterium ($\delta^2H$ from -94 to +510‰), as compared to that of the natural compound ($\delta^2H$ from -75 to -148‰). Presumably this enrichment of the nature-identical substance is due to the precursor material, especially when "synthetic" benzaldehyde is used (see below). However, in some cases even the global deuterium content may not be an absolute criterion, because the ranges of substances from different origins are overlapping. A more sophisticated positional hydrogen isotope analysis, e.g. by NMR, may be of advantage. However, corresponding data are not yet available. Finally the $\delta^{18}O$-value of cinnamic aldehyde is a promising approach for discrimination, as natural products may be enriched in $^{18}O$ compared to nature-identical cinnamon aldehydes *[249]*.

The isotopic analysis of *benzaldehyde*, bitter almond oil, has been of wide interest in recent years, hence many results are available for this substance (Table 6.5). The carbon isotope abundance is obviously not indicative for the origin of benzaldehyde, because the $\delta^{13}$C-values of nature-identical benzaldehyde (-25 to -28‰), basis toluene, are very close to and even overlapping those of natural benzaldehyde (-28 to -30.5‰) *[3, 261]*. On the other hand the average hydrogen isotope abundances and patterns seem to be more typical, as those for the nature-identical product are significantly, in some cases dramatically more positive than those for natural benzaldehyde. It turned out, that this deuterium enrichment is mainly concentrated in the carbonyl group and that it depends on the synthesis process used for the toluene oxidation. Obviously the catalytic oxidation of the starting material is accompanied by a large intramolecular kinetic isotope effect, while the chlorination to benzalchloride and its hydrolysis is not *[113]*. A partial positional deuterium isotope analysis by oxidative elimination of the hydrogen from the carbonyl group demonstrates the deuterium enrichment in this position. This could be confirmed by a more sophisticated NMR analysis, which even detected characteristic differences in other positions *[280]* (see 6.2.2.4.4).

Another source of inexpensive benzaldehyde, claimed to be natural, is aldol degradation of hydrated cassia oil (cinnamic aldehyde). This product, although prepared from a natural precursor, is not natural in the legal meaning, since its production process involves a chemical treatment of the precursor. It seems to be very difficult to differentiate the product derived from cassia oil from natural benzaldehyde, either on the basis of the mean deuterium or the $^{13}$C contents. The only possibility in this case seems to be positional isotopic analysis of deuterium by NMR *[280]*. Independently, in spite of the possible oxygen isotope exchange reactions *[281]*, oxygen isotope analysis for verification of the source of benzaldehyde seems to be promising. However, it is still in the first experimental stage *[250]*. In agreement with theoretical expectations (see 6.2.2.1) natural benzaldehyde shows $\delta^{18}$O-values of +19.3‰; nature-identical +14.9‰ (catalytic oxidation) or +6 to +7.5‰ (chlorination and hydrolysis), respectively. It is expected that the product from cassia oil will have $\delta^{18}$O-values close to the natural product, because the oxygen atom will equilibrate with water.

The $\delta^{13}$C-values of natural *anethole*, mainly the trans-isomer, originating from anise oil, star anise oil or fennel oil *[279]*, are between -26 and -33‰. This broad area of the $\delta^{13}$C-values may be due to the origin from different plant variants and local conditions of biosynthesis. Nature-identical anethole, which is prepared by isomerisation of estragole *[279, 282]*, has $\delta^{13}$C values with a range from -27.7 to -32.8‰. This is not very surprising, as estragol is often originating from the same natural source as anethole. The $^{13}$C-content of another nature-identical anethole (-30.7 to -32.1‰) may be due to its chemical synthesis from anisol and propionic acid derivatives or propenol, respectively *[279]*.

The hydrogen isotope abundance does also not permit a definite differentiation between natural and nature-identical anethole, as the product from synthesis ($\delta^2$H = -39 to -110‰) is only slightly enriched as compared to the natural one (-67 to -108‰). More positive values for nature-identical anethole may be found preferably for the product derived from estragole isomerisation as a consequence of the remarkable hydrogen isotope fractionation accompanying the H$^+$-abstraction from the methylene group in the side chain *[11]*. On the other hand, the low deuterium content of some nature-identical anetholes is caused by a remarkable depletion of $^2$H in position 1 of the side chain, probably going back to the precursor molecule itself or an isotope effect of the synthesis *[11, 283]*. The potential of an $^{18}$O-determination for the differentiation between natural and nature-identical anetholes has not been studied up to now.

Natural *eugenol* from clove leaf oil or from cinnamon oil has $\delta^{13}$C-values from -31.6 to -32.9‰ and $\delta^2$H-values from -89 to -116‰. As the natural product is inexpensive and available in large quantities, no nature-identical product is offered commercially *[279]*. Natural *isoeugenol* has been found to give the same range of $\delta^{13}$C- and $\delta^2$H-values (with a $\delta^{13}$C-value of -31.9‰ and a $\delta^2$H-value of -99‰), but as isoeugenol can be prepared by isomerisation of the sodium or potassium salt of eugenol, an isoeugenol produced by this isomerisation will not be different in the isotopic content from the natural isoeugenol. Nature-identical isoeugenol made from guajacol, however, should possess a different $\delta^{13}$C- and especially $\delta^2$H-value as compared to natural isoeugenol.

Results for other aromatic substances are compiled in Table 6.5, but are not discussed in detail, as the available data do not justify a discussion. *Methyl salicylate*, the main component of natural wintergreen oil is often used in cosmetic products. The quite negative $\delta^{13}$C-values of the natural product ($\delta^{13}$C-values from -32.6 to -33.7‰) are probably partially caused by the extreme depletion of the methyl groups (see 6.2.2.4.4) and permit perhaps a discrimination from the nature-identical product ($\delta^{13}$C-values from -27.7 to -30.8‰). The $\delta^2$H-values of natural methyl salicylate -130 to -163‰ *[3, 270]* are, with the exception of one sample (-56‰), in the range of other natural aromatic substances (e.g. benzaldehyde, cinnamic aldehyde, anethole). Nature-identical analogues show more positive $\delta^2$H-values.

For natural *benzyl acetate* a $\delta^{13}$C-value of -29.1‰ has been found, while -26.5‰ was measured for the nature-identical product. The $\delta^2$H-value of natural benzyl acetate (-113‰) is quite similar to that of other natural aromatic compounds, while for a nature-identical product a value of +352‰ has been found *[265]*. The reasons may be similar to those discussed for "synthetic" benzaldehyde.

The values measured for natural *methyl cinnamate* ($\delta^{13}$C-values between -25.7 and -30.6‰ and $\delta^2$H-values between -85 and -191‰, *[265, 271]*) are in line with those for other natural aromatic substances. The nature-identical substance seems to have similar $\delta^{13}$C- and partially very high $\delta^2$H-values.

A recent study of 2-phenylethanol and 2-phenylacetate by $^2$H-NMR made possible a discrimination between the natural and the "synthetic product". This would also have been possible by $^2$H-IRMS in the case of phenylacetate, but not for 2-phenylethanol *[284]*.

## 6.2.2.4.2 δ-Values of Aliphatic Substances

Non-isoprenoid aliphatic compounds of importance in flavourings and fragrances are mainly alcohols, esters, aldehydes and lactones. Their biogenesis in plants and microorganisms mostly starts from carbohydrates or amino acids. Some of them are directly correlated to pyruvate and acetyl-CoA, others originate indirectly from carbohydrates through amino acids. In any case, in the course of their synthesis, isotope effects and exchanges may contribute to their final δ-values. The non-statistical $^{13}$C-distribution in carbohydrates *[107, 285, 286]*, implying a relative enrichment of $^{13}$C in the positions 3 and 4, will lead to a relative depletion in all descendants preserving only positions 1, 2, 5 and 6 of sugars. This will mainly occur in the course of the pyruvate decarboxylase (PDC)- or the pyruvate dehydrogenase (PDH)- reactions. However, both reactions have been found to proceed with remarkable fractionation of $^{13}$C *[106, 287]*, resulting in an additional depletion of $^{13}$C in the final products as compared to the precursor carbohydrates. From the overlap of both parameters, we have to expect a global depletion of $^{13}$C by 3 to 6‰ for corresponding descendants relative to the carbohydrates from the same source, and additionally a pattern influenced by both factors. Concerning the hydrogen isotopes, a significant depletion of $^{2}$H is resulting from the phosphoenolpyruvate kinase reaction, whereby hydrogen is introduced from water with a high fractionation *[246, 288]*. In addition, the reduction steps during the formation of aliphatic substances introduce highly depleted hydrogen atoms into the products *[111, 246]*. As a consequence, natural aliphatic substances are depleted by about 150‰ compared to carbohydrates from the same source, with a natural range from -200 to -350‰ (Table 6.6). This is independent of the plants' photosynthesis type *[246]*. A biochemical systematic interpretation or prediction has recently been published *[111]*.

Systematics are also available for the $\delta^{18}$O-values of the compounds in question *[56]*: carboxyl and carbonyl functions in isotopic equilibrium with the surrounding water are, due to equilibrium isotope effects, enriched in $^{18}$O relative to this water by 19 and by 25 to 28‰, respectively. From here, the $\delta^{18}$O-values of natural alcohols, mostly descendants of carbonyl compounds, will have (maximally) similar $\delta^{18}$O-values, provided the precursors have attained isotopic equilibrium with water and their reduction has not been faster than their equilibration. Alcohols from addition of water to C=C double bonds or from exchange of halogen functions by OH groups, typical for synthetic alcohols, will have $\delta^{18}$O-values close to or even below that of the water, due to kinetic isotope effects. The few available results *[246, 289, 290]* seem to confirm this expectation. The $\delta^{18}$O-values of natural (and also synthetic) esters and lactones can be, especially in the carbonyl group, extremely high (up to 50‰), probably as a consequence of an intramolecular kinetic isotope effect on the activation of the carboxyl function.

**Table 6.6**: Range of $\delta^2H$- and $\delta^{13}C$-values of aliphatic compounds occuring in flavourings. Data include results from IRMS and from GC-C-IRMS measurements. Other $\delta$-values of some compounds are:

a. $\delta^{15}N$-value [‰]$_{Air-N2}$:
   Allylisothiocyanat: n. +1 to +11, n.i. –10 to -14 [291]; n. -5 to +10, n.i. –12 to +1 [292]; n. +1.7 to +17.0, n.i. –16.4 to +0.2 [293]
b. $\delta^{34}S$-value [‰]$_{V-CDT}$:
   Allylisothiocyanat: n. -17.1 to +12.4, n.i. -2.5 to +14.1 [293]
c. $\delta^{18}O$-value [‰]$_{V-SMOW}$:
   Ethanol: n. +16 to +28, n.i. ±0 to +5 [289]; n. +11.3, n.i. –2 to +5 [249]; n. +20.8 to +22.8, comm. +15.5 to +22.7 [208, 294]
   Isobutanol: n. +4.5, n.i +14.5 [249]
   Ethylbutyrate: n. average +24.8 [249]

| Substance | Status* | Origin/Source | $\delta^2H$ [‰]$_{V-SMOW}$ | $\delta^{13}C$ [‰]$_{V-PDB}$ | References |
|---|---|---|---|---|---|
| *Alcohols, acids and aldehydes* | | | | | |
| 1-Butanol | n. | pear (products) | -46 to -165 | -37.0 to -41.8 | [295] |
| | n. | pear brandy | -109 to -230 | -28.2 to -37.9 | [295] |
| | syn. | | -105 to -153 | -27.7 to -30.0 | [295] |
| Ethanol | n. | | -236 | n.d. | [246] |
| | n. | | -200 to -272 | -10.3 to -27.8 | [296] |
| | n. | various sources | -300 to -406 | -13.4 to -27.3 | [265, 297] |
| | n. | | -170 to -270 | -9.8 to -28.1 | [265, 289, 290] |
| | n. | tequila | n.d. | -12.1 to -14.8 | [208, 294] |
| | comm. | tequila | n.d. | -10.6 to -13.9 | [208, 294] |
| | n.i. | | -100 to -180 | -26.8 to -32.0 | [265] |
| | n.i. | | -110 to -269 | -23.4 to -34.0 | [289, 290] |
| | n.i. | | -117 to -140 | -25.0 to -32.0 | [296] |
| | n.i. | | -182 | -33.4 | [297] |
| Acetic acid | n. | fermentation | -200 to -400 | -12.0 to -30.0 | [265] |
| | n. | | n.d. | -20.3 to -22.7 | [298] |
| | n. | fermentation | n.d. | -10.9 to -25.6 | [299] |
| | n. | | n.d. | -25.9 to -28.5 | [300] |
| | n.i. | | n.d. | -26.9 to -28.1 | [298] |
| | n.i. | | -163 | -35.0 | [297] |
| | n.i. | | -120 to -180 | -16 to -31 | [265, 301] |
| | n.i. | | n.d. | -21.5 to -44.0 | [300] |
| Acetyl butyric acid | n. | | -286 | -20.3 | [265] |
| | n.i. | | -144 | -28.0 | [265] |
| Acetaldehyde | n. | | -140 to -266 | -12.1 to -21.8 | [3] |
| | n. | | -103 to -348 | -16.7 to -20.3 | [265] |
| | n.i. | | -68 to -83 | -26.3 to -30.7 | [3] |
| | n.i. | | -78 to -295** | -29.1 to -30.8 | [265] |
| Diacetyl | n. | | -294 | -25.7 | [265] |
| | n.i. | | -218 | -30.7 | [265] |
| Acetoin | n. | vinegar | n.d. | -26.5 to -27.9 | [302] |
| | n.i. | | +109 to -133 | -21.4 to -24.5 | [265] |
| 1-Hexanol | n. | pear (juice) | -134 to -212 | -36.9 to -42.7 | [295] |
| | | pear brandy | -80 to -230 | -34.1 to -39.1 | [295] |
| | n. | cactus pear | n.d. | -22.0 | [303] |
| | syn. | | -63 to -97 | -25.0 to -27.0 | [295] |
| Hexenol | n. | | -249 | -32.2 | [265] |

* n.d. = not determined, n. = natural, n.i. = nature-identical, syn. = synthetic, comm. = commercial
** acetaldehyde from catalytic oxidation of natural ethanol

| Substance | Status* | Origin/Source | $\delta^2H$ [‰]$_{V-SMOW}$ | $\delta^{13}C$ [‰]$_{V-PDB}$ | References |
|---|---|---|---|---|---|
| cis-3-Hexenol | n. | | -216 to -227 | -25.9 to -29.9 | [265] |
| | n. | | n.d. | -32 to -39 | [304] |
| | n. | | n.d. | -32.2 to -38.9 | [260] |
| | n.i. | Mentha arvensis | -178 | -27.4 | [265] |
| | n.i. | | n.d. | -27 to -28 | [304] |
| | n.i. | | n.d. | -27.8 | [260] |
| E-2-Hexenol | n. | cactus pear | n.d. | -21.8 | [303] |
| | | apple a.o. | -238 to -318 | n.d. | [305] |
| | syn. | | -41 to -131 | | [305] |
| Isobutanol | n. | | -293 | -22.4 | [265] |
| | n.i. | | -172 | -28.2 | [265] |
| E-2-Nonenol | n. | cactus pear | n.d. | -19.1 | [303] |
| E-2-Hexenal | n. | apple a.o. | -263 to -360 | n.d. | [305] |
| | syn. | | -14 to -109 | n.d. | [305] |
| Decanal | n. | orange | n.d. | -25.3 to -26.9 | [198, 306] |
| | n. | orange | n.d. | -25.7 to -27.0 | [307] |
| | n. | citrus fruits | -164 to -196 | n.d. | [305] |
| | n.i. | | n.d. | -27.6 to -30.0 | [198, 306] |
| | n.i. | | n.d. | -31.1 | [307] |
| | syn. | | -90 to -156 | n.d. | [307] |
| Dodecanal | n. | orange | n.d. | -27.1 to -28.2 | [198, 306] |
| | n.i. | | n.d. | -32.2 to -32.7 | [307] |
| Nonanal | n. | orange | n.d. | -27.3 to -28.6 | [198, 306] |
| | n.i. | | n.d. | -29.5 to -29.8 | [306, 307] |
| Octanal | n. | orange | n.d. | -27.4 to -29.2 | [198, 306] |
| | n. | orange | n.d. | -25.5 to -27.1 | [307] |
| | n.i. | | n.d. | -27.9 to -30.2 | [306] |
| *Esters* | | | | | |
| Butylacetate | n. | pear a. juice | -104 to -200 | -31.3 to -39.8 | [295] |
| | ? | pear | -213 to -262 | -27.2 to -28.7 | [295] |
| | syn. | | -131 to -152 | -29.7 to -31.3 | [295] |
| Amylacetate | n. | | -246 to -373 | -11.3 to -28.1 | [3] |
| | n.i. | | -89 to -145 | -26.5 to -31.5 | [3] |
| Butyl-2-methylbutanoate | n. | apple varieties | n.d. | -34.0 to -37.8 | [308] |
| | n.i. | | n.d. | -27.9 | [308] |
| Ethylacetate | n. | | -201 to -295 | -11.5 to -29.3 | [3] |
| | n. | | -231 to -295 | -11.5 to -29.3 | [265] |
| | n.i. | | +29 to -140 | -25.3 to -35.3 | [3] |
| | n.i. | | -141 to -175 | -29.8 to -32.3 | [265] |
| Ethylbutyrate | n. | | -273 | -14.8 | [265] |
| | n. | | -172 to -270 | -9.7 to -23.3 | [3] |
| | n. | orange juice | n.d. | -29.6 | [309] |
| | n. | microbiological | n.d. | -14.7 | [309] |
| | n.i. | | -188 | -27.2 | [265] |
| | n.i. | | -9 to -107 | -15.7 to -31.9 | [3] |
| Ethylcapronate | n. | | -194 to -267 | -27.3 to -32.2 | [3] |
| | n.i. | (mixtures) | -199 to -242 | -30.4 to -33.1 | [3] |
| Ethyl-2-methylbutyrate | n. | | -261 to -323 | -17.8 to -27.9 | [265] |
| | n.i. | | -144 | -25.9 | [265] |
| Ethyl-E,Z/E,E-2,4-decadienoate | n | pear and products | -187 to -236 | -34.3 to -41.9 | [295] |
| | ? | pear brandy | -193 to -239 | -36.7 to -40.7 | [295] |
| Ethylhexanoate | n. | pineapple | -128 to -165 | -12.8 to -22.9 | [207] |
| Ethyl-2-methylbutanoate | n. | apple varieties | n.d. | -30.9 to -35.0 | [308] |
| | n. | pineapple | -184 to -263 | -14.0 to -18.9 | [207] |
| | n.i. | | n.d. | -26.2 to -28.7 | [308] |
| | syn. | | -60 to -163 | -22.8 to -27.2 | [207] |

* n.d. = not determined, n. = natural, n.i. = nature-identical, syn. = synthetic, comm. = commercial

| Substance | Status* | Origin/Source | $\delta^2H$ [‰]$_{V\text{-SMOW}}$ | $\delta^{13}C$ [‰]$_{V\text{-PDB}}$ | References |
|---|---|---|---|---|---|
| Ethylpropionate | n. | | -252 to -270 | -15.4 to -26.5 | [265] |
| | n.i. | | -66 to -159 | -24.8 to -26.9 | [265] |
| Hexylacetate | n. | various sources | n.d. | -31 to -36 | [272] |
| | n. | pear | -48 to -157 | -29.7 to -38.3 | [295] |
| | n.i. | | n.d. | -27 to -34 | [272] |
| | syn. | | -83 to -93 | -26.2 to -26.7 | [295] |
| Methyl-E/Z-2,4-decadienoate | n. | pear and products | -219 to -255 | -36.1 to -39.3 | [272] |
| | n. | pear brandy | -215 to -258 | -35.0 to -39.3 | [272] |
| | syn. | | -131 to -142 | -30.6 | [272] |
| Methylhexanoate | n. | pineapple | -118 to -191 | -17.2 to -24.4 | [207] |
| | syn. | | -80 to -86 | -23.6 to -29.4 | [207] |
| Methyl-2-methylbutanoat | n. | pineapple | -208 to -234 | -17.7 to -19.7 | [207] |
| | syn. | | -49 to -107 | -26.6 to -31.8 | [207] |
| Hexyl-2-methylbutanoate | n. | apple varieties | n.d. | -36.7 to -39.2 | [308] |
| | n.i. | | n.d. | -26.7 to -27.0 | [308] |
| Isoamylacetate | n. | | -302 | -23.8 | [265] |
| | n.i. | | -126 | -28.7 | [265] |
| Isoamylbutyrate | n. | | -273 to -296 | -12.8 to -23.8 | [265] |
| | n.i. | | -107 | -28.1 | [265] |
| Propyl-2-methylbutanoate | n. | apple varieties | n.d. | -32.4 to -35.3 | [308] |
| | n.i. | | n.d. | -27.1 to -27.5 | [308] |
| *Lactones* | | | | | |
| δ-Decalactone | n. | various fruit | -171 to -228 | -33.5 to -37.9 | [310] |
| | syn. | | -171 | -28.2 | [310] |
| γ-Decalactone | n. | | -234 to -258 | -29.8 to -30.7 | [265] |
| | n. | | n.d. | -28.2 to -40.9 | [311] |
| | n. | | n.d. | -29.9 to -30.2 | [312] |
| | n. | apricot | n.d. | -33 to -40 | [272] |
| | n. | mango | n.d. | -36 to -40 | [272] |
| | n.i. | | -170 | -28.7 | [265] |
| | n.i. | | n.d. | -24.4 to -27.3 | [311] |
| | n.i. | | n.d. | -28.2 to -28.3 | [312] |
| | n.i. | | n.d. | -26 to -31 | [272] |
| γ-Decalactone | n. | various fruit | -160 to -206 | -34.6 to -38.4 | [310] |
| | syn. | | -151 to -184 | -27.4 to -28.3 | [310] |
| γ-Undecalacton | n.i. | | -116 to -146 | -27.4 to -28.1 | [291] |
| γ-Dodecalactone | n. | various souces | n.d. | -28 to -39 | [272] |
| | n.i. | | n.d. | -27.3 to -30.0 | [272] |
| *Furanes* | | | | | |
| Mesifuran | n. | strawberry | n.d. | -33.6 to -34.9 | [313, 314] |
| | n.i. | | n.d. | -39.2 to -39.7 | [313, 314] |
| Furaneol | n. | strawberry | n.d. | -23.7 to -26.8 | [313, 314] |
| | n. source | | n.d. | -18.5 to -27.2 | [313, 314] |
| | n. | pineapple | -141 to -195 | -20.9 to -29.4 | [207] |
| | n.i. | | n.d. | -27.8 to -39.0 | [313, 314] |
| | syn. | | -103 | -35.9 | [207] |
| *Aliphatic substances containing nitrogen or sulfur* | | | | | |
| Allylisothiocyanate | n. | | -55 to -157 | -27.9 to -29.0 | [291] |
| | n. | | n.d. | -25 to -32 | [292] |
| | n.i. | | -17 to -38 | -27.7 to -30.7 | [291] |
| | n.i. | | n.d. | -20 to -30 | [292] |
| Dimethylsulfide | n. | | -227 to -245 | -59.3 to -60.2 | [265] |
| | n.i. | | -170 | -46.2 | [265] |

* n.d. = not determined, n. = natural, n.i. = nature-identical, syn. = synthetic, comm. = commercial

## Alcohols, Aldehydes and Acids

The $^{13}$C-content and -pattern of the $C_2$-aliphates *ethanol, acetic acid* and *acetaldehyde* should reflect their direct origin from carbohydrates, especially, when they are main products. Most data are available for the $^{13}$C-content of ethanol in alcoholic beverages. As expected, this ethanol is depleted in $^{13}$C by about 1.6‰ as compared to the sugar from which it originates *[221, 315]*. Corresponding results are found for natural acetic acid. It is important to register that both products preserve the $^{13}$C-pattern from glucose ($\delta^{13}$C of C-1 by 2‰ more negative than that for C-2). Synthetic ethanol and acetic acid are remarkably depleted in $^{13}$C as compared to the natural products (range of -26 to -34‰ *[265, 289]*), and in most cases investigated, the pattern is reverse to that of the natural counterparts. Ethanol and acetic acid from $C_4$-plant material (corn, cane) have been found to possess $\delta^{13}$C-values of -10 to -15‰ *[9, 208, 289, 296]*. Any of these findings are also true for acetaldehyde, which is sometimes used for synthesis (-26 to -31‰) *[3, 265]*. Acetaldehyde obtained by catalytic oxidation of natural ethanol, which has, by law, to be called nature-identical, must have the same $\delta^{13}$C-values as natural acetaldehyde.

The $\delta^{13}$C-values of *higher alcohols and acids* (e.g. isobutanole, methylbutyric acid) are generally more negative in comparison to the source carbohydrates: the samples compiled in Table 6.6 are often produced from $C_4$-plant precursors. *Diacetyl* was found to possess a $\delta^{13}$C-value similar to that of acetic acid, whereas natural *acetoin* isolated from vinegar ranges from -26.5 to -27.9‰ *[302]*. The $^{13}$C-content of higher alcohols and acids obtained by chemical synthesis is close to the values for reported synthetic ethanol, therefore it is not a reliable criterion for a differentiation from natural products.

For *higher aldehydes* from citrus fruits (octanal, decanal) the $\delta^{13}$C-values have been found to be significantly different from those of the nature-identical analogues *[198, 306, 307]*. However, the number of tested nature-identical substances is too limited to take the $^{13}$C-content as a reliable criterion for origin assignment.

The $\delta^2$H-values of natural ethanol range from -170 to -270‰, depending on the origin of the precursor material, allowing a good discrimination of this material from the nature-identical product ($\delta^2$H-values from -110 to -180‰ *[289, 296]*), and its tentative assignment to geographical origins. The $\delta^2$H-values of acetic acid (measured as calcium salt, as the hydrogen atom of the carboxyl group is readily exchanging with the surrounding water hydrogen) are between -220 and -370‰ for natural acetic acid which is the same as for natural ethanol (methyl hydrogen, *[221, 296, 316]*). Nature-identical acetic acid was found to give $\delta^2$H-values from -100 to -180‰ *[265]*. Although a relatively higher $^2$H-abundance (range of $\delta^2$H from ±0 to -200‰ is typical for synthetic material as compared to natural alcohols, aldehydes and acids (range of $\delta^2$H from -200 to -400‰), it is advisable to determine in addition the $\delta^{13}$C- and even the $\delta^{18}$O-values of the compounds in order to obtain a reliable basis for the discrimination between nature-identical and natural material and to identify the geographical or botanical origin of the latter.

The $\delta^{18}$O-value of aliphatic alcohols and acids has so far only been investigated for a few examples, mainly for methodological reasons. However, the few results available underline that, on the basis of the above indicated isotopic correlations to the origin of the compounds, the $^{18}$O-content is obviously a potent means for the differentiation between natural and nature-identical material. As expected, results for natural ethanol indicate a $\delta^{18}$O-value between +15 and +30‰, while those for nature-identical products are by far less positive. A recent study has found correlations between the hydrogen and oxygen isotope abundances of ethanol and the fermentation water and the fermented carbohydrates and confirmed the expectations outlined before [private communication by G. Calderone (2005): Applications des modernes techniques isotopiques pour la caractérisation de produits alimentaires, PhD thesis, Université de Nantes].

**Esters of Aliphatic Acids and Alcohols**

For *esters* originating from the above mentioned alcohols or acids the $^{13}$C- and $^{2}$H-contents can give clues on the starting material, but do not permit an absolutely reliable differentiation of natural or nature-identical origin. In general the $\delta^2$H-values of esters are remarkably more positive for the nature-identical substances as compared to the natural products, as it was found already for alcohols and acids. The main problem in this context is provided by products from biotechnological or even natural origin when bound by a chemical process. In these cases the results of $\delta^{13}$C- and $\delta^2$H-values will be in between the range for pure natural and pure synthetic material. A reliable assignment of these esters can perhaps be done after hydrolysis and the $^{13}$C- and $^{18}$O-analysis of at least one of the products (see 6.2.2.4.4). While an exchange of $^{18}$O with the surrounding water must be taken into account for the carboxyl group *[82]*, the $\delta^{18}$O-value of the alcohol should be related to that of the original source of this part of the ester. However, only a few results confirming this supposition are available *[212]*.

As already indicated, a special problem with esters is their preparation from two natural precursor molecules by a chemical ester synthesis. Such products have to be labelled nature-identical. For an interesting positional $^2$H-NMR study on ethyl butyrate from enzymatic esterification of beet ethanol with butyric acid from milk see *[317]*. Another chance to detect a corresponding adulteration would be a positional carbon and oxygen isotope analysis of the ester components. Isotope effects on the esterification reaction in question seem to influence characteristically the δ-values of the atoms involved, and hence form a basis for the origin assignment of these compounds (for further details see 6.2.2.4.4).

Recently obtained results from $^2$H-NMR analysis of n-hexanol and (Z)-3-hexenol basically confirmed a lower $^2$H-content in natural compounds as compared to nature-identical ones, but in addition, it was even possible to detect differences between natural, "synthetic", biotechnologically produced and "hemi-synthetic" material (made from natural precursors) by means of the $^2$H-pattern analysis *[318, 319]*.

**Lactones**

The available data (Table 6.6) indicate that the $^2$H- and the $^{13}$C-contents of (long-chained) natural *lactones* can be quite more negative than those of the nature-identical ones *[265, 311, 312, 320]*. This is confirmed by $^2$H-NMR measurements on these compounds *[317, 321]*. It is obvious that natural lactones and esters with long-chained acids are, as descendants of fatty acids, distinctly depleted in $^{13}$C and $^2$H.

**Furanes**

Mosandl *[313, 314]* has found $\delta^{13}$C-values from -33.6 to -34.9‰ for *2,5-dimethyl-4-methoxy-2H-furan-3-one* ("*mesifuran*") and -23.7 to -26.8‰ for *2,5-dimethyl-4-hydroxy-2H-furan-3-one* ("*furaneol*") from strawberries. This difference, even found in products from the same source, suggests independent biosyntheses. Another natural sample of mesifuran showed a very positive $\delta^{13}$C-value (-18.5‰), which indicates its origin from a CAM-plant (pineapple). However, the recent data by Preston et al. *[207]* show that furaneol from this provenience can, but must not have quite positive $\delta^{13}$C-values. This and the seeming coincidence of the data for natural and nature-identical mesifurane show that the situation with these compounds is still not quite clear.

### 6.2.2.4.3 δ-Values of Isoprenoids

The natural precursors of any flavourings of this group are isopentenylpyrophosphate (IPP) and dimethylallyl pyrophosphate (DMAPP). In the classic pathway these are synthesised from three molecules acetyl/malonyl-CoA via mevalonic acid (MA). Recently a second pathway of the isoprenoid biosynthesis has been detected, in which IPP and DMAPP are synthesised from pyruvate and glyceraldehyde 3-phosphate via deoxyxylulose (DOX) 5-phosphate *[322-324]*.

The loss of the relatively $^{13}$C-enriched carbon atoms C-3 and C-4 of glucose *[107]*, and the kinetic isotope effect on the pyruvate-dehydrogenase reaction *[106]* in the biosynthesis of acetyl-CoA have for consequence that isoprenoids from the MA-pathway are depleted in $^{13}$C relative to carbohydrates from the same source by 4 to 6‰, whereas this $^{13}$C-depletion attains in products from the DOX-pathway only 2 to 4‰. $^{13}$C-patterns of the two types of isoprenoids, which would certainly be different, are still not yet available.

Correspondingly, the $^2$H-abundances and -patterns of representatives of the two groups are different: Due to differences in the mechanisms of and the kinetic isotope effects on the involved reactions, the average $\delta^2$H-values of MA-descendants range from -130 to -220‰, those for the DOX-descendants from -200 to -320‰. The deuterium patterns are very typical and characteristic and therefore fingerprints in the authenticity and origin assessment of the compounds (see 6.2.2.4.4 and Tables 6.12 and 6.14). The majority of isoprenoid flavourings are monoterpenes from higher plants, in some cases from bacteria, and they are all synthesised via the DOX pathway. Some sesquiterpenes and descendants (ionones) may also be originating from the MA-pathway.

Quite interesting and indicative in respect to the authentication of the compounds may also be their oxygen isotope characteristics. OH-groups of acyclic and of some monocyclic monoterpenes are introduced as OH⁻ from water, substituting the pyrophosphate functionality of the biosynthetic precursor. This hydrolysis is probably accompanied by a kinetic isotope effect, from where a relative low $\delta^{18}$O-value is to be expected. The $\delta^{18}$O-value of synthetic analogues will depend on the oxygen source and the process of their synthesis. In agreement with these expectations the $\delta^{18}$O-value of natural linalool ranges from -18 to +5.8‰, whereas that of the nature-identical product, probably synthesised from cis-pinane with $O_2$ in the presence of a radical initiator [279], is between +13.7 and +16.8‰ ([211], Table 6.7). The conversion of the alcoholic to a carbonyl group provides the possibility of a (partial) isotopic equilibration with water, hence an enhancement of the $\delta^{18}$O-value; an example is citral [56].

OH-groups in cyclic terpenes must contain oxygen from $O_2$, introduced by a monooxygenase reaction, hence have $\delta^{18}$O-values near +5‰; this is found for L-menthol [69]. Unexpectedly, also the natural cyclic ketones pulegone and menthone show relatively low $\delta^{18}$O-values [69]; perhaps these compounds have not had the time for an equilibration with water.

***Table 6.7:*** *Range of $\delta^2$H- and $\delta^{13}$C-values of isoprenoid compounds occuring in flavourings. Data include results from IRMS and from GC-C-IRMS measurements. Other δ-values of some compounds are:*

*$\delta^{18}$O-value [‰]$_{V\text{-}SMOW}$:*
*Citral: n. +13.4 to +19.7, n.i. +7 to + 16.8 [56]*
*L-Menthol: n. +5.9; menthone/ isomenthone: unknown +9.6; pulegone: n. +4.7[69]*
*Linalool: n. -18.0 to +5.8, n.i. +13.7 to + 16.8, comm. -10.4 to + 12.4 [211]*
*Linalylacetat: n. +19.1 to +32.2 n.i. +10.1 to + 12.2, comm. +16.2 to + 39.2 [211]*

| Substance | Status* | Origin/Source | $\delta^2$H [‰]$_{V\text{-}SMOW}$ | $\delta^{13}$C [‰]$_{V\text{-}PDB}$ | References |
|---|---|---|---|---|---|
| Monoterpenes, open-chained, monocyclic | | | | | |
| Geraniol | n. | lemon oil | n.d. | -27.3 to -29.3 | [325] |
| | n.i. | | -279 | -27.1 | [326] |
| Geranylacetate | n. | | -268 | -21.4 | [326] |
| | n. | lemon oil | n.d. | -29.0 to -30.0 | [325] |
| | n.i. | | -263 | -30.4 | [326] |
| Linalool | n. | various sources | -244 to-269 | n.d. | [118, 258, 266] |
| | n. | | -226 to -359 | -25.2 to -30.2 | [3] |
| | n. | | -160 to -270 | -24.6 to- 32.0 | [230] |
| | n. | | n.d. | -26.0 to -28.5 | [307, 327] |
| | n. | lavender, lavandin spike oil | n.d. | -25.1 to -31.9 | [328] |
| | n. | diff. ess. oils | -234 to -332 | n.d. | [305] |
| | n. | lavender oil | -241 to -274 | n.d. | [329] |
| | n. | lavender oil | -241 to -307 | -25.2 to -30.3 | [211] |
| | comm. | lavender oil | -190 to -294 | -24.8 to -33.0 | [211] |
| | comm. | lavender oil | -190 to -294 | n.d. | [329] |
| | syn. | | -185 to -209 | n.d. | [329] |
| | n.i. | | -185 to -209 | -24.9 to -28.9 | [211] |
| | n.i. | | -170 | n.d. | [118, 258, 266] |
| | n.i. | | -156 to -311 | -25.8 to -32.0 | [3] |
| | n.i. | | -162 to -215 | -27.0 to -29.2 | [230] |
| | n.i. | | -197 | -28.0 to -33.7 | [326] |
| | n.i. | | n.d. | -27.0 to -33.7 | [328] |
| | n.i. | | -207 to -301 | | [305] |

* n.d. = not determined, n. = natural, n.i. = nature-identical, syn. = synthetic, comm. = commercial

# Stable Isotope Ratio Analysis in Quality Control of Flavourings

| Substance | Status* | Origin/Source | $\delta^2$H [‰]$_{V\text{-SMOW}}$ | $\delta^{13}$C [‰]$_{V\text{-PDB}}$ | References |
|---|---|---|---|---|---|
| Linalylacetate | n. | various sources | -190 to -241 | -27.6 to -30.6 | [230] |
| | n. | Coriandrum sativum L. | n.d. | -27.2 to -29.8 | [307, 327] |
| | | Artemisia vulgaris L. | | | |
| | n. | diff. ess. Oils | -213 to -282 | n.d. | [305] |
| | n. | lavender oil | -238 to -276 | n.d. | [329] |
| | comm. | lavender oil | -187 to -274 | n.d. | [329] |
| | n. | lavender oil | -238 to -280 | -27.1 to -31.2 | [211] |
| | comm. | lavender oil | -187 to -274 | -23.8 to -32.5 | [211] |
| | n.i. | | -170 to -181 | -30.1 to -30.6 | [230] |
| | n.i. | | n.d. | -29.0 to -35.3 | [307, 327] |
| | n.i. | | -199 to -239 | | [305] |
| | syn. | | -173 to -197 | n.d. | [329] |
| Citral (Neral + Geranial) | n. | various sources | -251 to -276 | -11.6 to -27.0 | [118, 258, 266] |
| | n. | | -226 to -296 | -25.4 to -26.6 | [265] |
| | n. | lemon oil | n.d. | -10.4 to -26.6 | [325] |
| | n.i. | various sources | -174 | -27.6 | [118, 258, 266] |
| | n.i. | | -15 to -302 | -26.4 to -28.6 | [265] |
| Geranial | n. | diff. ess. oils | -216 to -297 | n.d. | [330] |
| | n.i. | | +38 to -177 | n.d. | [330] |
| Neral | n. | diff. ess. oils | -104 to -314 | n.d. | [330] |
| | n.i. | | +29 to -197 | n.d. | [330] |
| Myrcene | n. | Mentha piperita L. | -263 | n.d. | [118, 245] |
| | n. | various sources | n.d. | -24.2 to -25.8 | [307, 327] |
| Limonene | n. | lemon oil | n.d. | -26.6 to -28.5 | [325] |
| | n. | Mentha crispa L. | -264 | n.d. | [118] |
| | n. | | -215 to -250 | -26.1 to -28.5 | [326] |
| *Bicyclic monoterpenes* | | | | | |
| α-, β-pinene | n. | various sources | n.d. | -27.8 to -30.7 | [307, 327] |
| | n. | lemon oil | n.d. | -26.6 to -28.2 | [325] |
| | n. | | -280 to -310 | -27.8 to -28.0 | [326] |
| | n. | pine oil | n.d. | -28.5 | [118] |
| Menthol | n. | | n.d. | -28.1 | [307] |
| | n. | various sources | -358 to -394 | n.d. | [118, 266] |
| | n.i. | | -368 | -28.9 | [326] |
| | n.i. | | -196 to -250 | n.d. | [118, 266] |
| α-Terpineol | n. | | -264 to -286 | -27.5 to -27.8 | [265] |
| | n.i. | | -282 to -310 | -26.9 to -28.0 | [265] |
| Menthone | n. | Mentha piperita L. | -274 | n.d. | [118] |
| Carvone | n. | | n.d. | -27.2 to -28.5 | [307] |
| | n. | Mentha crispa L. | -276 | n.d. | [291] |
| | n. | oleum cervi | -232 to -238 | -26.3 to -26.5 | [265] |
| | n.i. | | -181 | n.d. | [307] |
| | n.i. | limonene | -203 to -205 | -27.9 to -28.8 | [326] |
| Pulegone | n. | Mentha pulegium L. | -238 | n.d. | [118] |
| | n. | Mentha pulegium L. | -339 | -29.3 | [326] |
| Safranal | n. | Crocus sativus L. | | -29.0 to -30.7 | [272, 331] |
| | n.i. | | | -23.7 to -25.0 | [272, 331] |
| *Higher terpenes and degradation products* | | | | | |
| α-Ionone | n. | biotechnological | -205 to -296 | -10.2 to -31.7 | [265] |
| | n. | | n.d. | -32.9 to -33.4 | [332, 333] |
| | n. | fermentation | n.d. | -8.6 to -9.1 | [332, 333] |
| | n. | raspberry | n.d. | -33 to -37 | [272] |
| | n.i. | | -55 to -172 | -26.6 to -28.1 | [265] |
| | n.i. | | n.d. | -24.3 to -27.1 | [332, 333] |
| | n.i. | | n.d. | -23 to -28 | [272] |
| β-ionone | n. | various sources | n.d. | -32 to -34 | [272] |
| | n.i. | | n.d. | -28.8 | [272] |
| Nootkatone | n. | | -160 | -30.9 | [265] |
| | n.i. | | -112 to -124 | -31.4 to -31.8 | [265] |

* n.d. = not determined, n. = natural, n.i. = nature-identical, syn. = synthetic, comm. = commercial

Monoterpenes of interest in flavourings can be open-chained or cyclic molecules. They can be more or less unsaturated, and they can contain various oxygen functions. Sometimes they may also be bound to non-terpene molecule parts. Among the compounds and their δ-values compiled in Table 6.7, typical representatives will be discussed in detail. Nature-identical isoprenoids are in this context compounds obtained from another natural isoprenoid precursor, because from a commercial point of view, a really synthetic product would not be of interest.

**Acyclic Monoterpenes and Their Esters**

*Geraniol* is a primary open-chained monoterpene, showing a relative positive $\delta^{13}$C-range from -27.3 to -29.3‰. As the $^{13}$C-content of natural *geranylacetate* should be not very different from that of the alcohol, the first product in Table 6.7 has obviously been obtained by way of biotechnology (partially) on a $C_4$-plant basis, whereas the product labelled nature-identical is presumably made from β-pinene and acetic anhydride. Proof is given for this supposition, as the acetyl residue of this product had an extremely negative $\delta^{13}$C-value (see also Table 6.9). The $\delta^2$H-values of the substances do not provide a basis for a reliable differentiation between natural and nature-identical geraniol and geranyl acetate. One can, however, expect that the deuterium distribution determined by $^2$H-NMR will be helpful. Yet up to now corresponding data have not been published.

*Linalool* and *linalylacetate* are very important substances among the monoterpenes for the flavouring industry. They can originate from lavender, bergamotte, petitgrain oils, geranium and coriander, but also from nature-identical origin *[230]*. As the latter are made on the basis of other natural compounds, one cannot expect large differences, especially in $^{13}$C and $^2$H *[3, 230]*. Great efforts have also been made to establish a reliable differentiation as to the (botanical) origins of the sample *[3, 230, 307, 327, 328]*. The situation is somewhat more encouraging for linalylacetate, where the natural substance has a significantly more positive $\delta^{13}$C-value than the nature-identical product *[230, 307, 327]*, probably due to the very negative $\delta^{13}$C-value of the acetyl group *[307, 327]* (see 6.2.2.4.4). However, even this difference will not be sufficient for a definitive origin assignment. The mean $\delta^2$H-values of natural linalool and linalylacetate are significantly lower than those of the synthetic counterparts (-297 and -196‰, respectively *[3]*), but in some cases contradictory results have also been reported *[3, 230]*. A multielement isotope analysis of linalool and linalyl acetate from lavender oils in combination with enantioselective MDGC-P-IRMS (see 6.2.2.2.2, Fig. 6.14) provides the optimal basis for a discrimination. Most indicative are the $\delta^{18}$O-values of the compounds, which are in the range as expected (see above) and the differences between the $\delta^{13}$C- and $\delta^{18}$O-values of linalool and the acetate moiety of linalyl acetate *[211]*. A three-dimensional display of the results (Fig. 6.18) demonstrates the successful application of the methodology.

*Citral* is a mixture of the cis- and trans-isomer aldehydes geranial and neral, which are very important as widespread components in various flavours. The origin assignment to the main natural sources, *Litsea cubeba* oil and lemon grass oil, respectively, can easily be performed, because lemon grass (*Cymbopogon citratus*) is a $C_4$-plant, providing $\delta^{13}$C-values of about -12‰, whereas $\delta^{13}$C-values around -26‰ are expected for citral from other sources (lemon, orange, *Litsea cubeba* oil) *[325]*. The differentiation of the $C_3$-originating natural compound from nature-identical products (pinene or petrochemical products as precursors) is more difficult (Table 6.7). However, the deuterium contents have been found to provide a reliable differentiation between natural $C_3$-plant and nature-identical citral ($\delta^2$H-values of ≤ -220‰ and -15 to -180‰, respectively), even though some samples, presumably made from pinene, can also have very negative $\delta^2$H-values (from -290 to -310‰).

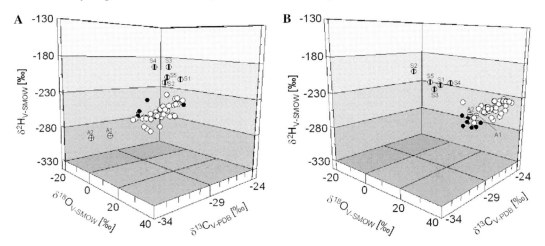

*Fig. 6.18:* (A) Multielement IRMS analysis of linalool from different origins: authentic (●), commercial (○), commercial non-authentic (⌀) and special aberrations (⊕). (B) Multielement IRMS analysis of linalyl acetate from different origins: authentic (●), commercial (○), commercial non-authentic (⌀) and special aberrations (⊕). Reproduced from reference [211]; with kind permission of Springer Verlag

## Cyclic Monoterpenes

*Limonene* is widespread in essential oils. The main source for (+)-limonene are citrus peel oils. From this source it is obtained in large quantities as a by-product of the orange juice production. Therefore nature-identical products (e.g. from pinene) are not of great importance. The $\delta^{13}$C-values of natural limonene from orange peel oil has been found to range from -26.1 to -28.5‰, and the reported $\delta^2$H-values are from -215 to -264‰, in agreement with results for other natural isoprenoids.

The bicyclic hydrocarbons *α-pinene* and *β-pinene* are not flavour compounds by themselves, but they are important as starting materials for the partial industrial synthesis of other isoprenoids, and they are produced in large quantities from turpentine oil. The $\delta^{13}$C-values of the compounds are somewhat more negative than those of other monocyclic isoprenoids like limonene, ranging from -26.6 to -30.7‰, and also their $\delta^2$H-values express a large deuterium depletion, covering a range of -280 to -310‰. According to $^2$H-NMR analysis (see 6.2.2.4.4) this depletion is mainly located at the pinene bridge, probably due to an isotope effect of the ring formation reaction *[230, 334, 335]*.

The $\delta^{13}$C-values of *menthol* are in the normal range and not suitable for a discrimination between natural and nature-identical origin. The same is true for *α-terpineol*. For both products, $^2$H-NMR measurement could perhaps give a reliable differentiation of the origin, but no data have been published on this topic until now.

The ketones *menthone, carvone* and *pulegone* are important starting materials for some herb and spice flavourings. Because of the high demand, menthone and carvone are often synthesised from other natural precursors (pinene, limonene). In spite of the fact that pinene is relatively depleted in $^{13}$C and $^2$H (see above), the differences between the nature-identical and the authentic material are not sufficient to be a reliable criterion for the assignment of its origin. Again here the $^2$H-NMR method and the $\delta^{18}$O-determination should be introduced as a means for a reliable determination of the origin.

*α-Ionone* is an important substance in raspberry flavour. It is a terpene-originating biological degradation product. In accordance with this origin the very negative $\delta^{13}$C-values of natural α-ionone are not unexpected (-31.0 to -33.4‰). On the other hand, α-ionone produced by fermentation, often on the basis of $C_4$-plant material (corn starch hydrolysate), is easily detected, even in mixtures, by $^{13}$C-GC-IRMS. However, compounds produced on the basis of $C_3$-plant precursors will not easily be identified by their $\delta^{13}$C-values. α-Ionone is available also as totally synthetic. The $\delta^{13}$C-values of this product have been found to be more positive than those of the natural substance (range from -24.3 to -28.1‰). The $\delta^2$H-values of natural and the fermentation-based ionones are remarkably more negative than those of the nature-identical counterparts (-205 to -296‰ and -55 to -172‰, respectively). Consequently a combination of $^{13}$C- and $^2$H-measurements should allow a secure proof of the origin of this substance (application of a multicompound-multielement-IRMS, see 6.2.2.4.4).

*Nootkatone* is a sesquiterpenoid carbonyl compound important in citrus (grapefruit) flavours. The $\delta^{13}$C-value (-30.9‰) is more negative by 6‰ than that of primary substances from the same origin *[336]*, but it is not sufficiently indicative for the discrimination from a nature-identical substance prepared from a hydrocarbon of orange oil ($\delta^{13}$C-values from -31.4 to -31.8‰). The deuterium content of natural nootkatone is somewhat more negative than that of the nature-identical product (-160‰ and -112 to -124‰, respectively). In both cases this value is unexpectedly positive for a natural isoprenoid, probably due to a hydrogen isotope effect accompanying the ring formation or oxidation. Again, further information is expected from $\delta^{18}$O-values measurements or a positional $^2$H-NMR analysis.

### 6.2.2.4.4 Intermolecular Isotopic Correlations and Isotopic Patterns

As already pointed out in chapter 6.2.2.1, isotope abundances, especially of primary products, are primarily fixed by those of their original elemental source. Within a "closed system", e.g. a plant, all secondary compounds must be isotopically correlated to the primary assimilation products (carbohydrates), and their isotope abundances or patterns can be characterised by moderate shifts from here in the course of their biosynthesis by isotope effects of the reactions involved. These will only become manifest in the case of metabolic branching and would, depending on material fluxes in different directions, lead to isotopically depleted compounds in one direction, and for compensation, because of an isotopic balance, to enriched products in the other *[33]*. The differences between individual compounds would depend on the magnitude of the isotope effects in question and reverse to the ratio of the metabolite fluxes in both directions. This will be true for complete molecules, as well as for molecule parts and even for defined positions within a molecule.

Most characteristic is the average difference of the $\delta^{13}C$-values between the main primary and secondary products of plants, cellulose and lignin ($\delta^{13}C$-difference 3.3 to 7.2‰; *[241]*). Similar reproducible differences are normally found between other primary and secondary products or even amongst secondary products, and this fact is therefore the basis for an intermolecular standardisation. In the case of molecules with parts originating from different pools or pathways, a δ-value difference will even become manifest between these moieties. A classic example is the quite constant $\delta^{13}C$-value difference of 5.5‰ between the sugar moiety (primary product) and many aglycons (secondary products) in glycosides *[242]*. In any case, the non-statistical $^{13}C$-distribution of primary products *[107]* will be transferred to all descendants, except for molecule positions which are directly concerned in reactions with isotope effects. In cases where these characteristic inter- or intramolecular correlations are not found, one might suspect that a compound from a foreign source had been added.

***Table 6.8:*** *$\delta^{13}C$-values ([‰]$_{V-PDB}$) of monoterpene flavouring compounds in lemon oils of different origins. Adapted from [325] with kind permission. Copyright [1993] Springer Verlag. All identified compounds are characterised by their relative amounts (concentration c [area %]) as calculated after the peak heights of the GC detector, and by their $\delta^{13}C$-values. The $\delta^{13}C$-values are the average of three measurements obtained by GC-C-IRMS*

| Compound | Italy | | Spain | | Argentina | |
|---|---|---|---|---|---|---|
| | $\delta^{13}C$[‰]$_{V-PDB}$ | c | $\delta^{13}C$[‰]$_{V-PDB}$ | c | $\delta^{13}C$[‰]$_{V-PDB}$ | c |
| β-Pinene | -27.71 | 13.37 | -27.41 | 11.07 | -28.17 | 12.81 |
| Limonene | -27.90 | 60.70 | -27.72 | 61.51 | -28.45 | 63.17 |
| γ-Terpinene | -29.62 | 11.01 | -29.27 | 10.44 | -30.38 | 9.33 |
| Nerol | -28.43 | 0.18 | -27.18 | 0.11 | -28.31 | 0.33 |
| Geraniol | -29.30 | 0.04 | -27.41 | 0.05 | -28.48 | 0.55 |
| Neral | -26.58 | 0.96 | -25.72 | 0.67 | -27.07 | 1.12 |
| Geranial | -25.98 | 1.42 | -25.36 | 0.99 | -26.60 | 1.79 |
| Neryl acetate | -30.19 | 0.91 | -29.53 | 0.70 | -30.58 | 0.64 |
| Geranyl acetate | -30.00 | 0.53 | -29.48 | 0.60 | -29.82 | 0.45 |

Intermolecular isotopic correlations are thus indicative for the authenticity of natural flavour mixtures. The method for their assessment is GC-C/P-IRMS (combustion/ reductive pyrolysis). As an early example for intermolecular isotope correlations, the result of a GC-C-IRMS analysis of the essential oil from *Coriandrum sativum* is given in Fig. 6.19 *[327]*. Further examples are corresponding analyses of oils from *Artemisia vulgaris* *[327]*, *Coriandrum sativum* *[337]* and various lemon oils (Table 6.8) *[325, 338, 339]*. In any of these cases typical correlations are found between the compounds of the same origin, even when their average δ-values may differ between products of different geographical origin. Thus it turns out that it is not the δ-value of the whole essential oil or of its main components by itself that is characteristic and indicative, but the intermolecular biogenetic and isotopic pattern, which is a fingerprint of the mixture.

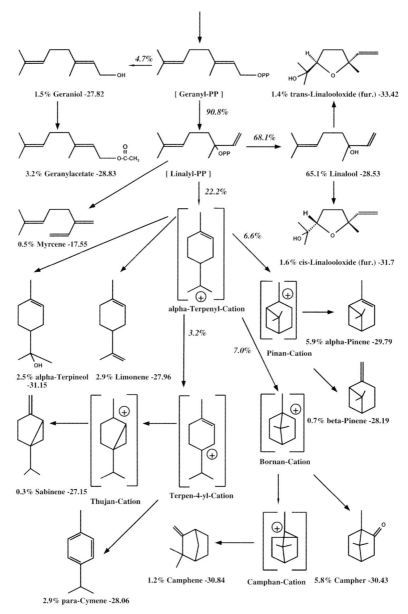

**Fig. 6.19:** *Scheme of isotopic correlations between monoterpenes from Coriandrum sativum L. All identified compounds are characterised by their relative amounts, as determined by GC, and by their δ$^{13}$C-values, obtained by GC-C-IRMS. The probable common precursors and intermediates are displayed in square brackets, and the calculated turnover rates in a given direction in % at the arrows. Modified with kind permission from [327]*

Most flavourings are complex mixtures of many compounds. As IRMS makes only sense with pure analytes, a strict purification of individual substances is indispensable. Therefore GC-IRMS has been further developed and optimised to multi-compound isotope ratio analysis by its coupling IRMS to capillary (c) and multidimensional (MD) gas chromatography (see 6.2.2.2.2). This methodology demands a strict intrinsic control and standardisation *[340]*; apart from the international standards (see Table 6.3) also secondary standards like the polyethylene foil IAEA-CH7 or the NBS22 oil are available from the IAEA in Vienna. However, as these substances are also not suitable for the direct standardisation of data from a coupled GC system for flavour isotope analysis, certificated tertiary laboratory standards for hydrogen have been developed by parallel analysis of flavour compounds by TC/EA-IRMS and MDGC-P-IRMS *[210]*.

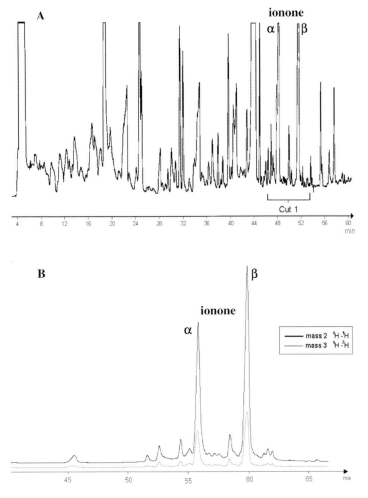

***Fig. 6.20:*** *Application of MDGC-P-IRMS (Fig. 6.14) for the isotope analysis of the most important components of a raspberry extract. Precolumn (A, FID) and main column (B, SIM detection) chromatogram. Reproduced from reference [210] with kind permission of the American Chemical Society*

The system by Sewenig et al. *[210]* as displayed in Fig. 6.14 has been used to isolate the two main compounds of a raspberry flavour, E-(α)- and E-(β)-ionone (Fig. 6.20) for carbon and hydrogen isotope analysis; a correlation diagram of the data from various authentic samples and extracts from brandy, yoghurts and syrups permitted the discrimination between products from natural and nature-identical origin.

A first step towards the analysis of isotopic patterns is the isotopic analysis of molecule parts, which must also have, in case of natural origin, intramolecular isotope correlations; in this case, pure compounds in sufficient amounts must be available. The performance of the molecule degradation depends on its structure, but in many cases a simple hydrolysis will be of practical value. Examples are compiled in Tables 6.9 and 6.10. They represent typical groups of compounds, as esters and glycosides, which can easily be hydrolysed, or a few other compounds which can be split in two parts by a suitable other method, and which can subsequently be converted into $CO_2$.

***Table 6.9:*** *Isotope distribution on different parts of esters and glycosides. Comparison of natural and nature-identical compounds. Data for the first seven compounds are $\delta^{13}C$ [‰]$_{V-PDB}$-values, for the last three compounds are $\delta^{18}O$ [‰]$_{V-SMOW}$-values. Data after [108] with kind permission and [341]*

| Compound | Molecular structure and molecule parts | | δ-value [‰]$_{Standard}$ | | | | | |
| --- | --- | --- | --- | --- | --- | --- | --- | --- |
| | | | Natural | | | Nature-identical | | |
| | A | B | A + B | A | B | A + B | A | B |
| Naringin | Aglycon | -Carbohydrate | -27.5 | -29.6 | -24.4 | | | |
| Hesperidin | Aglycon | -Carbohydrate | -26.5 | -28.3 | -23.3 | | | |
| Isoamylacetate | $C_5H_{11}$-O | -CO-$CH_3$ | -23.8[a] | -25.1 | -20.4 | -28.6 | -27.6 | -31.3 |
| cis-3-Hexenylacetate | $C_6H_{12}$-O | -CO-$CH_3$ | -34.8 | -35.5 | -32.7 | -32.9 | -28.0 | -47.6 |
| Linalylacetate | $C_{10}H_{18}$-O | -CO-$CH_3$ | -27.2 | -26.8 | -29.2 | -35.3 | -32.2 | -50.8 |
| Isoamylbutyrate | $C_5H_{11}$-O | -CO-$C_3H_7$ | -12.9 | -13.9 | -11.7 | -27.8 | -27.9 | -27.2 |
| Methylsalicylate | HO-$C_6H_4$-CO | -O-$CH_3$ | -32.6 | -30.8 | -45.5 | -27.7 | -27.5 | -29.1 |
| cis-3-Hexenylacetate | $C_6H_{12}$-O | -CO-$CH_3$ | +34.7 | +4.0 | +65.4 | +0.4 | -15.3 | +16.1 |
| Linalylacetate | $C_{10}H_{18}$-O | -CO-$CH_3$ | +19.8 | +8.0 | +31.6 | +15.6 | +9.6 | +21.6 |
| Isoamylbutyrate | $C_5H_{11}$-O | -CO-$C_3H_7$ | +22.9 | +7.7 | +38.1 | +21.7 | +1.9 | +41.5 |

[a] Probably of biotechnological origin based at least partially on a $C_4$-plant carbon source.

This is for example easily possible with the hydrolysis products of cyanogenic glycosides or glucosinolates, the cyanhydrines or mustard oils, respectively. The original α-C-atom of the precursor amino acid can be isolated from these compounds and converted into $CO_2$. In cyanogenic glycosides and glucosinolates from the amino acids Phe or Tyr, this position was enriched by about 12‰ in $^{13}C$ relative to the rest of the molecule *[242]*. In contrast, this position was depleted by approximately 6‰ in glycosinolates from homocysteine (sinalbin) *[307]*. The biochemical reason for this finding is probably understood, but not topic of this investigation. Finally, it has to be mentioned that for some terpenes fragmentations by ozone degradation may be possible *[334, 335]*, and that for intermediates of the glycolysis or citric acid cycle combinations of chemical and enzymatic degradation reactions can be used *[342]*.

*Table 6.10:* Carbon isotope distribution in parts of natural molecules and their nature-identical analogues. Data adapted from reference [108] with kind permission and unpublished results [307]. n.d. = not determined

| Compound | Molecular structure and molecule parts | | $\delta^{13}C$-value [‰]$_{V-PDB}$ | | | | | |
| --- | --- | --- | --- | --- | --- | --- | --- | --- |
| | | | Natural | | | Nature-identical | | |
| | A | B | Total | A | B | Total | A | B |
| Acetic acid | $CH_3$ | -COOH | -27.6 | -28.5 | -26.7 | -31.3 | -27.0 | -35.6 |
| Benzoic acid | $C_6H_5$ | -COOH | -29.8 | -29.7 | -29.8 | -28.9 | -27.9 | -34.9 |
| Cinnamic aldehyde | $C_6H_5$-CH | =CH-CHO | -28.4 | -28.8 | -26.8 | -29.3 | -28.1 | -33.4 |
| Cyanhydrine from amygdaline/ glucosinalbin | R-$C_6H_4$-CH(OH) | -CN | -30.6 | -32.3 | -18.8 | n.d. | n.d. | n.d. |
| Mustard oil from sinigrine | $CH_2$=CH-$CH_2$ | -N=C=S | -28.3 | -26.3 | -34.3 | -29.4 | -24.0 | -45.5 |

In any case it is important and advantageous to apply knowledge of (bio)chemical reaction mechanisms and/or implied isotope effects for the prediction or interpretation of isotopic patterns. For example in the case of the formation of esters, while remarkable hydrogen isotope effects are unlikely, the manifestation of isotope effects on carbon and oxygen atoms involved will depend on the relative amount of acid and alcohol applied for reaction [343]. Correspondingly, Rabiller et al. [344] did not find a significant difference on the deuterium abundance of the alcohol or acid moiety in the course of the esterification reaction, while on the other hand we found a dependence of $^{13}C$- and $^{18}O$-abundances on the conditions of esterification [343]. Therefore the corresponding findings will be the basis for the reconstruction of the synthesis of a given ester.

General aspects derived from these results are:

- Glycosides do preserve the $\delta^{13}C$-values of the primary products of the plant in their sugar moiety, and the $^{13}C$-depletion of the aglycon, normally an aromate, is -2,4 to -6.0‰ [242].

- Esters show similar average $\delta^{13}C$-values of acid and alcohol moieties, when these are biochemically closely related (e.g. from the same source). They can vary, when the two parts are from different (bio)chemical pathways. The $\delta^{18}O$-value of a given moiety of an ester is the more positive, the closer its metabolic connection is to carbohydrates. $\delta^{13}C$- and $\delta^{18}O$-values of both moieties depend in addition on the kind of esterification involved.

- Methyl groups of natural products are originating from S-adenosylmethionine, and they are generally depleted in $^{13}C$ by a few ‰ relative to other molecule parts [108, 238, 243, 244, 345]. However, also synthetic products can express a sometimes extreme depletion in this group, because its synthesis from $CH_3OH$, $CH_3I$ or $H_2N$-$CH_3$ may be accompanied by a large isotope discrimination. Hence the depletion of a methyl group by itself may not always be indicative.

Determinations of isotope patterns by SNIF-NMR® (see 6.2.2.3.2) are by far of more practical importance (for an early review see *[317]*). Although up to now this methodology is nearly exclusively limited to the elucidation of $^2$H-patterns, recent attempts for its extension to $^{13}$C-pattern determinations have been promising. Normally only pure substances are used for $^2$H-NMR measurements; $\geq 150$ mg sample and several hours' scans are needed. In the present context NMR spectral data of aromatic compounds, terpenes and some aliphatic compounds have been compiled.

The $^2$H-NMR spectra correspond qualitatively to the $^1$H-NMR spectra. Their signal assignment nomenclature follows their (descending) chemical shifts. The area of a given signal relative to the total signals' integral is the molar fraction $f_i$ of the corresponding isotopomer. The positional deuterium content $(D/H)_i$ in ppm (not to be confused with the chemical shift) can be calculated from $f_i$ by means the signal integral of an intrinsic standard (TMU = tetramethylurea) or by mass spectrometric determination of the average deuterium content of the compound $\overline{(D/H)}$ in question. For a better systematic presentation and comparison of the data of analogue compounds from different origins and syntheses the results are processed as canonical variables by a multifactorial calculation program *[346]*.

The most extensively investigated aromatic compounds are benzaldehyde (bitter almond oil) and vanillin. In agreement with the rule for descendants of the shikimic acid pathway as outlined in 6.2.2.1, the relative intensity of the $(D/H)_i$-values on the aromatic nucleus of the natural representatives of these compounds is $p > o \geq m$ and $m > o$, respectively. The signals of aromatic compounds with aliphatic side chains may not always be directly separated, but a separation will be attained after the oxidation of the side chain to a carboxyl function *[347]*. Natural benzaldeyde is obtained from amygdaline, a glycoside from bitter almonds, prune and other fruit kernels; it is furthermore produced by retroaldol reaction from cassia oil, main content cinnamaldehyde, and from toluene by catalytic oxidation or via benzalchloride. One of the first extensive studies on the product was performed by Hagedorn *[280]* (see Fig. 6.21 and Table 6.11). Whereas in this study, the signals for the p- and the m-isotopomers have not yet been completely separated, this has perfectly been performed in the work by Rémaud et al. *[348]*, also including data on cinnamon oil; this paper would be the optimal basis for everybody starting into the field.

Apart from some earlier papers (see Table 6.14) the most important research on vanillin originates from the same group *[277]*; this paper compiles the data of more than 140 samples originating from vanilla beans from lignin and guaiacol. It also includes data on the by-product of vanilla oil, p-hydroxybenzaldehyde and the average $\delta^{13}$C-values of both analytes. The results have recently been confirmed by an intensive additional study on all ingredients of vanilla beans from a limited defined area *[349]*. Further information on the authenticity of vanillin can be obtained, as already mentioned in 6.2.2.3.2, from the $^{13}$C-pattern of the compound *[238, 239]*. Even $^{17}$O-NMR data have been provided *[350]* but have not been confirmed by the $^{18}$O-pattern of vanillin obtained by degradation *[253]*.

**Table 6.11:** *Site-specific deuterium distribution in benzaldehyde by source. Adapted from [280] with kind permission, Copyright [1992] American Chemical Society. Values f [i] are the mole fractions of site-deuterated molecular species i. ADR (aromatic distribution ratio) = f [ortho]/f [meta + para]. The $\overline{(D/H)}$ are expressed in total parts per million [ppm] and are determined by standard combustion and IRMS*

| Source | Mole fraction of deuterium | | | | |
|---|---|---|---|---|---|
| | f [formyl] | f [ortho] | f [meta + para] | ADR | average ppm (±2) |
| synthetic/$C_6H_5CH_3$ / $O_2$ | 0.5449 | 0.1849 | 0.2702 | 0.6813 | 252 |
| synthetic/ $C_6H_5CH_3$ / $Cl_2$ | 0.1696 | 0.3326 | 0.4976 | 0.6681 | 151 |
| cassia $C_6H_5$-CH=CH-CHO | 0.1538 | 0.3630 | 0.4833 | 0.7518 | 142 |
| bitter almond oil | 0.1543 | 0.3117 | 0.5332 | 0.5906 | 140 |

**Table 6.12:** *Site-specific stable hydrogen isotope ratios of R(-)-linalool extracted from different botanical sources or prepared from fossil materials. Adapted from [230] with kind permission. Copyright [1992] American Chemical Society. The site-specific D/H-ratios are expressed in parts per million [ppm], position indication see Fig. 6.21*

| Botanical origin | $(D/H)_1$ | $(D/H)_2$ | $(D/H)_3$ | $(D/H)_4$ | $(D/H)_5$ | $(D/H)_6$ | $(D/H)_7$ | $(D/H)_8$ | $(D/H)_9$ | $(D/H)_{10}$ |
|---|---|---|---|---|---|---|---|---|---|---|
| lavender | 25.3 | 123.4 | 102.3 | 129.8 | 169.8 | 181.1 | 107.8 | 116.4 | 126.8 | 123.1 |
| spike lavender | 35.3 | 132.7 | 109.5 | 139.9 | 162.7 | 162.0 | 106.7 | 126.6 | 120.9 | 153.5 |
| bois de rose | 23.0 | 133.9 | 139.3 | 138.2 | 176.6 | 167.9 | 93.9 | 97.2 | 119.3 | 135.4 |
| bitter orange | 41.7 | 124.4 | 120.7 | 134.1 | 177.2 | 172.3 | 111.0 | 124.8 | 97.0 | 130.8 |
| bergamot | 30.4 | 135.2 | 123.3 | 142.6 | 168.6 | 172.4 | 120.9 | 117.4 | 104.4 | 126.3 |
| coriander | 48 | 132 | 150 | 141 | 176 | ≈165 | 92 | 128 | 120 | 130 |
| camphor | 20 | 132 | 123 | 141 | 173 | 181 | 89 | 140 | 91 | 125 |
| natural | 30.2 | 128.7 | 119.0 | 135.3 | 171.0 | 174 | 105.9 | 117.2 | 117.0 | 130.5 |
| synthetic | 106.8 | 102.6 | 126.5 | 89.9 | 153.7 | 128.4 | 135.0 | 131.9 | 140.3 | 140.0 |
| hemisynthetic | 26.4 | 138.6 | 167.2 | 132.0 | 162.8 | 172.7 | 104.9 | 88.7 | 122.1 | 138.6 |

Among the $^2$H-NMR data of isoprenoids practically exclusively those of monoterpenes are of interest in the present context. As already outlined before, these compounds are normally synthesised in plants via the deoxyxylulose-phosphate pathway and show therefore $^2$H-patterns in agreement with their common biosynthesis and its modulations *[111]*, e.g. in the case of R(-)-linalool, distinct deuterium depletions at position C-2 and C-9, corresponding to $(D/H)_1$ and $(D/H)_7$ in the NMR nomenclature but individual shifts from a common mean value, depending on the botanical origin *[230]* (see Fig. 6.22 and Tables 6.12 and 6.13).

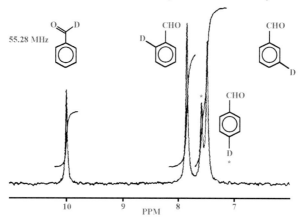

***Fig. 6.21:*** *Natural abundance 55.28 MHz $^2$H-NMR spectrum of benzaldehyde with indication of the signals for the different isotopomers. Results: see Table 6.11. Reprinted with kind permission from [280]. Copyright [1992] American Chemical Society*

A very extensive systematic review of 21 monoterpenes with geranilane, p-menthane and pinane/camphene skeletons has recently been published by Martin et al. *[351]*. Data on individual representatives are compiled, together with references on other compounds, in Table 6.14.

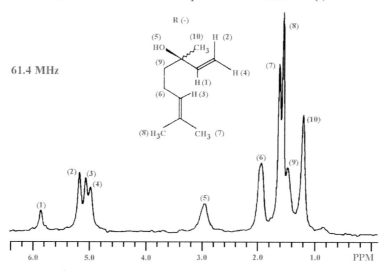

***Fig. 6.22:*** *61.4 MHz $^2$H-NMR spectrum of natural R(-)-linalool and assignment of the signals to individual positions or isotopomers. Results of relative amounts in products of different origin: see Table 6.12. Reprinted with kind permission from [230]. Copyright [1992] American Chemical Society*

*Table 6.13:* Site-specific stable hydrogen isotope analysis of linalylacetate by SNIF-NMR®. Adapted from [230] with kind permission. Copyright [1992] American Chemical Society. The overall D/H- and ($^{13}C/^{12}C$)-isotope ratios are determined by combustion and IRMS. $\delta^{13}C$-values are expressed in [‰] vs. V-PDB, the overall D/H- and the site-specific D/H-ratios in parts per million [ppm], position indication see Fig. 6.22; $(D/H)_5$ is the methyl group of the acetyl residue in this case

| Status | $\delta^{13}C$ | $\overline{(D/H)}$ | $(D/H)_1$ | $(D/H)_2$ | $(D/H)_3$ | $(D/H)_4$ | $(D/H)_5$ | $(D/H)_6$ | $(D/H)_7$ | $(D/H)_8$ | $(D/H)_9$ | $(D/H)_{10}$ |
|---|---|---|---|---|---|---|---|---|---|---|---|---|
| natural | -28.1 | 123.2 | 40.1 | 163.7 | 114.9 | 114.3 | 154.5 | 169.0 | 94.2 | 107.2 | 131.0 | 104 |
| synthetic | -30.4 | 128.4 | 105.3 | 77.8 | 140.4 | 114.7 | 206.2 | 126.9 | 89.9 | 134.7 | 131 | 119.8 |

*Table 6.14:* Authenticity and origin assignment of selected examples of flavouring compounds by deuterium-NMR. Further data on monoterpenes see [225, 227, 338, 351, 352]; for $^{13}C$-NMR see references [238, 239] and chapter 6.2.2.3.2., $^{18}O$-pattern of vanillin [253]

| Compound | | Deuterium pattern analysis | | |
|---|---|---|---|---|
| Name | Origin* | Aim | Result | Reference |
| Anethol | fennel, star anise, estragol, syn. | distinguish between different origins | botanical origin and environm. influences | [225, 283, 353] |
| Benzaldehyde | bitter almonds, cassia oil, syn. | to identify origin | possible, even analysis of mixtures | [280, 348] |
| 2-Phenylethanol | n. | distinguish n., syn. | possible | [284] |
| 2-Phenylethylacetate | syn. | syn. or biotechnol. Origin | possible | [284] |
| 2-Phenylacetate | n., syn. | Discrimination | possible | [347] |
| Eugenol | clove, bay | identify botan. origin | ? | [227] |
| Vanillin | *vanilla sp.*, lignin, synthesis | distinguish between various origins | discriminat. of nat., syn., semi-syn. origin | [11, 225, 353, 354, 355, 277, 349] |
| Methyl salicylate | wintergreen a. other essent. oils | distinction from syn. cpds. | possible | [356] |
| Salicin | *Salix sp.*, commerce | distinguish nat., syn. | possible | [364] |
| Coumarin | Tonka beans a. other nat. origin., syn. | distinguish nat., syn. | possible | [365] |
| Raspberry ketone | *Taxus baccata* and commercials | discriminat. nat., syn., biotechnol. | possible | [255] |
| Camphor | camphora tree, syn. from α-pinene | to identify semi-syn. products | possible | [351, 352, 357] |
| Carvone | spearmint, caraway, semisyn. limonene | to identify botanical origin | possible | [338, 351, 352] |

* n. = natural, n. i. = nature-identical, syn. = synthetic

|  Compound  |  | Deuterium pattern analysis |  |  |
|---|---|---|---|---|
| Name | Origin* | Aim | Result | Reference |
| Citral, Geraniol | n., syn. | distinguish between origins | possible | [227, 351, 352] |
| Limonene | orange | biosynthesis and optical isomers | possible | [358, 351, 352] |
| Linalool | lavender, orange, bergamot, coriander | to identify botanical origin | discrimination syn. a. botanical origin | [230, 317, 351, 352] |
| Linalylacetate | biological, syn. | assign origin | possible | [230, 351, 352] |
| α-Pinene | pinus sp. | biosynthesis and optical isomers | possible | [225, 359, 351, 352] |
| Acetic acid**, other Aliphates | n., syn. | distinguish natural and syn. origin | possible | [11, 354, 360, 361] |
| δ-Decalactone | n., syn. | origin assignment | possible | [362] |
| (R)-δ-Decanolide | n., syn. | origin assignment | possible | [321] |
| (Z)-3-Hexenol | n., syn. | distinguish n., syn. a. biotechnol. origin | possible | [319] |
| n-Hexanol | n., syn. | distinguish n. a. syn. | possible | [318] |
| Esters | n., syn. | distinction | possible | [344] |

\* n. = natural, n. i. = nature-identical, syn. = synthetic
\*\* also $^{13}$C-NMR

### 6.2.2.5 Perspectives and Future Aspects

Isotope ratio measurements entered quality control of flavouring compounds about 20 years ago, and today this methodology has been established as a routine test in a number of laboratories. The special field and the potential for isotope analysis is authenticity identification of flavourings and their components, mainly with regard to the differentiation between natural and nature-identical or synthetic products and to their assignment to botanical and climatic origins. To this end isotope analysis completes and enlarges the quality information provided by classic methods.

With this in mind this contribution has tried to show the state of the art and to outline the potential, but also the limits of the methods available and the information to be obtained. This will be of advantage and importance for producers, dealers and consumers, but also for adulterators, in so far as the latter try to conceive more and more sophisticated ways of blending. This in turn demands the adaptation and further development of the analytical methods. We believe that this will mainly concern methods which lead to the elucidation of inter- and intramolecular isotopic patterns, which will not and cannot be imitated, not only from a technical but also from a financial point of view. A very important progress of the last five years is that we have learned to understand and use the systematics of the isotopic patterns of natural compounds.

For certain aroma compounds carbon and hydrogen isotope analysis has meanwhile been supplemented by that of oxygen, nitrogen and/or sulphur stable isotopes. The substances that have been investigated up to now are allylisothiocyanate (mustard oil) using $^{15}$N- and $^{34}$S-analysis, dimethylsulphide (onion oil, garlic oil) using $^{34}$S-analysis *[293, 363]*, dimethylanthranilate (grapefruit oil) using $^{15}$N- *[247, 248]* and esters using $^{18}$O-determination *[343]*. Due to the increasing demand for "salty" flavours, which often contain nitrogen and/or sulphur atoms, the isotopic analysis of both elements in flavours will be more important in the future.

Without any doubt the most valuable development in mass spectrometry has been multi-compound/multi-isotope analysis. This implies the application of GC-C/P-IRMS to the on-line analysis, not only of carbon but also of other isotopes, preferably of hydrogen and oxygen, in the individual components of a mixture, and the use of the metabolic and isotopic correlations obtained from such an analysis. In the course of this chapter, the potential of (positional) oxygen isotope analysis has been emphasised several times and this will still be a challenge of the future. The advantage of GC-C/P-IRMS is its speed in performance and the very moderate demand on sample size and purity, and also its implication for automation. The information available can easily be correlated to that of other (classic) analyses. However, a disadvantage will be always that the data concern a global mean value for the whole molecule in question.

NMR has provided a lot of detailed information on individual compounds as a fingerprint and non-adulterable criterion for its (bio)synthetic and botanical origin. Especially here, the progress of the biochemical interpretation of the elucidated isotope patterns has enlarged the potential of this method. This will probably also be true for $^{13}$C- and perhaps $^{17}$O-NMR data, and therefore a great potential is seen in the exploitation and the development of this method as an additional tool in the quality analysis of flavourings. The disadvantage of NMR is still the demand for relative large samples, long measuring times and the need for standardisation by IRMS, but also here, enormous progress has been made recently.

The analytical scope applied to a given problem and the selection of the most suitable method will therefore be governed by the individual problem, and not least by its legal and commercial value but it will remain mainly a question of the personal experience and knowledge of the experimentalist. As already outlined, the investment for NMR equipment exceeds that for GC-C-IRMS combination by a factor of three to four, while the running costs, including personnel, are approximately the same for both methods. Since under these aspects not all possibilities can be applied for any one compound in its routine analysis, corresponding strategies and the optimum use of the methods available have to be developed. Therefore, in special cases, strategies of stepwise applications have to be adapted, as demonstrated for vanillin *[354]*, or for linalool *[211]* where the information to be obtained from each method is adapted to the corresponding possibilities of adulteration. In any case the natural variation of data for each sample has to be taken into account. For a definite statement the results of classic and isotope analysis have to be correlated *[230, 338]*. While each of these methods has its special area for providing optimum information, they are ultimately supplementary.

## REFERENCES

[1] H. Mommsen (1986): Archäometrie. Teubner Verlag, Stuttgart, pp. 202-236
[2] I. Karlén, I.U. Olsson, P. Kållberg, S. Kilicci (1964): Absolute determination of the activity of two $C^{14}$ dating standards. Arkiv för Geofysik 4, 465-471
[3] R.A. Culp, J.E. Noakes (1992): Determination of synthetic components in flavors by deuterium/hydrogen isotopic ratios. J. Agric. Food Chem. 40, 1892-1897
[4] P.G. Hoffman, M.C. Salb (1980): Radiocarbon ($^{14}C$) method for authenticating natural cinnamic aldehyde. J. Assoc. Off. Anal. Chem. 63, 1181-1183
[5] M.P. Neary, J.D. Spaulding, J.E. Noakes, R.A. Culp (1997): Tritium analysis of burn-derived water from natural and petroleum-derived products. J. Agric. Food Chem. 45, 2153-2157
[6] E.R. Schmid, I. Fogy, E. Kenndler (1978): Beitrag zur Unterscheidung von Gärungsessig und synthetischem Essig durch die Bestimmung der spezifischen $^3$H-Radioaktivität. Z. Lebensm. Unters. Forsch. 166, 221-224
[7] H.-L. Schmidt (1974): Analyse von stabil-isotop markierten Verbindungen. In: H. Simon (Hrsg.): Messung von radioaktiven und stabilen Isotopen, Band 2. Springer Verlag, Berlin, pp. 291-400
[8] H.-L. Schmidt, E. Schmelz (1980): Stabile Isotope in Chemie und Biowissenschaften. Chem. unserer Zeit 14, 25-34
[9] F.J. Winkler (1984): Application of natural abundance stable isotope mass spectrometry in food control. In: A. Frigerio, H. Milon (eds.): Chromatography and Mass Spectrometry in Nutrition Science and Food Safety. Elsevier Science Publishers, Amsterdam, pp. 173-190
[10] H.-L. Schmidt (1986): Food quality control and studies on human nutrition by mass spectrometric and nuclear magnetic resonance isotope ratio determination. Fresenius Z. Anal. Chem. 324, 760-766
[11] H.-O. Kalinowski (1988): Der Schrecken der Weinpanscher. Chem. unserer Zeit 22, 162-171
[12] R.C. Weast (ed.) (1977-1978): CRC Handbook of Chemistry and Physics, 58$^{th}$ edition. CRC Press Inc., Palm Beach, Florida, USA.
[13] R.A. Werner, W.A. Brand (2001): Referencing strategies and techniques in stable isotope ratio analysis. Rapid Comm. Mass Spectrom. 15, 501-519
[14] D.R. Lide (1999): Handbook of Chemistry and Physics. 80$^{th}$ Edition. CRC Press, Boca Raton, p. 1-10
[15] R. Gonfiantini (1978): Standards for stable isotope measurements in natural compounds. Nature 271, 534-536
[16] H. Craig (1957): Isotopic standards for carbon and oxygen and correction factors for mass-spectrometric analysis of carbon dioxide. Geochim. Cosmochim. Acta 12, 133-149
[17] T.B. Coplen (1996): New guidelines for reporting stable hydrogen, carbon, and oxygen isotope-ratio data. Geochim. Cosmochim. Acta 60, 3359-3360
[18] G. Junk, H.J. Svec (1958): The absolute abundance of the nitrogen isotopes in the atmosphere and compressed gas from various sources. Geochim. Cosmochim. Acta 14, 234-243
[19] A. Mariotti (1983): Atmospheric nitrogen is a reliable standard for natural $^{15}N$ abundance measurements. Nature 303, 685-687
[20] T.B. Coplen, H.R. Krouse, J.K. Böhlke (1992): Reporting of nitrogen-isotope abundances (Technical Report). Pure & Appl. Chem. 64, 907-908
[21] M.L. Jensen, N. Nakai. (1962): Sulfur isotope meteorite standards: Results and recommendations. In: M.L. Jensen (ed.): Biogeochemistry of Sulfur Isotopes. NSF Symposium, New Haven, Ct., pp. 30-35
[22] J. Macnamara, H.G. Thode (1950): Comparison of the isotopic constitution of terrestrial and meteoritic sulfur. Phys. Rev. Ser. II 78, 307-308
[23] J. Hoefs (1987): Stable Isotope Geochemistry. 3$^{rd}$ Edition. In: A. El Goresy, W. von Engelhardt, T. Hahn (eds.): Minerals and Rocks 9. Springer-Verlag, Berlin
[24] T.B. Coplen (1988): Normalization of oxygen and hydrogen isotope data. Chem. Geol. (Isot. Geosci. Sect.) 72, 293-297
[25] J. Koziet, A. Roßmann, G.J. Martin, P. Johnson (1995): Determination of the oxygen-18 and deuterium content of fruit and vegetable juice water. An european inter-laboratory comparison study. Anal. Chim. Acta 302, 29-37

[26]  K. Russe, S. Valkiers, P.D.P. Taylor (2004): Synthetic isotope mixtures for the calibration of isotope amount ratio measurements of carbon. Int. J. Mass Spectrom. 235, 255-262
[27]  H.A.J. Meijer, W.J. Li (1998): The use of electrolysis for accurate $\delta^{17}O$ and $\delta^{18}O$ isotope measurements in water. Isot. Environm. Health Stud. 34, 349-369
[28]  IAEA (1995): Reference and intercomparison materials for stable isotopes of light elements. IAEA-TECDOC-825. IAEA, Vienna. Summary of the meeting, pp. 7-11
[29]  T.B. Coplen, H.R. Krouse (1998): Sulphur isotope data consistency improved. Nature 392, 32
[30]  T. Ding, R. Bai, Y. Li, D. Wan, X. Zou, Q. Zhang (1999): Determination of the absolute $^{32}S/^{34}S$ ratio of IAEA-S-1 reference material and V-CDT sulfur isotope standard. Science in China Ser. D 42, 45-51 (English translation journal !)
[31]  T. Ding, S. Valkiers, H. Kipphardt, P. De Bievre, P.D.P. Taylor, R. Gonfiantini, R. Krouse (2001): Calibrated sulfur isotope abundance ratios of three IAEA sulfur isotope reference materials and V-CDT with a reassessment of the atomic weight of sulfur. Geochim. Cosmochim. Acta 65, 2433-2437
[32]  ISO 31-0: Quantities and Units, Part 0. General Principles, Subclause 2. International Organization for Standardization, Geneva (1992)
[33]  H.-L. Schmidt (2003): Fundamentals and systematics of the non-statistical distribution of isotopes in natural compounds. Naturwissenschaften 90, 537-552
[34]  W.G. Deuser, E.T. Degens (1967): Carbon isotope fractionation in the system $CO_2$ (gas) - $CO_2$ (aqueous) - $HCO_3^-$ (aqueous). Nature 215, 1033-1035
[35]  K. Emrich, D.H. Ehhalt, J.C. Vogel (1970): Carbon isotope fractionation during the precipitation of calcium carbonate. Earth Planet. Sci. Lett. 8, 363-371
[36]  J.F. Marlier, M.H. O'Leary (1984): Carbon kinetic isotope effects on the hydration of carbon dioxide and the dehydration of bicarbonate ion. J. Am. Chem. Soc. 106, 5054-5057
[37]  I.E. Keeley, D.R. Sandquist (1992): Carbon Freshwater Plant. Plant Cell Environ. 15, 1012-1035
[38]  P. Deines (1980): The isotopic composition of reduced organic carbon. In: P. Fritz and J.Ch. Fontes (eds.): Handbook of Environmental Isotope Geochemistry Vol. 1, The Terrestrial Environment, A, Chapt. 9. Elsevier Scientific Publishing Company, Amsterdam, pp. 329-406
[39]  J.R. Ehleringer, C.B. Field, Z. Lin, C. Kuo (1986): Leaf carbon isotope and mineral composition in subtropical plants along an irradiance cline. Oecologia 70, 520-526
[40]  J.R. Evans, T.D. Sharkey, J.A. Berry, G.D. Farquhar (1986): Carbon isotope discrimination measured concurrently with gas exchange to investigate $CO_2$ diffusion in leaves of higher plants. Aust. J. Plant Physiol. 13, 281-292
[41]  G.D. Farquhar, M.H. O'Leary, J.A. Berry (1982): On the relationship between carbon isotope discrimination and the intercellular carbon dioxide concentration in leaves. Aust. J. Plant Physiol. 9, 121-137
[42]  A.G. Condon, R.A. Richards, G.D. Farquhar (1987): Carbon isotope discrimination is positively correlated with grain yield and dry matter production in field-grown wheat. Crop Sci. 27, 996-1001
[43]  W.J.S. Downton, W.J.R. Grant, S.P. Robinson (1985): Photosynthetic and stomatal response of spinach leaves to salt stress. Plant Physiol. 77, 85-88
[44]  J.L. Araus, H.R. Brown, A. Febrero, J. Bort., M.D. Serret (1993): Ear photosynthesis, carbon isotope discrimination and the contribution of respiratory $CO_2$ to differences in grain mass in durum wheat. Plant Cell Environ. 16, 383-392
[45]  J.L. Araus, P. Santiveri, D. Bosch-Serra, C. Royo, I. Romagosa (1992): Carbon isotope ratios in ear parts of triticale. Plant Physiol. 100, 1033-1035
[46]  C. Körner, G.D. Farquhar, Z. Roksandic (1988): A global survey of carbon isotope discrimination in plants from high altitude. Oecologia 74, 623-632
[47]  C. Körner, G.D. Farquhar, S.C. Wong (1991): Carbon isotope discrimination by plants follows latitudinal and altitudinal trends. Oecologia 88, 30-40.
[48]  H.D. Freyer (1979): On the $^{13}C$ record in tree rings II. Registration of microenvironmental $CO_2$ and anomalous pollution effect. Tellus 31, 308-312
[49]  B. Martin, A. Bytnerowicz, Y.R. Thorstenson (1988): Effects of air pollutants on the composition of stable carbon isotopes, $\delta^{13}C$, of leaves and wood, and on leaf injury. Plant Physiol. 88, 218-223
[50]  B. Martin, E.K. Sutherland (1990): Air pollution in the past recorded in width and stable carbon isotope composition of annual growth rings of Douglas-fir. Plant Cell Environ. 13, 839-844

[51]  C.D. Keeling, W.G. Mook, P.P. Tans (1979): Recent trends in the $^{13}C/^{12}C$ ratio of atmospheric carbon dioxide. Nature 277, 121-123
[52]  J. Peñuelas, J. Azcón-Bieto (1992): Changes in leaf $\Delta^{13}C$ of herbarium plant species during the last 3 centuries of $CO_2$ increase. Plant Cell Environ. 15, 485-489
[53]  E. Medina, P. Minchin (1980): Stratification of $\delta^{13}C$ values of leaves in amazonian rain forests. Oecologia 45, 377-378
[54]  R. Park, S. Epstein (1961): Metabolic fractionation of $^{13}C$ and $^{12}C$ in plants. Plant Physiol. 36, 133-138
[55]  I. Levin, R. Grand, N.B. Trivett (1995): Long-term observations of atmospheric $CO_2$ and carbon isotopes at continental sites in Germany. Tellus 47B, 23-34
[56]  H.-L. Schmidt, R.A. Werner, A. Rossmann (2000): $^{18}O$ Pattern and biosynthesis of natural plant products. Phytochemistry 58, 9-32.
[57]  Y. Bottinga, H. Craig (1969): Oxygen isotope fractionation between $CO_2$ and water, and the isotopic composition of marine atmospheric $CO_2$. Earth Planet. Sci. Lett. 5, 285-295
[58]  P. Kroopnick, H. Craig (1972): Atmospheric oxygen: isotopic composition and solubility fractionation. Science 175, 54-55
[59]  P. Kroopnick, H. Craig (1976): Oxygen isotope fractionation in dissolved oxygen in the deep sea. Earth Planet. Sci. Lett. 32, 375-388
[60]  T.B. Coplen, J.A. Hopple, J.K. Böhlke, H.S. Peiser, S.E. Rieder, H.R. Krouse, K.J.R. Rosman, T. Ding, R.D. Vocke, K.M. Révész, A. Lamberty, P. Taylor, P. De Bièvre (2002): Compilation of minimum and maximum isotope ratios of selected elements in naturally occurring terrestrial materials and reagents. USGS Water-Resources Investigations Report 01-4222, USGS, Reston Virginia, pp. 39-40.
[61]  J.R. Gat (1980): The isotopes of hydrogen and oxygen in precipitation. In: P. Fritz, J.Ch. Fontes (eds.): Handbook of Environmental Isotope Geochemistry. Vol. 1, The Terrestrial Environment A, Chapt. 1. Elsevier Scientific Publishing Company, Amsterdam, pp. 21-47
[62]  J.R. Gat (1996): Oxygen and hydrogen isotopes in the hydrologic cycle. Ann. Rev. Earth Planet. Sci. 24: 225-262
[63]  International Atomic Energy Agency (1990): Environmental Isotope Data No. 9: World Survey of Isotope Concentration in Precipitation (1984-1987). Technical Reports Series No. 311, IAEA, Vienna
[64]  W. Dansgaard (1964): Stable isotopes in precipitation. Tellus 16, 436-468
[65]  H. Craig, L.I. Gordon (1965): Deuterium and oxygen-18 variation in the ocean and marine atmosphere. In: E. Tongiorgi (ed.): Stable Isotopes in Oceanographic Studies and Paleotemperatures. Lab. Geologia Nucleare, Pisa, pp. 9-130
[66]  G.J. Bowen, L.I. Wassenaar, K.A. Hobson (2005): Global application of stable hydrogen and oxygen isotopes to wildlife forensics. Oecologia 143, 337-348
[67]  H. Craig (1961): Isotopic variations in meteoric waters. Science 133, 1702-1703
[68]  H. Ziegler (1989): Hydrogen isotope fractionation in plant tissues. In: P.W. Rundel, J.R. Ehleringer, K.A. Nagy (eds.): Stable Isotopes in Ecological Research. Ecological Studies 68. Springer Verlag, New York, pp. 105-123
[69]  J. Bricout (1978): Recherches sur le fractionnement des isotopes stables de l'hydrogène et de l'oxygène dans quelques végétaux. Rev. Cytol. Biol. Végét. Bot. 1, 133-209
[70]  A. Ferhi, R. Létolle (1977): Transpiration and evaporation as the principal factors in oxygen isotope variations of organic matter in land plants. Physiol. Vég. 15, 363-370
[71]  C. Lesaint, L. Merlivat, J. Bricout, J.Ch. Fontes, R. Gautheret (1974): Sur la composition en isotopes stables de l'eau de la tomate et du mais. C.R. Acad. Sci. Paris, Sér. D 278, 2925-2930
[72]  H. Förstel (1985): Die natürliche Fraktionierung der stabilen Sauerstoff-Isotope als Indikator für Reinheit und Herkunft von Wein. Naturwissenschaften 72, 449-455
[73]  G.B. Allison, J.R. Gat, F.W.J. Leaney (1985): The relationship between deuterium and oxygen-18 delta values in leaf water. Chem. Geol. (Isot. Geosci. Sect.) 58, 145-156
[74]  S. Epstein, C.J. Yapp, J.H. Hall (1976): The determination of the D/H ratio of non-exchangeable hydrogen in cellulose extracted from aquatic and land plants. Earth Planet. Sci. Lett. 30, 241-252
[75]  S. Epstein, P. Thompson, C.J. Yapp (1977): Oxygen and hydrogen isotopic ratios in plant cellulose. Science 198, 1209-1215

[76]   B. Holbach, H. Förstel, H. Otteneder, H. Hützen (1994): Das Verhältnis der Stabilisotopen $^{18}$O and $^{16}$O zur Beurteilung von Auslandswein. Z. Lebensm. Unters. Forsch. 198, 223-229
[77]   M.F. Estep, T.C. Hoering (1980): Biogeochemistry of the stable hydrogen isotopes. Geochim. Cosmochim. Acta 44, 1197-1206
[78]   L.d.S.L. Sternberg (1989): Oxygen and hydrogen isotope ratios in plant cellulose: Mechanisms and Applications. In: P.W. Rundel, J.R. Ehleringer, K.A. Nagy (eds.): Stable Isotopes in Ecological Research. Ecological Studies 68. Springer Verlag, New York, pp. 124-141
[79]   K.A. Bonhoeffer (1934): Deuterium-Austausch in organischen Verbindungen. Z. Elektrochem. 40, 469-474
[80]   A. Schimmelmann, R.F. Miller, S.W. Leavitt (1993): Hydrogen isotopic exchange and stable isotope ratios in cellulose, wood, chitin, and amino compounds. In: P.K. Swart, K.C. Lohmann, J. McKenzie, S. Savin (eds.): Climate Change in Continental Isotopic Records. Geophysical Monograph 78, American Geophysical Union, Washington, pp. 367-374
[81]   D. Samuel, B.L. Silver (1965): Oxygen isotope exchange reactions of organic compounds. Advan. Phys. Org. Chem. 3, 123-186
[82]   K.W. Wedeking, J.M. Hayes (1983): Exchange of oxygen isotopes between water and organic material. Isot. Geosci. 1, 357-370
[83]   J. Dunbar, H.-L. Schmidt (1984): Measurement of the $^2$H/$^1$H ratios of the carbon bound hydrogen atoms in sugars. Fresenius Z. Anal. Chem. 317, 853-857
[84]   M.J. DeNiro (1981): The effects of different methods of preparing cellulose nitrate on the determination of the D/H ratios of non-exchangeable hydrogen of cellulose. Earth Planet. Sci. Lett. 54, 177-187
[85]   A. Roßmann, G. Haberhauer, S. Hölzl, P. Horn, F. Pichlmayer, S. Voerkelius (2000): The potential of multielement stable isotope analysis for regional origin assignment of butter. Eur. Food Res. Technol. 211: 32-40
[86]   F. Camin, K. Wietzerbin, A. Blanch Cordes, G. Haberhauer, M. Lees, G. Versini (2004): Application of multi-element stable isotope ratio analysis for the characterisation of French, Italian and Spanish cheese. J. Agric. Food Chem. 52, 6592-6601
[87]   M. Boner, H. Förstel (2004): Stable isotope variations as a tool to trace the authenticity of beef. Anal. Bioanal. Chem. 378, 301-310
[88]   O. Schmidt, J.M. Quilter, B. Bahar, A.P. Moloney, C.M. Scrimgeour, J.S. Begley, F.J. Monahan (2005): Inferring the origin and dietary history of beef from C, N, and S stable isotope ratio analysis. Food Chem. 91, 545-549
[89]   R.A. Werner, H.-L. Schmidt (2002): The in vivo nitrogen isotope discrimination among organic plant compounds. Phytochem. 61, 465-484
[90]   G.L. Sacks, T.J. Brenna (2005): $^{15}$N/$^{14}$N Position-specific isotopic analyses of poly-nitrogenous amino acids. Anal. Chem. 77, 1013-1019
[91]   L. Melander, W.H. Saunders (1980): Reaction Rates of Isotopic Molecules. Wiley, New York
[92]   Bigeleisen J. (1965): Chemistry of isotopes. Science 147, 463-471
[93]   W.W. Cleland, M.H. O'Leary, D.B. Northrop (eds.) (1977): Isotope effects on enzyme-catalyzed reactions, University Park Press, Baltimore
[94]   P.F. Cook (ed.) (1991): Enzyme Mechanism from Isotope Effects, CRC Press, Boca Raton
[95]   J.T. Christeller, W.A. Laing, J.H. Troughton (1976): Isotope discrimination by ribulose 1,5-diphosphate carboxylase. Plant Physiol. 57, 580-582
[96]   C.A. Roeske, M.H. O'Leary (1984): Carbon isotope effects on the enzyme-catalyzed carboxylation of ribulose bisphosphate. Biochem. 23, 6275-6284
[97]   F.J. Winkler, H. Kexel, C. Kranz, H.-L. Schmidt (1982): Parameters affecting the $^{13}$CO$_2$/$^{12}$CO$_2$ isotope discrimination of the ribulose-1,5-bisphosphate carboxylase reaction. In: H.-L. Schmidt, H. Förstel, K. Heinzinger (eds.): Stable Isotopes. Elsevier Scientific Publishing Company, Amsterdam, pp. 83-89
[98]   M.H. O'Leary (1981): Carbon isotope fractionation in plants. Phytochem. 20, 553-567
[99]   M.H. O'Leary, J.E. Rife, J.D. Slater (1981): Kinetic and isotope effect studies of maize phosphoenolpyruvate carboxylase. Biochem. 20, 7308-7314
[100]  F.J. Winkler, H.-L. Schmidt, E. Wirth, E. Latzko, B. Lenhardt, H. Ziegler (1983): Temperature, pH and enzyme-source dependence of the HCO$_3^-$-carbon isotope effect on the phosphoenolpyruvate carboxylase reaction. Physiol. Vég. 21, 889-895

[101] T. Whelan, W.M. Sackett, C.R. Benedict (1973): Enzymatic fractionation of carbon isotopes by phosphoenolpyruvate carboxylase from $C_4$ plants. Plant Physiol. 51, 1051-1054

[102] G.D. Farquhar (1983): On the nature of carbon isotope discrimination in $C_4$ species. Aust. J. Plant Physiol. 10, 205-226

[103] C.B. Osmond (1978): Crassulacean acid metabolism: A curiosity in context. Ann. Rev. Plant Physiol. 29, 379-414

[104] M.H. O'Leary (1988): Carbon isotopes in photosynthesis. Bioscience 38, 328-336

[105] J.C. Lerman, E. Deleens, A. Nato, A. Moyse (1974): Variation in the carbon isotope composition of a plant with Crassulacean Acid Metabolism. Plant Physiol. 53, 581-584

[106] E. Melzer, H.-L. Schmidt (1987): Carbon isotope effects on the pyruvate dehydrogenase reaction and their importance for relative carbon-13 depletion in lipids. J. Biol. Chem. 262, 8159-8164

[107] A. Roßmann, M. Butzenlechner, H.-L. Schmidt (1991): Evidence for a non-statistical carbon isotope distribution in natural glucose. Plant Physiol. 96, 609-614

[108] H.-L. Schmidt, M. Butzenlechner, A. Roßmann, S. Schwarz, H. Kexel, K. Kempe (1993): Inter- and intramolecular isotope correlations in organic compounds as a criterion for authenticity identification and origin assignment. Z. Lebensm. Unters. Forsch. 196, 105-110

[109] H.-L. Schmidt (1995): Isotopic patterns in natural products. Labels without labelling. In: J. Allen, R. Voges (eds.): Synthesis and Applications of Isotopes and Isotopically Labelled Compounds. Proc. 5th Int. Symp. Strasbourg 1994. Wiley, New York, pp. 869-875

[110] Y.-H. Luo, L. Sternberg, S. Suda, S. Kumazawa, A. Mitsui (1991): Extremely low D/H ratios of photoproduced hydrogen by cyanobacteria. Plant Cell Physiol. 32, 897-900

[111] H.-L. Schmidt, R.A. Werner, W. Eisenreich (2003): Systematics of $^2H$ patterns in natural compounds and its importance for the elucidation of biosynthetic pathways. Phytochem. Rev. 2, 61-85

[112] J. Bricout, J. Koziet (1985): Détection de l'addition de sucre dans les jus d'orange par analyse isotopique. Sci. Aliment. 5, 197-204

[113] M. Butzenlechner, A. Roßmann, H.-L. Schmidt (1989): Assignment of bitter almond oil to natural and synthetic sources by stable isotope ratio analysis. J. Agric. Food Chem. 37, 410-412

[114] L. Sternberg (1988): D/H ratios of environmental water recorded by D/H ratios of plant lipids. Nature 333, 59-61

[115] L. Sternberg, M.J. DeNiro, H. Ajie (1984): Stable hydrogen isotope ratios of saponifiable lipids and cellulose nitrate from CAM, $C_3$ and $C_4$ plants. Phytochem. 23, 2475-2477

[116] A.L. Sessions, T.W. Burgoyne, A. Schimmelmann, J.M. Hayes (1999): Fractionation of hydrogen isotopes in lipid biosynthesis. Org. Geochem. 30, 1193-1200

[117] Y. Chikaraishi, H. Naraoka, S.R. Poulson (2004): Hydrogen and carbon isotopic fractionation of lipid biosynthesis among terrestrial (C3, C4 and CAM) and aquatic plants. Phytochem. 65, 1369-1381

[118] J. Bricout, J. Koziet (1976): Détermination de l'origine des substances aromatiques par spectrographie de masse isotopique. Ann. Fals. Exp. Chim. 69, 845-855

[119] M. Byrn, M. Calvin (1966): Oxygen-18 exchange reactions of aldehydes and ketones. J. Am. Chem. Soc. 88, 1916-1922

[120] L. Sternberg, M.J. DeNiro (1983): Biogeochemical implications of the isotopic equilibrium fractionation factor between the oxygen atoms of acetone and water. Geochim. Cosmochim. Acta 47, 2271-2274

[121] L.W. Doner, H.O. Ajie, L. Sternberg, J.M. Milburn, M.J. DeNiro, K.B. Hicks (1987): Detecting sugar beet syrups in orange juice by D/H and $^{18}O/^{16}O$ analysis of sucrose. J. Agric. Food Chem. 35, 610-612

[122] M. Krummen, A.W. Hilkert, D. Juchelka, A. Duhr, H.-J. Schlüter, R. Pesch (2004): A new concept for isotope ratio monitoring liquid chromatography/mass spectrometry. Rapid Comm. Mass Spectrom. 18, 2260-2266

[123] M. Gensler, H.-L. Schmidt (1994): Isolation of the main organic acids from fruit juices and nectars for carbon isotope ratio measurements. Anal. Chim. Acta 299, 231-237

[124] B.H. Brown, S.J. Neill, R. Horgan (1986): Partial isotope fractionation during high-performance liquid chromatography of deuterium-labelled internal standards in plant hormone analysis: A cautionary note. Planta 167, 421-423

[125] P.Q. Baumann, D.B. Ebenstein, B.D. O'Rourke, K.S. Nair (1992): High-performance liquid chromatographic technique for non-derivatized leucine purification: Evidence for carbon isotope fractionation. J. Chromatogr. 573, 11-16

[126] R.J. Caimi, J.T. Brenna (1993): High-precision liquid chromatography-combustion isotope ratio mass spectrometry. Anal. Chem. 65, 3497-3500

[127] A. Barrie, J. Bricout, J. Koziet (1984): Gas chromatography - stable isotope ratio analysis at natural abundance levels. Biomed. Mass Spectrom. 11, 583-588

[128] A. Barrie, J.E. Davies, A.J. Park, C.T. Workman (1989): Continuous-flow stable isotope analysis for biologists. Spectrosc. Int. 1, 34-44 and Spectroscopy 4, 42-52

[129] R.A. Werner (2003): The online $^{18}O/^{16}O$ analysis: Development and application. Isot. Environ. Health Stud. 39, 85-104

[130] S. Epstein, T. Mayeda (1953): Variation of $O^{18}$ content of waters from natural sources. Geochim. Cosmochim Acta 4, 213-224

[131] T.B. Coplen, J.D. Wildman, J. Chen (1991): Improvements in the gaseous hydrogen-water equilibration technique for hydrogen isotope ratio analysis. Anal. Chem. 63, 910-912

[132] M. Gehre, H. Geilmann, J. Richter, R.A. Werner, W.A. Brand (2004): Continuous flow $^2H/^1H$ and and $^{18}O/^{16}O$ analysis of water samples with dual inlet precision. Rapid Comm. Mass Spectrom. 18, 2650-2660

[133] G. Gleixner, H.-L. Schmidt (1998): On-line determination of group-specific isotope ratios in model compounds and aquatic humic substances by coupling pyrolysis to GC-C-IRMS. ACS Symposium Series 707: 34-46

[134] T. Preston (1992): The measurement of stable isotope natural abundance variations. Plant Cell Environ. 15, 1091-1097

[135] W.A. Brand (2004): Mass spectrometer hardware for analyzing stable isotope ratios. In: P.A. de Groot (ed.): Handbook of Stable Isotope Analytical Techniques, Vol. I. Elsevier, Amsterdam, pp. 835-858.

[136] D. Rumble, T.C. Hoering, J.M. Palin (1993): Preparation of $SF_6$ for sulfur isotope analysis by laser heating sulfide minerals in the presence of $F_2$ gas. Geochim. Cosmochim. Acta 57, 4499-4512

[137] J. Santrock, S.A. Studley, J.M. Hayes (1985): Isotopic analyses based on the mass-spectrum of carbon-dioxide. Anal. Chem. 57, 1444-1448

[138] N.V. Grassineau, D.P. Mattey, D. Lowry (2001): Sulfur isotope analysis of sulfide and sulfate minerals by continuous flow-isotope ratio mass spectrometry. Anal. Chem. 72, 220-225

[139] B. Fry, S.R. Silva, C. Kendall, R.K. Anderson (2002): Oxygen isotope corrections for online $\delta^{34}S$ analysis. Rapid Comm. Mass Spectrom. 16, 854-858

[140] I. Kirshenbaum (1951): Physical Properties and Analysis of Heavy Water. McGraw-Hill Book Company, Inc., New York. Chapter 3: Isotopic analysis by the mass spectrometer, pp. 69ff

[141] I. Friedman (1953): Deuterium content of natural waters and other substances. Geochim. Cosmochim. Acta 4, 89-103

[142] A.L. Sessions, T.W. Burgoyne, J.M. Hayes (2001): Correction of $H_3^+$ contributions in hydrogen isotope ratio monitoring mass spectrometry. Anal. Chem. 73, 192-199

[143] Sessions AL, Burgoyne TW, Hayes JM (2001): Determination of the $H_3$ factor in hydrogen isotope ratio monitoring mass spectrometry. Anal. Chem. 73, 200-207

[144] B.W. Bromley, G.D. Hegeman, W. Meinschein (1982): A method for measuring natural abundance intramolecular stable carbon isotopic distributions in malic acid. Anal. Biochem. 126, 436-446

[145] M. Butzenlechner, S. Thimet, K. Kempe, H. Kexel, H.-L. Schmidt (1996): Inter- and intramolecular isotopic correlations in some cyanogenic glycosides and glucosinolates and their practical importance. Phytochem. 43, 585-592

[146] T.N. Corso, J.T. Brenna (1997): High-precision position-specific isotope analysis. Proc. Natl. Acad. Sci. 94, 1049-1053

[147] C.J. Wolyniak, G.L. Sacks, B.S. Pan, J.T. Brenna (2005): Carbon position-specific isotope analysis of alanine and phenylalanine analogues exhibiting nonideal pyrolytic fragmentation. Anal. Chem. 77, 1746-1752

[148] D.L. Buchanan, B.J. Corcoran (1959): Sealed tube combustion for the determination of carbon-14 and total carbon. Anal. Chem. 31, 1635-1638

[149] M.H. Engel, R.J. Maynard (1989): Preparation of organic matter for stable carbon isotope analysis by sealed tube combustion: A cautionary note. Anal. Chem. 61, 1996-1998

[150] F.J. Winkler, H.-L. Schmidt (1980): Einsatzmöglichkeiten der $^{13}$C-Isotopen-Massenspektrometrie in der Lebensmitteluntersuchung. Z. Lebensm. Unters. Forsch. 171, 85-94

[151] S.R. Boyd, C.T. Pillinger (1990): Determination of the abundance and isotope composition of nitrogen within organic compounds: A sealed tube technique for use with static vacuum mass spectrometers. Meas. Sci. Technol. 1, 1176-1183

[152] D.L. Eskew, P.M. Gresshoff, M. Doty, C. Mora C (1992): Sealed-tube combustion for natural $^{15}$N abundance estimation of $N_2$ fixation and application to supernodulating soybean mutants. Canad. J. Microbiol. 38, 598-603

[153] J. Bigeleisen, M.L. Perlman, H.C. Prosser (1952): Conversion of hydrogenic materials to hydrogen for isotopic analysis. Anal. Chem. 24, 1356-1357

[154] M.L. Coleman, T.J. Shepherd, J.J. Durham, J.E. Rouse, G.R. Moore (1982): Reduction of water with zinc for hydrogen isotope analysis. Anal. Chem. 54, 993-995

[155] A. Tanweer, G. Hut., J.O. Burgman (1988): Optimal conditions for the reduction of water to hydrogen by zinc for mass spectrometric analysis of the deuterium content. Chem. Geol. (Isot. Geosci. Sect.) 73, 199-203

[156] A. Tanweer (1990): Importance of clean metallic zinc for hydrogen isotope analysis. Anal. Chem. 62, 2158-2160

[157] A. Schimmelmann, M.J. DeNiro (1993): Preparation of organic and water hydrogen for stable isotope analysis: Effects due to reaction vessels and zinc reagent. Anal. Chem. 65, 789-792

[158] M. Gehre, R. Hoefling, P. Kowski, G. Strauch (1996): Sample preparation device for quantitative hydrogen isotope analysis using chromium metal. Anal. Chem. 68, 4414-4417

[159] D. Rittenberg, L. Ponticorvo (1956): A method for the determination of the $O^{18}$ concentration of the oxygen of organic compounds. Int. J. Appl. Radiat. Isot. 1, 208-214

[160] J. Dunbar, A.T. Wilson (1982): Re-evaluation of the $HgCl_2$ pyrolysis technique for oxygen isotope analysis. Int. J. Appl. Radiat. Isot. 34, 932-934

[161] M. Anbar, S. Guttmann (1959): Isotopic analysis of oxygen in inorganic compounds. Int. J. Appl. Radiat. Isot. 5, 233-235

[162] J.S. Lee (1962): Determination of $O^{18}$ concentration of sugars and glycogen. Anal. Chem. 34, 835-837

[163] P. Thompson, J. Gray (1977): Determination of $^{18}O/^{16}O$ ratios in compounds containing C, H and O. Int. J. Appl. Radiat. Isot. 28, 411-415

[164] C.A.M. Brenninkmeijer, W.G. Mook (1981): A batch process for direct conversion of organic oxygen and water to $CO_2$ for $^{18}O/^{16}O$ analysis. Int. J. Appl. Radiat. Isot. 32, 137-141

[165] C.E. Rees, B.D. Holt (1991): The isotopic analysis of sulphur and oxygen. In: H.R. Krouse, V.A. Grinenko (eds.): Stable Isotopes: Natural and Anthropgenic Sulphur in the Environment. Scope 43. John Wiley & Sons, Chichester, pp. 43-64.

[166] H.-L. Schmidt, A. Roßmann, R.A. Werner (1998): Stable isotope ratio analysis in quality control of flavourings. In: E. Ziegler, H. Ziegler (eds.): Flavourings - Production, Composition, Applications, Regulations. Wiley-VCH, Weinheim, New York, Chichester, pp. 539-594

[167] W. A. Brand (1996): High precision isotope ratio monitoring techniques in mass spectrometry. J. Mass Spectrom. 31, 225-235

[168] M. Minagawa, D.A. Winter, I.R. Kaplan (1984): Comparison of Kjeldahl and combustion methods for measurement of nitrogen isotope ratios in organic matter. Anal. Chem. 56, 1859-1861

[169] T. Preston, A. Barrie (1991): Recent progress in continuous flow isotope ratio mass spectrometry. Amer. Lab. 23, 32H-32L

[170] T. Preston, N.J.P. Owens (1983): Interfacing an automatic elemental analyser with an isotope ratio mass spectrometer : The potential for fully automated total nitrogen and nitrogen-15 analysis. Analyst 108, 971-977

[171] A. Barrie (1991): New methodologies in stable isotope analysis. In: Stable Isotopes in Plant Nutrition, Soil Fertility and Environmental Studies. IAEA, Vienna, pp. 3-25

[172] A. Giesemann, H.-J. Jäger, A.L. Norman, H.R. Krouse, W.A. Brand (1994): On-line sulfur-isotope determination using an elemental analyzer coupled to a mass spectrometer. Anal. Chem. 66, 2816-2819

[173] W. Meier-Augenstein (1999): Applied gas chromatography coupled to isotope ratio mass spectrometry. J. Chromatogr. A 842, 351-371

[174] J.T. Brenna, T.N. Corso, H.J. Tobias, R.J. Caimi (1997): High-precision continuous-flow isotope ratio mass spectrometry. Mass Spectrom. Rev. 16, 227-258
[175] M.E. Wieser, W. A. Brand (1999): Isotope Ratio Studies using Mass Spectrometry. In: J.C. Lindon, G.E. Tranter and J.L. Holmes (eds.): Encyclopedia of Spectroscopy and Spectrometry. Academic Press, New York, pp. 1072-1086
[176] R.M. Verkouteren (2005): Value assignment and uncertainty estimation of selected light stable isotope reference materials: RMs 8543-8545, RMs 8562-8564, and RM 8566. NIST Special Publication 260-149-2004 Edition.
[177] http://www.iaea.org/programmes/aqcs/news/mrm_3.htm
[178] U. Hener, W.A. Brand, A.W. Hilkert, D. Juchelka, A. Mosandl, F. Podebrad (1998): Simultaneous on-line analysis of $^{18}O/^{16}O$ and $^{13}C/^{12}C$ ratios of organic compounds using GC-pyrolysis-IRMS. Z. Lebensm. Unters. Forsch. A - Food Res. Technol. 206, 230-232
[179] D.A. Merritt, J.M. Hayes (1994): Nitrogen isotope analyses by isotope-ratio-monitoring gas chromatography/mass spectrometry. J. Am. Soc. Mass Spectrom. 5, 387-397
[180] W.A. Brand, A.R. Tegtmeyer, A. Hilkert (1994): Compound specific isotope analysis: extending toward $^{15}N/^{14}N$ and $^{18}O/^{16}O$. Org. Geochem. 21, 585-594
[181] J. Demmelmair, H.-L. Schmidt (1993): Precise $\delta^{13}C$-determination in the range of natural abundance of amino acids from protein hydrolysates by gas chromatography - isotope ratio mass spectrometry. Isotopenpraxis Environ. Health Stud. 29, 237-250
[182] G. Rieley (1994): Derivatization of organic compounds prior to gas chromatography-combustion-isotope ratio mass spectrometric analysis: Identification of isotope fractionation processes. Analyst 119, 915-919
[183] H.-L. Schmidt, R. Medina (1989): Patent GB 8921285.6.
[184] H.-L. Schmidt, K. Kempe, J. Demmelmair, R. Medina (1992): On-line gas chromatography and isotope ratio mass spectrometry of organic compounds in the range of natural abundances. In: T. Matsuo (ed.): Proc. Kyoto'92 Int. Conf. Biol. Mass Spectrom. Osaka University, Japan, pp. 382-383
[185] H.J. Tobias, K.J. Goodman, C.E. Blacken, J.T. Brenna (1995): High-precision D/H measurement from hydrogen gas and water by continous-flow isotope ratio mass spectrometry. Anal. Chem. 67, 2486-2492
[186] H.J. Tobias, J.T. Brenna (1996): Correction of ion source nonlinearities over a wide signal range in continous-flow isotope ratio mass spectrometry of water-derived hydrogen. Anal. Chem. 68, 2281-2286
[187] H.J. Tobias, J.T. Brenna (1996): High-precision D/H measurement from organic mixtures by gas chromatography continous-flow isotope ratio mass spectrometry using a palladium filter. Anal. Chem. 68, 3002-3007
[188] I.S. Begley, C.M. Scrimgeour (1996): On-line reduction of $H_2O$ for $\delta^2H$ and $\delta^{18}O$ measurement by continous-flow isotope ratio mass spectrometry. Rapid Comm. Mass Spectrom. 10, 969-973
[189] I.S. Begley, C.M. Scrimgeour (1997): High-precision $\delta^2H$ and $\delta^{18}O$ measurement for water and volatile organic compounds by continous-flow pyrolysis isotope ratio mass spectrometry. Anal. Chem. 69, 1530-1535
[190] R.A. Werner, B.E. Kornexl, A. Roßmann, H.-L. Schmidt (1996): On-line determination of $\delta^{18}O$ values of organic substances. Anal. Chim. Acta. 319, 159-164
[191] J. Koziet (1997): Isotope ratio mass spectrometric method for the on-line determination of oxygen-18 in organic matter. J. Mass Spectrom. 32, 103-108
[192] B.E. Kornexl, M. Gehre, R. Höfling, R.A. Werner (1999): On-line $\delta^{18}O$ measurement of organic and inorganic substances. Rapid Comm. Mass Spectrom. 13, 1685-1693
[193] T.W. Burgoyne, J.M. Hayes (1998): Quantitative production of H2 by pyrolysis of gas chromatographic effluents. Anal. Chem. 70, 5136-5141
[194] A.W. Hilkert, C.B. Douthill, H.J. Schlüter, W.A. Brand (1999): Isotope ratio monitoring gaschromatography/ mass spectrometry of D/H by high temperature conversion isotope ratio mass spectrometry. Rapid Comm. Mass Spectrom. 13, 1226-1230
[195] J.C. Fetzer, P.A. Bloxham, L.B. Rogers (1980): Gas chromatographic fractionations of the carbon and oxygen isotopes of carbon monoxide. Separ. Sci. Technol. 15, 49-56

[196]　A.T. Shepard, N.D. Danielson, R.E. Pauls, N.H. Mahle, P.J. Taylor, L.B. Rogers (1976): Gas-chromatographic fractionations of stable isotopes of carbon and oxygen in carbon dioxide. Separ. Sci. 11, 279-292

[197]　H. Illy, F. Botter (1980): Fractionnement des isotopes du carbone et de l´oxygène, obtenu par physisorption du dioxyde de carbone. J. Chim. Phys. 78, 17-22

[198]　R. Braunsdorf, U. Hener, G. Przibilla, S. Piecha, A. Mosandl (1993): Analytische und technologische Einflüsse auf das $^{13}C/^{12}C$-Isotopenverhältnis von Orangenöl-Komponenten. Z. Lebensm. Unters. Forsch. 197, 24-28

[199]　J. Mráz, P. Jheeta, A. Gescher, M.D. Threadgill (1993): Unusual deuterium isotope effect on the retention of formamides in gas-liquid chromatography. J. Chromatogr. 641, 194-198

[200]　Y. Benchekroun, S. Dautraix, M. Désage, J.L. Brazier (1997): Isotopic effects on retention times of caffeine and its metabolites 1,3,7-trimethyluric acid, theophylline, theobromine and paraxanthine. J. Chromatogr. 688B, 245-254

[201]　R.J. Caimi, J.T. Brenna (1997): Quantitative evaluation of carbon isotopic fractionation during reversed-phase high-performance liquid chromatography. J. Chromatogr. 757 A, 307-310

[202]　D.E. Matthews, J.M. Hayes (1978): Isotope-ratio-monitoring gas chromatography-mass spectrometry. Anal. Chem. 50, 1465-1473

[203]　S. Bilke, A. Mosandl (2002): Measurements by gas chromatography/pyrolysis/ mass spectrometry: fundamental conditions in $^{2}H/^{1}H$ isotope ratio analysis. Rapid Comm. Mass Spectrom. 16, 468-472

[204]　S. Sewenig, U. Hener, A. Mosandl (2003): Online determination of $^{2}H/^{1}H$ and $^{13}C/^{12}C$ isotope ratios of cinnamaldehyde from different sources using gas chromatography isotope ratio mass spectrometry. Eur Food Res Technol 217, 444-448

[205]　E. Richling, M. Appel, F. Heckel, K. Kahle, M. Kraus, C. Preston, W. Hümmer, P. Schreier (2006, in press): Flavor authenticity studies isotope ratio mass spectrometry – perspectives and limits. ACS Symposia Ser.

[206]　A. Mosandl (2004): Authenticity assessment: a permanent challenge in food flavor and essential oil analysis. J. Chromatogr. Sci. 42, 440-449

[207]　C. Preston, E. Richling, S. Elss, M. Appel, F. Heckel, A. Hartlieb, P. Schreier (2003): On-line gas chromatography combustion/pyrolysis isotope ratio mass spectrometry (HRGC-C/P-IRMS) of pineapple (*Ananas comosus* L. Merr.) volatiles. J. Agric. Food Chem. 51, 8027-8031

[208]　B.O. Aguilar-Cisneros, M.G. López, E. Richling, F. Heckel, P. Schreier (2002): Tequila authenticity assessment by headspace SPME-HRGC-IRMS analysis of $^{13}C/^{12}C$ and $^{18}O/^{16}O$ ratios of ethanol. J. Agric. Food Chem. 50, 7520-7523

[209]　D. Juchelka, T. Beck, U. Hener, F. Dettmar, A. Mosandl (1998): Multidimensional gas chromatography, online coupled with isotope ratio mass spectrometry (MDGC-IRMS): Progress in the analytical authentication of genuine flavour components. J. High Resolut. Chromatogr. 21, 145-151

[210]　S. Sewenig, D. Bullinger, U. Hener, A. Mosandl (2005): Comprehensive authentication of (E)-α(β)-ionone from raspberries, using constant flow MDGC-C/P-IRMS and enantio-MDGC-MS. J. Agric. Food Chem. 53, 838-845

[211]　J. Jung, S. Sewenig, U. Hener, A. Mosandl (2005): Comprehensive authenticity assessment of lavander oils using multielement/multicompound IRMS-analysis and enantioselective MDGC-MS. Eur. Food Res. Technol. 220, 232-237

[212]　C. Preston, E. Richling, K. Kahle, M. Kraus, W. Hümmer, M. Appel, F. Heckel, P. Schreier (2005): Ester synthesis: authenticity assessment by stable isotope analysis. In: T. Hofmann, M. Rothe, P. Schieberle (eds.): State-of-the-Art in Flavour Chemistry and Biology. DFA, Garching, pp. 358-362

[213]　C. Fuganti, G. Zucchi (1997): Recent studies on the biogeneration of flavours. Chim. Ind. 79, 745-750

[214]　Organikum - Organisch-chemisches Grundpraktikum (1986) 16. Auflage. VEB Deutscher Verlag der Wissenschaften, Berlin, pp. 414 ff

[215]　A.J. Birch, D.W. Cameron, Y. Harada, R.W. Rickards (1962): Studies in relation to biosynthesis. Part XXV. A preliminary study of the Antimycin-A complex. J. Chem. Soc., 303-305

[216]　K.D. Monson, J.M. Hayes (1982): Carbon isotopic fractionation in the biosynthesis of bacterial fatty acids. Ozonolysis of unsaturated fatty acids as a means of determining the intramolecular distribution of carbon isotopes. Geochim. Cosmochim. Acta. 46, 139-149

[217] W.G. Meinschein, G.G.L. Rinaldi, J.M. Hayes, D.A. Schoeller (1974): Intramolecular isotopic order in biologically produced acetic acid. Biomed. Mass Spectrom. 1, 172-174
[218] B.W. Bromley, G.D. Hegeman, W. Meinschein (1982): A method for measuring natural abundance intramolecular stable carbon isotopic distributions in malic acid. Anal. Biochem. 126, 436-446
[219] E. Melzer, M.H. O'Leary (1991): Aspartic acid synthesis in $C_3$ plants. Planta 185, 368-371
[220] P.H. Abelson, T.C. Hoering (1961): Carbon isotope fractionation in formation of amino acids by photosynthetic organisms. Proc. Natl. Acad. Sci. USA 47, 623-632
[221] A. Roßmann, H.-L. Schmidt (1989): Nachweis der Herkunft von Ethanol und der Zuckerung von Wein durch positionelle Wasserstoff- und Kohlenstoff-Isotopen-Verhältnismessung. Z. Lebensm. Unters. Forsch. 188, 434-438
[222] H. Günther (1992): NMR- Spektroskopie, 3. Auflage. G. Thieme Verlag, Stuttgart, p. 13
[223] G.J. Martin: Deuterium-NMR in the investigation of site-specific natural isotope fractionation (SNIF-NMR) - Automation of wine NMR. Laboratoire de RMN et Réactivité Chimique Université de Nantes - CNRS UA 472
[224] G. Martin (1992): SNIF-NMR - eine Methode zum Nachweis von Rübenzuckerzusatz in Fruchtsäften. Flüssiges Obst 59, 477-485
[225] M.L. Martin, G.J. Martin (1990): Deuterium NMR in the study of site-specific natural isotope fractionation (SNIF-NMR). In: P. Dietl, E. Fluck, H. Günther, R. Kosfeld, J. Seelig (eds.): NMR Basic Principles and Progress Vol. 23. Springer Verlag, Berlin, pp. 1-61
[226] French Patent 81-22710, European Patent 82-402-209-9, USP 85-4-550-082
[227] G.J. Martin, S. Hanneguelle, G. Remaud (1990): Authentification des arômes et parfums par résonance magnétique nucléaire et spectrométrie de masse de rapports isotopiques. Parfums Cosmét. Arômes 94, 95-109
[228] G.J. Martin, M.L. Martin, B.-L. Zhang (1992): Site-specific natural isotope fractionation of hydrogen in plant products studied by nuclear magnetic resonance. Plant Cell Environ. 15, 1037-1050
[229] F. Le Grand, G. George, S. Akoka (2005): How to reduce the experimental time in isotopic $^2$H NMR using the ERETIC method. J. Magnet. Res. 174, 171-176
[230] S. Hanneguelle, J.-N. Thibault, N. Naulet, G.J. Martin (1992): Authentication of essential oils containing linalool and linalyl acetate by isotopic methods. J. Agric. Food Chem. 40, 81-87
[231] S. Rezzi, C. Guillou, F. Reniero, V.M. Holland, S. Ghelli (2004): Natural abundance $^2$H-NMR spectroscopy. Application to food analysis. In P.A. de Groot (ed.), Handbook of Stable Isotope Analytical Techniques, Vol. I, Elsevier B.V., Amsterdam, pp. 103-121
[232] G.J. Martin, M.L. Martin (1988): The site-specific natural isotope fractionation-NMR method applied to the study of wines. In: H.F. Linskens, J.F. Jackson (eds.): Modern Methods of Plant Analysis, Vol. 6. Springer Verlag, Berlin, pp. 258-275
[233] C. Fauhl, R. Wittkowski (1996): On-line $^1$H-NMR to facilitate tube preparation in SNIF-NMR analysis. Z. Lebensm. Unters. Forsch. 203, 541-545
[234] P. Lesot, C. Aroulanda, I. Billault (2004): Exploring the analytical potential of NMR spectroscopy in chiral anisotropic media for the study of the natural abundance deuterium distribution in organic molecules. Anal. Chem. 76, 2827-2835
[235] E. Bengsch, J.Ph. Grivet, H.R. Schulten: French patent FR 2 530 026 (1982), European Patent 99 810 (1983)
[236] E. Bengsch, J.-Ph. Grivet (1981): Non-statistical label distribution in biosynthetic $^{13}$C enriched amino acids. Z. Naturforsch. 36b, 1289-1296
[237] V. Caer, M. Trierweiler, G.J. Martin, M.L. Martin (1991): Determination of site-specific carbon isotope ratios at natural abundance by carbon-13 nuclear magnetic resonance spectroscopy. Anal. Chem. 63, 2306-2313
[238] E. Tenailleau, P.Lancelin, R.J. Robins, S. Akoka (2004 a): NMR approach to the quantification of nonstatistical $^{13}$C distribution in natural products: vanillin. Anal. Chem. 76, 3818-3825
[239] E.J. Tenailleau, P.Lancelin, R.J. Robins, S. Akoka (2004 b): Authentication of the origin of vanillin using quantitative natural abundance $^{13}$C NMR. J. Agric. Food Chem. 52, 7782-7787
[240] E.M. Galimov (1985): The Biological Fractionation of Isotopes. Academic Press, Orlando, pp. 113-116
[241] R. Benner, M.L. Fogel, E.K. Sprague, R.E. Hodson (1987): Depletion of $^{13}$C in lignin and its implications for stable carbon isotope studies. Nature 329, 708-710

[242] M. Butzenlechner (1990): Isotopenverteilungen in Phenylpropan-Abkömmlingen, Spiegel ihrer Biosynthese und Grundlagen für ihre Herkunftsbestimmung. PhD thesis, Technische Universität München, Freising, Germany

[243] E.M. Galimov, L.A. Kodina, V.N. Generalova (1976): Experimental investigation of intra- and intermolecular isotopic effects in biogenic aromatic compounds. Geochem. Int. 1, 9-13

[244] D.A. Krueger, H.W. Krueger (1983): Carbon isotopes in vanillin and the detection of falsified "natural" vanillin. J. Agric. Food Chem. 31, 1265-1268

[245] J. Bricout, L. Merlivat, J. Koziet (1973): Fractionnement des isotopes de l'hydrogène au cours de la biosynthèse de certains constituants arômatiques des végétaux. C.R. Acad. Sc. Paris 277, Ser. D, 885-888

[246] J. Bricout (1979): Natural abundance levels of $^2$H and $^{18}$O in plant organic matter. In: E.R. Klein, P.D. Klein (eds.): Stable Isotopes. Proc. 3$^{rd}$ Int. Conf., Academic Press, New York, pp. 215-222

[247] K. Kempe, A. Roßmann, B.E. Kornexl, H.-L. Schmidt: unpublished results

[248] S. Faulhaber (1997): GC/IRMS analysis of mandarin essential oils. 1. $\delta^{13}C_{PDB}$ and $\delta^{15}N_{AIR}$ values of methyl N-methylanthranilate. J. Agric. Food Chem. 45, 2579-2583

[249] K. Kempe, A Roßmann, H.-L. Schmidt: unpublished results

[250] J. Demmelmair, A. Roßmann, H.-L. Schmidt: unpublished results

[251] C. Ruff, K. Hör, B. Weckerle, T. König, P. Schreier (2002): Authenticity assessment of estragole and methyl eugenol by on-line gas chromatography-isotope ratio mass spectrometry. J Agric. Food Chem. 50, 1028-1031

[252] C.A.M. Brenninkmeijer, W.G. Mook (1982): A new method for the determination of the $^{18}O/^{16}O$ ratio in organic compounds. In: H.-L. Schmidt, H. Förstel, K. Heinzinger (eds.): Stable Isotopes. Proc. 4$^{th}$ Int. Conf., Elsevier Scientific Publishing Company, Amsterdam, pp. 661-666

[253] G. Fronza, C. Fuganti, S. Serra, C. Guillou, F. Reniero (2001): The positional $\delta^{18}$O-value of extracted and synthetic vanillin. Helv. Chim. Acta 84, 351-359

[254] F.F. Bensaid, K.Witzerbin, G.J. Martin (2002): Authentication of natural vanilla flavourings: isotopic characterisation using degradation of vanillin into guaiacol. J. Agric. Food Chem. 50, 6271-6275

[255] G. Fronza, C. Fuganti, G. Pedrocchi-Fantoni, S. Serra, G. Zucchi (1999): Stable isotope characterization of raspberry ketone extracted from *Taxus baccata* and obtained by oxidation of the accompanying alcohol (betuligenol). J. Agric. Food Chem. 47, 1150-1155

[256] J. Bricout, J.Ch. Fontes, L. Merlivat (1974): Detection of synthetic vanillin in vanilla extracts by isotopic analysis. J. Assoc. Off. Anal. Chem. 57, 713-715

[257] P.G. Hoffmann, M.C. Salb (1979): Isolation and stable isotope ratio analysis of vanillin. J. Agric. Food Chem. 27, 352-355

[258] J. Bricout (1982): Possibilities of stable isotope analysis in the control of food products In: H.-L. Schmidt, H. Förstel, K. Heinzinger (eds.): Stable Isotopes. Proc. 4$^{th}$ Int. Conf., Elsevier Scientific Publishing Company, Amsterdam, pp. 483-493

[259] H.W. Krueger, R.H. Reesman (1982): Carbon isotope analyses in food technology. Mass Spectrom. Rev. 1, 205-236

[260] O. Bréas, F. Fourel, G.J. Martin (1994): $^{13}$C analysis of aromas and perfumes by a coupled GC-IRMS technique. The case of vanillin and leaf alcohol extracts. Analusis 22, 268-272

[261] R.A. Culp, J.E. Noakes (1990): Identification of isotopically manipulated cinnamic aldehyde and benzaldehyde. J. Agric. Food Chem. 38, 1249-1255

[262] C. Ruff, K. Hör, B. Weckerle, P. Schreier, T. König (2000): $^2$H/$^1$H ratio analysis of flavor compounds by on-line gas chromatography pyrolysis isotope ratio mass spectrometry (HRGC-P-IRMS): Benzaldehyde. J. High Resol. Chromatogr. 23, 357-358

[263] C. Ruff, K. Hör, B. Weckerle, T. König, P. Schreier (2000 a): Authentizitätskontrolle von Aromastoffen: Benzaldehyd in alkoholischen Getränken. Alkohol-Industrie 113, 144-145

[264] C. Ruff, K. Hör, B. Weckerle, T. König, P. Schreier (2000 b): Authentizitätskontrolle von Aromastoffen: Gaschromatographie-Isotopenverhältnis-Massenspektrometrie zur Bestimmung des $^2$H/$^1$H-Verhältnisses von Benzaldehyd in Lebensmitteln. Dtsch. Lebensm. Rdsch. 96, 243-246

[265] A. Roßmann, H.-L. Schmidt: unpublished results

[266] J. Bricout, J. Koziet (1978): Characterization of synthetic substances in food flavors by isotopic analysis. In: G. Charalambous, G.E. Inglett (eds.): Flavor of Foods and Beverages: Chemistry and Technology. Academic Press, New York, pp. 199-208

[267] M. Balabane, R. Letolle, J.-C. Bayle, M. Derbesy (1983): Determination du rapport $^{13}C/^{12}C$ et $^{2}H/^{1}H$ des differents anetholes. Parfums Cosmét. Arômes 49, 27-31

[268] M. Balabane, J.-C. Bayle, M. Derbesy (1984): Charactérisation isotopique ($^{13}C$, $^{2}H$) de l'origine naturelle ou de synthèse de l'anéthol. Analusis 12, 148-150

[269] S. Bilke, A. Mosandl (2002 a): $^{2}H/^{1}H$ and $^{13}C/^{12}C$ isotope ratios of *trans*-anethole using gas chromatography-isotope ratio mass spectrometry. J. Agric. Food Chem. 50, 3935-3937

[270] M. Butzenlechner, A. Roßmann, H.-L. Schmidt: unpublished results

[271] K. Fink, E. Richling, F. Heckel, P. Schreier (2004): Determination of $^{2}H/^{1}H$ and $^{13}C/^{12}C$ isotope ratios of (E)-methyl cinnamate from different sources using isotope ratio mass spectrometry. J. Agric. Food Chem. 52, 3065-3068

[272] H. Casabianca, J.-B. Graff, P. Jame, C. Perruchietti, M. Chastrette (1995): Application of hyphenated techniques to the chromatographic authentication of flavours in food products and perfumes. J. High Resol. Chromatogr. 18, 279-285

[273] D.A. Krueger, H.W. Krueger (1985): Detection of fraudulent vanillin labelled with $^{13}C$ in the carbonyl carbon. J. Agric. Food Chem. 33, 323-325

[274] H. Ziegler, C.B. Osmond, W. Stichler, P. Trimborn (1976): Hydrogen isotope discrimination in higher plants: Correlations with photosynthetic pathway and environment. Planta 128, 85-92

[275] L. Sternberg, M.J. DeNiro (1983): Isotopic composition of cellulose from $C_3$, $C_4$, and CAM plants growing near one another. Science 220, 947-949

[276] B. Fayet, C. Fraysse, C. Tisse, I. Pouliquen, M. Guerere, G. Lesgards (1995): Analyse isotopique par GC-SMRI du carbone 13 de la vanille dans des crèmes glacées. Analusis 23, 451-453

[277] G. Remaud, Y.-L. Martin, G.G. Martin, G.J. Martin (1997): Detection of sophisticated adulterations of natural vanilla flavors and extracts: Application of the SNIF-NMR method to vanillin and p-hydroxybenzaldehyde. J. Agric. Food Chem. 45, 859-866

[278] A. Kaunzinger, D. Juchelka, A. Mosandl (1997): Progress in the authenticity assessment of Vanilla. 1. Initiation of authenticity profiles. J. Agric. Food Chem. 45, 1752-1757

[279] K. Bauer, D. Garbe, H. Surburg (1990): Common Fragrance and Flavor Materials. Preparation, Properties and Uses, 2$^{nd}$ revised edition. VCH, Weinheim

[280] M.L. Hagedorn (1992): Differentiation of natural and synthetic benzaldehydes by $^{2}H$ nuclear magnetic resonance. J. Agric. Food Chem. 40, 634-637

[281] M. Senkus, W.G. Brown (1938): Oxygen exchange reactions of benzaldehyde and some other compounds. J. Org. Chem. 2, 569-573

[282] Ullmanns Encyklopädie der technischen Chemie Bd.14, 3. Auflage (1963). Urban & Schwarzenberg, München - Berlin, pp. 760-761

[283] G.J. Martin, M.L. Martin, F. Mabon, J. Bricout (1982): A new method for the identification of the origin of natural products. Quantitative $^{2}H$-NMR at the natural abundance level applied to the characterization of anetholes. J. Am. Chem. Soc. 104, 2658-2659

[284] G. Fronza, C. Fuganti, P. Graselli, S. Servi, G. Zucchi, M. Barbeni, M. Cisero (1995): Natural abundance $^{2}H$ nuclear magnetic resonance study of the origin of 2-phenylethanol and 2-phenylethylacetate. J. Agric. Food Chem. 43, 439-443

[285] C.M. Scrimgeour, W.M. Bennet, A.A. Connacher (1988): A convenient method of screening glucose for $^{13}C:^{12}C$ ratio for use in stable isotope tracer studies. Biomed. Environ. Mass Spectrom. 17, 265-266

[286] G. Gleixner, H.-J. Danier, R.A. Werner, H.-L. Schmidt (1993): Correlations between the $^{13}C$ content of primary and secondary plant products in different cell compartments and that in decomposing basidiomycetes. Plant Physiol. 102, 1287-1290

[287] M.J. DeNiro, S. Epstein (1977): Mechanism of carbon isotope fractionation associated with lipid synthesis. Science 197, 261-263

[288] L. Ponticorvo (1968): Some observations on isotope fractionations of deuterium in nature. PhD thesis, Columbia University, New York, NY

[289] K. Misselhorn, H. Brückner, M. Müßig-Zufika, W. Grafahrend (1983): Nachweis des Rohstoffs bei hochrektifiziertem Alkohol. Branntweinwirtschaft 123, 162-170

[290] K. Misselhorn, W. Grafahrend (1990): Rohstoffnachweis bei hochgereinigtem Alkohol. Branntweinwirtschaft 130, 70-73

[291] A. Roßmann, K. Kempe, B.E. Kornexl, H.-L. Schmidt: unpublished results

[292]  G. Remaud (1993): Technical improvements of SNIF-NMR application to mustard oil. In: 2nd European Symposium Food Authenticity - Isotope Analysis and Other Advanced Analytical Techniques, 20-22 October 1993, Eurofins, Nantes (France)

[293]  G.S. Remaud, Y.-L. Martin, G.G. Martin, N. Naulet, G.J. Martin (1997): Authentication of mustard oils by combined stable isotope analysis (SNIF-NMR and IRMS). J. Agric. Food Chem. 45, 1844-1848

[294]  C. Bauer-Christoph, N. Christoph, B.O. Aguilar-Cisneros, M. G. López, E. Richling, A. Roßmann, P. Schreier (2003): Authentication of tequila by gas chromatography and stable isotope ratio analyses. Europ Food Res Technol 217, 438-443

[295]  K. Kahle, C. Preston, E. Richling, F. Heckel, P. Schreier (2005): On-line gas chromatography combustion/pyrolysis isotope ratio mass spectrometry (HRGC-C/P-IRMS) of major volatiles from pear fruit (*Pyrus communis*) and pear products. Food Chem. 91, 449-455

[296]  P. Rauschenbach, H. Simon, W. Stichler, H. Moser (1979): Vergleich der Deuterium- und Kohlenstoff-13-Gehalte in Fermentations- und Syntheseethanol. Z. Naturforsch. 34c, 1-4

[297]  G. Remaud, C. Guillou, C. Vallet, G.J. Martin (1992): A coupled NMR and MS isotopic method for the authentification of natural vinegars. Fresenius J. Anal. Chem. 342, 457-461

[298]  E.R. Schmid, I. Fogy, P. Schwarz (1978): Beitrag zur Unterscheidung von Gärungsessig und synthetischem Säureessig durch die massenspektrometrische Bestimmung des $^{13}C/^{12}C$-Isotopenverhältnisses. Z. Lebensm. Unters. Forsch. 166, 89-92

[299]  D.A. Krueger (1992): Stable carbon isotope ratio method for detection of corn-derived acetic acid in apple cider vinegar: Collaborative study. J. Assoc. Off. Anal. Chem. 75, 725-728

[300]  C. Nubling, B. Fayet, C. Tisse, M. Guerere, G. Lesgards (1989): Contribution a la reconnaissance de l´origine de l´acide acetique dans les vinaigres d´alcool. Ann. Fals. Exp. Chim. 82, 385-392

[301]  D.A. Krueger, H.W. Krueger (1984): Comparison of two methods for determining intramolecular $^{13}C/^{12}C$ ratios of acetic acid. Biomed. Mass Spectrom. 11, 472-474

[302]  G. Rinaldi, W.G. Meinschein, J.M. Hayes (1974): Intramolecular carbon isotopic distribution in biologically produced acetoin. Biomed. Mass Spectrom. 1, 415-417

[303]  B. Weckerle, R. Bastl-Borrmann, E. Richling, K. Hör, C. Ruff, P. Schreier (2001). Cactus pear (*Opuntia ficus indica*) flavour constituents – chiral evaluation (MDGC-MS) and isotope ratio (HRGC-IRMS) analysis. Flav. Fragr. J. 16, 360-363

[304]  M. Lees, O. Bréas, G.J. Martin (1993): Potential applications of GC-C-IRMS in the field of flavours and essential oils - An overview. In: 2nd European Symposium Food Authenticity - Isotope Analysis and Other Advanced Analytical Techniques, 20-22 October 1993, Eurofins, Nantes, France

[305]  K. Hör, C. Ruff, B. Weckerle, T. König, P. Schreier (2001a): Flavor authenticity studies by $^2H/^1H$ ratio determination using on-line gas chromatography pyrolysis isotope ratio mass spectrometry. J. Agric. Food Chem. 49, 21-25

[306]  R. Braunsdorf, U. Hener, A. Mosandl (1992): Analytische Differenzierung zwischen natürlich gewachsenen, fermentativ erzeugten und synthetischen (naturidentischen) Aromastoffen. II. Mitt.: GC-C-IRMS-Analyse aromarelevanter Aldehyde - Grundlagen und Anwendungsbeispiele. Z. Lebensm. Unters. Forsch. 194, 426-430

[307]  K. Kempe, M. Kohnen: unpublished results

[308]  V. Karl, A. Dietrich, A. Mosandl (1994): Gas chromatography - isotope ratio mass spectrometry measurements of some carboxylic esters from different apple varieties. Phytochem. Anal. 5, 32-37

[309]  B. Byrne, K.J. Wengenroth, D.A. Krueger (1986): Determination of adulterated natural ethyl butyrate by carbon isotopes. J. Agric. Food Chem. 34, 736-738

[310]  H. Tamura, M. Appel, E. Richling, P. Schreier (2005): Authenticity assessment of γ- and δ-decalactone from *Prunus* fruits by on-line gas chromatography combustion/pyrolysis isotope ratio mass spectrometry (HRGC-C/P-IRMS). J. Agric. Food Chem. 53, 5397-5401

[311]  A. Bernreuther, J. Koziet, P. Brunerie, G. Krammer, N. Christoph, P. Schreier (1990): Chirospecific capillary gaschromatography (HRGC) and on-line HRGC-isotope ratio mass spectrometry of γ-decalactone from various sources. Z. Lebensm. Unters. Forsch. 191, 299-301

[312]  A. Mosandl, U. Hener, H.-G. Schmarr, M. Rautenschlein (1990): Chirospecific flavor analysis by means of enantioselective gas chromatography, coupled on-line with isotope ratio mass spectrometry. J. High Resol. Chromatogr. 13, 528-531

[313] A. Mosandl (1992): Analytical origin control of flavours and fragrances. In: H. Woidich, G. Buchbauer (eds.): Proceedings on 12$^{th}$ Int. Congr. Flav. Fragr. Ess. Oils, Vienna , 4. to 8. Oct., pp. 164-176

[314] A. Mosandl (1992): Echtheitskontrolle natürlicher Duft- und Aromastoffe. Merck Kontakte (Darmstadt) 3, 38-48

[315] A. Roßmann, H.-L. Schmidt, F. Reniero, G. Versini, I. Moussa, M.H. Merle (1996): Stable carbon isotope content in ethanol from EC data bank wines from Italy, France and Germany. Z. Lebensm. Unters. Forsch. 203, 293-301

[316] G.J. Martin, M.L. Martin, F. Mabon, M.-J. Michon (1982): Identification of the origin of natural alcohols by natural abundance hydrogen-2 nuclear magnetic resonance. Anal. Chem. 54, 2380-2382

[317] G.J. Martin, S. Hanneguelle, G. Remaud (1991): Application of site-specific natural isotope fractionation studied by nuclear magnetic resonance (SNIF-NMR) to the detection of flavour and fragrance adulteration. Ital. J. Food Sci. 3, 191-213

[318] G. Fronza, C. Fuganti, G. Zucchi, M. Barbeni, M. Cisero (1996): Natural abundance $^2$H nuclear magnetic resonance study of the origin of n-hexanol. J. Agric. Food Chem. 44, 887-891

[319] M. Barbeni, M. Cisero, C. Fuganti (1997): Natural abundance $^2$H nuclear magnetic resonance study of the origin of (Z)-3-hexenol. J. Agric. Food Chem. 45, 237-241

[320] S. Nitz, H. Kollmannsberger, B. Weinreich, F. Drawert (1991): Enantiomeric distribution and $^{13}$C/$^{12}$C isotope ratio determination of γ-lactones: appropriate methods for the differentiation between natural and non-natural flavours? J. Chromatogr. 557, 187-197

[321] G. Fronza, C. Fuganti, P. Grasselli, M. Barbeni, M. Cisero (1993): Natural abundance $^2$H nuclear magnetic resonance study of the origin of (R)-δ-decanolide. J. Agric. Food Chem. 41, 235-239

[322] M. Rohmer, M. Knani, P. Simonin, B. Sutter, H. Sahm (1993): Isoprenoid biosynthesis in bacteria: a novel pathway for the early steps leading to isopentenyl diphosphate. Biochem. J. 295, 517-524.

[323] M. Rohmer (1999): A mevalonate-independent route to isopentenyl diphosphate. In: D. Cane (ed.): Comprehensive Natural Product Chemistry, Vol. 2, Pergamon Press, Oxford, pp. 45-67

[324] F. Rohdich, K. Kis, A. Bacher, W. Eisenreich (2001): The non-mevalonate pathway of isoprenoids: genes, enzymes and intermediates. Curr. Opin. Chem. Biol. 5, 535-540

[325] R. Braunsdorf, U. Hener, S. Stein, A. Mosandl (1993): Comprehensive cGC-IRMS analysis in the authenticity control of flavours and essential oils. Part I: Lemon oil. Z. Lebensm. Unters. Forsch. 197, 137-141

[326] A. Roßmann, S. Schwarz, H.-L. Schmidt: unpublished results

[327] K. Kempe, M. Kohnen (1994): δ$^{13}$C-values of monoterpenes in essential oils as a mirror of their biosynthesis. Isotopenpraxis Environ. Health Stud. 30, 205-211

[328] U. Hener, R. Braunsdorf, P. Kreis, A. Dietrich, B. Maas, E. Euler, B. Schlag, A. Mosandl (1992): Chiral compounds of essential oils. X: The role of linalool in the origin evaluation of essential oils. Chem. Mikrobiol. Technol. Lebensm. 14, 129-133

[329] S. Bilke, A. Mosandl (2002 b): Authenticity assessment of lavender oils using GC-P-IRMS: $^2$H/$^1$H isotope ratios of linalool and linalylacetate. Eur Food Res Technol 214, 532-535

[330] K. Hör, C. Ruff, B. Weckerle, T. König, P. Schreier (2001b): $^2$H/$^1$H ratio analysis of flavor compounds by on-line gas chromatography-pyrolysis-isotope ratio mass spectrometry (HRGC-P-IRMS): Citral. Flav. Fragr. J. 16, 344-348

[331] D. Semiond, S. Dautraix, M. Desage, R. Majdalani, H. Casabianca, J.L. Brazier (1996): Identification and isotopic analysis of safranal from supercritical fluid extraction and alcoholic extracts of saffron. Anal. Lett. 29, 1027-1039

[332] R. Braunsdorf, U. Hener, D. Lehmann, A. Mosandl (1991): Analytische Differenzierung zwischen natürlich gewachsenen, fermentativ erzeugten und synthetischen (naturidentischen) Aromastoffen. Mitt. I: Herkunftsspezifische Analyse des (E)-α(β)-Ionons. Dt. Lebensm.-Rund. 87, 277-280

[333] A. Mosandl (1992): Capillary gas chromatography in quality assessment of flavours and fragrances. J. Chromatogr. 624, 267-292

[334] S. Schwarz, H.-L. Schmidt: unpublished

[335] S. Schwarz (1991): Intramolekulare Isotopenverteilung bei Isoprenoiden: Biosynthetische Ursachen und praktische Möglichkeiten zur Bestimmung ihrer natürlichen Herkunft bzw. Naturbelassenheit. PhD thesis, Technische Universität München, Freising, Germany

[336]  A. Roßmann, W. Rieth, H.-L. Schmidt (1990): Möglichkeiten und Ergebnisse der Kombination von Messungen der Verhältnisse stabiler Wasserstoff- und Kohlenstoff-Isotope mit Resultaten konventioneller Analysen (RSK-Werte) zum Nachweis des Zuckerzusatzes zu Fruchtsäften. Z. Lebensm. Unters. Forsch. 191, 259-264
[337]  C. Frank, A. Dietrich, U. Kremer, A. Mosandl (1995): GC-IRMS in the authenticity control of the essential oil of *Coriandrum sativum* L.. J. Agric. Food Chem. 43, 1634-1637
[338]  G.J. Martin (1993): The multi-site and multi-component approach for the stable isotope analysis of aromas and essential oils. In: 2$^{nd}$ European Symposium Food Authenticity - Isotope Analysis and Other Advanced Analytical Techniques, 20-22 October 1993, Eurofins, Nantes France
[339]  A. Mosandl, R. Braunsdorf, A. Dietrich, B. Faber, V. Karl, T. Köpke, D. Lehmann, B. Maas (1994): Recent development in the authenticity control of flavours and fragrances. In: H. Maarse, D.G. van der Heij (eds.): Trends in Flavour Research. Elsevier Science B.V., Amsterdam, pp. 89-98
[340]  Mosandl (2004): Authentizitätsbewertung von Aromastoffen mittels enantio-GC und Isotopen-MS. Mitt. Lebensm. Hyg. 95, 618-631
[341]  H.-L. Schmidt, D. Weber, A. Rossmann, R.A. Werner (1999): The potential of intermolecular and intramolecular isotopic correlations for authenticity control, in R. Teranishi, E.L.Wick, I. Hornstein (edts), Flavor Chemistry, Thirty Years of Progress, Kluver Academic/Plenum Press, New York 1999, pp. 55-61
[342]  H.-L. Schmidt: unpublished results
[343]  R.A. Werner, K. Kempe, H.-L. Schmidt (1995): Determination of $\delta^{13}$C- and $\delta^{18}$O-values of esters and their hydrolysis products for the evaluation of their origin (Abstract). Isotopenpraxis Environ. Health Stud. 31, 274
[344]  C. Rabiller, F. Mazé, F. Mabon, G.J. Martin (1991): Fractionnements isotopiques d´origine enzymatique et chimique dans la préparation d´esters. Analusis 19, 18-22
[345]  E. Deléens, N. Schwebel-Dugué, A. Tremolières (1984): Carbon isotope composition of lipidic classes isolated from tobacco leaves. FEBS Lett. 178, 55-58
[346]  O.A. Güell, J.A. Holcombe (1990): Analytical applications of Monte Carlo Techniques. J. Anal. Chem. 62, 529-542
[347]  J. Aleu, G. Fronza, C. Fuganti, S. Serra, C. Fauhl, C. Guillou, F. Reniero (2002): Differentiation of natural and synthetic phenylacetic acids by $^2$H NMR of the derived benzoic acids. Eur. Food. Res. Technol. 214, 63-66
[348]  G. Remaud, A.A. Debon, Y.-L. Martin, G.G. Martin, G.J. Martin (1997) : Authentication of bitter almond oil and cinnamon oil: application of the SNIF-NMR method to benzaldehyde. J. Agric. Food Chem. 45, 4042-4048
[349]  T.V. John, E. Jamin (2004): Chemical investigation and authenticity of Indian vanilla beans. J. Agric. Food Chem. 52, 7644-7650
[350]  G.J. Martin, M.L. Martin (1999): Thirty years of Flavor NMR, in R. Teranishi, E.L.Wick, I. Hornstein (edts), Flavor Chemistry, Thirty Years of Progress, Kluver Academic/Plenum Press, New York 1999, pp.19-30
[351]  G.J. Martin, S. Lavine-Hanneguelle, F. Mabon, M.L. Martin (2004): The fellowship of natural abundance $^2$H-isotopomers of monoterpenes. Phytochem. 65, 2815-2831
[352]  G.J. Martin (1995): Multisite and multicomponent approach for the stable-isotope analysis of aromas and essential oils, in R.L. Rouseff, M.M. Leahy (eds.): Fruit Flavors: Biogenesis, Characterization and Authentication. Developed from a symposium sponsored by the Division of Agricultural and Food Chemistry at the 206th National Meeting of the American Chemical Society, Chicago, Illinois, August 22-27, 1993. ACS Symposium Series, Nr. 596, Am. Chem. Soc. Washington, pp. 79-93
[353]  G.J. Martin, N. Naulet (1988): Precision, accuracy and referencing of isotope ratios determined by nuclear magnetic resonance. Fresenius Z. Anal. Chem. 332, 648-651
[354]  G. Martin, G. Remaud, G.J. Martin (1993): Isotopic methods for control of natural flavours authenticity. Flav. Fragr. J. 8, 97-107
[355]  C. Maubert, C. Guérin, F. Mabon, G.J. Martin (1988): Détermination de l'origine de la vanilline par analyse multidimensionnelle du fractionnement isotopique naturel spécifique de l'hydrogène. Analusis 16, 434-439

[356]  F. Le Grand, G. George, S. Akoka (2005): Natural abundance $^2$H-ERETIC-NMR authentication of the origin of methyl salicylate. J. Agric. Food Chem. 53, 5125-5129
[357]  D.M. Grant, J. Curtis, W.R. Croasmun, D.K. Dalling, F.W. Wehrli, S. Wehrli (1982): NMR determination of site-specific deuterium isotope effects. J. Am. Chem. Soc. 104, 4492-4494
[358]  M.F. Leopold, W.W. Epstein, D.M. Grant (1988): Natural abundance deuterium NMR as a novel probe of monoterpene biosynthesis: limonene. J. Am. Chem. Soc. 110, 616-617
[359]  G.J. Martin, P. Janvier, S. Akoka, F. Mabon, J. Jurczak (1986): A relation between the site-specific natural deuterium contents in α-pinenes and their optical activity. Tetrahedron Lett. 27, 2855-2858
[360]  C. Vallet, M. Arendt, G.J. Martin (1988): Site specific isotope fractionation of hydrogen in the oxydation of ethanol into acetic acid. Application to vinegars. Biotechnol. Techn. 2, 83-88
[361]  C. Guillou, G. Remaud, G.J. Martin (1991): Application of deuterium NMR and isotopic analysis to the characterization of foods and beverages. Trend. Food Sci. Technol. 2, 85-89
[362]  M. Barbeni, P. Cabella, M. Cisero (1994): Biogeneration of δ-decalactone and $^2$H-NMR study of its origin. In: H. Maarse, D.G. van der Heij (eds.): Trends in Flavour Research. Elsevier Science B.V., Amsterdam, pp. 487-491
[363]  A. Roßmann, F. Pichlmayer: unpublished results
[364]  E. Brenna, G. Fronza, C. Fuganti, F.G. Gatti, M. Pinciroli, S. Serra (2004): Differentiation of extractive and synthetic salicin: The $^2$H aromatic pattern of natural 2-hydroxybenzyl alcohol. J. Agric. Food Chem. 52, 7747-7751
[365]  E. Brenna, G. Fronza, C. Fuganti, F.G. Gatti, V. Grande, S. Serra, C. Guillou, F. Reniero, F. Serra (2005): Stable isotope characterization of the *ortho*-oxygenated phenylpropanoids: coumarin and melilotol. J. Agric. Food Chem. 53, 9383-9388

## 6.2.3 Enantioselective Analysis
*Armin Mosandl*

Chiral discrimination has been recognized as one of the most important principles in biological activity and also odour perception [1-9]. Besides enantioselective biogenesis, the evaluation of chirality in the origin control of flavourings and fragrances has to be discussed with regard to some fundamental conditions.

### 6.2.3.1 Chiral Resolution and Chromatographic Behaviour of Enantiomers

Resolution (Rs) is defined as the separation of two peaks in terms of their average peak width at half height (I) or at a base width of $4\sigma$ (II) [10].

$$Rs = 1.177 \cdot \frac{\Delta t_R}{W_{h1} + W_{h2}} \quad (I)$$

$$Rs = 2 \cdot \frac{\Delta t_R}{W_{b1} + W_{b2}} \quad (II)$$

$\Delta t_R$ : absolute difference in retention time of the two peaks 1(2)
$W_{h1(2)}$ : peak width of 1(2) at half height
$W_{b1(2)}$ : peak width of 1(2) at base ($4\sigma$)

In the case of enantiomeric pairs the term chiral resolution (cRs) is used.

As outlined in Table 6.15, 100 % separation occurs, if chiral resolution cRs = 2.50; in practice cRs $\geq$ 1.50 (99.73% separation) is defined as a "baseline" resolution [10]. Thus, optimal chiral resolution (cRs $\geq$ 1.50) should be achieved using a suitable chiral stationary phase in enantioselective capillary gaschromatography (enantio-cGC).

*Table 6.15: Separation as a function of cRs (From ref. [10])*

| cRs | Separation (%) | $\Delta tR$ (in $\sigma$) |
|---|---|---|
| 0.5 | 68.28 | 2 |
| 0.9 | 92.82 | 3.6 |
| 1.0 | 95.44 | 4 |
| 1.1 | 97.22 | 4.4 |
| 1.25 | 98.76 | 5 |
| 1.50 | 99.73 | 6 |
| 2.0 | 99.99 | 8 |
| 2.50 | 100.00 | 10 |

Although separation factor $\alpha$ has the greatest impact on peak resolution, an accurate determination of high enantiomeric excess (ee-values) as shown in Fig. 6.23 [11] is only possible, in the case of highly resolved enantiomer separation.

# Enantioselective Analysis

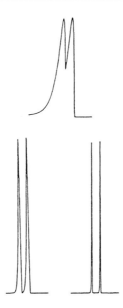

*Fig. 6.23:* Chiral separations with identical separation factors ($\alpha$) and different chiral resolutions (cRs) (From ref. [11])

At the present state of knowledge, the mechanisms of GC enantiomer separation have not been elucidated. Unusual chromatographic behaviour and reversal of the elution order of enantiomers have been observed. Consequently, the usefulness of a given chiral stationary phase as well as the order of elution of separated enantiomers cannot be predicted. References of definite chirality are essential to identify the separated isomers, no matter whether directly stereoanalyzed with chiral stationary phases or via derivatized stereoisomers [12-19].

## 6.2.3.2 Sample Clean-up

Optimum chiral resolution without any racemization and reliable interpretation of chromatographic behaviour of enantiomers have to be considered as the first targets in enantioselective analysis (section 6.2.3.1.). Even if a universal recommendation on sample clean-up cannot be given, depending on the complexity of samples to be analyzed, preseparation procedures of highest efficiency are needed if a reliable stereodifferentiation should be achieved.

The demanding challenge is to yield chiral volatiles of highest chemical purity, ready for the direct resolution into their mirror images by *GC*-techniques. This ideal may be realized by preseparation techniques in nonchiral media (chiral discriminations excluded), using high performance thin layer chromatography (HPTLC) *[20, 21]* or high performance liquid chromatography (HPLC), off-line *[22, 23]* or on-line *[24, 25]* coupled with enantio-cGC.

In particular, enantioselective multidimensional gas chromatography (enantio-MDGC) with the combination of a non-chiral precolumn and a chiral main column

*[26, 27]* has been demonstrated as a powerful method to the direct stereoanalysis of chiral volatiles without any further clean-up or derivatization procedures *[28-50]*.

A schematic diagram of enantio-MDGC, well proved in quality assurance and origin control of flavourings and fragrances is outlined in Fig. 6.24.

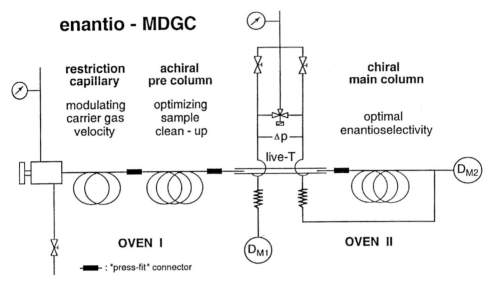

*Fig. 6.24:* Schematic diagram of enantioselective multidimensional gas chromatography, according to ref. *[28]*

A double-oven system with two independent temperature controls, two detectors (DM1, DM2) and a "live switching" coupling piece, is used. At optimum pneumatic adjustment of the MDGC-system definite fractions, eluted from the precolumn, are selectively transferred onto the chiral main column (heart-cutting technique).

At first the chromatographic conditions of the chiral main column must be optimized carefully. Capillary columns coated with chiral stationary phases of suitable enantioselectivities are used as main columns. Chiral resolutions are commonly achieved isothermally or by low temperature programming rates, starting at least 20°C below precolumn temperature. Precolumns are chosen with respect to (i) the versatility of application, to (ii) the direct injection of high sample volumes and (iii) with respect to the requisite time of analysis.

At optimized operating conditions, uncoated and deactivated restriction capillaries are installed between injector and precolumn by means of simple "press-fit" connectors to reduce carrier gas velocity within the precolumn. By means of such a column combination suitable pre- and main columns may be easily exchanged and adopted to optimal efficiency *[28]*.

### 6.2.3.3 Detection Systems

If optimum chiral resolution and high efficiency sample clean-up, the claims of first priority in enantioselective analysis (sections 6.2.3.1. and 6.2.3.2.), are realized rather simple detection systems, such as flame ionization detectors (FID) are suitably used.

# Enantioselective Analysis

The ideal detector is universal yet selective, sensitive and structurally informative. Mass spectrometry (MS) currently provides the closest approach to this ideal. The combination of multi-dimensional gas chromatography with high resolution MS or mass-selective detectors in the single ion monitoring (SIM)-mode is currently the most potent analytical tool in enantioselective analysis of chiral compounds in complex mixtures [29]. Nevertheless, it must be pointed out that the application of structure specific detection systems like MS [51] or Fourier transform infrared (FT-IR) [52] cannot save the fundamental challenges to optimum (chiral) resolutions and effective sample clean-up [53].

The effectiveness of mass selective detection in the SIM(MIM) mode and selected wavelength chromatograms (SWC) in FT-IR-detection depends on efficient sample clean-up. In raw extracts the risk of co-eluting substances which remain indiscriminative from chiral volatiles increases with the complexity of flavouring and fragrance extracts to be analyzed.

*Table 6.16:* Mass-selective detection of 2-methylbutanoic acid (esters) and analogous mass fragments (from ref. [53])

| m/e = 85 | | m/e = 88 | |
|---|---|---|---|
| $[C_5H_9O]^+$ | 2-methylbutanoic acid (ester) | $[C_4H_8O_2]^+$ | methyl 2-methylbutanoate |
| $[C_4H_5O_2]^+$ | 4-alkylsubstituted γ-lactones | $[C_4H_8O_2]^+$ | McLafferty-rearrangement: ethyl esters |
| $[C_5H_9O]^+$ | pentanoic acid (ester) butylketones | **m/e = 101** | |
| $[C_5H_9O]^+$ | 2-substituted pyrane derivates | $[C_5H_9O_2]^+$ | methyl 2-methylbutanoate |
| $[C_6H_{13}]^+$ | fragment from alkyl residues ≥ $C_6$ | $[C_5H_9O_2]^+$ | β,γ-cleavage: ethyl esters |

As outlined in Table 6.16, the mass-selective detection of 2-methylbutanoic acid (esters) may interfere with the γ-lacton fragment (m/e = 85) and/or analogous mass fragments from different classes of compounds. As to be seen from Table 6.16, even the selectivity of simultaneous multiple ion detection of 2-methylbutanoic acid (ester) fragments (m/e = 85, 88, 101) in the MIM mode remains inconclusive, if the sample clean-up procedure was insufficient.

Similar considerations have to be taken into account, if FT-IR detection in the selected wave length chromatogram (SWC) mode shall be used (Table 6.17).

Consequently, high efficiency sample clean-up remains an essential condition to chirality evaluation and legally binding interpretation of authenticity assessment of flavourings [53].

**Table 6.17:** Half-height widths of CO-valence absorption of different carbonyl compounds (from ref. [53])

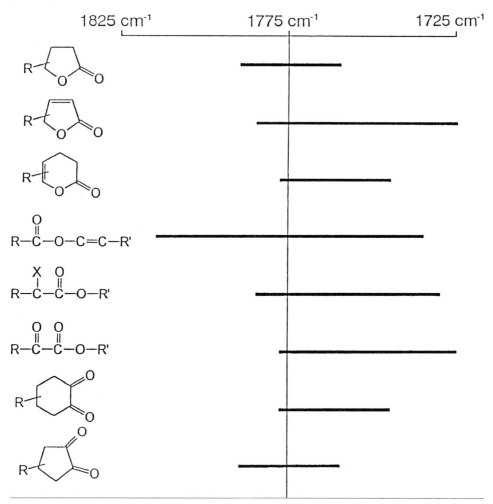

### 6.2.3.4 Stereodifferentiation and Quantification

The total amount of characteristic flavouring constituents and their relative distribution in aroma extracts have been included in official legal assessments for many years, using internal standards for quantification and referring to the merits of comprehensive stock-taking of volatile constituents from complex flavouring and fragrance extracts during the first era of flavour research [54]. The measurements are precise and reproducible, but with respect to the validation of the origin of flavourings and essential oils their usefulness is rather limited [49].

Enantioselective differentiations of optically and sensorially active compounds point out new possibilities in structure-function relationships and biogenesis of chiral volatiles [49, 22].

Enantiomeric purity and enantiomeric excess (ee) are usual terms used in the determination of enantiomers. Enantiomeric purity is defined as the measured ratio (expressed as a percentage) of the detected enantiomers, whereas ee-values describe the relative difference of the separated enantiomers (expressed as a percentage). Usually quantifications are given in ee-values, but one should note, that convincing results can be concluded only for baseline-resolved enantiomers (cRs ≥ 1.50). Exact calculations of partially resolved mirror images, as frequently happened in the current literature, remain unintelligible in view of differences in sensory qualities and odour thresholds of enantiomers Fig. 6.25, [1-9].

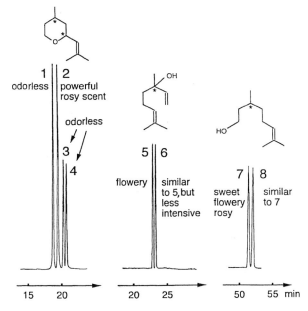

*Fig. 6.25: Enantioselective cGC separation and odor impression by enantiomer eluate sniffing of cis rose oxides 1(2R, 4S), 2(2S, 4R) and trans rose oxides 3(2R, 4R), 4(2S, 4S) (column B); program: 60°C/5 min isothermal// 95°C isothermal; linalool 5(R), 6(S) (column A); program: 60°C/5 min isothermal// 80°C/5 min isothermal/1.2°C/min/200°C; citronellol 7(S), 8(R) (column A); program: 60°C/5 min isothermal//70°C/10 min isothermal/1.0°C/min/200°C (from ref. [109])*

Partially separated enantiomeric pairs (cRs = 0.9 - cRs = 1.50) should be calculated approximately as the ratio of enantiomers (expressed as a percentage).

Approximately quantifications in terms of enantiomeric ratios are also more conclusive than ee-values, if concentrations are too low for precise calculations [50].

### 6.2.3.5 Limitations

Three types of limitations have to be accepted in enantioselective analysis:
- racemates of natural origin, generated by non-enzymatic reactions (autoxidation, photooxidation, etc.)
- racemization during processing or storage of foodstuffs, if structural features of chiral compounds are sensitive

– blending of natural and synthetic chiral flavouring compounds.

Nevertheless the systematic evaluation of origin specific enantiomeric ratios has proved to be a valuable criterion for differentiating natural flavouring compounds from those of synthetic origin.

### 6.2.3.6 Analysis of Individual Classes of Compounds

First success in enantioselective flavouring analysis was achieved by chromatographic separations of diastereomeric derivatives. In spite of limited sensitivity and frequently laborious work-up conditions, these methods revealed reliable insight into enantiomeric distribution of $\gamma(\delta)$ lactones and other chiral fruit flavouring compounds, as reviewed previously [55].

A real breakthrough in enantioselective analysis occurred, since enantioselective cGC became available. Three classes of chiral stationary phases are of special interest:

- amino acid derivatives, chemically bonded to polysiloxanes [56, 57].
- optically active metal chelates [58]
- modified cyclodextrins [11, 14, 59-64]

In particular, since 1988 selectively alkylated/acylated $\alpha$-, $\beta$- and $\gamma$-cyclodextrins have been synthesized, serving as chiral stationary phases in enantioselective cGC (Fig. 6.26).

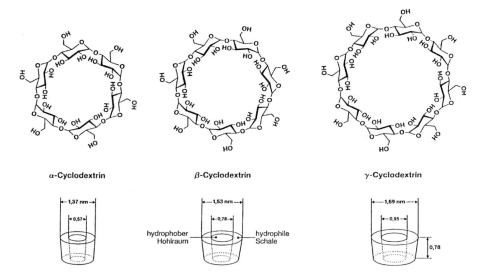

*Fig. 6.26:* Chemical structure and conical shape of cyclodextrins (from ref. [116])

Owing to the high melting points of pure cyclodextrins and many of its derivatives, the chromatographic efficiency of such chiral stationary phases was poor. Improved performance was achieved by König and co-workers [14] and Armstrong et al. [60], who introduced cyclodextrin derivatives which are liquid or waxy at room temperature. As an alternative approach, Schurig and co-workers [61] diluted high melting cyclodextrin derivatives with polysiloxane stationary phases to obtain chiral selectivity below the melting point of the pure cyclodextrin. Dilution of high melting point cyclodextrin derivatives lowers the minimum GC-working temperature. Meanwhile the influence of polysiloxane solvents to the enantioselectivity and column efficiency is well documented *[65-67]*.

Although the mechanisms of chiral recognition and molecular inclusion have not been elucidated, modified cyclodextrins have been reported as versatile chiral phases with a wide range of applications. The chiral resolution of many cyclic monoterpenes *[11]* and investigations of the enantiomeric distribution of α-pinene, β-pinene and limonene from foods, drugs and essential oils have been reported in the literature *[49, 33-35, 37]*. Also *cis/trans* α-irones, highly responsible fragrance compounds of the iris oil *[68]*, filbertone [(E)-5-methyl-2-hepten-4-one] the character impact compound of hazelnuts *[69, 115]*, damascone *[70]* and geosmin *[68]* have been separated into their mirror images *[68-70, 115]*. Even polar molecules such as 1,2-ketols, 1,2-glycols and free carboxylic acids were successfully stereoanalyzed *[17]*.

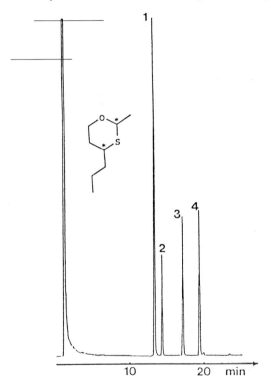

***Fig. 6.27:*** *Stereoselective differentiation of 2-methyl-4-propyl-1,3-oxathiane. Elution order:* **(1)** *2R, 4S (enriched);* **(2)** *2S, 4R;* **(3)** *2R, 4R;* **(4)** *2S, 4S (from ref. [64])*

## Chiral Sulphur Compounds

3-Methylthiohexanol and *cis/trans*-2-methyl-4-propyl-1,3-oxathiane have been reported by Winter et al. *[71]* to play important roles in the delicate flavour of the yellow passionfruit, in spite of their occurrence in the extreme trace levels. Using complexation GC, the stereoselective analysis of all the four stereoisomers of 2-methyl-4-propyl-1,3-oxathiane was achieved. This was the first direct stereochemical analysis of an essential chiral trace compound from fruit flavours *[12]*. While first attempts to the chirality evaluation of this potent odorant from yellow passionfruits were described *[72]*, the direct stereoselective analysis of natural 1,3-oxathianes, 3-mercapto-hexyl acetate and butanoate now has been realized *[73]*. Modified cyclodextrins *[64, 68]*, using enantioselective cGC serve as promising alternatives (Fig. 6.27 - 6.29).

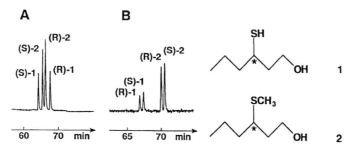

*Fig. 6.28: Stereodifferentiation of 3-mercaptohexanol (1) and 3-methylthiohexanol (2) on heptakis(2,3-di-O-acetyl-6-O-tert-butyldimethylsilyl)-β-cyclodextrin (DIAC-6-TBDMS-β-CD) (A) and heptakis(2,3-di-O)-methyl-6-O-tert-butyldimethylsilyl)-β-cyclodextrin (DIME- 6- TBDMS-β-CD) (B) (from ref. [110])*

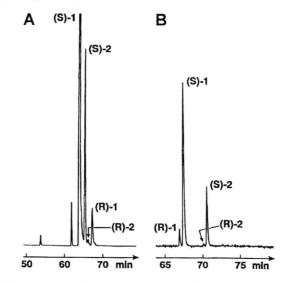

*Fig. 6.29: Stereodifferentiation of 3-mercaptohexanol (1) and 3-methylthiohexanol (2) of yellow passion fruits, using different GC-systems (A, B) (from ref. [110])*

## γ-Lactones

Chiral γ-lactones are important compounds of many fruits and give strawberries, peaches, apricots and many other fruits their characteristic and distinctive notes *[74]*. Albrecht and Tressl *[75]* investigated the biogenetic sequence of γ-decalactone. These results indicate that (E)-3,4-epoxydecanoic acid, formed from (E)-3-decenoyl-CoA, an intermediate of the β-oxidation of linoleic acid, is the genuine precursor in the biosynthesis of γ-decalactone.

# Enantioselective Analysis

Owing to their common enzymatic pathways, chiral aroma compounds from fruits and other natural sources should be characterized by origin specific enantiomeric ratios. Indeed, from freshly harvested strawberries γ-decalactone and γ-dodecalactone have been detected with high enantiomeric purity favouring the (4R)-configurated γ-lactones; γ-$C_{10}$(4R; >98 % ee), γ-$C_{12}$(4R; >99 % ee) [28]. On the other hand, racemic γ-decalactone in aroma relevant amounts has not yet been observed in fruits. Thus, the detection of racemic γ-decalactone from fruit-containing food indicates the addition of "nature-identical" γ-C10-lactone.

Furthermore, enantio-MDGC, employing heart-cutting techniques from DB-1701 as the preseparation column on to heptakis (3-O-acetyl-2,6-di-O-pentyl)-β-cyclodextrin as the chiral main column, was described by Mosandl et al. [28] as a powerful tool in the direct enantiomer separation of chiral γ-lactones from complex matrices without any further clean-up or derivatization procedures.

Comprehensive data on natural γ-lactones from fruits have been reported by Nitz et al. [29], Bernreuther et al. [30] and Guichard et al. [31].

Comparative investigations of γ-lactones from apricots of different varieties and cultivars [76] in connection with corresponding enantioselective analysis [31] indicate some important conclusions:

- If present at all, odd numbered γ-lactones occur only in trace amounts.
- The most abundant γ-lactones are the even numbered homologues γ-$C_6$, γ-$C_8$, γ-$C_{10}$, γ-$C_{12}$.
- The enantiomeric distribution of chiral γ-lactone homologues in fruits has been demonstrated to increase in favour of the 4R- configured lactones with increasing length of the alkyl side chain [29-31, 46].

It is also interesting to note the detection of dihydroactinidiolide as the first γ-lactone racemate of natural origin, generated in ripening apricots [31]. This surprising fact also has been confirmed for dihydroactinidiolide from raspberries [68, 115].

With regard to chirality evaluation as an indicator for the genuineness of natural flavourings and fragrances only chiral volatiles of high enantiomeric purity and characteristic and small ranges of ee-values should be validated in relation to their total amounts (Table 4). Furthermore, it is remarkable to pay more attention to the *odour activity values (OAV)* of compounds analyzed [77, 78].

$OAV = c/a$   concentration:   c [µg/kg]
odour threshold:   a [µg/kg]

Even if comprehensive data about OAV-values are not yet available or existing values may differ significantly, owing to methodological differences, OAV-values will become more and more important. In this context γ-hexalactone is mentioned as an unsuitable example. In spite of its relatively large amounts, γ-$C_6$ lactone is useless for the origin assessment of flavours, owing to the wide range of ee values detected (Table 6.18). Furthermore, a high odour threshold (a = 13,000 µg/kg) [79] and unspecific odour quality [8] reveals γ-$C_6$ lactone as insignificant to the odour impression of natural fruit flavouring [31]. On the other hand the even numbered γ-lactones $C_8$ (a = 95 µg/kg [79]), $C_{10}$ (a = 88 µg/kg [79]; 11µg/kg [80]; 1 µg/kg [81]) and $C_{12}$

(a = 7 µg/kg [80]), in particular γ-decalactone, have proved to be useful indicators of naturalness for many fruit flavourings and fragrances, if their genuine amount occurs in the aroma-relevant range (OAV >1).

Of course, chiral compounds occurring in trace amounts and far below their odour activity values in foods investigated, e. g. γ-decalactone in raspberries and gooseberries [82], have to be neglected.

In this context it should be mentioned that γ-lactones do not occur as genuine constituents of coconuts [83]. Thus, it seems to be rather clear, that aroma-relevant amounts of γ-lactones in coconut products are of non-natural origin and there is no reason to look for their stereoanalysis in the low ppb-range [82].

On the other hand, chiral δ-lactones from coconuts, in particular δ-octalactone and δ-decalactone, have been evaluated as suitable indicators of naturalness [36].

## δ-Lactones

Likewise, chiral δ-lactones are known as characteristic flavouring compounds of fruits and dairy products [74].Their stereodifferentiation was achieved with modified γ-cyclodextrin by König et al. [84] and their chromatographic behaviour interpreted by coinjection with enantiopure references, as described by Palm et al. [36]. Using enantio-MDGC and the column combination OV 1701/octakis(3-O-butyryl-2,6-di-O-pentyl)-γ-cyclodextrin the simultaneous stereodifferentiation of all aroma-relevant 4(5) alkylsubstituted γ(δ)-lactones has been reported recently [42, 85] (Fig. 6.30).

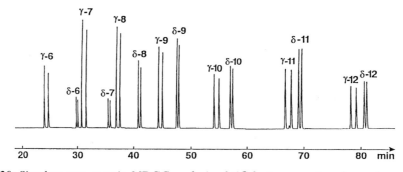

*Fig. 6.30:* Simultaneous enantio-MDGC analysis of γ(δ) lactones, main column chromatogram of references (from ref. [85])

Peaches, apricots and greengages contain γ(δ)-lactones with characteristic enantiomeric ratios in favour of R-configured lactones, proving reliable indicators of genuineness during all stages of fruit processing (Fig. 6.31).

Whereas aroma-relevant amounts of γ-lactones from raspberries are not detectable, their δ-$C_8$, δ-$C_{10}$-lactones are enantiopure S-enantiomers [42]. In cheddar cheese-aroma the enantiomeric purity of (5R)-δ-lactones ($C_{10}$-$C_{14}$) increases with increasing side chain length [68].

The enantioselective analysis of massoilactone (dec-2-en-5-olide), a coconut-typical δ-lactone of natural origin, has been reported by Bernreuther et al. [47] and Werkhoff et al. [68], revealing (-)-massoilactone of high enantiomeric purity (5R; >99% ee).

# Enantioselective Analysis

**Table 6.18:** Distribution of γ-lactone homologues from different apricot cultivars (from ref. [31])

A = Relative amount of each γ-lactone (peak heights were measured on the β-cyclodextrin chromatogram and for each γ-lactone the sum of the two enantiomer peaks was divided by the sum of all the lactones peaks.) R = Relative amount of the (R)-enantiomers [HR/(HR + HS)] in percent, where HR and HS are the heights of the peak for the (R)- and (S)-enantiomers, respectively].

| Cultivar | $\gamma$-$C_6$ A(%) | $\gamma$-$C_6$ R(%) | $\gamma$-$C_7$ A(%) | $\gamma$-$C_7$ R(%) | $\gamma$-$C_8$ A(%) | $\gamma$-$C_8$ R(%) | $\gamma$-$C_9$ A(%) | $\gamma$-$C_9$ R(%) | $\gamma$-$C_{10}$ A(%) | $\gamma$-$C_{10}$ R(%) | $\gamma$-$C_{11}$ A(%) | $\gamma$-$C_{11}$ R(%) | $\gamma$-$C_{12}$ A(%) | $\gamma$-$C_{12}$ R(%) |
|---|---|---|---|---|---|---|---|---|---|---|---|---|---|---|
| Bergeron 1986 | 26 | 88 | 1 | 89 | 13 | 93 | 1 | 83 | 57 | 96 | tr[a] | – | 2 | 98 |
| Bergeron 1985 | 21 | 80 | 0.8 | 92 | 9 | 92 | 1.2 | 89 | 58 | 95 | tr | – | 10 | >99 |
| Polonais 1986 | 18 | 96 | 0.6 | 81 | 17 | 92 | 1.4 | 76 | 62 | 95 | tr | – | 1 | >99 |
| Moniqui 1986 | 15 | 91 | 0.3 | 81 | 10 | 90 | 0.7 | 83 | 71 | 95 | tr | – | 3 | 99 |
| Palsteyn 1986 | 49 | 82 | 2.5 | 80 | 25 | 90 | tr | – | 23 | 92 | tr | – | 0.5 | >99 |
| PrecoceThyrinthe 1986 | 87 | 99 | 1 | 81 | 7 | 90 | 0.2 | 64 | 4 | 91 | tr | – | 0.8 | 93 |
| Rouge du Roussillon 1988 | 52 | 71 | 0.8 | 83 | 20 | 87 | 0.6 | 88 | 26 | 97 | tr | – | 0.6 | >99 |
| Rouge du Roussillon 1986 | 53 | 71 | 1.6 | 90 | 19 | 88 | 0.4 | 77 | 22 | 98 | tr | – | 4 | 99 |
| Rouge du Roussillon 1985 | 29 | 61[b] | tr | – | 17 | 86 | 1 | 91 | 42 | 96 | tr | – | 11 | >99 |

[a] Traces of; γ-lactones not quantified.
[b] This low value was confirmed after isolation of this γ-hexalactone from the complex extract by high-performance liquid chromatography [SiO2 5 μm; 1.5 ml/min pentane-diethyl ether (70:30) RI detection]. The eluate in the interval 15-16 min was concentrated and analysed by multi-dimensional GC; cf. separation conditions.

Even the higher homologues of δ-lactones ($C_{13}$-$C_{18}$) were resolved into their enantiomers, using a thin film capillary coated with modified β-cyclodextrin [86] (Fig. 6.32).

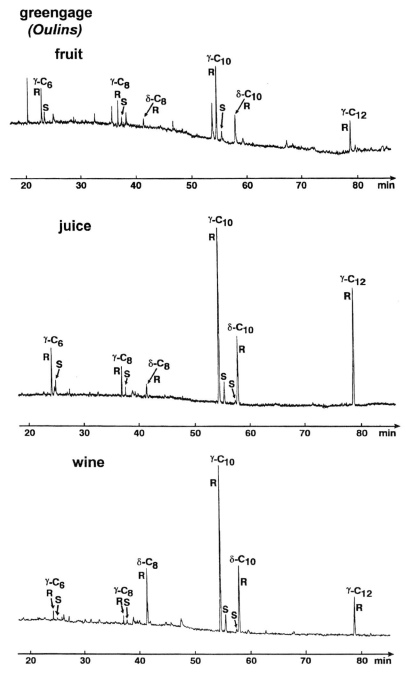

***Fig. 6.31:*** *Simultaneous enantio-MDGC analysis of γ (δ) lactones from greengages (main column separation) (from ref. [85])*

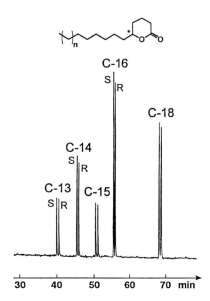

***Fig. 6.32:*** *Enantioselective differentiation of δ-lactones $C_{13}$ to $C_{18}$ (order of elution: (S) (I), (R) (II) for $C_{13}$, $C_{14}$ and $C_{16}$); chromatographic conditions: 24 m x 0.23 mm i.d. glass open tubular column coated with 50% DIAC-6-TBDMS-β-CD in PS 255, df = 0.05 μm; carrier gas H2, 50 cm/sec; temperature programme: 120°C//15 min isothermal//1.5°C/min to 220°C (from ref. [86])*

## 2-Alkylcarbonic Acids (Esters)

2-Alkylcarbonic acids have been separated into their enantiomers without any derivatization and their sequence of elution was assigned by co-injection with enantio pure references [17]. Latest results on stereoselective flavour evaluation revealed characteristic sensory properties for all the enantiomers of 2-alkylbranched acids, esters and corresponding alcohols. Tremendous differences between the mirror images of 2-methylbutanoic acid have been found. While the R-enantiomer exhibits a penetrating, cheesy-sweaty odour, the S-enantiomer emits a pleasant sweet and fine fruity note [87, 88]. All commercially available homologues of 2-methylbutanoic acid esters and 2-methylbutyl acetate are simultaneously stereoanalyzed, using heptakis (2,3-di-O-methyl-6-O-tert-butyldimethylsilyl)-β-cyclodextrin (DIME-β-CD) in PS 268 as the chiral stationary phase [88] (Fig. 6.33, Table 6.19).

***Table 6.19:*** *Sensory properties of the esters shown in chromatogram of Fig. 6.33 (from ref. [88])*

| Compound | (R)-Enantiomer | (S)-Enantiomer |
|---|---|---|
| 2-Methylbutanoates | | |
| Methyl (**1**) | sweet, fruity, apple-like | sweet, fruity, apple-like; considerably stronger than the (R) enantiomer |
| Ethyl (**2**) | sweet, fruity, reminiscent of apples | substantially stronger than the (R) enantiomer; sweet, fruity, full-ripe apple |

| Compound | (R)-Enantiomer | (S)-Enantiomer |
|---|---|---|
| Propyl (**4**)[b] | weak, unspecific | intensive, full-ripe apple-note; not as strong as (S)-2 |
| Isobutyl (**5**) | nearly odourless | nearly odourless |
| Isopentyl (**6**)[b] | nearly odourless | nearly odourless |
| Pentyl (**7**) | nearly odourless | nearly odourless |
| Hexyl (**8**) | nearly odourless | nearly odourless |
| 2-Methylpentanoates acetate (**3**) | nearly odourless | nearly odourless |
| 2-Methylpentanoates | | |
| Methyl (**9**) | nearly odourless | fruity sweet, apple-like |
| Ethyl (**10**)[b] | fruity sweet, apple-like | very intensive, fruity sweet, very pleasant apple-note; highest aroma quality of the investigated substances |
| Acid | | |
| 2-Methylpent-anoic acid (**11**) | weak, sweaty fruity | sweaty fruity |

[a]  Conditions: flow rate, 2.6 mL/min; split, 30 mL/min; injection volume; 1 µL 0.1% solution of the racemate; subsequent 1:1 splitting between FID and sniffing detector.
[b]  Elution order deduced;

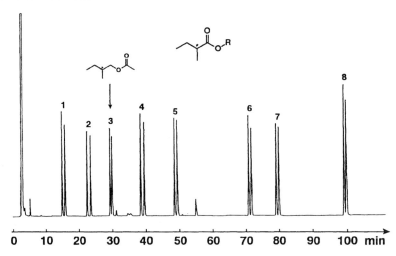

*Fig. 6.33:* FID chromatogram of the enantioselctive analysis of methyl 2-methylbutanoate (**1**), ethyl 2-methylbutanoate (**2**), 2-methylbutyl acetate (**3**), propyl 2-methylbutanoate (**4**), isobutyl 2-methylbutanoate (**5**), isopentyl 2-methylbutanoate (**6**), pentyl 2-methylbutanoate (**7**) and hexyl 2-methylbutanoate (**8**), simultaneously analysed by sniffing GC-see Table 6.19 (from ref. [88])

From the analytical point of view it is worth noting the biogenetic pathway of 2-methylbutanoic acid starting from isoleucine [(2S)-amino-(3S)-methylpentanoic acid]. The S-configuration of the precursor is expected to remain; but also enzymatic racemization (by enolization of the intermediate 2-oxo-3-methylpentanoic acid) is known from the literature [89].

Appropriate analytical techniques without any racemization were developed and applied to apples and apple containing foods. In all investigated food the S-enantiomer has been identified with high enantiomeric purity. Thus, the addition of "nature-identical", synthetic 2-methylbutanoic acid racemate is easily detected [39].

The direct (but not baseline resolved) stereodifferentiation of ethyl 2-methylbutanoate, a well known impact flavouring compound of the apple aroma, was reported simultaneously by Takeoka et al. [90] and Mosandl et al. [17]. Meanwhile further enantioselective procedures have been developed, indicating (S)-ethyl 2-methylbutanoate as the unique antipode from natural flavourings [39, 91] and its impressive and pleasant apple note at extreme dilution has been recognized previously [87, 88]. As the latest advance in this field the first simultaneous stereoanalysis of 2-methylbutanoic acid, its methyl and ethyl esters and its corresponding alcohol 2-methylbutanol was realized, using selectivity adjusted enantio-MDGC with perethylated β-cyclodextrin as the chiral stationary phase. All the four chiral compounds were detected as enantiopure S-enantiomers from apples. The application of this technique to a commercially available apple flavouring concentrate reveals the adulteration by means of methyl and ethyl ester racemates [45].

## Chiral Monoterpenoids from Mentha-Species

The essential oils of the Mentha species are known to be valuable ingredients of pharmaceutical and cosmetic preparations. Menthone, isomenthone, menthol and menthyl acetate, the most important Mentha oil constituents, are defined by their amounts as substantial parameters of mentha oil quality. Even off-line HPTLC/enantio-GC coupling has proved to be a valuable method in authenticity control of mint and peppermint oils. After preseparating by HPTLC (silica gel; dichloromethane) the menthyl acetate fraction is isolated and directly injected for stereodifferentiation with permethyl β-cyclodextrin. Racemic menthyl acetate was detected by Kreis et al. as a frequently used adulteration of Mentha oils [20].

Two-dimensional GC in the direct enantiomer separation of menthone, isomenthone and menthol with Ni(HFC)2 as the chiral main column has been reported by Bicchi et al. [92]. Werkhoff et al. [68] isolated these compounds from peppermint oils before stereoanalysis with permethylated β-cyclodextrin.

The simultaneous optical resolution of menthone, isomenthone, menthol as well as menthyl acetate has already been achieved using a column combination of three different columns, coated with modified cyclodextrins. All four enantiomeric pairs of these compounds were directly stereoanalyzed. Appropriate dilutions of mint and peppermint oils were analyzed by a single chromatographic run and without any preseparation [93].

As reported by Faber et al., genuine mint and peppermint oils contain enantiopure (1R)-configured monoterpenoids [94] (Fig. 6.34).

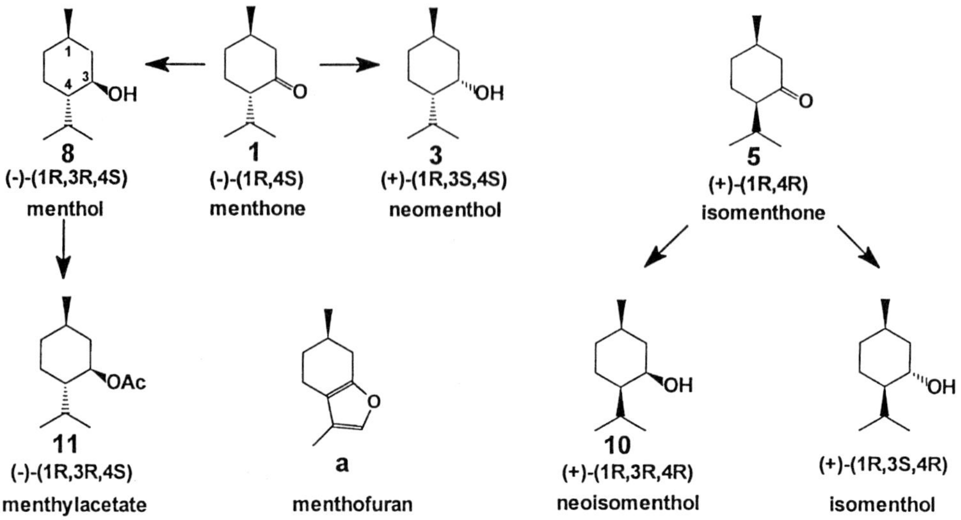

*Fig. 6.34: Chiral monoterpenoids from Mentha-species (from ref. [94])*

By means of enantio-MDGC analysis using the column combination SE 52 / DIME-β-CD in PS 268 the simultaneous stereodifferentiation of menthone (**1**, **2**) neomenthol (**3**, **4**), isomenthone (**5**, **6**) menthol (**7**, **8**), neoisomenthol (**9**, **10**) and menthyl acetate (**11**, **12**) has been verified and applied to various peppermint oils (Fig. 6.35; Fig. 6.36).

The occurrence of the strange (1S)-enantiomers of (+)-menthylacetate (**12**) and (+)-menthol (**7**) in commercially available mint oil samples has to be conceived as an adulteration with compounds of strange biological origin or coming from chemical synthesis.

## Lavandula Oil Constituents

Lavender oils are steam distilled oils from Lavandula angustifolia MILLER. For many years the adulteration of lavender oils by means of synthetic linalool and/or linalyl acetate has been easily revealed, by detecting the synthetic by-products dihydrolinalool and/or dehydrolinalool. But this detection fails, if these unnatural by-products are not present.

Chirality evaluation of linalyl acetate and linalool have been introduced as new and substantial criteria in the authenticity control of lavender oils. In particular, linalyl acetate from genuine lavender oils has high enantiomeric purity favouring the R-configuration, irrespective of the Lavandula species, and storage or work-up conditions *[32]*. Using 2,3-di-O-acylated-6-O-silylated cyclodextrins as a new generation of chiral stationary phases in enantio-cGC, most of chiral compounds of Lavandula oil, are stereoanalyzed simultaneously (Fig. 6.37; Fig. 6.38).

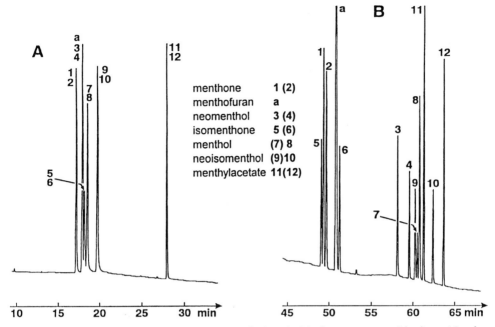

*Fig. 6.35:* Simultaneous enantio-MDGC analysis of chiral monoterpenoids from Mentha-species, pre column (A), main column (B) (from ref. [94])

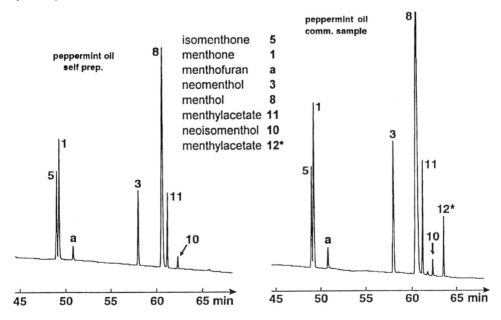

*Fig. 6.36:* Enantioselective MDGC analysis of peppermint oils (main column separation); peppermint oil, self prepared (left); adulterated commercial sample (right) (from ref. [94])

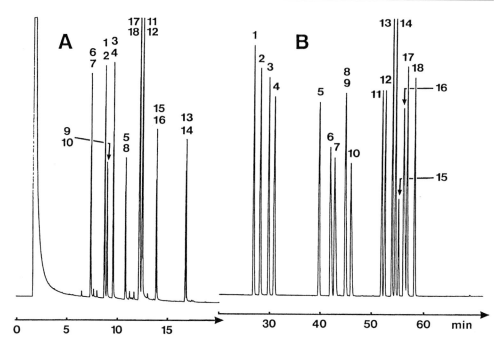

*Fig. 6.37:* Simultaneous stereoanalysis of Lavandula oil constituents, using enantioselective MDGC (standard mixture). (A) Preseparation of racemic compounds; unresolved enantiomeric pairs of octan-3-ol (**6**, **7**), trans-linalool oxide (**1**, **2**), oct-1-en-3-ol (**9**, **10**), cis-linalool oxide (**3**, **4**), camphor (**5**, **8**), linalool (**17**, **18**), linalyl acetate (**11**, **12**), terpinen-4-ol (**15**, **16**), lavandulol (**13**, **14**). (B) Chiral resolution of enantiomeric pairs, transferred from the precolumn: trans-linalool oxide: **1** (2S, 5S), **2** (2R, 5R); cis-linalool oxide: **3** (2R, 5S), **4** (2S, 5R); camphor: **5** (1S), **8** (1R); octan-3-ol: **6** (R), **7** (S); oct-1-en-3-ol: **9** (S), **10** (R); linalyl acetate: **11** (R), **12** (S); lavandulol: **13** (R), **14** (S); terpinen-4-ol: **15** (R), **16** (S); linalool: **17** (R), **18** (S) (from ref. [111])

The enantiomeric pairs of trans (**1**, **2**) linalool oxides (fur.) and cis (**3**, **4**) linalool oxides (fur.), camphor (**5**, **8**), octan-3-ol (**6**, **7**), oct-1-en-3-ol (**9**, **10**), linalyl acetate (**11**, **12**), lavandulol (**13**, **14**), terpinen-4-ol (**15**, **16**) and linalool (**17**, **18**) are analyzed by a single chromatographic run without any preseparation.

Investigations on authentic essential oils of Lavandula species are summarized in Table 6.20, detecting the chiral monoterpenoids with high enantiomeric purities as characteristic constituents of genuine lavender oils: Besides R-linalyl acetate (**11**) and R-linalool (**17**), the chiral main compounds of genuine lavender oils 2R-trans-(**2**), 2R-cis-(**3**)-linalool oxide (fur.), (R)-lavandulol (**13**), as well as S-terpinen-4-ol (**16**) are detected [111].

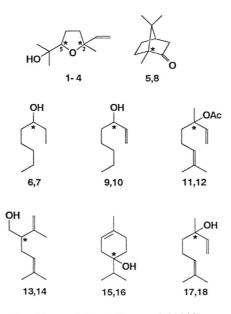

*Fig. 6.38:* Chiral compounds of Lavandula oil (from ref. [111])

*Table 6.20:* Enantiomeric distributions of chiral volatiles, investigated from authentic samples of Lavandula oils (from ref. [111])

| Sample no. | trans-Linalool oxide | | cis-Linalool oxide | | Linalyl acetate | | Lavandulol | | Terpinen- 4-ol | | Linalool | |
|---|---|---|---|---|---|---|---|---|---|---|---|---|
| | (2S,5S) | (2R,5R) | (2R,5S) | (2S,5R) | (R) | (S) | (R) | (S) | (R) | (S) | (R) | (S) |
| | 1 | 2 | 3 | 4 | 11 | 12 | 13 | 14 | 15 | 16 | 17 | 18 |
| 1 | 13.6 | 86.4 | 88.5 | 11.5 | >99 | <1 | 93.2 | 6.8 | — | — | 97.4 | 2.6 |
| 2 | — | — | — | — | >99 | <1 | 98.5 | 1.5 | 2.2 | 97.8 | 94.5 | 5.5 |
| 3 | 11.4 | 88.6 | 86.0 | 14.0 | >99 | <1 | 98.0 | 2.0 | — | — | 96.6 | 3.4 |
| 4 | 15.4 | 84.6 | 86.7 | 13.3 | >99 | <1 | 89.8 | 10.2 | 2.0 | 98.0 | 95.1 | 4.9 |
| 5 | 9.8 | 90.2 | 91.5 | 8.5 | >99 | <1 | >99 | <1 | 5.8 | 94.2 | 97.5 | 2.5 |
| 6 | 3.9 | 96.1 | 91.5 | 8.5 | >99 | <1 | >99 | <1 | 1.8 | 98.2 | 97.3 | 2.7 |
| 7 | 4.2 | 95.8 | 92.9 | 7.1 | >99 | <1 | 98.3 | 1.7 | 2.0 | 98.0 | 96.9 | 3.1 |
| 8 | <5 | >95 | >95 | <5 | >99 | <1 | 98.7 | 1.3 | 1.7 | 98.3 | 98.2 | 1.8 |
| 9 | 14.6 | 85.4 | 90.0 | 10.0 | >99 | <1 | >99 | <1 | 1.9 | 98.1 | 96.1 | 3.9 |
| 10 | 12.5 | 87.5 | 95.8 | 4.2 | >99 | <1 | >99 | <1 | 1.6 | 98.4 | 97.1 | 2.9 |
| 11 | — | — | — | — | >99 | <1 | >99 | <1 | 1.9 | 98.1 | 95.2 | 4.8 |
| 12 | 23.3 | 76.7 | 82.9 | 17.1 | 98.8 | 1.2 | >99 | <1 | 1.9 | 98.1 | 97.2 | 2.8 |
| 13 | 13.8 | 86.2 | 89.3 | 10.7 | >99 | <1 | 96.2 | 3.8 | 10.9 | 89.1 | 95.1 | 4.9 |

## Authenticity Control of Linalool

The enantiomeric distribution of linalool from Flores Lavandulae in relation to different sample work-up conditions is shown in Table 6.21. Some important facts can be concluded:

*Table 6.21:* Enantiomeric distribution of linalool from Flores Lavandulae in relation to different sample work-up conditions (from ref. [97])

| distillation time [min] | yield of essential oil [ml] | %-vol. | %-rel. | distillation stock ph-value | enantiomeric ratio linalool* R : S [%] | |
|---|---|---|---|---|---|---|
| *1. hydrodistillation* | | | | | | |
| 5 | 0.30 | 1.5 | 41 | 4.95 | 96.0 | 4.0 |
| 10 | 0.46 | 2.3 | 63 | 4.95 | 95.4 | 4.6 |
| 20 | 0.54 | 2.7 | 74 | 4.93 | 94.4 | 5.6 |
| 40 | 0.61 | 3.1 | 84 | 4.92 | 93.3 | 6.7 |
| 60 | 0.68 | 3.4 | 93 | 4.90 | 92.3 | 7.7 |
| 90 | 0.70 | 3.5 | 96 | 4.90 | 91.5 | 8.5 |
| 120 | 0.73 | 3.7 | 100 | 4.90 | 91.8 | 8.2 |
| *2. hydrodistillation (copper plates added)* | | | | | | |
| 20 | 0.52 | 2.6 | 71 | 4.92 | 94.4 | 5.6 |
| 60 | 0.67 | 3.4 | 92 | 4.91 | 91.9 | 8.1 |
| *3. steam distillation* | | | | | | |
| 5 | 0.57 | 2.9 | 78 | – | 98.4 | 1.6 |
| 10 | 0.61 | 3.1 | 84 | – | 98.1 | 1.9 |
| 20 | 0.64 | 3.2 | 88 | – | 98.0 | 2.0 |
| 30 | 0.67 | 3.4 | 92 | – | 98.0 | 2.0 |
| *4. diethyl ether extract* | | | | | | |
| 5 | – | – | – | – | 98.3 | 1.7 |
| 300 | – | – | – | – | 98.3 | 1.7 |

* linalyl acetate: enantiomeric purity, R > 99 % in any case.

- Steam distillation (30 min) corresponds with good manufacturing practice (GMP) in the processing of lavender oil and is without any influence on the genuine chirality of linalool, as to conclude by comparison with the results on linalool enantiomeric ratio from the diethyl ether extraction.

  Since (S)(+)-linalool in amounts up to 8 % may be formed during unusual and extremely prolonged times of hydrodistillation (2 h) it has to be accepted as the maximum from technologically induced partial racemization. Nevertheless (R)(-) linalool remains a reliable indicator of genuine lavender oils, produced under the GMP-conditions of steam distillation. Higher amounts (> 15%) of (S)(+) linalool have to be interpreted as an addition of synthetic linalool-racemate.

Also, the addition of copper plates in order to simulate the application of cupric vessels during lavender oil production does not influence the chiral stability of linalool and linalyl acetate.

– Even if (R)(–)-linalyl acetate, the genuine main component of lavender oils, is hydrolyzed in considerable amounts during prolonged procedures of hydrodistillation, the absolute configuration of the ester remains unchanged to be a reliable indicator of genuineness.

In summary, scope and limitations of linalool's enantiomeric ratio in the authenticity control of flavourings and fragrances are obvious:

– Owing to its liability in acidic medium, enantioselective analysis of linalool (and its cyclic monoterpenoic derivatives) is not suitable as parameter in the origin assessment of acidic foods like fruits, fruit-juices, wines etc.

– Consequently, also genuine (R)(-)-linalool from natural extracts gets lost, if these extracts are used as ingredients of acidic formulations.

– Nevertheless, under the conditions of *good manufacturing practice (GMP)* the chirality evaluation of linalool has proved to be a reliable indicator in the authenticity control of lavender, bergamot, and orange oils. This fact will be of considerable importance in the quality assurance for the flavour and fragrance industry.

*GMP-quality*

| Lavender oil | (R)(–) linalool (> 94%) [95] |
| Bergamot oil | (R)(–) linalool (> 99.5 %) [96] |
| Orange oil | (S)(+) linalool (> 92 %) [97] |

– If to extend these experiences to the chirality evaluation of linalool from other natural substrates, generally attention must be paid to the GMP-conditions as well as to the pH-values of the substrates investigated.

Comparative investigations on the enantiomeric purity of linalool and linalyl acetate in the essential oils of Lavandula angustifolia MILLER leads to quick and simple techniques of gas chromatographic chirality evaluation of lavender oil compounds. Their usefulness in the authenticity control of lavender oil compounds is demonstrated in Fig. 6.39 [95].

**Chiral Buchu Leaf Oil Compounds**

Owing to their characteristic minty-fruity notes, reminiscent of blackcurrants, buchu leaf oils are well appreciated in the composition of flavourings and fragrances. 8-Mercapto-p-menthan-3-one and its thiolacetate have been described by Sundt et al. [98] and simultaneously by Lamparsky et al. [99] as impact flavouring compounds of the flavour of cassis. They have so far not been identified in blackcurrant. All four stereoisomers of 8-mercapto-p-menthan-3-one exhibit distinct and characteristic sensory impressions [100].

The (1R)-diastereomers emit a more intensive odour than the (1S)-configured compounds, but only the later diastereomers have been recognized to be peculiar to buchu leaf oil and reminiscent of blackcurrant. An enantio-MDGC method was developed, detecting (1S)-configured menthone (**3**), isomenthone (**6**), (1S)-pulegone (**8**), (1S)-configured thiols (**9**, **12**) and (1S)-configured thiolacetates (**14**, **15**) as enantiopure chiral sulphur compounds from genuine buchu leaf oil *[101]* (Fig. 6.41).

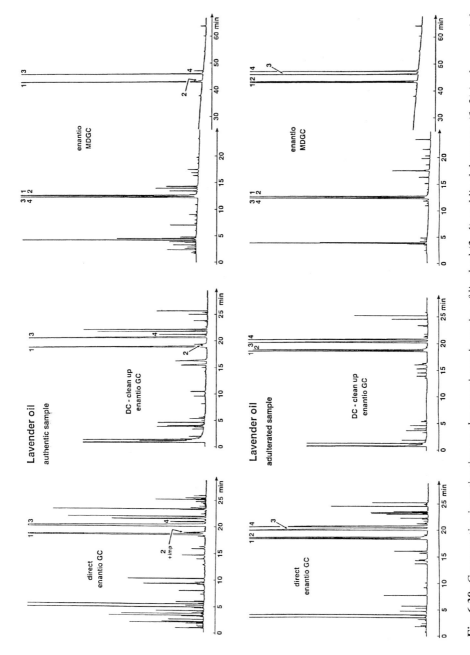

***Fig. 6.39:*** *Comparative investigation on the enantiomeric purity of linalool (**3**, **4**) and linalyl acetate (**1**, **2**) in the essential oils of lavender. Authentic sample (upper trace); adulterated sample (basic trace) (from ref. [95])*

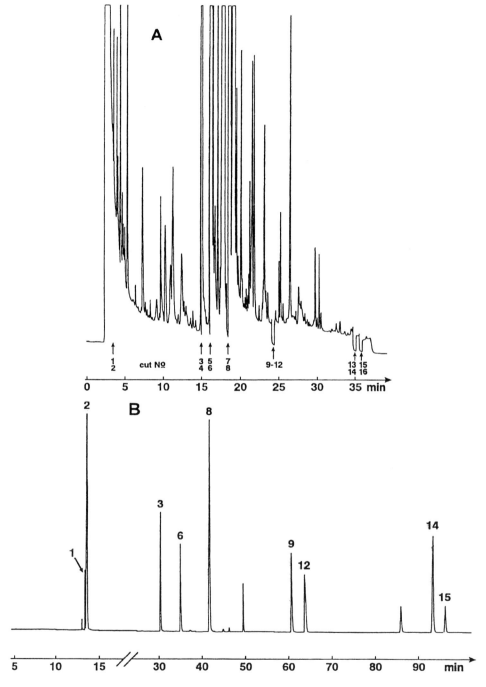

*Fig. 6.40:* Enantio-MDGC analysis of laboratory prepared buchu leaf oil. A: pre-column separation; B: main column separation; limonene: **1** (S), **2** (R); menthone: **3** (1S, 4R); isomenthone: **6** (1S, 4S); pulegone: **8** (S); cis-3-oxo-p-menthane-8-thiol: **9** (1S, 4R); trans-3-oxo-p-menthane-8-thiol: **12** (1S, 4S); trans-3-oxo-p-menthane-8-thiol acetate: **14** (1S, 4S); cis-3-oxo-p-menthan-8-thiol acetate: **15** (1S, 4R) (from ref. [101])

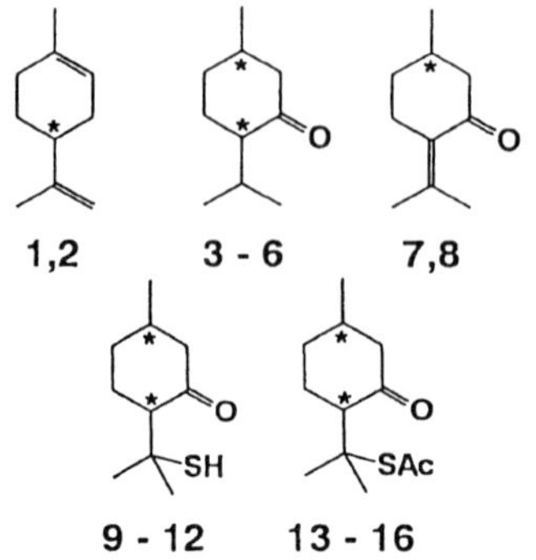

*Fig. 6.41: Chiral compounds of buchu leaf oil (from ref. [101])*

8-Mercapto-p-menthan-3-one is prepared by reacting pulegone with hydrogen sulfide, yielding racemic cis/trans diastereomers in the case of racemic pulegone or strange 1R-configurated diastereomers, when (R)(+)-pulegone from Pennyroyal oil (Mentha pulegium L.) was used as starting material (Fig. 6.42). In any case, these products can exactly be differentiated from genuine buchu leaf oil compounds by enantio-MDGC analysis (Fig. 6.40).

From a legal point of view it is interesting to note that up to now (1R)-configured stereoisomers of 3-oxo-p-methane-8-thiol have not been detected in nature. Therefore, those stereoisomers have to be classified as "artificial" flavourings in the sense of order regulations by the European Union.

### 3-(2H)-Furanones

Cyclopentenolones with a planar vicinal enol-oxo configuration are known to be powerful aroma active substances with distinct caramel notes. By methylation of the enolic function, this flavour impression is changed drastically to a sweet, mildew, and mouldy odour in the case of 2,5-dimethyl-4-methoxy-3-[2H]-furanone (**2**). This so-called "mesifurane" as well as "pineapple ketone" (**1**) were stereodifferentiated with modified cyclodextrin *[103]*. Although (**1**) and (**2**) can be stereoanalyzed without any racemization, both compounds were detected in strawberries, pineapples, grapes and wines as racemates (Fig. 6.43).

# Enantioselective Analysis

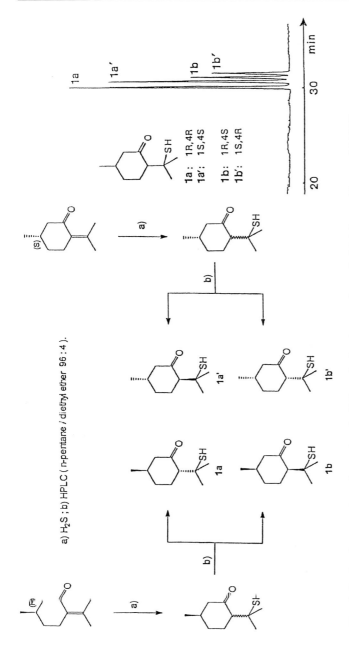

***Fig. 6.42**: Synthesis and configurational cGC-analysis of p-menthane-8-thiol-3-one (from ref. [102])*

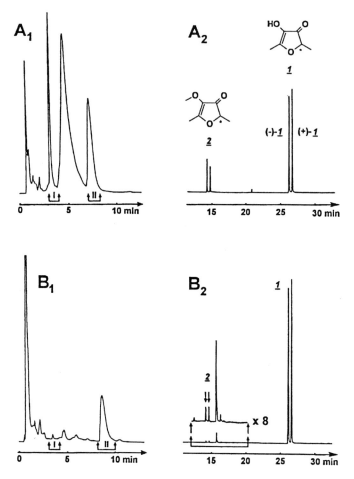

*Fig. 6.43:* Stereodifferentiation of 2,5-dimethyl-4-hydroxy-3[2H]-furanone **(1)** and its methyl-ether **(2)** from strawberries (A) and pineapples (B). A1/B1: HPLC-chromatogram of raw extracts. A2/B2: enantio-cGC analysis of fractions I, II from raw extracts (from ref. [62])

Obviously optically active furanones are equilibrated by keto-enol tautomerism in acidic media. Thus, enantioselective analysis cannot be applied to assign the origin of chiral 3[2H]-furanones [62]. This can only be expected by stable isotope measurements [104, 105].

### 6.2.3.7 Special Measurements

In chapter 6.2.2. isotope ratio mass spectrometry (IRMS) has been reported as a powerful tool in the origin assessment of food compounds.

In comparison to classical IRMS methods, the instrumental configuration of cGC-IRMS combines the precision of IRMS to the high purification effect of cGC-separation, with large savings on laborious sample clean-up procedures.

## Internal Isotopic Standard (i-IST)

Under the conditions of accurate sample clean-up, real high resolution (base line: Rs ≥ 1.5) and quantitative peak evaluation the chromatographic procedures HPLC, HPTLC, and preparative high resolution segment chromatography (PHS) are proved to be reliable methods in concentrating natural flavour compounds from complex matrices. Thus cGC-IRMS may be used as a suitable tool in evaluating precise δ13CPDB-measurements. But unfortunately, the usability of IRMS-analysis in the authenticity control of flavouring and fragrance compounds is limited, as most plants cultivated for human nutrition belong to the group of C3-plants yielding δ13CPDB-values partially overlapping with those of synthetic substances from fossil sources (or those of CAM-plants). To overcome this principal limitation the use of suitable internal isotopic standards (i-IST) was introduced [112]. By this way, the main influence of isotope discrimination, caused by CO2-fixation during photosynthesis is eliminated and only the influence of enzymatic reactions during secondary biogenetic pathways is investigated [113].

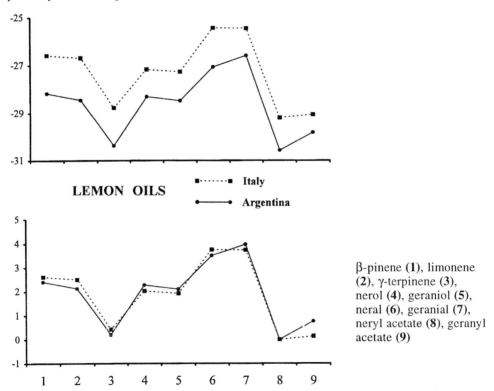

*Fig. 6.44:* $\delta^{13}C$-*fingerprint of lemon oil compounds of different provenance (upper trace); same samples, calculated versus neryl acetate* (**8**) *as an internal isotopic standard (i-IST) at bottom (from ref. [113])*

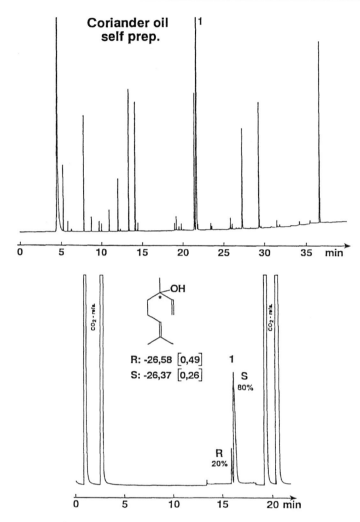

*Fig. 6.45:* Enantio-IRMS analysis of linalool from coriander seed oil, self prepared (from ref. [106])

In order to choose a suitable i-IST, the following aspects have to be taken into account:

- The substance selected as an i-IST should be a characteristic genuine compound of less importance in view of sensorial relevance.
- The compound must be available in sufficient amounts and free of isotopic discrimination during sample clean-up.
- The selected compound should be biogenetically related to the compounds under investigation.
- Its chemical inertness during storage and/or technical processes is mandatory.
- The compound selected for i-IST, must not be a legally allowed additive.

# Enantioselective Analysis

In Fig. 6.44 characteristic monoterpenes and monoterpenoids from lemon oils are compared. By introducing the i-IST method the individual influences of provenance are eliminated.

## Enantioselective cGC-IRMS

On-line coupling of enantio cGC with IRMS (enantio cGC-IRMS) is one of the latest developments in originspecific analysis of flavouring and fragrance compounds. Enantio-IRMS detects enantiomers of the same source with identical $\delta^{13}C$ levels. As outlined in Fig. 6.45, linalool from Coriander oil is detected as an R(20%) : S(80%) enantiomeric ratio with identical isotope levels of the enantiomers.

Enantio-IRMS also offers a direct method to detect conclusively a blend of enantio-pure chiral flavour compounds with synthetic racemates. An origin-specific enantiomeric ratio may be imitated but not yet detectable, neither by enantioselective analysis nor by IRMS-measurements. However, in the case of enantio-IRMS a simulated origin-specific enantiomeric distribution is proved by different $\delta^{13}C$-levels of the detected enantiomers.

In Fig. 6.46 the experiments to reveal the origin of linalool from a commercially available spike oil are compared *[107]*. While enantioselective analysis detects the R(80%) : S(20%) enantiomeric ratio, indicating a blend with synthetic racemate, the amount of synthetic racemate cannot be calculated, owing to a conceivable partial racemization of linalool during the processing of Lavandula oils. By means of enantio-IRMS investigation the blend of linalool from different origin is proved [R(-26.1 ‰), S(-30.6‰)], whereas simple IRMS-analysis [R+S(-27.0‰)] even simulates the occurrence of a genuine compound, due to the overlapping ranges of $\delta^{13}C$-values of substances from $C_3$-plants with those prepared from fossil sources.

## Origin specific analysis of (E)-α(β)-ionone

(E)-α-ionone from raspberries and many other natural sources occurs as enantio-pure (R)(+)-enantiomer *[70, 108]*. Therefore, "nature-identical" E-α-ionone racemate and enantiopure (R)(+)-E-α-ionone from genuine natural sources are detected unambiguously (Fig. 6.47; Fig. 6.48)

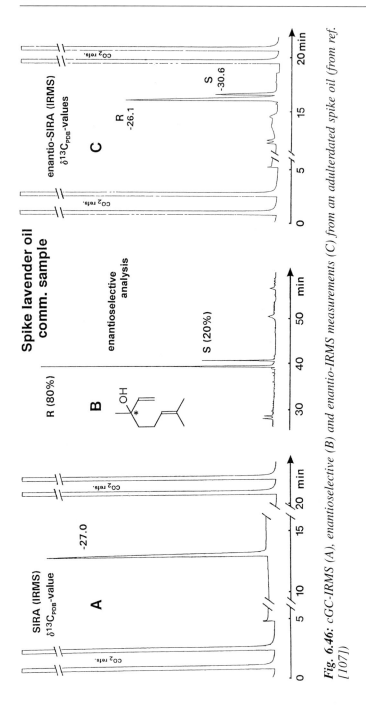

**Fig. 6.46:** cGC-IRMS (A), enantioselective (B) and enantio-IRMS measurements (C) from an adulterdated spike oil (from ref. [107])

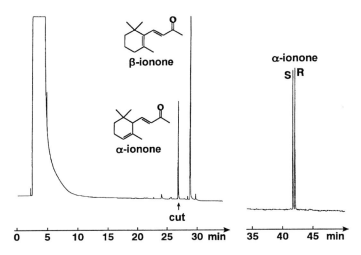

*Fig. 6.47:* Enantio-MDGC analysis of α-ionone. Pre-column separation α(β) ionone (left); enantiomer separation of racemic α-ionone standard (right) (from ref. [85, 114])

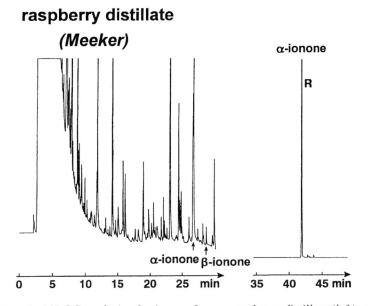

*Fig. 6.48:* Enantio-MDGC analysis of α-ionone from a raspberry distillate (left); enantiopure (R)-α-ionone detected (right) (from ref. [85, 114])

Using stable isotope mass spectrometry (E)-α- and (E)-β-ionone from raspberries are well differentiated from their synthetic analogues and from a commercially available product, which was declared to be of natural origin [108].

As natural α-ionone exclusively prefers the (R)-configuration, the commercial sample (declaration "fermentative") has, therefore, to be interpreted as a "nature-identical" compound, synthesized with compounds of natural origin (Table 6.22).

**Table 6.22:** $^{13}C_{PDB}$ values of (E)-α(β)-ionone from different origins (from ref. [108])

| Origin | (E)-α-Ionone | | | | (E)-β-ionone |
|---|---|---|---|---|---|
| | (R)-(+)- (%) | (S)-(–)- (%) | $\delta^{13}C_{PDB}$ (s)[a] | | $\delta^{13}C_{PDB}$ (s)[a] |
| Synthetic A | 50 | 50 | -24.33 | (0.16) | n.d.[b] |
| Synthetic B | 50 | 50 | -26.69 | (0.10) | n.d. |
| Synthetic C | n.d. | n.d. | n.d. | | -28.63 (0.22) |
| Synthetic D | 50 | 50 | -27.05 | (0.15) | -25.59 (0.09) |
| Raspberry (fruit) | > 99.9 | < 0.1 | -32.86 | (0.59) | -33.41 (0.37) |
| Raspberry (mash) | > 99.0 | < 0.1 | -33.43 | (0.38) | -31.41 (0.38) |
| Commercial[c] | 50 | 50 | -9.12 | (0.06) | -8.57 (0.16) |

[a] Standard deviations (s) are given in parentheses (n= 5)
[b] n.d.: Not detectable
[c] sample declaration "fermentative"

### 6.2.3.8 Authenticity Assessment – Latest Developments

In recent years, enantioselective chromatography was extended to semipreparative separations and now simulated moving-bed technology (SMB) has become available for the resolution of some synthetic racemates in preparative scale. In the pharmaceutical industry this progress is of considerable interest with respect to enantiopure drug applications.

Under the condition of large amounts of chiral stationary phases with reproducible batch-to-batch properties and economical feasibility of the SMB process, this new technique might also be promising for large-scale continuous separations of synthetic flavour and fragrance racemates [117], and perhaps for delivering enantiopure compounds. However, from the legal point of view, one should note these compounds are of synthetic origin and are judged as "nature-identical" but not as "natural" flavourings.

Consequently, in the case of enantiopure compounds, generated by the SMB process, the expressiveness of enantioselective analysis and the conclusions regarding the authentication of the enantiomeric analytes might be seriously hampered or even impossible. Bringing together both online techniques (enantioselective analysis and isotope ratio mass spectrometry) will be the key to overcoming the difficulties in the legal interpretation of enantiomers from chromatographic racemate separation.

Presently, online cGC-IRMS determination of $\delta^{13}C$ and $\delta^{2}H$ values of flavour compounds are established [107, 118-123] and, further on, $\delta^{18}O$ measurements using online cGC-IRMS procedures will be available. Finally, enantio-MDGC combined with multielement- and multicomponent IRMS methods will open the door to comprehensive authenticity assessment of natural flavour and fragrance compounds in the near future.

So far, general conditions for the official monographs of essential oils are recommended as follows:

– extraction by steam distillation, according to GMP conditions

- evaluation of characteristic components by ring tests, using authentic drug materials
- quantification by cGC/MS, using internal standards
- authenticity assessment, using enantio-MDGC/MS or (enantio)-MDGC-IRMS, including multielement ($\delta^{13}$C, $\delta^2$H, $\delta^{18}$O, $\delta^{15}$N) and multicomponent analysis.

## Conclusions

Enantioselective capillary gas chromatography and (enantio)c-GC-IRMS measurements have been demonstrated as sophisticated methods, which will strongly accelerate insights into the origin of flavourings and fragrances. Continous advances in analytical origin assignment will be adopted for quality assurance in the flavour industry and also should be reflected in legal regulations as a consequence.

## REFERENCES

[1] G.F. Russell and J.I. Hills. Odor differences between enantiomeric isomers. Science, 172, 1043 (1971).
[2] L. Friedman and J.G. Miller. Odor incongruity and chirality. Science, 172, 1044 (1971).
[3] A. Mosandl and G. Heusinger. 1,3-Oxathianes, chiral fruit flavour compounds. Liebigs Ann. Chem., 1185-1191 (1985).
[4] G. Ohloff. Chemistry of odor stimuli. Experientia, 42, 271-279 (1986).
[5] C. Günther and A. Mosandl. Stereoisomere Aromastoffe XII: 3-Methyl-4-octanolid-"Quercuslacton, Whiskylacton"-Struktur und Eigenschaften der Stereoisomeren. Liebigs Ann. Chem., 2112-2122 (1986).
[6] A. Mosandl, G. Heusinger and M. Gessner. Analytical and sensory differentiation of 1-octen-3-ol enantiomers. J. Agric. Food Chem., 34, 119-122 (1986).
[7] A. Mosandl and M. Gessner. Stereoisomeric flavour compounds XXIII: δ-lactone flavour compounds - structure and properties of the enantiomers. Z. Lebensm. Unters. Forsch., 187, 40-44 (1988).
[8] A. Mosandl and C. Günther. Stereoisomeric flavor compounds XX: Structure and properties of γ-lactone enantiomers. J. Agric. Food Chem., 37, 413-418 (1989).
[9] G. Ohloff, Riechstoffe und Geruchssinn - Die molekulare Welt der Düfte, Springer Verlag, Berlin, 1990.
[10] P. Sandra. Resolution - definition and nomenclature. J. High Resolut. Chromatogr., 12, 82-86 (1989).
[11] W.A. König. Modifizierte Cyclodextrine als chirale Trennphasen in der Gaschromatographie. Kontakte (Darmstadt), 1990, (2), 3-14.
[12] A. Mosandl, G. Heusinger, D. Wistuba and V. Schurig. Analysis of a chiral aroma compound by complexation gas chromatography. Z. Lebensm. Unters. Forsch., 179, 385-386 (1984).
[13] W. Deyer, M. Gessner, G. Heusinger, G. Singer and A. Mosandl. Stereoisomeric flavor compounds XIII: Critical remarks on the interpretation of chiral stereodifferentiation in gas chromatography. J. Chromatogr., 366, 385-390, (1986).
[14] W.A. König, S. Lutz, P. Mischnick-Lübbecke, B. Brassat and G. Wenz. Cyclodextrins as chiral stationary phases in capillary gas chromatography. I. Pentylated α-cyclodextrin. J. Chromatogr. 447, 193-197 (1988).
[15] A. Mosandl and U. Hagenauer-Hener. Stereoisomeric flavor compounds XXVI: HRGC-Analysis of chiral 1,3-dioxolanes. J. High Resolut. Chromatogr. Chromatogr. Comm., 11, 744-749 (1988).
[16] W.A. König, D. Icheln and I. Hardt. Unusual Retention behaviour of methyl lactate and methyl 2-hydroxybutyrate enantiomers on a modified cyclodextrin. J. High Resolut. Chromatogr., 14, 694-695 (1991).

[17] A. Mosandl, K. Rettinger, K. Fischer, V. Schubert, H.-G. Schmarr and B. Maas. Stereoisomeric Flavor Compounds XLI: New applications of permethylated β-cyclodextrin phase in chiral CGC analysis. J. High Resolut. Chromatogr., 13, 382-385 (1990).

[18] W.A. König, S. Lutz, C. Colberg, N. Schmidt, G. Wenz, E. von der Bey, A. Mosandl, C. Günther and A. Kustermann. Cyclodextrins as chiral stationary phases in capillary gas chromatography, Part III: Hexakis (3-0-acetyl-2,6-di-0-pentyl)-α-cyclodextrin. J. High Resolut. Chromatogr. Chromatogr. Comm., 11, 621-625 (1988).

[19] A. Mosandl, M. Gessner, C. Günther, W. Deger and G. Singer. Stereoisomeric flavor compounds XIV: (S)-O-Acetyl lactylchloride - a versatile chiral auxiliary in stereodifferentiation of enantiomeric flavor components. J. High Resolut. Chromatogr. Chromatogr. Comm., 10, 67-70 (1987).

[20] P. Kreis, A. Mosandl and H.-G. Schmarr. Mentha-Öle, enantioselektive Analyse des Menthylacetats zur Qualitätsbeurteilung von Mentha-Ölen. Dtsch. Apoth. Ztg., 130, 2579-2581(1990).

[21] P. Kreis, D. Juchelka, C. Motz and A. Mosandl. Chirale Inhaltsstoffe ätherischer Öle IX: Stereodifferenzierung von Borneol, Isoborneol und Bornylacetat. Dtsch. Apoth. Ztg., 131, 1984-1987 (1991).

[22] A. Mosandl and A. Kustermann. Stereoisomere Aromastoffe XXX: HRGC-Analyse chiraler γ-Lactone aus Getränken und Fruchtzubereitungen. Z. Lebensm. Unters. Forsch., 189, 212-215 (1989).

[23] A. Mosandl, A. Kustermann, U. Palm, H.-P. Dorau and W.A. König. Stereoisomeric flavour compounds XXVIII: Direct chirospecific HRGC-analysis of natural γ-lactones. Z. Lebensm. Unters. Forsch., 188, 517-520 (1989).

[24] H.-G. Schmarr, A. Mosandl and K. Grob. Stereoisomeric Flavour Compounds XXXVIII: Direct chirospecific analysis of γ-lactones using on-line coupled HPLC- HRGC with a chiral separation column. Chromatographia, 29, 125-130 (1990).

[25] A. Artho and K. Grob. Nachweis der Aromatisierung von Lebensmitteln mit γ-Lactonen. Wie identisch mit der Natur müssen „naturidentische" Aromen sein?. Mitt. Gebiete Lebensm. Hyg., 81, 544-558 (1990).

[26] C. Wang, H. Frank, G. Wang, L. Zhou, E. Beyer and P. Lu. Determination of amino acid enantiomers by two-column gas chromatography with valveless column switching. J. Chromatogr., 262, 352-359 (1983).

[27] G. Schomburg, H. Husmann, E. Hübinger and W.A. König. Multidimensional Capillary Gas Chromatography - Enantiomeric separations of selected cuts using a chiral second column. J. High Resolut. Chromatogr. Chromatogr. Commun., 7, 404-410 (1984).

[28] A. Mosandl, U. Hener, U. Hagenauer-Hener and A. Kustermann. Stereoisomeric flavor compounds XXXII: Direct enantiomer separation of chiral γ-lactones from food and beverages by multidimensional gas chromatography. J. High Resolut. Chromatogr., 12, 532-536 (1989).

[29] S. Nitz, H. Kollmannsberger and F. Drawert. Determination of Non Natural Flavours in Sparkling Fruit Wines. 1. Rapid Method for the Resolution of Enantiomeric γ-Lactones by Multidimensional GC. Chem. Mikrobiol. Technol. Lebensm., 12, 75-80 (1989).

[30] A. Bernreuther, N. Christoph and P. Schreier. Determination of the enantiomeric composition of γ-lactones in complex natural matrices using multidimensional capillary gas chromatography. J. Chromatogr., 481, 363-367 (1989).

[31] E. Guichard, A. Kustermann and A. Mosandl. Chiral flavour compounds from apricots - distribution of γ-lactone enantiomers and stereodifferentiation of dihydroactinidiolide using multidimensional gas chromatography. J. Chromatogr., 498, 396-401 (1990).

[32] A. Mosandl and V. Schubert. Stereoisomere Aromastoffe XXXIX: Chirale Inhaltsstoffe ätherischer Öle I. Stereodifferenzierung des Linalylacetats - ein neuer Weg zur Qualitätsbeurteilung des Lavendelöls. Z. Lebensm. Unters. Forsch., 190, 506-510 (1990).

[33] P. Kreis, U. Hener and A. Mosandl. Chirale Inhaltsstoffe ätherischer Öle III: Stereodifferenzierung von α-Pinen, β-Pinen und Limonen in ätherischen Ölen, Drogen und Fertigarzneimitteln. Dtsch. Apoth. Ztg., 130, 985-988 (1990).

[34] U. Hener, P. Kreis and A. Mosandl. Chiral compounds of essential oils IV. Enantiomer distribution of α-pinene, β-pinene and limonene in essential oils and extracts; Part 2: Perfumes and cosmetics. Flav. Fragr. J., 5, 201-204 (1990).

[35] U. Hener, P. Kreis and A. Mosandl. Enantiomer distribution of α-pinene, β-pinene and limonene in essential oils and extracts; Part 3: oils for alcoholic beverages and seasoning. Flav. Fragr. J., 6, 109-111 (1991).

[36] U. Palm, C. Askari, U. Hener, E. Jakob, C. Mandler, M. Geßner, A. Mosandl, W.A. König, P. Evers and R. Krebber. Stereoisomere Aromastoffe XLVII: Direkte chirospezifische HRGC-Analyse natürlicher δ-Lactone. Z. Lebensm. Unters. Forsch., 192, 209-213 (1991).

[37] P. Kreis, U. Hener and A. Mosandl. Chirale Inhaltsstoffe ätherischer Öle V: Enantiomerenverteilung von α-Pinen, β-Pinen und Limonen in ätherischen Ölen und Extrakten; Citrusfrüchte und citrushaltige Getränke. Dtsch. Lebensm. Rundsch., 87, 8-11 (1991).

[38] A. Mosandl, K. Fischer, U. Hener, P. Kreis, K. Rettinger, V. Schubert and H.-G. Schmarr. Stereoisomeric Flavor Compounds. XLVIII Chirospecific Analysis of Natural Flavors and Essential Oils, Using Multidimensional Gas Chromatography. J. Agric. Food Chem., 39, 1131-1134 (1991).

[39] A. Mosandl, K. Rettinger, B. Weber and D. Henn. Untersuchungen zur Enantiomerenverteilung von 2-Methylbuttersäure in Früchten und anderen Lebensmitteln mittels multidimensionaler Gaschromatographie (MDGC). Dtsch. Lebensm.Rundsch., 86, 375-379 (1990).

[40] A. Hollnagel, E.-M. Menzel and A. Mosandl. Chiral Aroma Compounds of Sherry II. Direct Enantiomer Separation of Solerone and Solerole. Z. Lebensm. Unters. Forsch., 193, 234-236 (1991).

[41] V. Schubert, R. Diener and A. Mosandl. Enantioselective Multidimensional Gas Chromatography of some Secondary Alcohols and their Acetates from Banana. Z. f. Naturforsch., 46c, 33-36 (1991).

[42] D. Lehmann, C. Askari, D. Henn, F. Dettmar, U. Hener and A. Mosandl. Simultane Stereodifferenzierung chiraler γ- und δ-Lactone. Dtsch. Lebensm. Rundsch., 87, 75-77 (1991).

[43] C. Askari, U. Hener, H.-G. Schmarr, A. Rapp and A. Mosandl. Stereodifferentiation of some chiral monoterpenes, using multidimensional gas chromatography. Fresenius J. Anal. Chem., 340, 768-772 (1991).

[44] V. Schubert and A. Mosandl. Stereoisomeric Flavour Compounds LIII: Chiral Compounds of Essential Oils VIII; Stereodifferentiation of linalool using multidimensional gas chromatography. Phytochem. Anal., 2, 171-174 (1991).

[45] V. Karl, H.-G. Schmarr and A. Mosandl. Simultaneous stereoanalysis of 2-alkylbranched acids, -esters, and -alcohols, using selectivity adjusted column systsem in multidimensional gas chromatography. J. Chromatogr., 587, 347-350 (1991).

[46] A. Bernreuther, J. Koziet, P. Brunerie, G. Krammer, N. Christoph and P. Schreier. Chirospecific capillary gaschromatography (HRGC) and on-line HRGC-isotope ratio mass spectrometry of γ-decalactone from various sources. Z. Lebensm. Unters. Forsch., 191, 299-301 (1990).

[47] A. Bernreuther, V. Lander, M. Huffer and P. Schreier. Enantioselective Analysis of Dec-2-en-5-olide (Massoilactone) from Natural Sources by Multidimensional Capillary Gas Chromatography. Flav. Fragr. J., 5, 71-73 (1990).

[48] E. Guichard, A. Mosandl, A. Hollnagel, A. Latrasse and R. Henry. Chiral γ-lactones from Fusarium Poae: Enantiomeric ratios and sensory differentiation of cis-6-γ-dodecenolactone enantiomers. Z. Lebensm. Unters. Forsch., 193, 26-31 (1991).

[49] A. Mosandl, U. Hener, P. Kreis and H.-G. Schmarr. Chiral compounds of essential oils II. Enantiomer distribution of α-pinene, β-pinene and limonene in essential oils and extracts; Part 1, Rutaceae and gramineae. Flav. Fragr. J., 5, 193-199 (1990).

[50] K. Rettinger, V. Karl, H.-G. Schmarr, F. Dettmar, U. Hener and A. Mosandl. Chirospecific analysis of 2-alkylbranched alcohols, -acids, and -esters; chirality evaluation of 2-methylbutanoates from apples and pineapples. Phytochem. Anal., 2, 184-188 (1991).

[51] K. Haase-Aschoff, I. Haase-Aschoff and H. Laub. Chirale Analyse und Bewertung der 2-Methylbuttersäure und ihrer Ester in Früchten und Fruchtprodukten. Lebensmittelchemie, 45, 107-109 (1991).

[52] G. Full, A. Bernreuther, G. Krammer and P. Schreier. On-line-HRGC-FTIR zur direkten Enantiodifferenzierung von Aromastoffen aus komplexen Matrices. Labo, 1991, 30-34.

[53] U. Hener and A. Mosandl. Zur Herkunftsbeurteilung von Aromen. Dtsch. Lebensm. Rundsch. 89, 307-312 (1993).

[54] H. Maarse and C. A. Visscher „Volatile Compounds in Food - Qualitative and quantitative Data. Vol. I-III, 6th Edition TNO-CIVO Food Analysis Institute Zeist, 1989

[55] A. Mosandl. Chirality in flavor chemistry - recent developments in synthesis and analysis. Food Rev. Int., 4, 1-43 (1988).

[56] H. Frank, G. J. Nicholson and E. Bayer. Rapid Gas chromatographic separation of amino acid enantiomers with a novel chiral stationary phase. J. Chromatogr. Sci., 15, 174-176 (1977).

[57] W.A. König, I. Benecke and S. Sievers. New results in the gaschromatographic separation of enantiomers of hydroxy acids and carbohydrates. J. Chromatogr., 217, 71-79 (1981).

[58] V. Schurig and W. Bürkle. Extending the Scope of Enantiomer Resolution by Complexation Gas Chromatography. J. Am. Chem. Soc., 104, 7573-7580 (1982).

[59] Z. Juvancz, G. Alexander and J. Szejtli. Permethylated β-Cyclodextrin as Stationary Phase in Capillary Gas Chromatography. J. High Resolut. Chromatogr. Chromatogr. Commun., 10, 105-107 (1987).

[60] D.W. Armstrong, C.-D. Chang and W.Y. Li. Relevance of Enantiomeric Separations in Food and Beverage Analyses. J. Agric. Food Chem., 38, 1674-1677 (1990).

[61] V. Schurig and H.-P. Nowotny. Separation of enantiomers on diluted permethylated b-cyclodextrin by high resolution gas chromatography. J. Chromatogr., 441, 155-163 (1988).

[13] H.-P. Nowotny, D. Schmalzing, D. Wistuba and V. Schurig. Extending the Scope of Enantiomer Separation on Diluted Methylated β-Cyclodextrin Derivatives by High Resolution Gas Chromatography. J. High Resolut. Chromatogr., 12, 383-393 (1989).

[13] V. Schurig and H.-P. Nowotny. Gaschromatographische Enantiomerentrennung an Cyclodextrinderivaten. Angew. Chem., 102, 969-1108 (1990).

[62] G. Bruche, H.-G. Schmarr, A. Bauer, A. Mosandl, A. Rapp and L. Engel. Stereoisomere Aromastoffe LI: Stereodifferenzierung chiraler Furanone - Möglichkeiten und Grenzen der herkunftsspezifischen Aromastoff-Analyse. Z. Lebensm. Unters. Forsch., 193, 115-118 (1991).

[63] H.-G. Schmarr, A. Mosandl and A. Kaunzinger. Influence of Derivatization on the Chiral Selectivity of Cyclodextrins: Alkylated/Acylated Cyclodextrins and γ/δ-Lactones as an Example. J. Microcol. Sep. 3, 395-402 (1991).

[64] A. Dietrich, B. Maas, V. Karl, P. Kreis, D. Lehmann, B. Weber and A. Mosandl. Stereoisomeric Flavor Compounds, Part LV: Stereodifferentiation of some chiral volatiles on heptakis (2,3-di-O-acetyl-6-O-tert-butyl-dimethylsilyl)-β-cyclodextrin. J. High Resolut. Chromatogr., 15, 176-179 (1992).

[65] H.-G. Schmarr, A. Mosandl, H.-P. Neukom and K. Grob. Modified Cyclodextrins as Stationary Phases for Capillary GC; Consequences of Dilution in Polysiloxanes. J. High Resolut. Chromatogr., 14, 207-210 (1991).

[66] A. Dietrich, B. Maas, G. Brand, V. Karl, A. Kaunzinger and A. Mosandl. Stereoisomeric Flavor Compounds, Part: LX. Diluted modified cyclodextrins as chiral stationary phases - the influence of the polysiloxane solvent. J. High Resolut. Chromatogr. 15, 769-772 (1992).

[67] A. Dietrich, B. Maas and A. Mosandl. Diluted modified cyclodextrins as chiral cGC phases, the influence of the polysiloxane solvents to the enantioselectivity and column efficiency. J. Microcol. Sep., 6, 33-42 (1994).

[68] P. Werkhoff, S. Brennecke and W. Bretschneider. Fortschritte bei der chirospezifischen Analyse natürlicher Riech- und Aromastoffe. Chem. Mikrobiol. Technol. Lebensm., 13, 129-152 (1991).

[69] M. Güntert, R. Emberger, R. Hopp, M. Köpsel, W. Silberzahn and P. Werkhoff. Chirospecific analysis in flavor and essential oil chemistry Part A. Filbertone - the character impact compound of hazel-nuts. Z. Lebensm. Unters. Forsch., 192, 108-110 (1991).

[70] P. Werkhoff, W. Bretschneider, M. Güntert, R. Hopp and H. Surburg. Chirospecific analysis in flavor and essential oil chemistry Part B. Direct enantiomer resolution of trans-α-ionone and trans-α-damascone by inclusion gas chromatography. Z. Lebensm. Unters. Forsch., 192, 111-115 (1991).

[71] M. Winter, A. Furrer, B. Willhalm and W. Thommen. Identification and Synthesis of two New Organic Sulfur Compounds from the Yellow Passion Fruit (Passiflora edulis f. flavicarpa). Helv. Chim. Acta, 59, 1613-1620 (1976).

[72] G. Singer, G. Heusinger, O. Fröhlich, P. Schreier and A. Mosandl. Chirality evaluation of 2-methyl-4-propyl-1,3-oxathiane from the yellow passion fruit. J. Agric. Food Chem., 34, 1029-1033 (1986).

[73] B. Weber, Dissertation, Universität Frankfurt/Main, 1996.

[74] J.A. Maga. Lactones in Foods. Crit. Rev. Food Sci. Nutrit., 8, 1-56 (1976).

[75]   W. Albrecht and R. Tressl. Studies on the Biosynthesis of γ-Decalactone in Sporobolomyces odorus. Z. Naturforsch., 45c, 207-216 (1990).
[76]   E. Guichard and M. Souty. Comparison of the relative quantities of aroma compounds found in fresh apricot (Prunus armeniaca) from six different varieties. Z. Lebensm. Unters. Forsch., 186, 301-107 (1988).
[77]   W. Grosch. Analyse von Aromastoffen. Chem. Unserer Zeit, 24, 82-89 (1990).
[78]   I. Blank and W. Grosch. Evaluation of potent odorants in dill seed and dill herb (Anethum graveolens L.) by aroma extract dilution analysis. J. Food Sci., 56, 63-67 (1991).
[79]   T.J. Siek, J.A. Albin, L.A. Sather and R.C. Lindsay. Comparison of flavor thresholds of aliphatic lactones with those of fatty acids, esters, aldehydes, alcohols and ketones. J. Dairy Sci., 54, 1-4 (1971).
[80]   J.G. Leffingwell and D. Leffingwell. GRAS Flavor Chemicals - detection threshold. Perfum. Flavourist, 16, 1-19 (1991).
[81]   N. Fischer and F.J. Hammerschmidt. A contribution to the analysis of fresh strawberry flavour. Chem. Mikrobiol. Technol. Lebensm., 14, 141-148 (1992).
[82]   S. Nitz, H. Kollmannsberger, B. Weinreich and F. Drawert. Enantiomeric distribution and 13C/12C isotope ratio determination of γ-lactones: appropriate methods for the differentiation between natural and non-natural flavours? J. Chromatogr., 557, 187-197 (1991).
[83]   R. Eberhardt, H. Woidich and H. Pfannhauser, in P. Schreier (Editor), Flavour '81, de Gruyter, Berlin, 1981. Analysis of Natural and Artifical Coconut Flavouring in Beverages. p. 377-383
[84]   W.A. König, R. Krebber and P. Mischnick. Cyclodextrins as Chiral Stationary Phases in Capillary Gas Chromatography. Part V: Octakis (3-O-butyryl-2,6-di-O-pentyl)-γ-cyclodextrin. J. High Resolut. Chromatogr., 12, 732-738 (1989).
[85]   D. Lehmann, A. Dietrich, S. Schmidt, H. Dietrich and A. Mosandl. Stereodifferenzierung von γ(δ) Lactonen und (E)-α-Ionon verschiedener Früchte und ihrer Verarbeitungsprodukte. Z. Lebensm. Unters. Forsch., 196, 207-213 (1993).
[86]   B. Maas, A. Dietrich, V. Karl, A. Kaunzinger, D. Lehmann, T. Köpke and A. Mosandl. Tert-Butyldimethylsilyl-substituted cyclodextrin derivatives as versatile chiral stationary phases in capillary GC. J. Microcol. Sep., 5, 421-427 (1993).
[87]   K. Rettinger, C. Burschka, P. Scheeben, H. Fuchs and A. Mosandl. Chiral 2-alkyl-branched acids, esters and alcohols. Preparation and stereospecific flavour evaluation. Tetrahedron Asymm., 2, 965-968 (1991).
[88]   V. Karl, A. Dietrich and A. Mosandl. Stereoisomeric Flavour Compounds LXIV: The chirospecific analysis of 2-alkyl-branched flavour compounds from headspace extracts and their sensory evaluation. Phytochem. Anal., 4, 158-164 (1993).
[89]   O.A. Mamer, S.S. Tjoa, Ch. R. Sriver and G.A. Klassen. Demonstration of a new mammalian isoleucine catabolic pathway yielding an (R)-Series of metabolites. Biochem. J., 160, 417-426 (1976); and literature cited therein.
[90]   G. Takeoka, R.A. Flath, T.R. Mon, R.G. Buttery, R. Teranishi, M. Güntert, R. Lautamo and J. Szejtli. Further Applications of Permethylated β-Cyclodextrin Capillary Gas Chromatographic Columns. J. High Resolut. Chromatogr., 13, 202-206 (1990).
[91]   K. Rettinger, B. Weber and A. Mosandl. Stereoisomere Aromastoffe XLII: 2-Methylbuttersäure-Enantiomerenverteilung in Äpfeln und apfelhaltigen Lebensmitteln. Z. Lebensm. Unters. Forsch., 191, 265-268 (1990).
[92]   C. Bicchi and A. Pisciotta. Use of two-dimensional gas chromatography in the direct enantiomer separation of chiral essential oil components. J. Chromatogr., 508, 341-348 (1990).
[93]   C. Askari, P. Kreis, A. Mosandl and H.-G. Schmarr. Chiral Compounds of Essential Oils VII: Quality Evaluation of Mentha Oils, using enantioselective CGC-Analysis of Monoterpenoid Constituents. Arch. Pharm., 325, 35-39 (1992).
[94]   B. Faber, A. Dietrich and A. Mosandl. Chiral compounds of essential oils. Part XV. Stereodifferentiation of characteristic compounds of Mentha species by multi-dimensional gas chromatography. J. Chromatogr., 666, 161-165 (1994).
[95]   P. Kreis, A. Dietrich, D. Juchelka and A. Mosandl. Methodenvergleich zur Stereodifferenzierung von Linalool und Linalylacetat in ätherischen Ölen von Lavandula angustifolia MILLER. Pharm. Ztg. Wiss., 138, 149-155 (1993).

[96]  A. Cotroneo, I. Stagno d'Alcontres and A. Trozzi. On the genuineness of citrus essential oils. Part XXXIV. Detection of added reconstituted bergamot oil in genuine bergamot essential oil by high resolution gaschromatography with chiral capillary columns. Flavour Fragr. J., 7, 15-17 (1992).

[97]  P. Kreis. R. Braunsdorf, A. Dietrich, U. Hener, B. Maas and A. Mosandl, in: Progress in Flavour Precursor Studies, Analysis, Generation, Biotechnology (P. Schreier, P. Winterhalter, eds). Allured Publ. Corp. 1993. Enantioselective analysis of linalool - scope and limitations, pp. 77-82.

[98]  E. Sundt, B. Willhalm, R. Chappaz and G. Ohloff. Das organoleptische Prinzip von Cassis-Flavor im Buccublätter"l. Helv. Chim. Acta, 54, 1801-1805 (1971).

[99]  D. Lamparsky and P. Schudel. p-Menthane-8-thiol-3-one, a new component of Buchu leaf oil. Tetrahedron Lett. 36, 3323-3326 (1971).

[100] T. Köpke and A. Mosandl. Stereoisomere Aromastoffe LIV. 8-Mercapto-p-menthan-3-on, Reindarstellung und chirospezifische Analyse der Stereoisomeren. Z. Lebensm. Unters. Forsch., 194, 372-376 (1992).

[101] T. Köpke, A. Dietrich and A. Mosandl. Chiral compounds of essential oils XIV: Simultaneous stereoanalysis of Buchu leaf oil compounds. Phytochem. Anal., 5, 61-67 (1994).

[102] A. Mosandl, T. Köpke and W. Bensch. Stereoisomeric Flavour compounds. LXII. Structure elucidation of 8-mercapto-p-menthan-3-one isomers. Tetrahedron Asymm., 4, 651-654 (1993).

[103] A. Mosandl, G. Bruche, C. Askari and H.-G. Schmarr. Stereoisomeric Flavor Compounds XLIV: Enantioselective analysis of some important flavor molecules. J. High Resolut. Chromatogr., 13, 660-662 (1990).

[104] A. Barrie, J. Bricout and J. Koziet. Gas Chromatography - Stable Isotope Ratio Analysis at Natural Abundance Levels. Biomed. Mass Spectrom., 11, 583-588 (1984).

[105] M. Rautenschlein, K. Habfast and W. Brand, in: Stable isotopes in paediatric, nutritional and metabolic research; ed. T.E. Chapman, R. Berger, D.J. Reyngoud, A. Okken, Intercept, Andover, Hampshire, UK 1990. High-Precision Measurement of 13C/12C ratios by on-line combustion of GC eluates and isotope Ratio Mass Spectrometry. p. 133-148.

[106] A. Mosandl, R. Braunsdorf, G. Bruche, A. Dietrich, U. Hener, V. Karl, T. Köpke, P. Kreis, D. Lehmann and B. Maas; in: Fruit flavors - biogenesis, characterization and authentication. ACS Symp. Ser., 596, 94-112 (1995). New methods to assess authenticity of natural flavours and essential oils.

[107] U. Hener, R. Braunsdorf, P. Kreis., A. Dietrich, B. Maas, E. Euler, B. Schlag, and A. Mosandl, Chiral compounds of essential oils X: The role of linalool in the origin evaluation of essential oils, Chem. Mikrobiol. Technol. Lebensm. 14, 129-133 (1992).

[108] R. Braunsdorf, U. Hener, D. Lehmann and A. Mosandl. Analytische Differenzierung zwischen natürlich gewachsenen, fermentativ erzeugten und synthetischen (naturidentischen) Aromastoffen I: Herkunftsspezifische Analyse des (E)-α(β)-Ionons. Dtsch. Lebensm. Rundsch., 87, 277-280 (1991).

[109] B. Maas, A. Dietrich, and A. Mosandl. Enantioselective capillary gaschromatography-olfactometry in essential oil analysis. Naturwiss., 80, 470-472 (1993).

[110] B. Weber, A. Dietrich, B. Maas, A. Marx, J. Olk, and A. Mosandl. Stereoisomeric Flavour Compounds LXVI: Enantiomeric distribution of the chiral sulfur-containing alcohols in yellow and purple passion fruits. Z. Lebensm. Unters. Forsch., 199, 48-50 (1994).

[111] P. Kreis and A. Mosandl. Chiral compounds of essential oils XI: Simultaneous stereoanalysis of Lavandula oil constituents. Flavour Fragr. J., 7, 187-193 (1992)

[112] R. Braunsdorf, U. Hener, S. Stein, and A. Mosandl. Comprehensive cGC-IRMS analysis in the authenticity control of flavours and essential oils. Part I: Lemon Oil. Z. Lebensm. Unters. Forsch., 197, 137-141 (1993).

[113] R. Braunsdorf. Herkunftsspezifische Aromastoffanalytik mittels Isotopenmassenspektrometrie. Dissertation, Universität Frankfurt/M., 1994.

[114] D. Lehmann. Neue Ergebnisse zur herkunftsspezifischen Aromastoffanalyse. Dissertation, Universität Frankfurt/M., 1994.

[115] P. Werkhoff, S. Brennecke, W. Bretschneider, M. Güntert, R. Hopp and H. Surburg. Chirospecific analysis in essential oil, fragrance and flavor research. Z. Lebensm. Unters. Forsch., 196, 307-328 (1993).

[116] K. Cabrera and G. Schwinn. Cyclodextrine als chirale Selektoren für die Enantiomerentrennung in der HPLC (Teil 1): Dynamische Belegung von RP-Phasen. Kontakte (Darmstadt), 1989, (3), 3-8.

[117] M. Juza, M. Mazzotti and M. Morbidelli. Simulated moving-bed chromatography and its application to chirotechnology. TIBTECH 18, 108-118 (2000).
[118] S. Bilke and A. Mosandl. Measurements by gas chromatography / pyrolysis / mass spectrometry: fundamental conditions in $^2$H/ $^1$H isotope ratio analysis. Rapid Commun. Mass Spectrom., 16, 468-472 (2002).
[119] S. Bilke and A. Mosandl. Authenticity assessment of lavender oil using GC-P-IRMS: $^2$H/ $^1$H-ratios of linalool and linalyl acetate. Eur. Food Res. Technol., 214, 532-535 (2002).
[120] S. Bilke and A. Mosandl. $^2$H/ $^1$H-and $^{13}$C/ $^{12}$C isotope ratios of trans-anethole using gas chromatography–isotope ratio mass spectrometry. J. Agric. Food Chem., 50, 3935-3937 (2002).
[121] S. Sewenig, U. Hener, and A. Mosandl. Online determination of $^2$H/$^1$H and $^{13}$C/$^{12}$C isotope ratios of cinnamaldehyde from different sources using gas chromatography isotope ratio mass spectrometry. Eur. Food. Res. Technol., 217, 444-448 (2003).
[122] J. Jung, S. Sewenig, U. Hener, and A. Mosandl. Comprehensive authenticity assessment of lavender oils using multielement/ multicomponent IRMS-analysis and enantioselective MDGC-MS. Eur. Food Res. Technol., 220, 232-237 (2005).
[123] S. Sewenig, D. Bullinger, U. Hener, and A. Mosandl. Comprehensive authentication of (E)-α(β)-ionone from raspberries, using constant flow MDGC-C/P-IRMS and enantio-MDGC/MS. J. Agric. Food Chem., 53, 838-844 (2005).
[124] H.-L. Schmidt, R. A. Werner,, A. Roßmann, A. Mosandl, P. Schreier in: Flavourings: Production, Composition, Application, Regulations (H. Ziegler, ed.), 602-663. Wiley-VCH, Weinheim 2006.

## 6.2.4 Key Odorants of Food Identified by Aroma Analysis

*Werner Grosch*

### 6.2.4.1 Introduction

In numerous studies the volatile fractions of food have been analysed to identify their components. Due to the progress in instrumental analysis, particularly gas chromatography (GC) and mass spectrometry (MS), about 7600 different volatile substances have been identified in more than 400 food products up to 2004 *[1]*.

Among all the volatile compounds, only a limited number are important for aroma. According to a proposal by Rothe and Thomas *[2]* only those compounds actually contribute to aroma whose concentration in food exceeds their odour thresholds. To estimate the importance of a volatile compound for the aroma of a particular food, the ratio of concentration to its odour threshold was calculated. This value was denoted aroma value *[2]*, odour unit *[3]* or odour activity value (OAV) *[4]*. In the following the latter term is used.

The knowledge that not all of the volatiles (e.g. more than 800 in roasted coffee) *[5]* that occur in a food contribute to its aroma was the rationale for changing the methodology of analysis. Since 1984, when the procedure for charm analysis was published *[4]*, techniques have been developed that focus on the identification of compounds contributing to the aroma with higher OAV.

It is the objective of this chapter to provide a survey of the state of art in the methodology of aroma analysis (6.2.4.2) and to discuss important results which were obtained by the application of these methods (6.2.4.3). The data listed in 6.2.4.3 will give a first orientation on the composition of important food aromas.

### 6.2.4.2 Aroma Analysis

#### 6.2.4.2.1 Outline of the procedure

To detect the odour-active volatiles, Fuller et al. *[6]* described a system for the sniffing of GC effluents which was improved and applied to food samples by Dravnieks and O'Donnell *[7]*. The new technique, named GC olfactometry (GCO), was the starting point for the development of a systematic approach to the identification of the compounds which cause food aromas. As summarised in Table 6.23 the analytical procedure consists of screening for key odorants by special GCO techniques, quantification and calculation of OAVs as well as aroma-recombination studies. During the last decade these steps have been critically reviewed by Acree *[8]*, Blank *[9]*, Grosch *[10, 11]*, Mistry et al. *[12]* and Schieberle *[13]*.

*Table 6.23:* Steps in aroma analysis

| Step | |
|---|---|
| 1 | Preparation of a concentrate smelling like the food sample |
| 2 | Separation of the concentrate by GC and localisation of key-odorants by charm analysis or AEDA |
| 3 | Detection of highly volatile key odorants by gas chromatography-olfactometry of headspace samples (GCOH) |
| 4 | Enrichment and identification of key odorants |
| 5 | Quantification of key odorants and calculation of their OAVs |
| 6 | Preparation of a synthetic blend (aroma model) of the key odorants on the basis of the quantitative data obtained in step 5. Critical comparison of the aroma profile of the synthetic blend with that of the original |
| 7 | Comparison of the overall odour of the aroma model with that of models in which one or more components are omitted (omission experiments) |

### 6.2.4.2.2 Screening for Key Odorants

**Medium and Lower Boiling Odorants**

After the preparation of an aroma concentrate as detailed in *[11]*, dilution experiments are performed (Table 6.23). As reviewed by Acree *[8]* and Grosch *[11]*, two techniques – charm analysis and aroma extract dilution analysis (AEDA) – are used to screen the potent, medium and lower volatile odorants on which the identification experiments are then focused. In both procedures, an extract obtained from the food is diluted, and each dilution is analysed by GCO. This procedure is performed until no odorants are perceivable by GCO.

In the case of AEDA, which is mostly applied *[11]*, the result is expressed as flavour dilution (FD) factor. The FD factor is the ratio of the concentration of the odorant in the initial extract in which the odour is still detectable by GCO *[14, 15]*. Consequently, the FD factor is a relative measure and is proportional to the OAV of the compound in air. As an example, the analysis of the aroma of the baguette crust *[16]* will be discussed. After separation of the acidic fraction, the neutral/basic volatiles were investigated by AEDA. Results listed in Table 6.24 reveal 21 odorants in the FD factor range 32-512, of which 2-acetyl-1-pyrroline (no. 1), 2-ethyl-3,5-dimethylpyrazine (no. 10) and (E)-2-nonenal (no. 17) showed the highest FD factors.

Charm analysis constructs chromatographic peaks, the areas of which are proportional to the amount of the odorant in the aroma concentrate *[4]*. The difference between the two methods is that charm analysis measures the dilution value over the entire time the compounds elute, whereas AEDA simply determines the maximum dilution value *[8]*.

The Osme method *[17]* and olfactometry global analysis (OGA) *[18]* constitute GCO techniques which are rarely applied. However, a critical comparison of AEDA, Osme and OGA has indicated that the three methods agree in the evaluation of the most potent odorants [19]. A further method – aroma extract concentration analysis (AECA) – is the reverse of AEDA [20, 21]. It starts with the undiluted extract which then is concentrated stepwise and after each step an aliquot is analysed by GCO [20]. This procedure avoids an underestimation of the FD factors of reactive odorants which are degraded in AEDA.

*Table 6.24:* AEDA of baguette crust volatiles (neutral/basic compounds) [16]

| No. | Compound | Odour quality[a] | FD factor[b] |
|---|---|---|---|
| 1 | 2-Acetyl-1-pyrroline | Roasty | 512 |
| 2 | Dimethyltrisulphide | Cabbage-like | 64 |
| 3 | 1-Octen-3-one | Mushroom-like | 128 |
| 4 | 2-Methyl-3-ethylpyrazine | Earthy | 32 |
| 5 | Trimethylpyrazine | Earthy | 32 |
| 6 | 2-Acetyl-2-thiazole | Roasty | 32 |
| 7 | Acetylpyrazine | Sweet, roasty | 128 |
| 8 | Phenylacetaldehyde | Honey-like | 128 |
| 9 | 3-Ethyl-2,5-dimethylpyrazine | Earthy | 64 |
| 10 | 2-Ethyl-3,5-dimethylpyrazine | Earthy | 512 |
| 11 | 2-Acetyl-2-thiazoline | Roasty | 128 |
| 12 | Nonanal | Soapy | 32 |
| 13 | 2-Phenylethanol | Honey-like | 128 |
| 14 | (Z)-2-Nonenal | Green | 256 |
| 15 | (E,Z)-2,6-Nonadienal | Cucumber-like | 64 |
| 16 | 2,3-Diethyl-5-methylpyrazine | Earthy | 128 |
| 17 | (E)-2-Nonenal | Green, fatty | 512 |
| 18 | (E,E)-2,4-Nonadienal | Fatty | 256 |
| 19 | (E,Z)-2,4-Decadienal | Deep-fried | 128 |
| 20 | (E,E)-2,4-Decadienal | Deep-fried | 256 |
| 21 | trans-4,5-Epoxy-(E)-2-decenal | Metallic | 128 |

[a] Odour quality perceived at the sniffing port.
[b] FD factor on capillary SE-54.

## Highly Volatile Odorants

During preparation of the aroma concentrate (step 1 in Table 6.23), the highly volatile odorants are lost or masked in the gas chromatogram by the solvent peak. To overcome this limitation, the screening for key odorants has to be completed by GCO of static headspace samples (GCOH) [22, 23].

In the example of baguette crust (Table 6.25), analysis started with a headspace volume of 20 mL and 13 odorants were revealed by GCO. The headspace sample was then reduced in a series of steps to determine the most potent, highly volatile odorants. GCOH of volumes of 10 and 2.5 mL indicated only 10 and 7 odorants, respectively (Table 6.25). After reduction to 0.2 mL, only 2,3-butanedione was found. According to this experiment, 2,3-butanedione was the most potent highly volatile odorant of baguette crust. A comparison of Tables 6.24 and 6.25 shows that some odorants were detected by both GCOH and AEDA (e.g. 1-octen-3-one, dimethyltrisulphide, 2-acetyl-1-pyrroline).

*Table 6.25: GCO of headspace samples of the baguette crust [16]*

| No. | Compound | Odour quality | Volume[a] (mL) | FD factor[b] |
|---|---|---|---|---|
| 1 | 2,3-Butanedione | Buttery | 0.2 | 100 |
| 2 | 1-Octen-3-one | Mushroom-like | 0.5 | 40 |
| 3 | Methional | Boiled potato | 1.0 | 20 |
| 4 | Methylpropanal | Malty | 2.5 | 8 |
| 5 | 3-Methylbutanal | Malty | 2.5 | 8 |
| 6 | Dimethyltrisulphide | Cabbage-like | 2.5 | 8 |
| 7 | 2-Acetyl-1-pyrroline | Roasty | 2.5 | 8 |
| 8 | 2-Methylbutanal | Malty | 10 | 2 |
| 9 | 2-Ethyl-3,5-dimethylpyrazine | Earthy | 10 | 2 |
| 10 | Hexanal | Green, fatty | 10 | 2 |
| 11 | 2,3-Pentanedione | Buttery | 20 | 1 |
| 12 | 2-Acetylfuran | Smoky | 20 | 1 |
| 13 | (E)-2-Nonenal | Green, fatty | 20 | 1 |

[a] Lowest headspace volume required to perceive the odorant at the sniffing port.
[b] The highest headspace volume (20 mL) was equated to a FD factor of one. The FD factors of the other odorants were calculated on this basis.

### 6.2.4.2.3 Identification

The results obtained in the screening procedures are not corrected for losses of odorants during the extraction and concentration procedures. For this reason, the identification experiments should be focused not only on compounds with the highest FD factors, but also on those perceived at lower dilutions [15]. Furthermore, it is necessary to enrich odorants which do not appear as peaks in the gas chromatogram or are concealed by large peaks of odourless volatiles. Enrichment (step 4 in Table 6.23) is performed by column chromatography [24, 25] or with multidimensional gas chromatography of MDGC [26]. In MDGC, the aroma extract is separated on a polar pre-column, then a section of the effluent containing the analyte is cryofocused with liquid nitrogen and subsequently transferred to an non-polar main column which is combined with a mass spectrometer (MS) and a sniffing port.

In the identification experiment, it is necessary to compare by GCO the odour quality of the analyte with that of an authentic sample at approximately equal levels. Only when, in addition to GC and MS data, the sensorial properties are correspondent, the analyte, which has been perceived by GCO in the volatile fraction, is correctly identified. All of the odorants reported in the following section were identified according to this procedure.

### 6.2.4.2.4 Quantitative Analysis and Calculation of OAVs

**Introduction**

Charm analysis and AEDA both constitute screening procedures, as the results are not corrected for the losses of odorants during the isolation procedure. Furthermore, in GCO, the odorants are completely volatilised and then evaluated by sniffing, whereas the volatility of the aroma compounds in foods depends on their solubility in the aqueous and/or oily phase as well as their binding to non-volatile food constituents. To elucidate which of the compounds revealed in the dilution experiments contribute with high OAV to the aroma, quantification of the odorants with higher FD factors and calculation of their OAVs are the next steps in the analytical procedure.

**Quantitative Analysis**

The odorants occurring in foods differ strongly in concentration, volatility, and reactivity as well as in binding to food matrix. Consequently, it is difficult to quantify them precisely in the usual way by applying internal standards that differ significantly in chemical nature from the analytes [11, 15, 27]. The results are wrong because the losses of standard and analytes during extraction, work-up and GC separation are different. The losses during the analytical procedure do not affect the results when stable isotopomers of the analytes are used as internal standards in stable isotope dilution assays (SIDAs).

In SIDAs the majority of the odorants used as internal standards are labelled with deuterium [11, 13]. However, during the quantification procedure some deuterated odorants might undergo deuterium-protium exchange, which would falsify the results. Examples are furaneol [28] and sotolon [29], which are consequently labelled instead with carbon-13.

The precision of SIDA has been confirmed in model experiments [29, 30]. Although after clean-up, the yields of some analytes were lower than 10%, the results of quantification were correct as the standards showed equal losses.

**OAV**

To approach the situation in food, OAVs are calculated on the basis of odour threshold values which have been estimated in a medium that predominates in the food, e.g. water, oil, starch. As an example the OAVs of the key odorants of baguette crust are listed in Table 6.26. The highest OAVs were found for the roasty smelling 2-acetyl-1-pyrroline (no. 7), followed by furaneol (no. 20), 2,3-butanedione (no. 2), (E)-2-nonenal (no. 13), 1-octen-3-one (no. 9) and methional (no. 6). It is assumed that these odorants contribute strongly to the aroma of baguette crust.

*Table 6.26: Concentration of odorants in the crust of a baguette and their OAV [16]*

| No. | Compound | Concentration (μg/kg) | Odour threshold[a] (μg/kg) | OAV[b] |
|---|---|---|---|---|
| 1 | Methylpropanal | 1733 | 56 | 31 |
| 2 | 2,3-Butanedione | 1516 | 6.5 | 233 |
| 3 | 3-Methylbutanal | 926 | 32 | 13 |
| 4 | 2-Methylbutanal | 1147 | 53 | 22 |
| 5 | Hexanal | 239 | 30 | 8 |
| 6 | Methional | 26.6 | 0.27 | 99 |
| 7 | 2-Acetyl-1-pyrroline | 16.7 | 0.0073 | 2288 |
| 8 | Dimethylsulphide | 1.9 | 0.086 | 22 |
| 9 | 1-Octen-3-one | 4.9 | 0.044 | 111 |
| 10 | 2-Ethyl-3,5-dimethylpyrazine | 2.5 | 0.17 | 15 |
| 11 | (Z)-2-Nonenal | 0.4 | 0.036 | 11 |
| 12 | 2,3-Diethyl-5-methylpyrazine | 0.4 | 0.14 | 3 |
| 13 | (E)-2-Nonenal | 64.4 | 0.53 | 122 |
| 14 | (E,Z)-2,4-Decadienal | 14.2 | 1.81 | 8 |
| 15 | (E,E)-2,4-Decadienal | 33.5 | 2.7 | 12 |
| 16 | 4-Vinylguaiacol | 1071 | 18 | 60 |
| 17 | trans-4,5-Epoxy-(E)-2-decenal | 1.8 | 0.19 | 9 |
| 18 | Acetic acid[c] | 271000 | 31140 | 9 |
| 19 | 2-/3-Methylbutyric acid[c] | 1129 | 24 | 47 |
| 20 | Furaneol[c] | 6040 | 13 | 465 |

[a] Odour threshold in starch.
[b] The OAV is calculated by dividing the concentration by the odour threshold of the compound.
[c] Odorants nos. 18-20 showed the highest FD factors in AEDA of the acidic fraction [16].

Leffingwell and Leffingwell [31] have pointed out that the calculation of OAV is also helpful for flavourists. As an example they cited the unstable 2-acetyl-1-pyrroline, responsible for the popcorn-like note in the odour profile of cooked fragrant rice [32], which can be replaced in an aroma reconstruction by a similarly smelling, stable compound. 2-Acetylthiazole and acetylpyrazine were selected, and on the basis of their odour threshold values in water, the amounts were calculated which would reach the OAV of 2-acetyl-1-pyrroline in cooked fragrant rice [32]. Two flavours were mixed in which 2-acetyl-1-pyrroline had been replaced either by 2-acetylthiazole or acetylpyrazine. Sensory experiments indicated that the sample containing 2-acetylthiazole provided a very good reconstituted rice flavour. Leffingwell and Leffingwell [31] concluded that "the odour unit (OAV) concept is an effective and inexpensive tool for simplifying the otherwise difficult problem of converting a complex flavour analysis into a 'practical' flavour system".

### 6.2.4.2.5 Aroma Model

On the basis of qualitative and quantitative data, aroma models are prepared to check whether the results of aroma analyses are correct (step 6 in Table 6.23). In case of a liquid food, the preparation of an aroma model is simple. The odorants, in concentrations found in the food, are dissolved in water, aqueous ethanol, or an odourless edible oil (examples in 6.2.4.3). Difficulties arise in the case of solid foods, as it is generally not possible to reproduce the composition and distribution of the non-volatile components as well as the distribution of odorants in the food matrix. However, the aroma model becomes independent from the matrix when it is identical with the composition of the odorants in the air above the food *[11]*. Using this method an aroma model for baguette crust was prepared *[33, 34]*. Its composition (Table 6.27) was derived from the concentration of the odorants in the baguette headspace *[34]*. Table 6.28 indicates that the odour profiles of the baguette and its aroma model agreed very well. Consequently, the volatiles listed in Table 6.27 are the key odorants of freshly baked baguette. A further example showing an aroma simulation on the basis of the odorant composition in the headspace was performed with roasted coffee *[35]*. Also in this case the odour of the model was very similar to that of the original. An advantage of headspace aroma models is that the concentrations of the odorants smelled by the assessor are identical with those above the food.

***Table 6.27:*** *Composition of a model for the headspace aroma of a baguette [33, 34]*

| No. | Compound | Concentration$^a$ (ng/L air) |
| --- | --- | --- |
| 1 | Methylpropanal | 830 |
| 2 | 2-Methylbutanal | 320 |
| 3 | 3-Methylbutanal | 150 |
| 4 | Methional | 6.6 |
| 5 | Dimethylsulphide | 4.2 |
| 6 | 2,3-Butanedione | 980 |
| 7 | 2-Acetyl-1-pyrroline | 3.7 |
| 8 | Hexanal | 216 |
| 9 | 1-Octen-3-one | 6.7 |
| 10 | (E)-2-Nonenal | 28 |
| 11 | (E,E)-2,4-Decadienal | 7.8 |

$^a$ Concentration in the olfactometer *[33, 34]*.

*Table 6.28:* Headspace odour profiles of a baguette and the corresponding aroma model [34]

| Attribute | Odour intensity[a] | |
|---|---|---|
| | **Baguette** | **Model** |
| Roasty | 1.8 | 1.7 |
| Malty | 1.9 | 2.0 |
| Sweet | 1.2 | 1.1 |
| Fatty | 1.0 | 0.9 |

[a] The intensity of the attributes was scored by 5 assessors on the scale 0 (absent) to 3 (strong). Mean values are presented in the table.

#### 6.2.4.2.6 Omission Experiments

In the dilution experiments, on which the aroma models are based, the odour impact of the volatiles is evaluated after separation by GC (cf. 6.2.4.2.2). Perceptual interactions of odorants, which in most cases were characterised by inhibition and suppression *[8, 36]*, are excluded by this procedure. Therefore, the question as to which compound among the volatiles evaluated by the dilution experiments actually contributes to the aroma has to be answered by omission experiments (step 7 in Table 6.23). Examples of such experiments are given in 6.2.4.3.

### 6.2.4.3 Applications

#### 6.2.4.3.1 Introduction

Since 1985, the methodology discussed in the preceding sections has been applied to the aroma analysis of a large number of foods. The compounds which were identified and their concentrations, as far as measured with a precise method like SIDA, are discussed in the following section. Additionally, as they are of potential commercial interest, the OAVs are listed in the tables.

#### 6.2.4.3.2 Meat

**Beef and Pork**

Boiled beef and pork differ in their aromas. Beef smells more intensely meaty, sweet-caramel-like and malty, whereas pork is stronger in sulphurous and fatty odour notes *[21, 37]*. According to Table 6.29, the pronounced odour notes of beef are caused by high concentration of furaneol (no. 1), 2-furfurylthiol (no. 2), 3-mercapto-2-pentanone (no. 3) and 2-methyl-3-furanthiol (no. 6). Omission experiments confirmed that these volatiles and in addition octanal, nonanal, (E,E)-2,4-decadienal, are the key odorants of boiled beef *[21]*. The higher concentration of the caramel-like smelling furaneol (no. 1) in beef than in pork is due to higher levels of its precursors glucose-6-phosphate and fructose-6-phosphate *[37]*.

*Table 6.29:* Key odorants of boiled beef and pork [38][a]

| No. | Odorant | Concentration (µg/kg)[b] | | Odour activity value (OAV)[c] | |
|---|---|---|---|---|---|
| | | Beef[d] | Pork[d] | Beef | Pork |
| 1 | 4-Hydroxy-2,5-dimethyl-3(2H)-furanone | 9075 | 2170 | 908 | 217 |
| 2 | 2-Furfurylthiol | 29 | 9.5 | 2900 | 950 |
| 3 | 3-Mercapto-2-pentanone | 69 | 66 | 99 | 94 |
| 4 | Methanethiol | 311 | 278 | 1555 | 1390 |
| 5 | Octanal | 382 | 154 | 546 | 220 |
| 6 | 2-Methyl-3-furanthiol | 24 | 9.1 | 3429 | 1300 |
| 7 | Nonanal | 1262 | 643 | 1262 | 643 |
| 8 | Methional | 36 | 11 | 180 | 55 |
| 9 | (E,E)-2,4-Decadienal | 27 | 7.4 | 135 | 37 |
| 10 | (E,Z)-2,4-Decadienal | n.a.[e] | 7.2 | n.a. | 180 |
| 11 | 12-Methyltridecanal | 962 | n.a. | 9620 | n.a. |
| 12 | Dimethylsulphide | 105 | n.a. | 350 | n.a. |
| 13 | (Z)-2-Nonenal | 6.2 | 1.4 | 310 | 70 |
| 14 | (E)-2-Nonenal | 32 | 15 | 128 | 60 |
| 15 | Acetaldehyde | 1817 | 3953 | 182 | 395 |
| 16 | 1-Octen-3-one | 9.4 | 4.8 | 188 | 96 |
| 17 | Methylpropanal | 117 | 90 | 167 | 29 |
| 18 | Butyric acid | 7074 | 17200 | 2.6 | 6.3 |
| 19 | 3-Methylbutanal | 26 | 27 | 74 | 77 |

[a] Forerib from beef and shoulder from pork were trimmed of all excess fat and samples (400 g) in water (300 mL) were boiled in a pressure cooker ($8 \times 10^4$ Pa, 116°C) for 45 min.
[b] Values are related to cooked meat.
[c] OAV: concentration/nasal odour threshold in water.
[d] Fat content of the analysed samples: 8.8%, beef; 1.7%, pork.
[e] n.a.: not analysed because of minor importance during AEDA and/or GCOH.

Sulphurous and fatty odour notes predominated in the odour profile of boiled pork. It is assumed [37, 38] that the sulphurous (e.g. methanethiol) and fatty odorants (e.g. octanal, nonanal) are clearly perceptible in pork due to the much lower concentration of furaneol (Table 6.29). Although when compared to pork, the levels of octanal and nonanal were twice as high in beef (Table 6.29) the fatty odour note was weak. In particular the fourfold higher concentration of furaneol might have reduced the intensity of the fatty odour.

Omission experiments [21] indicated that 12-methyltridecanal (no. 11) did not belong to the key odorants of boiled beef in spite of its high OAV (Table 6.29). Most likely its smell was suppressed by other odorants. This is in contrast to the juice released during stewing of lean beef, where it was one of the essential odorants and also of importance for the mouth feeling *[39]*. No. 11 originates from microorganisms present in the rumen *[40]*. There are indications that small amounts of no. 11 are absorbed by the animal and transported to the muscular tissue where it is incorporated into plasmalogens. During a longer cooking period no. 11 is liberated by hydrolysis *[41]*. Consequently, in beef that is cooked for a short period, e.g. pan-fried for only 5 to 10 minutes, the amount of liberated no. 11 is so small that it is not detectable in AEDA *[42]*.

## Chicken

(E,E)-2,4-Decadienal (no. 16 in Table 6.30) proposed by Pippen and Nonaka *[43]* as character-impact odorant of cooked chicken belonged together with methanethiol (no. 4) to the volatiles showing the highest OAVs in a study on boiled chicken aroma. Also in the skin of but not in the meat of roasted chicken, decadienal was the most potent odorant due to its high OAV (Table 6.30). Decadienal is formed by autoxidation of linoleic acid *[46]*. In boiled chicken and in the skin of roasted chicken, the formation of 2,4-decadienal is favoured by the high fat content (see legend of Table 6.30) and by its high content of linoleic acid which is up to 10 times higher than in beef and pork *[47]*. On the other hand, the lean meat remaining after separation of the skin from roasted chicken provides only traces of 2,4-decadienal and the nonenals (nos. 12-14) both with linoleic acid as their precursor. On the basis of high OAV, the roasty-malty note in the odour profile of roasted chicken is caused by the Strecker aldehydes (nos. 19-21) (Table 6.30), 2-acetyl-1-pyrroline (no. 23) and the pyrazines (nos. 25 and 26).

*Table 6.30:* Key odorants of boiled and roasted chicken [38, 44, 45]

| No. | Compound | Concentration (µg/kg)[a] | | | OAV[b] | | |
|---|---|---|---|---|---|---|---|
| | | Boiled chicken[c] | Roasted chicken[d] | | Boiled chicken[c] | Roasted chicken[d] | |
| | | | Meat | Skin | | Meat | Skin |
| 1 | 2-Furfurylthiol | 1.1 | 0.1 | 1.9 | 110 | 10 | 190 |
| 2 | 3-Mercapto-2-pentanone | 35 | 29 | 27 | 50 | 41 | 21 |
| 3 | 2-Methyl-3-furanthiol | 2.0 | 0.4 | 4.1 | 286 | 57 | 585 |
| 4 | Methanethiol | 1456 | 202 | 164 | 7280 | 1010 | 820 |
| 5 | Hydrogensulphide | +[e] | 280 | n.a. | – | 29 | – |
| 6 | Methional | 15 | 53 | 97 | 74 | 265 | 485 |
| 7 | Dimethyltrisulphide | 0.8 | <0.03 | 0.5 | 133 | <5 | 83 |
| 8 | Acetaldehyde | 4700 | 3815 | 3287 | 470 | 382 | 329 |
| 9 | Hexanal | 1596 | 283 | 893 | 355 | 63 | 198 |
| 10 | Octanal | 148 | 190 | 535 | 211 | 271 | 764 |

| No. | Compound | Concentration (µg/kg)[a] | | | OAV[b] | | |
|---|---|---|---|---|---|---|---|
| | | Boiled chicken[c] | Roasted chicken[d] | | Boiled chicken[c] | Roasted chicken[d] | |
| | | | Meat | Skin | | Meat | Skin |
| 11 | Nonanal | +[f] | 534 | 832 | – | 534 | 832 |
| 12 | (Z)-2-Nonenal | n.a. | 5.5 | 10.5 | – | 275 | 525 |
| 13 | (E)-2-Nonenal | 75 | 23 | 147 | 300 | 92 | 588 |
| 14 | (E,E)-2,4-Nonadienal | 57 | 2.3 | 24 | 570 | 23 | 240 |
| 15 | (E,Z)-2,6-Nonadienal | n.a. | 3.1 | 6.0 | – | 155 | 300 |
| 16 | (E,E)-2,4-Decadienal | 302 | 11 | 711 | 1510 | 55 | 3555 |
| 17 | (E,Z)-2,4-Decadienal | 31 | n.a. | n.a. | 775 | – | – |
| 18 | 1-Octen-3-one | 5.7 | 7.2 | 10.8 | 114 | 144 | 216 |
| 19 | Methylpropanal | 39 | 83 | 538 | 56 | 119 | 769 |
| 20 | 2-Methylbutanal | n.a. | 8 | 455 | – | 6 | 350 |
| 21 | 3-Methylbutanal | 5.0 | 17 | 668 | 14 | 49 | 1670 |
| 22 | 2-Acetyl-2-thiazoline | n.a. | 2.6 | 5.8 | – | 2.6 | 6 |
| 23 | 2-Acetyl-1-thiazoline | n.a. | 0.2 | 2.9 | – | 2 | 29 |
| 24 | 2-Propionyl-1-pyrroline | n.a. | n.a. | 0.8 | – | – | 8 |
| 25 | 2-Ethyl-3,5-dimethylpyrazine | n.a. | n.a. | 4.3 | – | – | 27 |
| 26 | 2,3-Diethyl-5-methylpyrazine | n.a. | n.a. | 2.5 | – | – | 28 |
| 27 | 4-Hydroxy-2,5-dimethyl-3(2H)-furanone | 1232 | 50 | 395 | 123 | 5 | 40 |
| 28 | 3-Hydroxy-2,5-dimethyl-2(5H)-furanone | 4.5 | 0.6 | 1.9 | 15 | 2 | 6 |

[a] Values are related to cooked meat.
[b] OAV: concentration/nasal odour threshold in water.
[c] Fowl (1.4 kg without innards, fat content 20%) was boiled for 30 min in a pressure cooker at 116°C and $8 \times 10^4$ Pa.
[d] Broiler fowl (1.5 kg without innards) was roasted in a frying pan containing coconut oil (15 g) for 1 h at 180°C. Skin (fat content 47%) and meat (fat content 1%) of the breast were separately analysed.
[e] n.a.: not analysed because of minor importance during AEDA and/or GCOH.
[f] According to GCOH the compound contributes to the aroma.

### 6.2.4.3.3 Fish

The mild green-metallic-mushroom aroma of fresh fish is formed by enzymatic-oxidative degradation of polyunsaturated fatty acids (PUFA) with the participation of lipoxygenases of varying specificity [48]. Consequently, oxidation products of PUFA were detected as the most important odorants of raw salmon (Table 6.31). In particular (Z)-1,5-octadien-3-one (no. 10), (E,Z)-2,6-nonadienal (no. 12), propanal (no. 2) and acetaldehyde (no. 1) are significant contributors to the aroma. During boiling the concentration of nos. 1, 2 and 10 decreases somewhat (Table 6.31).

*Table 6.31:* Odorants of raw and boiled salmon (Salmo salar) and of boiled cod (Gadus morhua) [49]

| No. | Odorant | Concentration (µg/kg)[a] | | | OAV[b] | | |
|---|---|---|---|---|---|---|---|
| | | Salmon | | Cod boiled | Salmon | | Cod boiled |
| | | raw | boiled | | raw | boiled | |
| 1 | Acetaldehyde | 3700 | 2300 | 1300 | 370 | 230 | 130 |
| 2 | Propanal | 3500 | 1700 | n.a. | 500 | 240 | – |
| 3 | 2,3-Butanedione | 57 | 52 | 200 | 11 | 10 | 40 |
| 4 | 2,3-Pentanedione | 141 | 234 | 86 | 28 | 47 | 17 |
| 5 | Hexanal | 35 | 58 | 115 | 3 | 6 | 11 |
| 6 | (Z)-3-Hexenal | 2.6 | 3.9 | 1.3 | 87 | 130 | 43 |
| 7 | (Z)-4-Heptenal | 3.0 | 6.0 | 1.6 | 50 | 100 | 27 |
| 8 | Methional | 3.0 | 8.0 | 11 | 75 | 200 | 275 |
| 9 | 1-Octen-3-one | 0.5 | 0.4 | 0.7 | 50 | 40 | 70 |
| 10 | (Z)-1,5-Octadien-3-one | 0.4 | 0.3 | 0.1 | 1000 | 750 | 250 |
| 11 | (Z,Z)-3,6-Nonadienal | n.a. | 5.7 | 1.3 | – | 114 | 26 |
| 12 | (E,Z)-2,6-Nonadienal | 9.3 | 9.7 | 35 | 465 | 485 | 175 |
| 13 | (E)-2-Nonenal | 2.0 | 2.7 | n.a. | 25 | 34 | – |
| 14 | (E,E)-2,4-Nondienal | 2.2 | 2.6 | 3.2 | 37 | 43 | 53 |
| 15 | (E,E)-2,4-Decadienal | 4.8 | 6.0 | 3.5 | 160 | 200 | 117 |
| 16 | Methanethiol | n.a. | n.a. | 100 | – | – | 50 |
| 17 | Methylpropanal | n.a. | n.a. | 27 | – | – | 39 |
| 18 | 2-Methylbutanal | n.a. | n.a. | 20 | – | – | 22 |
| 19 | 3-Methylbutanal | n.a. | n.a. | 51 | – | – | 204 |
| 20 | Dimethylsulphide | n.a. | n.a. | 77 | – | – | 39 |
| 21 | Dimethyltrisulphide | n.a. | n.a. | 0.15 | – | – | 19 |

[a] Amount in raw or boiled fish.
[b] OAV: concentration/nasal odour threshold in water.
  n.a.: not analysed because of minor of importance during AEDA and/or GCOH.

In contrast to salmon, cod is a lean fish as its lipid content amounts to less than 1%. Consequently, the concentration of octadienone (no. 10) in boiled cod is only one third of that in boiled salmon (Table 6.31). However, as proven in a model experiment *[50]*, this low concentration is enough to generate a fishy smell. Furthermore, this experiment revealed that addition of the amount of methional (no. 8) occurring in cod (Table 6.31) enhances the fishy odour. A further difference between boiled salmon and boiled cod is the contribution of methanethiol (no. 16) and of the Strecker aldehydes (nos. 17-19) to the aroma of the latter (Table 6.31).

### 6.2.4.3.4 Cheese

**Swiss Cheese (Emmental)**

The potent odorants of Swiss cheese have been evaluated by the methods reported in 6.2.4.2 *[51-53]*. Table 6.32 shows the identified neutral odorants. Furanones nos. 7 and 8 are of special interest as there are indications that – in addition to sweet tasting calcium and magnesium propionate *[54]* – they contribute to the sweet and caramel-like note which has been perceived in the flavour profile of emmental *[55]*. Further experiments revealed that lactic acid bacteria cause the production of furanone no. 7 in Swiss cheese *[56]*.

*Table 6.32:* Neutral odorants of Swiss cheese (emmental) [51, 52]

| No. | Compound | Concentration (µg/kg) | OAV[a] |
|---|---|---|---|
| 1 | 2,3-Butanedione | 98 | 10 |
| 2 | 3-Methylbutanal | 64 | 12 |
| 3 | Ethyl butanoate | 25 | 1 |
| 4 | Ethyl 3-methylbutanoate | 0.9 | 9 |
| 5 | Ethyl hexanoate | 62 | 2 |
| 6 | Methional | 33 | 165 |
| 7 | 4-Hydroxy-2,5-dimethyl-3(*2H*)-furanone | 2035 | 81 |
| 8 | 5-Ethyl-4-hydroxy-2-methyl-3(*2H*)-furanone | 141 | 24 |
| 9 | δ-Decalactone | 927 | 2 |

[a] OAV: concentration odour threshold in sunflower oil.

To verify whether the volatiles listed in Table 6.32 are actually the key odorants, an aroma model was prepared by using an unripened cheese (UC) as base *[53]*. The odorants and in addition the compounds showing high taste activity values *[54]* were quantified in UC and in Swiss cheese *[53]*. The differences in the concentration of these compounds in both samples were calculated, and, accordingly, the compounds were dissolved in water and/or sunflower oil and then added to freeze-dried UC. The flavour model obtained agreed in colour, pH, water, protein and fat content with grated Swiss cheese, only the texture was more grainy *[53]*.

The odour profile differed somewhat because the nutty, sweet and pungent odour notes were more intense in the model and the buttery note was weaker. However, the taste profile of the model matched that of the corresponding Swiss cheese sample very well *[53]*. Omission experiments supported the conclusion that methional (no. 6 in Table

6.32), the furanones nos. 7 and 8, acetic acid, propionic acid, lactic acid, succinic acid, glutamic acid, sodium, potassium, calcium, magnesium, ammonium, phosphate and chloride are the character impact odour and taste compounds of Swiss cheese. The contribution of 2,3-butanedione and δ-decalactone to the flavour of Swiss cheese remains open, because the concentration of these odorants in UC is relatively high.

**Swiss Gruyere Cheese**

Aroma models were prepared for gruyere by applying the methods reported for emmental [57, 58]. The aroma of the models was similar to that of the original cheeses. Consequently, it was concluded that 2-/3-methylbutanal, methional, dimethyltrisulphide, phenylacetaldehyde, 2-ethyl-3,5-dimethylpyrazine, methanethiol, as well as acetic, propionic, butyric, 3-methylbutyric and phenylacetic acids are the key odorants of gruyere.

Compared to emmental cheese, the aroma of gruyere cheese is more intensely sweaty and less caramel-like [55]. The composition of the key odorants reflects these differences as the concentrations of the caramel-like smelling furanones are lower and that of methylbutyric acid is higher [57, 58].

*Table 6.33: Odorants of camembert cheese [59-61]*

| No. | Compound | Concentration (µg/kg)[a] | OAV[b] |
|---|---|---|---|
| 1 | 2,3-Butanedione | 110 | 11 |
| 2 | 3-Methylbutanal | 142 | 11 |
| 3 | Methional | 125 | 625 |
| 4 | 1-Octen-3-ol | 130 | 4 |
| 5 | 1-Octen-3-one | 2.1 | <1 |
| 6 | Phenylethyl acetate | 320 | 2 |
| 7 | Dimethyltrisulphide | 10.1 | 4 |
| 8 | Methanethiol | 260 | 4333 |
| 9 | Dimethylsulphide | 250 | 208 |
| 10 | Acetaldehyde | 25 | 114 |
| 11 | Methylen-bis(methysulphide) | 360 | 12 |
| 12 | 2-Acetyl-1-pyrroline | 3.2 | 32 |
| 13 | 2-Phenylethanol | 130 | <1 |
| 14 | Hexanal | 124 | 1 |
| 15 | δ-Decalactone | 1080 | 3 |

[a] Concentration levels are related to fresh weight.
[b] OAV: concentration/odour threshold in oil.

**Camembert**

The flavour profile of camembert was simulated on basis of the odorants presented in Table 6.33 and by using important taste compounds [59-61]. As in emmental and gruyere, the butter-like aroma note is perceivable but its intensity is lower than in unripened cheese as odorants formed during ripening become evident.

Among the odorants listed in Table 6.33, 1-octen-3-ol (no. 4), the character impact aroma compound of mushrooms *[62]*, is also responsible for the characteristic mushroom-like note of camembert, which is intensified by 1-octen-3-one (no. 5). Although the concentration of this ketone is much lower than that of the alcohol, it can be aroma-active in cheese because its odour threshold is 100 times lower than that of the alcohol *[60]*. Methanethiol, methional, dimethylsulphide, dimethyltrisulphide and methylene-bis(methylsulphide) generate the sulphurous odour note, whereas phenylethyl acetate is responsible for the floral odour note *[61]*.

### 6.2.4.3.5 Edible Fats and Oils

**Butter**

The three compounds presented in Table 6.34 are the key odorants of butter *[63]*. A comparison of the odour profiles of five samples of butter (Table 6.35) with the results of quantitative analysis (Table 6.34) show that the concentrations of these three odorants, which were found in samples 1 and 2, produce an intensive butter aroma. In samples 3 and 5, the concentration of 2,3-butanedione is too low and, therefore, the buttery odour quality is weak. In sample 4, the excessively high butyric acid concentration stimulates a rancid off-flavour.

*Table 6.34: Key odorants of five samples of butter [63]*

| Compound | Sample 1 | | Sample 2 | | Sample 3 | | Sample 4 | | Sample 5 | |
|---|---|---|---|---|---|---|---|---|---|---|
| | C[a] | OAV[b] | C | OAV | C | OAV | C | OAV | C | OAV |
| 2,3-Butandione | 0.62 | 138 | 0.34 | 76 | 0.11 | 24 | 0.32 | 71 | <0.01 | <2 |
| (R)-δ-Decalactone | 5.0 | 42 | 4.91 | 41 | 3.06 | 26 | 2.15 | 18 | 3.8 | 32 |
| Butyric acid | 4.48 | 33 | 3.63 | 27 | 2.66 | 20 | 94.5 | 704 | 2.58 | 19 |

[a] C: concentration in mg/kg.
[b] OAV: concentration/nasal odour threshold in sunflower oil.

*Table 6.35: Odour profile of butter samples [63]*

| No. | Sample | Odour | |
|---|---|---|---|
| | | Quality | Intensity[a] |
| 1 | Sour cream | Buttery, creamy, sweet | 3 |
| 2 | Sour cream | Buttery, creamy | 2-3 |
| 3 | Sour cream | Slightly buttery, mild sour | 1-2 |
| 4 | Sour cream | Rancid-like, butyric acid | 3 |
| 5 | Sweet cream | Slightly sour, mild | 1 |

[a] Evaluation: 1, weak; 2, medium; 3, strong.

## Olive Oil

The key odorants of two olive oils, I and S, with very different odour profiles (Table 6.36) are listed in Table 6.37. The apple-like and green odour qualities, which are characteristic for oil I, are caused by aldehydes nos. 5 to 7 and 15 (Table 6.37). Their concentration is higher in oil I than in S. The very potent odorant no. 10 occurred only in oil S. As a result of its high OAV and its blackcurrant-like odour quality, it is the character impact odorant of this oil [64, 65].

***Table 6.36:*** *Odour profiles of virgin olive oils and their aroma models* [a] *[64]*

| Odour attribute[b] | Oil I | Model I | Oil S | Model S |
|---|---|---|---|---|
| Pungent | 2.1 | 2.1 | 1.5 | 1.7 |
| Fruity | 1.7 | 2.0 | 2.2 | 1.8 |
| Green, leaf-like | 1.6 | 1.4 | 1.2 | 1.2 |
| Fatty | 1.4 | 1.3 | 1.2 | 1.3 |
| Apple-like | 2.4 | 2.5 | 0 | 0 |
| Black currant-like | 0 | 0 | 2.8 | 2.9 |
| Similarity[c] |  | 2.6 |  | 2.7 |

[a] The oil samples originated from Italy (I) and Spain (S).
[b] The intensity of the attributes was nasally scored on the scale 0 (absent) to 3 (strong).
[c] The similarity of the overall odour was evaluated on the scale 0 (no similarity) to 3 (identical with the original).

Aroma models prepared on the basis of the quantitative data shown in Table 6.37 agreed very well with the original oil samples (Table 6.36). The similarity scores amounted to 2.6 (oil I) and 2.7 (oil S), respectively. In these experiments [64] an odourless plant oil was used as the solvent for the odorants. Reduction of the aroma model for oil I to only seven odorants (nos. 3, 6, 7, 12, 16, 17, 19) lowered the similarity score to 2.2 but the characteristic overall odour remained was still preserved. In the case of oil S, a mixture containing only odorants nos. 1, 2, 8, 9 and 10 did not differ in the aroma from that of the complete aroma model. This result indicates that the other compounds quantified in oil S (Table 6.37) are not important for the aroma.

**Table 6.37:** Key odorants of two olive oils with different odour profiles[a] [64, 65]

| No. | Compound | Concentration[b] | | OAV[c] | |
|---|---|---|---|---|---|
| | | I | S | I | S |
| 1 | 3-Methylbutanal | 62 | 102 | 12 | 19 |
| 2 | 2-Methylbutanal | n.a. | 70 | – | 32 |
| 3 | 1-Penten-3-one | 26 | n.a. | 36 | – |
| 4 | Ethyl isobutyrate | 1.4 | 4.2 | 1.1 | 3.4 |
| 5 | Hexanal | 1770 | 137 | 5.9 | <1 |
| 6 | (Z)-3-Hexenal | 36 | n.a. | 21 | – |
| 7 | (E)-2-Hexenal | 6770 | n.a. | 16 | – |
| 8 | Ethyl 2-methylbutyrate | 2.1 | 14 | 8.1 | 55 |
| 9 | Ethyl 3-methylbutyrate | n.a. | 5.3 | – | 8.5 |
| 10 | 4-Methoxy-2-methyl-2-butanethiol | n.a. | 4.3 | – | 253 |
| 11 | Octanal | 382 | 99 | 7 | 2 |
| 12 | (Z)-3-Hexenyl acetate | 2250 | n.a. | 11 | – |
| 13 | Guaiacol | 28 | 7.5 | 1.8 | <1 |
| 14 | Ethyl cyclohexylcarboxylate | 0.54 | 0.36 | 3.4 | 2.2 |
| 15 | (Z)-2-Nonenal | 28 | 1.3 | 6.3 | <1 |
| 16 | (E,Z)-2,4-Decadienal | 255 | 9.7 | 26 | <1 |
| 17 | (E,E)-2,4-Decadienal | 422 | 127 | <1 | 5.1 |
| 18 | trans-4,5-Epoxy-(E)-2-decenal | 32 | 22 | 17 | 15 |
| 19 | Acetic acid | 6830 | 1840 | 55 | 15 |
| 20 | 2-/3-Methylbutyric acid | 81 | 32 | 3.7 | 1.5 |
| 21 | Acetaldehyde | 587 | 410 | 2668 | 1864 |
| 22 | Propanal | 409 | 75 | 44 | 8 |
| 23 | 2-Phenylethanol | 843 | – | 4 | – |

[a] The oil samples originated from Italy (I) and Spain (S).
[b] Values in µg/kg.
[c] OAV: concentration/nasal odour threshold in oil.
  n.a.: not analysed.

#### 6.2.4.3.6 Fruits

**Orange**

The aroma of the most important citrus fruit, the orange, has been analysed in detail *[66, 67]*. The potent odorants identified in the freshly pressed juice of the cultivar Valencia late are shown in Table 6.38. On the basis of high OAV it is expected that esters nos. 12-14 as well as (Z)-3-hexenal (no. 3), acetaldehyde (no. 1) and (R)-limonene are especially important for the aroma of orange juice.

*Table 6.38:* Key odorants of juice of Valencia late oranges [67]

| No. | Compound | Concentration[a] | OAV[b] |
|---|---|---|---|
| 1 | Acetaldehyde | 8305 | 332 |
| 2 | Hexanal | 197 | 19 |
| 3 | (Z)-3-Hexenal | 187 | 747 |
| 4 | Octanal | 25 | 3 |
| 5 | 1-Octen-3-one | 4.1 | 4 |
| 6 | Nonanal | 13 | 3 |
| 7 | Methional | 0.4 | <1 |
| 8 | Decanal | 45 | 9 |
| 9 | (E)-2-Nonenal | 0.6 | <1 |
| 10 | (E,E)-2,4-Decadienal | 1.2 | 6 |
| 11 | trans-4,5-Epoxy-(E)-2-decenal | 4.3 | 36 |
| 12 | Ethyl 2-methylpropanoate | 8.8 | 440 |
| 13 | Ethyl butanoate | 1192 | 1192 |
| 14 | Ethyl (S)-2-methylbutanoate | 48 | 8000 |
| 15 | Ethyl hexanoate | 63 | 13 |
| 16 | Ethyl (R)-3-hydroxyhexanoate | 1136 | 4 |
| 17 | (R)-α-Pinene | 308 | 62 |
| 18 | Myrcene | 594 | 42 |
| 19 | (R)-Limonene | 85598 | 228 |
| 20 | 2-Methylbutanol | 270 | <1 |
| 21 | 3-Methylbutanol | 639 | <1 |
| 22 | (S)-Linalool | 81 | 13 |
| 23 | Vanillin | 67 | 3 |

[a] Concentration: μg/kg fresh weight.
[b] OAV: concentration/odour threshold in water.

An aroma model was very similar to the aroma of the juice when it contained 0.1% fat in addition to the volatiles listed in Table 6.38 *[67]*. The terpene-like odour quality was decreased and the fruity note was enhanced. Omission experiments (Table 6.39) revealed that only the absence of acetaldehyde (no. 1) and (R)-limonene (no. 19) was detectable with high significance (a = 0.1) by either nasal or retronasal evaluation. Retronasally an aroma difference between the complete model and the model, in which one odorant was lacking, was detectable when (Z)-3-hexenal (no. 3), 1-octen-3-one (no. 5) or trans-4,5-epoxy-(E)-2-decenal (no. 11) were omitted (Table 6.39).

*Table 6.39:* Odour differences between the complete aroma model and models containing various compounds less [67]

| Odorant(s) omitted[a] | Significance $\alpha^b$ (%) | |
|---|---|---|
| | Orthonasal | Retronasall |
| Acetaldehyde | 1 | 0.1 |
| (Z)-3-Hexenal | n.d. | 1 |
| 1-Octen-3-one | n.d. | 5 |
| Decanal | 5 | n.d. |
| trans-4,5-Epoxy-(E)-2-decenal | n.d. | 1 |
| (R)-Limonene | 0.1 | 1 |
| Terpene hydrocarbons | 0.1 | 0.1 |
| Esters | 0.1 | 0.1 |
| Aldehydes | 0.1 | 0.1 |
| Esters nos. 12, 14 and 15 | 5 | 5 |
| Myrcene, (R)-α-pinene | n.d. | n.d. |

[a] All odorants listed were present in the complete aroma model.
[b] Determination of the odour difference by triangle test and calculation of the significance α according to [69]; 0.1 highest significance; 5 lowest significance.
 n.d., not detectable.

The odour intensities of volatiles showing similar odour qualities are partially additive [68]. To substantiate such additive effects, three groups of odorants (terpene hydrocarbons, esters or aldehydes) were omitted from the aroma model for orange juice. For all groups, a significant difference from the complete model was observed (Table 6.39). Omission of esters nos. 12, 14 and 15 with ethyl butanoate (no. 13) still present was clearly detectable. This indicates that the fruity quality in the odour profile is enhanced by additive effects. In contrast, no difference was perceivable when (R)-α-pinene (no. 17) and myrcene (no. 18) were omitted. The concentration of the odorants in juice differs depending on the variety. Thus, the weaker citrus note of Navel oranges compared with the above discussed variety Valencia late is due to a 70% lower content of (R)-limonene [67].

Mandarin oranges, e.g. clementines, differ from oranges in the occurrence of the sinensals, ethyl cinnnamate, ethylphenyl acetate, 3-sec-butyl-2-methoxyprazine and dodecanal as intense odorants [70]. On the other hand, many odorants detected in orange juice were also found in mandarins but differed in their concentration. Especially, the concentration of ethyl butanoate, ethyl 2-methylbutanoate and limonene, but also of α-pinene was high in orange juice compared to clementines.

## Grapefruit

Analysis of grapefruit juice *[71]* gave high OAV for 4-mercapto-4-methylpentan-2-one (MMP), ethyl (S)-2-methylbutanoate, (Z)-3-hexenal, ethyl butanoate, decanal, ethyl methylpropanoate and 1-p-menthen-8-thiol (MTO, probably the R-enantiomer). The aroma of a solution containing 19 odorants in the levels found in grapefruits was very similar to the natural aroma of this fruit *[71]*.

The concentrations of the two sulphur compounds, MMP and MTO, in grapefruit juices were 0.8 and 0.01 µg/kg, respectively *[71]*. Omission experiments indicate that the grapefruit-like odour was lacking when MMP was absent. MTO, which occurs in even lower concentration in oranges, contributes to the aroma but its impact is not so typical when compared to that of MMP.

*Table 6.40: Odorants of the apple varieties Elstar and Cox Orange [72]*

| No. | Compound | Elstar | | Cox Orange | |
|---|---|---|---|---|---|
| | | C[a] | OAV[b] | C | OAV |
| 1 | (E)-β-Damascenone | 1.4 | 1813 | 0.99 | 1320 |
| 2 | Methyl 2-methylbutanoate | 0.2 | <1 | 1.8 | 7 |
| 3 | Ethyl butanoate | 0.7 | 7 | 0.3 | 3 |
| 4 | Hexyl acetate | 5595 | 112 | 1500 | 30 |
| 5 | Butyl acetate | 4640 | 93 | 1595 | 32 |
| 6 | 2-Methylbutyl acetate | 240 | 48 | 217 | 43 |
| 7 | Hexanal | 85 | 19 | 48 | 11 |
| 8 | (E)-2-Hexenal | 77 | 2 | 114 | 2 |
| 9 | (Z)-3-Hexenal | 6.4 | 25 | 30 | 120 |
| 10 | (Z)-2-Nonenal | 8.8 | 440 | 1.1 | 55 |
| 11 | Butanol | 4860 | 10 | 975 | 2 |
| 12 | Hexanol | 1390 | 3 | 350 | <1 |
| 13 | Linalool | 9.3 | 19 | 4.5 | 9 |

[a] C: concentration in µg/ kg fresh weight.
[b] OAV: concentration/nasal odour threshold in water.

## Apple

The potent odorants identified in two apple varieties with fruity/green (Elstar) and fruity/sweet/aromatic (Cox Orange) odour are shown in Table 6.40. The fruity note in the aroma profile of both varieties is produced by acetic acid esters nos. 4-6. Ethyl esters, which on molar basis are more aroma-active than acetates *[72, 73]* and which predominate in some other fruits, e.g. olives (Table 6.37) and orange (Table 6.38), are of minor importance for the aroma of apples. Hexanal (no. 7), (Z)-3-hexenal (no. 9) and (Z)-2-nonenal (no. 10) are responsible for the green/apple-like note. (E)-β-damascenone (no. 1), which smells like cooked apples, has the highest OAV in both varieties due to its very low odour threshold. However, the odour impact of no. 1 might be strongly reduced by the interaction with other odorants *[74]*. Eugenol and (E)-anethol contribute to the aniseed-like note which is characteristic especially for the aroma of Cox Orange *[72]*.

**Strawberries**

The concentrations and OAV of odorants detected in strawberry juice are given in Table 6.41. To confirm that these compounds are actually the key aroma compounds, they were dissolved in a model juice matrix of pH 3.5 *[75]*. Sensory evaluations showed that the intensities of the fruity, sweet and green odour notes of the model were nearly as high as those of the natural juice. Only the fresh strawberry note was somewhat lower in the model.

*Table 6.41: Odorants of fresh strawberry juice [75]*

| No. | Compound | Concentration (µg/kg) | OAV[a] | N[b] |
|---|---|---|---|---|
| 1 | 4-Hydroxy-2,5-dimethyl-3(2H)-furanone | 16239 | 1624 | 6 |
| 2 | (Z)-3-Hexenal | 333 | 1332 | 5 |
| 3 | Methylbutanoate | 4957 | 991 | 4 |
| 4 | Ethyl butanoate | 410 | 410 | 4 |
| 5 | Ethyl 2-/3-methylbutanoate | 7 | 70 | 4 |
| 6 | Methyl 2-/3-methylbutanoate | 48 | 192 | 3 |
| 7 | Acetic acid | 74513 | 1 | 3 |
| 8 | 2,3-Butanedione | 1292 | 431 | 3 |
| 9 | Butanoic acid | 1831 | <1 | 2 |
| 10 | 2-/3-Methylbutanoic acid | 2237 | 4 | 2 |
| 11 | Ethyl methylpropanoate | 43 | 430 | 2 |
| 12 | 4-Methoxy-2,5-dimethyl-3(2H)-furanone | 20 | <1 | 1 |

[a] OAV: concentration/odour threshold in water at pH 3.5.
[b] In triangle tests the aroma model containing the 12 odorants was compared by six trained panellists with models lacking in one. N: number of panellists detecting an odour difference.

Omission of furaneol (no. 1 in Table 6.41) led to a very significant aroma defect which was perceived by all six members of the panel. The model smelt green and fruity. When (Z)-3-hexenal (no. 2) was missing, the caramel-like/sweetish note of furaneol predominated *[75]*. In contrast, a lack of odorants nos. 9-12 was only detected by one or two panellists, respectively, indicating a lower aroma impact of these substances.

### 6.2.4.3.7 Vegetables

**Potatoes, French Fries**

The volatiles that have been established in dilution experiments *[76]* as important for the aroma of boiled potatoes are listed in Table 6.42. The potato aroma note can be reproduced with an aqueous solution (pH 6) of pyrazines nos. 4 and 5, methional (no. 2), methanethiol (no. 11) and dimethylsulphide (no. 12) in the concentrations given in Table 6.42 *[77]*. Despite its smell of boiled potatoes, methional (no. 11) is not important for the aroma (see below). In the drying process applied for producing a granulate, the concentration of the pyrazines nos. 4 and 5 decrease, and therefore, the intensity of the potato note also decreases *[77]*.

*Table 6.42:* Odorants of boiled potatoes [76, 77]

| No. | Compound | Concentration[a] | OAV[b] |
|---|---|---|---|
| 1 | Hexanal | 102 | 3 |
| 2 | Methional | 65 | 241 |
| 3 | Dimethyltrisulphide | 1.0 | 12 |
| 4 | 2,3-Diethyl-5-methylpyrazine | 0.17 | 1 |
| 5 | 3-Isobutyl-2-methoxypyrazine | 0.07 | 14[c] |
| 6 | (E,E)-2,4-Decadienal | 7.3 | 3 |
| 7 | trans-4,5-Epoxy-(E)-2-decenal | 58 | 305 |
| 8 | 4-Hydroxy-2,5-dimethyl-3(2H)-furanone | 67 | 5 |
| 9 | 3-Hydroxy-4,5-dimethyl-2(5H)-furanone | 2.2 | 1 |
| 10 | Vanillin | 1001 | 218 |
| 11 | Methanethiol | 15.4 | 77[c] |
| 12 | Dimethylsulphide | 8.8 | 29[c] |

[a] Concentration in µg/kg fresh weight.
[b] OAV: concentration/odour threshold in starch.
[c] OAV on the basis of the odour threshold in water.

The volatiles detailed in Table 6.43 were identified in French fries prepared in palm oil *[78, 79]*. The flavour profile of a solution of these compounds in odourless sunflower oil was close to that of the original (Table 6.44). The greatest intensity differences were found for the caramel and malty notes when the model was retronasally examined. Variation of the composition of the aroma model indicated the odorants that are essential for the characteristic aroma *[79]*. An extract of the results is shown in Table 6.45.

*Table 6.43: Odorants of French fries prepared in palm oil [79]*

| No. | Compound | Concentration[a] | OAV[b] |
|---|---|---|---|
| 1 | 2-Ethyl-3,5-dimethylpyrazine | 41.9 | 19 |
| 2 | 3-Ethyl-2,5-dimethylpyrazine | 592 | 10 |
| 3 | 2,3-Diethyl-5-methylpyrazine | 41.4 | 83 |
| 4 | 2-Ethenyl-3-ethyl-5-methylpyrazine | 5.4 | 11 |
| 5 | 3-Isobutyl-2-methoxypyrazine | 8.6 | 11 |
| 6 | 1-Octen-3-one | 3.9 | <1 |
| 7 | (Z)-2-Nonenal | 15.7 | 3 |
| 8 | (E)-2-Nonenal | 138 | <1 |
| 9 | (E,Z)-2,4-Decadienal | 1533 | 383 |
| 10 | (E,E)-2,4-Decadienal | 6340 | 35 |
| 11 | trans-4,5-Epoxy-(E)-2-decenal | 771 | 592 |
| 12 | 4-Hydroxy-2,5-dimethyl-3(2H)-furanone | 2778 | 111 |
| 13 | 3-Hydroxy-4,5-dimethyl-2(5H)-furanone | 5.2 | 26 |
| 14 | Methylpropanal | 5912 | 1739 |
| 15 | 2-Methylbutanal | 10599 | 1059 |
| 16 | 3-Methylbutanal | 2716 | 503 |
| 17 | 2,3-Butandione | 306 | 31 |
| 18 | Methional | 783 | 3915 |
| 19 | Methanethiol | 1240 | 20667 |

[a] Concentration in µg/kg French fries.
[b] OAV: concentration/odour threshold in oil.

*Table 6.44: Flavour profiles of French fries (FF) and the corresponding aroma model [79]*

| Attribute | Intensity[a] | | | |
|---|---|---|---|---|
| | Nasal | | Retronasal | |
| | FF | Model | FF | Model |
| Earthy | 1.8 | 1.9 | 1.8 | 1.7 |
| Deep fried, fatty | 2.7 | 2.5 | 2.4 | 2.6 |
| Boiled potato | 1.6 | 1.7 | 2.3 | 2.0 |
| Caramel | 1.5 | 1.3 | 0.8 | 1.2 |
| Malty | 1.6 | 1.4 | 1.6 | 1.2 |

[a] The intensity of the attributes was nasally and retronasally scored on the scale 0 (absent) to 3 (strong).

**Table 6.45:** *Odour of the aroma model for French fries as affected by the absence of one or more odorants [79]*

| Exp. | Odorant omitted in the model[a] | Number[b] | Intensity of the odour difference[c] |
|---|---|---|---|
| 1 | Methanethiol | 5 | 2.5 |
| 2 | (E,Z)-2,4-decadienal and (E,E)-2,4-decadienal | 5 | 2.5 |
| 3 | 2-Ethyl-3,5-dimethylpyrazine and 3-ethyl-2,5-dimethylpyrazine | 4 | 1.5 |
| 4 | 2,3-Diethyl-5-methylpyrazine and 2-ethenyl-3-ethyl-5-methylpyrazine | 0 | |
| 5 | Methylpropanal, 2-methylbutanal and 3-methylbutanal | 4 | 0.5 |
| 6 | trans-4,5-Epoxy-(E)-2-decenal | 4 | 0.5 |
| 7 | 1-Octen-3-one, (Z)-2-nonenal and (E)-2-nonenal | 1 | 0.5 |
| 8 | Methional | 0 | |

[a] Models lacking in one or more components were each compared to the model containing the complete set of 19 odorants listed in Table 6.43.
[b] Number of five assessors detecting an odour difference in triangle test.
[c] Odour difference between complete and reduced models; rating scale: 0 (no deviation) to 3 (strong deviation).

In exp. 1, the model without methanethiol differed clearly from the complete model indicating that this thiol is indispensable for the aroma of French fries. In exp. 2, the judges agreed that the decadienal stereoisomers had a great impact on the aroma and were responsible for the deep-fried, fatty odour note. Of the four pyrazines, the absence of 2-ethyl-3,5-dimethylpyrazine and 3-ethyl-2,5-dimethylpyrazine was clearly recognised in exp. 3, but the absence of the two pyrazines, 2,3-diethyl-5-methyl- and 2-ethenyl-3-ethyl-5-methylpyrazine, was not noted in exp. 4. 2,3-Diethyl-5-methylpyrazine, therefore, did not play a role in the aroma of French fries, although its OAV was much higher than the OAV of the other three pyrazines (Table 6.43). Absence of the three Strecker aldehydes and of 4,5-epoxy-(E)-2-decenal changed the aroma of the model in exps. 5 and 6, respectively, but the deviation in the odour intensity and quality was only small. In accordance with their low OAV, the carbonyl compounds which were lacking in exp. 7, did not significantly contribute to the aroma. The result of exp. 8 was very surprising. Although methional was the odorant showing the second highest OAV of all of the volatiles occurring in French fries (Table 6.45), the aroma of the model in which this odorant was absent did not differ from that of the complete model. This confirms the observation reported above that methional, despite its smell of boiled potatoes, is not important for this odour quality in processed potatoes.

## Tomatoes

The key odorants of fresh tomatoes and tomato paste are compared in Table 6.46. (Z)-3-hexenal (no. 1), β-damascenone (no. 2), epoxy-decenal isomers nos. 3 and 5, β-ionone (no. 4), 1-octen-3-one (no. 5), 2,4-decadienal isomers nos. 11 and 12 as well as 1-penten-3-one (no. 13) are the most important odorants of fresh tomatoes *[80]*.

***Table 6.46:*** *Odorants of fresh tomatoes (I) and tomato paste (II) [24, 80, 82]*

| No. | Compound | I | | II | |
|---|---|---|---|---|---|
| | | $C^a$ | $OAV^b$ | $C^c$ | $OAV^b$ |
| 1 | (Z)-3-Hexenal | 9700 | 3.9E+4 | <1 | – |
| 2 | (E)-β-Damascenone | 3 | 1500 | 5.7 | 2850 |
| 3 | trans-4,5-Epoxy-(E)-2-decenal | 610 | 5080 | n.a. | – |
| 4 | β-Ionone | 15 | 2140 | n.a. | – |
| 5 | 1-Octen-3-one | 5 | 1000 | n.a. | – |
| 6 | cis-4,5-Epoxy-(E)-2-decenal | 90 | 750$^d$ | n.a. | – |
| 7 | Hexanal | 1840 | 409 | n.a. | – |
| 8 | 3-Methylbutanal | 38 | 190 | 38 | 190 |
| 9 | Methional | 1 | 5 | 26 | 130 |
| 10 | Phenylacetaldehyde | 120 | 30 | n.a. | – |
| 11 | (E,Z)-2,4-Decadienal | 38 | 540$^d$ | n.a. | – |
| 12 | (E,E)-2,4-Decadienal | 23 | 330 | n.a. | – |
| 13 | 1-Penten-3-one | 600 | 600 | n.a. | – |
| 14 | 2-Phenylethanol | 300 | 7 | 1280 | 28 |
| 15 | Dimethylsulphide | n.a. | – | 1390 | 1390 |
| 16 | Linalool | n.a. | – | 5.3 | 3.5 |
| 17 | 2-/3-Methylbutyric acid | n.a. | – | 1840 | 2.5 |
| 18 | 4-Hydroxy-2,5-dimethyl-3(2H)-furanone | 220 | 22 | 690 | 69 |
| 19 | 5-Ethyl-4-hydroxy-2-methyl-3(2H)-furanone | n.a. | – | 31 | 6 |
| 20 | 3-Hydroxy-4,5-dimethyl-2(5H)-furanone | n.a. | – | 17 | 215 |
| 21 | Eugenol | <1 | – | 95 | 95 |
| 22 | Vinylguaiacol | n.a. | – | 80 | 16 |
| 23 | Vanillin | n.a. | – | 240 | 8 |
| 24 | Methylpropanal | n.a. | – | 28 | 40 |
| 25 | Acetic acid | n.a. | – | 218000 | 10 |

[a] The concentration C (μg/kg fresh weight) was quantified with a conventional method, the accuracy of which had been examined *[80]*.
[b] OAV = Concentration/odour threshold in water.
[c] The concentration C (μg/kg paste) was quantified using SIDA (6.2.4.2).
[d] Value based on the threshold of the other isomer.

The aroma is changed during the production of tomato paste *[81, 82]*. This is mainly caused by losses of (Z)-3-hexenal (no. 1) and the odorants nos. 3-7 and nos. 10-13. Additionally, dimethylsulphide (no. 15) is formed and the concentration of β-damascenone (no. 2), methional (no. 9) and furaneol (no. 18) increased (Table 6.46).

Two aroma reconstitution experiments have been performed for tomato paste. In the first experiment *[81]* a synthetic mixture containing the odorants nos. 2, 8, 9, 15, 17 (3-isomer), 21 and 1-nitro-2-phenylethane was used. The concentration of the first six odorants was in the range shown in Table 6.46. In the second experiment *[82]* the aroma model contained the 15 odorants in the concentration equal to those reported in Table 6.46. A comparison indicated that the aroma of this mixture was very close to the aroma of the original. Obviously, 1-nitro-2-phenylethane does not contribute to the aroma of tomato paste.

### 6.2.4.3.8 Spices

**Pepper**

Pepper is commercially available as black and white pepper. While black pepper is harvested before full maturity is reached and then dried, white pepper is the kernel of the ripe fruit after removal of the pulp. The flavour of white pepper is milder than that of the black pepper.

The odorants of black pepper are listed in Table 6.47. Their OAVs are calculated on the basis of odour threshold values in starch because at app. 45% it is the major component of the kernel *[83]*. As a result of its low odour threshold, linalool (no. 11) shows the highest OAV, followed by limonenes nos. 9a and 9b, (-)-β-pinene (no. 6a) and (S)-α-phellandrene (no. 8b).

The aroma of black pepper is approached with a synthetic mixture of the odorants listed in Table 6.47 *[83]*. Omission tests indicated α- and β-pinene (nos. 4 and 6), myrcene (no. 7), (S)-α-phellandrene (no. 8b), limonene (no. 9), linalool (no. 11), butyric acid (no. 12), 3-methylbutyric acid (no. 14) and the Strecker aldehydes nos. 1-3 as key odorants.

White pepper contains the same typical odorants as black pepper, but in lower concentrations, e.g. the monoterpene fraction amounts only to 40-60% *[83]*.

**Bell Pepper**

The odorants of dry sweet bell pepper powder have been analysed in detail *[84, 85]*. Omission experiments established ethyl methylpropanoate, methyl 2- and 3-methylbutanoate, ethyl 2- and 3-methylbutanoate, 3-isopropyl-, 3-isobutyl- and 3-sec-butyl-2-methoxypyrazine as character impact odorants.

**Dill**

AEDA and sensory studies show that (S)-α-phellandrene in combination with dill ether generate the aroma of dill herb *[86, 87]*. An essential odorant of dill fruits is (S)-carvone, which smells like caraway *[86]*.

*Table 6.47:* Odorants of black pepper [83]

| No. | Compound | Concentration[a] | OAV[b] |
|---|---|---|---|
| 1 | Methylpropanal[c] | 1.03 | 18 |
| 2 | 2-Methylbutanal[c] | 1.99 | 37 |
| 3 | 3-Methylbutanal[c] | 4.18 | 130 |
| 4a | (-)-α-Pinene[c] | 2070 | 610 |
| 4b | (+)-α-Pinene[c] | 486 | 230 |
| 5a | (-)-Sabinene[c] | 4470 | 89 |
| 5b | (+)-Sabinene | 285 | 45 |
| 6a | (-)-ß-Pinene | 3950 | 1400 |
| 6b | (+)-ß-Pinene | 298 | 140 |
| 7 | Myrcene[c] | 870 | 460 |
| 8a | R-(-)-α-Phellandrene | 227 | 160 |
| 8b | S-(+)-α-Phellandrene[c] | 1390 | 1300 |
| 9a | S-(-)-Limonene[c] | 4000 | 1400 |
| 9b | R-(+)-Limonene[c] | 3280 | 1800 |
| 10 | 1,8-Cineol[c] | 22.4 | 270 |
| 11 | (±)-Linalool[c] | 231 | 3300 |
| 12 | Butyric acid[c] | 26.9 | 270 |
| 13 | 2-Methylbutyric acid | 1.28 | 53 |
| 14 | 3-Methylbutyric acid[c] | 2.99 | 540 |

[a] Concentration: mg/kg.
[b] OAV: concentration/odour threshold in starch.
[c] Compound present in the aroma model for black pepper [83].

## Parsley

The most important odorants of parsley leaves are listed in Table 6.48. An aroma model formulated on the basis of the quantitative data (Table 6.48) was described as clearly parsley-like [88]. Differences between the odour profile of the model and that of parsley leaves were observed for the spicy and the green-grassy notes which were stronger and weaker, respectively, in the model. The parsley-like character of the aroma model was completely lost when p-mentha-1,3,8-triene (no. 1 in Table 6.48) and myrcene (no. 2) were omitted [88].

*Table 6.48: Odorants of parsley leaves [88]*

| No. | Compound | Concentration[a] | OAV[b] |
|---|---|---|---|
| 1 | p-Mentha-1,3,8-triene | 393.3 | 26219 |
| 2 | Myrcene | 18.0 | 1284 |
| 3 | 2-Isopropyl-3-methoxypyrazine | 0.0014 | 350 |
| 4 | 2-sec-Butyl-3-methoxypyrazine | 0.0077 | 2567 |
| 5 | Myristicin | 57.8 | 1927 |
| 6 | 1-Octen-3-one | 0.003 | 60 |
| 7 | (Z)-1,5-Octadien-3-one | 0.001 | 833 |
| 8 | Linalool | 0.694 | 116 |
| 9 | (E,E)-2,4-Decadienal | 1.058 | 5290 |
| 10 | (Z)-6-Decenal | 5.898 | 17347 |
| 11 | Methanethiol[c] | 0.26 | 1300 |
| 12 | (Z)-3-Hexenal | 0.2 | 800 |
| 13 | p-Methylacetophenone[c] | 0.624 | 26 |
| 14 | (Z)-3-Hexenyl acetate | 0.164 | 21 |
| 15 | (Z)-3-Hexenol | 0.433 | 11 |
| 16 | ß-Phellandrene[c] | 204.2 | 5672 |
| 17 | 1-Isopropenyl-4-methylbenzene[c] | 15.9 | 187 |

[a] Concentration: mg/kg fresh weight.
[b] OAV: concentration/odour threshold in water.
[c] Compound lacking in the aroma model for parsley [88].

## Basil

Fresh basil has a strong green, flowery, clove- and pepper-like aroma [89] which is generated by the compounds given in Table 6.49. An aroma model prepared on the basis of these data was in good agreement in the odour profile with that of basil [89]. Omission experiments show that eugenol (no. 6), (Z)-3-hexenal (no. 1), α-pinene (no. 10), 4-mercapto-4-methylpentan-2-one (no. 3), linalool (no. 4) and 1,8-cineol (no. 2) make the largest contributions to the aroma.

*Table 6.49:* Odorants of fresh basil leaves [89]

| No. | Compound | Concentration[a] | OAV[b] |
|---|---|---|---|
| 1 | (Z)-3-Hexenal | 12.4 | 413000 |
| 2 | 1,8-Cineol | 64 | 246000 |
| 3 | 4-Mercapto-4-methylpentan-2-one | 0.01 | 83300 |
| 4 | Linalool | 60.2 | 40000 |
| 5 | 4-Allyl-1,2-dimethoxybenzene | 495 | 9900 |
| 6 | Eugenol | 89.0 | 8900 |
| 7 | Wine lactone | 0.0034 | 425 |
| 8 | Methyl cinnamate | 2.6 | 235 |
| 9 | Estragol | 1.2 | 165 |
| 10 | α-Pinene | 1.8 | 90 |
| 11 | Decanal | 0.039 | 6 |

[a] Concentration: mg/kg fresh weight.
[b] OAV: Concentration/odour threshold in water.

### 6.2.4.3.9 Roasted Seeds

**Hazelnuts**

5-Methyl-(E)-2-hepten-3-one (filbertone) has been identified as character impact odorant of roasted hazelnuts [90-92]. Its odour threshold is extremely low: 5 ng/kg (water as solvent).

In a model experiment, simulating the roasting process, the concentration of filbertone increased at 180°C from 1.4 to 600 µg/kg in 9 min and to 1150 µg/kg in 15 min. Oil from unroasted nuts contains less than 10 µg/kg [93].

**Popcorn**

The compounds listed in Table 6.50 cause the typical smell of popcorn. As a result of its very high OAV, acetyltetrahydropyridine (no. 1) is the character impact odorant followed by 2-acetyl-1-pyrroline (no. 2). The reverse was found for the roasty note of baguette crust [95]. The causes for this difference have been clarified by model studies [96].

*Table 6.50:* Odorants in fresh hot-air popped corn [94]

| No. | Compound | Concentration[a] | OAV[b] |
|---|---|---|---|
| 1 | 2-Acetyltetrahydropyridine | 437 | 7283 |
| 2 | 2-Acetyl-1-pyrroline | 24 | 1200 |
| 3 | 2-Propionyl-1-pyrroline | 17 | 850 |
| 4 | Acetylpyrazine | 8 | 20 |

[a] Concentration: µg/kg.
[b] OAV: concentration/odour threshold in water.

## Sesame

Roasting of the rather odourless sesame seeds generates an intense aroma characterised by roasty, burnt, meat-like or sulphurous odour notes. Several studies have been performed to identify the odorants responsible for these notes in roasted white and black sesame *[97-99]*. The results obtained for moderately roasted white sesame are presented in Table 6.51 as an example.

*Table 6.51:* Odorants of roasted white sesame [99]

| No. | Compound | Concentration[a] | OAV[b] |
|---|---|---|---|
| 1 | 2-Acetyl-1-pyrroline | 30 | 300 |
| 2 | 2-Furfurylthiol | 54 | 135 |
| 3 | 2-Phenylethylthiol | 6 | 120 |
| 4 | 4-Hydroxy-2,5-dimethyl-3(2*H*)-furanone | 2511 | 50 |
| 5 | 2-Ethyl-3,5-dimethylpyrazine | 53 | 18 |
| 6 | 2-Methoxyphenol | 269 | 14 |
| 7 | 2-Pentylpyridine | 19 | 4 |
| 8 | Acetylpyrazine | 26 | 3 |
| 9 | 4-Vinyl-2-methoxyphenol | 72 | 1 |
| 10 | (E,E)-2,4-Decadienal | 89 | <1 |

[a] Concentration: µg/kg.
[b] OAV: concentration/odour threshold in sunflower oil.

On the basis of high OAV, 2-acetyl-1-pyrroline (no. 1), 2-furfurylthiol (no. 2), 2-phenylethylthiol (no. 3) and furaneol (no. 4) are the most important contributors to the overall roasty, caramel-like aroma of the moderately roasted sesame. The two thiols nos. 2 and 3, but not the unstable 2-acetyl-1-pyrroline (no. 1), were also identified as key odorants of white and black sesame seeds which had been longer roasted and which elicited intense burnt or even rubbery odour notes *[97, 98]*.

### 6.2.4.3.10 Beverages

**Roasted Coffee (reviews in *[5, 100]*)**

The aroma of coffee is described as roasty/earthy/sulphurous, sweet/caramel-like and smoky/phenol-like *[101]*. It can be largely approximated with 28 odorants in the concentrations present in Table 6.52.

*Table 6.52:* Concentrations and OAV of potent odorants in ground, medium roasted Arabica coffee from Colombia: yields of odorants in the production of the beverage [101, 102]

| No. | Compound | Concentration (mg/kg) | OAV[a] | Yield[b] (%) |
|---|---|---|---|---|
| 1 | 3-Methyl-2-butenthiol | 0.0086 | 2.9 E+5 | 85 |
| 2 | 2-Methyl-3-furanthiol | 0.068 | 9.7 E+3 | 34 |
| 3 | 2-Furfurylthiol | 1.68 | 1.7 E+5 | 19 |
| 4 | Methional | 0.228 | 1.1 E+3 | 74 |
| 5 | 3-Mercapto-3-methylbutyl formate | 0.077 | 2.2 E+4 | 81 |
| 6 | Methanethiol | 4.70 | 2.4 E+4 | 72 |
| 7 | Dimethyltrisulphide | 0.028 | 2.8 E+3 | n.a. |
| 8 | 2-Ethyl-3,5-dimethylpyrazine | 0.249 | 1.6 E+3 | 79 |
| 9 | 2-Ethenyl-3,5-dimethylpyrazine | 0.052 | 5.2 E+2 | 35 |
| 10 | 2,3-Diethyl-5-methylpyrazine | 0.073 | 8.1 E+2 | 67 |
| 11 | 2-Ethenyl-3-ethyl-5-methylpyrazine | 0.018 | 1.8 E+2 | 25 |
| 12 | 2-Isobutyl-3-methoxypyrazine | 0.099 | 1.9 E+4 | 23 |
| 13 | 4-Hydroxy-2,5-dimethyl-3(*2H*)-furanone | 112 | 1.1 E+4 | 95 |
| 14 | 2-Ethyl-4-hydroxy-5-methyl-3(*2H*)-furanone | 16.8 | 1.4 E+4 | 93 |
| 15 | 3-Hydroxy-4,5-dimethyl-2(*5H*)-furanone | 1.36 | 6.8 E+1 | 78 |
| 16 | 5-Ethyl-3-hydroxy-4-methyl-2(*5H*)-furanone | 0.104 | 1.4 E+1 | n.a. |
| 17 | Guaiacol | 3.04 | 1.2 E+3 | 65 |
| 18 | 4-Ethylguaiacol | 1.42 | 2.8 E+1 | 49 |
| 19 | 4-Vinylguaiacol | 55.2 | 2.8 E+3 | 30 |
| 20 | Vanillin | 3.41 | 1.4 E+2 | 95 |
| 21 | (E)-β-Damascenone | 0.222 | 3.0 E+5 | 11 |
| 22 | Acetaldehyde | 139 | 1.4 E+4 | 73 |
| 23 | Propanal | 17.4 | 1.7 E+3 | n.a. |
| 24 | Methylpropanal | 32.3 | 4.6 E+4 | 59 |
| 25 | 2-Methylbutanal | 20.7 | 1.1 E+4 | 62 |
| 26 | 3-Methylbutanal | 18.6 | 4.7 E+4 | 62 |
| 27 | 2,3-Butanedione | 48.4 | 3.2 E+3 | 79 |
| 28 | 2,3-Pentanedione | 34.0 | 1.1 E+3 | 85 |

[a] OAV: concentration/odour threshold in water.
[b] Yield of the odorant in the preparation of the beverage (1 L) by percolation of coffee powder (54 g) with water (ca. 90°C).
n.a.: not analysed.

Omission experiments *[101]* confirmed earlier assumptions *[103]* that 2-furfurylthiol (no. 3) is the outstanding odorant among the aroma compounds of coffee. In addition, it was shown that 4-vinylguaiacol (no. 19), pyrazines nos. 8-10, furanones nos. 13-16 and carbonyl compounds nos. 22-26 had the greatest impact on the coffee aroma.

The aroma is changed when a brew is prepared from ground coffee. Caramel-like, buttery and phenolic notes become more intense. AEDA shows that this change in the aroma profile is caused by a shift in the concentrations *[102]*. As detailed in Table 6.52, the polar odorants are preferentially extracted by hot water leading to yields higher than 80% for S-compounds nos. 1 and 4, furanones nos. 13 and 14, vanillin (no. 20) and pentanedione (no. 28). On the other hand, the yield of the character impact odorant of ground coffee, 2-furfurylthiol, is with 19% relatively low.

## Wine

The odorants and the taste compounds of wines which are produced from the grape cultivars Gewürztraminer and Scheurebe have been identified *[104, 105]*. Synthetic mixtures containing the odorants listed in Table 6.53 and the taste compounds were prepared. A sensory study indicated that the aromas of these mixtures and the corresponding wine samples agreed very well *[105, 106]*. The two cis-rose oxide diastereomers (nos. 4 and 5) and 4-mercapto-4-methylpentan-2-one (no. 23) are the cultivar-specific odorants in Gewürztraminer and Scheurebe, respectively (Table 6.53). In addition, 3-methylbutyl acetate (no. 2), ethyl hexanoate (no. 3), ethyl oxtanoate (no. 6), geraniol (no. 8), ethyl isobutanoate (no. 12) and linalool (no. 14) exhibit high but different OAV in the two cultivars of wine.

Ethanol has a strong effect on the wine bouquet as the odour threshold of many volatiles increases in its presence, e.g. the threshold of ethyl 2- and 3-methylbutanoate by a factor of 100 *[106]*.

## Green Tea

The odorants of green tea, powder and brew, are detailed in Table 6.54. 3-Methyl-2,4-nonanedione (no. 3) is of special interest among the odorants showing high OAV. It smells green and strawy and was detected at first in soybean oil showing a light-induced off-flavour (Table 6.55). Precursors of no. 3 are furanoid fatty acids *[23, 107]*.

*Table 6.53:* Odorants of the wines Gewürztraminer and Scheurebe [104, 105]

| No. | Compound | Gewürztraminer | | Scheurebe | |
|---|---|---|---|---|---|
| | | $C^a$ | $OAV^b$ | $C^a$ | $OAV^b$ |
| 1 | Acetaldehyde | 1.86 | 4 | 1.97 | 4 |
| 2 | 3-Methylbutyl acetate | 2.9 | 97 | 1.45 | 48 |
| 3 | Ethyl hexanoate | 0.49 | 98 | 0.28 | 56 |
| 4 | (2S,4R)-Rose oxide | 0.015 | 105 | 0.003 | 15 |
| 5 | (2R,4S)-Rose oxide | 0.006 | | | |
| 6 | Ethyl octanoate | 0.63 | 315 | 0.27 | 135 |
| 7 | (E)-β-Damascenone | 0.00083 | 17 | 0.00098 | 20 |
| 8 | Geraniol | 0.221 | 7 | 0.038 | 1 |
| 9 | 3-Hydroxy-4,5-dimethyl-2(5H)-furanone | 0.0054 | 1 | 0.0033 | 1 |
| 10 | Wine lactone | 0.0001 | 10 | 0.0001 | 10 |
| 11 | Ethanol | 90000 | | 90000 | |
| 12 | Ethyl isobutanoate | 0.15 | 10 | 0.48 | 32 |
| 13 | Ethyl butanoate | 0.21 | 11 | 0.184 | 9 |
| 14 | Linalool | 0.175 | 12 | 0.307 | 20 |
| 15 | Ethyl acetate | 63.5 | 8 | 22.5 | 3 |
| 16 | 1,2-Diethoxyethane | 0.375 | 8 | n.a. | |
| 17 | 2,3-Butanedione | 0.15 | 2 | 0.18 | 2 |
| 18 | Ethyl 2-methylbutanoate | 0.0044 | 4 | 0.0045 | 5 |
| 19 | Ethyl 3-methylbutanoate | 0.0036 | 1 | 0.0027 | 1 |
| 20 | Methylpropanol | 52 | 1 | 108 | 3 |
| 21 | 3-Methylbutanol | 128 | 4 | 109 | 4 |
| 22 | Dimethyltrisulphide | 0.00025 | 1 | 0.00009 | <1 |
| 23 | 4-Mercapto-4-methylpentan-2-one | <0.00001 | <17 | 0.0004 | 667 |
| 24 | (3-Methylthio)-1-propanol | 1.415 | 3 | 1.040 | 2 |
| 25 | Hexanoic acid | 3.23 | 1 | 2.47 | 1 |
| 26 | 2-Phenylethanol | 18 | 2 | 21.6 | 2 |
| 27 | Ethyl trans-cinnamate | 0.002 | 2 | 0.023 | 2 |
| 28 | Eugenol | 0.0054 | 1 | 0.0005 | <1 |
| 29 | (Z)-6-Dodecenoic acid-γ-lactone | 0.00027 | 3 | 0.00014 | 1 |
| 30 | Sulphur dioxide | 7.3 | 2 | 30 | 10 |

[a] C: concentration (mg/kg).
[b] OAV: concentration/odour threshold in water/ethanol (9+1, ww).

*Table 6.54:* Odorants of green tea powder and brew [107]

| No. | Compound | Powder | | Brew[a] | |
|---|---|---|---|---|---|
| | | C[b] | OAV[c] | C[b] | OAV[d] |
| 1 | (Z)-1,5-Octadien-3-one | 1.8 | 36 | 0.012 | 30 |
| 2 | 3-Hydroxy-4,5-dimethyl-2(5H)-furanone | 49 | 38 | 0.6 | 8 |
| 3 | 3-Methyl-2,4-nonanedione | 83 | 55 | 0.56 | 28 |
| 4 | (Z)-4-Heptenal | 112 | 37 | 0.63 | 11 |
| 5 | (Z)-3-Hexenal | 101 | 51 | 0.28 | 9 |
| 6 | (E,Z)-2,6-Nonadienal | 61 | 14 | 0.48 | 24 |
| 7 | 1-Octen-3-one | 6 | 6 | 0.03 | 3 |
| 8 | (E,E)-2,4-Decadienal | 127 | 17 | 0.9 | 18 |
| 9 | (E)-β-Damascenone | 9 | 45 | 0.01 | 10 |

[a] Brew (1 kg) was obtained from the powder (10 g).
[b] C: concentration in μg/kg.
[c] OAV: concentration/odour threshold in cellulose.
[d] OAV: concentration/odour threshold in water.

A comparison of the concentration levels in Table 6.54 on the basis of equal amounts of dry matter indicates that in most cases only a portion of the odorants is extracted during preparation of the brew, e.g. 67% of octadien-3-one (no. 1) and dione no. 3, 56% of (Z)-4-heptenal (no. 4), 28% of (Z)-3-hexenal (no. 5) and 11% of β-damascenone (no. 9). However, sotolon (no. 2) is an exception. Its concentration increased during brewing of the green tea beverage.

### 6.2.4.3.11 Aroma Changes

Characterisation of aroma changes during processing or storage of foods is a further application of AEDA and SIDA [108]. Increased levels of odorants that negatively affect the flavour of food, the loss of odorants that have a positive impact on flavour, and the appearance of new odorants causing off-flavours are all revealed by a comparative AEDA of fresh and deteriorated samples. The examples listed in Table 6.55 show that several potent odorants that contribute to undesirable aroma changes have been identified by this method. These volatiles can be used as indicators for studying the development of the corresponding off-flavour.

*Table 6.55:* Aroma changes investigated by AEDA

| Food | Off-flavour/aroma change | Odorant(s) involved | Reference |
|---|---|---|---|
| Trout | Fatty, fishy | (Z)-3-Hexenal, (Z,Z)-2,6-nonadienal | *[109]* |
| Butter oil | Cardboard-like | (E)-2-Nonenal, (Z)-2-nonenal | *[110]* |
| Soybean oil | Beany-strawy | 3-Methyl-2,4-nonanedione, 2,3-butanedione | *[82, 111]* |
| White bread crust | Aroma staling | Loss of 2-acetyl-1-pyrroline | *[32]* |
| Meat, cooked and refrigerated | Warmed-over flavour | Loss of furaneol, formation of hexanal and epoxydecenal | *[44, 113]* |
| Parsley, dry | Hay-like | 3-Methyl-2,4-nonanedione | *[114]* |
| Pepper, black | Musty, mouldy | 2,3-Diethyl-5-methylpyrazine, 2-Isopropyl-3-methoxypyrazine | *[83]* |
| Pepper, white | Faecal smell | Skatole, p-cresol | *[83]* |
| Coffee brew | Aroma staling | Loss of odour-active thiols | *[115]* |
| Gruyere cheese | Potato-like | Increase of methional | *[57, 58]* |
| Beer | Aroma staling | Increase of esters, methional and phenylacetaldehyde | *[116]* |
| Buttermilk | Metallic | (E,Z)-2,6-Nonadienol | *[117]* |
| Milk powder | Musty, wet popcorn | 2-Aminoacetophenone | *[118]* |
| Orange juice | Changes during processing | Loss of (Z)-3-hexenal, acetaldehyde, increase of carvone | *[119]* |
| Mandarin orange | Changes during processing | Increase of 4-vinyl-2-methoxyphenol and furaneol | *[70]* |

## References

[1] Nijssen, B., van Ingen-Visscher, K., Donders, J. (2004) Volatile Compounds in Food. 8.1. Qualitative and Quantitative Data (VCF data base: www.voeding.tno.nl). TNO Nutrition and Food Research Institute, Zeist, The Netherlands

[2] Rothe, M., Thomas, B. (1963) Aroma substances of bread (in German). Z. Lebensm. Unters. Forsch. 119, 302-310

[3] Guadagni, D.G., Buttery, R.G., Harris, J. (1966) Odour intensities of components. J. Sci. Food Agric. 17, 142-144

[4] Acree, T.E., Barnard, J., Cunningham, D.G. (1984) A procedure for the sensory analysis of gas chromatographic effluents. Food Chem. 14, 273-286

[5] Grosch, W. (1998) Flavour of coffee. A review. Nahrung/Food 42, 344-350

[6] Fuller, G.H., Steltenkamp, G.A., Tisserand, G.A. (1964) The gas chromatography with human sensor: Perfumer model. Ann. NY Acad. Sci. 116, 711-724

[7] Dravnieks, A., O'Donnell, A. (1971) Principles and some techniques of high-resolution headspace analysis. J. Agric. Food Chem. 19, 1049-1056

[8] Acree, T.E. (1993) Bioassays for flavor. In Acree, T.E., Teranishi, R. (eds.) Flavor Science. Sensible Principles and Techniques. ACS Professional Reference Book. American Chemical Society, Washington DC, pp. 1-20

[9]   Blank, I. (1997) Gas chromatography-olfactometry in food aroma analysis. In Marsili, R. (ed.) Techniques for Analyzing Food Aroma. Marcel Dekker, New York, pp. 293-329
[10]  Grosch, W. (2001) Evaluation of the key odorants of foods by dilution experiments, aroma models and omission. Chem. Senses 26, 533-545
[11]  Grosch, W. (2004) Aroma compounds. In Nollett, L. (ed.) Handbook of Food Analysis, 2nd edition, Vol. I. Marcel Dekker, New York, pp. 717-746
[12]  Mistry, B.S., Reineccius, T., Olson, L.K. (1997) Gas chromatography-olfactometry for the determination of key odorants in foods. In Marsili, R. (ed.) Techniques for Analyzing Food Aroma. Marcel Dekker, New York, pp. 265-291
[13]  Schieberle, P. (1996) New developments on methods for analysis of volatile flavor compounds and their precursors. In Goankar, A.G. (ed.) Characterization of Food-emerging Methods. Elsevier, Amsterdam, pp. 403-433
[14]  Ullrich, F., Grosch, W. (1987) Identification of the most intense flavour compounds formed during autoxidation of linoleic acid. Z. Lebensm. Unters. Forsch. 184, 277-282
[15]  Grosch, W. (1993) Detection of potent odorants in foods by aroma extract dilution analysis. Trends Food Sci. Technol. 4, 68-73
[16]  Zehentbauer, G., Grosch, W. (1998) Crust aroma of baguettes. I. Key odorants of baguettes prepared in two different ways. J. Cereal Sci. 28, 81-92
[17]  McDaniel, M.R., Miranda-Lopez, R., Watson, B.T., Micheals, N.J., Libbey, L.M. (1990) Pinot noir aroma: a sensory/gas chromatographic approach. In Charalambous, G. (ed.) Flavours and Off-flavours. Elsevier Science, Amsterdam, pp. 23-25
[18]  Van Ruth, S.M., Roozen, J.P. (1994) Gas chromatography/sniffing port analysis and sensory evaluation of commercially dried bell peppers (*capsicum annuum*) after rehydration. Food Chem. 51, 165-170
[19]  LeGuen, S., Prost, C., Demaimay, M. (2000) Critical comparison of three olfactometric methods for the identification of the most potent odorants in cooked mussels (*mytilus edulis*). J. Agric. Food Chem. 48, 1307-1314
[20]  Kerscher, R., Grosch, W. (1997) Comparative evaluation of boiled beef aroma by aroma extract dilution and concentration analysis. Z. Lebensm. Unters. Forsch. 204, 3-6
[21]  Grosch, W., Kerscher, R., Kubickova, J., Jagella, T. (2001) Aroma extract dilution analysis versus aroma extract concentration analysis. In Leland, J.V., Schieberle, P., Buettner, A., Acree, T. (eds.) Gas Chromatography Olfactometry: The State of the Art. ACS Symposium Series 782, pp. 138-147
[22]  Holscher, W., Steinhart, H. (1992) Investigation of roasted coffee freshness with an improved headspace technique. Z. Lebensm. Unters. Forsch. 195, 33-38
[23]  Guth, H., Grosch, W. (1993) Identification of potent odorants in static headspace samples of green and black tea powders on the basis of aroma extract dilution analysis (AEDA). Flavour Fragrance J. 8, 173-178
[24]  Schieberle, P., Grosch, W. (1987) Quantitative analysis of aroma compounds in wheat and rye bread crusts using a stable isotope dilution assay. J. Agric. Food Chem. 35, 252-257
[25]  Blank, I., Sen, A., Grosch, W. (1992) Potent odorants of the roasted powder and brew of Arabica coffee. Z. Lebensm. Unters. Forsch. 195, 239-245
[26]  Weber, B., Maas, B., Mosandl, A. (1995) Stereoisomeric distribution of some chiral sulfur-containing trace components of yellow passion fruits. J. Agric. Food Chem. 42, 2438-2441
[27]  Blank, I., Milo, C., Lin, J., Fay, L.B. (1999) Quantification of aroma-impact components by isotope dilution assay – Recent developments. In Teranishi, R., Wick, E.L., Hornstein, I. (eds.) Flavor Chemistry. Thirty Years of Progress. Kluwer Academic/Plenum, New York, pp. 63-74
[28]  Rychlik, M., Grosch, W. (1996) Identification and quantification of potent odorants formed by toasting of wheat bread. Lebensm. Wiss. Technol. 29, 515-525
[29]  Blank, I., Schieberle, P., Grosch, W. (1992) Quantification of the flavour compounds 3-hydroxy-4,5-dimethyl-2(5H)-furanone and 5-ethyl-3-hydroxy-4-methyl-2(5H)-furanone by a stable isotope dilution assay. In Schreier, P., Winterhalter, P. (eds.) Progress in Flavor Precursors Studies. Allured Publishing, Carol Stream, pp. 103-109
[30]  Guth, H., Grosch, W. (1990) Deterioration of soya-bean oil: quantification of primary flavour compounds using a stable isotope dilution assay. Lebensm. Wiss. Technol. 23, 513-522

[31] Leffingwell, J.C., Leffingwell, D. (1991) GRAS flavour chemicals: detection thresholds. Perfumer & Flavorist 16, 1, 3-4, 6-8, 10, 13-14, 16-19

[32] Buttery, R.G., Turnbaugh, J.G., Ling, L.C. (1988) Contribution of volatiles to rice aroma. J. Agric. Food Chem. 36, 1006-1009

[33] Zehentbauer, G., Grosch, W. (1997) Apparatus for quantitative headspace analysis of the characteristic odorants of baguettes. Z. Lebensm. Unters. Forsch. 205 262-267

[34] Grosch, W. (1997) Flavour of white bread. In Kruse, H.-P., Rothe, M. (eds.) Proceedings of the 5th Wartburg Aroma Symposium, University Potsdam, pp. 179-191

[35] Mayer, F., Grosch, W. (2001) Aroma simulation on the basis of the odorant composition of roasted coffee headspace. Flavour Fragrance J. 16, 180-190

[36] Laing, D.G. (1988) Coding of chemosensory stimulus mixtures. Ann. NY Acad. Sci. 510, 61-66

[37] Kerscher, R. (2000) Aroma differences as affected by the animal species (in German). Thesis, Technical University of Munich

[38] Kerscher, R., Grosch, W. (2000) Comparison of the aromas of cooked beef, pork and chicken. In Schieberle, P., Engel, K.-H. (eds.) Frontiers of Flavor Sicence. Deutsche Forschungsanstalt für Lebensmittelchemie, Garching, pp. 17-20

[39] Guth, H., Grosch, W. (1994) Identification of the character impact odorants of stewed beef juice by instrumental analysis and sensory studies. J. Agric. Food Chem. 42, 2862-2866

[40] Kerscher, R., Nürnberg, K., Voigt, J., Schieberle, P., Grosch, W. (2000) Occurrence of 12-methyltridecanal in microorganisms and physiological samples isolated from beef. J. Agric. Food Chem. 48, 2387-2390

[41] Guth, H., Grosch, W. (1993) 12-Methyltridecanal, a species-specific odorant of stewed beef. Lebensm. Wiss. Technol. 26, 171-177

[42] Cerny, C., Grosch, W. (1992) Quantification of character-impact odour compounds of roasted beef. Z. Lebensm. Unters. Forsch. 196, 417-422

[43] Pippen, E.L., Nonaka, M. (1960) Volatile carbonyl compounds of cooked chicken. II. Compounds volatized with steam during cooking. Food Res. 25, 764-769

[44] Kerler, J., Grosch, W. (1997) Character impact odorants of boiled chicken. Z. Lebensm. Unters. Forsch. 205, 232-238

[45] Kerscher, R., Grosch, W. (1998) Quantification of 2-methyl-3-furanthiol, 2-furfurylthiol, 3-mercapto-2-pentanone, and 2-mercapto-3-pentanone in heated meat. J. Agric. Food Chem. 46, 1954-1958

[46] Grosch, W. (1987) Reactions of hydroperoxides. Products of low molecular weight. In Chan, H.W.-S. (ed.) Autoxidation of Lipids. Academic Press, pp. 95-139

[47] Scherz, H., Senser, F. (2000) Food Composition and Nutrition Tables. Medpharm Scientific Publishers, Stuttgart

[48] Josephson, D.B., Lindsay, R.C. (1986) Enzymatic generation of volatile aroma compounds from fresh fish. In Parliament, T.H., Croteau, R. (eds.) Biogeneration of Aromas. American Chemical Society, Washington, DC, pp. 201-219

[49] Milo, C., Grosch, W. (1996) Changes in the odorants of boiled salmon and cod as affected by the storage of the raw material. J. Agric. Food Chem. 44, 2366-2371

[50] Masanetz, C., Guth, H., Grosch, W. (1998) Fishy and hay-like off-flavours of dry spinach. Z. Lebensm. Unters. Forsch. A 206, 108-113

[51] Preininger, M., Rychlik, M., Grosch, W. (1994) Potent odorants of the neutral volatile fraction of Swiss cheese (Emmental). In Maarse, H., van der Heij, D.G. (eds.) Trends in Flavour Research. Elsevier Science Publishers, Amsterdam, pp. 267-270

[52] Preininger, M., Grosch, W. (1994) Evaluation of key odorants of the neutral volatiles of Emmentaler cheese by calculation of odour activity values. Lebensm. Wiss. Technol. 27, 237-244

[53] Preininger, M., Warmke, R., Grosch, W. (1996) Identification of the character impact flavour compounds of Swiss cheese by sensory studies of models. Z. Lebensm. Unters. Forsch. 202, 30-34

[54] Warmke, R., Belitz, H.-D., Grosch, W. (1996) Evaluation of taste compounds of Swiss cheese (Emmentaler). Z. Lebensm. Unters. Forsch. 203, 230-235

[55] Muir, D.D., Hunter, E.A., Banks, J.M., Horne, D.S. (1995) Sensory properties of hard cheese. Identification of key attributes. Int. Dairy J. 5, 157-177

[56] Preininger, M., Grosch, W. (1995) Determination of 4-hydroxy-2,5-dimethyl-3(2H)-furanone (HDMF) in cultures of bacteria. Z. Lebensm. Unters. Forsch. 201, 97-98

[57] Rychlik, M., Bosset, J.O. (2001) Flavour and off-flavour compounds of Swiss Gruyere cheese. Evaluation of potent odorants. Int. Dairy J. 11, 895-901
[58] Rychlik, M., Bosset, J.O. (2001) Flavour and off-flavour compounds of Swiss Gruyere cheese. Identification of key odorants by quantitative instrumental and sensory studies. Int. Dairy J. 11, 903-910
[59] Kubickova, J., Grosch, W. (1997) Evaluation of potent odorants of Camembert cheese by dilution and concentration techniques. Int. Dairy J. 7, 65-70
[60] Kubickova, J., Grosch, W. (1998) Quantification of potent odorants in Camembert cheese and calculation of their odour activity values. Int. Dairy J. 8, 17-23
[61] Kubickova, J., Grosch, W. (1998) Evaluation of flavour compounds of Camembert cheese. Int. Dairy J. 8, 11-16
[62] Fischer, K.-H., Grosch, W. (1987) Volatile Compounds of importance in the aroma of mushrooms (*psalliota bispora*). Lebensm. Wiss. Technol. 20, 233-236
[63] Schieberle, P., Gassenmeier, K., Guth, H., Sen, A., Grosch, W. (1993) Character impact odour compounds of different kinds of butter. Lebensm. Wiss. Technol. 26, 347-356
[64] Reiners, J., Grosch, W. (1998) Odorants of virgin olive oils with different flavor profiles. J. Agric. Food Chem. 46, 2754-2763
[65] Guth, H., Grosch, W. (1991) A comparative study of the potent odorants of different virgin olive oils. Fat Sci. Technol. 93, 335-339
[66] Hinterholzer, A., Schieberle, P. (1998) Identification of the most odor-active volatiles in fresh, hand-extracted juice of Valencia late oranges by odor dilution techniques. Flavour Fragrance J. 13, 49-55
[67] Buettner, A., Schieberle, P. (2001) Evaluation of aroma differences between hand-squeezed juices from Valencia late and Navel oranges by quantitation of key odorants and flavor reconstitution experiments. J. Agric. Food Chem. 49, 2387-2394
[68] Meilgaard, C. (1975) Aroma volatiles in beer: purification, flavour, threshold and interaction. In Drawert F. (ed.) Geruch- und Geschmacksstoffe. Verlag Hans Carl, Nürnberg, pp. 211-254
[69] Jellinek, G. (1985) In Morton, I.D. (ed.) Sensory Evaluation of Food: Theory and Practice, Series in Food Science and Technology. Ellis Horwood, Chichester, UK.
[70] Schieberle, P., Mestres, M., Buettner, A. (2003) Characterization of aroma compounds in fresh and processed mandarin oranges. In Cadwallader, K.R., Weenen, H. (eds.) Freshness and Shelf Life of Foods. ACS Symposium Series 836, pp. 162-174
[71] Buettner, A., Schieberle, P. (2001) Evaluation of key aroma compounds in hand-squeezed grapefruit juice (*citrus paradisi* Macfayden) by quantitation and flavour reconstitution experiments. J. Agric. Food Chem. 49, 1358-1363
[72] Fuhrmann, E., Grosch, W. (2002) Character impact odorants of the apple cultivars Elstar and Cox Orange. Nahrung/Food 46, 187-193
[73] Grosch, W. (2000) Specificity of the human nose. In Schieberle, P., Engel, K.-H. (eds.) Frontiers of Flavour Science. Deutsche Forschungsanstalt für Lebensmittelchemie, Garching, pp. 213-219
[74] Grosch, W. (1998) Which compounds occurring in heated food are preferred by the olfactory organ (In German). Lebensmittelchemie 52, 143-146
[75] Schieberle, P., Hofmann, T. (1997) Evaluation of the character impact odorants in fresh strawberry juice by quantitative measurements and sensory studies on model mixtures. J. Agric. Food Chem. 45, 227-232
[76] Mutti, B., Grosch, W. (1999) Potent odorants of boiled potatoes. Nahrung 43, 302-306
[77] Grosch, W. (1999) Aroma of boiled potatoes, dried potatoes and French fries (in German). Kartoffelbau 50, 362-364
[78] Wagner, R., Grosch, W. (1997) Evaluation of potent odorants of French fries. Lebensm. Wiss. Technol. 30, 164-169
[79] Wagner, R., Grosch, W. (1998) Key odorants of French fries. J. Am. Oil Chem. Soc. 75, 1385-1392
[80] Mayer, F., Takeoka, G., Buttery, R., Nam, Y., Naim, M., Bezman, Y., Rabinowitch, H. (2003) Aroma of fresh field tomatoes. In Cadwallader, K.R., Weenen, H. (eds.) Freshness and Shelf Life of Foods. ACS Symposium Series 836, pp. 144-161
[81] Buttery, R.G. (1993) Quantitative and sensory aspects of flavour of tomato and other vegetables and fruits. In Acree, T.E., Teranishi, R. (eds.) Flavor Science. Sensible Principles and Techniques. ACS Professional Reference Book. American Chemical Society, Washington, DC, pp. 259-286

[82]   Guth, H., Grosch, W. (1999) Evaluation of important odorants by dilution techniques. In Teranishi, R., Wick, E.L., Hornstein, I. (eds.) Flavor Chemistry. Thirty Years of Progress. Kluwer Academic Plenum Publishers, New York, pp. 377-386

[83]   Jagella, T., Grosch, W. (1999) Flavour and off-flavour compounds of black and white pepper I-III. Eur. Food Res. Technol. 209, 16-21, 22-26, 27-31

[84]   Zimmermann, M., Schieberle, P. (2000) Important odorants of sweet bell pepper powder (*capsicum annuum cv. annuum*). Differences between samples of Hungarian and Moroccan origin. Eur. Food Res. Technol. 211, 175-180

[85]   Schieberle, P., Zimmermann, M. (2001) Characterization of important aroma compounds of bell pepper powder from Hungary and Morocco (in German). In Gewürz- und Heilpflanzen. Deutsche Gesellschaft für Qualitätsforschung (Pflanzliche Nahrungsmittel) e.V., Proceedings der XXXVI Vortragstagung, pp. 141-151

[86]   Blank, I., Grosch, W. (1991) Evaluation of potent odorants in dill seed and dill herb (*anethum graveolens* L.) by aroma extract dilution analysis. J. Food Sci. 56, 63-67

[87]   Blank, I., Sen, A., Grosch, W. (1992) Sensory study on the character-impact flavour compound of dill herb (*anethum graveolens* L.). Food Chem. 43, 337-343

[88]   Masanetz, C., Grosch, W. (1998) Key odorants of parsley leaves (*petroselinum crispum* [Mill.] Nym.ssp. *crispum*) by odour-activity values. Flavour Fragrance J. 13, 115-124

[89]   Guth, H., Murgoci, A.-M. (1997) Identification of the key odorants of basil (*ocimum basilicum* L.). Effect of different drying procedures on the overall flavour. In Kruse, H.-P., Rothe, M. (eds.) Flavour Perception. Aroma Evaluation. University Potsdam, pp. 233-242

[90]   Kinlin, T.E., Muralidhara, R., Pittet, A.O., Sanderson, A., Waldradt, J.P. (1972) Volatiles components of roasted filberts. J. Agric. Food Chem. 20, 1021-1028

[91]   Güntert, M., Emberger, R., Hopp, R., Köpsel, M., Silberzahn, W., Werkhoff, P. (1991) Chirospecific analysis in flavour and essential oil chemistry. Part A. Filbertone: the character impact compound of hazelnuts. Z. Lebensm. Unters. Forsch. 192, 108-110

[92]   Matsui, T., Guth, H., Grosch, W. (1998) A comparative study of potent odorants in peanut, hazelnut, and pumpkin seed oils on the basis of aroma extract dilution analysis (AEDA) and gas chromatography-olfactometry of headspace samples (GCOH). Fett/Lipid 100, 51-56

[93]   Pfnuer, P., Matsui, T., Grosch, W., Guth, H., Hofmann, T., Schieberle, P. (1999) Development of a stable isotope dilution assay for the quantification of 5-methyl-(E)-2-hepten-3-one: application to hazelnut oils and hazelnuts. J. Agric. Food Chem. 47, 2944-2947

[94]   Schieberle, P. (1995) Quantitation of important roast-smelling odorants in popcorn by stable isotope dilution assays and model studies on flavour formation during popping. J. Agric. Food Chem. 43, 2442-2448

[95]   Schieberle, P. (1990) Occurrence and formation of roast-smelling odorants in wheat bread crust and popcorn. In Bressière, Y., Thomas, A.F. (eds.) Flavour Science and Technology. John Wiley, Chichester, pp. 105-108

[96]   Hofmann, T., Schieberle, P. (1998) 2-Oxopropanal, hydroxy-2-propanone, and 1-pyrroline: important intermediates in the generation of the roast-smelling food flavor compounds 2-acetyl-1-pyrroline and 2-acetyltetrahydropyridine. J. Agric. Food Chem. 46, 2270-2277

[97]   Schieberle, P. (1993) Studies on the flavour of roasted white sesame seeds. In Schreier, P., Winterhalter, P. (eds.) Progress in Flavour Precursor Studies. Allured Publishing, Carol Stream, IL, pp. 341-360

[98]   Schieberle, P. (1994) Important odorants in roasted white and black sesame seeds. In Kuhihara, K., Suzuki, N., Ogawa, H. (eds.) Olfaction and Taste, ISOT XI. Springer, Tokyo, pp. 263-267

[99]   Schieberle, P. (1996) Odour-active compounds in moderately roasted sesame. Food Chem. 55, 145-152

[100]  Grosch, W. (2001) Chemistry III: Volatile compounds. In Clarke, R.J., Vitzthum, O.G. (eds.) Coffee. Recent Developments. Blackwell Science, Oxford, pp. 68-89

[101]  Czerny, M:, Mayer, F., Grosch, W. (1999) Sensory study on the character impact odorants of roasted Arabica coffee. J. Agric. Food Chem. 47, 695-699

[102]  Mayer, F., Czerny, M., Grosch, W. (2000) Sensory study on the character impact aroma compounds of coffee beverage. Eur. Food Res. Technol. 211, 272-276

[103]  Reichstein, R., Staudinger, H. (1955) The aroma of coffee. Parfum. Essent. Oil. Res. 46, 86-88

[104]  Guth, H. (1997) Identification of character impact odorants of different white wine varieties. J. Agric. Food Chem., 45, 3022-3026
[105]  Guth, H. (1997) Quantitation and sensory studies of character impact odorants of different white wine varieties. J. Agric. Food Chem. 45, 3027-3032
[106]  Guth, H. (1998) Comparison of different white wine varieties in odor profiles by instrumental analysis and sensory studies. In Waterhouse, A.L., Ebeler, S.E. (eds.) Chemistry of Wine Flavor, ACS Symposium Series 714, pp. 39-52
[107]  Guth, H., Grosch, W. (1993) Furanoid fatty acids as precursors of a key aroma compound of green tea. In Schreier, P., Winterhalter, P. (eds.) Progress in Flavour Precursor Studies. Allured Publishing, Carol Stream, pp. 401-407
[108]  Maarse, H., Grosch, W. (1996) Analysis of taints and off-flavours. In Saxby, M.J. (ed.) Food Taints and Off-Flavours, 2nd ed. Blackie Academic & Professional, London, pp. 72-106
[109]  Milo, C., Grosch, W. (1993) Changes in the odorants of boiled trout (*salmo fario*) as affected by the storage of the raw material. J. Agric. Food Chem. 41, 2076-2081
[110]  Widder, S., Grosch, W. (1994) Study on the cardboard off-flavour formed in butter oil. Z. Lebensm. Unters. Forsch. 198, 297-301
[111]  Guth, H., Grosch, W. (1989) 3-Methylnonane-2,4-dione: an intense odour compound formed during flavour reversion of soya-bean oil. Fat Sci. Technol. 91, 225-230
[112]  Schieberle, P. Grosch, W. (1992) Changes in the concentrations of potent crust odorants during storage of white bread. Flavour Fragrance J. 7, 213-218
[113]  Kerler, J., Grosch, W. (1996) Odorants contributing to warmed-over flavor (WOF) of refrigerated cooked beef. J. Food Sci. 61, 1271-1284
[114]  Masanetz, C., Grosch, W. (1998) Hay-like off-flavour of dry parsley. Z. Lebensm. Unters. Forsch. 206, 114-120
[115]  Hofmann, T., Schieberle, P. (2004) Influence of meladonins on the aroma staling of coffee beverage. In Shihidi, F., Weerasinghe, D.K. (eds.) Neutraceutical Beverages. Chemistry, Nutrition, and Health Effects. ACS Symposium Series 871, pp. 200-215
[116]  Komarek, D., Schieberle, P. (2003) Changes in key beer aroma compounds during natural beer aging. In Cadwallader, K.R., Weenen, H. (eds.) Freshness and Shelf Life of Foods. ACS Symposium Series 836, pp. 70-79
[117]  Schieberle, P., Heiler, C: (1997) Influence of processing and storage on the formation of the metallic smelling (*E,Z*)-2,6-nonadienol in buttermilk. In Kruse, H.-P., Rothe, M. (eds.) Flavour Perception. Aroma Evaluation. University Potsdam, pp. 213-220
[118]  Preininger, M., Ullrich, F. (2001) Trace compound analysis for off-flavor characterization of micromilled milk powder. In Leland, J.V., Schieberle, P., Buettner, A. Acree, T.E. (eds.) Gas Chromatography-Olfactometry, ACS Symposium Series 782,pp. 46-61
[119]  Buettner, A., Schieberle, P. (2001) Application of a comparative aroma extract dilution analysis to monitor changes in orange juice aroma compounds during processing. In Leland, J.V., Schieberle, P., Buettner, A. Acree, T.E. (eds.) Gas Chromatography-Olfactometry, ACS Symposium Series 782, pp. 33-45

## 6.3 Microbiological Testing

*Heinz-Jürgen Lögtenbörger*

Microbial attack can cause spoilage or impair the quality of any foodstuff, including flavourings and flavouring components, provided that the preconditions for growth are met. Only the control of microbial growth can prevent such quality defects. Fungi, which taxonomically speaking also include yeasts, and bacteria are very significant for foodstuffs *[1-4]*. Other microorganisms such as protozoa, algae, cyanobacteria and also viruses are only of secondary importance *[5]*.

Microorganisms are living creatures which cannot be seen by the human eye. They become visible only when greatly magnified under a microscope, or they may be cultured on special nutrient media to such an extent that they become macroscopically visible *[6]*. The typical shape of some microorganisms in microscope are shown in Fig. 6.49-6.51.

Microorganisms may have both positive and negative effects on foodstuffs. They may cause spoilage of foodstuffs and some, for example the typical food poisoning pathogens (*Salmonella*, *Staphylococcus aureus* etc.), may even poison it. Moulds too may form toxic substances *[7-9]*.

Microorganisms are, however, also used to produce and finish foodstuffs and to preserve them by fermentation (c.f. chapter 2.2 and 3.2.1.1.2). The production of high quality beer or wine would be inconceivable without yeasts. Cured sausages with their typical lactic acid starter cultures and sometimes with mould cultures on their surface also depend on microorganisms.

Finally, microorganisms also produce ingredients for foodstuffs, such as in citric acid fermentation in which citric acid is obtained from molasses using a mould.

The growth of microorganisms on foodstuffs depends on factors which are endogenous and exogenous to the product. Endogenous factors are determined by the nature of the foodstuff itself. They include water activity, which is related to the relative humidity of air and osmotic pressure, pH, the nutrient content of the foodstuffs, further growth factors, coupled with antimicrobial compounds and protective microbiological structures in the foodstuff. Exogenous factors comprise environmental influences acting upon the foodstuff, such as temperature, humidity and the composition of the surrounding atmosphere. Depending on the various processing and production methods used, the endogenous and exogenous factors may be altered in such a way that microbial growth does or does not proceed. Processing methods have thus in recent times been cited as the third factor *[10]*.

As a rule, distilled flavourings, natural and nature-identical flavourings are strictly speaking insensitive to microbiological action. They frequently contain alcohols and an increased proportion of oil, or they are synthesised from chemicals which act as antimicrobial compounds. In these flavourings, the endogenous factors do not allow microbial growth. It should, however, be mentioned that the dormant form of spore-

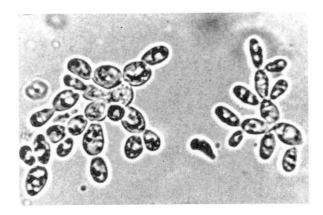

*Fig. 6.49:* Yeast

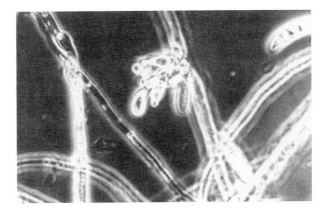

*Fig 6.50:* Mould

*Fig. 6.51:* Bacteria

forming bacteria may survive in alcohol and, when the flavouring is used in a very small quantity in the finished product, the alcohol content may be reduced to such a level that the spores can germinate if the conditions for microbial growth are met.

However, the development of ever more complicated flavouring systems containing natural ingredients may also result in microbiological problems [11].

Fruit juice concentrates or bases for the soft drink industry consist mainly of natural ingredients. Due to their composition and the presence of further endogenous factors, the possible range of spoilage microbes is, however, greatly restricted. Only moulds, yeasts, lactic acid and acetic acid bacteria may develop in these products [12-15]. If beverage bases are preserved, growth is generally greatly reduced as well as many microorganisms are killed. However, in manufacturing the finished product, the concentration of the preservative is reduced to such an extent that microorganisms may grow. Unpreserved fruit juice concentrates and beverage bases are highly susceptible to microbiological action and appropriate changes must thus be made to exogenous factors or to the processing methods if they are to keep for an extended period. Finished drinks made from these raw materials can be split into two groups - carbonated and non-carbonated drinks. Due to the low oxygen content, the range of spoilage microbes in carbonated drinks is restricted to fermentable yeasts and lactic acid bacteria [16-17]. Lactic acid bacteria are very demanding in terms of nutrient content growth factors and pH value. Consequently they may grow only under relatively mild conditions in drinks with a high content of nutrients and growth factors, such as isotonic drinks with their high vitamin and mineral content (see Figures 6.52 and 6.53).

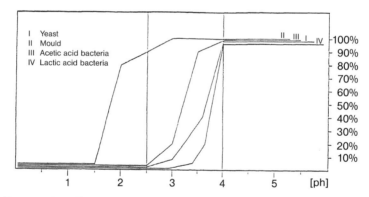

*Fig. 6.52: Danger of spoilage in non-carbonated soft drinks*

Fruit juices and nectars, and sometimes also drinks with low juice content, undergo heat treatment (pasteurisation, high-temperature short-time pasteurisation, hot filling) before, during or immediately after filling of the finished drink. This ensures that the finished drink no longer contains any microorganisms causing spoilage [18]. Cold-filled soft drinks require exceptionally clean production equipment in order to avoid reinfection with spoilage microbes [19-21].

As a result of enhanced quality awareness and standards, great efforts are currently being made to ensure the prevention of economic damage, due to the multiplication of bacteria in foodstuffs, through the application of selected processes.

# Microbiological Testing

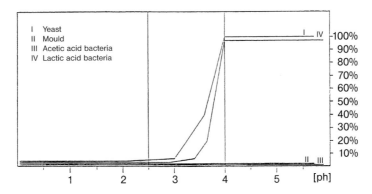

*Fig. 6.53: Danger of spoilage in carbonated soft drinks*

The individual raw materials as well as the flavouring ingredients can be treated according to the pertinent legislation. However, the final products must also be microbiologically protected.

Different processes are applied to achieve this purpose, depending on the foodstuff involved. The technical possibilities available are pasteurisation and sterilisation processes, sterile filtration and the application of modified atmosphere packaging (MAP). The foodstuffs are chemically treated with preservatives. The application of ionising radiation is also permitted in some countries.

However, such processes could also have a negative influence on the sensory quality of the foodstuffs. For example, specific foodstuffs cannot be heat-treated without causing deterioration in quality. The taste of the product will be impaired if the content of preservatives is increased. The treatment of the final product cannot guarantee absolute protection from bacterial multiplication. Resistant bacteria are able to survive heat treatment and microaerophile microorganisms even continue to grow in MAP. Microorganisms have been identified which are able to metabolise preservatives and therefore cannot be prevented from multiplying. The so-called 'hurdle technology' is frequently used, i.e. a combination of different processes is employed in order to obtain the highest possible degree of protection (e.g. pasteurisation and preservation) [22-24].

Due to their low water content, powder flavourings are generally not subject to microbiological spoilage. However, depending on the raw materials used, they may very well be contaminated with microbes. Spray or belt drying or similar drying processes do not kill all microorganisms in their entirety. As a rule, drying reduces the microbial count by 1-2 powers of ten. An important feature of powder flavourings is that after rehydration it is possible microorganisms can grow again. Freeze drying is also used in the production of starter cultures for various purposes in the food industry. In this process, suspensions of microorganisms with an elevated microbial count are carefully preserved through freeze drying and they are prepared in such a way that they may be easily and safely used by the end user.

Spices and spice preparations may be contaminated microbiologically to a greater or lesser degree depending on production process (e.g. fermentation). The degree of contamination depends on their preservation, which is subject to differing legislation. In some countries, spices may be fumigated with ethylene oxide, in others sterilization with ionising radiation is permitted. At present, neither of these processes is permitted in Germany, and under certain conditions spices there may have higher microbial contamination. Spices with high microbiological standards can only be obtained by appropriate raw material selection. The raw spices themselves are generally protected from spoilage by their low water content. Only their usage in food presents again favourable conditions for the growth of microorganisms and subsequent spoilage of the product. Typical examples are the production of cured or boiled sausages. Methods for microbial reduction such as salting or scalding yield only a partial reduction in the number of microbes present. Microbes may thus survive and grow in the finished product and lead to more serious quality problems. It is particularly with spices, due to their origin and the hygiene conditions in the country of origin, that pathogens may be expected. Hence special attention must be given to the microbiological testing of spices and spice products [25].

Table 6.56 shows the different microbiological tests recommended for various flavourings and flavouring systems.

The Table provides a general guide for the microbiological testing of flavourings and some finished products. The scope of microbiological testing varies from case to case and it depends on product composition, processing conditions, the manner in which the product is to be used and, in certain circumstances, on the legal requirements of different countries.

A specially equipped laboratory is necessary for the microbiological testing. Staff must have the appropriate professional knowledge; staff without basic microbiological training must to be equipped the necessary skills. Most microorganisms usually are not hazardous to health. However all materials, especially the detection media, should none the less be treated as potentially hazardous and they should be handled with appropriate caution. All contaminated equipment, nutrient media etc. must be sterilized before disposal. It is absolutely essential to wear protective clothing. Overalls should be changed or removed when leaving the laboratory.

The following equipment and materials are essential requirements for a basic microbiological testing laboratory:

- autoclave to sterilize nutrient media and equipment
- microbiologically clean working area; for certain work a laminar flow bench
- sterile Petri dishes, pipettes, Drigalski spatulae, inoculation loops
- Bunsen burner
- gas burner and cartridge
- incubator with different temperature settings
- microscope
- refrigerator
- pH meter
- water baths
- magnetic stirrer with hotplate
- pressure cooker
- soap and hand disinfectant.

# Microbiological Testing

Table 6.56: Microbiological tests recommended for various flavourings

| Flavourings | Microorganisms | | | | | | Comments |
|---|---|---|---|---|---|---|---|
| | Total microbial count | E. coli/coliforms | Yeasts/moulds | Acetic acid/lactic acid bacteria | Salmonella | Staph. aureus | Enterococci | |
| Bases and concentrates for the soft drinks industry | (+) | - | + | + | - | - | - | For products with pH > 4, total microbial count as initial guide. |
| Flavourings based on natural products and fruitpreparations (e.g. compounds, pastes) | + | (+) | + | + | (+) | (+) | (+) | E. coli/coliforms at pH ≥ 5.0. Salmonella, Staph. aureus and enterococci in neutral systems and for products based on animal raw materials (egg products, milk products). |
| Natural flavourings containing no or little alcohol (max. 5% v/v) (e.g. flavourings for the baking industry) | + | + | + | + | - | - | - | |
| Purely plant based powder flavourings | + | + | (+) | (+) | - | - | - | Yeasts, moulds, lactic and acetic acid bacteria only if fruit components are contained. |
| Powder flavourings containing raw materials of animal origin | + | + | (+) | (+) | + | (+) | (+) | Additionally yeasts, moulds, lactic and acetic acid bacteria if fruit components are contained. |
| Spices | + | + | + | - | + | (+) | - | |
| Flavourings produced by biotechnological processes | + | + | (+) | (+) | (+) | (+) | (+) | Depending on production process and intended use. |

If pathogens bacteria are also being investigated the micro laboratory, it must be separated from other laboratory (safety airlock). For the handling of certain pathogens (e.g. *Salmonella*, *Staph. aureus* etc.) a special permit is required and a certified supervisor and specially trained staff must perform such work. In Germany, the Federal Law on Epidemics governs such operations [26].

Preparation for microbiological testing generally comprises sterilization of the sampling container and the working material. Sampling should be done under sterile conditions as possible, i.e. by thorough flame sterilization of the sampling taps or spoons and any other equipment with a gas burner. Caution is required with powder products or products containing alcohol as they can ignite readily. A naked flame should not be used for such work.

Subsequently, the necessary nutrient media must be prepared and sterilized. Nutrient media are solutions of nutrients in gel or liquid form which are selected to provide optimum growth conditions for the microbes to be detected. One can chose from complete nutrient media (e.g. plate count agar) which detect the complete range of bacteria or selective nutrient media, the composition of which is arranged such that only very specific groups of microbes will grow and particular forms may be recognised (e.g. coliform bacteria, lactic acid bacteria). Such media usually contain inhibitors for the non-selected microbes and an excess of growth factors for those to be detected. Selective nutrient media sometimes also contain special substances (e.g. indicators), which highlight the presence of certain microorganisms by means of colour reactions. In general, the nutrient media are supplied as powders which need to be rehydrated usually with distilled water. However, wide range of ready-to-use nutrient media are, also commercially available. Once prepared, sterilized and ready for use, the nutrient media must be kept cool and protected from light.

After sampling, the products are prepared in the laboratory. The sample preparation is illustrated below with two simple methods for determining the total microbial count:

If the sample is expected to have a low microbial count, it is used undiluted. In the event of higher microbial counts, decimal dilution series in sterilized physiological saline are used. The degree of dilution must be taken into account when interpreting the results.

*Method 1: Koch plate casting procedure*

The sample material or dilutions are placed in a sterile Petri dish and then mixed with liquefied nutrient agar which is kept at 45°C (e.g. plate count or standard I agar). After solidification of the agar, the Petri dish is usually incubated for 48 hours at 37°C.

*Method 2: Plating procedure*

The liquid sample or dilution is evenly spread directly on to solidified agar in a Petri dish using a Drigalski spatula. The sample is then incubated at 37°C.

After incubation, all developed microbes are counted and if necessary multiplied by the respective degree of dilution. They are recorded as the total microbial count per reference quantity.

Other methods and their exact description should be taken from specialist literature *[27-30]*.

## REFERENCES

[1] Kreger-van Rij, N.J.W. (Ed.), The yeasts: a taxonomic study, Amsterdam, Elsevier Science, 1984
[2] Müller, G., Weber, H., Mikrobiologie der Lebensmittel, Hamburg, Behr's Verlag, 1996
[3] Krieg, N.R., Holt, J.G. (Eds.), Bergey's Manuel of Systematic Bacteriology, Volume 1, 9th ed. Baltimore, Williams & Williams, 1994
[4] Sneath, P., Mair, N., Sharpe, M., Holt, J.G. (Eds.), Bergey's Manual of Systematic Bacteriology – Volume 2, Baltimore, Williams & Williams, 1986
[5] Holzapfel, W. (Ed.), Lexikon Lebensmittelmikrobiologie und -hygiene, Hamburg, Behr's Verlag, 2002
[6] Difco Laboratories (Eds.), Difco Manual, Detroit, Michigan, 1984
[7] Müller, G., Mikrobiologie pflanzlicher Lebensmittel, Darmstadt, Steinkopf-Verlag, 1988
[8] Zickrick, K., Mikrobiologie tierischer Lebensmittel, Leipzig Fachbuchverlag, 1986
[9] Samson, R.A., Hoeckstra, E.S., Frisvad, J.C., Filtenborg, O. (Eds.), Introduction to food-and airborne fungi, Centralbureau voor Schimmelculturen, Utrecht, 2000
[10] Sinell, H.-J., Einführung in die Lebensmittelhygiene, Berlin, Verlag P. Parey, 2003
[11] Dittrich, H.-H., Getränketechnik Nr. 6, 198 ff (1990)
[12] Dittrich, H.-H. (Ed.), Mikrobiologie der Lebensmittel-Getränke, Hamburg, Behr's Verlag, 1993
[13] Back, W., Brauwelt $\underline{122}$, 357 ff (1982)
[14] Back, W., Brauwelt $\underline{121}$, 43 ff (1981)
[15] Back, W., Confructa-Studie Nr. V/86 (1986)
[16] Back, W., Farbatlas und Handbuch der Getränkebiologie, Band 1, Nürnberg, Verlag Hans Carl, 1994
[17] Back, W., Farbatlas und Handbuch der Getränkebiologie, Band 2, Nürnberg, Verlag Hans Carl, 2000
[18] Back, W., Der Mineralbrunnen $\underline{42}$, 244 ff (1991)
[19] Bäuml, J., Getränketechnik, Nr. 5, 209 ff (1988)
[20] Wallhäußer, K.-H., Praxis der Sterilisation-Desinfektion-Konservierung-Keimidentifizierung-Betriebshygiene, Stuttgart, Thieme-Verlag, 1995
[21] Schriftenreihe der Schweizerischen Gesellschaft für Lebensmittelhygiene, Betriebshygiene, Heft 5, Industriestraße 12, CH-8157 Dielsdorf, 1977
[22] Casas, E., Ancos, B., Valderama, M.J., Cano, P., Peinado, J.M., International Journal of Food Microbiology $\underline{94}$, 93-96 (2004)
[23] Liewen, M.B., Marth, E.H., Journal of Food Protection $\underline{48}$, 364-375 (1985)
[24] Hocking, A.D., Pitt, J.J., Foodspoilage II Heat-resistant fungi, CSIRO Food Res. Quart. 44 (1984)
[25] Rosenberger, A., Werner, H., Fleischwirtschaft $\underline{73}$, 830 ff (1993)
[26] Bundesministerium für Gesundheit und Soziale Sicherung (Eds.), Infektionsschutzgesetz (IfSG), Gesetz zur Verhütung und Bekämpfung von Infektionskrankheiten beim Menschen § 44 ff, Stand 2001
[27] Becker, B., Baumgart, I., Mikrobiologische Untersuchung von Lebensmitteln, Loseblattsammlung, Hamburg, Behr's Verlag, 2002
[28] Bundesinstitut für gesundheitlichen Verbraucherschutz und Veterinärmedizin (Eds.), Amtliche Sammlung und Untersuchungsverfahren nach § 35 LMBG, aktualisierte Loseblattsammlung, Berlin/Wien/Zürich, Beuth Verlag, 2004
[29] Food and Drug Administration, Bacteriological Analytical Manual, Washington, 1998
[30] International Federation of Fruit Juice Procedures, Microbiology Working Group (Eds.), Microbiological Methods, Loseblattsammlung, Eigenverlag, 2004

# 7 Legislation / Toxicology

*Uwe-Jens Salzer*

## 7.1 Introduction / Definitions

In general each country has its own legislative rules from history and tradition. The legislator is the parliament which passes a bill or enacts a law, i.e. the "food act" which may authorise the government (or a governmental agency) to execute the law via detailed rules i.e. regulations, decrees and/or orders. Besides these some countries have guiding principles on which industry, consumers, science, and food control have to agree. These guiding principles may be listed in a national "codex alimentarius" for all intents and purposes. They have the force of law. Finally industry may agree on guidelines or standards for certain food products.

Because of general interest, the definitions of I.O.F.I. [1] in its Code of Practice should be mentioned at this point:

- *Flavouring Substance*
  Defined chemical component with flavouring properties, not intended to be consumed as such, e.g.: vanillin, ethylvanillin, citral.

There are three different categories of flavouring substances, these are:

- *Natural Flavouring Substance*
  Chemically defined substance used for its flavouring properties not intended to be consumed as such and suitable for human consumption at the level used, which is obtained by appropriate physical, microbiological or enzymatic processes from a foodstuff or material of vegetable or animal origin either as such or after processing by food preparation processes. Salts of natural flavouring substances with the following cations $NH_4^+$, $Na^+$, $K^+$, $Ca^{2+}$ and $Fe^{3+}$ or the anions $Cl^-$, $SO_4^{2-}$ and $CO_3^{2-}$ are classified as natural flavouring substances.

- Nature-Identical Flavouring Substance
  Flavouring substance obtained by synthesis or isolated through chemical processes from a natural aromatic raw material and chemically identical to a substance present in natural products intended for human consumption, either processed or not. Salts of nature-identical flavouring substances with the following cations $NH_4^+$, $Na^+$, $K^+$, $Ca^{2+}$ and $Fe^{3+}$ or the anions $Cl^-$, $SO_4^{2-}$ and $CO_2^{2-}$ are classified as nature-identical flavouring substances. e.g.: vanillin, citral (obtained by chemical synthesis or from oil of lemongrass through the bisulfite derivative).

- *Artificial Flavouring Substance*
  Flavouring substance, not yet identified in a natural product intended for human consumption, either processed or not, e.g.: ethylvanillin.

The flavourist may combine many of these substances to create a

- *Flavouring*
  Concentrated preparation, with or without flavour adjuncts, used to impart flavour with the exception of only salty, sweet or acid tastes. It is not intended to be consumed as such. e.g.: flavouring for confectionery.

But he/she needs some more ingredients like

- *Flavour Adjuncts*
   Food additives and food ingredients necessary for the production, storage and application of flavourings as far as they are non-functional in the finished food

and

- *Flavour Enhancers*
   Substance with little or no odour at the level used, the primary purpose of which is to increase the flavour effect of certain food components well beyond any flavour contributed directly by the substance itself. e.g.: monosodium glutamate

Instead of chemically defined substances he/she may also use a

- *Natural Flavour Concentrate*
   A preparation used for its flavouring properties, not intended to be consumed as such and suitable for human consumption at the level used, which is obtained by appropriate physical, microbiological or enzymatic processes from a foodstuff or material of vegetable or animal origin, either as such or after processing by food preparation processes.

Or he/she will not only create a blended flavouring, but a

- *Process Flavouring*
   A product or a mixture prepared for its flavouring properties and produced from ingredients or mixtures of ingredients which are themselves permitted for use in foodstuffs, or are present naturally in foodstuffs, or are permitted for use in process flavourings, by a process for the preparation of foods for human consumption. Flavour adjuncts may be added. This definition does not apply to flavouring extracts, processed natural food substances or mixtures of flavouring substances.

or a

- *Smoke Flavouring*
   A concentrated preparation, not obtained from smoked food materials, used for the purpose of imparting a smoke type flavour to foodstuffs. Flavour adjuncts may be added.

Besides the flavour industry the Codex Committee on Food Additives and Contaminants CCFAC [2] has elaborated definitions which are:

- *Natural Flavours and Natural Flavouring Substances* are preparations and single substances respectively, acceptable for human consumption, obtained exclusively by physical, microbiological or enzymatic processes from material of vegetable or animal origin either in the raw state or after processing for human consumption by traditional food preparation processes (including drying, roasting and fermentation) [3].

- *Nature-Identical Flavouring Substances* are substances chemically isolated from aromatic raw materials or obtained synthetically; they are chemically identical to substances present in natural products intended for human consumption, either processed or not [4].

- *Artificial Flavouring Substances* are those substances which have not yet been identified in natural products intended for human consumption, either processed or not [4].

- *Adjuncts* are foodstuffs and food additives which are essential in the manufacture and use of flavourings [3].

Many countries possess a "food act" which, as its prime purpose, intends to protect the consumer

- against damage to health
- against misleading statements and deception about the quality of the offered food.

These food acts normally are supplemented by regulations as mentioned before. In this way food additives and flavourings are regulated. *Food additive* means

any substance not normally consumed as a food in itself and not normally used as a characteristic ingredient of food whether or not it has nutritive value, the intentional addition of which to food for a technological purpose in the manufacture, processing, preparation, treatment, packaging, transport or storage of such food results, or may be reasonably expected to result, in it or its by-products becoming directly or indirectly a component of such foods [5].

Flavourings are either food additives, i.e. in the USA see 7.5.1, or treated differently, i.e. in the European Union where they have a directive on their own, see 7.4.2.

Food additives are regulated via a "positive list", i.e. only those additives may be used which appear on a list of permitted substances. Flavourings are regulated either in this way, e.g. in the USA, or differently in some European countries where all natural and nature-identical flavouring substances may be used with the exception of those which are forbidden, a so-called "negative list". In a so-called "mixed list system" some flavouring substances (i.e. artificial ones) are regulated via a positive list and the others via a negative list (German flavour regulation).

Both systems have advantages and disadvantages. The positive system may appear safer for the consumer because only permitted flavouring substances may be used. On the other side natural flavouring substances and their nature-identical counterparts have been used via flavouring preparations since ancient times and mankind has had a long experience with substances which may be forbidden or restricted in use (e.g. safrol in nutmeg and other herbs and spices). Ultimately, it is not possible to distinguish in a final food whether a flavouring substance has been added as such or via a food ingredient (e.g. juice concentrate).

We also need to consider the so-called "Blue Book" on flavouring substances of the Council of Europe which does contain several lists of flavouring substances and source materials in different categories, see 7.3.

## 7.2 Toxicological Considerations

There are different opinions on the necessary scope of safety evaluations of flavouring substances. The following facts need to be borne in mind:

- The number of known flavouring materials is much larger than that of all other food additives combined.
- The levels at which flavouring materials occur, or are added, are relatively low. Their flavour impact limits the risk of an incidental overdose by making the food unpalatable.
- The vast majority of flavouring materials occur widely in traditional foods. They are not "new".
- The chemical structure of flavouring materials is generally of the type that may be expected to occur in foods as a result of biogenetic processes.

These facts have led to a number of safety evaluations of and regulatory approaches to flavouring materials [6].

Especially the relatively low exposure level has been considered by several authors:

- In 1967, Frawley [7] analysed the results of over 200 chronic feeding studies in rodents and found that apart from a few exceptions (e.g. heavy metals, pesticides), there were no adverse effects observed at dietary levels below 100 ppm. Considering a safety factor of 1000 to cover any such exceptions, as well as considering the limited nature of his data base, he arrived at a figure of no-concern of 0.1 ppm (which translates to somewhere between 1 and 5 µg/kg body weight/day in consumers).
- Rulis [8] analysed an entirely different data base (159 substances) and came to a similar conclusion, i.e. there is essentially no probable toxicity produced in rodents at doses below 1 mg/kg body weight/day (i.e. 1 µg/kg/day in consumers, benefiting from a safety factor of 1000).
- These two studies did not consider potential carcinogenesis. Rulis [9], however, has proposed a "threshold of regulation" using the TD50 data base of Gold and co-workers. The TD50 values determined for 343 carcinogens (by oral route in rodents) in this data base are widely dispersed but show a logarithmic normal distribution. Rulis has been able to demonstrate that it is possible to transform this into a distribution of the level of exposure which corresponds to a life-time risk of one in a million. Only 15 % of the carcinogens (i.e. the most potent), produce a $10^{-6}$ risk of cancer when consumed at levels above 50 ppt in the diet. If it is assumed that 10 % of substances of unknown toxicity are carcinogens, it can be calculated that there is a 95 %

probability of not exceeding the 10-6 risk for chemicals consumed at dietary levels of 1 ppb.

The approaches outlined above, all treat prospective chemicals as potential equals. They do not take account of any other data which may have been available at the time of evaluation, even though this may be influential in deciding which exposure levels do not have toxicological significance.

There is one approach which considers an additional factor: historical consumption as a natural component of foods by Stofberg *[10]*. In this model, flavouring materials with maximum added levels in food of 5 ppm, industrial use as flavouring ingredients and with a total consumption in the USA of less than 10 kg/year and "Consumption Ratios" exceeding 10, would be considered to be of minimal significance in this context. The criteria of 10 kg/year corresponds to 36.5 ng/kg/day in persons consuming 10 times the national average (assuming that 10 kg represents only 60 % of real industrial use and assuming a body weight of 50 kg). This is close to the estimates of a life-time carcinogenic risk of one in a million (see above) and well below the exposure estimates for insignificant risk of systemic toxicity. The Consumption Ratio offers some epidemiological reassurances but more importantly, points to the futility of trying to reduce human exposure by regulating intentional addition to the diet through the use of industrially produced flavourings.

In response to suggestions by JECFA in 1967 *[11]* which already suggested lowering the priority for evaluation of substances for which *per capita* consumption was below 3.65 mg/year, a system of priority setting was developed *[12]* which was accepted by JECFA and the Council of Europe Expert Committee on Flavouring. This is specifically designed to rank the priorities for evaluating different flavouring substances. One can not dissociate, however, the recognition that a substance needs less urgent attention from the concept that its perceived risk to health (under current exposure conditions) is of lower significance.

The JECFA method adds the refinement of considering structural aspects which are known to lead to increased toxicity, including carcinogenicity. The structural-alert considerations of FDA's "Red Book" and FEMA's "Decision Tree" could be integrated into a system for defining insignificant exposure just in the same way as they are used in the JECFA approach for determining priorities. When the exposure to a substance has been estimated, the Priority Ranking can be used as an indication of significance in the context of a regulatory threshold *[13]*.

In this line JECFA recommended during its meeting in Rome, 14-23 Febr. 1995, *[79]* a procedure proposed by I.C.Monroe should be applied to the evaluation of flavouring agents. The Procedure incorporates a series of criteria designed to provide a means of evaluation such agents in a consistent and timely manner. The criteria, which have been drawn up in the light of the principles for the safety evaluation of flavouring substances, take account of available information on intake from current uses, structure-activity relationships, and metabolism and toxicity data. They incorporate procedures outlined in the report of the Committee *[12]*. The use of these criteria provides a means of ranking flavouring substances in terms of concern over potential inherent toxicity and provides guidance on the nature and extent of the data required to perform a safety evaluation *[80]*.

The first and most important activity in evaluating the safety of flavouring substances was initiated in 1960 in the USA where the Flavor and Extract Manufacturers' Association (FEMA) organised a panel of expert scientists for this purpose.

The evaluation of the safety of flavourings, as practised by the panel, involves the simultaneous application of a series of criteria, every one of which must be considered in arriving at a determination of safety-in-use. No single criterion is used to the exclusion of others or is expected to require absolute, independent justification. Instead, all must be mutually supportive and unanimously agreed upon for a conclusion of safety to be reached. These criteria are:

- chemical identity,
- structure and purity,
- natural occurrence of the chemical in food,
- concentration of the chemical in food and in the total diet,
- toxicological evidence in animals and man and
- metabolic fate in mammals.

Details about the procedure used and the criteria employed have been published by Woods and Doull *[14]*, and an update was given by Schrankel in 2004 *[81]*.

The data used for the safety evaluation of GRAS substances have been published in the form of Scientific Literature Reviews (SLR) prepared by FEMA originally under contract with the FDA.

The SLRs cover groups of substances with related chemical structures and, therefore, presumably, related metabolic and toxicological properties. The titles of these SLRs together with their order number are listed in Flavor and Fragrance Materials 2004 *[15]*. This publication contains for each GRAS substance an alpha numeric code to the SLR. The SLRs are available either in hard copy or on microfiche from the Customer Service National Technical Information Services (NTIS), 5285 Port Royal Road, Springfield, VA 22161 (USA).

Another approach has been taken by the EU Scientific Committee on Food (SCF) in the Guidelines for the evaluation of flavourings for use in foodstuffs: 1. Chemically defined flavouring substances *[16]*. In the view of the Scientific Committee for Food there is no reason to expect that occurrence in nature is any guarantee of safety and therefore there is no toxicological reason why these categories should be treated differently from artificial ones. The Committee considers it necessary from the point of view of consumer protection to establish measures of acceptability of those flavouring substances presently in use in food. According to the SCF, factors having a bearing on the safety in use of flavourings may be considered in the following categories:

(1) Toxicological and other biological properties of the flavouring principle(s).

(2) Toxicological and other biological properties of components other than the flavouring principles in the flavouring. These may be different if a given flavouring is derived from natural sources or prepared synthetically. Special care shall be taken to avoid toxic substances which are known to derive from a specific source of method of synthesis.

(3) Quantity of flavouring consumed. This, in turn, depends on the amount occurring in food, the number of different foods in which it may be used and the frequency with which these foods are consumed.

In two Annexes to the cited paper [16] detailed principles for the evaluation of flavouring substances have been set.

Subsequently the EU Commission and Council have agreed on a Regulation laying down a community procedure for flavouring substances used in foodstuffs [17], where it is stated in the Annex:

Flavouring substances can be approved provided that:

– they present no risk to the health of the consumer, in accordance with the scientific assessment

– and their use does not mislead the consumer

– they must be subjected to appropriate toxicological evaluation.

## 7.3 FAO/WHO and Council of Europe

Both organisations are international bodies, confederations of nations which are mentioned here because they have scientific bodies for the evaluation of substances like flavouring substances. Since both organisations do not have the power to enact laws they can only make recommendations. But many smaller countries who do not have these scientific possibilities have adopted the recommendations into their own law.

The Food and Agriculture Organization of the United Nations (FAO) and the World Health Organization (WHO) created the Joint FAO/WHO Expert Committee on Food Additives (JECFA) back in 1957. This panel has evaluated food additives since then. According to the definitions of the Codex Alimentarius (Commission) for food additives, flavouring substances come under this category. JECFA has so far evaluated 41 substances and established acceptable daily intakes (ADI) for 35. At present new flavouring substances on the agenda of JECFA have been selected, see 7.2.

The Council of Europe was established on 5.5.1949. Within this confederation of European states a "Partial Agreement in the Social and Public Health Field" was agreed upon under Resolution (59)23 on 16.11.1959 by the following states: Belgium, France, Germany, Italy, Luxembourg, Netherlands and the United Kingdom. Later Austria, Denmark, Ireland, Sweden and Switzerland acceded to the activities of it. A committee of experts on the health control of foodstuffs was established within the Public Health Committee of this agreement. A subsidiary body of this committee is the committee of experts on flavouring substances (CEFS). This committee published the first edition of the "Flavouring Substances and Natural sources of Flavourings", the so-called "Blue Book", in 1970. Recently Volume I of the 4th edition has been published [18].

The 3rd edition contains a list of botanical and zoological source materials, subdivided in fruits and vegetables (N1), plants including herbs, spices and seasonings the use of which is acceptable (N2), plants with long history of use which are temporary acceptable (N3) and plants which are used, but which cannot be classified owing to insufficient information (N4).

The use of certain source materials for flavourings may be limited by the presence of an active principle the limit of which in the final product should not be exceeded. These limits for active principles, (e.g. ß-asarone, quassine, quinine, safrole, etc.) have been used for many national regulations and for the EC Flavour Directive [19].

The defined flavouring substances are divided into List 1 (No. 1 – 761), flavouring substances which may be added to foodstuffs without hazard to public health, and into List 2 (No. 2001 – 2353) flavouring substances which may be added temporarily to foodstuffs without hazard to public health. Specific limits in foods and beverages are proposed for many substances; in some instances these limits are considerably lower than the average maximum use levels of the substances, as determined by inquiries in the USA. In the 4th edition the flavouring substances are combined into a single list arranged according to their chemical structure. The substances are coded into categories A and B corresponding to lists I and II of the 3rd edition based on information available:

    A  flavouring substances which may be used in foodstuffs (referred to as List I substances in the 3rd edition);

    B  flavouring substances for which further information is required before the Committee of Experts is able to offer a firm opinion on their safety-in-use (referred to as List II substances in the 3rd edition). These substances can be used provisionally in foodstuffs.

No mention is made of the toxicological principles and procedures which have been used by the experts in the evaluation process. Some flavouring substances of the 3rd edition do not appear in the 4th edition or they have been "down graded" to category B without any explanation. Subsequently the recommendations of the "Blue Book" are only of limited value when compared to those of FEXPAN [14].

Furthermore, CEFS has published guidelines for smoke flavourings [20a], thermal process flavourings [20b] and flavouring preparations produced by enzymatic or microbiological processes [20c]. It is discussing plant tissue cultures for the production of flavourings.

## 7.4 European Union

### 7.4.1 General

The European Economic Communities (EEC) began with the treaty of Rome on 28.3.1957 signed by Belgium, France, Germany, Italy, Luxembourg and the Netherlands. In 1972 the United Kingdom, Ireland and Denmark joined the communities and 15 years later Greece, Spain and Portugal. The treaty of Maastricht 1992 which became effective on 1.11.1993 changed the name of the confederation to "European Union". Between the twelve EU and the six EFTA (European Free Trade Agreement)-countries Austria, Finland, Iceland, Norway, Sweden, and Switzerland another agreement was reached which led to the foundation of the European Economic Area

(EEA) on 1.1.1994. The importance of this event in this context is that the EFTA countries agreed to adopt the EU food legislation in their own law. In the meantime Austria, Finland and Sweden have joined the European Union on 1.1.1995 and on 1.5.2004, the Czech Republic, Cyprus, Estonia, Hungary, Latvia, Lithuania, Malta, Slovakia and Slovenia.

The EU has the following legislative bodies: The Commission and the Council of the 15 member states, both established in Brussels, and the European Parliament which holds its meetings in Strasbourg.

The Commission gets assistance from several advisory committees, the most important in this connection is the Scientific Committee on Foods, SFC, established in 1969 (69/414/EEC) and confirmed 1995 (95/273/EC) which is composed of 20 highly qualified scientists and which gives opinion at the request of the Commission on any problem concerning public health resulting from the consumption of foodstuffs [21a+b].

The SCF was replaced in 2002 by the European Food Safety Authority (EFSA) established through the General Regulation on Foodstuffs: (EC) No. 178/2002 of 28 January 2002 laying down the general principles and requirements of food law, establishing the EFSA and laying down procedures in matters of food safety [74]. This can be considered as the "EU General Food Law". EFSA with its place of business and legal seat in Parma, Italy, consists of an Advisory Forum for Food Safety with delegates from the Member States, a Scientific Committee SCC and 8 scientific panels; to be mentioned in this context is the Panel AFC which deals with food additives, flavourings, processing aids and materials in contact with food.

Since the legislative procedure of the EU is rather complicated it will be refrained from giving details, except some general remarks: Only the Commission can submit proposals for new rules. Any such proposal has to go through the Council of Ministers and the European Parliament (EP) before the rule is finally published either via the "consultation procedure" (one reading in the EP), the "co-operation procedure" (two readings in the EP) or, since 1.11.1993, the "co-decision procedure" (two readings in the EP and if EP rejects a common position by establishing a conciliation committee).

Such rules can be directives which are binding only for the member states. They have to be implemented into their own national law. Or they can be regulations which are binding for any subject to law in the EU. And finally it can be a decision which is binding for only the addressee(s) of it.

A lot of vertical (product-related) and horizontal rules have been published within food legislation since 1962. In 1988 the Commission decided to concentrate on horizontal rules which up to now deal with labelling, materials and articles in contact with foodstuffs, additives, flavourings, processing aids, pesticide residues, and official control of foodstuffs.

## 7.4.2 Flavour Legislation

According to the general food law [74] flavourings fall within the definition of "food". They are regulated by the Flavour Directive 88/388/EEC which has been published in 1988 *[19]* after many years of discussion and consultation. It puts in order the categorising of flavouring ingredients, (Art. 1), the labelling of flavourings (Art. 9), and purity criteria and maximum limits for certain active principles (Art. 4). It does not have provisions concerning flavouring substances and other flavouring ingredients (Art. 5) and adjuncts for flavourings (Art. 6). Specific rules for these purposes will have to follow. The Flavour Directive has been amended by Directive 91/71/EEC *[22]* which regulates the labelling of flavourings for the end consumer.

These directives have been implemented by the member states in the following national laws (examples):

- *Austria*: Österr. Lebensmittelbuch, III. Auflage, Kapitel A9 "Aromen" 3.11.93 (Mitt. Österr. Sanitätsverwaltung, Heft 12/1993)

- *Belgium*: 24.1.90 Arrêté royal relatif aux arômes destinés à être utilisés dans les denrées alimentaires – Moniteur Belge 24.04.90, 7634-7637; modified by 3.3.92 Arrête royal – Moniteur Belge 15.04.92, 8465-8466

- *Denmark*: Bekendtgørelse om aromaer m.m., der ma anvendes til Levnedsmidler – Sundhedsministeriets bekendtgørelse nr. 560 af 1. august 1990; revised in nr. 529 af 9. juni 1992

- *Finland*: Nr. 522 Förordning om aromen och deras tillverkningsämnen 12.6.92 (statens tryckericentral, Helsingfors 1992, 1607-1609)

- *France*: 1. Décret no 91-366 du 11 avril 1991 relatif aux arômes destinés â être employés dans les denrées alimentaires (Journal officiel de la Republique Francaise 17. avril 1991, 5044-5045) modified by Décret no 92-814 du 17 août 1992 (Journal officiel de la Republique Francaise 22 août 1992, 11473)
  2. Arrêté du 11 Juillet 1991 relatif à l'etablissement de critêres generaux de qualité et de pureté pour les arômes alimentaires (Journal officiel de la Republique Francaise 2 août 1991, 10271-10272).

- *Germany*: Verordnung zur Änderung der Aromenverordnung und anderer lebensmittelrechtlicher Verordnungen 29.10.91 (Bundesgesetzblatt Teil I Nr. 61, 2045-2050).

- *Greece* (Translation): Number 1437/89 Approval of the amendment of the article 44 of the Food Code in compliance with the directive nr. 88/388/EEC (Official Gazette of the Greek Republic, Athens, 17. September 1990, 2nd issue, edition Nr. 599).

- *Ireland*: European Communities (Flavourings for use in foodstuffs for human consumption) Regulations, 1992 – 30. Jan. 1992 (Statutory Instruments S.I. Nr. 22 of 1992)

- *Italy*: Decreto Legislativo 25 gennaio 1992, n. 107 – Attuazione delle direttive 88/388/CEE e 91/71/CEE relative agli aromi destinati ad essere impiegati nei prodotti alimentari ed ai materiali di base per la loro preparazione (Supplimento ordinario alla GAZZETTA UFFICIALE, Serie generale – n. 39, 17-2-1992)

- *Luxembourg*: 1. Règlement grand-ducal du 20 décembre 1990 relatif aux arômes destinés à être employeés dans les denrées alimentaires (Luxembourg Memorial A, 31. Dec. 1990, 1544-1548) 2. Règlement ministériel du 21 avril 1992 concernant l'etiquetage des aromes destinés à être vendus au consommateur final (Luxembourg Memorial A, No. 27, 8 mai 1992, 905)

- *Netherlands*: 1. Besluit van 4 februari 1992, houdende Warenwetbesluit Aroma's (Staatsblad van het Koninkrijk der Nederlanden, Jaargang 1992, Nr. 95) 2. Warenwetregeling Stoffen in aroma's (Staatscourant 51, 12 maart 1992)

- *Portugal*: Portario no 620/90 de 2 de Agosto (Diario da Republica – 1 Serie Nr. 178 – 3-8-1990, 3171-3173)

- *Spain*: 1. Real Decreto 1477/1990, de 2 de novembre, por el que se aprueba la Reglementación Técnico-Sanitaria de los aromas que se utilizan en los productos alimenticions y de los materiales de base para su produción (BOE num. 280, Jueves 22 novembre 1990, 34604-34611) 2. Real Decreto 1320/1992, de 30 de octobre, por el que se modifica el Real Decreto 1477/1990, (BOE num. 279, Viernes 20 novembre 1992, 39385)

- *Sweden*: Statens Livsmedelsverks kungörelse om ändring i Kungörelsen (SLV FS 1993:34) med föreskrifter och allmänna råd om aromer m.m. (SLV FS 1996: 1-1996-03-25)

- *United Kingdom*: The Flavourings in Food Regulations 1992, 24th August 1992 (Statutory Instruments 1992 no. 1971).

A major revision of the EC-Directive 88/388 is underway since end of 2002. Among other things that are under discussion, it will be changed to a regulation. A final draft is expected for the end of 2006.

Together with the Flavour Directive a Council Decision 88/389/EEC *[23]* has been published asking the Commission to establish an inventory of source materials for flavourings, flavouring substances (see Part 3.2.1.2. of this book), and source materials and reaction conditions for process flavourings and smoke flavourings. All of these have been compiled by EFFA *[24]* and handed over to the Commission. The Commission has not published them.

In amending the Flavour Directive, Article 5, the Commission has prepared a draft "Proposal for a council directive establishing a list of flavouring substances used in foodstuffs" following the SCF guidelines for the evaluation of flavourings for use in foodstuffs: I. Chemically defined flavouring substances *[16]*.

The final Regulation (EC) No. 2232/96 laying down a Community procedure for flavouring substances used or intended for use in or on foodstuffs was published in 1996 *[17]*.

This regulation will end in a positive list of flavouring substances to be used in foodstuffs in a stepwise procedure as follows: According to Article 3 the Member States had to notify to the Commission a list of the flavouring substances which may be used in foodstuffs marketed on their territory. On the basis of the notification the flavouring substances have been entered in a ‚register' *[82]*. The register can be amended, the intellectual property rights of the flavour manufacturers shall be protected. In Article 4 details are put down for a programme for the evaluation of these flavouring substances resulting in Regulation (EC) No. 1565/2000 laying down the measures necessary for the adoption of an evaluation programme in application of Reg. 2232/96 *[83]*. That programme defines the order of priorities according to which the flavouring substances are to be examined and the flavouring substances which are to be subject of the scientific cooperation. According to Article 5 the (positive) list of flavouring substances shall be evaluated and adopted within 5 years according to Regulation (EC) 622/2002 *[84]*. The flavouring substances in the register have been grouped according to their chemical structure for easier evaluation procedure; for details see chapter 3.2.1.2. The deadline for final evaluation was in 2004, but the resulting list will not be published before 2008 as the evaluation takes more time than expected. Until April 2005 only the following groups (see 3.2.1.2) had been evaluated: 1+2 (2003), 3+6 (2004) and 7, 9+11 (2005) that means 7 out of 34 (= 20%). New additions to that list may be authorized following evaluation.

The general criteria for the use of flavouring substances in the Annex to this regulation have been mentioned above, see 7.2.

In Article 5 of the Flavour Directive *[19]* it is said: "The Council, acting in accordance with the procedure laid down in Article 100a of the Treaty, shall adopt 1. Appropriate provisions concerning: "… source materials used for the production of smoke flavourings or process flavourings, and the reaction conditions under which they are prepared". The EU legislature has complied with this provision and finalised Regulation (EC) No. 2065/2003 on smoke flavourings *[85]*. Smoke flavourings have been regulated first of all flavouring sources, as there have been different legal and administration rules for evaluation and approval of smoke flavourings in the EU. In Germany, Denmark and Sweden they had to be approved, in all other Member States they could be sold without any restriction. The Regulation puts in order in 21 articles production, safety requirements, approval, traceability, and control measures. The production starts with condensing or collecting the smoke in liquid, so-called "primary products". These will be used for making different smoke flavourings. The toxicological assessment shall concentrate on the primary products. A procedure for safety evaluation and approval of smoke flavourings has been established in this Regulation on base of these opinions. In order to get an approval the applicant has to give detailed information about the production method, use, specification, etc. (see 3.2.4.2). According to Article 10, the EU Commission shall establish a Community list of approved "primary products". The approval is valid according to Article 9 for ten years and may be prolonged by application. Former national regulations will be annulled after 16.06.2005. Annex I indicates conditions for production of smoke condensates ("primary products") and Annex II the necessary information for scientific evaluation of these products. The necessary data are contained in the "SCF report on smoke flavourings" *[25]* which is based on the Council of Europe guidelines *[20a]*. It contains technical and toxicological considerations, categorising smoke flavourings and listing principles of toxicological evaluation.

Another concept had to be prepared in completion of Directive 88/388/EEC: *[19]*, Article 6, in accordance with the procedure set out in Article 10:1. The list of substances or materials authorised in the Community as:

- additives necessary for the storage and use of flavourings
- products used for dissolving and diluting flavourings, etc.

The legislator has answered this demand with Directive 2003/114/EC, amending Directive 95/2/EC *[29]*. For details on food additives see 7.4.3. According to the amending Directive, generally authorised food additives of Annex I may be used *quantum satis* (without limitations) in flavourings; furthermore some additives may be used with limitations, i.e. solvents, preservatives, antioxidants, phosphates, polysorbates and E 416 Karaya gum, E 459 β-Cyclodextrin, E 553 Silicon dioxide, and E 900 Dimethyl polysiloxane. The levels of food additives present in flavourings should be the minimum required to achieve the intended purpose. In addition, consumers should be guaranteed correct, adequate and non-misleading information and the additive should be of no health hazard. It does not have a technological function in the foodstuff, otherwise it is a mixture "flavouring + food additive" which has to be labelled accordingly. This applies also to colours and sweeteners which are regarded as always having a technological function in the aromatised food. Food manufacturers should be informed about the concentration of all additives in flavourings in order to enable them to comply with Community legislation. The Directive has to be implemented into national law of the Member States by July 2005.

Finally some words about labelling of flavourings in the list of ingredients of the final food which is regulated by the Labelling Directive 79/112/EEC. This was amended by Directive 91/72/EEC *[26]* which inserted a new Annex III stating that flavouring shall be designated either by the word "flavouring(s)" or by a more specific name or description of the flavouring. According to Article 9 of the Flavour Directive *[19]*, a flavouring is "natural" if its aromatising components consist only of natural flavouring substances and/or flavouring preparations. A "natural apple flavour" may consist exclusively or almost exclusively of components from apple. "Almost exclusively" means rounding off the taste with components from other natural sources is possible, in this case, for example, with pear.

## 7.4.3 Food Additive Legislation

In general a food additive may be permitted in the EU if the European Food Safety Authority *[74]* has found that it is harmless to the health of consumers and that it has an established ADI (acceptable daily intake). Besides this, there must be a technological need for the substance.

Presently food additives have been permitted in the EU by 3 directives in general or for special food use. The additives are divided in the following categories: sweeteners, colours and other. Specific criteria of purity for sweeteners and colours have been published in 1995 (95/31/EC last amended by 2004/46/EC, and 95/45/EC, last amended by 2004/47/EC), the other ones in 1996 (96/77/EC, last amended by 2004/45/EC).

The directive on colouring matters was the first directive ever adopted and published by the EU commission *[27a]*, now overruled by *[27b]*. Preservatives had been regulated by Directive 64/54/EEC *[28a]* with 24 amendments. The directive on antioxidants followed six years later *[28b]*. The category of food additives, the emulsifiers, stabilisers, thickeners and gelling agents had been regulated by Directive 74/329 *[28c]*. These three categories are now regulated by the European Parliament and Council Directive 95/2/EC of February, 20th, 1995, amended the fifth time by Directive 2003/114/EC, another amendment being underway *[29]*. Rules on sweeteners are laid down in 94/35/EC of 30.6.94, last amended by 2003/115/EC of 22.12.2003 *[30]*.

General conditions for use in foodstuffs and approval of these substances have been regulated by the so-called Framework Directive *[31]*. Annex I of this directive lists the categories of additives for which individual additives and related conditions of use have to be specified, whilst Annex II contains the general criteria governing the inclusion of such additives in the lists. The directive also lays down general labelling requirements. A safeguard clause allows Member States to provisionally suspend or restrict the use of specific additives if they perceive danger to human health, even though such additives do conform to the directive.

An amendment has been adopted by the Council on 16.6.1994 and published on 10.9.1994 allowing national derogations for "traditional" food products *[75]*.

A revision of the "Framework Directive" by a regulation has been under discussion in Brussels since 2002 and there will be a regulation on enzymes.

### 7.4.4 Other Rules Concerning Flavourings

Some rules of the EU are significant for flavourings and therefore they must be mentioned in this context.

First and most important is the Directive on Extraction Solvents *[71]*. It contains lists of solvents which may be used during the processing of raw materials for foodstuffs, food components or food ingredients. The maximum residue limits and conditions of use are specified for a number of them. Annex I consists of those extraction solvents which may be used in compliance with food manufacturing practice for all purposes, Annex II and III lists those for which conditions are specified. Annex II lists solvents for food products and Annex III lists those for flavouring preparations from natural raw materials. The directive has been amended in 1992 deleting cyclohexane and isobutane from Annex III. A second amendment, however, has added cyclohexane again (Directive 94/52/EC of 7.12.94). For the complete list of permitted extraction solvents for flavouring ingredients see chapter 3.3.1, extraction solvents.

In flavouring preparations and natural flavouring ingredients you may find pesticide residues, contamination by radioactive fallout and other contaminants. For pesticide residues in foodstuffs of plant origin there are three directives, 76/895 last amended by Regulation (EC) No. 807/2003), 86/362/EEC (last amended by Directive 2004/61/EC) and 90/642/EEC (last amended by Directive 2004/95/EC) and in foodstuffs of animal origin the Directive 86/363/EEC (last amended by Directive 2004/61/EC). Furthermore, there is Directive 91/414/EEC (last amended by Directive 2004/99/EC) concerning the placing of plant protection products on the market. All of these directives will be replaced step by step by Regulation (EC) No. 393/2005 on maximum residue levels of pesticides in food and feed of plant and animal origin and amending Directive 91/414/EEC *[73]*. In analysing flavouring ingredients for these residues the concentration process has to be kept in mind.

Other contaminants of concern are mycotoxins, dioxins, and 3-monochloropropane-1,2-diol (3-MCPD) *[86]*.

After the 'Chernobyl-accident' many food products had been contaminated by radioactivity. A first regulation 1707/86 had established limits for Cs 134 and 137 in Bq/kg. This regulation has been replaced by Council Regulation (Euratom) No 3954/87 *[87]*. This regulation is of special importance for raw materials originating from Eastern Europe countries.

Genetic engineering has also to be mentioned. At first all novel food products have been regulated by Directive 90/220/EEC. Later the Regulation (EC) No. 1829/2001 on genetically modified food and feed was issued covering both food additives and flavourings *[88]*.

Last but not least, products causing allergic reactions have to be mentioned. The Directive 2003/89/EC amending Directive 2000/13/EC relating to labelling of foodstuffs *[89]* demands clear labelling of the following ingredients: cereals containing gluten, crustaceans, eggs, fish, peanuts, soybeans, milk, nuts, celery, mustard, sesame seeds and products thereof, and finally sulphur dioxide and sulphites at concentrations of > 10 mg/kg. Flavourings are included in the labelling.

Finally quite another subject is to be mentioned. It is not a food regulation, but derives from the EC legislation on the classification and labelling of those chemicals which have to be regarded as dangerous substances *[32]*. If chemicals (that may be flavouring ingredients as well) have been classified according to this rule as dangerous they have to be labelled as is shown in Figure 7.1.

Flavouring ingredients may be flammable, toxic, corrosive, irritant, harmful, and/or dangerous for the environment.

The nature of the special risks attributed to dangerous substances has to be indicated by so-called R-phrases (i.e. R 21 Flammable). Safety/precautions/stock keeping/cleaning/first aid advice etc. concerning dangerous substances have to be labelled by so-called S-phrases (i.e. S 35 Keep away from acids). In 1988 this classification and labelling of dangerous substances has been developed further to cover "dangerous preparations" *[33]* defining those mixtures which contain at least one dangerous substance. Foodstuffs are exempted according to Article 1(3) but only "in a finished stage intended for the final consumer". Therefore food ingredients like flavourings fall under this legislation. The mixture may be tested by physical methods whether it is flammable etc. or by toxicological tests whether it is harmful etc. The latter procedure may be replaced by a mathematical method compounding the relative risks of the single ingredients (by percentage). For details the directive has to be consulted or a practical handbook issued by the EEC Commission *[34]*. This Directive has been implemented in national law by the Member States as shown in Table 7.1.

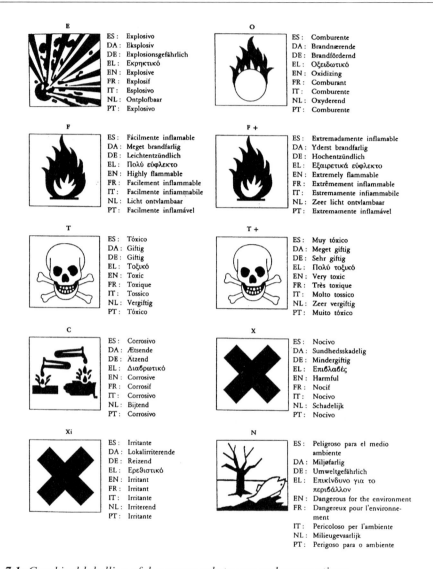

*Fig. 7.1:* Graphical labelling of dangerous substances and preparations

*Table 7.1: Implementation of Directive 88/379/EEC*

| Member State | Date | Enacted | Transition period until |
|---|---|---|---|
| EEC | 07.06.88 | to be implemented until 07.06.91 | 07.06.92 |
| Austria | - | - | - |
| Belgium (Gazette No.97) | 17.05.93 | | |
| Denmark (Bekendtgørelse No.586) | 08.08.91 | 21.08.91 | 08.06.92 |
| Finland | - | - | - |
| France (Arrêté) | 21.02.90 | 01.06.91 | 07.06.92 |
| Germany (BGBL I Nr. 57) | 30.10.93 | 26.10.93 | 30.04.94 |
| Greece (Gazette No.567B) | 06.09.90 | ? | ? |
| Ireland | - | - | - |
| Italy (Decreto No.46) | 29.02.92 | 28.02.93 | ((31.12.92)) |
| Luxembourg | - | - | - |
| Netherlands (Besluit No.534) | 08.09.91 | 01.01.92 | 08.06.92 |
| Portugal (Decreto-Lei No.120/92) | 01.06.92 | | 01.01.94 |
| Spain (Real Decreto 1078/1993) | 10.03.95 | 10.03.95 | |
| Sweden | - | - | - |
| UK | - | - | - |

## 7.5 America

### 7.5.1 Introduction

Inspired by the great economic success of the European Union (formerly European Economic Communities – EEC) other countries have made plans to form agreements or unions bearing some resemblance to the EU. In November 1993 the "North American Free Trade Agreement – NAFTA" was agreed upon between Canada, Mexico and the United States of America. It became effective as of 1.1.94. It is only a free trade agreement, but nevertheless the first step of an alliance. Therefore, in the next subchapter these three countries will be dealt with together, in the order USA, Canada, Mexico.

On 29.11.1991 another agreement was signed between Argentina, Brazil, Paraguay and Uruguay, the "Creation Letter" for the so-called "MERCOSUR". It started fully operating on 1.1.1995. Chile joined 1.1.1996. MERCOSUR goes further than NAFTA and it plans to harmonise legislation similar to the EU. In the area of food legislation the countries have set up various technical groups which prepare standards on additives, labelling, packaging, weights and measures etc. The member countries are committed to adopt these standards. Details for flavourings will be given in subchapter 7.5.3. Besides the MERCOSUR countries, Colombia will be mentioned.

## 7.5.2 The NAFTA Countries

### USA

The Federal Food, Drug, and Cosmetic Act (FFDCA) of 1938 of the United States of America was completed in 1958 by the Food Additives Amendment. This amendment established the requirement of "safety", which led to a major change in the Food and Drug Administration's (FDA) approach to its activities. The US Congress applied the term "safe" as the criterion for action, thereby supplementing, but not replacing, the use of the term "adulteration". The result was a shift in focus – from an approach which emphasised evaluation of the food in its entirety to one which emphasised evaluation of individual food components.

The Congress described the term "safety" as "reasonable certainty that no harm will result from the proposed use of an additive", recognising that it is not possible to prove that no harm will result under any conceivable circumstances. However, it was generally interpreted to mean that if an additive is safe, there is no risk associated with it. Accordingly, the amendment directed FDA to consider consumption patterns, the cumulative effect of the ingredient in the diet of man or animals, and appropriate safety factors.

Under the Food Additives Amendment, FDA's regulatory activities became divided between continuing the past practices which relied on the prohibitions against adulteration under Section 402 and the even more elaborate procedures under the new provisions of the Food, Drug and Cosmetic Act in Section 409.

Congress, furthermore, changed the traditional requirement that FDA prove adulteration, by establishing the then novel approach that industry must prove safety. Therefore, FDA was able to place the data-collection burden on the petitioners. FDA became a reviewer, a specialised evaluator.

According to Section 201 (s) of the FFDCA as amended

- The term "food additive" means any substance the intended use of which results or may reasonably be expected to result, directly or indirectly, in its becoming a component or otherwise affecting the characteristics of any food (including any substance intended for use in producing, manufacturing, packing, processing, preparing, treating, packaging, transporting, or holding food; and including any source of radiation intended for any such use), if such substance is not generally recognised, among experts qualified by scientific training and experience to evaluate its safety, as having been adequately shown through scientific procedures (or, in the case of a substance used in food prior to January 1, 1958, through either scientific procedures or experience based on common use in food) to be safe under the conditions of its intended use; except that such term does not include

    (1) a pesticide chemical in or on a raw agricultural commodity; or

    (2) a pesticide chemical to the extent that it is intended for use or is used in the production, storage, or transportation of any raw agricultural commodity; or

    (3) a colour additive; or

(4) any substance used in accordance with a sanction or approval granted prior to the enactment of this paragraph pursuant to this Act, the Poultry Products Inspection Act (21 U.S.C. 451 and the following) or the Meat Inspection Act of March 4, 1907 (34 Stat. 1260), as amended and extended (21 U.S.C. 71 and the following); or

(5) a new animal drug.

The safety of food additives is regulated in section 409(a):

- A food additive shall, with respect to any particular use or intended use of such additives, be deemed to be unsafe for the purposes of the application of clause (2)(C) of section 402(a), unless -

 there is in effect, and it and its use or intended use are in conformity with, a regulation issued under this section prescribing the conditions under which such additive may be safely used.

While such a regulation relating to a food additive is in effect, a food shall not, by reason of bearing or containing such an additive in accordance with the regulation, be considered adulterated.

If the Food Additives Amendment had required that all food ingredients in use in 1958 be evaluated as food additives, the marketing of the entire food supply would have been severely disrupted. To prevent that from occurring, the US Congress incorporated into the amendment two grandfather clauses. One provided that if a substance had a "prior sanction" (a specific approval of its use prior to the amendment), it would not be considered a food additive and would not necessitate a petition and regulation for its use. A prior-sanctioned ingredient would continue to be subject to the "adulteration" limitations of Section 409. The second grandfather clause defined a category of substances which would not be considered food additives on the basis of general recognition of safety by the scientific community. The "generally recognised as safe" (GRAS) status would derive from demonstrable common use in food prior to 1958 or from scientific data which demonstrated its safety. Rather than to be subjected strictly to the limitations of the "adulteration" provisions or the same requirements that are applied to food additives, GRAS ingredients are subjected to very ambiguous standards of "general recognition of safety" [35].

Permitted food additives are listed in the CFR Title 21 Part 170 180, substances generally recognised as safe in Part 182-184. Because it is not so easy to find a given substance in the CFR the "Food Chemical News Guide" [36] is recommended. It lists all substances in alphabetical order.

Flavourings fall under the definition of food additives. They therefore need either an approval or an evaluation as GRAS. For better understanding see Table 7.2 and [90].

In response to the Food Additives Amendment, the Flavor and Extract Manufacturers' Association (FEMA) organized in 1960 a Panel of Expert Scientists (FEXPAN), not affiliated with the flavour industry, but with special competencies in toxicology, pharmacology and biochemistry. The Panel evaluated the "GRAS status" of the many hundreds of organic chemicals contained in natural and artificial foods and flavourings. The FDA was kept informed of this activity of FEMA. Subsequent to the release in 1965 of the first GRAS list of some 1100 flavouring substances and their use levels, the FDA temporarily waived bulk labelling requirements and adopted virtually the entire list.

The Panel has further evaluated the safety of flavouring substances. The results of these studies have been reported in a series of papers in "Food Technology". The publications 'GRAS 3' to 'GRAS 21' are covering substances with FEMA numbers 2001 to 4068 and are reporting names, synonyms and average maximum use levels in foods. The Panel used these use levels as basis for its judgement that the substances are "generally recognised as safe" for their intended uses. These figures do not have the meaning of limits or restrictions.

The principles and rationales adopted in evaluating the safety of present or proposed flavouring substances to determine whether or not such ingredients can be regarded as GRAS have been described in the above cited reports and summarised in a paper by L.A. Woods and J.Doull [14] as mentioned before.

Of similar importance is the labelling of flavourings and flavoured food products.

Label regulations relating to use of flavourings in foods are described in § 101.22 of Title 21 of the Code of Federal Regulations.

If the flavouring consists of one ingredient, it must be declared by its common or usual name. If the flavouring consists of two or more ingredients, the label may either declare each ingredient by a common or usual name or state "All flavor ingredients contained in this product are approved for use in a regulation of the FDA or are listed as generally recognized as safe on a reliable published association list". Non-flavouring ingredients like solvents, carriers, flavour enhancers etc. have to be listed separately.

The label of a food to which a flavour is added may state "natural flavor", "artificial flavor" or any combination thereof, as the case may be. Foods which contain "both a characterising flavor from the product whose flavor is simulated and other natural flavors which simulates, resembles or reinforces the characterizing flavor" must be labelled "with other natural flavor" (WONF).

Any incidental additives which are present in the food at insignificant levels and which do not have any technical or functional effect in that food are exempt from food labelling. This is the case for most of the non-flavouring ingredients of a flavouring with the exception of e.g. salt and monosodium glutamate.

For completeness here are the definitions for natural and artificial flavourings:

- The term "natural flavor" or "natural flavoring" means the essential oil, oleoresin, essence or extractive, protein hydrolysate, distillate, or any product of roasting, heating or enzymolysis, which contains the flavoring constituents derived from a spice, fruit or fruit juice, vegetable or vegetable juice, edible yeast, herb, bark, bud, root, leaf or similar plant material, meat, seafood, poultry, eggs, dairy products, or fermentation products thereof, whose significant function in food is flavoring rather than nutritional. Natural flavors include the natural essence or extractives obtained from plants listed in §§

182.10, 182.20, 182.40, and 182.50 and part 184 of this chapter, and the substances listed in § 172.510 of this chapter.

- The term "artificial flavor" or "artificial flavoring" means any substance, the function of which is to impart flavor, which is not derived from a spice, fruit or fruit juice, vegetable or vegetable juice, edible yeast, herb, bark, bud, root, leaf or similar plant material, meat, fish, poultry, eggs, dairy products, or fermentation products thereof. Artificial flavor includes the substances listed in §§ 172.515(b) and 182.60 of this chapter except where these are derived from natural sources *[37]*.

*Table 7.2: Permitted flavourings in the USA (FDA-CFR = Code of Federal regulations, Title 21)*

| | Flavourings | | | |
|---|---|---|---|---|
| | Approved Food Additives | | Safe by recognition of experts | |
| FDA-CFR | 172.510 | Natural flavoring substances and natural substances used in conjunction with flavors. | 182.10 | Spices and other natural seasonings and flavorings. |
| | 172.515 | Synthetic flavoring substances and adjuvants. | 182.20 | Essential oils, oleoresins (solvent-free) and natural extractives (including distillates) |
| | 172.520 | Cocoa with dioctyl sodium sulfoccinate for manufacturing. | 182.40 | Natural extractives (solvent-free) used in conjunction with spices, seasonings, and flavorings. |
| | 172.530 | Disodium guanylate. | | |
| | 172.535 | Disodium inosinate. | 182.50 | Certain other spices, seasonings, essential oils, oleoresins, and natural extracts. |
| | 172.540 | DL-Alanine. | | |
| | 172.560 | Modified hop extract. | | |
| | 172.575 | Quinine. | | |
| | 172.580 | Safrole-free extract of sassafras. | 182.60 | Synthetic flavoring substances and Adjuvants. |
| | 172.585 | Sugar beet extract flavour base. | | |
| | 72.590 | Yeast-malt sprout extract. | | |
| Private | | | FEMA Expert Panel | |
| | | | Other Expert Panels | |

For further details see "Flavor labelling in the United States" by K.J.Bauer in *[38]*.

In accordance with the EU, the US Congress has passed the Food Allergen Labelling and Consumer Protection Act of 2004 (FALCPA) to be enforced on 1.1.2006. Eight products possibly causing allergenic reactions have to be labelled (also in flavourings): milk, eggs, fish, crustaceans, tree nuts (almonds, pecan and walnuts), wheat, peanuts, and soybeans. Exceptions are e.g. refined edible oils *[91]*.

## Canada

After the USA, Canada is the second most important country in North America. Like other countries, Canada does have a Food and Drug Act *[39]* which consists of 4 parts. Details are governed by Food and Drug Regulations. Also these regulations consist of several parts. In Part B – Foods – Division 1 is a general one, Division 10 deals with Flavouring Preparations and Division 16 with Food Additives. Sweeteners are regulated in Part E.

Division 10 defines the following terms:

- *Extract or Essence* (B 10.003) shall be a solution in ethyl alcohol, glycerol, propylene glycol or any combination of these, of sapid or odorous principles, or both, derived from the plant after which the flavouring extract or essence is named, and may contain water, a sweetening agent, food colour and a Class II preservative or Class IV preservative.

- *Artificial or Imitation Extract/Essence* (B 10.004) shall be a flavouring extract or essence except that the flavour principles shall be derived in whole, or in part, from sources other than the aromatic plant after which it is named, and if such extract or essence is defined in these Regulations, the flavour strength of the artificial or imitation extract or essence shall be not less than that of the extract of essence.

- *Flavour* (B 10.005)

    (a) shall be a preparation, other than a flavouring preparation described in section B 10.003 of sapid or odorous principles, or both, derived from the aromatic plant after which the flavour is named;

    (b) may contain a sweetening agent, food colour, Class II preservative, Class IV preservative or emulsifying agent; and

    (c) may have added to it the following liquids only:

        (i) water;

        (ii) any of, or any combination of, the following: benzyl triacetate, glyceryl tributyrate, isopropyl alcohol, monoglycerides and diglycerides, 1,2-propylene glycol or triethyl citrate;

        (iii) edible vegetable oil; and

        (iv) brominated vegetable oil, sucrose acetate isobutyrate or mixtures thereof, when such flavour is used in citrus-flavoured or spruce-flavoured beverages.

- *Artificial or Imitation Flavour* (B 10.006) shall be a flavour except that the flavouring principles may be derived in whole or in part from sources other than the aromatic plant after which it is named, and if such flavour is defined in these Regulations, the flavouring strength of the artificial or imitation flavour shall be not less than that of the flavour.

- *Extract / Essence / Flavour Naturally Fortified* (B 10.007) shall be an extract, essence or flavour derived from the named fruit to which other natural extractives have been added and 51 per cent of the flavouring strength shall be derived from the named fruit.

Special definitions exist for Almond, Anise and Celery Seed Extracts (B 10.009 – 10.011). According to B 10.008 the word "artificial" or "imitation" shall be an integral part of the name of such flavouring preparations in identical type.

Furthermore you find two definitions in Division 1 where the list of ingredients is defined. According to the amendments of Oct. 29,1992 [40] are

- *Flavour* (B 01.010(3)(b), table item 4) one or more substances prepared for their flavouring properties and produced from animal or vegetable raw materials or from food constituents derived solely from animal or vegetable raw materials.

- *Artificial (Imitation / Simulated) Flavour* (B 01.010(3)(b), table item 5) one or more substances prepared for their flavouring properties and derived in whole or in part from components obtained by chemical synthesis.

According to item 13 in the same table a flavouring labelled with the name of the plant or animal source followed by the word "flavour" has to be obtained solely from the plant or animal source after which the flavour is named.

As is easily seen these definitions are not congruent and somewhat confused and outdated. A revision has been under discussion for several years. In practice the word "imitation" is not used anymore. It has been replaced by the word "artificial" instead. The combination "natural and artificial" is similar to the one employed in the USA.

*The labelling for natural flavourings in practice is as follows:*

- *Natural "Orange" Flavour*
  for a flavouring which contains only flavouring ingredients derived from "Orange"

- *Natural Orange Flavour*
  for a flavouring in which 51 % of the naturally fortified perceived flavouring is derived from "Orange"

- *Natural Flavour Compound, Naturally fortified ("Orange" Type)*
  for a flavouring which may, but does not have to contain flavouring ingredients from "Orange".

As a consequence the Canadian Food Inspection Agency (CFIA) wants to change the labelling rules of the Food and Drug Regulations and published a discussion paper in January 2003 titled "Clarifying the labelling rules for highlighted ingredients, flavourings and sensory characteristic descriptions" [92].

As far as the labelling of the finished food is concerned, the ingredients and components of flavouring preparations, artificial flavouring preparations, spice mixtures, seasoning or herb mixtures etc. are not required to be declared on the label of the finished foods (B 01.009(2)). The following ingredients of preparations have, however, to be shown in the list of ingredients: salt, glutamic acid or its salts, hydrolysed plant protein, aspartame, potassium chloride and any ingredient that performs a function in, or has any effect on, that food (B 01.009(3)), which is in accordance with B 01.010(3), see above.

Food Additives are defined in B 01.001 as any substance, including any source of radiation, the use of which results, or may reasonably be expected to result in it or its by-products becoming a part of or affecting the characteristics of a food, but does not include:

(a) any nutritive material that is used, recognised, or commonly sold as an article or ingredient of food,

(b) vitamins, mineral nutrients and amino acids, other than those listed in the tables to Division 16,

(c) spices, seasonings, flavouring preparations, essential oils, oleoresins and natural extractives,

(d) agricultural chemicals, other than those listed in the tables to Division 16,

(e) food packaging materials and components thereof, and

(f) drugs recommended for administration to animals that may be consumed as food.

They are regulated in Division 16 by a list of permitted substances and they are grouped together as shown in Table 7.3.

**Mexico**

Food legislation in Mexico is described easily. A general food law exists. It was published on 18.1.1988 *[41]* and it regulates food in general, hygiene rules for manufacturing and equipment, food standards and food additives, including flavourings. Food additives are handled in "titulo noveno, articulo 657-701", page 71-79. They are classified in 21 categories, starting with flavour enhancers in Category I and ending with flavourings, Category XXI. A list of single substances or products is published in each category. These lists may be amended by the Health Ministry if necessary. Articulo 688 deals with flavourings in the following way:

| | | |
|---|---|---|
| I | | Essential oils |
| II | | Non-natural concentrates with essential oils |
| III | | Natural essences |
| IV | | Concentrates from essential oils and fruits |
| V | | Fruit juice concentrates |
| VI | | Artificial bases |
| VII | | Artificial essences |
| VIII | | Artificial concentrates |
| IX | | Artificial concentrates with fruit juice |
| X | | Extracts and distillates. |

Some flavouring ingredients are forbidden for weaning food, i.e. Coumarin, 4-Ethoxyphenyl Urea, Nitrobenzene, Safrol, Thujone etc. (articulo 701).

*Table 7.3: Permitted groups of food additives in Canada*

| Table No. | Title |
|---|---|
| I | Food additives that may be used as anticaking agents |
| II | Food additives that may be used as bleaching maturing and dough conditioning agents |
| III | Food additives that may be used as colouring agents |
| IV | Food additives that may be used as emulsifying, gelling, stabilising, and thickening agents |
| V | Food additives that may be used as food enzymes |
| VI | Food additives that may be used as firming agents |
| VII | Food additives that may be used as glazing and polishing agents |
| VIII | Miscellaneous food additives |
| IX | Revoked by P.C. 1977-2550 of September 15, 1977 |
| X | Food additives that may be used as pH adjusting agents, acid-reacting materials and water correcting agents |
| XI (I) | Food additives that may be used as class I preservatives |
| (II) | Food additives that may be used as class II preservatives |
| (III) | Food additives that may be used as class III preservatives |
| (IV) | Food additives that may be used as class IV preservatives |
| XII | Food additives that may be used as sequestering agents |
| XIII | Food additives that may be used as starch modifying agents |
| XIV | Food additives that may be used as yeast foods |
| XV | Food additives that may be used as carrier or extraction solvents |

## 7.5.3 South American Countries

### Brazil

In 1977 a flavouring regulation was published as "Resulacão Nr.22/76" *[42]*. It had been put together according to recommendations of I.O.F.I. *[1]*.

Definitions are given for flavouring (1.1.), natural flavouring raw material (1.2.), natural flavouring product (1.3.), natural flavouring substance (1.4.), nature-identical flavouring substance (1.5.) and artificial flavouring substance (1.6). Flavourings are classified as natural flavouring (2.1.), reinforced natural flavouring (2.2.), reconstituted flavouring (2.3.), imitation flavouring (2.4.) and artificial flavouring (2.5.).

Reinforcing means that limited quantities can be used of natural or nature-identical substances already present in the product which is to be reinforced (2.2.).

Imitation flavourings contain the same substances as reinforced flavourings, but in unlimited amounts, or only natural and nature-identical substances existing in the product which is imitated, or natural raw materials from the product which is imitated and any other natural product or substance or nature-identical substance (2.4.). The artificial flavouring contains either an artificial substance or a natural or nature-identical substance not existing in the flavouring which gives the name (2.5.).

Designations for reinforced flavourings are "aroma natural reforcado de ..." or "aroma reforcado de ..." (3.5.), for reconstituted flavourings "aroma reconstitudo de ..." or "aroma de ... reconstitudo" (3.6.). Natural flavourings without defined flavour are called "natural composto" and they can be given fantasy names (3.7.). Optional designations may indicate the specific intended use (3.8.).

Raw materials, flavourings and flavouring substances included in officially adopted lists from FDA, FEMA, Council of Europe, or others, can be used, provided that the respective use restrictions are respected.

The label of the flavouring has to mention the designation, the application and quantity to be used (if there are use restrictions), the nature and percentage of the solvent used (when there is a limit).

When the purpose of the flavouring is to impart a definite taste, the label of the foodstuff must indicate whether natural or reinforced flavourings are used "sabor natural de ...", "sabor de..."; for reconstituted flavourings "sabor reconstituido de ...", for imitation flavourings "sabor imitacao de ...", for artificial flavourings "sabor artificial de ...".

When the purpose is to enhance the natural flavour, or to impart to the foodstuff a non-specific flavour, the label statement should be: "Contém aromatizante natural de ...", "Contém aromatizante de ...", "Contém aromatizante natural composto", "Contém aromatizante natural reforcado de ...", "Contém aromatizante reconstitudo de ...", "Contém aromatizante imitacao de ...", "Aromatizado artificialmente".

This flavouring regulation will be replaced by the MERCOSUR Res. No.46/93 *[43]*: After a definition (1.) flavourings are classified into

    2.1.    Natural ingredients
    2.1.1.  essential oils
    2.1.2.  extracts
    2.1.3.  resins

    2.2.    Synthetic ingredients
    2.2.1.  nature-identical
    2.2.2.  artificial

    2.3.    Flavouring blends
    2.4.    Process flavourings (with Anexo 1)
    2.5.    Smoke flavourings (with Anexo 2)

Flavourings can be dry, liquid or paste-like (3.).

In part 5 the authorised lists of ingredients are mentioned. According to 5.1. all materials published in "Flavor and Fragrance Materials" *[15]* may be used with due consideration for restrictions of certain substances in part 7 and 8. Part 7 is identical with Annex II of the EC Flavour Directive *[19]*. Furthermore (5.3.) the flavouring ingredients listed by FAO/WHO, Council of Europe *[18]*, FDA and F.E.M.A. (see *[36]*) are authorised for use. A list of botanicals is contained in Anexo 3 according to 5.2.

Lists of additives have been expanded on in part 6:

- 6.1. Solvents and Carriers
- 6.2. Antioxidants
- 6.3. Sequestrants
- 6.4. Preservatives
- 6.5. Emulsifiers and Stabilizers
- 6.6. Weighting Agents
- 6.7. Acids, Bases and Salts
- 6.8. Anticaking Agents
- 6.9. Extraction Solvents

The member states have adopted this Regulation so that it became effective on 1.1.1995.

Food additives have been defined in MERCOSUR Res. No. 04/93 (44a), Anexo A. It contains 22 categories. Single substances are compiled in MERCOSUR Res. No. 19/93 *[44b]*. These two regulations also became effective on 1.1.1995.

**Argentina**

The "Codigo alimentario Argentiono" (updated issue 1977) contains the following regulations for flavourings:

Bitter substances are defined (Art. 1292); negative lists (Art. 1293) and a positive list with 11 bitter substances (Art. 1294) are given.

Flavouring additives are defined (Art. 1298); they comprise the following categories:

(1) natural essences

(2) extracts

(3) balsams, oleoresins and gums

(4) substances isolated from essential oils or extracts

(5) synthetic or artificial chemical flavouring substances (Art. 1299).

The definitions for extracts make a distinction between liquid extracts and dry extracts: concretes, resinoids and dry purified extracts or absolutes (Art. 1306). Descriptions of 7 extracts are given (Art. 1307).

The list of isolated substances contains 8 items with specifications (Art. 1309). The names, but no specifications, are given for 78 authorised synthetic flavouring substances (Art. 1311).

Natural essences, extracts and artificial essences can be dissolved in water, ethyl alcohol, glycerine, propylene glycol, edible fats and oils, liquid vaseline; labelled "emulsion" (Art. 1315). Powdered or granulated flavourings can be prepared with sugar, starches, gums, calcium silicate (max. 2% in the finished food); they are labelled "polvo" (Art. 1316).

Essences and extracts of 9 plants as well as safrole, coumarin and hydroxycoumarin, thujone and pinocamphone are explicitly prohibited (Art. 1320). Also forbidden is the use of salicyl aldehyde, hydrocarbons, pyridine derivatives, nitro compounds and organic nitrites (Art. 1321).

This regulation was replaced in the same way as in Brazil by the MERCOSUR rule *[43]*.

**Chile**

In the food regulations *[45]* the articles 250-252 deal with flavourings. According to Art. 250 the denominations "esencia" (essence) or "aceites esenciales" (essential oils) and "extractos aromaticos" (flavouring extracts) can be used for inoffensive natural or artificial preparations.

Definitions are given in Art. 251 for:

    a) "Esencias y extractos naturales" (natural essences and extracts)

    b) "Esencias y extractos artificiales" (artificial essences and extracts)

    c) "Aceite esencial natural" (natural essential oil) without addition of alcohol, other solvents or colorants, "extracto natural" (natural extract) obtained exclusively from the name giving natural product, containing an approved solvent, "extracto alcoholico" (alcoholic extract) is a synonym for an extract, containing one or more natural flavouring substances dissolved in alcohol or in another approved solvent.

    d) The designations "savor de ..." (with ... flavouring), "sabor del principio aromatico de ..." (with the flavour of the aromatic principle ...) or "extracto sintetico de ..." (synthetic extract) refer to a natural raw material, but mean chemical products obtained by synthesis.

    e) The designations "espiritu", "espiritu destilado", "aroma destilado" or "espiritu destilado alcoholico" (distilled spirits) apply to products obtained by distillation of essential oils in the presence of ethyl alcohol.

    f) "Infusion" is obtained by maceration of a natural raw material in water, alcohol of their mixtures.

    g) "Agua de ..." is the aqueous phase obtained by vapor distillation of essential oils.

    h) "Esencia soluble de ..." (soluble essence of ...) applies to the constituents of an essential oil treated in a way that in alcoholic solution or as a colloid is soluble in water.

A negative list of prohibited ingredients is contained in Article 252:

a) harmful substances

b) alcohols (except ethanol), chloroform, acetone, pyridine bases, nitrobenzene, acids (except those especially approved by the National Health Service), ethyl chlorides, bromide and iodide, ethyl or amyl nitrites, toxic denaturants

c) essential oils from wormwood varieties, rue, cade oil and turpentine

d) sweeteners, colorants and preserving agents which are not authorised

e) trace metals in higher amounts than those authorised.

Lists of food additives (preserving agents, antioxidants, thickening agents etc.) have been published as well as a special list of flavourings, flavour enhancers and flavouring adjuncts, comprising the following substances [46]:

Salt
Saccharose, glucose, glucono-delta-lactone, lactose, maltose, honey, sorbitol, mannitol
Saccharine (only with special authorisation labelling)
Milk powder
Natural spices and their extracts
Authorised essences and authorised active principles
Hydrolysed proteins
Monosodium glutamate
Maltol
Acetic, citric, lactic, tartaric acids and their Na salts,
Na and Ca-inosinates, Na and Ca-guanylates
Na-dioctylsulfosuccinate.
Propylene glycol is mentioned as authorised solvent.

This regulation will be replaced by the MERCOSUR rule [43].

Modified lists of food additives have been published in 1990.

**Colombia**

Food additives and flavourings are regulated by Decree 2106 [47a]. Food additives are classified in 21 categories; Article 3, paragraph 6 gives the following definitions for flavourings:

6. *Flavourings:* The substance or mixture of substances with odorous and flavouring properties, capable to confer or intensify the aroma or taste of foodstuffs.

Flavourings may be:

6.1. *Natural flavouring:* Consisting of simple substances, chemically defined, or of compounds extracted from vegetables, or sometimes animal substances, natural or manufactured, suitable for human consumption, exclusively by physical means.

6.2. *Nature-identical flavourings:* Aromatic substances, obtained by organic synthesis, or by adequate chemical processes, and the chemical structure of which is identical with the isolated active principle of natural flavourings, or food products suitable for human consumption.

6.3. *Artificial flavourings:* Aromatic substances, chemically defined which are not found in natural products, and which are obtained by organic synthesis, and compounded or not with natural flavourings or nature-identical flavourings.

The Health Ministry has to establish and to update lists of additives (Art. 5, 6, 7 and 8).

For the transport and sale of additives, reference is made to the Decree 2333 *[47b]* (Art. 15, 16).

This Decree regulates the hygiene conditions for manufacturing and distribution of foodstuffs. According to Art. 2.1. substances added to improve organoleptic characteristics are classified as foodstuffs. Every foodstuff has to obtain a sanitary register (Art. 129); natural foodstuffs not submitted to a transformation process (like fruit, vegetables, etc.) are exempt from licensing (Art. 131).

# 7.6 Asia

## 7.6.1 General

Some words on the largest continent, Asia, which consists of so many different areas.

First to be mentioned is the area of the former Soviet Union with the G.U.S. countries. In all these countries food legislation is in transition from the former centralist Soviet rules to national rules. Subsequently not much more is known so far. Russia is eager to bring its Food Law and Food Standards (GOST) in line with European rules and Codex guidelines.

The "Middle East" will be described in chapter 7.6.2. The Arabian countries Jordan, Iraq, Sudan and the members of the Gulf Cooperation Council (GCC) Bahrain, Quatar, Kuwait, Osman, Saudi Arabia and the United Arabic Emirates have founded the Arabian Standardisation and Metrology Organisation (ASMO). Naturally Israel is no member.

The third major block is the "Far East" with many large countries which are becoming more and more important for the world wide economy. Japan, China, India, Indonesia, South Korea, Malaysia, Singapore, Taiwan, and Thailand will be described briefly, always under the aspect of food additives and flavouring regulations.

## 7.6.2 The "Middle East"

With the exception of Israel all countries are Arabian speaking countries. They also have the Muslim religion in common. Muslim dietary laws are based on the Quran, see chapter 7.8.

In addition to these religious rules modern food standards have been accepted in most of the countries. Fortunately, the countries of the Arabian League have agreed upon the ASMO, located in Amman (Jordan) *[48]*. Based on the standards of the Codex Alimentarius Commission of FAO/WHO *[2]* the ASMO had modified CAC standards or developed new ones where necessary. The members of the Arabic League are not bound to adopt these standards, but in practice all food standards of the Arabian countries are based or influenced by ASMO.

The GCC came to a special agreement which is the "Standardization & Metrology Organisation for G.C.C. Countries" *[49]*. In contrary to ASMO, this organisation aims at identical law for all products in all GCC countries. These standards combine ASMO standards and traditional ones of the Arabian countries.

As far as food additives are concerned ASMO has adopted the CAC list of harmless substances with amendments. Usage details are regulated within the product standards. Furthermore, there is a larger number of standards for spices, but only two on essential oils.

Flavourings are not mentioned, but there does exist a Saudi Standard 951/1995 "Flavours permitted for use in foodstuffs". The Standard gives definitions for natural flavourings (obtained from vegetable or animal raw materials), nature-identical and artificial flavouring substances and flavour enhancers.

Thirteen biological active substances in natural flavourings are limited . No artificial flavouring substances except those mentioned can be used. This table contains 727 substances from the list in the US CFR. These substances are artificial as well as nature-identical. The fact that some important ingredients are missing will have no legal consequences since nature-identical flavouring substances are not restricted and can be used according to good manufacturing practices.

The following label information shall be declared in Arabic on the containers: common name and code number.

## 7.6.3 The "Far East"

Within this area Japan will be dealt with first. The other countries will follow in alphabetic order.

### Japan

As in most countries the basis for the food regulation is the "Food Sanitation Law" (FSL) *[50]*. It is a framework law which empowers the Ministry of Health and Welfare (MHW) to lay down detailed regulations.

In Article 2-2 of the FSL additives are defined: "In this Law, the term "additive" means substances to be used in or on food, in the process of the manufacturing of food or for the purpose of the processing or preserving of food, by adding, mixing, infiltrating, or other means."

Article 6 establishes a "positive system" for food additives: "No person shall sell, or manufacture, import, process, use, store or display with intent to sell any additive (excluding any natural flavoring agent and any substance which is generally provided for eating or drinking as a food and which is used as a food additive) or any preparation or food that contains such additive, unless the Minister of Health and Welfare designates it as not injurious to human health based upon the opinion of the Food Sanitation Council."

Appendix 1-4 contain lists of food additives, divided into synthetic and natural ones. Standards for manufacturing, using and labelling of additives according to article 7 and 11 of the FSL are compiled in Appendix 7 *[50]*. The principles for the labelling requirements are as follows:

- All food additives shall be declared on pre-packaged processed food with exceptions of carry-over and processing aids.

- Food additives used for the purpose of eight categories, such as antioxidant, preservative, food colour etc. shall be declared in both, substance name and category name.

- When the collective name, such as leavening agent, flavouring, etc. is more familiar for the general public, the collective name may be used in place of a substance name.

- The ingredients and their percentage in preparations shall be declared, except for flavourings.

- All non-chemically synthesised food additives, except those already listed, have to be notified.

- The term "natural" is not allowed for labelling since this expression is not used in the list of food additives. The only approved term is chemical synthetics.

"Guidelines for Designation of Food Additives and for Revision of Standards for Use of Food Additives" have been published 1996 *[51]*. These Guidelines describe not only the procedures used to apply for designation of chemically synthesized food additives and for establishment of standards for use, but also the scope of accompanying documentation for these applications, such as safety evaluation results, and the generally used methods for studies which are required to prepare the documentaion.

Although there is considerable discrepancy between the regulations of the other countries and the Japanese standards, the regulations of Japan reflect the historical background of provisions of the original Food Sanitation Law and it will take time to get harmonisation to international guidelines.

The Information Service Department of the External Trade Organisation (JETRO) published "Flavoring agents as food additives" in 2003 *[93]*. It contains four tables: (1) Synthetic flavoring agents with 78 substances and (2) 18 groups with similar chemical structures, i.e. ethers, esters, ketones, etc., (3) Substances that have been confirmed as the result of an investigation conducted by the Japan Flavor & Fragrance Materials Association in the 2000 Health Sciences Research Program and (4) Substances additionally confirmed in 2003.

The labelling of flavours can be summarised as follows:

- Flavourings are declared by the collective name "flavours" (in Japanese "Kohryo").
- When one of the "other than chemical synthetic" flavourings of the list of 577 substances is used, its name together with the word "flavour" is declared: e.g. Orange Flavour, Vanilla Flavour. In cases where mixtures of several ingredients are used, the declaration is only "Flavour".
- The carry-over principle applies for antioxidants, solvents, emulsifiers, etc. used in water-soluble, oil-soluble, emulsion and powdered flavourings.
- Preparations of flavouring and food additives, whereby the food additive has a function in the food have to be declared, e.g. Flavour + Food Colour.

**China**

The People's Republic of China used Codex Alimentarius standards as model in designing its General Standard for the Labeling of Foods, which went into effect on 1.2.95.

The standard, promulgated by the State Bureau of Technical Supervision, adheres closely to the 1991 Codex General Standards for Pre-packed Foods. However, the standard includes certain unique requirements:

- the language used on food labels must be standard Chinese;
- Chinese Pinyin (Latinized phonetic transcription) can be used at the same time, but the spelling must be correct, and should not be in bigger letters than the corresponding Chinese characters; and
- ethic or foreign languages can be used at the same time, but there must be certain close correlations with the Chinese characters.

Other state standards for food additives have been published in 1995: The "Hygienic Standards for Food Additives in Use" cover antiseptic agents, antioxidants, chlorophoric agents, bleaching agents, acidifying agents, coagulants, porous agents, thickeners, defoaming agents, sweeteners, colourants, emulsifiers, quality enhancing agents, anticoagulants, perfumes and others. Each additive is listed by category, name, application, maximum dose, and additional "remarks" as needed.

A translation dated 8.5.1986 of the Chinese flavouring regulation states:

"All those flavouring material that have been suggested for approval by WHO or of which ADI values have been established or that have been approved by either two organisations such as FEMA, CoE, IOFI can be used in China. In case a test is required, generally an acute toxicological test is sufficient, and the evaluation will be made according to published data."

However, a list of flavouring ingredients and where they can be used was compiled, resulting in a "new public health standard" GB 2760-86 (18.9.1986). The list comprises 121 natural flavourings, 257 synthetic flavouring substances, and 118 temporary permitted flavourings (e.g. nutmeg oil, onion oil, ethyl vanillin, allyl hexanoate, etc.).

Since then China is focusing on two issues:

(1) further rapid expansion of their positive list of flavourings, and
(2) completing a positive list of flavouring substances for import/export control.

Consequently Chinese authorities indicated in 1995 that they will accept flavouring substances of the US-FEMA-GRAS list for the time being and think about "GRAS-ing" additional natural products currently in use.

Rather than adopting the entire list by legislation, it is preferred that each ingredient be formally submitted to the Food Additives Committee for review and addition to the Chinese list of flavour ingredients. It appears that the review procedure is more of a formality, which leads to the development of ingredient specifications prior to inclusion on the Chinese positive list.

In 2003 the State Food and Drug Administration (SFDA) was set up as a new agency with far-reaching powers to control the food and drug market. It will group food and drug safety under a single umbrella and remove responsibility from the Ministry of Public Health (MPH) and State Administration for Industry [94].

**India**

The Government of India enacted a Central legislation called the "Prevention of Food Adulteration Act" (PFA) which came into effect from 1st June 1955. This Act repealed all laws existing at that time in Indian States concerning food adulteration, a new edition was published 1981.

The Act makes provisions for prevention of adulteration of food and lays down that no person shall manufacture for sale, store, sell or distribute any adulterated or misbranded food which is not in accordance with the conditions of the licence or any article which contravenes the provision of the Act or Rules. The standards on various articles of food are specified in the rules. In the case of foods which have not been standardised in the PFA rules, it is necessary that a list of ingredients used in the product in their descending order of composition is given on the label.

The Central Committee for Food Standards (CCFS) under the Directorate General of Health Service, Ministry of Health and Family Welfare, is responsible for the operation of the PFA Act. Provisions of this Act are mandatory and contravention of the rules leads to both fine and imprisonment.

The CCFS has laid down a list of permitted additives along with their limits in various foodstuffs. The Bureau of Indian Standards has laid down the specifications for identity and purity of these additives. As of writing this summary Indian Standards have been formulated on about 75 additives. Most of these have been based on the work done by the joint FAO/WHO Expert Committee on Food Additives (JECFA) [52a,b].

An amendment of the Prevention of Food Adulteration Rules has been published in 1988 [53]. A new rule 64 BBB on the use of menthol has been inserted, limiting menthol in confectionery to 0.1 %. In addition to the prohibition of the use of coumarin, dihydrocoumarin and tonkabean, also ß-asarone and cinnamyl anthranilate are banned in foods (Rule 63A).

*General labelling requirements for food:*

All declarations must be in English or Hindi in Devnagri script and surrounded by a line. The word "Synthetic" must be the same size as the product name.

Unless otherwise provided in the rules, every package containing food must bear a label on which the following is specified:

(1) the name, trade name or description of the food

(2) ingredients used in descending order of composition by weight or volume as the case may be. For substances falling in the respective classes and appearing in the list of additives permitted in food generally the following class titles may be used: Antioxidants, bleaching agents, emulsifying and stabilising agents, preservatives, edible colours, edible flavours and edible gums.

(3) name and complete address of manufacturer or importer or vendor or packer.

(4) the net weight or number or measure or volume or content as required.

(5) a distinctive batch number or lot number or code number being preceded by the words "Batch No", "Batch", "Lot number" or "Lot".

(6) the month and year in which the commodity is manufactured or pre-packed.

As the PFA does not allow innovation and the enforcement of the law is a major problem, a new food law is underway. It will ensure safety and quality for consumers, bring about greater transparency, self-regulation, harmonisation of standards and rules, and give a boost to the country's agriculture processing industry *[95]*.

## Indonesia

All imported foods must be registered. There are no standards of composition for foodstuffs.

The regulations on additives list the additives under their group names and any restrictions regarding their use in foods and levels of use *[66]*. The following classes of additives are listed: Antioxidants, anti-caking agents, acids, bases buffers, enzymes, bleaching and dough conditioners, nutrients, preservatives, emulsifiers and stabilisers, thickeners, colours, flavours and aromatic agents, sequestrants and miscellaneous agents.

The artificial sweeteners, saccharin and cyclamate, and their sodium and calcium salts, may be used in low-calorie foods and saccharin is permitted in certain low-calorie soft drinks. If manufacturers wish to use an additive which is not listed as permitted and is not specifically prohibited, they may obtain permission from the Minister of Health.

Certain colours and the following substances are prohibited in foods: Borax, boric acid, brominated vegetable oils, chloramphenicol, potassium chlorate, diethyl pyrocarbonate, nitrofuranzone, dulcin (4-ethoxyphenylurea) and salicylic acid and its salts.

According to a Regulation of the Ministry of Health on Food Additives [54], there are only 89 flavouring substances and 4 flavour enhancers listed. However, this does not mean that other flavourings are not allowed. In order to use these one has to apply to the Director General of Drug and Food Control. Applications must be accompanied by scientific data supporting their safety and a list of countries in which they are permitted. Natural flavourings are regulated as in Japan.

*General labelling requirements for food [67]*

The following information must be given on the label in Indonesian or English. The size of letters and numbers should be minimum 0.75mm and their colour should be in sufficient contrast to the background colour.

This information should be written on the principal or main part of the label:

(1) The name of the foodstuff, which can be a name laid down by law or an appropriate description.

(2) The net contents in metric units. The net contents of imported foods may be declared in the measures valid in their country of origin.

(3) The registration number.

This information may be shown on any part of the label:

(4) List of ingredients in descending order of proportion, including components of ingredients. Additives may be listed by their generic names, for example antioxidants, preservatives, emulsifiers, seasoning and colouring substances.

(5) The name, address and country of origin of the manufacturing company must be given.

(6) The production code.

**Korea (South)**

Base of the Korean food legislation is the "Food Sanitation Act" of 1962 in the version of 1986, with several modifications since. It applies to all food, drinks, additive, and synthetic compounds, for locally produced as well as imported products. Rules for additive materials used for adding, mixing, or permeating in or upon food in any way in the process of manufacturing, processing or preserving the food – are found in Article 6 which determines the criteria for manufacture and the use of food additives. For those foods and additives not identified by a criteria as described above, the processor may submit his own criteria and standards which may then be recognised through inspection by a food sanitation inspection agency.

According to Article 11 the Ministry of Health and Welfare has established a list of food additives and their standards. An English translation of parts of the Korean Food Additive Code of December 1996, which is proposed for adoption on June 1, 1997 has been transmitted to the Committee on Sanitary and Phytosanitary Measures of the World Trade Organisation WTO (98 pages).

The Code mentions 382 synthetic additives, comprizing 69 individual flavouring substances and the following 18 chemical groups: ketones, lactones, aromatic aldehydes, aromatic alcohols, esters, ethers, isothiocyanates, indole and its derivatives, fatty acids, aliphatic aldehydes, aliphatic alcohol, aliphatic hydrogen carbide aldehydes, (translation error, probably hydrocarbons), thio alcohols, thio ethers, terpene hydrocarbons, phenols, phenolethers, furfural and its derivatives.

In addition to 156 natural additives, items No. 157 gives a definition and lists 271 source materials for natural flavourings comprising 248 botanical (e.g. sassafras, woodruff, etc.) and 1 animal species. Item 272 of this list are Other natural flavouring substances: flavouring substances that are manufactured/processed using appropriate food ingredients according to Korean Food Code Chapter 3, Common Standards & Specifications in general, 3. Required conditions for raw materials.

**Malaysia**

In 1983 the Food Law Nr. 281/1983 was proclaimed as the first unilateral law in this field for all Malaysian States. It is a frame work like general food laws in other countries governing food control, further forbidding food adulteration and danger against consumers' health.

More important are the Food Regulations Nr. 281/85 dated 26.9.85, entry into force on 1.10.1986. They combine horizontal standards about food additives and contaminants, labelling, analytical methods, nutritional ingredients, and irradiation. Food additives are regulated in Article 19 to 25 with annexes, flavouring ingredients are included in article 22 together with Annex 8.

The addition of food additives to food is prohibited, except as otherwise permitted by these Regulations. There are lists of permitted preservatives, colours, flavours, flavour enhancers, antioxidants, nutrient supplements, and food conditioners, which include emulsifiers, stabilisers, anti-foaming agents, thickeners, modified starches, gelling agents, acidity regulators, enzymes, solvents and anticaking agents. Additives may only be added to foods where specified.

"Flavouring Substance" means any substance that, when added to food, is capable of imparting flavour to that food, and the definition includes spices.

The permitted flavouring substances are:

- the spices specified
- the flavouring substances naturally present in raw fruit or spices as specified and,
- the substances specified (in the 8th Schedule).

The list in Schedule 8 is quite similar to the list of synthetic flavouring substances in the regulation CFR § 172.515 of the USA.

The following spices or substances naturally present in spices are prohibited in food: cade oil, calamus and its derivatives, coumarin, pyroligneous acid, safrole and isosafrole, sassafras oil. Standards for 23 spices are given.

The main difficulty with this Malaysian legislation was the following: In the USA, the list of the Federal Register is not a positive list since the use of any other GRAS substance is also legally permitted and there is no reason to restrict flavouring substances to those mentioned on the list. However, in the Malaysian Food Regulations 1985 the list had the meaning of a positive list and no other substance was legally permitted. Since it was practically impossible to control whether the Malaysian list was respected the authorities had to accept the cooperation of the industry.

The result was the Food (Amendment) Regulations 1988 (P.U.(A) 162/88) in which a new article 22 was issued as follows:

- Definitions for natural and nature-identical flavouring substances are given. All flavouring substances shall be permitted with the exception of the prohibited substances (Table I to the 8th Schedule, see below) and those imported without written approval of the Director.
- A flavouring substance can only be imported if it has been certified as safe and suitable for use in food by the authorities of the country of origin or manufacture and if its importation has been approved by the Director.
- Substances in Table 7.4 I to the 8th Schedule (see below) are limited. If more than one of these substances is present the amount of each shall be such that when expressed as a percentage of the amount permitted singly the sum of the combined percentages does not exceed one hundred.

*Table 7.4: I = restricted and II = prohibited flavouring substances in Malaysian Food Regulations, Article 22, 8th Schedule*

| I | II |
|---|---|
| Agaric Acid | Aloin |
| Total Hydrocyanic Acid | Berberine |
| Pulegone | Beta-Azarone |
| Quassin | Cade Oil |
| Calamus Oil | Cocaine |
| Thujones | Coumarin |
| | Diethylene Glycol |
| | Diethylene Glycol Monoethyl Ether |
| | Hypericine |
| | Nitrobenzene |
| | Pyroligenous Acid |
| | Safrole and Isosafrole |
| | Santonin |
| | Sassafras Oil |

The 8th Schedule was substituted by a new one, mentioning prohibited substances in Table 7.4 II and limited substances in Table 7.4 I. The long list copied from the US regulation has been deleted. The 8th Schedule allows the addition of preservatives to essences and flavouring emulsions:

800 mg/kg sulphur dioxide,

350 mg/kg benzoic acid, 800 mg/kg sorbid acid.

In this way the positive system of flavouring ingredients was changed to a mixed list system.

*General labelling requirements for food*

Imported food may be labelled in Bahasa Malaysian or English and may include translation in any other language.

The following are required:

(1) Appropriate designation or description of food containing the common name of its principal ingredients.

(2) Declare "mixed (insert appropriate designation of food)" or "blended (insert appropriate designation of food)" for standardised mixed or blended foods.

(3) State "CONTAINS (state whether beef or pork, or its derivatives, or lard, as applicable)" or similar words when present, immediately below designation of the food.

(4) State "CONTAINS ALCOHOL" or similar where added, in capital bold-faced lettering of sans-serif character, min 6 point, immediately below designation of food.

(5) Declare two or more ingredients in descending order of proportion by weight and the proportion declared where required by the regulations.

(6) Declare edible fat or oil together with common name of animal or vegetable from which derived.

(7) State presence of food additives in form "contains permitted (state type of relevant food additive)".

(8) State minimum net weight or volume or number of contents of package.

(9) Declare name and business address of manufacturer or packer or the agent or owner of the rights of manufacture or packing for food imported or manufactured or packed locally, and the name and business address of the importer in Malaysia and the name of the country of origin of the food.

(10) EXPIRY DATE or EXP DATE (state date as day, month, year) USE BY (state date as day, month, year) or CONSUME BY or CONS BY (state date as day, month, year). The "date of minimum durability", i.e. the date until which the food kept in accordance with stated storage conditions will retain specific qualities for which tacit or express claim has been made, must be in the following form: BEST BEFORE or BEST BEF (state date as day, month, year). Necessary storage directions must be stated. Date marking must be in capital bold-faced lettering or sans-serif character, min 6 point.

(11) Strength, weight or quantity of ingredients or food components may be declared as "per cent", meaning per cent by weight.

**Philippines**

In the absence of other regulations, products complying with USA regulations should be acceptable. There are regulatory guidelines concerning food additives dated 25 May 1984 *[68]*, which include a list of permitted additives and their recommended levels of use in certain food products based on USA regulations. The following groups of additives are included: acidulants, anti-caking agent, antioxidants, colours, dough conditioners, emulsifiers, flavour enhancers, humectants, nutrient supplements, pH control agents, preservatives, raising agents, sequestrants, stabilisers and thickeners, surface- finishing agents, sweeteners, those used as processing aids and other miscellaneous additives. Quantities of two or more substances which produce the same technological effect should be reduced proportionately. Food standards may also restrict the use of additives.

General labelling requirements for food *[69]*

The labels of all pre-packaged foods must bear the following information:

(1) Name of the food: indicating its true nature in a specific form. Must give additional statements of packing medium, form or style, condition or type of treatment unless form is discernible. Name must be in bold type letters on principal display panel and of a size related to the most prominent printed matter.

(2) List of ingredients: in descending order or proportion on principal display panel or information panel. Only optional ingredients required on standardised foods unless otherwise specified. Declare added water unless part of an ingredient, e.g. brine, syrup. Use specific names except for spices, flavours and colours. Declare flavours as "Natural flavor(s)", "Nature-identical flavor(s)" or "Flavor(s)" and "Artificial flavor(s)". Declare combinations of natural and nature-identical flavours as such or as "Flavor(s)". Declare artificial smoke flavours as "Artificial flavor" or "Artificial smoke flavor". No reference to the product having been smoked or having a true smoked flavour may be made. Declare colours by their common name or as "Food color(s)" or "Color(s)" if derived from or identical to substances derived from plant materials and as "Artificial color(s)" for coal-tar dyes or other synthetic chemical compounds. Indicate specific name of vegetable oil used. Declare mixtures as "vegetable oil (name of oils in decreasing order of proportion)" of "blend of vegetable oil (name of possible oil blend)". Declare food additives by common name or class name indicating functional category. Following class names may be used except when otherwise stated in a food standard:

| | |
|---|---|
| Anti-caking agent(s) | Glazing agent(s) |
| Acidulant(s)/ food acid(s) | Humectant(s) |
| pH-Control agent(s) | Leavening agent(s) |
| Emulsifier(s) | Preservative(s) |

Firming agent(s)  
Flavour enhancer(s)  
Flour treatment agent(s)  
Bleaching agent(s)  
Dough conditioner(s)  
Maturingagent(s)  
Declare following by common name:  
Sodium chloride/salt  
Sodium nitrite  
Sodium/potassium nitrate  
Monosodium glutamate/MSG/Vetsin  
specific name for non-nutritive sweeteners  

Antimicrobial agent(s)  
Antioxidant(s)  
Stabiliser(s)/thickener(s)  
Modified starch(es)  
Vegetable gum(s)  
Sequestrant(s)  

In dehydrated food products declare acetic acid or sodium diacetate as such or as acidulant(s); do not use terms vinegar or vinegar powder. Processing aids and additives carried over from ingredients at levels less than required to achieve a technological function need not be declared.

(3) Net contents: must be declared using metric system of measurement or "SI" units on principal display panel or information panel in lines parallel to base of package.

(4) Name and address of manufacturer, packer of distributor.

(5) Lot identification code: must be embossed or permanently marked on individual packages or containers.

(6) Open-date marking: is required.

(7) Alcoholic beverages: must indicate alcohol content as percentage or proof units on principal display panel.

(8) Language: must be English of Philipino or any major dialect or a combination of these. Imported foods may carry corresponding English translations on labels.

## Singapore

Singapore has a modern and very clear food legislation. The "Sale of Food Act as of 1.5.1973, revised in 1985 and 2002, secures wholesomeness and purity of food and fixes standards for the same; it prevents the sale or other disposition, or the use of articles dangerous or injurious to health and it provides regulation of food establishments. Details are regulated by the Food Regulations 1988: Part I contains definitions, part II the administration, part III food additives and part IV standards and labelling. Only food additives permitted in part IV may be used. There are lists of permitted emulsifiers and stabilisers, colours, antioxidants, preservatives, anti-caking agents, and miscellaneous food additives. There are also regulations on flavours, flavour enhancers, humectants, sequestrants, nutrients, solvents for flavours and mineral hydrocarbons. Saccharin is the only permitted artificial sweetener; however, any food containing it may only be imported or sold under licence.

Of special interest for the flavouring industry is the Regulation 22:

(1) In these Regulations, "flavouring agent" means any wholesome substance that when added or applied to food is capable of imparting taste or odour, or both, to a food.

(2) Permitted solvents for flavourings are diethyl ether, ethyl acetate, ethyl alcohol, glycerol, isopropyl alcohol, propylene glycol and water. Flavourings may also be used as emulsions prepared with permitted emulsifiers.

(3) Natural flavouring agents shall include natural flavouring essences, spices and condiments.

(4) Natural flavouring essences or extracts may contain permitted solvents, sweetening agents, colourants or chemical preservatives.

(5) The use of coumarin, tonka bean, safrole, sassafras oil, dihydrosafrole, isosafrole, agaric acid, nitrobenzene, dulcamara, pennyroyal oil, oil of tansy, rue oil, birch tar oil, cade oil, volatile bitter almond oil containing hydrocyanic acid, and male fern as flavouring agents is prohibited.

(6) Articles of food may have in them natural flavouring agents as specified in these Regulations.

(7) Synthetic flavouring essences or extracts shall include any artificial flavouring or imitation flavouring which may resemble the sapid or odoriferous principles of an aromatic plant.

(8) Synthetic flavouring essences or extracts shall not contain any of the prohibited substances *[see (5)]*.

Also of interest is Regulation 23:

(1) In this Regulation, "flavour enhancer" means any substance which is capable of enhancing or improving the flavour of food, but does not include any sauce, gravy, gravy mix, soup mix, spice, or condiment.

(2) No person shall import, sell, advertise, manufacture, consign or deliver any flavour enhancer for use in food intended for human consumption other than:

(a) ethyl maltol,
(b) monosodium salt of L-glutamic acid,
(c) sodium and calcium salts of guanylic and inosinic acid,
(d) L-cystine.

Tolerance limits for solvents residues in spice oleoresins are given in Regulation 36-3 (h):

| | |
|---|---|
| trichloroethylene | 30 ppm |
| methylene chloride | 30 ppm |
| ethylene dichloride | 30 ppm |
| hexane | 25 ppm |

Special standards are given for some essences of meat, almond, ginger etc.

*General labelling requirements for food*

The following label statements should be in English and clearly legible in a prominent position on the label. Items 1, 2 and 3 should be in printed letters min 1.5 mm in height.

(1) The appropriate designation.

(2) A list of ingredients in descending order of proportion must be given using their common or usual names or appropriate designations. If an ingredient consists of two or more constituents, it is not necessary to name the original ingredient provided the constituents are listed separately. The generic term for the declaration of flavourings is "natural/artificial flavouring essence".

(3) The minimum quantity of food in the container or wrapper, expressed in terms either of net weight or measure in metric units, should be declared, although non-metric units may also be shown.

(4) On imported foods, the name and address of the local importer and the name of the country of origin must be given.

(5) Foods containing artificial sweetening agents (does not include sugar, carbohydrates or polyhydric alcohols) must declare "This (name of food) contains the artificial sweetening agent (name of artificial sweetening agent)".

## Taiwan

A "Health Food Control Act" was promulgated on February 3, 1999 and enacted. Furthermore the "Law Governing Food Sanitation" promulgated on January 28, 1975, amended last on February 9, 2000, is in place. In Article 3, the term "food additive" is defined. It includes materials added to food for the purpose of flavouring. The use of food additives in food products and the maximum quantity that may be used shall meet the regulation set by the central agency-in-charge. According to Article 14 food additives should not be manufactured, imported or exported unless inspected, tested, registered and authorised by the authority . According to this basic food law the registration of imported flavourings is required.

In application of the Food Law an official list of food additives, their scope of use and maximum use levels has been issued in 1976. The additives are classified according to their function, maximum use levels are indicated for certain substances, 18 functional groups are distinguished. On 22nd September 1982, with regards to the above registration of flavourings, the three Directives A., B., and C. were issued:

A. Explanation in Inspection-Registration of Flavours for Importing

B. Guide for Application for Importation of Natural Food Flavours

C. Guide for Application for Inspection Registration of Food Flavours for Importation

A food and health management manual for the application and dosage of additives for food was issued by the National Health Administration in May 1986. This document contains lists of additives grouped according to categories, the foods in which these additives can be used, as well as dosage guidelines and restrictions.

A list of 89 flavouring substances and 13 groups of substances without limitations is given (similar to the Japanese list). A restrictive list with 14 limited substances mentions e.g. β-asarone 0.1 ppm, coumarin in beverages 2 ppm, pulegone 100 ppm, quassine 5 ppm, quinine 85 ppm, safrole 1 ppm, α- and β-thujones 0.5 ppm etc. Propylene glycol and glycerol are mentioned as carrier solvents.

The import of food additives, including flavouring essences, is subject to prior registration, which is usually effected on behalf of the exporter by a local firm, and an agency agreement or power of attorney authorising the firm to register the product for the exporter would be required.

A possible address for registration is: The Board of Foreign Trade, 1 Hukou St, Taipei, Taiwan.

**Thailand**

Base of the Thai Food Legislation is the Food Law of 8.5.1979 together with "Announcements" of the Health Ministry about labelling, food additives and contaminants, pre-packed foods etc. Furthermore Thai Industrial Standards (TIS) have been elaborated.

The Ministry of Public Health has issued Announcement No. 84 B.E. 2527 about Food Additives on December 25, 1984 (which became effective from September 19, 1985) as per the following important impositions:

- Food Additives are regarded as Special-Controlled Food, the food manufacturers or importers must apply for registration of the food recipe.

- Other types of Food Additives which are not specified in this announcement must have the quality or standard, packaging and storage according to the approval of (Thai) FDA. This also covers the mixing of 2 or more types of Food Additives.

- Food Additives must be used according to their objectives, names, kind of foods, and highest quantity permitted in food, or according to (Thai) FDA's approval.

- Method of analysis used for registration purposes must follow the imposition in the last section of the Announcement No. 84 B.E. 2527. For other Food Additives which are not specified in this announcement the specifications of quality or standard, packaging, and storage must be submitted to the (Thai) FDA for approval, including a sample for technical analysis.

- The ingredient listing of Food Additives on labels must be arranged as per Announcement No. 68 B.E. 2525. Regarding materials not permitted in food the latest list has been published on 4.2.1994 [72] comprising e.g. brominated vegetable oil, salicylic acid, coumarin, dihydrocoumarin, methanol, and diethylene glycol.

The labelling of flavourings is regulated by Announcement No. 120 of the Health Ministry [55]. Natural, nature-identical and synthetic flavourings are distinguished.

Labels of flavourings sold directly to consumers must be in Thai, imported flavourings which are not for sale to consumers can also be labelled in English.

The following information has to be shown on the label:

- name of the flavour with the appropriate classification "natural flavour", "nature-identical flavour" or "synthetic flavour"
- net content according to the metric system
- name of producer and country of origin
- day, month and year of production or repacking, indicating clearly whether "production" or "repacking".

When importing flavourings into Thailand it is required to submit a certification by the government of the exporting country covering the following three subjects:

- the flavouring has been approved for use in foods in the exporting country,
- the flavouring has been approved for sale in the exporting country
- the flavouring is formulated as the product sold in the exporting country.

*General labelling requirements for food*

The following statements in the Thai language, with or without another language, should be included on the labels of food for retail sale:

(1) Commercial name of the food and type or kind of food.

(2) Food ingredient registration number.

(3) Name and address of producer or packer, and country of origin for imported food.

(4) Net quantity according to the metric system.

(5) Important ingredients in percentage weight by approximation, unless the percentage by weight is required by the FDA, in decreasing order of quantity.

(6) Day, month and year of production and day, month and year of expiration for food which cannot be stored for more than 90 days; month and year of production or day month and year of expiration for food which can be stored for more than 90 days; day, month and year of expiration for food notified by the FDA. The words "Produced" or "Expiration" must be used.

As of 1.7.2003 Good Manufacturing Practice (GMP) standards are mandatory throughout Thailand [96].

## 7.7 South Africa, Australia and New Zealand

**South Africa**

Although there is no legislation for flavourings in South Africa the onus is on the flavouring manufacturer not to supply a product which could be injurious to the public health. Therefore flavourings which conform to the national legislation of an exporting country would be acceptable in South Africa according to the Foodstuffs, Cosmetics and Disinfectants Act *[56]*.

The use of smoke in foodstuffs is regulated in the Preservatives and Antioxidants Regulation. Here it is mentioned that where the process of smoking is applied or where a smoke solution is added, the smoke or smoke solutions shall be derived from wood or ligneous vegetable matter in the natural state. Smoke or smoke solutions derived from wood or ligneous vegetable matter which has been impregnated, coloured, gummed, painted, coated or treated in any manner liable to impart substances harmful to human health are not permissible.

**Australia and New Zealand**

The standard for flavourings and flavour enhancers is to be found in the Australia New Zealand Food Standards Code *[57]*.

Chapter 1: General Food Standards, Part 1.3 Substances Added to Food, Standard 1.3.1 Food additives. The purpose of which is any substance not normally consumed as a food in itself and not normally used as an ingredient of food, but which is intentionally added to a food to achieve one or more of the technological functions specified in Schedule 5. It or its by-products may remain in the food. Food additives are distinguishable from processing aids (Standard 1.3.3) and vitamins and minerals (Standard 1.3.2).

The Standard regulates the use of food additives in the production and processing of food. A food additive may only be added to food where expressly permitted in this standard. Additives can only be added to food in order to achieve an identified technological function according to Good Manufacturing Practice.

Flavouring substances are mentioned under No. 11 of the Standard. Permitted are those which are either (a) listed in at least one of the following publications (i) *Food Technology*, generally recognised as safe (GRAS) lists of flavouring substances published by the FEMA (see 7.5.2); or (ii) *Flavouring Substances and Natural Sources of Flavourings,* Council of Europe; or US *Code of Federal Regulations,* 21 CFR Part 172.515; or (b) a substance that is a single chemical entity obtained by physical, microbiological, enzymatic, synthetic or chemical processes, from material of vegetable or animal origin either in its raw state or after processing by traditional preparation processes including drying, roasting, and fermentation.

*Labelling*

Food Label

The label of a food containing a flavouring must either include the statement "Flavour added" or the statement of ingredients must include the phrase "Flavours".

*Flavouring Label*

The labels used on the packages of flavourings must include one of the following or similar statements:

"Natural (description of flavouring) flavouring"

"Nature-identical (description of flavouring) flavouring"

"Artificial (description of flavouring) flavouring"

Smoke flavourings must include the statement "smoke flavouring" on the label. Vanilla oleoresin and vanilla essence must be declared as such.

The word "flavour" or "flavouring" may be replaced by the word "essence" or "extract" and the words "nature-identical" may be replaced by the word "imitation".

## 7.8 Religious Dietary Rules

### 7.8.1 Introduction

Because of the growing importance for the food flavouring business religious dietary rules of Jews and Muslims are mentioned. Increasingly pre-packed food show signs or certificates that they are "kosher" and/or "halal" approved. The meaning of these words will be explained.

### 7.8.2 "Kosher"

The kosher dietary laws "kashruth" are observed to varying degrees by members of the Jewish faith. Because there are almost 6 million Jews living in the USA at this time and almost 400 000 in Canada and in addition Seventh-Day Adventists and Muslims who purchase specific kosher foods to meet their religious needs, "kosher" has become an important issue in food production in the USA. From there it has been spreading to the European food industry.

The system of kosher laws is part of a larger religious and historical context in Judaism and must be considered as such. It is contained in the Bible, particularly in the first Five Books of Moses which constitute the "Torah". It has been translated into religious practice from there. Kosher laws are not "health laws" as such. Many of them may make sense from a health point of view especially historically, but that has not been their justification.

There are four major concepts in the kosher laws:

(1) Plants

Man was initially a vegetarian in the Garden of Eden. In practical terms all plant materials are kosher.

(2) Flesh and Blood

Apparently man is permitted to eat flesh because of his desire for it.

(3) Milk and Meat

The third major concept of the kosher laws concerns the separation of milk and meat.

(4) Acceptable Species of Animal

The fourth major concept includes the designation of which specific species of animals are fit to eat, and which of those must be made kosher ("koshered").

The mammals that are acceptable for a kosher meat meal both chew their cud and are cloven- (or split-) hoofed. We note that these animals are all herbivores. The following are the general characteristics of kosher birds: They (1) are not "birds of prey" like owl, kite, falcon etc., (2) do not have a front toe, (3) have a craw (and a double-skin stomach, easily separated), and (4) catch food thrown into the air, then place it on the ground and tear it with their bill before eating. There are also laws for fish and seafood. Fish with fins and removable scales are acceptable. This includes fish with cycloic and octenoid scales but excludes those fish with placoid and gadoid scales, such as shark. All shellfish (molluscan and crustacean) are prohibited.

Kosher slaughter involves the use of a very sharp knife with a quick single cut by hand of the carotid arteries, jugular veins, and windpipe of a live animal by a trained religious slaughterer ("shochet") to permit maximal removal of blood. Kosher slaughter of poultry must also be done by hand.

Additional cuts are made on the accepted bird to permit proper effectiveness of the ritual salting and soaking process. There are also special cutting procedures for beef and lamb. The koshering process may change the organoleptic properties of meat; many customers are convinced that it enhances the quality of kosher poultry and roasts.

As a result of these concepts all foods and the equipment used to prepare them must be defined as belonging to one of four categories: meat (Yiddish: fleishig), dairy (Yiddish: milchig), neutral (Hebrew: parve or pareve), or unacceptable (Yiddish: traif). Parve products can be used with either milk or meat foods. Rabbinical supervision of food production ensures the initial correct classification of products into one of these four categories.

There are special requirements concerning the equipment used to prepare kosher foods. These must be observed if the plant produces exclusively kosher products or a combination of kosher and non-kosher products. In many cases products that are "pareve" can be made on equipment that was used to produce either dairy or meat products. However, products produced on such equipment should not be mixed.

Materials that are rabbinically defined as either nonporous (e.g. glass) or porous but easily desorbable (e.g. stainless steel and other metals) can be koshered using one of several approved processes. Different rabbis have their own criteria for determining which materials desorb foods easily and thus can be cleaned by an appropriate koshering process. The process selected depends on the history of the items used, particularly with respect to heat treatments.

Kosher laws have consequences which extend far beyond the worlds of meat, poultry, and fish. The inclusion of food additives and flavourings, increased food processing, and the use of more sophisticated machinery have made the observance of kashruth more complex than ever. To ensure that proper handling has taken place, rabbinical supervision is required, and a system of marking and of certifying materials is used. [58-60].

As far as flavourings are concerned it can be said that pure synthetic flavouring ingredients are without problems for kosher products. However, most flavourings contain natural ingredients or are derived from natural source materials.

In the following some examples of areas of concern in the production of kosher flavourings are discussed:

- Alcohol is an important food ingredient that is often used as a carrier for various flavourings and for the production of vinegar. In general grape products and related products must be produced completely by Jews in order to be acceptable as kosher. Once these are heated (e.g. pasteurised), however, or alcohol coming from other sources (i.e. potatoes) may be handled or manufactured by non-Jews.

- Eggs must come from kosher birds. Chicken eggs are kosher if there are no blood spots. Industrial egg products might not be kosher if no special precautions are taken to avoid eggs with blood spots to be processed. Eggs are generally considered "pareve".

- Gelatine can be produced from the skin and bones of beef animals (and the skin of pork). If produced from kosher-slaughtered beef and subsequently handled in an appropriately kosher fashion, the product would be kosher. Bones and skins are considered "pareve" or neutral; therefore, gelatine can be used with meat or dairy products.

- Lactose and caseinate, whey powder etc. are dairy products and should not be used in flavourings intended to flavour meat containing products. The same problem arises with lactic acid.

- Emulsifiers and other functional ingredients, such as sodium or magnesium stearate, mono- and diglycerides, glycerol, polysorbates, and monostearates, can be derived from plant or animal sources. Products of animal origin may only be used if derived from kosher slaughtered beef animals etc. Therefore it is recommended to use synthetic products.

- Hydrolysed (Enzyme-Digested) Protein are often used as flavouring components and/or functional ingredients in many food products. The use of dairy proteins as a source of hydrolysed protein would define any kosher product using this ingredient as dairy. Hydrolysed proteins from animal sources would have to be produced as kosher to be acceptable. Only hydrolysed protein of plant origin can be used without any problems.

Obviously, processing plants must make specific efforts to comply with all these regulations and inspections. "Mixed" plants must scrupulously separate "kosher" and "traif" (or in some cases, meat and dairy) materials. Generally, the procedure for koshering equipment is to let it stand for 24 hours and then run boiling water through it until it overflows. The complete procedures for kosher production equipment cannot be covered within the space allotted here. A rabbi involved in kosher certification should be consulted for further details.

There is a large number of individual rabbis and rabbinical organisations who provide certification services. It is recommended in general to use American ones because their certificates are recognised all over the world. The KASHRUT Magazine *[61]* provides information and addresses, i.e. those of the "Circle U" *[62]* or "Circle K" *[63]* organisation and others. In the UK the Rabbinical Council (Beth Din) in London is well known.

### 7.8.3 "Halal"

Halal is a Quranic word meaning lawful or permitted which indicates it belongs to the Islam, the world's second largest religion. The basic guidance in Islam about the food laws is revealed in the Quran (the divine book) from Allah (the Creator) to Muhammad (the Prophet) for all mankind. The food laws are explained and put into practice through the Sunnah (the life, actions, and teachings of Prophet Muhammad) as recorded in the Hadith (the compilation of the traditions of Prophet Muhammad). These laws are strictly observed by Muslims of all ethnic and geographic origins. It has been estimated that there are 1,8 billion Muslim consumers spread over more than 112 countries throughout the world.

There are 11 generally accepted principles pertaining to halal (permitted) and haram (prohibited) in Islam for providing guidance to Muslims in their customary practices. These "laws" are binding to the faithful and must be observed at all times.

A general guideline for use of the term "halal" was under discussion within the FAO/WHO Codex Alimentarius Committee on Food Labelling. The guideline has been adopted at the 22nd session of the Codex Alimentarius Commission on June 23-28, 1997.

General Quranic guidance dictates that all foods are halal except those that are specifically mentioned as haram (unlawful or prohibited). The Islamic dietary laws may be divided according to the basic three food groups:

- All products derived from plants are lawful for Muslim consumption except when hazardous or fermented. Alcoholic beverages such as wine, beer, and hard liquors are strictly prohibited. Foods containing added amounts of alcohol are also prohibited because such foods by definition become impure.
- The milk group is permitted at all times.
- The meat group may be divided into two groups: marine animals and land animals. All fresh- and saltwater animals are permitted except when poisonous. Among the land animals, only pigs and boars have been specifically prohibited by name. The prohibition by species also includes birds that have claws, such as falcons, vultures, and eagles. ("Birds of prey", see kosher), carnivorous animals with claws and fans such as lions and tigers, and finally animals that live both on land and in water such as frogs and crocodiles.

All meat must be slaughtered. Slaughter is necessary because it drains the blood from the animal. Blood drinking is prohibited. There are strict laws guiding the slaughtering of animals. Any Muslim having reached puberty is allowed to slaughter after saying the name of Allah and facing Mecca. Meat slaughtered by people of the Jewish or Christian faith (People of the Book) may also be eaten.

Biotechnology and bioengineering techniques have started to reshape the food supply picture, and questions are being raised about the permissibility of foods produced using these techniques. Islam is a viable religion for all times, and such issues are being reviewed on a case-by-case basis by the Muslim scholars. *[64-65]*

Two organisations in the UK are to be mentioned which give assistance *[77a+b]*.

Consequences for the flavouring industry are that pork and pork products (gelatine etc.) as well as alcohol from any source is strictly forbidden.

### 7.8.4 Comparison of Kosher and Halal Requirements

Both "rules" prohibit pork and some "birds of prey" for human consumption. Halal flavourings must not contain any alcohol. As far as kosher is concerned, alcohol is permitted. This has to be kosher, coming from grapes it has to be manufactured by Jews.

Animals have to be slaughtered before being fit for human consumption, the slaughtering processes are "regulated" by different rules for each religion. Muslims may eat kosher meat but Jews not halal one. Furthermore, according to kosher dictary laws meat and dairy products must be eaten separately.

Biotechnology and genetic engineering open up new opportunities for expanding the food supply. At the same time, these new technologies may create some difficulties for the religious scholars in making halal and/or kosher determinations of food ingredients, food products, food materials, and even modified species of animals and plants.

Leaders in both communities are reviewing the concepts and practices in food biotechnology and genetic engineering, in order to evaluate properly their implication for foods. Details are given in *[70]*.

### REFERENCES

[1]   I.O.F.I.: International Organization of the Flavour Industry; 8, Rue Charles-Humbert; CH-1205 Genève; Switzerland
[2]   Codex Alimentarius Commission of FAO and WHO; Via delle Terme di Caracalla; I-00100 Roma; Italy
[3]   CCFAC Alinorm 87/12A (1987)
[4]   Codex Guide to the Safe Use of Food Additives (Second Series) CAC/FAL 5-1979
[5]   Council Directive of 21 December 1988 on the approximation of the laws of the Member States concerning food additives authorised for use in foodstuffs intended for human consumption (89/107/EEC), Official Journal of the European Countries Nr. L 40/27 – 11.2.89
[6]   Safety Evaluation and Regulation of Flavouring Substances – J. Stofberg – Perf.Flav.$\underline{8}$ (Aug./Sept.), 53-62 (1983)
[7]   Scientific Evidence and Common Sense as a Basis for Food- Packaging Regulations – J.P. Frawley – Fd.Cosmet.Toxicol. (1967), 293-308

[8]  Safety Assurance Margins for Food Additives Currently in Use – A.M. Rulis – Reg. Toxicol. Pharmacol. 7, 160-168 (1987)

[9]  Establishing a Threshold of Regulation – A.M. Rulis in: Risk Assessment in Setting National Priorities, ed. by J.J. Bonin + D.E. Stevenson (Plenum Publ.Corp., 1989)

[10] Beyond the GRAS Lists – The Minimal Risk of Additional Nature Identical Flavoring Materials – J. Stofberg – Perf.Flav.12 (Oct./Nov.), 19-22 (1987)

[11] 1th Report of the Joint FAO/WHO Expert Committee of Food Additives (WHO, Geneva, 1968), page 14

[12] A Flavor Priority Ranking System / Acceptance and Internationalisation – O. Easterday et al. in Food Safety Assessment, Chapter 16 (American Chemical Society, 1992)

[13] Toxicologically Insignificant Exposure to Chemically-Defined Substances – P. Cadby – Private communication

[14] GRAS Evaluation of Flavouring Substances by the Expert Panel of FEMA – L.A. Woods and J. Doull – Reg.Tox.Pharmacol. 14,, 48-58 (1991)

[15] Flavor and Fragrance Materials 1997 (Allured Publ.Co., Wheaton IL/USA)

[16] Commission of the EC, Directorate-General for internal market and industrial affairs III/C/1, December 1991

[17] Regulation (EC) No. 2232/96 of the European Parliament and of the Council Laying down a Community procedure for flavouring substances used or intended for use in or on foodstuffs, Official Journal of the European Communities No. L 299/1-4-23.11.96

[18] First Edition Strasbourg 1970, 2nd ed. 1973, 3rd ed. 1981,4th ed. (Volume I) 1992

[19] Council Directive of 22 June 1988 on the approximation of the laws of the member states relating to flavourings for use in foodstuffs and to source materials for their production (88/388/EEC), Official Journal of the European Communities No. L 184/61 – 15.07.88.

[20a] Council of Europe guidelines concerning the transmission of flavour of smoke to food (Council of Europe Press, 1992)

[20b] Guideline for safety evaluation of thermal process flavourings; (Council of Europe Publishing, 1995)

[20c] Council of Europe guidelines for flavouring preparations produced by enzymatic or microbiological processes; (Council of Europe Press, 1994)

[21a] The Scientific Committee for Food: Procedures and Program for European Food Safety – K.A. v.d. Heijden – Food Technol. 46, 102-106 (1992)

[21b] Commission decision of 6.7.95 relating to the institution of a Scientific Committee for Food (95/273/EC), Official Journal for the European Communities No. L 167/22-23 – 18.7.95

[22] Commission Directive of 16. January 1991 completing Council Directive 88/388/EEC (91/71/EEC), Official Journal of the European Communities Nr. L 42/25 – 15.2.91

[23] Council Decision of 22 June 1988 on the establishment by the Commission, of an inventory of the source materials and substances used in the preparation of flavourings (88/389/EEC), Official Journal of the European Communities No. L 184/67 – 15.07.88.

[24] EFFA: European Flavour and Fragrance Association; Square Marie Louise 49; B-1040 Bruxelles, Belgium; until 1991 known as: Bureau de Liaison des Syndicats Europeen (CEE) des Produits Aromatiques

[25] Commission of the European Communities Directorate-General Industry II/E/1 CS-FLAV-49-FINAL, 30. July 1993

[26] Commission Directive of 16 January 1991 amending Council Directive 79/112/EEC in respect of the designation of flavourings in the list of ingredients on the labels of foodstuffs (91/72/EEC), Official Journal of the European Communities No. L 42/27 – 15.2.1991

[27a] Council Directive of 23 October 1962 on the approximation of the rules of the Member States concerning the colouring matters authorised for use in foodstuffs intended for human consumption, Official Journal of the European Communities No. 115, 2645/62 – 11.11.1962

[27b] European Parliament and Council Directive 94/36/EC of 30 June 1994 on colours for use in foodstuffs, Official Journal of the European Communities No. L 237/13-29-10.9.94

[28a] Council Directive of 5 November 1963 on the approximation of the laws of the Member States concerning the preservatives authorised for use in foodstuffs intended for human consumption (64/54/EEC), Official Journal of the European Communities No. 12, 161/64 – 27.1.1964

| | |
|---|---|
| [28b] | Council Directive of 13 July 1970 on the approximation of the laws of the Member States concerning the antioxidants authorised for use in foodstuffs intended for human consumption (70/357/EEC), Official Journal of the European Communities No. L 157, 31 – 18.7.1970 |
| [28c] | Council Directive of 18 June 1974 on the approximation of the laws of the Member States relating to emulsifiers, stabilizers, thickeners and gelling agents for use in foodstuffs (74/329/EEC), Official Journal of the European Communities No. L 189, 1 – 12.7.1974 |
| [29] | European Parliament and Council Directive No. 95/2/EC of 20 February 1995 on food additives other than colours and sweeteners, Official Journal of the European Communities No. L 61/1-40- 18.3.95 with Corrigendum No. L 248/60 – 14.10.95, last amended by Directive 2003/114/EC of the European Parliament and of the Council of 23 December 2003, OJ of the EC No. L 24/58-64 – 29.1.2004 |
| [30] | European Parliament and Council Directive 94/35/EC of 30 June 1994 on sweeteners for use in foodstuffs, Official Journal of the European Communities No. L 237/3 – 12 – 10.9.94, last amended by Directive 2003/115/EC of 22.12.2003, OJ of the EC L 24/65-71 of 29.1.2004 |
| [31] | Council Directive of 21 December 1988 on the approximation of the laws of the Member States concerning food additives authorized for use in foodstuffs intended for human consumption (89/107/EEC), Official Journal of the European Communities No. L 40, 27 – 21.12.1988 |
| [32] | Council Directive of 27 June 1967 on the approximation of laws, regulations and administrative provisions relating to the classification, packaging and labelling of dangerous substances (67/548/EEC), Official Journal of the European Communities No. 196/1 – 16.8.1967 |
| [33] | Council Directive of 7 June 1968 on the approximation of the laws, regulations and administrative provisions of the Member States relating to the classification, packaging and labelling of dangerous preparations (88/379/EEC), Official Journal of the European Communities No. L 187/14 – 16.7.1988 |
| [34] | Classification and labelling of dangerous preparations (Directive 88/379/EEC) / Practical Handbook – Commission of the European Communities, Directorate-General Internal Market and Industrial Affairs (Luxembourg: Office for Official Publications of the EC, 1992), ISBN 92-826-4231-3 |
| [35] | Regulation the Safety of Food – R.D. Middlekauff – Food Technol.43, 296-307(1989) |
| [36] | The Food Chemical News Guide, 1341 G Street N.W., Washington, D.C. 20005 (USA) |
| [37] | Code of Federal Regulations Title 21 / Food and Drugs, § 101.22(a)(3) and § 101.22(a)(1) (US Government Printing Office, Washington, reprinted every year) |
| [38] | Flavor Labeling in the United States – K.J. Bauer – Dragoco Report 38 (1), 5-19(1993) and in Source Book of Flavours, edited by G. Reineccius (Chapman & Hall Ltd., London SE1 8HN, 2nd Ed. 1993), page 860-875 |
| [39] | An Act Respecting Food, Drugs, Cosmetics and Therapeutic Devices – Short Title: This Act may be cited as the Food and Drugs Act. |
| [40] | Canada Gazette Part II, Vol. 126, No. 24, page 4478 |
| [41] | "Reglameto de la ley general de salud en materia de control sanitario de actividades, establecimientos, productos y servicios" – (Mexican) Diario Oficial de la Federacion 18.1.1988, Tomo CD XII No. 11, Primera Sección, page 1-128; Segunda Sección, page 1-30 |
| [42] | Resolucao No. 22/76 (Retificado pela Resolucao No. 40/76 – Publicado no D.O.U. – Secao I – Parte I – 26-07-77) – (Brazilian) Diario Oficial, 16.3.1977, page 3026-3028 |
| [43] | MERCOSUR/GMC/RES No 46/93; VISTO: El Art. 13 del Tratado de Asunción, el Art. 10 de la Decisión No. 4/91 des Consejo del Mercado Somún y la Recomendación No. 36/93 del Subgrupo de Trabajo No. 3 "Normas Tecnicas" – not yet published |
| [44a] | MERCOSUR/GMC/RES. No. 04/93, VISTO: El Tratado de Asunción suscrito el 26 de marzo de 1991 y lo acordado en el Subgrupo de Trabajo No. III – Normas Técnicas |
| [44b] | MERCOSUR/GMC/RES 19/93; VISTO: El art. 13 del Tratado de Asunción, el art. 10 de la Decisión No. 4/91 del Consejo del Mercado Común y la Recomendación No. 25/93 del Subgrupo de Trabajo No. 3 "Normas Técnicas"- not yet published |
| [45] | "Reglamento sanitario de los alimentos", approved by Order No. 377, 12.8.1960. Modifications published in (Chilean) Diario Oficial No. 26805, 29.7.1967. |
| [46] | Order No. 327, 30.6.75 – (Chilean) Diario Oficial No. 29353, 13.1.1976 |
| [47a] | Colombian Ministry of Health: Decree 2106, 26.7.1983 |
| [47b] | Decree 2333, 2.8.1982 |

[48]  Arab Organization for Standardization and Metrology (ASMO). P.O. Box 926161, Amman/Jordan
[49]  Standardization & Metrology Organization for Golf Cooperation Council Countries, P.O. Box 85245, Riyadh 11691/Saudi Arabia
[50]  Food Sanitation Law: Law No. 233, 24.12.47 Last Amendment: Law No. 55, 30.5.2003 (Publ. by Japan Food Additives Ass. Chuo-Kun, Tokyo 103, Japan)
[51]  Guidelines for Designation of Food Additives and for Revision of Standards for Use of Food Additives (Publ. by Japan Food Additives Ass. Chuo-Kun, Tokyo 103, Japan, 20.6.96)
[52a] India: Present Situation of Food Law "Standardization in the Field of Processed Food" – by the "Bureau of Indian Standards" – Alimentalex / Intern. Food Law Review Nr. 4 (December 1990)
[52b] India: Standardization in the Field of Processed Food – S.A. Govindan – Alimentalex / Intern. Food Law Review Nr.7 (June 1992)
[53]  GSR 454(E) of April 15, 1988 – Gazette of India: Extraordinary (Part II – Sec. 3(i))
[54]  Republic of Indonesia, The Ministry of Health – Regulation No. 722/Men.Kes/Per/IX/88 on Food Additives: Section X. Flavour, Flavour Enhancer (1988/1990)
[55]  Thai Government Gazette No. 106, Section 43 DD. 21 March 1993, B.E. 2532
[56]  Foodstuffs, Cosmetics and Disinfectants Act, 1972 (Act 54 of 1972), last amended by Regulation 747 of 17 August 2001, http://www.polity.org.za/html/govdocs/regulations/2001/reg0747.html
[57]  Australia New Zealand Food Standards Code (up to and including Amendment 75): Chapter 1 – General, Food Standards, Chapter 2 – Food Product Standards, Chapter 3 – Food Safety Standards, Chapter 4 - Primary Production Standards, http://www.foodstandards.gov.au/foodstandardscode/
[58]  An Introduction to the Kosher Dietary Laws for Food Scientists and Food Processors – J.M. Regenstein + C.E. Regenstein – Food Technol 33, 89-99 (1979)
[59]  The Kosher Dietary Laws and their Implementation in the Food Industry – J.M. Regenstein + C.E. Regenstein – Food Technol. 42, 86-94 (1988)
[60]  Current Issues in Kosher Foods – J.M. Regenstein + C.E. Regenstein – Trends in Food Science & Technology 1991, 50-54
[61]  P.O. Box 204, Brooklyn N.Y. 11204 (USA)
[62]  Union of Orthodox Jewish Congregations of America ; 116 East 27th Street, New York, N.Y. 10016 (USA)
[63]  Organized Kashrut Laboratories, P.O. Box 218, Brooklyn, N.Y. 11204 (USA)
[64]  An Introduction to Moslem Dietary Laws – S. Twaigery and D. Spillman – Food Technol. 43, 88-90 (1989)
[65]  Islamic Food Laws: Philosophical Basis and Practical Implications – M.M. Chaudry – Food Technol. 46, 92-93 + 104 (1992)
[66]  Permitted additional substances for foods [in Indonesia] – Regulation No. 235/Men. Kes/Per/VI/79
[67]  Labelling and advertising of foods [in Indonesia] – Regulation No. 79/Men. Kes/Per/III/1978
[68]  [Philippines] Regulatory Guidelines concerning Food Additives. Administrative Order No. 88 – As.1984.
[69]  Rules and Regulations Governing the Labelling of Pre-packaged Food Products Distributed in the Philippines. Administrative Order No. 88 – Bs. 1984.
[70]  Implication of biotechnology and genetic engineering for kosher and halal foods – M.M. Chaudry and J.M. Regenstein – Trends Food Sci & Techn. 5, 165-168 (1994)
[71]  Council Directive of 13 June 1988 on the approximation of the Laws of the Member States on extraction solvents used in the production of foodstuffs and food ingredients, Official Journal of the European Communities No. L 157/28 – 33 – 24.6.88
[72]  Thai Government Gazette No. 111, 9 March 1994, B.E. 2536, Notification No. 151 (28 Dec.1993)
[73]  Regulation (EC) No. 396/2005 of the European Parliament and of the Council of 23 February 2005 on maximum residue levels of pesticides in or on food and feed of plant and animal origin and amending Council Directive 91/414/EEC, Official Journal of the European Communities L 70/1-16 of 6.3.2005
[74]  Regulation (EC) No. 178/2002 of the European Parliament and of the Council of 28 January 2002 laying down the general principles and requirements of food law, establishing the European Food Safety Authority and laying down procedures in matters of food safety, Official Journal of the European Communities L 31/1-24 – 1.2.2002 – amended by Regulation (EC) No. 1642/2003 of 22 July 2003, OJ of the EC L 245/4-6 of 29.9.2003

[75] Directive 94/34/EEC of the European Parliament and the Council of 30 June 1994 Modifying Directive 89/107/EEC Concerning Food Additives Authorized for use in Foodstuffs Intended for Human Consumption – Official Journal of the European Communities No. L 237/1-2 – 10.9.94
[76] Alinorm 97/22, page 32-34 Draft General Guidelines for Use of the Term "halal"
[77a] Halal Food Authority; P.O. Box 279 ; London WC1H OHZ
[77b] Halal Food Board; 517 Moseley Road; Birmingham B12 9BX
[78] State Standard of the People's Republic of China [GB 7718-94]: General Standard for the Labelling of Foods. Promulgated on February, 4th, 1994, implemented since February, 1th, 1995
[79] WHO Technical Report Series 859 – Evaluation of Certain Food Additives and Contaminants – 44. Report of the Joint FAO/WHO Expert Commitee on Food Additives (World Health Organisation, Geneva, 1995) ISBN 92-4-120859-7, page 2-3
[80] Toxicological databases and the concept of thresholds of toxicological concern as used by the JECFA for the safety evaluation of flavouring agents – A. G. Renwick – Toxicology Letters $\underline{149}$(1-3), 223-334 (2004)
[81] Safety evaluation of food flavourings – K. R. Schrankel – Toxicology $\underline{198}$(1-3), 203-211 (2004)
[82] Regulation (EC) No. 2232/96 of the European Parliament and of the Council of 28 October 1996 laying down a Community procedure for defined chemical flavouring substances used or intended for use in or on foodstuffs, amended by Decision 1999/217/EC of 23 February 1999 adopting a register of flavouring substances used in or on foodstuffs, Official Journal of the European Communities L 84 of 27.3.1999, last modified by Decision 2004/357/EC of 20 April 2004, OJ of the EC L 113/28 of 20.4.2004
[83] Commission Regulation (EC) No. 1565/2000 of 18 July 2000 laying down the measures necessary for the adoption of an evaluation programme in application of Regulation (EC) No. 2232/96, Official Journal of the European Communities L 180/8-16 of 19.7.2000
[84] Commission Regulation (EC) No. 622/2002 of 11 April 2002 establishing deadlines for the submission of information for the evaluation of chemical defined flavouring substances used in or on foodstuffs, Official Journal of the European Communities L 95/10-11 of 12.4.2002
[85] Regulation (EC) No. 2065/2003 of the European Parliament and of the Council of 10 November 2003 on smoke flavourings used or intended for use in or on foods, Official Journal of the European Communities L 309 of 26.11.2003
[86] Commission Regulation (EC) No. 466/2001 of 8 March 2001 setting maximum levels for certain contaminants in foodstuffs, Official Journal of the European Communities L 77/1-13 of 16.3.2001, amended by Regulation (EC) No. 2375/2001 of the Council of 29 November 2001, Official Journal of the EC L 321/1-5 of 6.12.2001
[87] Council Regulation (Euratom) No. 3954/87 of 22 December 1987 laying down maximum permitted levels of radioactive contamination of foodstuffs and of feedingstuffs following a nuclear accident or any other case of radiological emergency, Official Journal L 371 of 30.12.1987, amended by Regulation 2218/89 of 18 July 1989, OJ L 211 of 22.7.1989
[88] Regulation (EC) No. 1829/2003 of the European Parliament and of the Council of 22 September 2003 on genetically modified food and feed, Official Journal of the European Communities L 268/1-28 of 18.10.2003
[89] Directive 2000/13/EC of the European Parliament and of the Council of 20 March 2000 on the approximation of the laws of the Member States relating to labelling, presentation and advertising of foodstuffs, Official Journal of the European Communities L 109/29 of 6.5.2000, completed by Directive 2003/89/EC of 10 November 2003 amending Directive 200/13/EC as regards indication of the ingredients present in foodstuffs
[90] http://www.cfsan.fda.gov/~dms/grasguid.html
[91] World Food Law monthly, April 2004 No. 70, 14, and Sept. 2004 No.75, 1-2
[92] http://www.inspection.gc.ca/english/bureau/inform/20030116disc.shtml
[93] www.jetro.go.jp/se/e/standards_regulation/flavor2003aug-e.pdf
[94] World Food Law monthly, April 2003 No. 58, 17
[95] World Food Law monthly, January 2003 No. 55, 17
[96] The Nation (Thailand) 6.5.2003

# Index

a
AAS, atom absorption spectrometry  587
Absolue  248
Acetaldehyde  279
Acetic acid and acetates  277, 385
Acetoin  162
5-Acetyl-2,3-dihydro-1,4-thiazine  279, 281
N-Acetylglycine (NAG)  558
2-Acetyl-1-pyrrolin  281
2-Acetyltetrahydropyridine  281
Acrylamide  292
Activity coefficient  25, 72
Adenosine monophosphate, AMP  356
Adenosine triphosphate, ATP  358
Adjuncts  640
Affinity  24
Agar-agar  443
Agitated extraction tower  42
Alanine  278f, 287, 290
Alapyridaine  287
Alcoholic beverages  487-514
Alfa-Laval extractor  38
Algae  124, 744
Alginate  439
Allspice  241
Almond  306, 412
Amadori rearrangement  277
– compounds  276-279, 287
– degradation of A. compounds  277f
Ames test  291
Amino acids  268, 278, 459
– interactions with free A.  449f
AMP, adenosine monophosphate  356
Amplification column  76
Amylopectin  439
Amylose  443
Analytical methods  587-743
Angelica  216
Angostura bark  216
Animal fats  543
Anise  217, 624
Anthocyanins  473
Anti-aging  7
Antioxidants  373-376
Apple  166f, 412, 679
– aroma  723
– juice  167f, 171ff
Apricot  413
Aquavit  493f
Aqueous essences  187

Arabian Standardisation and Metrology Organization, ASMO  784
Arachidonic acid  283f
ARD extractor  43f
Argentina, legislation  781
Aroma  576, 578
– analysis  704-743
– changes  737
– chemicals  5, 140
– compound  140
– reconstitution  729
– recovery  166, 175, 177
Aroma extract concentration analysis, AECA  706
Aroma extract dilution analysis, AEDA  705
Aroma model  710
– baguette crust  710
– camembert cheese  717
– grapefruit juice  722f
– gruyere cheese  717
– olive oils  718ff
– orange  720ff
– reconstitution  729
– roasted coffee  710
– strawberry juice  724
– Swiss cheese  716
– wine  735
Aroma value, definition  704
Aronia  470
ARP, Amadori rearrangement products  277
Arrack  491
Artificial compounds  159
Artificial flavouring substance  139, 158, 755, 757
Artificial flavourings  775
Artificial sweeteners  470, 476, 522
Ascorbic acid  374, 472
Ascorbyl palmitate  374
Asia, legislation  784-800
ASMO, Arabian Standardisation and Metrology Organization  784
Asparagine  233
Asparagus  431
Assessor  579f
Atom absorption spectrometry, AAS  588
Atomisation  98, 307
Atomiser  98f
– centrifugal a.  99
– nozzle a.  99
ATP, adenosine triphosphate  358

Australia, legislation   801
Authenticity control   619, 684f
Autoxidation   282ff
Azeotrope distillation   81
Azeotrope, maximum   73, 81
Azeotrope, minimum   73, 81

b
Bacteria   109, 145, 744
Baguette crust, aroma   705
Baked goods   531
Banana   413
– puree   174
Bärenfang   503
Base notes   429, 557
Basil   218
– aroma   731f
Beef   263, 287f, 559
– aroma   711ff
– broth   287
– boiled   711ff
Beer   345f, 474, 506-511
– flavour   270
– types   510
Benzaldehyde   1, 158, 444, 602, 623f
Benzene in beverages   386
Benzo[a]pyrene   290
Benzoates   382ff
Benzoic acid in foods   373
Benzpyrene   311f
Bergamot oil   191
Berl saddles   93
Berries, juice   171
Beta-carotene   473
Beverage emulsions   471
Beverage industry   4
Beverages   333, 466, 473ff, 733ff
BHA, butyl-hydroxyanisole   373
BHT, butyl-hydroxytoluene   373
Billberry   417
Binodal curve   28
Binomial distribution   580
Biocatalysts   123, 148, 263
Bioreactors   120
– cell culture reactors   129
Biotechnological processes   120-134, 142, 260
Biotechnology   120, 260
Biotransformation   121, 143ff
– by microorganisms   145
– of flavour precursors   128
Biscuits   290, 531

Bitter   578
– almond oil   219, 623
– blockers   11
– orange oil   204
Bitters   495
Black currant   413
– buds   219
Black mustard   236
Blackberry   418
Blended flavourings   391-435
Blessed thistle   219
Blood orange
– oil   204
– essence oil   207
Blue Book   758
Blueberry   417
Bollmann extractor   21
Bone-stock   557
Bonito   353
Bonotto extractor   22
Borage   220
Bouillon   552, 558ff
– cubes   549
Brambles   418
Brandies   488ff
Brazil, legislation   779ff
Bread   274, 282, 292, 327ff, 531
Brewing   345, 506ff
– process   509
Bromelain   335
Brominated vegetable oil, BVO   776
Brown extractor   166f
Bubble
– flow   57
– gum   522
– b.-cap plates   93
Buchu leaf oil   220, 685ff
Buddhist conform   292
Bunspice   533
2,3-Butanedione   277, 279, 707
Butter   284, 718
Butterfat   265
Butyric acid   161, 385
BVO, brominated vegetable oil   776

c
$C_3$-plants   606
$C_4$-plants   606
Cakes   531
Calamus   62
Callus cultures   130, 263
Calories   7, 465
Camembert cheese   717

Camomile 61f
CAM-plants (crassulaceam acid metabolism) 606
Canada, legislation 775ff
Cape Gooseberry 418
Capers 220
Caramel 289, 521f
– colours 474
Caraway 221
Carbohydrate-based replacer 456f
Carbon Dioxide, $CO_2$ 50, 52
Carbonyl fragments 277
Cardamom 60, 221
Cardboard odour 283
Carmine 473
Carrier
– bound enzymes 128f
– distillation 79
– material for spray drying 97
– matrix for freeze dried products 114f
– permitted c. solvents 317-321
– permitted c~s 317-321
– sugars as c~s for flavours 438
Carrot seed oil 222
Carry-through effect 376
Carvone 162, 636
β-Caryophyllene epoxide 163
Cascarilla 222
Cassia cinnamon 224, 556
Cayenne pepper 242
CCFAC, Codex Committee on Food Additives and Contaminants 160, 756
CEFS, Committee of Experts on Flavouring Substances 761
Celery 223
Cell cultures 121, 271
– flavour manufacturing from c. c. 271
– immobilized cultures 143f
– plant c. c. 130, 261, 271
– plant tissue cultures 143
– reactors 129
– suspension cultures 143
– tuber melanosporum cells 272
Cell tissue cultures 143, 271
Cellulose 439
Centaury 223
Centrifugal
– atomizer 99f
– evaporator 88f
– extractor 37
Centrifuge 23, 167, 170, 177, 180ff
– decanter c. 180ff
Cereal malt 335

Ceylon cinnamon 224
Character impact compounds, CIC 396, 411ff
Charm analysis 705f, 708
Cheese 282, 292
– aroma 716-718
– enzymatic modified c. 264, 349
– flavour 269
– fresh c. products 545
– ripening 348f
Chemaesthetic stimulation 579
Chemical groups 161, 790
Chemical ionisation, CI 595
Chemical potential 71
Cherimoya 418f
Cherry 413f
Chervil 223
Chewing gum 522f
Chicken 263, 268, 284, 288, 551, 559
– aroma 713f
– boiled c. aroma 713
– roasted c. 713
Chico 418
Chile, legislation 782f
Chill proofing enzymes 346
China, legislation 787
Chinese Gooseberry 422
Chinese Restaurant Syndrome 356, 554
Chips 307
Chiral
– resolution, cRs 664
– separation 665
– stationary phases 696
– sulphur compounds 686
Chlorophyll 272, 474
Chloroplasts 272
3-Chloro-1,2-propandiol 292
Chocolate 282, 289f, 426, 526ff, 537
Cholesterol 7, 453
Chromatography 590ff
– absorption c. 591
– affinity c. 591
– c. of terpenes 190
– ion exchange c. 591
Chymosin 335
CI, chemical ionisation 595
CIC, character impact components 396, 411ff
Cinchona bark 223
1,8-Cineol 163
Cinnamic aldehyde 141, 158, 163, 623
Cinnamon 224f, 624
Citral 161, 414, 635

Citrus
- aqueous essences    187
- aurantifolia    196f, 414
- aurantium    191f, 204
- essence oil    187
- juices    167, 176ff
- latifolia    196f
- limon    194f, 414
- oils    187-211
- paradisi    192f, 415
- peel oil    167f, 176, 187, 468, 635
- recovery, taste oil    187
- reticulata    199f, 415
- sinensis    201f, 414
Clary sage    225
Clear soft drinks    468f
Cloudberry    424
Clouding soft drinks    469f
Cloudy concentrates    470
Cloves    225
$CO_2$, Carbon Dioxide    50-65
- fixation    606
Cocoa    274, 287, 290, 426, 498
- liqueur    498
Coconut    415, 674
Code of Federal Regulations    159
Cofactor    149, 151
Coffee    282, 289, 425, 498
- freeze dried    114
- liqueur    498
- brew    737
- roasted    733
- - aroma model, omission test    733
- - aroma reconstitution    733
Cognac oil    226
Cola    433
Cold pressed oil    187
Colour shade    472ff
- red    473
Colours    466
Colours for soft drinks    472
Colombia, legislation    682f
Column characteristics
- loading range    40
- throughput maximum    39
Column fillings    93
- Berl saddles    93
- Grating rings    93
- Intralox rings    93
- Pall rings    93
- Raschig rings    93
Columns
- amplification c.    76ff
- pulsed packed c.    40
- Scheibel c.    41
- sieve plate c.    91
- static sieve tray c.    40
- stripping c.    76f
- tray c.    93
Combustion    608
Compressed tablets    524
Confectioneries    515-530
Consumption ratio    759
Contacting, stage-wise or differential    32
Contaminants    386
Continuous distillation    75ff
Continuous phase    26
Continuous rectification    76, 90f
Convenience food    7, 138, 274, 549ff
Cooler    505
Cooling table    517, 522
Coriander    226, 693
Cornmint oil    240
Coumarin    158
Council of Europe    759
Countercurrent distillation    90
Countercurrent rectification    20
Cow shed smell    545
Crackers    307
Cranberry    417
Creatinine    291f
Cross-current extraction    30
Culinary aroma    561ff
Culinary products    274, 549
Curacao    497
Curculin    369
Custard apple    418
Cut-back juice    177
Cyclodextrins    348, 443, 670
Cyclopentenolones    317
Cyclotene®    290, 366
Cysteine    279, 284, 288f

d
Da Vinci Principle    10
DAD, diode array detection    592
Dairy products    542-552
Dalton's law    71, 81
Dangerous preparations    769
Dangerous substances    769
DATEM, diacetyl tartaric acid esters of monoglycerides    328f
Davana oil    227
De novo biosynthesis    143f
DEAE-cellulose    271
2,4(E,E)-Decadienal    283

# Index

2,4(E,Z)-Decadienal 283
Decaffeination 51
γ-Decalactone 162
Decanal 283
Decanter 22
2(E)-Decenal 283
Dehydrated convenience food 549-572
Dehydration 109, 382
Dense gas extraction 50ff
Deoxyglycosones 276f
Deterpenisation 189
Deuterium spectrum 616
Dextrins 473
Diacetyl 162
Dichloropropanols 264
Dielectric constant 25
Difference point or pole 33
Digeration 513
Digestion 18
Diketopiperazines 287
Dill 227, 730
Diluent 26
4,5-Dimethylthiazol 285
Diode array detection, DAD 592
Dipole moment 25
Discontinuous rectification 91
Disodium 5'-guanylate 353
Disodium 5'-inosinate 353
Dispersed phase 26f, 35f
Dispersion mode 41
Distillation 66-96
– azeotrope d. 81f
– carrier d. 79f
– continuous d. 83f
– countercurrent d. 90
– distillative separation of terpenes 190
– extractive d. 82f
– high vacuum d. 87f
Distillation-extraction process 30
Distilled lime oil 188
Distilled spirits 487
Distribution coefficient, K 24
Distribution curve 29
2,6(E,Z)-Dodecadienal 284
Double jacket 85
Dough 531
Dry product 556
Drying 109-119
– chamber 110
– methods 109ff
– spray d. 97ff
– thermal d. 109f
– vacuum d. 109

DUO tests 582
DUO-TRIO tests 582
Durian 424
Dust explosion 103

e
EC Directive 88/383 765
EC Flavour Directive 762
ECD, electron capturing detector 594
ECP, ethyl-2-cyclopentene-2-ol-1-one 366
EFFA, European Flavour and Flagrance Association 765
Effervescent tablets 525
EFSA, European Food Safety Authority 160, 763
Egg 289, 804
Egg yolk 284, 289
EI, electron impact 595
Elderberry 470
Electric discharge treatment 20
Electrochemical-detection 592
Electron impact, EI 595
EMC, enzyme modified cheese 264, 349
Emulsifiers 322-330, 456, 539
Emulsion stabiliser 330
Emulsions 470
Enantio cGC 664
Enantio-IRMS 692f
Enantio-MDGC, enantioselective multidimensional GC 665f
Enantiomeric excess, ee 669
Enantiomeric purity 669
Enantioselective analysis 664-703
Enantioselective cGC-IRMS 693
Encapsulation 97ff, 532, 565
Energy drinks 468
Enhancement 557
Enhancer 11
Entrainers 54, 83
Enzymatic activities 261
Enzymatic reactions 261f
Enzyme
– classification 148f, 336
– production 121, 337
– reactors 123
Enzyme modified cheese, EMC 264, 349
Enzymes 148f, 335-350
– carbohydrases 343
– endopeptidases 339
– exopeptidases 339
– for brewing 345f
– for cheese ripening 348f

- for clarification of fruit juices 173f, 179, 347
- hemicellulases 266, 338, 342
- in the baking industry 342ff
- industrial e. 337ff
- lipases 264f, 340
- lipoxygenases 121, 341
- origin of e. 337
- oxidoreductases 336, 341
- pectinesterases 266
- pectintranseliminase 266
- pectolytic e. 266
- polygalacturonases 266
- proteases 107, 263, 266, 339
- proteinases 342
- -glycosidases 267

Epithelium 576
Equation, Clausius-Clapeyron 74
Equation, Gibbs-Duhem 73
Equation, Langmuir-Knudsen 87
Equation, Prausnitz 73
Equation, Redlich-Kwong 73
Equilibria, ideal-nonideal 71
Equilibrium curve 27, 32
Equilibrium phase diagrams 73
Equipment, continuous distillation 83
Equipment, countercurrent distillation 90
Essence oils 187
Essential oils 5, 137, 453, 456
Ethnic foods 12
Ethyl butanoate 161
Ethyl maltol 362
N-Ethyl-p-menthane-3-carboxamide 163
Ethyl vanillin 159, 368, 528
Ethylene oxide 748
Eucalyptus oil 228
Eugenol 58f, 163
European Food Safety Authority (EFSA) 160, 763
European Union, legislation 762-771
Evaporator 83-89, 166-178
- ALFA-LAVAL e. 167, 175
- centrifugal e. 81
- circular e. 86f
- falling-film e. 83f
- forced circulation evaporator 85f
- LUWA e. 86
- rotatory e. 85
- SAKO e. 86
- Sambay thin-film e. 79
- TASTE e. 166, 174
- thin-film e. 86

Exposure level 758

Extract 23
Extracting agent 82
Extraction 17-48
- batteries 36f
- battery with decanter 22f
- countercurrent e. 20ff
- cross current e. 30ff
- e. tower with mechanical agitation 40
- e. tower without mechanical agitation 39
- liquid-liquid e. 24
- maceration 18
- membrane e. 178f, 271
- multistage e. 30-35
- single-stage e. 27-30
- selectivity 26
- solid-liquid e. 17f
- solvents 26, 314ff
- soxhlet e. 20
- stages 23
- tower 38-46
- water e. 170
- with $CO_2$ 50ff

Extractive concentration of citrus oils 190
Extractive distillation 82
Extractive methods 190
Extractors 20-46
- citrus juice e., Brown 166f
- citrus juice e., FMC 166f, 177
- juice e. 166ff

Extrusion 291

f
Falling film evaporator 83f
FAO, Food and Agriculture Organization of the United Nations 761f
Fat 7
Fat degradation 274
Fat replacer 12, 455ff
- interactions of volatiles with f.-r. 455-459
Fatty acids 282
FD factor 705
FDA, Food and Drug Administration 771
FEMA, Flavour and Extract Manufactures' Association 160, 760
Fennel 229, 624
Fenske method 76
Fenugreek 229
Fermentation 121, 145, 260
- in-situ-f. 121
- microbial f. 267f
- of cell cultures 129
- spontaneous f. 121f

# Index

- submerged f. 125f
- surface f. 124
Fermenter 125f
Ferulic acid 286
FEXPAN, Panel of Expert Scientists 160, 773
FFDCA, Federal Food, Drug and Cosmetic Act 771
FID, flame ionization detector 593f, 666
Filbert 416
Filtration 171, 173, 178ff
Fine bakery products 533
Fining aids 173
Fish 263, 288
- aroma 714f
- boiled 715
- fresh 714
Fitness 6, 479
Fixed bed column reactor 271
Fixes 551f
Flame ionization detector, FID 593f, 666
Flash distillation 66
Flavour
- adjunct 756
- analysis 378
- binding and release 437f
- body 557ff
- chemical 140
- definition 138, 576
- dilution factor 705
- enhancer 286
- formulated f.s 5
- fruit f. 410-425
- generation by fermentation 121ff
- industry 1-13, 137f
- intrinsic f. 557
- legislation 763-767
- manufacturing from cell cultures 271f
- manufacturers of f. 3
- migration 9
- modifier 351-372
- potentiator 351
- precursors 282, 285
- preservation by drying 114
- profile 378ff
- pyramid 570
- research 137
- substance 140
Flavour dilution factor, FD factor 705
Flavour & Fragrance Industry 1-13, 138
Flavour & Fragrance Market 2
Flavour extracts 260
- by enzymatic reactions 140

- by microbial processes 140, 145
Flavouring
- agent 140
- definition 138, 755
- permitted f. ingredients 138
- preparations based on biotechnology 260-273
- preparations by enzymatic reactions 261ff
- preparations from microbial fermentation 267f
- preparations 141
- substance 140ff, 158, 755
Flavourings
- alcoholic f. 430
- applification of f. 403f
- artificial f. 158-165
- blended f. 391-435
- commercially available f. 152ff
- dairy f. 430f
- fermented f. 429ff
- for bakery products 531-534
- for beverages 466-514
- for confectioneries 515-530
- for dairy products 542-548
- for ice-cream 535-541
- in fruit preparations 545f
- inventory of f. 160
- natural f. 140ff, 152, 755f
- natural-identical f. 158-165
- of dehydrated convenience food 549-570
- process f. 138ff, 274-297
- smoke f. 138ff, 298-313
- vegetable f. 431f
Flocculation aids 173
Flotation 171, 173
Flow conductance 69
Flow regime 74
Fluid bed dryers 525
Fluorescence detection 591
FMC extractor 166ff
FNU value 467
Fondants 518f
Food additive 757, 772
Food additive legislation 767f
Food and Agriculture Organization of the United Nations, FAO 761f
Food categories 161
Food emulsifier 323f
Food industry 7, 533f
Food proteins 330
Food-Minus 8

Food-Plus 8
Forced circulation evaporator 85
Formic acid 277, 491
Fortification 7
Fourier transform infrared, FT-IR 667
Fragrances 4
Free excess enthalpy 73
Freeze
− concentration 112f, 177
− dried coffee 114
− dried herbs 116
− dried mushrooms 115
− dried products 112-117
− drying 109-119
− shelf life of f. dried products 116
Freezing 110
Freezing point 110
French fries 724ff
− omission test 724
Freshness 7
FreshNote process 180
Fructose syrups 348
Fructose 282, 348, 367
Fruit
− aroma, analysis 720ff
− flavours 270, 410ff
− preparations 545
− purees 174
− teas 469
− wines 505
Fruit juice 166-186, 469, 746
− clarification 173, 179f, 347
− concentrates 166-186, 469, 746
− freshly squeezed 171
− liqueurs 496f
− production 174
FUFOSE, Functional Food Science in Europe 479
Fugacity coefficient 72f
Functional drinks 478-482
− definitions 478ff
Functional food 7, 479f
Fungi 145, 744
Furaneol® 163, 366
Furanones 281, 688
Furfural 279, 454
Furfuryl mercaptan 163
2-Furfurylthiol 280

g
Galanga 229
Garlic 230
Gas chromatography, GC 589, 593
Gas chromtography-olfactometry (GCO) 704
GC enantiomer separation 665
GC, gas chromatography 589, 593
GCC; Golf Cooperation Council 784
GC-FTIR, GC-Fourier transform infrared spectroscopy 596
GC-IRMS, GC- isotope ratio mass spectrometry 608
GC-MS, GC-mass spectrometry 595
GCO , GC olfactometry 704
− headspace , GCOH 706f
Gelling 334
Genetic engineering 131-133
− genetically engineered products 133
Gentian 230
− spirit 492
Geranial 161, 188
Geraniol 161
Geranium-aroma 387
German purity law 271
Gibbs' excess enthalpy 25
Ginger 62, 231
Glucose 276ff, 438, 475
− syrup 348, 475
Glutamate 352ff, 578
Glutamic acid 288, 353
Gluten 343f
Gluten free 8
Glycemic 8, 476
Glycemic index 7
Glyceraldehyde 288
Glycerol 287-290
Glycine 276-278, 282, 290
Glycolaldehyde 277
Glycosidically bound flavouring substances 152
GMP, guanosine monophosphate 357
Gooseberry 418
Graesser contactor 44f
Grape 415f
− juice 166f
− extract 473
Grapefruit 192-194, 415
− aqueous essence 193
− aroma 722f
− essence oil 193
− juice 173, 723
− oil 192
GRAS, generally recognized as safe 292, 773
Grating rings 95
Gravies 371

Index                                                                 819

Green   410, 474
Green flavour complexes   272
Green notes   141ff
Grenadine   418
Grenco process   178
Grill flavours   306
Gruyere cheese   717
Guaiacol   286, 368
Guanosine monophosphate, GMP   286, 356f
Guar gum   439, 459
Guava   422
Gum arabic   331, 471

h
Halal   292, 802
Haptic sensations   351, 575
Hazelnut   416, 732
– roasted   529, 732
HDMF, 4-hydroxy-2,5-dimethyl-3(2H) furanone   366
Headspace analysis   706f
Health   7, 455
Health claim concepts   479f
Health food   7
Health food development   480
Health Organisation Joint Expert Committee on Food Additives (JECFA)   160, 356
Healthy nutrition   478-483
Heat
– conductivity   70
– denatured proteins   445
– exchanger   85
– generation   70f
– transfer   70
– transfer coefficient   70
Heating pump   92
Heavy metal contamination   587
Height equivalent of a theoretical plate, HETP   74
Height equivalent of a theoretical stage, HETS   33
Height of a transfer unit, HTU   34
Height of diffusion rate, HDU   35
2(E)-Heptenal   283
Herbs   214-261
HETP, height equivalent of theoretical plate   74
HETS, high equivalent of theoretical stage   33
Heteroazeotrope   80, 82
Hexanal   161, 283
(Z)-3-Hexen-1-ol   161

2(E)-Hexenal   284
Heyns   276
Hibiscus extract   469
High boilings   516ff
High fibre   7
High vacuum distillation   87
HLB, hydrophile-lipophile balance   324
HMF, 4-hydroxy-5-methyl-3(2H)furanone   366
Homogenizers   268
Honey liqueur   503
Horseradish   231
Hotfill   171, 178, 746
HPLC, high performance liquid chromatography   590ff, 665
HPLC-MS, HPLC-mass spectrometry   592f
HPTLC, high performance thin layer chromatography   665
Human senses   422, 575, 579
HVP, hydrolyzed vegetable proteins   264, 490ff
Hydrocolloids   453f
Hydrogen bridge binding   424f
Hydrogen sulphide   279, 285, 288f
Hydrolysed vegetable protein   284, 288, 292
Hydrolyzed vegetable proteins, HVP   264
Hydrophile-lipophile balance, HLB   324
Hydrophobic interactions   445ff
4-Hydroxy-2,5-dimethyl-3(2H)-furanone   281f, 289f
4-Hydroxy-5-methyl-3(2H)-furanone   279, 366f
Hydroxyproline   290
Hyssop   232

i
Ice-cream   535-541
Iced teas   469
ICP-AES, inductively coupled plasma-atomic emission spectrometry   588
IMF's, intermediate moisture foods   381f
Immobilized cultures   143
IMP, inosine monophosphate   356, 558
India, legislation   788
Indonesia, legislation   789
Inductively coupled plasma-atomic emission spectrometry, ICP-AES   588
Inorganic salts   454ff
Inosine monophosphate, IMP   286, 356, 558
In-situ-generation of flavours   121
Intalox rings   93
Intelligent flavours   11

Interactions of volatiles
- and food ingredients 437
- with amino acids 449-450
- with carbohydrates 438-445
- with complex systems and foodstuff 459-462
- with fat-replacers 455-459
- with lipids 450-453
- with proteins 445-449
Interface area 74
Interfacial area per unit volume 26
Interfacial tension 26
Intermediate moisture foods, IMF's, 381f
Intermolecular forces 25, 71
International isotopic standard, i-IST 691f
International Organization of the Flavour Industry, IOFI 138f,159, 755
Inventory for flavouring substances 160
IOFI, International Organization of the Flavour Industry 138f,159, 755
Ionising radiation 747
IQ mutagens 291
Irregular column packings 93
IRMS, isotope ratio mass spectrometry 609
Isoeugenol 163, 368, 625
Isotonic beverages 746
Isotope
- abundance 608
- analysis 613
- discrimination 606
- effects 603
- ratio mass spectrometry, IRMS 609
Isotopic correlations 637ff
Isotopic pattern 606, 637ff
Isotopomer 602

j
Jabuticaba 418
Japan, legislation 785f
JECFA, Joint Expert Committee on Food Additives 160, 759, 761
Jellies and gums 520f
JET COOKER 520
Juice
- extraction 167-171
- extractors 166
- oil 176
Juices 9, 171-173
- clarified j. 171ff, 175
- cloudy concentrates 470
- concentrates 173ff, 469
- natural cloudy j. 171
- reconstituted j. 178

Juniper berries 232
Juniper-based spirits 490f

k
Karr column 42
Key lime 197ff
Key odorants 704ff
- screening 705
Kinetic gas theory 68
Kitchen aids 549-572
Kiwi 422
Koch plate casting procedure 750
Koiji 124, 337
Kokumi 11
Korea (South), legislation 790f
Kosher 292, 802
Kühni extractor 43

l
Labelling directive 79/110/EEC 767
Lactic acid starter cutures 744
Lactic acid 162, 287, 804
Lactones 630, 672ff
Lamb 288, 803
Lamellar phase 284
Langmuir-Knudsen equation 87
Laurel 233
Lavandula oils 680ff, 693
Law, Boyle and Mariotte 67
Law, Dalton's 71
Law, Raoult's 71
Law, Trouton's 70
LC Taste™ 160
Leakage rate 83
Lecithine 275
Legislation and Toxicology 753-810
Lemon 194-196, 414
- essence oil 195
- juice 173, 414
- oil 194, 691
- grass oil 635
Leucine 278, 290
Life-style 9
Light taste 545
Likens-Nickerson apparatus 30f
Lime oil 196, 415
Linalool type 218
Linalool 162
Linoleic acid 284
Linolenic acid 284
Lipid oxidation 276, 282ff
Lipid-based fat-replacers 456
Lipids 450

Index 821

– interactions of volatiles with l.   450-453
Lipoxygenase   282
Liqueurs   487, 496-504
Liquid-liquid
– continuous countercurrent extraction   32
– extraction   24ff
– extraction tower   38
– multi-stage extraction   30
– single-stage extraction   27
Liquid-solid extraction   17, 589
Liquorice   233f
Litsea cubeba oil   565
Loading diagram   32
Lovage   234
Low caffeine   7
Low nicotine   7
Low-oxygen system   102f
Lulo   417
Lutein   473
LUWA evaporator   88
Lychee   422

m
Macapuno   415
Maceration   18ff, 513
Maillard products   263, 276-282
Maillard reaction   274, 276-282, 287
Malaysia, legislation   791ff
Malt   274, 276, 289
Maltodextrins   101, 319, 443
Maltol   162, 289f, 362ff
Mandarin   199-203, 415
– essence oil   201
– juice   173
– oil   199ff
– oranges, aroma   722
Mango   423
Manufacturers   3
Marjoram   60, 235
Market   8
Market share   2
Marshmallow   521
Masking agents   11
Mass selective detector, MSD   595
Mass spectrometry, MS   595, 667
Mass transfer coefficient   35, 74
Mass transfer rate   26, 74
Massive chocolate products   512
Massoia oil   248
Massoilactone   674
Maximum azeotrope   81f
Mayonnaise   549
McCabe-Thiele diagrams   75, 76

MCP, 3-methyl-2-cyclopentene-2-ol-1-one   366f
MDGC, multidimensional gas chromatography   594, 612
Mead   505
Meat   274f, 427f
– cooked   561
– extract   557
– flavour   263f, 561
– snacks   307
Meat aromas, analysis   711
Meat flavour   274, 288f
Melanoidins   278
Melon   423
Membrane contactor   44f
Membrane extraction   271
Membrane processes   178ff
Mentha species   679f
1-Menthen-8-thiol   160
Menthol   60, 162
Menthyl lactate   162
3-Mercapto-2-butanone   279
3-Mercapto-2-methylpentan-1-ol   276, 285f
3-Mercato-2-pentanone   279
MERCOSUR   771
Meristema   130
Methional   279
Methionine   288
2-Methylbutanal   279
3-Methylbutanal   279
2-Methyl butyrate   161
Methyl chavicol type   219
4-Methyl-5-ethylthiazol   285
2-Methyl-3-furanthiol   279
Methyl ketones from butterfat   265
2-Methylpropanal   279
Methyl salicylate   158, 625
4-Methylthiazol   285
4-Methyl-5-vinylthiazol   285
Mexico, legislation   778
Microbial growth   744f
Microbial processes   139,260
Microbiological testing   744-751
Microemulsion   289
Microfiltration   179f
Micro-organisms   120, 145ff, 261ff, 744ff
– water activity and growth of m. in food   377ff
Microwave system   289, 547
Microwave heating   291
Microwave   290
Middle notes   429, 559

Milk
- defects   545
- fat   538
- rice   545
- Solids Non Fat, MSNF   538
Minimum azeotrope   73, 81
Minimum reflux ratio   77f
Minimum solvent ratio   32
Mint oils   240f, 679
Miraculin, Mirlin®   369
Miscella   17, 21f
Miscibility   24
- gap   28, 79f
Mixed-blend fat-replacer   456f
Mixer-settler   35, 41
Mixtures, thermodynamic fundamentals   71-83
Modified cyclodextrins   670
Molar evaporation enthalpy   70
Monosodium glutamate, MSG   353ff, 552f, 564
Mould   744
Mouthfeel   351, 455, 475f
MS, mass spectrometry   595, 667
MSD, Mass selective detector   595
MSDI, Maximised Survey-Derived Daily Intake   161
MSG, monosodium glutamate   353ff, 552f
- hypersensitivity   356
MSNF, Milk Solids Non Fat   538
Mugwort   235f
Multifunctionality   10f
Multistage distillation   67
Murcott tangerine oil   200f
Mushroom flavours   270
Mustard   226

n
NAFTA   771
Nanostructure   284, 289
Naranjilla   417
Naringin   163
Nasal-smelling   576f
Nasopharynx   576f
Natural
- abundance   644
- extracts   5
- flavour concentrate   756
- flavouring   775
- flavouring substance   140ff, 152, 756f
 - commercially available n. f. s.   152ff
- meat flavours   263f
- origin   619

Nature-identical   1, 158ff
Nature-identical flavouring substance   139, 158, 755
Near water drinks   468
Negative List   757
Neral   161
NMR, nuclear magnetic resonance   614ff
2,6(E,Z)-Nonadienal   284
Non-alcoholic beverages   466
Nonanal   283
2(E)-Nonenal   283
3(Z)-Nonenal   283
Nootkatone   162
Novel food   476
NPD, nitrogen-phosphorus detector   594
Number of a transfer unit, NTU   34
Nut   284
Nutmeg   236
Nutrition   7

o
OAV, odour activity values   160, 673
1,5(Z)-Octadien-3-one   284
Octanal   161, 283
2(E)-Octenal   283
Odorants, identification   707
Odorants, quantification   708
Odorants, stable isotope dilution assay (SIDA)   708
Odour   140f, 380f
- activity values, OAV   673
- threshold   436f
Odour activity value (OAV)   160, 673
- calculation   708
- definition   704
Off-flavours   121, 289
- produced by lipoxygenases   121
- produced by peroxidases   121
- produced by proteases   121
Oldshue-Rushton extractor   43
Oleic acid   283
Olfactometry global analysis   706
Olfactory
- epithelium   437, 576
- organ   437
- perception   576
Oligosaccharides   348
Olive oil   718ff
Omission test   711
Onion   237, 276, 285f
Operating line   76
Oral / pulmonary cavity   437, 576
Orange   203-209, 414

- aroma 720ff
- essence oil 203
- juice 173, 414
- oil 203, 414
- wash pulp, OWP 170
- water phase 208
Oregano 237
Organic 7
Organoleptic 575
- quality control, OQC 576
- quality testing, OQT 576
Ornithine 281
Osme method 706
Osmotic drying 382
Ostwald's triangle diagram 28
OWP, orange wash pulp 170
Oxidation taste 545
2-Oxopropanal 277, 281

p
Packings
- irregular p. 93
- regular p. 93
PAH, polycyclic aromatic hydrocarbons 290f
Paired comparison test 582
Pall rings 93
Panelists 575
Panettone 533
Panned work 525f
Papain 335f
Papaya 424
Papillae 578
Paprika 241
- Oleoresin 375
Parsley 238, 730
Partition coefficient 24
Passiflora edulis 424
Passion fruit 173, 424, 672
Pasteurisation 746
Pastries 531
Pathogens (salmonella, staphylococcus aureus) 744
Paw Paw 424
PCH, polycyclic hydrocarbons 312
Peach 416
Peanuts 306
Pear 416
- juice 171, 173, 179
Pecans 306
Pectin 439, 481
2-Pentylpyridine 286
Pepper 239, 729

Pepper, black 729
Peppermint oil 240f, 679
Peptides 275
Percolation 20f, 514
Perilla aldehyde 162
Permeation 18
Persian lime 197
Pesticide residues 768
Petitgrain oils 188, 209
Phase transfer 74
Phenolic compounds 298f, 374, 454
Phenyl acetaldehyde 158, 279
Phenylalanine 276, 279, 290
Philippines, legislation 794f
Phospholipids 282, 289
Photosynthesis 272
Physical processes 17-119, 141f
Pimento 241
Pineapple 416
- juice 166f, 173
- ketone 688
Plait point 28
Plant
- homogenates 142
- tissue 266f
- tissue cultures 143
Plate efficiency number 33
Plate exchange efficiency 75
Plating procedure 633
Plum 424
Podbielniak extraktor 37
Polarity 18
Polycyclic aromatic hydrocarbons, PAH 290f
Polycyclic hydrocarbons, PCH 312
Polyglycerol polyricineolate 330
Polyhydric alcohols 325, 476
Polymethylsiloxane 275
Polysaccharides 267, 439
Pomegranate 418
Popcorn 732
Pork 263, 288, 711ff
Positive List 757
Potato 432
- starch 440f
Potatoes, boiled 724ff
Precipitates 473
Precolumns 666
Preservation by drying 109
Preservatives 377-387
- Food Additives Directive 382
Presses 168
- belt press 168

- Bucher press   168
- hydraulic press   168
- one sieve belt press   170
Pressure range   67f, 110
Prickly Pear   425
Process flavourings   274-297, 756
Process flavours   559
Processing of citrus oils   188
Proline   281, 288, 290
Propionic acid   385
Propyl gallate   373
Propylene glycol   290, 382
Protein-based fat-replacer   456f
Proteins   263, 445ff
- anhydrous p.   447
- heat denatured p.   446
- interactions of volatiles with p.   445-449
- native p.   445
Protozoa   124, 744
Pulp wash   170, 173
Pulsed extraction tower   41
Pumping speed   68
Pumps, for vacuum generation   68-69
- liquid-seal pumps   69
- pumping speed   68
Purine alkaloids   454ff
Pyroligneous acid   290
1-Pyrroline   281
Pyrolysis   289, 291
Pyruvic acid   162

q
QC, quality control   575
Quadronics extractor   38
Quality control, QC   575
Quasi-equilibrium   18
Quince   425

r
Radioactive isotope   602
Radioactivity, regulations   769
Raffinate   24
Raisin   415
Rancidity   283
Raoult's law   71
Raschig rings   93
Raspberry   416f, 695
Ratio of enantiomers   669
RDC extractor   43f
Reaction flavours   559
Reactor   290f
Reboiler evaporator   85
Receiver phase   24

Reciprocating plate extraction tower   42
Recovery oils   187
Rectification   74-79, 90-94
- continuous r.   76ff, 91
- discontinuous r.   91
- partly continuous r.   92
- semi-continuous r.   90
Red pepper   242f
Reductone   277f
Reflux ratio, Vr   76
- minimum r. r.   77f
Regular column packings   93
Rehydration   114, 747
Relative volatility factor   71
Release phase   24
Religious dietary rules   802-810
Rennet   335, 348
Resolution, Rs   664
Retro-nasal smelling   422, 577
Reverse osmosis, RO   180
Rhamnose   282, 290, 366
Riboflavin   472
Ribonucleic acid   268
Ribonucleotides   356ff, 558
Robatel extractor   38
Roquefort flavour   269
Rosemary   242, 374
Rotary evaporator   85
Rotocell extractor   21
R-phrases   769
Rum   491
RVP, relative vapour pressure   452
RZE extractor   43f

s
Safety evaluations of flavouring substances   758
Safety of food additives   772
Saffron   243
Safranal   162
Sage   244, 374
SAKO evaporator   86
Salt replacers   554
Saltiness   552ff
Salty snacks   9
Sambay thin film evaporator   86
Sample preparation methods   589f
Santalyl acetate   162
Sauces   551f, 570
Savory   244
Savoury   570
- process flavourings   288ff
- pyramid   557

Savoury flavour   288ff, 549
– pyramid   557
– taste   552f
SCF, Scientific Committee on Food   760
Schiff base   276
Screw-conveyor extractor   22
SDE, simultaneous distillation-extraction   589
Seafood   263, 304ff
Seasonings   264, 570
Seaweed extracts   331
Selected wavelength chromatograms, SWC   667
Selectivity of extraction solvents   29, 50, 52ff
Senses   576ff
Sensory   575f
– analysis   575-586
– quality control, SQC   576
– quality testing, SQT   576
Sensory expertise   12
Separation factor   664
Sesame seeds   286
Sesame, roasted   732f
SFE, supercritical fluid extraction   50-65
Shiitake   353
Sieve plates   91, 93
Significance level   580
Significance table   582ff
SIM, selected ion monitoring   595, 667
Simultaneous distillation/extraction   589
Singapore, legislation   796ff
Single stage spray drying   99
Sintering   568
SIRA, stable isotope ratio analysis   602-663
Smoke
– components   291, 303
– condensates   310
– generation   309
– solutions   289
Smoke flavourings   298-313, 756
– application on various foods   304ff
– dry s. f.   292
– liquid s. f.   289
– preparations   309f
smoked
– colour   289
– food   289
Snack foods   306
SNIF-NMR®, site-specific natural fragmenta-tion-NMR   617
Sodium chloride   362, 454, 578
Soft drinks   466-486

– based on clear juices   469
– based on emulsiones   470ff
– based on flavour   468
– based on plant extract   469
– circle   466f
– energy drinks   468
– ingredients for s. d.   467ff
– with high juice content   469f
– with low juice content   470
Solid-liquid absolute countercurrent extraction   22
Solid-liquid discontinuous countercurrent extraction   22
Solid-liquid extraction   17f, 589
Solid-liquid relative continuous counter-current extraction   21
Solvents
– carrier s.   317-321
– dense gases as s.   50-65
– extraction s.   314ff, 768
– evaluation   50f
– miscibility   24, 27
– polarity   18
– properties   18, 50, 52
– selectivity   50
Sorbic acid and sorbates   385
Sorbic acid derived off-notes   387
Soups   563, 566
Sour dough   123
Sour milk products   544ff
Soursop   420
South Africa, legislation   800
Soxhlet-extraction   20, 589
Soy nuts   307
Soy sauce   124, 282, 292, 569
Spearmint oil   241
Spices   214-261, 729ff
Spike oil   694
Spirits   487-496
Spore-forming bacteria   745
Sport drinks   476
Spray chilling   104, 327
Spray drying   97-108, 565
– low oxygen s. d. system   103
– single stage s. d.   99
Spreads   329
Stabilisers   330-335, 538
Stability   330, 408ff, 472, 483
Stable emulsion   322
Stable isotope
– measurements   619
– pattern analysis   614
– ratio analysis, SIRA   602-663

– ~s   604
Star anise   217, 624
Starch   439
– complexing   326
– modified starches   471
Starfruit   425
Starter cultures   121
Statistical analysis   583
Steam distillation   81f
Stereodifferentiation   668f
Sterilization   125, 750
Stoke's law   470
Stollen   533
Strawberry   417, 537, 724
Strecker acid   279
Strecker aldehydes   278ff
Strecker degradation   276, 278
Stripping column   76
Sublimation   110
Succinic acid   162
Sucrose   475
Sugar   475
Sugar apple   424
Sugar confectionery   9
Sugar substitutes   475f
Sulfurol   285
Sulphur containing flavouring substances   289
Sulphur dioxide   336
Supercritical fluid extraction, SFE   50-65
Surface tension   18
Suspension cultures   143
SWC, selected wavelength chromatograms   667
Sweet basil   218
Sweet marjoram   235
Sweet orange oil   205f
Sweetener   466, 475, 536
– production   348
Sweeteners   475f
Sweetsop   420
Synergism   361
Synergistic effects   361f
Syrups   475

t
Tactile sensations   351, 575
Taiwan, legislation   798
TAMDI (Theoretical Added Maximum Daily Intake)   161
Tangerine oil   199
Tarragon   245
Taste   140, 380, 575

– activity value (TAV)   160
– compounds   286ff, 735
– dilution analysis (TDA)   558
– enhancement   557ff
– enhancers   263, 558
– on the tongue   578
TAV, taste activity value   160
TBHQ, t-butyl-hydroxy quinone   373
Tea   427
– liqueurs   498
Tea drinks   469
Tea, green   736
$\alpha$-Terpineol   162
$\alpha$-Terpinyl acetate   162
Thailand, legislation   799f
Theobromine   287
Theoretical extraction stages   23, 26, 29
Theoretical plate number   74
Thermodynamic equilibrium fundamentals   25
Thermolysis   268
Thiamine degradation   284f
Thiamine   284f, 288
Thiazole   285
Thickening   334
Thin-film evaporator   86
Think-drinks   11
Thyme   245
Thymol   163
TIC, total ion chromatogram   595
Tie line   28
Tissue cultures   261
Tocopherols   374
Toffee   521f
Tomato
– fruit   431f
– puree   175
– paste   727f
– aroma   727f
Tongue   577
Top notes   556f
Top Ten flavour houses   2
Total ion chromatogram, TIC   595
Toxicology and Legislation   753-810
– toxicological considerations   250, 758-761
– toxicological tests   291
Tray column   93
Trehalose   476
Triangular diagram   28
Triangle test   582
Trickle flow   57
2,4,7(E,Z,Z)-Tridecatrienal   284

# Index

Trigeminal stimuli 578f
Triglycerides 329
TRIO test 582
Trouton's law 70
Truffle 272
Turbidity 468
Turkish delight 521
Turmeric 246
Tutti frutti 433

u
Ultrafiltration 173, 179
Ultrasound 18
– treatment 18
Umami 287f, 292, 353, 358f, 552ff
2-Undecanone 161
2(E)-Undecenal 283
UNIFAC method 25
USA, legislation 771ff
UV/VIS detection 591
UV/VIS spectroscopy 587

v
Vacuum generation 68f
Valencene 164
Valine 290
Values of natural products 605ff, 619ff
Vanilla 247, 432, 533
Vanilla flavour from cell cultures 272
Vanillin 1, 158, 163, 286, 368f, 528
– synthesis 164
Vapour-liquid loading 74
Vapour pressure curves 67, 80
Vapour pressure, partial 71
Veal 288
Vegetable 288
Vegetarian 7
Vermouth 505

Vinylguaiacol 286
Virial coefficient 73
Viruses, bacteriophages 744
Vitamins 7, 468
Vodka 493

w
Walnuts 306
Warmed over flavour, WOF 283
Washing 190
Water activity $a_w$ 377, 744
Water extracted solids, WESO 170
Water extraction 170
Weighting agents 470
Wellness 7
WESO, water extracted solids 170
Wheat gluten 287, 516
Wheat gluten hydrolysate 287
White mustard 236
WHO, World Health Organization 761
Wine 503f, 735f
Wine-like beverages 505f
Wintergreen oil 625
WOF, warmed over flavour 283
WONF, with other natural flavours 775
Wood pyrolysis 309
World Health Organization, WHO 761
Wormwood 247

x
Xylose 279, 284, 287-289

y
Yeast 123, 145
– autolysates 268f
– extracts 268, 557, 559, 568f
Yellow 472f
Yoghurt cultures 123

**Related Titles**

Otto L. Piringer, Albert L. Baner (Eds.)
**Plastic Packaging Materials
for Food and Pharmaceuticals**
2007
ISBN 3-527-31455-5

Horst Surburg, Johannes Panten
**Common Fragrance and Flavor Materials**
*Preparation, Properties and Uses*
5th, Completely Revised and Enlarged Edition
2006
ISBN 3-527-30364-2

James G. Brennan (Ed.)
**Food Processing Handbook**
2005
ISBN 3-527-30719-2

Jean-Nicolas Wintgens (Ed.)
**Coffee: Growing, Processing, Sustainable Production**
*A Guidebook for Growers, Processors, Traders, and Researchers*
2004
ISBN 3-527-30731-1

Wolfgang Aehler (Ed.)
**Enzymes in Industry**
*Production and Applications*
2003
ISBN 3-527-29592-5

Georg-Wilhelm Oetjen, Peter Haseley
**Freeze-Drying**
2004
ISBN 3-527-30620-X